Failure Characteristics of Deep Rocks

Failure Characteristics of Deep Rocks

Guest Editors

Zhenyu Han
Diyuan Li
Xin Cai

MDPI

Basel • Beijing • Wuhan • Barcelona • Belgrade • Novi Sad • Cluj • Manchester

Guest Editors
Zhenyu Han
School of Resources and Safety Engineering
Central South University
Changsha
China

Diyuan Li
School of Resources and Safety Engineering
Central South University
Changsha
China

Xin Cai
School of Resources and Safety Engineering
Central South University
Changsha
China

Editorial Office
MDPI AG
Grosspeteranlage 5
4052 Basel, Switzerland

This is a reprint of the Special Issue, published open access by the journal *Minerals* (ISSN 2075-163X), freely accessible at: https://www.mdpi.com/journal/minerals/special_issues/FCDR.

For citation purposes, cite each article independently as indicated on the article page online and as indicated below:

Lastname, A.A.; Lastname, B.B. Article Title. *Journal Name* **Year**, *Volume Number*, Page Range.

ISBN 978-3-7258-3241-5 (Hbk)
ISBN 978-3-7258-3242-2 (PDF)
https://doi.org/10.3390/books978-3-7258-3242-2

Contents

Preface

The following reprint of the Special Issue *Failure Characteristics of Deep Rocks* explores the complex mechanisms of rock failure in deep geological settings. This collection was inspired by the increasing demand for understanding the behavior of deep rocks subjected to extreme stress, high temperatures, and coupled environmental factors. These insights are critical for advancing applications in mining, tunneling, geothermal energy extraction, and carbon storage, where challenges such as rockbursts, instability, and structural failure pose significant risks.

The scope of this reprint spans theoretical analyses, experimental studies, numerical simulations, and field investigations, addressing topics such as mechanical properties, crack propagation, coupled thermal–hydraulic–mechanical interactions, and practical strategies for enhancing underground excavation safety. The aim is to provide a comprehensive resource for researchers and practitioners focused on deep rock mechanics, offering new knowledge to improve safety, efficiency, and sustainability in engineering practices.

This work was made possible by the contributions of leading experts in the field, who have shared valuable findings from their creative research. The editors extend their gratitude to the authors for their dedication and to the peer reviewers for their critical insights, which have significantly improved the quality of the articles.

This reprint is addressed to a broad audience, including academic researchers, industry professionals, and students seeking to deepen their understanding of deep rock failure mechanics. It serves as both a reference and a foundation for future investigations into the fundamental processes and engineering applications of rock mechanics.

The editors would like to acknowledge the support of the journal's editorial team and the institutions that facilitated this work, ensuring the successful compilation of this reprint. We hope that this reprint will inspire further research and collaboration in addressing the challenges posed by deep geological environments.

Zhenyu Han, Diyuan Li, and Xin Cai
Guest Editors

minerals

Article

A Study on the Dynamic Strength Deterioration Mechanism of Frozen Red Sandstone at Low Temperatures

Yang Yang [1], Niannian Zhang [2] and Jianguo Wang [3,*]

1 School of Civil and Resources Engineering, University of Science and Technology Beijing, Beijing 100083, China; yangyang@ustb.edu.cn
2 Yunnan Superstar Certified Safety Engineer Office Co., Ltd., Kunming 650106, China; znn88916@163.com
3 Faculty of Land Resources Engineering, Kunming University of Science and Technology, Kunming 650093, China
* Correspondence: wangjg0831@163.com

Abstract: In this study, the dynamic mechanical properties of red sandstone at low temperatures were studied by performing SHPB dynamic impact tests. According to damage and energy theories, the influences of different low temperatures on the dynamic strength, damage variable, and energy dissipation of red sandstone were analyzed. Combined with a fracture morphology analysis, the deterioration mechanism of the dynamic mechanical strength of red sandstone was deduced at lower negative temperatures. The research results showed that lower negative temperatures (<−30 °C) caused "frostbite" in red sandstone, which resulted in a sharp reduction in the macroscopic, dynamic mechanical strength of rock under high strain. Transient engineering disasters can easily occur under such a dynamic disturbance. According to the fracture morphology analysis, low temperatures generated a large number of cracks at the interface between the components of red sandstone. The plastic deformation ability of the crack tip was poor, and stability loss and expansion under high strain rate were readily achieved, resulting in low-stress brittle failure. However, due to the complex mineral composition of the cementitious materials, they were more susceptible to low temperature. Therefore, under the double action of dynamic load and low temperatures, it was found that damage occurred in the cementitious materials first, and then fracture of the red sandstone as a whole resulted.

Keywords: frozen red sandstone; dynamic strength deterioration; damage variable; fracture morphology

Citation: Yang, Y.; Zhang, N.; Wang, J. A Study on the Dynamic Strength Deterioration Mechanism of Frozen Red Sandstone at Low Temperatures. *Minerals* **2021**, *11*, 1300. https://doi.org/10.3390/min11121300

Academic Editor: Stephen E. Laubach

Received: 9 October 2021
Accepted: 18 November 2021
Published: 23 November 2021

1. Introduction

The mechanical response of rock at low temperature is significantly different from that at room temperature [1,2]. Especially in the case of lower negative temperature, uniaxial compressive strength, tensile strength, and elastic modulus of rock increase significantly with a decrease in negative temperature [3–5]. In the meantime, the internal friction angle and cohesion also increase [6–10], and the overall static performance of rock is enhanced. However, negative temperatures can also make rocks more brittle, and failure strain is reduced. Meanwhile, rocks also bear huge ice swelling forces brought by water-ice phase transformation, which can lead to sudden engineering disasters occurring more easily in geotechnical engineering in cold areas. In addition, in cold climate regions, rocks are subjected to repeated processes of freeze–thaw weathering and, consequently, the integrity of those rocks gradually deteriorates [11,12]. Therefore, it is key to solve the problem of mineral resource exploitation and geotechnical engineering construction in cold regions to reveal and understand the mechanical properties of rocks, especially the dynamic properties under low negative temperatures.

However, as compared with the numerous studies in the field of rock statics [13–19], dynamic studies are relatively scarce, and most of them have focused on −20 °C temperature conditions, which are not low enough to fully reflect the influence of negative temperature changes on rock dynamic characteristics, let alone to explore the dynamic

disaster mechanism of geotechnical engineering in cold regions. According to the above problem, in the present study, a dynamic impact test of water-saturated frozen red sandstone is carried out by using a split Hopkinson pressure bar (SHPB) experimental system to study the dynamic mechanical properties of low-temperature frozen red sandstone under compression load. The range of test temperatures is from 25 °C to −40 °C, which can comprehensively reflect the influence of low temperatures on rock dynamic behaviors. In addition, combined with scanning electron microscopy, the microscopic fracture characteristics of frozen rock are accurately characterized by fracture morphology analysis. On this basis, the effects of low temperatures and water-ice phase transformation on dynamic mechanical strength and deformation characteristics of rock are studied.

2. Dynamic Impact Test of Frozen Red Sandstone

2.1. Specimen Preparation and Corresponding Parameters

The red sandstones were taken from the Hongqinghe Coal Mine. According to the test standard recommended by the Society of Rock Dynamics, cylindrical specimens with ϕ 74 × 37 mm were prepared (see Figure 1), ensuring that the end face and axial direction of the specimens met the requirements of flatness and straightness, respectively (see Figure 2). The prepared red sandstone specimens were saturated or dried and their basic physical parameters were measured (see Table 1), and then the freezing treatment and comparison group tests were conducted.

Figure 1. Some of the prepared rock specimens.

Figure 2. The SHPB experimental system.

Table 1. Basic physical parameters of red sandstone.

Rock Type	Velocity of Longitudinal Wave m·s^{-1}		Dry Density kg·m^{-3}	Saturated Density kg·m^{-3}	Saturation Rate %
	Dry	Saturated			
Red sandstone	2780	3651	2292	2371	3.46

2.2. Experimental Scheme

The strain rates that correspond to engineering projects, such as mechanical or blasting excavation of frozen vertical shaft, vibration loading of geotechnical engineering in cold areas, impact loading, and seismic loading of liquefied natural gas and LPG reservoir are mostly in the range of 10~1000, which belong to medium and high strain rates, and their dynamic mechanical properties can be tested through a Hopkinson experimental system. Therefore, this study is based on the dynamic impact test, and an SHPB experimental system is used to simulate the dynamic disturbance of cryogenic frozen rock during engineering construction or service.

The preparation of low-temperature frozen rock specimens was accomplished in a high- and low-temperature testing machine. After dropping to the specified temperature, the specimens needed to be stabilized in a low-temperature chamber for at least 24 h to ensure consistency of the temperatures inside and outside the specimens. For the SHPB dynamic impact test under high strain rate, the gas pressure was set at 0.72 MPa. The average strain rate of red sandstone is 101 s^{-1}. According to freezing temperature, the specimens were divided into the following seven groups (at least three specimens in each group): the 25 °C Group, −5 °C Group, −10 °C Group, −15 °C Group, −20 °C Group, −30 °C Group, and −40 °C Group. During the experiment, the specimens needed to be quickly transferred to the Hopkinson experimental device from the freezing chamber; the transfer time should be less than 30 s. Due to the poor thermal conductivity of rock and the short transfer time, we considered that the overall temperature of the rock specimens were not affected by the external room temperature (25 °C) during the entire experimental process. In addition to the dynamic impact test on saturated rock, a contrast test on a dry specimen group was also conducted to analyze the influence of water and the water-ice phase transformation on the mechanical properties of rock at high strain rates.

3. Temperature Effect on Mechanical Properties of Red Sandstone under High Strain Rate

3.1. Analysis of Dynamic Mechanical Properties

The dynamic compressive strength and failure strain of water-saturated red sandstone at different temperatures were obtained according to the measured voltage value, and the curves of corresponding characteristic parameters changing with temperature were drawn, as shown in Figures 3 and 4.

The results of the low-temperature rock statics tests show that the uniaxial and triaxial compressive strengths of rock increase with a decrease in temperature at negative temperatures [13–15]. However, under high strain rate, we found that low negative temperatures caused "frostbite" effects in rocks, that is, large numbers of microcracks and intermedium spaces were produced in rocks due to the differences in shrinkage rates and degrees of solid ice, minerals, and other media under low negative temperatures. These secondary defects were prominent in the deterioration of rocks under high strain rate, resulting in a decrease in dynamic compressive strength of the rocks. These results are different from the results obtained from static and quasi-static tests, in which the strength increases as the temperature decreases. As can be seen from Figure 5, the dynamic compressive strength of water-saturated red sandstone decreases from 122 to 86 MPa in the temperature range from −30 °C to −40 °C, the reduction ratio of which is about 41.68%, and the rock dynamic performance is greatly degraded. The dynamic compressive strength of red sandstone at −10 °C, −15 °C, −20 °C, and −30 °C increases by 14.87%, 16.87%, 22.84%, and 23.97%, respectively, as compared with that at −5 °C. These results indicate that a negative temperature from −5 °C to −30 °C is not enough to generate a large number of secondary defects on the interfaces of different media. On the contrary, the rock shrinks when it encounters cold, and the contact between various media is closer, thus, the overall bearing capacity of the rock is enhanced. The variations in the failure strain of red sandstone reflect that with a decrease in temperature (from 25 °C to −10 °C), the brittleness of the rock gradually increases, and it is easy to go from small deformation to failure directly, and therefore the

failure strain value gradually decreases. However, with a further decrease in temperature, secondary defects caused by the lower negative temperature begin to appear in the rock. These defects further develop and expand under the impact load, resulting in an increase in rock failure strain.

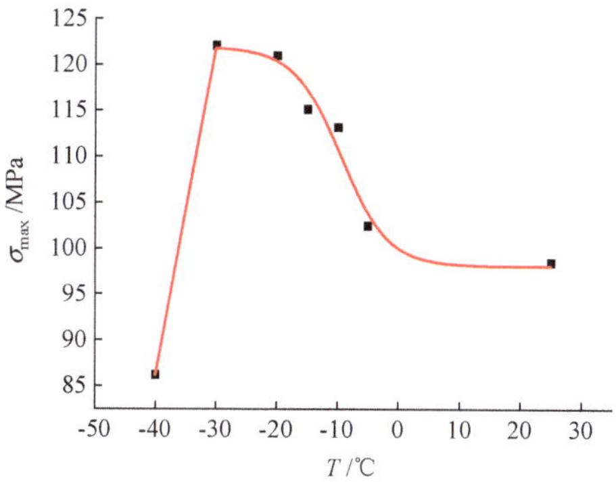

Figure 3. Dynamic compressive strength of water-saturated red sandstone at different temperatures.

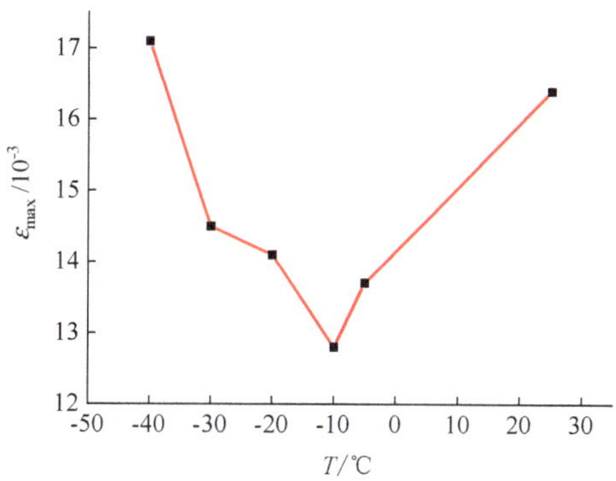

Figure 4. Failure strain change of water-saturated red sandstone at different temperatures.

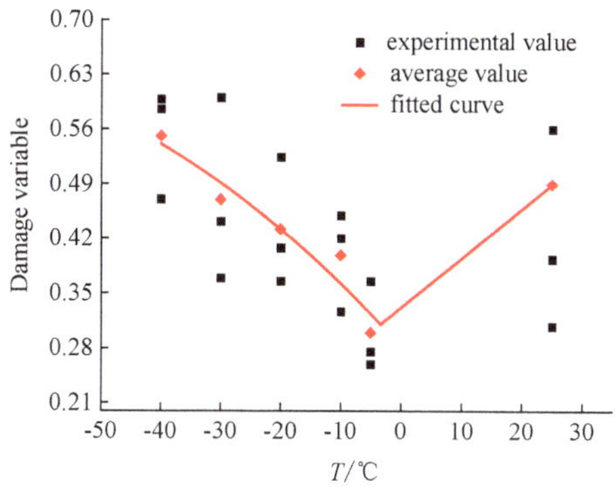

Figure 5. Relationship between damage variable and temperature in the dynamic impact test.

The above analysis of the dynamic mechanical property curve shows that "frostbite" occurs in red sandstone when the temperature drops to −30 °C, that is, the dynamic mechanical properties of red sandstone deteriorate. The reason is that a large number of microcracks occur in negative-temperature-treated rocks under high strain rate loading.

These microcracks have little effect on rock strength properties under low strain rate or static or quasi-static test conditions, but their effect can be amplified by the high strain rate, resulting in deterioration of rock mechanical properties and a sharp reduction in dynamic mechanical strength.

3.2. *Influence of Negative Temperature on Damage Variables*

The formula for calculating the total dissipated energy density of rock specimens under impact load is as follows:

$$w_d = \frac{W_L}{V} \quad (1)$$

where w_d is the total dissipated energy density, V is the specimen volume, and W_L is the total dissipated energy of rock impact failure.

The area enclosed by the stress-strain curve of rock in the SHPB dynamic impact test is considered to be the total absorption energy density u of rock deformation and failure:

$$u = \int \sigma d\varepsilon \quad (2)$$

In order to study the influences of different temperatures on rock dynamic damage, the damage variable d represented by energy concept is introduced [20,21]:

$$d = \frac{w_d}{u} \quad (3)$$

where w_d and u represent the total dissipated energy density and total absorbed energy density of rock deformation and failure under impact load, respectively.

According to Equations (1)–(3), the damage variables of red sandstone specimens under impact load are calculated, as shown in Table 2, along with corresponding temperature parameters.

According to general damage rules, it is considered that d = 1.0 indicates material destruction, but in many cases, when the damage variable value is far less than 1.0, the material may also be destroyed, such as rock material under dynamic loading at negative temperatures. In this study, energy dissipation is used to define the damage variable, and the larger the damage variable value is, the higher the damage and breakage degree of the corresponding rock specimen is. It can be observed from the damage variables in Table 2 that there are significant differences in damage and breakage degrees of red sandstone under the same impact load, which is caused by temperature change, or more specifically, the different negative temperatures encountered by the rock in the saturated state. Although the damage variables in Table 2 represent the damage and failure of rock under impact load, their change trends at different temperatures can still reflect the influence of negative temperature change on rock damage and failure. Figure 5 shows the variation of damage variable and temperature under dynamic impact of red sandstone. It can be seen from the figure that the value of the damage variable d is the minimum at −5 °C, and the degree of rock breakage is low at −5 °C (Figure 6a). When the temperature drops from 25 °C to −5 °C, the value of the damage variable decreases sharply, the dissipated energy decreases, and the failure mode of rock changes from severe fracture to bulk splitting. In the temperature range from −5 °C to −40 °C, the damage variable increases gradually as the temperature decreases, the dissipated energy increases gradually, and the degree of breakage gradually intensifies, as shown in Figure 6a–e. As compared with 25 °C, the damage variable of −30 °C and −40 °C increases by 11.9% and 30.9%, respectively. The dissipated energy at −30 °C and −40 °C is also higher than that at 25 °C; correspondingly, the degree of rock breakage is more serious. In the temperature range from −5 °C to −30 °C, the damage variable value d of red sandstone is smaller than that at 25 °C.

Table 2. The measured damage variables at different temperatures in the dynamic impact test.

Serial Number	T °C	U J·m^{-3}	w_d J·m^{-3}	d	
				Experimental Value	Average Value
1	25	2,026,020	798,014.6	0.39	0.42
2		1,824,890	1,022,309.0	0.56	
3		2,170,870	669,018.2	0.31	
4	−5	2,308,220	635,319.7	0.28	0.30
5		2,309,290	599,024.3	0.26	
6		2,036,210	744,266.1	0.37	
7	−10	1,843,130	774,945.0	0.42	0.40
8		1,829,330	597,333.3	0.33	
9		1,554,370	698,912.0	0.45	
10	−20	1,803,090	945,189.0	0.52	0.43
11		2,121,560	775,307.4	0.37	
12		2,225,180	907,565.0	0.41	
13	−30	2,046,070	904,847.4	0.44	0.47
14		2,196,160	810,938.4	0.37	
15		1,944,590	1,167,188.0	0.60	
16	−40	2,000,980	941,746.6	0.47	0.55
17		1,754,530	1,049,546.0	0.60	
18		1,658,340	970,915.8	0.59	

(a) (b) (c) (d) (e)

Figure 6. Failure modes of red sandstone at different negative temperatures. (**a**) −5 °C, **b**) −10 °C, (**c**) −20 °C, **d**) −30 °C, (**e**) −40 °C.

According to Section 3.1, the change from normal temperature to negative temperatures within the range from 25 °C to −30 °C improves the impact resistance of water-saturated frozen red sandstone, and thus the dynamic compressive strength is significantly improved. However, with a decrease in negative temperature (from −5 to −30 °C), red sandstone tends to be brittle. In detail, under the same impact load, the failure strain value becomes smaller, the dissipated energy increases, and the failure degree of rock gradually intensifies. Under temperatures lower than −30 °C, the dissipated energy and damage variables continue to increase, but the dynamic compressive strength of rock specimens decreases sharply, and the impact failure resistance decreases significantly, inducing deterioration of the dynamic mechanical properties at lower negative temperatures.

The values of damage variable of water-saturated red sandstone under dynamic impact do not change continuously in the temperature range from 25 to −5 °C, but their variation in the range from −5 °C to −40 °C has regularity, thus, the values of damage variable and temperature in this negative temperature range can be fitted as the following Equation (4):

$$d = -5.04739\mathrm{e}^{-5}T^2 - 0.00846T + 0.28345 \\ (R^2 = 0.95738),\ T \in [-40, -5] \tag{4}$$

3.3. Macroscopic Failure Pattern of Red Sandstone at Negative Temperatures

As shown in Figure 6, under the same impact pressure loading, the water-saturated red sandstone under different negative temperature conditions shows different failure modes. Firstly, at −5 °C, the red sandstone fractures into several large pieces under the action of stress waves, and the failure surfaces are mostly inclined at 90° angle with obvious tensile traces. From −10 °C to −20 °C, the degree of rock fragmentation intensifies, and a large number of fragments and a small number of conical structures appear in addition to the existence of tensile splitting bodies. From −30 °C to −40 °C, the rock is further broken, and the broken body is mainly a small cone body with strong shear action. In conclusion, the breakage degree of red sandstone at high strain rate gradually intensifies with a decrease in negative temperature, and the failure mode also changes under different temperatures, from the initial tension failure to the shear failure.

According to the analyses of Sections 3.1 and 3.2, it can be inferred that the intervention of initial negative temperature (−5 °C) leads to the transformation of pore water in sandstone into ice. Ice can effectively fill and bond the pores and microfissures in sandstone, and enhance the adhesion between mineral particles. Therefore, the rock integrity is enhanced, the effective bearing area is increased, and the dynamic strength is also improved. At this point, the sandstone is more prone to tensile failure along the loading direction under impact. When the negative temperature further decreases (from −10 to −30 °C), the water-saturated frozen red sandstone shows a shrinking trend as a whole, and the interlocking between mineral particles and solid ice is closer. The dynamic mechanical strength of rock is greatly improved. In this condition, negative temperature is equivalent to applying precompression stress to the rock specimen, thus, the rock specimen has a certain ability to resist radial expansion deformation. However, it is still sensitive to radial shrinkage deformation (and becomes more sensitive with a decrease in negative temperature), and therefore the number of fragments under the action of reverse tensile stress wave continuously increases. There are large numbers of microcracks and weak structures produced in the rock, and these weak structures converge into a large number of shear cracks at the mesoscopic level, causing the rock to show shear failure and disintegration and resulting in a large number of conical fracture fragments [15–18]. Table 3 shows the macroscopic failure characteristics of red sandstone at different temperatures.

Table 3. Macroscopic failure characteristics of red sandstone at different temperatures.

T	−5 °C	−10~−30 °C	−40 °C
Characteristics of macroscopic failure	Damage mode is given priority with the tension damage. The shape of broken body is mainly the columnar splitting structure.	The failure mode changes from tensile failure to shear failure. The number of columnar splitting structures decrease, while the number of lamellar spallation structures increase.	Shear failure is the main failure mode, and the shape of the broken body is mainly the lamellar and conical structure.
Change law of dynamic strength	Gradually Increasing →		Dramatically reducing →
Cause analysis	The water-ice phase transformation occurs, the bond between mineral particles increases, the rock integrity is enhanced, and the dynamic strength is improved. Sandstone with enhanced brittleness and integrity is more prone to tensile failure along the loading direction under impact load.	With a decrease in negative temperature, the red sandstone shrinks as a whole, and the interlacing between mineral particles and solid ice becomes closer, and the dynamic mechanical strength increases significantly. The resistance of rock to radial contraction is further reduced, and the rock fracture is lamellar structure under the action of reverse tensile wave.	The shrinkage rate and amplitude of each component material in sandstone are quite different when it is cooled. A large number of cracks are formed at the contact of the components, and the plastic deformation ability at the crack tip is poor. Low stress brittle failure occurs easily under negative temperature and high strain rate loading, and the strength drops sharply.

4. Analysis of Mesoscopic Fracture Morphology

The fracture morphology of water-saturated red sandstone at different temperatures is observed in order to explore the deterioration mechanism of dynamic mechanical properties of rock at lower negative temperatures.

Red sandstone is a multiphase body composed of mineral particles and cementation, and there are a lot of medium voids inside. Under the action of negative temperature, the rock shrinks, some voids close, and the connection and occlusion between the components become closer. As compared with a normal temperature fracture, the roughness of the microscopic fracture is lower, the mineral particles can be seen but the three-dimensional sense is lower, and the integration with the surrounding cements and other matrix components is significant (see Figure 7a–c, ×150 represents a magnification of 150 times). The rock is more compact and has higher strength as a whole. However, at lower negative temperatures, due to the great difference in shrinkage rate and amplitude among mineral particles, ice media, cementation components, and other media, the separation phenomenon between components is very apparent, and the three-dimensional sense of fracture morphology is enhanced. The mineral particles are prominent, and the cementation and matrix components are broken in disorder (see Figure 7d–f). At this time, a large number of pores, voids, and microcracks are generated at the interfaces of the media (see Figure 8). Under dynamic load, these defective structures and microcracks are easy to develop rapidly and even expand unstably, resulting in component separation, particle stripping, and other fracture phenomena under shear action.

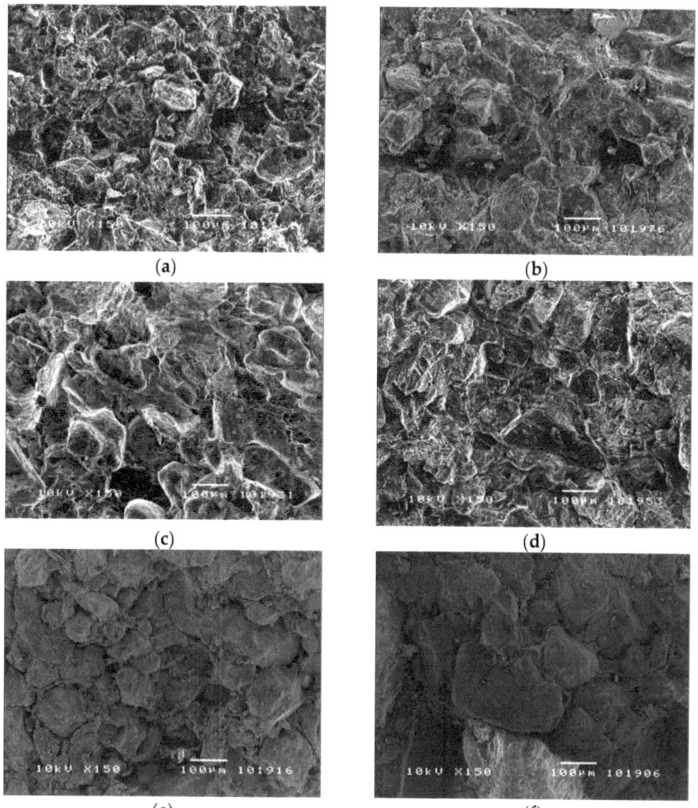

Figure 7. Fracture morphology of red sandstone at different temperatures. (**a**) 25 °C red sandstone, ×150, (**b**) −5 °C red sandstone, ×150, (**c**) −10 °C red sandstone, ×150, (**d**) −20 °C red sandstone, ×150, (**e**) −30 °C red sandstone, ×150, (**f**) −40 °C red sandstone, ×150.

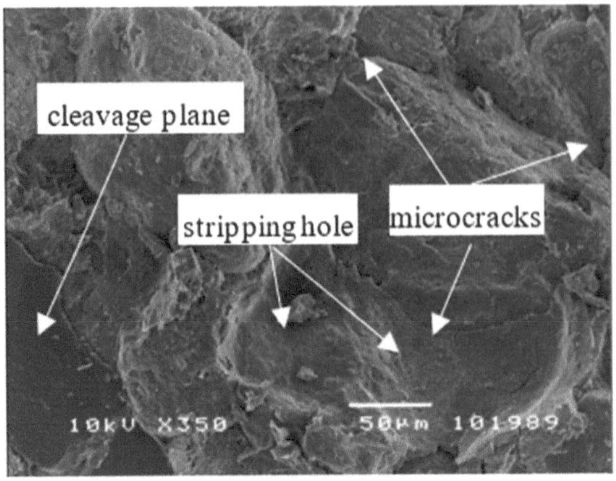

(**a**)

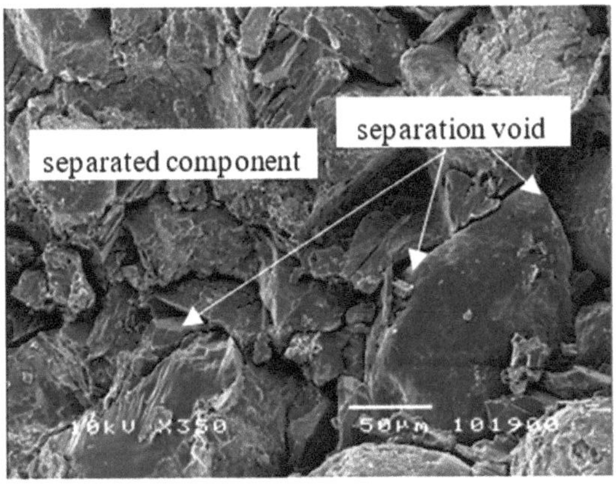

(**b**)

Figure 8. Local microscopic fracture morphology. (**a**) −40 °C red sandstone, ×350, (**b**) −30 °C red sandstone, ×350.

According to the observation of fracture morphology in Figure 8 (×350 represents a magnification of 350 times), it can be found that the fracture modes of saturated frozen red sandstone are mainly cement fracture and brittle fracture. Among them, due to the complex mineral composition of the cementitious materials, they were more susceptible to low temperature. Therefore, under the double action of dynamic load and low temperatures, it was found that the damage occurred at the cementitious materials first, and then caused the fracture of the red sandstone as a whole.

5. Discussion

5.1. Influence of Water-Ice Phase Transformation on Dynamic Strength of Red Sandstone

According to the comparison of Figures 3 and 9, it is found that as the temperature drops to negative values, the dynamic compressive strength of the two red sandstones in the comparison group increases under cold contraction, but there is a large difference between them. The dynamic compressive strength of the dry red sandstone increases by 14%, while that of the water-saturated red sandstone only increases by 4%, with a difference of 10%. Therefore, it can be inferred that water-ice phase transformation has a certain degree of deterioration effect in the strength of saturated rock under high strain rate conditions.

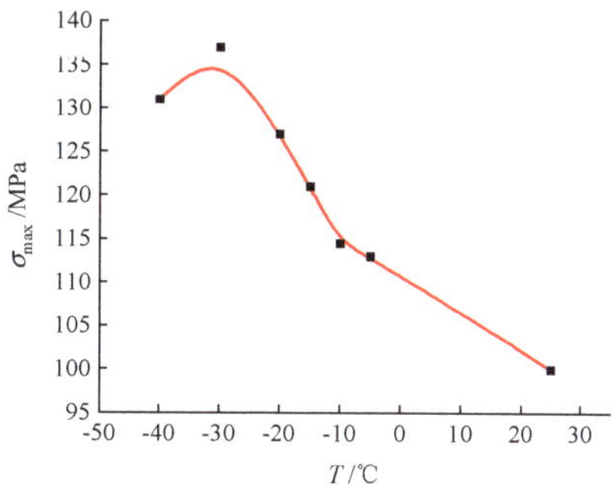

Figure 9. Dynamic compressive strength of dry red sandstone at different temperatures.

The phenomenon, in which pore water transfers from liquid to solid, increases the water volume by 9%, which can produce ice expansion force within the rock. According to related studies [1,2], under the condition of no drainage and consolidation, the ice expansion force in water-saturated rock is up to 200 MPa. These ice expansion forces can produce microcracks in the rock, and can also expand the original cracks. Referring to the above research conclusions and combined with the present test, it can be determined that the ice expansion force in water-saturated red sandstone leads to a large number of microcracks in the rock. These microcracks can reduce the dynamic mechanical strength of water-saturated red sandstone as compared with dry red sandstone. The deterioration of strength can be obviously detected only by comparison, because the macroscopic strength of the overall contraction and compaction of the two rocks increases under the condition of negative temperature.

5.2. Fracture Mode of Frozen Red Sandstone

In the field of mesofracture, the order of energy dissipation of fracture modes under the same conditions is as follows: cement fracture $\leq$ intergranular fracture $\leq$ transgranular fracture $\leq$ quasi cleavage fracture [22,23]. According to the analysis of Sections 3 and 4, it is found that dynamic strength and dissipated energy correspond to the fragmentation and fracture mode, as shown in Table 4.

Table 4. Fracture model vs. dissipated energy.

T °C	Fracture Mode	Dynamic Strength MPa	Dissipated Energy W_L J
25	Cement fracture, intergranular fracture	98.51	147.63
−5	Cement fracture	102.47	109.10
−10	Cement fracture	113.16	131.63
−20	Cement fracture	121.01	145.05
−30	Cement fracture, intergranular fracture	122.12	158.78
−40	Cement fracture, transgranular fracture	86.23	169.13

As shown in Table 4, the fracture mode of red sandstone at various temperatures under the same impact load has significant differences, which further induces differences in the failure mode and fragmentation of rock, and the dissipated energy required for rock fragmentation, thus, affecting the macroscopic mechanical strength of rock.

According to the change of fracture mode and fracture morphology, it can be inferred that a low negative temperature (in this series of tests, it means the temperature after dropping to −30 °C) will cause "frostbite" in red sandstone under high strain rate, that is, the dynamic compressive strength and bearing capacity of red sandstone will decrease, which is called "frostbite effect" under high strain rate. Many researchers [3–8] have

conducted numerous negative temperature loading tests on rocks under static or quasi-static conditions. However, the results have shown that even when the temperature dropped to −160 °C, there was no "frostbite" phenomenon of strength decline that occurred, which indicated that the occurrence of frostbite was related to the loading method, and the frostbite effect occurred only in the combination of low negative temperature and high strain rate loading.

To some extent, the water-ice phase transformation will deteriorate the dynamic strength of water-saturated red sandstone, which does not occur in the static load test, indicating that the rock bearing capacity under high strain rate is more sensitive to microcracks, microvoids, and other defect structures. However, in static or quasi-static cases, due to a long loading time and low strain rate, the rock has a relatively long compaction stage (or crack closure stage), and it is not so sensitive to the emergence of microdefect structure, which is reflected in the macroscopic view that its strength does not decrease significantly. This insensitivity to microdefect structure becomes more prominent when the temperature is lower than −30 °C. Under high strain rate loading, the dynamic compressive strength of red sandstone decreases sharply after −30 °C, while its static strength continues to increase, and even increases faster at lower negative temperatures [3,4]. In the present experiment, it is observed that the dynamic mechanical strength of red sandstone decreases after −30 °C, because the overall property of red sandstone tends to be brittle after this temperature, and different materials (such as mineral particles, cemented materials, and solid ice) in the rock have great differences in shrinkage rate and shrinkage amplitude when they are cooled. Therefore, a large number of secondary defects such as microvores and microcracks are produced at the contact interface of the components. These secondary defects have poor plastic deformation ability under the action of negative temperature. Under the loading of high strain rate, low stress brittle failure often occurs directly without any deformation, thus, leading to a sharp decline in the strength and bearing capacity of red sandstone. Under static or quasi-static loading, negative temperature will also generate secondary defects in rocks [13–15]. However, because under negative temperature rocks have enough time for defects to close during a longer loading time, even if there is local microscopic fracture, a lower negative temperature can make the rock shrink and tightly restrain its development. Therefore, it is impossible for high strain rate instantaneous loading to occur (under high strain rate loading in low temperature brittle media, cracks are usually quite unstable and difficult to suppress).

The strength deterioration of red sandstone after −30 °C is because different materials (such as mineral particles, cemented materials, and solid ice) in the rock have great differences in shrinkage rate and amplitude when they are cooled, and a large number of cracks are generated at the interface between the constituent materials. As compared with mineral particles, cement is more susceptible to negative temperatures due to its complex composition. Therefore, under the coupling action of dynamic load and negative temperatures, cement is usually destroyed first, which leads to the overall fracture of red sandstone.

6. Conclusions

In this study, a dynamic impact experiment on low-temperature frozen red sandstone was carried out to study the temperature effect on the dynamic mechanical properties of red sandstone under high strain rates. On the basis of existing damage and energy theories, the effects of different low temperatures on the strength, damage variables, and energy dissipation of red sandstone were analyzed. Combined with the fracture morphology analysis, the deterioration mechanism of the dynamic mechanical strength of red sandstone at lower negative temperatures was deduced.

(1) A change in negative temperature significantly affects the dynamic mechanical properties of red sandstone, which is equivalent to applying precompression stress to red sandstone. Although the rock tends to be brittle as a whole, its ability to resist tensile

stress is enhanced. Therefore, within the range from 25 °C to −30 °C, its dynamic compressive strength increases gradually with a decrease in temperature.

(2) Lower negative temperatures lead to "frostbite" and deterioration of dynamic mechanical properties of red sandstone under high strain rate loading. The test results show that the dynamic compressive strength of red sandstone decreases sharply after −30 °C, which is different from the results obtained in static or quasi-static tests at −30 °C.

(3) At high strain rates, the fracture degree of red sandstone gradually intensifies with a decrease in negative temperature (from −5 to −40 °C), and the failure mode of the stone gradually transitions from the initial negative temperature tension failure to the slip shear failure at a lower negative temperature, which is consistent with the transformation of the internal structural fracture mode at the mesoscopic level.

Author Contributions: Data curation, Y.Y., N.Z. and J.W.; Validation, N.Z.; Writing—original draft, Y.Y., N.Z. and J.W.; Writing—review and editing, Y.Y. and J.W. All authors have read and agreed to the published version of the manuscript.

Funding: This research was funded by the Key project of the National Natural Science Foundation of China, grant number 51934001.

Data Availability Statement: The experimental data used to support the findings of this study are included within the article.

Acknowledgments: The authors thank the China University of Mining and Technology (Beijing), for providing instruments to conduct the research.

Conflicts of Interest: The authors declare that they have no conflict of interest. The funders had no role in the design of the study; in the collection, analyses, or interpretation of data; in the writing of the manuscript, and in the decision to publish the results.

References

1. Liu, Q.S.; Huang, S.B.; Kang, Y.S.; Cui, X. Advance and review on freezing-thawing damage of fractured rock. *Chin. J. Rock Mech. Eng.* **2015**, *34*, 452–471.
2. Chen, W.Z.; Tan, X.J.; Yu, H.D.; Yuan, K.; Li, S. Advance and review on thermo-hydro-mechanical characteristics of rock mass under condition of low temperature and freeze-thaw cycles. *Chin. J. Rock Mech. Eng.* **2011**, *30*, 1318–1336.
3. Aoki, K.; Hibiya, K.; Yoshida, T. Storage of refrigerated liquefied gases in rock caverns: Characteristics of rock under very low temperatures. *Tunnell. Undergr. Space Technol.* **1990**, *5*, 319–325. [CrossRef]
4. Yamabe, T.; Neaupane, K.M. Determination of some thermo-mechanical properties of Sirahama sandstone under subzero temperature condition. *Int. J. Rock Mech. Min. Sci.* **2001**, *38*, 1029–1034. [CrossRef]
5. Dwivedi, R.D.; Soni, A.K.; Goel, R.K.; Dube, A.K. Fracture toughness of rocks under sub-zero temperature conditions. *Int. J. Rock Mech. Min. Sci.* **2000**, *37*, 1267–1275. [CrossRef]
6. Tang, M.M.; Wang, Z.Y.; Sun, Y.L.; Ba, J. Experimental study of mechanical properties of granite under low temperature. *Chin. J. Rock Mech. Eng.* **2010**, *29*, 787–794.
7. Xi, J.M.; Yang, G.S.; Pang, L.; Lv, X.-T.; Liu, F.-L. Experimental study on basic mechanical behaviors of sandy mudstone under low freezing temperature. *J. China Coal Soc.* **2014**, *39*, 1262–1268.
8. Xu, G.M.; Liu, Q.S.; Peng, W.W.; Chang, X.X. Experimental study on basic mechanical behaviors of rocks under low temperatures. *Chin. J. Rock Mech. Eng.* **2006**, *25*, 2502–2508.
9. Yang, G.S.; Lv, X.T. Experimental study on the sandy mudstone mechanical properties of shaft sidewalls under the frozen conditions. *J. Min. Saf. Eng.* **2012**, *29*, 492–496.
10. Yang, G.S.; Xi, J.M.; Li, H.J.; Cheng, L.; Lv, X. Experimental study on the mechanical properties of soft rock of coal mine shaft sidewalls under the frozen conditions. *Chin. J. Undergr. Space Eng.* **2012**, *8*, 690–697.
11. Lei, W.; Zhi, J.W.; Abbas, T.; Quansheng, L.; Huai, L. Deterioration of dynamic mechanical properties of granite due to freeze-thaw weathering: Considering the effects of moisture conditions. *Cold Reg. Sci. Technol.* **2020**, *176*, 103092.
12. Ke, B.; Zhou, K.; Xu, C.; Hongwei, D.; Li, J.; Bin, F. Dynamic mechanical property deterioration model of sandstone caused by freeze-thaw weathering. *Rock Mech. Rock Eng.* **2018**, *51*, 2791–2804. [CrossRef]
13. Shan, R.L.; Yang, H.; Zhang, J.X.; Guo, Z.; Dinghong, Z. Mechanical properties of saturated red sandstone of Meilinmiao mine under loading and unloading at negative temperatures. *J. Min. Saf. Eng.* **2016**, *33*, 924–931.
14. Shan, R.L.; Yang, H.; Guo, Z.M.; Liu, X.; Song, L. Experimental study of strength characters of saturated red sandstone on negative temperature under triaxial compression. *Chin. J. Rock Mech. Eng.* **2014**, *33*, 3657–3664.

15. Shan, R.L.; Zhang, L.; Yang, H.; Zhang, j.X.; Guo, Z.M. Experimental study of freeze-thaw properties of saturated red sandstone. *J. China Univ. Min. Technol.* **2016**, *45*, 923–929.
16. Li, H.J. Experimental Study on Rock Mechanical Properties under Freezing Conditions. Ph.D. Thesis, Xi'an University of Science and Technology, Xi'an, China, 2009.
17. Cheng, L. Experimental Study on Rock Mechanical Properties under Freezing Conditions and Application in Project. Ph.D. Thesis, Xi'an University of Science and Technology, Xi'an, China, 2009.
18. Xi, J.M.; Yang, G.S.; Dong, X.H. Effect of freezing temperature on mechanical properties of sandy mudstone. *J. Chang'an Univ. (Nat. Sci. Ed.)* **2014**, *34*, 92–97.
19. Li, Y.P.; Wang, Z.Y. Uniaxial compressive mechanical properties of rock at low temperature. *J. Univ. Sci. Technol. Beijing* **2011**, *33*, 671–675.
20. Xu, J.Y.; Fan, J.S.; Lü X, C. *Dynamic Mechanical Properties of Rock with the Confining Pressure*; Northwestern Polytechnical University Press: Xi'an, China, 2012.
21. Xie, H.P.; Gao, F.; Zhou, H.W.; Zuo, J.P. Fractal fracture and fragmentation in rocks. *J. Disasaster Prev. Mitig. Eng.* **2003**, *23*, 1–9.
22. Qian, Y. Experimental Study on Blasting Fragment-Size of Joints Rock Mass based on Fractal Theory. Ph.D. Thesis, Wuhan University of Technology, Wuhan, China, 2005.
23. Wen, M.; Xu, J.Y.; Wang, H.Y.; Fang, X.Y.; Zheng, G.H. Fractography analysis of sandstone failure under low temperature-dynamic loading coupling effects. *Chin. J. Rock Mech. Eng.* **2017**, *36*, 3822–3830.

minerals

MDPI

Article

Physical Simulation Test on Surrounding Rock Deformation of Roof Rockburst in Continuous Tunneling Roadway

Yaobin Shi [1,2,3], Yicheng Ye [1,2,3], Nanyan Hu [1,3,*], Yu Jiao [1] and Xianhua Wang [4]

1 School of Resource and Environmental Engineering, Wuhan University of Science and Technology, Wuhan 430081, China; shiyaobin@wust.edu.cn (Y.S.); yeyicheng@wust.edu.cn (Y.Y.); jiaoyu00wust@163.com (Y.J.)
2 State Environmental Protection Key Laboratory of Mineral Metallurgical Resources Utilization and Pollution Control, Wuhan University of Science and Technology, Wuhan 430081, China
3 Hubei Key Laboratory for Efficient Utilization and Agglomeration of Metallurgic Mineral Resources, Wuhan University of Science and Technology, Wuhan 430081, China
4 Sinosteel Wuhan Safety and Environmental Protection Research Institute Co., Ltd., Wuhan 430081, China; ssriwangxh@163.com
* Correspondence: hunanyan@wust.edu.cn

Abstract: To study the occurrence process, as well as the temporal and spatial evolution laws, of rockburst disasters, the roof deformation of continuous heading roadways during rockburst was studied through a physical similarity simulation test with a high similarity ratio and low strength. The deformation and failure evolution law of the roadway roof in the process of rockburst were analyzed by using detection systems, including a strain acquisition system and a high-power digital micro-imaging system. The results show that the rockburst of the roadway roof can be divided into four stages: equilibrium, debris ejection, stable failure, and complete failure stage. According to the stress state of a I–II composite crack, the theoretical buckling failure strength of the surrounding rock is determined as 1.43 times the tensile strength. The flexural failure strength of a vanadium-bearing shale is 1.29–1.76 times its compressive strength. With continuous advancement in the mining time, the internal expansion energy of the roadway roof-surrounding rock in the equilibrium stage continuously accumulates. The fracture network continuously increases, developing to the stable failure stage, with bending deformation, accompanied by continuous particle ejection until the cumulative stress in the failure stage increases, and the tensile state of the rock surrounding the roof expands radially into deep rock. A microscopic damage study in similar material demonstrated that the deformation of the roadway roof is non-uniform and uncoordinated. In the four stages, the storage deformation of the rock surrounding the roadway roof changes from small accumulation to continuous deformation, to the left (or deep rock). Finally, the roadway roof-surrounding rock becomes completely tensioned. The research results presented in this study provide a reference for the prediction and control of rockburst in practical engineering.

Keywords: rockburst; roadway; roof-surrounding rock; physical similarity simulation test; digital image processing (DIP) technology

Citation: Shi, Y.; Ye, Y.; Hu, N.; Jiao, Y.; Wang, X. Physical Simulation Test on Surrounding Rock Deformation of Roof Rockburst in Continuous Tunneling Roadway. *Minerals* **2021**, *11*, 1335. https://doi.org/10.3390/min11121335

Academic Editors: Diyuan Li, Zhenyu Han, Xin Cai and Shijie Xie

Received: 24 October 2021
Accepted: 24 November 2021
Published: 29 November 2021

1. Introduction

In underground mining, roadway excavation causes the rock mass of the surrounding rock surface to change from a three-dimensional stress state to an approximately one- or two-dimensional stress state, resulting in stress readjustment and local concentration [1–3]. When the elastic strain energy stored in the rock mass exceeds the rock mass strength, the excess strain energy will be released violently, resulting in rockburst accidents, which pose serious threats to the safety of personnel [4–6].

In recent years, the frequency of rockbursts in mining underground and space engineering has increased significantly. The factors influencing rockburst are complex, mainly

including the excavation mode, surrounding rock stress state, rock mass structure, and its performance [7–11]. Scholars have deeply studied the incubation process, occurrence mechanism, and prevention technology of rockburst and have discussed the rockburst characteristics of rock masses in different environments [12,13]. In the study of rockburst mechanism, strength theory, stiffness theory, impact tendency theory, energy theory, instability theory, and fracture damage theory have been proposed [14]. The classification research has shown that there are more than five types of rockburst [15], and rock roofs with different thicknesses have distinct failure patterns under the complex stress [16]. Rockbursts can be divided into horizontal stress type, vertical stress type and mixed stress type, as well as burst ejection type, flake-spalling type and tunnel wall collapse type. Research has shown that rockbursts mostly occur in hard rock masses under high-stress environments, and that the severity is closely related to the stress concentration of the surrounding rock, which affects the ratio of rock-buckling failure strength ($\sigma_{\theta max}$) and rock uniaxial compressive strength, one of the important criteria for rockburst [17–19].

At present, research into rockburst has mainly been conducted through indoor rock mechanics testing, numerical analysis, field monitoring, and physical simulation testing. Physical similarity simulation testing is a main and important laboratory experimental method to study the development, expansion, rockburst, and final failure of rock fractures [20,21]. Jiang Lishuai et al. [22] have established two physical models utilizing similar materials and studied the influence of the advancing direction on the mining effect caused by a fault. Rockbursts affected by faults with different mining directions were compared and analyzed, and the correlation among advancing direction, strata behaviors, and rockburst induction was studied. Zhou Nan et al. [23] the used experimentally measured strain–stress curves of the backfill body and similarity theory in order to design and employ four experimental models for physical simulation. A non-contact strain measurement system and pressure sensors were used to monitor the deformation of the overlying strata and changes in abutment stress in front of the face during mining of the models for varying roof-controlled backfilling ratios. Chen Bing et al. [24] developed rock-like materials for simulating faults and surrounding rocks and carried out simulation experiments on the activation process of different types of analogical faults. They found that the failure process of rock-like samples with analogical concealed faults and analogical conduction faults could be divided into three stages, including the crack initiation, crack generation, and propagation stages. Li Shugang et al. [25] have carried out uniaxial compression experiments with different loading rates using a DYD-10 electronic universal testing machine for physical similarity simulation testing in order to study the influence of fracture evolution characteristics of rock-like materials under different strain rates. Their results show that, with an increase in the strain rate, the peak strength of the rock-like specimen increased, and the main fracture gradually developed in the direction parallel to the maximum principal stress, which led to the failure form of the specimen evolving from shear failure to tensile failure. Ma Dan et al. [26] studied non-Darcy hydraulic properties and the deformation behaviors of granular gangues through laboratorial, theoretical, and in situ aspects. They conducted a series of compression and seepage tests on granular gangues under the variables original grain size grade (GSG) and stress rate. They found that the fractal dimension reveals more obvious increases in the samples with the higher original GSGs and lower stress rates. Additionally, an improved model was established to predict the permeability evolution by the fractal dimension in their research.

Therefore, using physical simulation models bearing similarity to the field geometry and environment to simulate the rockburst process can more accurately reflect the dynamic rockburst development process. However, the occurrence process and temporal and spatial evolution law of rockbursts in high-stress roadways need to be further studied, as they have important reference significance for eliminating rockburst disasters and improving rockburst prediction and control.

In this paper, based on a rockburst-like material test featuring low strength and high brittleness in a gently inclined multi-layer deposit, a large-scale physically similar simula-

tion test model was constructed. The strip-mining method was adopted, and the strain characteristics and laws of roadway-surrounding rock under different mining intensity were analyzed using XL2101g multi-point high-speed whole process controlled static strain acquisition system and digital image processing (DIP) technology. The occurrence process and evolution law of rockburst in the rock surrounding the roadway—comprising a gently inclined multi-layer deposit—was studied, and the buckling failure strength and uniaxial compressive strength of the rock mass were determined, thus providing a scientific basis for revealing the temporal and spatial evolution law of rockburst.

2. Materials and Methods

2.1. Engineering Background

We studied the mining project of the Shanghengshan multi-layer shale deposit, which has 12 ore bodies in the ore section. The ore bodies are produced in layers, with good continuity and a simple shape. The ore-bearing rocks are mainly carbonaceous shale, siliceous shale, and vanadium-bearing shale, followed by siliceous rock. The ore body has a dip of 150–220° and an inclination of 5–25°. The ore length is 615–952 m, the thickness is 0.75–7.27 m, the thickness variation coefficient is 37.07–64.59%, and the inclined extension depth is 103–223 m.

The brittle rock shale with obvious rockburst tendency was chosen as the test material to establish the roadway rockburst model [27]. The physical and mechanical parameters of shale materials were obtained by uniaxial compression and Brazilian splitting tests, as detailed in Table 1.

Table 1. Physical and mechanical parameters of rock mass in Shanghengshan Mine.

Rock Formation	Density /(kg/m^3)	Porosity /%	Compressive Strength /MPa	Tensile Strength /MPa	Elastic Modulus /GPa	Poisson's Ratio
Vanadium-bearing shale	2 482.53	14.33	76.69	12.56	58.80	0.21
Carbonaceous shale	2 429.86	18.61	49.45	7.42	50.80	0.20
Siliceous shale	2 564.72	16.07	112.63	17.24	59.80	0.21

2.2. Sample Preparation of Similar Materials

Considering the balanced equation, geometric equation, physical equation, stress boundary conditions, and displacement boundary conditions of the prototype and model, rock-like materials and rock materials followed the similarity criteria. In the design of the physical similarity stimulation test [28], we determined the geometric physical quantity similarity ratio as $C_l = 100$ and mechanical similarity ratios as $C_\rho = 1.5$ and $C_\sigma = 150$. The dimensionless parameters, such as Poisson's ratio, the porosity, and internal friction angle of similar materials were selected with the similarity constant of 1. Meanwhile, the geometric similarity constant and mechanical similarity constant met the following requirements.

$$C_\sigma/(C_\rho C_l) = 1 \tag{1}$$

Cylindrical standard samples with a diameter of 50 mm and a height of 100 mm were selected for samples of similar materials. Portland cement and gypsum were selected as the cementing materials, ordinary river sand as the filling material, and glycerol as the cementing agent. The proportion test of similar materials had been previously given in the literature [28]. The proportions and physical and mechanical parameters of similar materials in each mine and rock stratum of the multi-layer vanadium-bearing shale deposit were determined, as shown in Table 2, and the actual similarity constants of the similar materials are shown in Table 3.

2.3. Test Equipment and Methods

A self-developed adjustable similar simulation experimental device system [29] was adopted for the test, and the layered pressing method is used for model construction.

We determined the amount of similar materials for different rock layers of the physical similarity model according to the similarity ratio. We filled and tamped, layer by layer, from bottom to top, and used mica powder between the different rock layers (thickness 2–3 mm) to build a 4500 mm × 2200 mm × 400 mm (length × height × width) physical similarity simulation model, as shown in Figure 1.

Table 2. Similar material selection and mechanical parameters.

Rock Formation	ρ /(kg/m^3)	σ_c /MPa	σ_t /kPa	n /%	E /GPa	ν	c /kPa	ϕ /(°)
Siliceous shale	1 704.41	0.75	114.87	16.21	0.399 1	0.213 3	71.218	41.087
Vanadium-bearing shale	1 661.74	0.51	84.01	14.46	0.394 4	0.210 0	61.891	40.275
Carbonaceous shale	1 620.98	0.33	49.42	18.56	0.336 8	0.203 3	56.748	40.106

Table 3. Similar constants of similar materials.

Rock Formation	Density	Compressive Strength	Tensile Strength	Elastic Modulus	Poisson's Ratio	Cohesion Internal	Friction Angle	$C_{\sigma l}(C_\rho \cdot C_l)$
Siliceous shale	1.505	150.173	150.082 7	149.837	0.985	150.243	0.998	1.002 0≈1
Vanadium-bearing shale	1.494	150.373	149.506 0	149.087	1.000	150.264	1.003	1.000 3≈1
Carbonaceous shale	1.499	149.848	150.141 6	150.831	0.984	149.785	0.997	0.993 5≈1

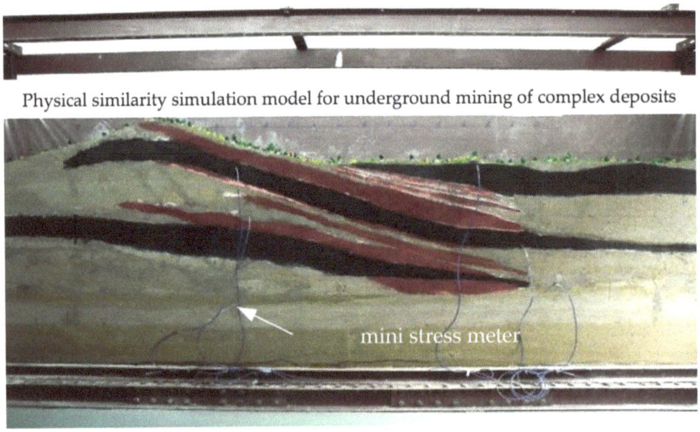

Figure 1. Physical similarity simulation test model of gently inclined multi-layer deposit.

In the physical similarity simulation test, only the influence of rock self-weight stress was considered, and the influence of stress in other directions was not considered, such as in the radial direction.

In the physical similarity simulation test model, the XL2101G multi-point high-speed whole process controlled static strain acquisition system was used to monitor the vertical strain of the surrounding rock. The layout of sampling points for vertical strain is shown in Figure 2. In addition, a mini stress meter was used to detect the rock stratum pressure. The sample and instrument were consistent with the ambient temperature before the test, and we ensured that the laboratory was closed during the test. The mining of a certain sub rock stratum was carried out, and the layout of the ore room adopted a "mining one by one" approach in the simulated mining test [30]. The design width of pillar and stope was 15 m. To simulate the explosion process of the rock surrounding the roadway during roadway excavation, continuous mining without backfilling was selected for the test; that is, while maintaining the consistency of roadway section shape and size and the spacing distance between two adjacent roadways, the elastic strain energy of the simulated roadway-surrounding rock continues to accumulate and increase.

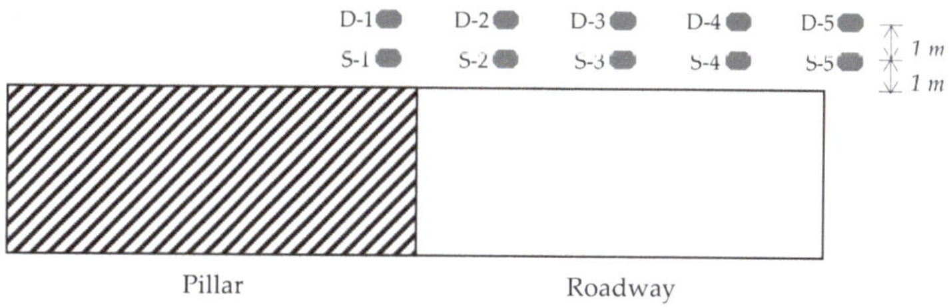

Figure 2. Layout of sampling points of roadway roof.

Roadway excavation causes the stress of the rock mass in the surrounding rock surface to change from a three-dimensional stress state to an approximately one- or two-dimensional stress state, resulting in stress readjustment and local concentration. When the elastic strain energy stored in the rock mass exceeds the rock mass strength, the excess strain energy is released violently, resulting in a rockburst accident [31–33]. To simulate the rockburst failure phenomenon of a roadway excavation, the 300 times digital micro-imaging system was used in the test. The roadway roof cracks were captured by the digital micro camera, the images of cracks were numerically processed using Origin software (9.4.2.380), and the roadway roof deformation and failure process (DIP) was analyzed at the microscopic level.

Taking the initial step excavation roadway (roadway 1#) as the sample, the evolution process of roadway 1# roof rockburst in continuous excavation was analyzed. The schematic diagrams of the test prototype and the mining are shown in Figure 3 [28]. The actual situation of Sections 1 and 2 of the mining stope is shown in Figure 4.

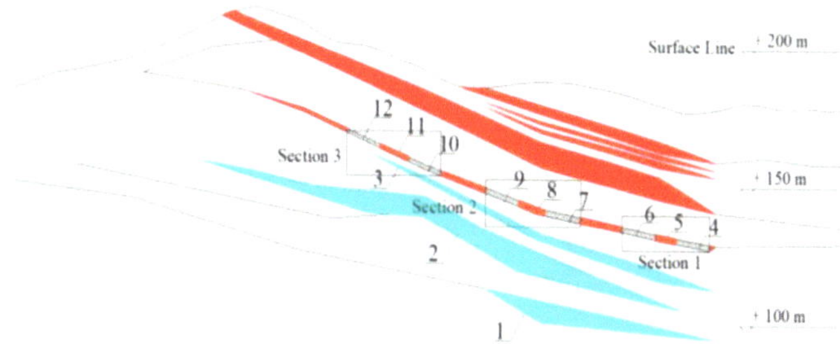

Figure 3. Schematic diagram of physical simulation test. Legend:1—Ore deposit V1; 2—Ore deposit V2 and V3; 3—Ore deposit V1; 4—Roadway1#; 5—Pillar 1#; 6—Roadway2#; 7—Roadway3#; 8—Pillar 2#; 9—Roadway4#; 10—Roadway5#; 11—Pillar 3#; 12—Roadway6#.

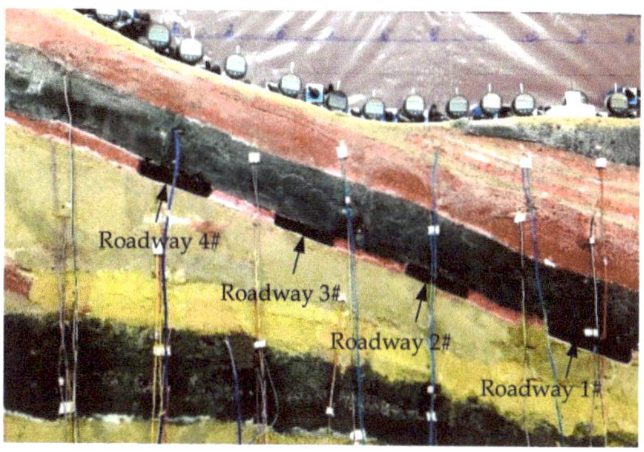

Figure 4. Status of similar simulation in mining Sections 1 and 2.

3. Stress Analysis of Roadway-Surrounding Rock Affected by Adjacent Face Mining

The main reason for the brittle fracture and even instability failure of rock is crack propagation and instability propagation throughout the rock [34,35]. The failure and fracturing of rock are essentially the same process; that is, the crack propagation in the surrounding rock leads to the final failure of rock, and the crack often shows the characteristics of mode I, mode II, or a composite mode [36,37].

3.1. Mechanical Model

Based on the theory of elastic–plastic mechanics [38,39], considering that the mining strip thickness of the deposit is smaller than the inclined length (W_s) of the working face and the strip pillar (W_p) interval, the working face can be calculated as an inclined crack, and the calculation and analysis model are shown in Figure 5.

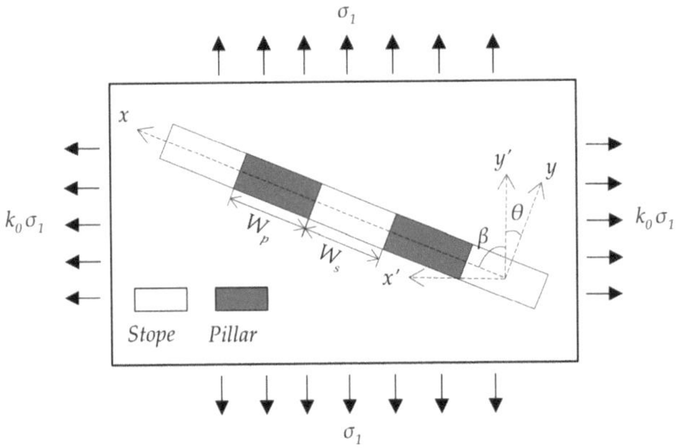

Figure 5. Plane stress mechanical model of roadway and pillar.

It was assumed that there is a penetrating oblique crack (strip width) with a length of W_s in the infinite plate, and the edge is subjected to a uniformly distributed biaxial tensile pressure σ_1 and $k_0\sigma_1$ (where k_0 is the lateral pressure coefficient), where the vertical stress is $\sigma_1 = \gamma H$(γ is the unit weight of ore and rock; H is the buried depth). Considering the working face direction and σ_1 the included angle of the action direction is β, and the edge is not under force.

3.2. Stress Function of Mode I and Mode II Crack Tip Affected by Mining in Adjacent Working Face

Cracks in the state of open mode I and sliding mode II were considered. For the two-dimensional elastic problems of mode I and mode II cracks, the Westergaard complex stress functions [40–42] for the tips of mode I and II cracks affected by the mining of adjacent working faces are given as Equations (2) and (3), respectively.

$$\left.\begin{aligned} \sigma_x &= \mathrm{Re}Z_{\mathrm{I}}(z) - y\mathrm{Im}Z_{\mathrm{I}}'(z) \\ \sigma_y &= \mathrm{Re}Z_{\mathrm{I}}(z) + y\mathrm{Im}Z_{\mathrm{I}}'(z) \\ \tau_{xy} &= -y\mathrm{Re}Z_{\mathrm{I}}'(z) \end{aligned}\right\} \tag{2}$$

$$\left.\begin{aligned} \sigma_x &= 2\mathrm{Im}Z_{\Pi}(z) + y\mathrm{Re}Z_{\Pi}'(z) \\ \sigma_y &= -y\mathrm{Re}Z_{\Pi}'(z) \\ \tau_{xy} &= \mathrm{Re}Z_{\Pi}(z) - y\mathrm{Im}Z_{\Pi}'(z) \end{aligned}\right\} \tag{3}$$

The tensor expression can be expressed by the following formula:

$$\sigma_{ij,\mathrm{I}} = \frac{K_{\mathrm{I}}}{\sqrt{2\pi r}} f_{ij,\mathrm{I}}(\theta) \tag{4}$$

$$\sigma_{ij,\Pi} = \frac{K_{\Pi}}{\sqrt{2\pi r}} f_{ij,\Pi}(\theta) \tag{5}$$

where K_I and K_{II} are the stress intensity factors of mode I and mode II cracks, respectively, θ is the crack angle, and $W_p = 2r$.

For the problem of mode I and mode II cracks in a an infinite plate, the stress intensity factor is defined as follows:

$$K_{\mathrm{I}} = \lim_{|\zeta|=0} Z_{\mathrm{I}}(\zeta) \cdot \sqrt{2\pi\zeta} \tag{6}$$

$$K_{\Pi} = \lim_{|\zeta|=0} Z_{\Pi}(\zeta) \cdot \sqrt{2\pi\zeta} \tag{7}$$

where, ζ is a new coordinate, with polar coordinates used to represent $\zeta = re^{i\theta}$.

When subjected to two-dimensional uniform tensile stress, the analytical function of the complex stress function of the through crack on the x-axis can be written as follows:

$$Z_{\mathrm{I}}(\zeta) = \frac{\sigma(\zeta + a)}{\sqrt{\zeta(\zeta + 2a)}} \tag{8}$$

$$Z_{\Pi}(\zeta) = \frac{\tau(\zeta + a)}{\sqrt{\zeta(\zeta + 2a)}} \tag{9}$$

where τ is the uniform shear stress of the mode II crack problem (the main difference from the mode I crack problem), and $W_s = 2a$.

The stress intensity factor at the tip of an inclined crack on an infinitely thin plate is the product of the horizontal crack stress intensity factor and the correction coefficient, which is $F_I = k_0 \cdot cos^2\beta + sin^2\beta$ (where $F_{II} = (1-k_0)\ cos\beta \cdot sin\beta$). At the same time, considering the mutual influence between adjacent working faces, the stress at multiple inclined crack tips with certain spacing is the product of the stress at a single inclined crack tip and the correction coefficient, as shown in Equation (10) [43]. Therefore, bringing in Equations (6) and (7), the following formula is obtained:

$$\alpha = \left[\frac{W_s + W_p}{\pi \frac{W_s}{2}} tg \frac{\pi \frac{W_s}{2}}{W_s + W_p} \right]^{1/2} \tag{10}$$

$$K_{\mathrm{I}} = \alpha \cdot F_{\mathrm{I}} \cdot \sigma \sqrt{\pi a} \tag{11}$$

$$K_{\Pi} = \alpha \cdot F_{\Pi} \cdot \sigma \cdot \sqrt{\pi a} \tag{12}$$

Bringing Equations (11) and (12) into Equations (4) and (5), respectively, we have:

$$\sigma_{ij,\mathrm{I}} = \frac{1}{\sqrt{2}} \cdot \alpha \cdot F_{\mathrm{I}} \cdot \gamma H \cdot \sqrt{\frac{W_s}{W_p}} f_{ij,\mathrm{I}}(\theta) \tag{13}$$

$$K_{\Pi} = \alpha \cdot F_{\Pi} \cdot \sigma \cdot \sqrt{\pi a} \tag{14}$$

3.3. Stress Function at Crack Tip of I–II Composite Mode

The stress field at the tip of a mode I–II composite crack can be transformed into a superposition of mode I crack and mode II crack solutions; that is, the stress components near the tip of mode I and mode II cracks, as expressed in Equations (13) and (14), respectively, are superimposed. Therefore, the expression of the stress component at the crack tip under the of I–II composite mode is as follows:

$$\sigma_x = \frac{1}{\sqrt{2}} \cdot \gamma H \cdot \sqrt{\frac{W_s}{W_p}} \cdot \alpha \cdot \left[F_{\mathrm{I}} \cdot \cos\frac{\theta}{2} (1 - \sin\frac{\theta}{2} \sin\frac{3\theta}{2}) - F_{\Pi} \cdot \sin\frac{\theta}{2} (2 + \cos\frac{\theta}{2} \cos\frac{3\theta}{2}) \right] \tag{15}$$

$$\sigma_y = \frac{1}{\sqrt{2}} \cdot \gamma H \cdot \sqrt{\frac{W_s}{W_p}} \cdot \alpha \cdot \left[F_{\mathrm{I}} \cdot \cos\frac{\theta}{2}\left(1+\sin\frac{\theta}{2}\sin\frac{3\theta}{2}\right) + F_{\mathrm{II}} \cdot \sin\frac{\theta}{2}\cos\frac{\theta}{2}\cos\frac{3\theta}{2} \right] \tag{16}$$

For the plane strain problem, the maximum principal stress is as follows:

$$\begin{aligned}
\sigma_{1,2} &= \frac{\sigma_x+\sigma_y}{2} \pm \sqrt{\left(\frac{\sigma_x-\sigma_y}{2}\right)^2 + \tau_{xy}^2} \\
&= \frac{1}{2\sqrt{2}} \cdot \gamma H \cdot \sqrt{\frac{W_s}{W_p}} \cdot \alpha \cdot \left[F_{\mathrm{I}} \cdot \cos\frac{\theta}{2} - F_{\mathrm{II}} \cdot \sin\frac{\theta}{2}\cos\frac{\theta}{2}\cos\frac{3\theta}{2} \right] \\
&\pm \left\{ \left[F_{\mathrm{I}} \cdot \cos\frac{\theta}{2}\sin\frac{\theta}{2} - F_{\mathrm{II}} \cdot \sin\frac{\theta}{2} \right]^2 + \left[F_{\mathrm{I}} \cdot \sin\frac{\theta}{2}\cos\frac{\theta}{2}\cos\frac{3\theta}{2} + F_{\mathrm{II}} \cdot \cos\frac{\theta}{2}\left(1-\sin\frac{\theta}{2}\sin\frac{3\theta}{2}\right) \right]^2 \right\}^{1/2}
\end{aligned} \tag{17}$$

For roadway-surrounding rock, we calculated $\sigma_{1,2} = 17.97 \pm 0$ Mpa.

According to the stress condition of a I–II composite mode crack, the crack is mainly affected by tensile and shear stress, and calculations show that $\sigma_{tmax} = 1.43\sigma_t$.

4. Results

4.1. Analysis of Roadway Roof Deformation

As shown in Figure 6, the cumulative stress state of the surrounding rock near the roadway roof was the same as that in the deep part of the roadway roof, while the stress concentration of the surrounding rock near the roadway roof was more obvious than that in the deeper part. The stress state of the surrounding rock on both sides of the roadway roof changed from a compression to a tension state, and the central monitoring point continued to be in a tension state. This indicates that maintenance of the roadway roof needs to focus on the rock ejection in the center of the roof in future mining work. The maximum strain or tensile position of the surrounding rock near the roadway roof is on the left side (cheek) of the working face, and the strain difference of the left-most monitoring point increased with the mining time step, consistent with the actual engineering phenomenon. This reflects that the monitoring point at the shallower part of the roadway roof along the rock stratum trend has greater pushing force than other parts. It can be concluded that the stress concentration of the surrounding rock is closely related to the strike of the rock stratum until rockburst. In other words, rockburst is closely related to the direction of tectonic stress [44,45].

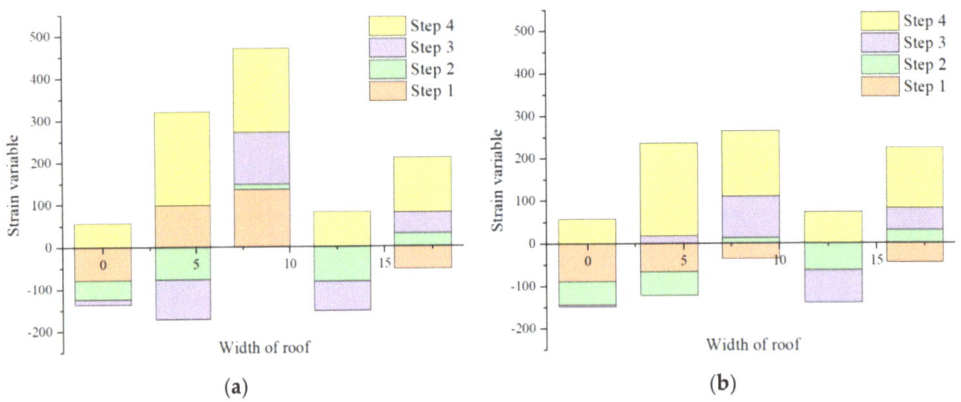

Figure 6. Strain of surrounding rock of roadway roof in different mining time step: (**a**) Monitoring points near roadway roof and (**b**) Monitoring points in the deep part of roadway roof.

In addition to geological factors, mining conditions and construction procedures are also very important factors affecting rockburst. The stress state of deep monitoring points of the roadway roof changed from a compression to a tension state with the advance of mining time. In the process of continuous mining, the roof of the adjacent roadway and the deep monitoring points of the roadway roof were in tension at the fourth stope step. This

indicates that with the continuous increase in the number of adjacent stopes, the dynamic load of the roadway roof and the surrounding rock stress of the first stope adjusted rapidly, which can easily lead to the occurrence of rockburst accidents.

4.2. Deformation and Failure Process of Roadway Roof

The data were graphically processed using Origin software, as shown in Figure 7. The analysis showed that the failure of the roadway roof could be roughly divided into equilibrium, debris ejection, stable failure, and complete failure stages.

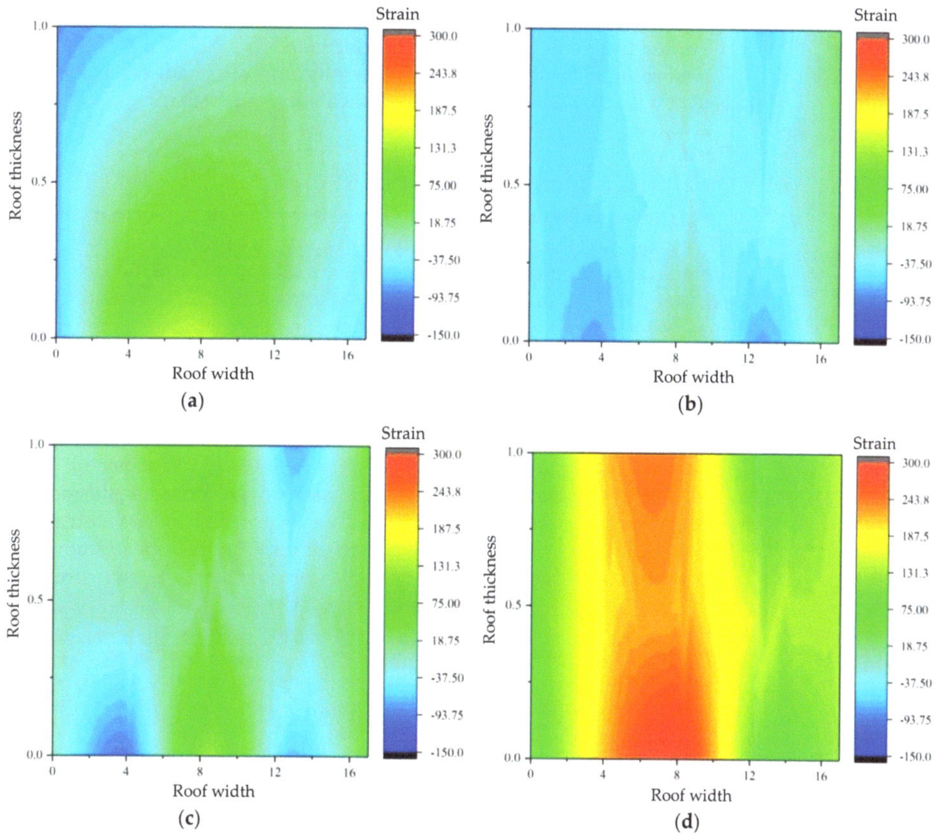

Figure 7. Deformation and failure process of roadway-surrounding rock: (**a**) First mining step; (**b**) Second mining step; (**c**) Third mining step; (**d**) Fourth mining step.

After the first mining step of roadway excavation, the overall deformation of the roof was small. The deformation of the roadway roof was pulled along the roadway centerline and distributed symmetrically. The surrounding rock near the roadway roof was in a certain tensile state, and the roadway roof was in a certain stable state.

With the continuous expansion of mining time step, in the second mining step, the roadway 1# midline roof was stretched, and deformed, the effects of which extended to the deep rock level. Meanwhile, adjacent monitoring points and the roof on both sides of the roadway were compressed to a certain extent. The roadway roof was in a non-uniform and uncoordinated deformation state, debris fell intermittently on the inner wall of the roadway, and roadway roof entered the debris ejection stage.

In the third mining time step, the overall tensile deformation of the roadway 1# continued to increase, during which the deep monitoring point on the left side of the

roadway (1.5 m from the roadway) changed from a compression to tensile state, and the compressive strain of the monitoring point near the roadway (0.5 m from the roadway) accumulated. The debris falling from the roadway roof continued to increase, and the roadway roof entered the debris ejection stage.

In the fourth mining time step, the roadway 1# roof was in a fully tensioned state, and the deformation increased sharply. The maximum tensile strain accumulation area shifted to the left side of the roadway, and the a tension state of the monitoring points near the roadway was obvious, compared with the deep monitoring points. On the model, it can be seen that the rock stratum on the surface of the model had formed fractures, the fracture of the roadway roof had a large expansion degree, and the expansion and connection of different fracture points formed a dense fracture network. Cracking damage had occurred on the left inner wall of the roadway and adjacent pillars.

4.3. Temporal and Spatial Evolution Law of Microscopic Damage and Failure in Roadway Roof Rockburst Process

Although the XL12101G multi-point high-speed whole process controlled static strain acquisition system could observe the macro deformation of the roof of the continuous excavation roadway, the damage in the rock mass was not well reflected using this system. Digital image processing (DIP) technology can accurately characterize the microscopic damage and failure law in rock [46,47]. By analyzing the temporal and spatial evolution law of rock damage fracture extension, expansion and nucleation, the microscopic damage process of the roof slate rockburst in the continuous excavation roadway was studied using DIP technology.

DIP technology studies the temporal and spatial evolution law of microscopic rock damage and failure and shows the mathematical significance of microscopic rock damage. The images were collected by a 300 X digital microscopic imaging system, and the micro damage data of the sampled images were analyzed using the MATLAB software (7.10).

In the process of determining the micro damage data of the rock mass, it was necessary to determine the cross-scale transformation threshold from the collected image macro data to the micro damage micro data [48].

The gray value calculation method was used to determine the micro data of rock mass fractures. Through normalization of the sampling data from the five monitoring points in the first mining time step, the micro data of rock mass fractures were accumulated and calculated, respectively, according to $F = (0.05, 0.10, 0.15, \ldots, 0.90, 0.95)$. Combined with the polynomial fitting of fracture and non-fracture data, $F = 0.35$ was determined as the cross-scale transformation threshold from macro to micro data of rock mass damage, as shown in Figure 8.

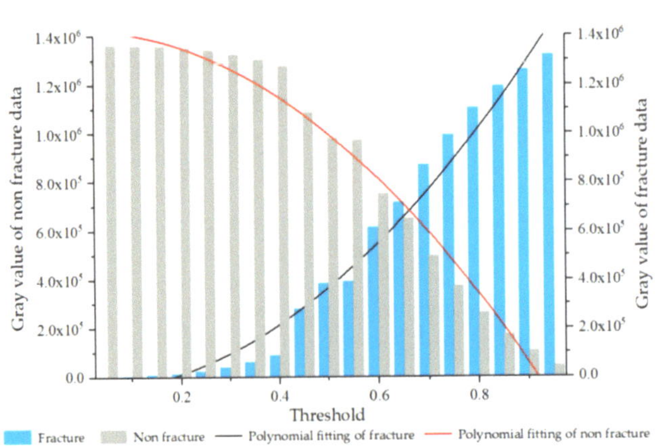

Figure 8. Determination of cross-scale transformation threshold from macro data to micro data of rock mass damage.

Taking the roof monitoring point S-3 of the roadway centerline as an example, microscopic image data collected at the sampling points in the first to fourth mining time steps are shown in Figure 9.

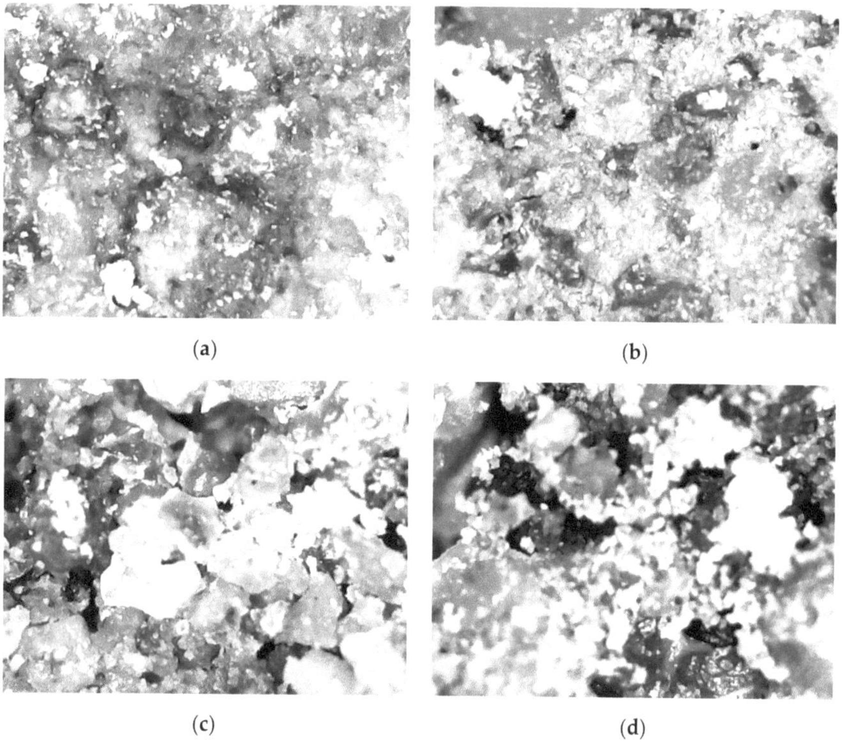

Figure 9. Images were acquired at different time steps at S-3 sampling points (300X): (**a**) First mining step; (**b**) Second mining step; (**c**) Third mining step; (**d**) Fourth mining step.

The microscopic images were converted into corresponding gray images and the Itophat through MATLAB to obtain the corresponding binary image, where the gray threshold change was used to realize the binarization of the binary image (Figure 10).

The binary images of monitoring points were obtained by numerical analysis. Black pixels in the image represent the cracked areas, while white pixel represent non-cracked areas. The data analysis showed that with the continuous advancement of mining time steps, the rock fractures in the S-3 roof increased, and the rock micro damage occurred through local crack occurrence and propagation, damage compaction, crack nucleation and extension and fracture network formation. After the second time step mining, the in situ stress in the post-mining rock promoted the compaction and closure of micro cracks in the surrounding rock, and the development and penetration of cracks was limited, which reduced the binary data of cracks; however, the cracks gathered and nucleated in the rock in the third time step. The expansion speed of rock fractures become faster, and the phenomenon of fracture nucleation in the new area was prominent. The reason for this is that, with the continuous progress of mining, the potential energy accumulated in the equilibrium stage increases significantly. After entering the debris ejection stage, the energy continues to accumulate and, at the stable failure stage, the energy release is concentrated, the process of fracture nucleation, and extension is accelerated. Finally, the fracture network is dense in the complete failure stage.

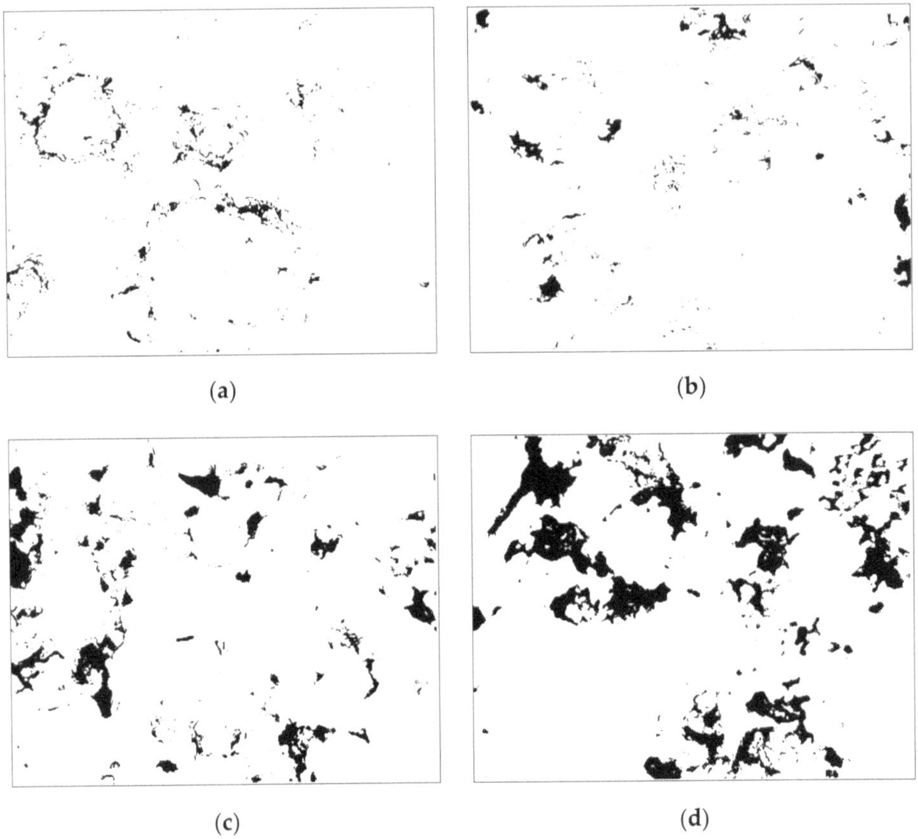

Figure 10. Gray binary image of S-3 sampling points (300X): (a) First mining step; (b) Second mining step; (c) Third mining step; (d) Fourth mining step.

According to the binary data, the density damage of S-3 sampling points in the first to fourth time steps was 5.75%, 2.64%, 5.98%, and 6.55%, respectively. The sampling points went through the process of crack expansion, compaction, and re-expansion. The gray level of the binary data of the sampling points was divided into high and low levels, as shown in Table 4. With a higher proportion of low gray level, more cracks are derived.

Table 4. Gray level distribution at different time steps of S-3 sampling points.

Time Step	Low Gray Level/ Level 0–85	Medium Gray Level/Level 86–170	High Gray Level/ Level 170–256
First	0.93%	44.97%	54.10%
Second	1.84%	34.41%	63.75%
Third	5.37%	35.85%	58.78%
Fourth	11.71%	37.01%	51.28%

We analyzed the sampling data of the roof of the continuous driving roadway 1 # to determine the rock density damage according to the binarization diagram of sampling points at different time steps. By analyzing the micro damage change, we could determine the density damage increment (△F) of the roof rock in different mining steps, as shown in Figure 11.

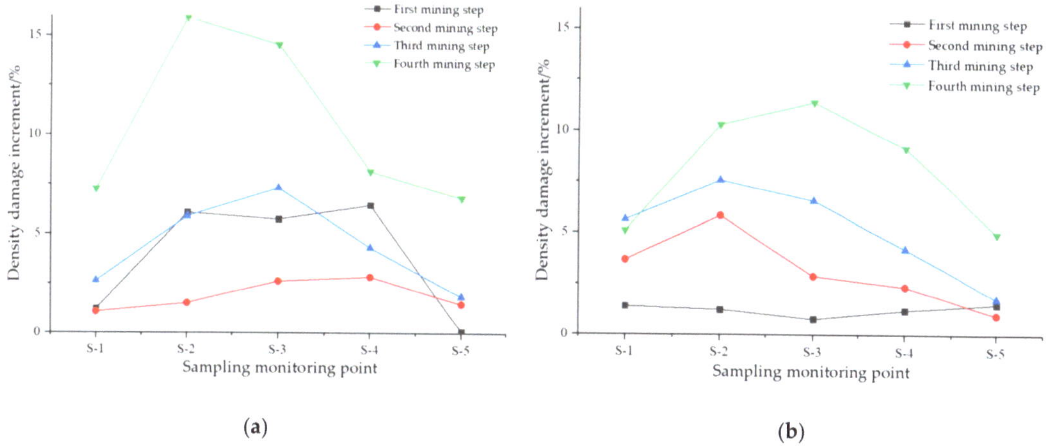

Figure 11. Density damage of roadway roof in (**a**) near roadway roof sampling points and (**b**) deeper sampling points.

The rock density damage of the roadway roof continued to increase with the advance of mining time, and the rock density damage increment at deeper sampling points continued to increase as a whole. In the equilibrium stage, the load of the rock surrounding the roadway roof was low, the rock density damage was symmetrical, and the density damage of surrounding rock near the roadway roof was high. In the stage of debris ejection, the density damage increment of the rock surrounding the roadway roof was less than that in the previous step, and the rock mass was in the stage of compaction and density. The density damage of the surrounding rock of the deep roadway roof continued to increase. In the stable failure stage, the density damage of the rock surrounding the roadway roof on the same vertical line was similar overall, whereas the density damage increment of left surrounding rock was larger than that of right surrounding rock. In the stage of complete failure, the density damage increment of the rock surrounding the roadway roof increased greatly; it was 1.5–4.9 times higher than that in the initial time step (abnormal points were eliminated). The density damage increment of the surrounding rock of the deep roadway roof was centrally symmetrical and evenly distributed, increasing by 2.5–7.2 times, when compared with the initial time step (abnormal points were eliminated), which explains the continuous extension process of the failure state of the roadway roof from the adjacent surrounding rock to the deep surrounding rock.

Using the density damage increment research method, the temporal and spatial evolution law of micro damage and failure in the process of roadway roof rockburst was analyzed, and the cumulative process of roadway roof deformation and temporal and spatial laws of failure were explained. The corresponding conclusions are similar to those drawn in Section 2.1.

4.4. *Deformation and Failure Characteristics of Roadway Roof*

Assuming that the surrounding rock of the rectangular roadway is regarded as a homogeneous and isotropic continuous medium, the conformal variation can be used to map the outside of a rectangular roadway to the unit circle through the mapping function $Z = \omega(\xi)$, as shown in Figure 12. Based on the deformation characteristics of rock and the theory of elasticity, the mechanical constitutive equation of the volume deformation of the roadway-surrounding rock is determined, with the following partial tensor form [49]:

$$\begin{cases} \left(D+\frac{k_1}{\eta}\right)\sigma_{rs} = \left[(k_1+k_2)D+\frac{k_1k_2}{\eta}\right]\varepsilon_{rs} \\ \sigma = 3k\varepsilon \end{cases} \tag{18}$$

where, D is the differential operator for time (t), $D^n = \partial^n/\partial t^n$. σ_{rs} is the partial stress tensor, ε_{rs} is the strain partial tensor, k_1 is the shear deformation modulus of viscous components of the surrounding rock, k_2 is the shear deformation modulus of the elastic components of the surrounding rock, and η is the viscosity coefficient of the viscous component of the surrounding rock.

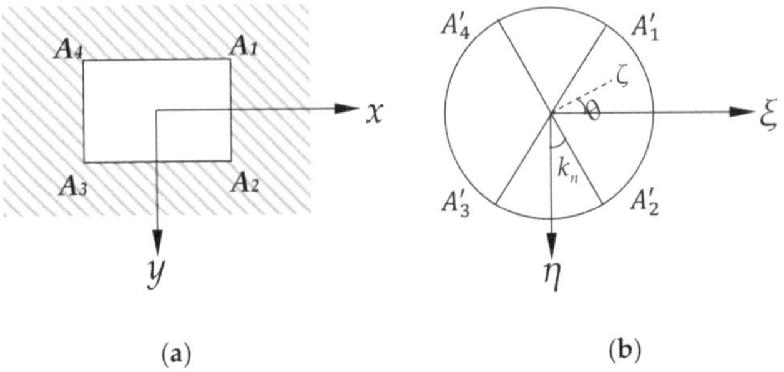

Figure 12. Mapping of Z plane rectangle (**a**) to ζ plane unit circle (**b**).

It can be found that the stress tensor of roadway-surrounding rock is positively correlated with the strain tensor and is not related whatsoever with the process of reaching this state.

The stress accumulation increment of the rock surrounding the roadway roof is shown in Figure 13. With the initial excavation, the shear stress of the rock surrounding the roadway roof gradually concentrated, and the stored energy is small. After entering the debris ejection stage, the strain energy in the surrounding rock accumulated continuously, small particles were ejected from the surrounding rock, the ejected particle size gradually increased, and there was no obvious buckling deformation of the roof. However, the high-power microscopic imaging system showed that tensile crack failure had begun to occur in the surrounding rock, forming new tensile cracks, which continued to expand and nucleate. When it developed to the stable failure stage, under the further action of tangential stress, the elastic strain energy accumulating in the surrounding rock of the roof adjacent to the roadway continued to increase, the deformation state of the surrounding rock in the deep part of the roof around the roadway centerline obviously changed, and the local bending deformation of the surrounding rock was accompanied by continuous particle ejection. The high-power microscopic imaging system showed that the surrounding rock fractures expanded intensively, forming a very obvious fracture network (or fracture family). When mining in the fourth step, the cumulative stress in the surrounding rock roof surged, the tensile state of the roadway centerline roof expanded to the deep rock along the radial direction, the cumulative deformation curve formed a V-shaped rockburst zone, and the deformation deflection tensor of monitoring points near the corner of the low-level (high burial depth) roof changed from a compression to a tension state.

According to the Hoek brown rock mass strength failure criterion [50]:

$$\sigma_1 = \sigma_3 + \sigma_{\text{ci}} \left(m_b \frac{\sigma_3}{\sigma_{cj}} + s \right)^{0.5} \tag{19}$$

where, σ_1 is the maximum principal stress when the rock mass is damaged(MPa), σ_3 is the minimum principal stress in the case of rock mass failure(MPa), σ_{ci} is the rock block strength(MPa), and m_b and s are dimensionless parameters representing the properties of rock materials, which are taken as 10 and 1, respectively, in this study.

Considering that the inner wall of the roadway after excavation showed $\sigma_3 = 0$, the roadway buckling failure satisfies $\sigma_{\theta max} = \sigma_1 = \sigma_{ci}$; that is, in theory, when the tangential stress of surrounding rock around the roadway reaches the uniaxial compressive strength,

the roadway will be damaged. However, in our experimental study, we found that the buckling failure strength of the roof from the first to third time steps was $\sigma_{\theta max} \approx$ (1.29–1.76) σ_{ci}, as shown in Table 5. This is also consistent with previous research results at home and abroad. The results of the analysis are due to the influence of the size effect and the strong effect on the roof after roadway excavation [35,51,52].

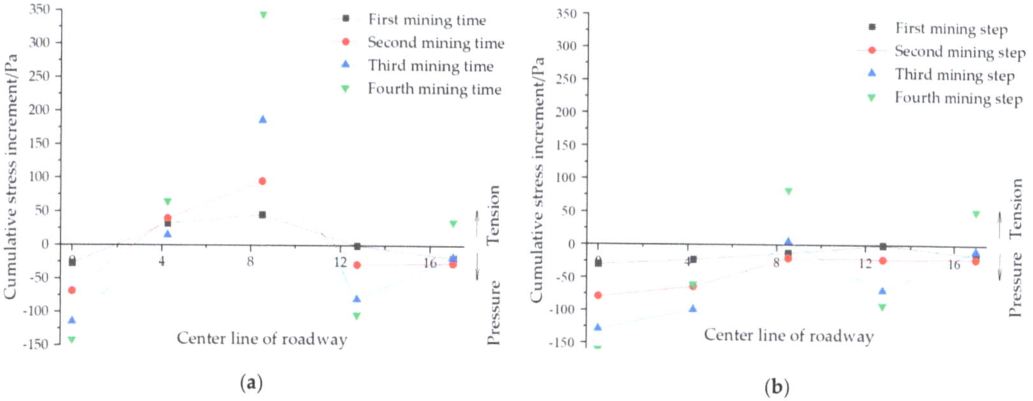

Figure 13. Cumulative stress increment of roadway roof-surrounding rock: (**a**) Near roadway roof sampling points and (**b**) deeper sampling points.

Table 5. Buckling strength coefficient of roadway roof-surrounding rock.

#	D-1	D-2	D-3	D-4	D-5
Peak load/Pa	255.630 5	225.563 7	146.756 4	195.853 1	147.642 0
$\sigma_{\theta max}/\sigma_c$	1.57	1.29	1.69	1.35	1.59
#	S-1	S-2	S-3	S-4	S-5
Peak load/Pa	241.386 9	40.3965 8	187.357 7	206.746 4	153.157 1
$\sigma_{\theta max}/\sigma_c$	1.68	1.31	1.76	1.44	1.63

5. Discussion

In this study, special emphasis was put on developing simple theoretical tools that could rapidly assess the stability of the surrounding rock for a continuous driving roadway. Such simplicity and efficiency are crucial for practicing engineers, in particular, during indoor physical similarity simulation and rock mechanics test design. In the laboratory test, some parameters of rock materials were idealized, ignoring uncertain factors such as rock hydrogeology, natural weathering, and disturbance. In view of the surrounding rock deformation and failure process of underground engineering under the action of the overburden of self-weight and support pressure in geotechnical engineering, the following assumptions were made [53]:

1. The rock material studied is a brittle material, which can be considered to have linear elasticity and can be studied using the relevant theories of linear elastic fracture mechanics;
2. The rock material was considered to be isotropic;
3. The deformation and failure process of the surrounding rock is quasi-static and isothermal;
4. In the micro research of rock, the cracks studied are regarded as ideal cracks, and the crack size is much smaller than the rock mass size. Ignoring the influence of finite size on the calculation results, it is assumed that the rock mass size is infinite.

Whether the judgment of the initial in situ stress field is reliable, and whether the selection of rock mass parameters is reasonable, will directly affect the economy, reliability, and safety of engineering design and construction [54]. E. T. Brown and E. Hoek [55] summarized the measurement results of in situ stress in different regions of the world in

1978 and summarized the vertical stress in countries all over the world, σ_v, and changes with the depth, H, with the fitting formula as follows.

$$\sigma_v = 0.027H \tag{20}$$

The ratio of average horizontal in situ stress to vertical in situ stress was obtained, and the relationship is as follows.

$$\frac{100}{H} + 0.30 \leq k_0 \leq \frac{1500}{H} + 0.50 \tag{21}$$

In this paper, the regression analysis equation of Zhao D.A. for the ratio of average horizontal in situ stress to vertical in situ stress of sedimentary rocks in China was selected [56], as is shown in Figure 14.

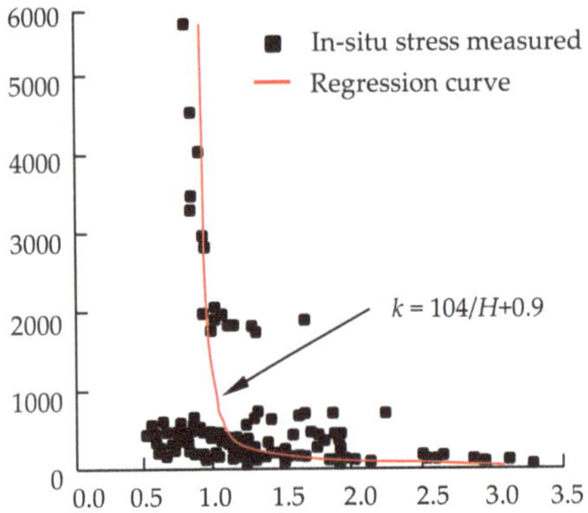

Figure 14. Variation law of average horizontal in situ stress and vertical in situ stress ratio with the depth of sedimentary rocks in China.

At the same time, the comparative analysis of the rock flexural failure strength coefficient, as calculated in Section 2, showed that there was a certain gap between the rock flexural failure strength coefficient obtained by theoretical analysis and the maximum flexural failure strength coefficient calculated by physical similarity simulation experiment. Affected by the characteristics of multi-layer strata, mining disturbances, and monitoring of time effects, the compaction of the roadway-surrounding rock roof occurred to a certain extent, and its endurance strength was enhanced in the process of the physical similarity simulation test. Moreover, due to the large geometric similarity selected in the physical similarity simulation test, there must be a certain error between the stress moment of the roof of the overhead roadway and the actual project.

Correspondingly, from the experimental results, the average flexural failure strength coefficient of 10 sampling monitoring points was 1.531, which differed by 7.01% from the rock flexural failure strength obtained by theoretical analysis. The theoretical flexural failure strength coefficient was still within the range of the experimental research results, and the error met the requirements of test regulations. Therefore, we believe that the comparison results can still reasonably approximate the theoretical analysis results and reflect the final results of rock deformation and failure.

6. Conclusions

In this study, we used a physical similarity simulation test to carry out a roof rockburst simulation test considering underground roadway excavation. The incubation process of rockburst was studied using XL2101g multi-point high-speed whole process controlled static strain acquisition system and digital micro-imaging system. The evolution law of roof

deformation during roadway excavation was preliminarily revealed. Our main conclusions are as follows.

1. Failure of the roadway roof can be divided into four stages. In the equilibrium stage, the expansion energy in the roof accumulates continuously, accompanied by the continuous increase in the fracture network. At the stage of stability failure, with the deep surrounding rock of the roof around the roadway centerline as the center, the local bending deformation of the surrounding rock of the roof appears, accompanied by continuous particle ejection. Finally, in the failure stage, the roof stress increases accumulatively, the tensile state of the roof of the roadway centerline expands to the deep rock along the radial direction, the cumulative deformation curve forms a V-shaped rockburst zone, and the surrounding rock deformation is obvious.

2. The deformation of roadway roof showed non-uniform and uncoordinated deformation. In the equilibrium stage, the stored deformation energy of the roadway roof is small, and the roof deformation is centrosymmetric. After entering the clastic stage, the strain energy of the surrounding rock accumulates continuously, small particle ejection occurs in the surrounding rock, and the strain energy of the surrounding rock near the corner of roadway increases continuously. In the stable breaking stage, the surrounding rock around the roadway centerline is deformed obviously, and local buckling deformation occurs, forming an obvious fracture network (or fracture cluster). In the failure stage, the stress concentration of the roadway roof shifts to the left side of the roadway, and the roadway is fully tensioned as a whole.

3. The rock surrounding the roadway roof is mainly affected by tensile stress and shear stress. According to the stress state of a I-II composite crack, it was determined that the theoretical buckling failure strength of the roadway roof-surrounding rock is 1.43 σ_t. The results of the physical similarity simulation model test show that the theoretical buckling failure strength of the rock surrounding the roadway roof is (1.29–1.76) σ_{ci}. The analysis results are due to the influence of the size effect and the strong effect on the roof after roadway excavation. We expect that the roof will be affected by size and strength effects after roadway excavation. The theoretical flexural failure strength coefficient was still within the range of the experimental research results, and the error met the requirements of test regulations.

Author Contributions: Conceptualization, Y.Y., Y.S., and X.W.; methodology, N.H., Y.Y. and Y.S.; software, Y.S.; validation, Y.S. and Y.J.; formal analysis, N.H., Y.Y. and Y.S.; investigation, N.H. and Y.S.; resources, Y.Y. and N.H.; data curation, Y.S. and Y.J.; writing—original draft preparation, Y.S., Y.Y., and N.H.; writing—review and editing, N.H., Y.S. and Y.Y.; visualization, Y.S.; supervision, N.H. and Y.Y.; project administration, Y.Y. and X.W.; funding acquisition, Y.Y. and Y.S. All authors have read and agreed to the published version of the manuscript.

Funding: This research was funded by the Natural Science Foundation of Hubei Province (No.2020CFB123), Scientific Research Project of Hubei Provincial Department of Education (No.Q20201109), Key Research and Development Plan of Hubei Province (No. 2020BCA082), and Innovation and entrepreneurship training program for college students of WUST (No. 20ZB012).

Institutional Review Board Statement: Not applicable.

Informed Consent Statement: Not applicable.

Data Availability Statement: Not applicable.

Conflicts of Interest: The authors declare no conflict of interest.

References

1. Li, T.; Gong, H.; Xu, G.L.; Kong, D.Z. Study on the influence of in situ stress distribution on the stability of roadway surrounding rock. *Adv. Civ. Eng.* **2021**, *10*, 3570523. [CrossRef]
2. Yu, Y.; Lu, J.F.; Chen, D.C.; Pan, Y.X.; Zhao, X.Q.; Zhang, L.Y. Study on the Stability Principle of Mechanical Structure of Roadway with Composite Roof. *Minerals* **2021**, *11*, 1003. [CrossRef]
3. Zhao, G.P.; Yang, S.L. Analytical solutions for rock stress around square tunnels using complex variable theory. *Int. J. Rock Mech. Min. Sci.* **2018**, *106*, 74–83. [CrossRef]

4. Wang, J.; Apel, D.B.; Pu, Y.Y.; Hall, R.; Wei, C.; Sepehri, M. Numerical modeling for rockbursts: A state-of-the-art review. *J. Rock Mech. Geotech. Eng.* **2020**, *13*, 457–478. [CrossRef]
5. Li, Z.L.; He, X.Q.; Dou, L.M.; Song, D.Z.; Wang, G.F. Numerical investigation of load shedding and rockburst reduction effects of top-coal caving mining in thick coal seams. *Int. J. Rock Mech. Min. Sci.* **2018**, *110*, 266–278. [CrossRef]
6. Shadabfar, M.; Gokdemir, C.; Zhou, M.L.; Kordestani, H.; Muho, E.V. Estimation of Damage Induced by Single-Hole Rock Blasting: A Review on Analytical, Numerical, and Experimental Solutions. *Energies* **2020**, *14*, 29. [CrossRef]
7. Mu, W.Q.; Li, L.C.; Chen, D.Z.; Wang, S.X.; Xiao, F.K. Long-term deformation and control structure of rheological tunnels based on numerical simulation and on-site monitoring. *Eng. Fail. Anal.* **2020**, *118*, 104928. [CrossRef]
8. Tahmasebinia, F.; Zhang, C.G.; Canbulat, I.; Vardar, O.; Saydam, S. Numerical and analytical simulation of the structural behaviour of fully grouted cable bolts under impulsive loading. *Int. J. Min. Sci. Technol.* **2018**, *28*, 807–811. [CrossRef]
9. Brink, S.; Dorfling, C.; Aldrich, C. An acoustic sensor for prediction of the structural stability of rock. *Int. J. Rock Mech. Min. Sci.* **2016**, *85*, 187–191. [CrossRef]
10. Fang, Q.C.; Shang, L.; Shang, Y.H.; Zhao, Y.; Ou, N.Y. Study on the Reinforcement Measures and Control Effect of the Surrounding Rock Stability Based on the Shield Tunneling Under Overpass Structure. *J. Eng. Sci. Technol. Rev.* **2016**, *9*, 131–138. [CrossRef]
11. Sainoki, A.; Maina, D.; Schwartzkopff, A.K.; Obara, Y.; Karakus, M. Impact of the intermediate stress component in a plastic potential function on rock mass stability around a sequentially excavated large underground cavity. *Int. J. Rock Mech. Min. Sci.* **2018**, *127*, 104223. [CrossRef]
12. Zhao, D.S. Retrospection on the development of rock mass mechanics and the summary of some unsolved centennial problems. *Chin. J. Rock Mech. Eng.* **2021**, *40*, 1297–1336.
13. Ahmad, M.; Hu, J.L.; HadzimaNyarko, M.; Ahmad, F.; Tang, X.W.; Rahman, Z.U.; Nawaz, A.; Abrar, M. Rockburst Hazard Prediction in Underground Projects Using Two Intelligent Classification Techniques: A Comparative Study. *Symmetry* **2021**, *13*, 632. [CrossRef]
14. Guo, L.; Li, X.B.; Yan, X.M. Research progress and development tendency of rockburst. *Min. Technol.* **2006**, *6*, 16–20.
15. Kaiser, P.K.; Cai, M. Design for rock support system under rockburst condition. *J. Rock Mech. Geotech. Eng.* **2012**, *4*, 215–227. [CrossRef]
16. Ma, D.; Wang, J.J.; Cai, X.; Ma, X.T.; Zhang, J.X.; Zhou, Z.L.; Tao, M. Effects of height/diameter ratio on failure and damage properties of granite under coupled bending and splitting deformation. *Eng. Fract. Mech.* **2019**, *220*, 106640. [CrossRef]
17. Du, H.; Song, D.Q.; Chen, Z.; Guo, Z.Z. Experimental study of the influence of structural planes on the mechanical properties of sandstone specimens under cyclic dynamic disturbance. *Energy Sci. Eng.* **2020**, *8*, 4043–4063. [CrossRef]
18. Pei, F.; Ji, H.G.; Zhao, J.W.; Geng, J.M.; Kong, H.L. Energy Evolution and AE Failure Precursory Characteristics of Rocks with Different Rockburst Proneness. *Adv. Civ. Eng.* **2020**, *9*, 8877901. [CrossRef]
19. Aslannezhad, M.; Keshavarz, A.; Kalantariasl, A.; Aslannezhad, M.; Keshavarz, A.; Kalantariasl, A. Evaluation of mechanical, chemical, and thermal effects on wellbore stability using different rock failure criteria. *J. Nat. Gas. Sci. Eng.* **2020**, *78*, 103278. [CrossRef]
20. Li, X.B.; He, W.R.; Xu, Z.H.; Gu, F. Study on Law of Overlying Strata Breakage and Migration in Downward Mining of Extremely Close Coal Seams by Physical Similarity Simulation. *Adv. Civ. Eng.* **2020**, *2020*, 2898971. [CrossRef]
21. Makeiev, S.; Andreiev, S.; Ryzhov, H. The study of the process of physic-chemical destruction of coal by the method of physical modeling. International Conference Essays of Mining Science and Practice. *E3S Web Conf. EDP Sci.* **2019**, *109*, 00054. [CrossRef]
22. Jiang, L.S.; Wang, P.; Zheng, P.Q.; Luan, H.J.; Zhang, C.; Corradi, M. Influence of Different Advancing Directions on Mining Effect Caused by a Fault. *Adv. Civ. Eng.* **2019**, *2019*, 7306850. [CrossRef]
23. Zhou, N.; Zhang, J.X.; Yan, H.; Li, M.; Wang, M. Deformation Behavior of Hard Roofs in Solid Backfill Coal Mining Using Physical Models. *Energies* **2017**, *10*, 557. [CrossRef]
24. Chen, B.; Zhang, S.C.; Li, Y.Y.; Li, Z.K.; Zhou, H.J. Physical simulation study of crack propagation and instability information discrimination of rock-like materials with faults. *Arab. J. Geosci.* **2020**, *13*, 1–14. [CrossRef]
25. Li, S.G.; Du, X.H.; Zhao, P.x.; Xiao, P.; Lin, H.F.; Shuang, H.Q. Experimental study on crack evolution characteristics of rock-like materials under different strain rates. *J. Geophys. Eng.* **2018**, *15*, 2071–2078.
26. Ma, D.; Zhang, J.X.; Duan, H.Y.; Huang, Y.L.; Li, M.; Sun, Q.; Zhou, N. Reutilization of gangue wastes in underground backfilling mining: Overburden aquifer protection. *Chemosphere* **2021**, *264*, 128400. [CrossRef]
27. Shi, Y.B.; Ye, Y.C.; Hu, N.Y.; Huang, X.; Wang, X.H. Experiments on Material Proportions for Similar Materials with High Similarity Ratio and Low Strength in Multilayer Shale Deposits. *Appl. Sci.* **2021**, *11*, 9620. [CrossRef]
28. Ye, Y.C.; Shi, Y.B.; Wang, Q.H.; Yao, N.; Lu, F.; Yue, Z. Test model research on low strength similar material of Shanghengshan multilayer shale deposit. *Rock Soil Mech.* **2014**, *35*, 114–120.
29. Shi, Y.B.; Ye, Y.C.; Wang, Q.H.; Yue, Z.; Lu, F.; Xia, Z.Q. Pillar deformation characteristics for multilayer gently-dipping deposits by strip mining and mining intensity simulation. *Min. Metall. Eng.* **2015**, *35*, 4–9.
30. Ye, Y.C.; Shi, Y.B.; Wang, Q.H.; Liu, Y.Z.; Yao, N.; Lu, F. Experimental study of deformation of wall rock and stoping sequence in mining gently inclined and multilayer deposits by backfill mining in Shanghengshan. *J. Min. Saf. Eng.* **2015**, *32*, 407–413.
31. Hu, X.C.; Su, G.S.; Chen, G.Y.; Mei, S.M.; Feng, X.T.; Mei, G.X.; Huang, X.H. Experiment on rockburst process of borehole and its acoustic emission characteristics. *Eng. Geol.* **2019**, *52*, 783–802. [CrossRef]
32. Tian, C.L.; Yang, X.L.; Sun, H.T.; Liu, Y.B.; Hu, Q.T. Experimental study on the overburden movement and stress evolution in multi-seam mining with residual pillars. *Energy Sci. Eng.* **2019**, *7*, 3095–3110. [CrossRef]

33. Hu, G.J.; Yang, T.H.; Zhou, J.R.; Yu, Q.L.; Xie, L.K.; Liu, H.L.; Zhao, Y.; Taheri, A. Mechanism of Surrounding Rock Failure and Crack Evolution Rules in Branched Pillar Recovery. *Minerals* **2017**, *7*, 96. [CrossRef]
34. Zhu, Z.M.; Gao, W.T.; Wan, D.Y.; Wang, M.; Shu, Y.; Niu, L.L. Numerical Study of Fracture Characteristics of Deep Granite Induced by Blast Stress Wave. *Shock Vib.* **2021**, *2021*, 2579569. [CrossRef]
35. Zhang, S.; Wang, H.Y.; Li, X.J.; Zhang, X.L.; An, D.C.; Yu, B.X. Experimental study on development characteristics and size effect of rock fracture process zone. *Eng. Fract. Mech.* **2020**, *241*, 107377. [CrossRef]
36. Cao, K.W.; Khan, N.M.; Liu, W.; Hussain, S.; Zhu, Y.G.; Cao, Z.T.; Bian, Y.C. Prediction Model of Dilatancy Stress Based on Brittle Rock: A Case Study of Sandstone. *Arab. J. Sci. Eng.* **2021**, *46*, 2164–2176. [CrossRef]
37. Wang, M.; Zhu, Z.M.; Wang, X. The growth of mixed-mode I/II crack under impacting loads. *Chin. J. Rock Mech. Eng.* **2016**, *35*, 1323–1332.
38. Justo, J.; Castro, J.; Cicero, S.; Sanchez, C.M.; Husillos, R. Notch effect on the fracture of several rocks: Application of the Theory of Critical Distances. *Theor. Appl. Fract. Mech.* **2017**, *90*, 251–258. [CrossRef]
39. Wang, X.; Su, O. Specific energy analysis of rock cutting based on fracture mechanics: A case study using a conical pick on sandstone. *Eng. Fract. Mech.* **2019**, *213*, 197–205. [CrossRef]
40. Shi, Y.B.; Ye, Y.C.; Liu, Y.Z.; Wang, Q. H Analysis for the breadth of the pillar-damaged mining strip based on I-II mixed crack mode. *J. Saf. Environ.* **2017**, *17*, 888–895.
41. Dumont, N.A.; Mamani, E.Y.; Cardoso, M.L. A boundary element implementation for fracture mechanics problems using generalised Westergaard stress functions. *Eur. J. Comput. Mech.* **2018**, *27*, 401–424. [CrossRef]
42. Deng, J.L.; Yang, P.; Dong, Q.; Yan, X. Elasto-Plastic Fracture Analysis of Finite-Width Cracked Stiffened Plate. *Appl. Mech. Mater.* **2014**, *496*, 1052–1057. [CrossRef]
43. Paliwal, B.; Hammi, Y.; Chandler, M.; Moser, R.D.; Horstemeyer, M.F. A three-invariant cap-viscoplastic rate-dependent-damage model for cementitious materials with return mapping integration in Haigh-Westergaard coordinate space. *Int. J. Solids Struct.* **2020**, *182*, 77–99. [CrossRef]
44. Lv, X.; Wang, S.Z.; Qiu, Y.; Liu, X.X.; Gong, F.Q. Study on the Influence of Tectonic Stress on Stability of Horseshoe-Shaped Tunnels. *Adv. Mater. Sci. Eng.* **2021**, *2021*, 5566142. [CrossRef]
45. Zhang, Z.Q.; Chen, F.F.; Li, N.; He, M.M.; Moayedi, H. Influence of Fault on the Surrounding Rock Stability for a Mining Tunnel: Distance and Tectonic Stress. *Adv. Civ. Eng.* **2019**, *2019*, 2054938. [CrossRef]
46. Azarafza, M.; Koçkar, M.K.; Faramarzi, L. Spacing and block volume estimation in discontinuous rock masses using image processing technique: A case study. *Environ. Earth Sci.* **2021**, *80*, 1–13. [CrossRef]
47. Liu, Q.S.; Wang, Z.W. Review of numerical modeling based on digital image processing for rock mechanics applications. *Chin. J. Rock Mech. Eng.* **2020**, *39*, 3286–3296.
48. Sun, B. Adaptive multi-scale beam lattice method for competitive trans-scale crack growth simulation of heterogeneous concrete-like materials. *Int. J. Fract.* **2021**, *228*, 85–101. [CrossRef]
49. Li, M.; Mao, X.B. Based on the complex variable functions of rectangular roadway surrounding rock stress and deformation viscoelastic analysis. *Chin. Q. Mech.* **2011**, *32*, 195–202.
50. Lee, Y.K.; Pietruszczak, S. Limit equilibrium analysis incorporating the generalized hoek–brown criterion. *Rock Mech. Rock Eng.* **2021**, *54*, 4407–4418. [CrossRef]
51. Ban, H.X.; Peng, Z.L.; Fang, D.N.; Yao, Y.; Chen, S.H. A modified conventional theory of mechanism-based strain gradient plasticity considering both size and damage effects. *Int. J. Solids Struct.* **2020**, *202*, 384–397. [CrossRef]
52. Hütter, G.; Sab, K.; Forest, S. Kinematics and constitutive relations in the stress-gradient theory: Interpretation by homogenization. *Int. J. Solids Struct.* **2020**, *193*, 90–97. [CrossRef]
53. Anderson, T.L. *Fracture Mechanics: Fundamentals and Applications*; CRC Press: Boca Raton, FL, USA, 2017; pp. 136–157.
54. Girard, J.; Silber, R.E.; Mohiuddin, A.; Chen, H.Y.; Karato, S. Development of a Stress Sensor for In-Situ High-Pressure Deformation Experiments Using Radial X-Ray Diffraction. *Minerals* **2020**, *10*, 166. [CrossRef]
55. Brown, E.T.; Hoek, E. Technical note trends in relationships between measured in-situ stress and depth. *Int. J. Rock Mech. Min. Sci. Geomech. Abstr.* **1978**, *15*, 211–215. [CrossRef]
56. Zhao, D.; Chen, Z.M.; Cai, X.L.; Li, S.G. Analysis of distribution rule of geostress in china. *Chin. J. Rock Mech. Eng.* **2007**, *26*, 1265–1271.

minerals

Article

Experimental Study on the Short-Term Uniaxial Creep Characteristics of Sandstone-Coal Composite Samples

Dawei Yin [1], Feng Wang [1,*], Jicheng Zhang [1], Faxin Li [1], Chun Zhu [2] and Fan Feng [1]

[1] State Key Laboratory of Mine Disaster Prevention and Control, Shandong University of Science and Technology, Qingdao 266590, China; yindawei@sdust.edu.cn (D.Y.); 202081010096@sdust.edu.cn (J.Z.); lifaxin@sdust.edu.cn (F.L.); fengfan0213@sdust.edu.cn (F.F.)

[2] School of Earth Sciences and Engineering, Hohai University, Nanjing 210098, China; zhu.chun@hhu.edu.cn

* Correspondence: wangfeng@sdust.edu.cn; Tel.: +86-137-8985-9661

Abstract: In this investigation, the uniaxial short-term creep tests with multi-step loading were conducted on the sandstone-coal composite samples, and the characteristics of creep strength, creep deformation, acoustic emission (AE), and creep failure of composite samples were studied, respectively. The creep strength of the composite sample decreased with the stress-level duration, which was mainly determined by the coal and influenced by the interactions with the sandstone. The creep deformation and damage of sandstone weakened the deformation and damage accumulation within the coal, resulting in the larger strength for the composite sample compared with the pure coal sample. The axial creep strain of composite sample generally increased with the stress-level or the stress-level duration under same conditions. The AE characteristics of composite sample were related to the creep strain rate, the stress level, the stress level duration, and the local failure or fracture during creep loading. The micro or macro failure and fracture within the composite sample caused the rise in the axial creep strains and the frequency and intensity of AE signals, especially the macro failure and fracture. The creep failures of composite samples mainly occurred within the coal with the splitting ejection failure accompanied by the local shear failure, and no obvious failures were found within the sandstone. The coal in the composite sample became more broken with the stress-level duration.

Keywords: sandstone-coal composite samples; creep strength; creep deformation; acoustic emission characteristics; creep failure; short-term uniaxial creep loading

Citation: Yin, D.; Wang, F.; Zhang, J.; Li, F.; Zhu, C.; Feng, F. Experimental Study on the Short-Term Uniaxial Creep Characteristics of Sandstone-Coal Composite Samples. *Minerals* **2021**, *11*, 1398. https://doi.org/10.3390/min11121398

Academic Editor: Gianvito Scaringi

Received: 15 October 2021
Accepted: 8 December 2021
Published: 10 December 2021

1. Introduction

Coal is the foundation of energy, and the major resource in China, which effectively promotes the development of economy and society nationally [1–3]. In order to ensure the safety, high efficiency and green mining of coal resources in China, different types and functions of coal pillars have been left around the mining area [4–8], such as the strip coal pillar, the waterproof coal pillar, the section coal pillar, etc. A considerable number of left coal pillars, as the permanent coal pillars, bear loads for a long time to support the overlying strata and to control the surface subsidence or collapse [9–14]. The long-term stability of these coal pillars determines the safety and stability of overlying strata, surface buildings and ecological environment, etc. And their failure and instability can induce the overlying strata caving and surface collapsing in China [7], for instance, the failure and instability of left coal pillars in the 402 panel of Majiliang Coal Mine (Figure 1a) caused the formation of an oval surface collapse area with about 79,103 m^2; the failure and instability of left coal pillars in the 404 panel of Wajinwan Coal Mine (Figure 1b) induced the formation of an surface collapse area with about 163,000 m^2, and resulted in 18 deaths and 19 injuries. Therefore, it is important to investigate the long-term stability of these left coal pillars.

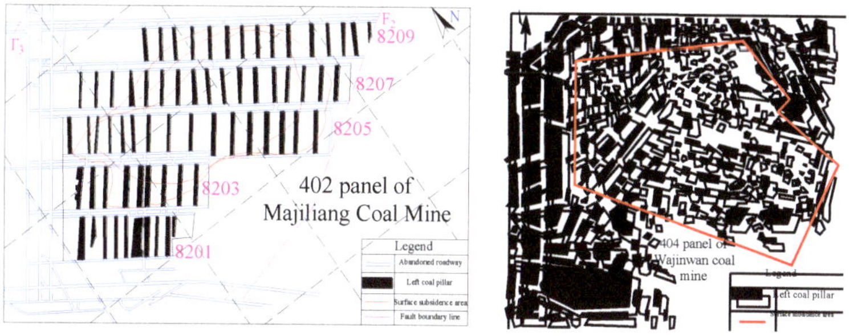

(a) (b)

Figure 1. Surface collapsing induced by the failure and instability of left coal pillars in China [4]: (**a**) Surface collapsing induced by the failure and instability of left coal pillars in 402 panel of Majiliang Coal Mine; (**b**) Surface collapsing induced by the failure and instability of left coal pillars in 404 panel of Wajinwan Coal Mine.

Many investigations have achieved that the long-term bearing failure and instability of the coal pillar are a mechanical behavior gradually developing from the edge to its interior [4,7–12]. These are not only related to the properties of coal pillar itself, but also affected by the composite structure consisting of coal pillar and surrounding rock layers [6–36]. In other words, the instability of coal pillar is the result of the overall failure and instability of composite structure under overlying strata loads. Therefore, the long-term bearing characteristics and failure mechanisms of coal pillar should be investigated considering the composite structure. Generally, these composite structures were simplified to the rock–coal, coal–rock, and rock–coal–rock composite samples with a bonded or frictional interface to study their mechanical behaviors in previous experimental and numerical simulation investigations [12–36]. These studies focused on the effects of the height ratio of coal to rock [12,22,31–33], the interface between rock and coal [11–14,19], the composite model [36], the loading conditions (uniaxial loading, triaxial loading, cyclic loading and unloading, and true triaxial loading) [14,21,24,25,27,34,35], the defects within the coal or rock, and the mechanical properties (strength, stiffness, lithology, etc.) of rock or coal [19,20,26] on the strength, deformation, energy evolution, electromagnetic radiation characteristics [28,29], acoustic emission (AE) characteristics, and failure properties of composite samples. These above achievements are of great significance to understand mechanical characteristics of the composite structure consisting of coal pillar and surrounding rock layers.

The creep tests on rock or coal samples are very important methods to understand the long-term bearing characteristics and failure mechanisms of coal or rock mass [37–46], and these investigations mainly focused on pure coal or rock samples. Fatemeh and Mark [37] carried out the creep experiments on clay- and carbonate-rich shale samples with short-term (4 h) and long-term (4 weeks) periods, respectively, the results showed that the shale samples follow the same creep trend through time, regardless of the loading history. Cao et al. [38] proposed a nonlinear damage creep constitutive model describing the creep deformation of soft rock. Maranini and Brignoli [39] investigated the creep behavior of a porous chalk, named Pietra Liccese, and proposed an elasto-viscoplastic model describing the slow irreversible deformation in time of the rock surrounding underground openings. Yang et al. [40] analyzed the influences of confining pressure and pore pressure on short-term mechanical behavior of red sandstone samples, and the creep contribution to rock deformation increased with the pore pressure, and the samples showed significant time-dependent effect at higher deviatoric stresses. Yang et al. [41] studied the evolution characteristics of deformation field for red sandstone during the creep loading using the digital image correlation technique. Zong et al. [42] discussed the effects of the confining pressure and damage on short-term mechanical behavior of fractured sandstone with different degrees of damage. Yang et al. [43] and Cao et al. [44] analyzed the AE characteristics of coal samples under creep loading, respectively. Chen et al. [45] and Wu et al. [46] proposed

the creep model for coal samples and the nonlinear creep damage model for salt rock, respectively. These above achievements are of great significance to understand the creep behavior of engineering coal or rock mass

However, the creep investigations that focused on the rock–coal, coal–rock, and rock–coal–rock composite samples have been scant. The creep tests can be classified as long-term creep tests and short-term creep tests based on the stress level duration, but there are no definite time requirements for this classification. Short-term creep tests typically have a stress level duration of several hours or days, whereas long-term creep tests can have a stress level duration of several weeks or even years. A large number of investigations have shown that the creep parameters measured by the short-term creep tests can give a reasonably understanding of the long-term behavior of the rock and help avoid time-consuming [37,40,44–47]. In this investigation, we conducted the short-term uniaxial multi-step loading creep tests with different stress-level durations on the sandstone-coal composite samples, and the characteristics of creep strength, creep deformation, AE, and creep failure of composite samples were studied, respectively.

2. Experimental Procedure

2.1. Sample Preparation

In this investigation, the rock in the composite sample was the sandstone. Sandstone and coal blocks were taken from the Daizhuang Coal Mine in Jining City of China. The sandstone was from the immediate roof. The heights of both the rock and coal layers were 50 mm in the composite samples with a diameter of 50 mm and a height of 100 mm. Firstly, the sandstone and coal blocks were drilled into the cylindrical samples with a diameter of 50 mm. Then, the cylindrical samples were cut into 50 mm height by a stone-sawing machine. Then both ends of cylinder samples were flattened, and smoothed to meet the experimental requirements, respectively, by a stone grinding machine. Finally, the cylinder coal and sandstone samples with a diameter of 50 mm and a height of 50 mm were bonded with superglue to make the 100 mm high sandstone-coal composite samples. Nine composite samples were well prepared, three of which (Figure 2a) were used to test their mechanical properties under uniaxial loading, and the other composite samples (Figure 2b) were subjected to the short-term uniaxial creep tests.

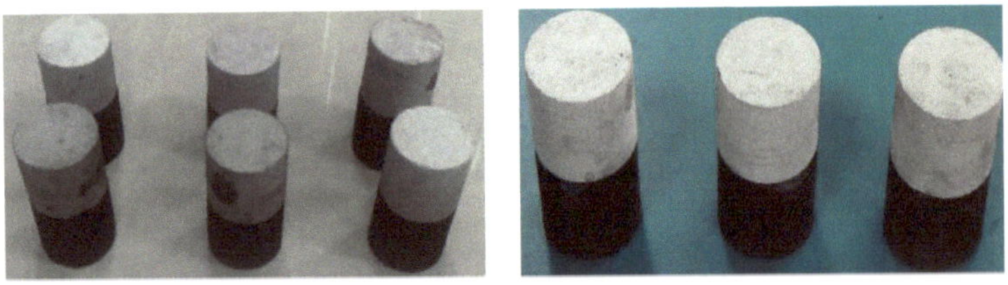

(a) (b)

Figure 2. Sandstone-coal composite samples: (**a**) Sandstone-coal composite samples for creep tests; (**b**) Sandstone-coal composite samples for uniaxial compression tests.

2.2. Test System

The testing system for short-term uniaxial creep with multi-step loading on sandstone-coal composite samples is shown in Figure 3, including a loading system, an AE monitoring system, and a digital video camera (DVC) (Sony, Tokyo, Japan). During each test, they were synchronized to have the same time stamp in order to facilitate analyses of test results.

An AG-X250 servo-controlled testing system (Shimadzu, Kyoto, Japan) was used to perform the short-term uniaxial creep tests on the composite samples [12–14,20]. The AE event was selected to analyze the AE characteristics of composite samples during creep loading, which was monitored using a PCI-2 AE system (MISTRAS, Princeton, NJ, USA) by MISTRAS. The AE sensor was a R3α with a diameter of 19 mm and a height of 22 mm.

Its resonant frequency was 29 kHz, and the range of frequency was 20–180 kHz. The main amplifier gain, threshold, floating threshold, and sampling frequency of the AE monitoring system were 40 dB, 45 dB, 6 dB, and 1 MHz, respectively. The Vaseline was applied between the sample surface and AE sensor in order to improve the coupling conditions. One AE sensor was fixed to the sandstone surface with the adhesive tape. Before the tests, the AE sensor was subjected to a pencil-lead breaking test proposed by ASTM (American Society for Testing and Materials) to ensure that the amplitude measured was above 90 dB. A SONY portable DVC was used to capture the stage points in progressive failure processes of composite samples during creep loading. Generally, the laboratory temperature affected the creep characteristics of samples, and the creep deformation increased with the decrease of the laboratory temperature. The laboratory temperature was set to 24 °C to reduce the temperature effects during creep tests by running the air conditioning.

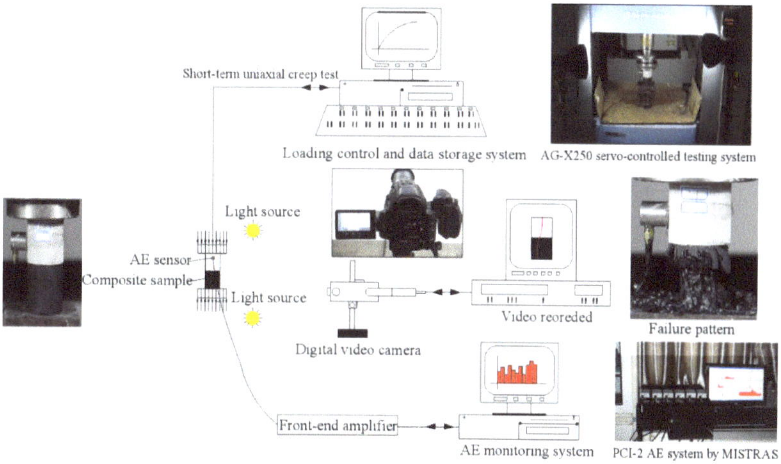

Figure 3. Testing system for short-term uniaxial creep on sandstone-coal composite samples (revised from Yin et al. [20]).

2.3. Test Scheme

Before the short-term uniaxial creep testing, uniaxial compressive tests were carried out on three sandstone-coal composite samples (labeled as RC-1, RC-2 and RC-3) to obtain their basic mechanical properties, using the AG-X250 servo-controlled testing system under the displacement-controlled mode at a loading rate of 0.01 mm/s. Figure 4 shows uniaxial stress-strain curves of composite samples; their uniaxial mechanical properties are listed in Table 1.

Table 1. Mechanical properties of sandstone-coal composite samples under uniaxial loading.

Sample Type	Sample No.	Uniaxial Compressive Strength (UCS)/MPa	Peak Strain	Elastic Modulus/GPa (at 40–60% of Peak Stress)
Sandstone-coal composite sample	RC-1	14.36	0.00985	2.11
	RC-2	14.08	0.00775	2.52
	RC-3	13.99	0.00853	2.44
	Average	14.14	0.00871	2.36

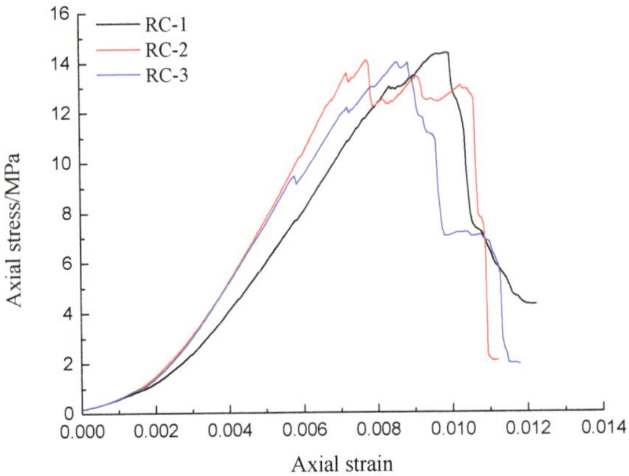

Figure 4. Uniaxial stress-strain curves of sandstone-coal composite samples.

In Table 1, the average UCS, peak strain and elastic modulus of sandstone-coal composite samples were 14.14 MPa, 0.00871, and 2.36 GPa, respectively. Generally, the creep tests were carried out by methods of single-stage loading, multi-stage loading, and multi-stage loading and unloading in previous investigations on the creep characteristics of pure coal or rock samples [37–46]. In this investigation, the short-term uniaxial creep tests with multi-stage loading were conducted on sandstone-coal composite samples. According to the results of uniaxial compression tests, the axial stresses of 50%, 60%, 70%, 80%, 85%, 90%, 95%, … … of the average UCS (σ) were gradually applied on composite samples until they underwent the failure, as shown in Figure 5. The corresponding axial stresses were 7.07 MPa, 8.48 MPa, 9.90 MPa, 11.31 MPa, 12.02 MPa, 12.73 MPa, 13.44 MPa, … … , respectively. The stress level durations (S) were set to 2 h and 6 h, respectively. The axial stress was firstly applied on the composite samples from 0 kN to the 13.88 kN (the first stress level) at a loading rate of 0.005 kN/s, and the loading rate between two adjacent stress levels was also set to 0.005 kN/s. The test data acquisition interval between two adjacent stress level was set to 0.5 s, and the data acquisition interval during each stress level was set to 1 s.

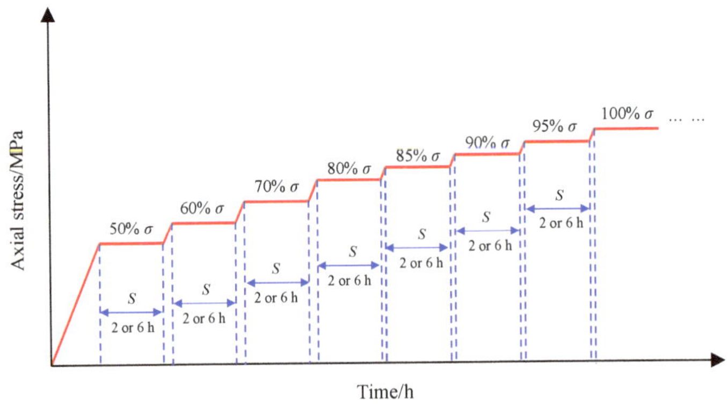

Figure 5. Creep test scheme for sandstone-coal composite samples.

3. Creep Strength Characteristics of Composite Samples

3.1. Creep Strengths of Composite Samples

The creep strength σ_{rc} of the composite sample was given in Equation (1),

$$\sigma_{rc} = \sigma_{n-1} + (\sigma_n - \sigma_{n-1})\frac{t}{S} \tag{1}$$

where, σ_n was the stress level leading to failure; σ_{n-1} was the stress level before σ_n, t was the loading duration at the stress level of σ_n; S was the horizontal loading duration of each level of stress.

Table 2 shows the values of σ_{rc}, σ_n, σ_{n-1}, t and creep coefficients (ratio of σ_{rc} and σ) of composite samples. Among them, the stress level durations corresponding to Groups A and B composite samples were 2 h and 6 h, respectively. The values of σ_{rc} and creep coefficients for composite samples were compared in Figure 6a,b, respectively.

Table 2. σ_n, σ_{n-1}, t and creep coefficients of sandstone-coal composite samples.

Sample Type	Sample No.	S/h	σ_n/MPa	σ_{n-1}/MPa	t/s	σ_{rc}/MPa	Creep Coefficient
Sandstone-coal composite sample	A-1	2	13.44	12.73	380	12.77	90.24%
	A-2		14.14	13.44	782	13.52	95.47%
	A-3		13.44	12.73	1035	12.83	90.88%
	Average value		—	—	—	13.05	92.20%
	B-1	6	13.44	12.73	2700	12.82	90.66%
	B-2		12.02	11.31	3435	11.42	80.76%
	B-3		11.31	9.90	455	9.93	70.23%
	Average value		—	—	—	11.39	80.55%

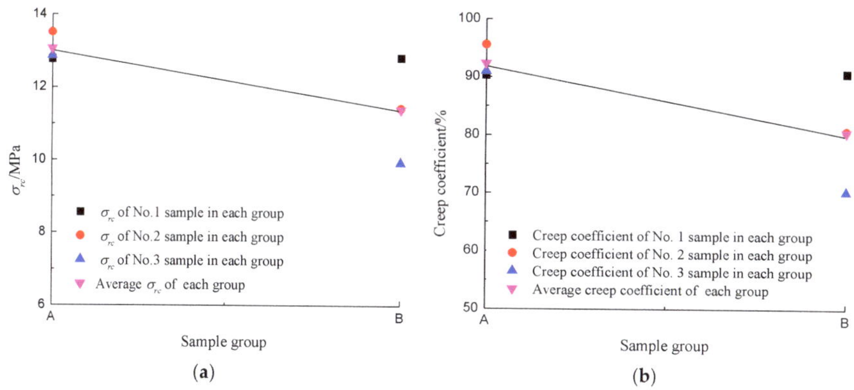

Figure 6. Comparisons of σ_{rc} and creep coefficients for composite samples: (a) σ_{rc} comparison of composite samples; (b) Creep coefficient comparison of composite samples.

In Table 2 and Figure 6, σ_{rc} and creep coefficients of composite samples varied with the stress level duration. The average σ_{rc} of Group A composite samples was 13.05 MPa with an average creep coefficient of 92.20%; whereas the average σ_{rc} of Group B composite samples was 11.39 MPa, and their average creep coefficient was 80.55%. These illustrated that as S increased from 2 h to 6 h, σ_{rc} and creep coefficients of composite samples gradually decreased, which were consistent with previous investigations on creep characteristics of pure coal or rock samples. Compared with Group A composite samples, both average σ_{rc} and creep coefficient of Group B composite samples decreased by 12.45%.

3.2. Analyses on Creep Strength Characteristics of Composite Samples

The failure properties could well reflect the strength characteristics of the sample under loading. After the short-term uniaxial creep loading, the failures of composite samples mainly occurred within the coal, and no obvious failures were found in the sandstone (in Section 6). These illustrated that the failure of coal determined the overall instability of the composite sample, i.e., the creep strength of coal played a major in the creep strength of the composite sample. In order to analyze the relationship between creep strengths of pure coal samples and composite samples, the short-term uniaxial creep tests with multi-stage loading were conducted on pure coal samples (labeled as PC-1 and PC-2) with a diameter of 50 mm and a height of 50 mm using the AG-X250 servo-controlled testing system (Figure 7). And the test scheme was same as that of composite samples.

Figure 7. Short-term uniaxial creep tests on pure coal samples.

Figure 8a,b show the creep curves of PC-1 and PC-2 pure coal samples, respectively, and the values of their creep strengths (σ_{pc}), σ_n, σ_{n-1} and t are given in Table 3. In Table 3, the values of σ_{pc} for PC-1 and PC-2 pure coal samples were 12.46 MPa and 11.39 MPa, respectively, which were different from σ_{rc} of composite samples. σ_{pc} and σ_{rc} are compared in Figure 9.

Table 3. σ_n, σ_{n-1} and t and creep coefficients of pure coal samples.

Sample Type	Sample No.	σ_n/MPa	σ_{n-1}/MPa	t/s	σ_{pc}/MPa	T/h
Pure coal sample	PC-1	12.73	12.02	4481	12.46	2
	PC-2	12.02	11.31	2480	11.39	6

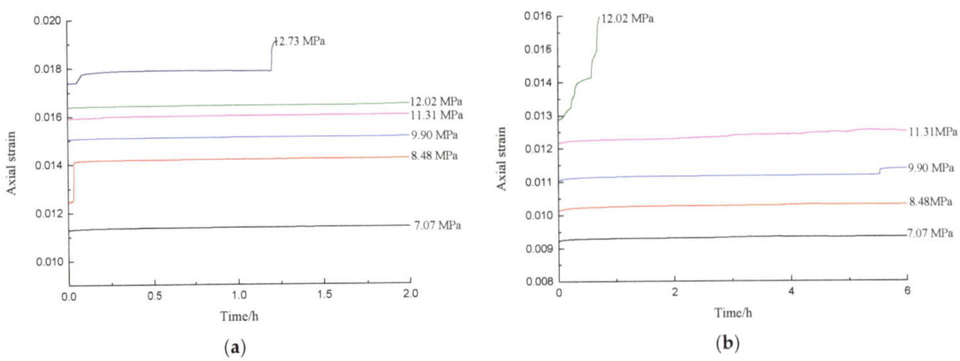

Figure 8. Creep curves of pure coal samples: (**a**) Creep curve of PC-1 pure coal sample; (**b**) Creep curve of PC-2 pure coal sample.

For the pure coal sample, under the actions of gradual stress levels, it was incurred creep deformations and damages. When the accumulative creep deformations (damages) reached its creep deformations (damages) limit, pure coal sample would fail. However, the sandstone-coal composite sample was a combined body, under the actions of gradual stress levels, the rock and coal in the composite sample were both incurred creep deformations and damages. The average UCS of sandstone (72.89 MPa) was 4.82 times larger than that of coal (15.11 MPa). The coal in the composite sample would occur creep failure first, inducing the overall instability and failure of the composite sample. While the sandstone was still at the linear elastic stage due to its larger strength. Before the creep failure of coal, the sandstone has had creep deformations occur, consuming some applied mechanical work as strain energy in it. Therefore, under the certain external input energy, the creep deformation and damage in coal of the composite sample was smaller than that in pure coal sample. The creep deformation of sandstone limited the accumulations of creep deformation (damage) in coal, and which would be destroyed at a higher input energy level compared with the pure coal sample. Therefore, the creep strengths of composite samples were larger than that of pure coal samples. In Figure 9a, compared with σ_{pc} of PC-1 pure coal sample, the values of σ_{rc} for A-1, A-2, and A-3 composite samples increased by 2.49%, 8.51%, and 2.97%, respectively. In Figure 9b, compared with the σ_{pc} of PC-2 pure coal sample, the values of σ_{rc} for B-1, and B-2 composite samples increased by 2.49%, 8.51%, and 2.97%, respectively. However, σ_{rc} of B-3 composite sample decreased by 12.82% compared with σ_{pc} of PC-2 pure coal sample. This was mainly because that the initial damages in coal were relatively larger that of B-1, B-2 composite samples and PC-2 pure coal sample.

It is important to point out that the sandstone was at the linear elastic stage when the coal occurred the creep failure. The creep failure of coal could cause the rebound deformation of sandstone, and strain energy in it would release rapidly. Partial strain energy was used to overcome the damping of sandstone to restore its the initial state. The other strain energy exerted work on the coal to promote its further creep failure. This process appeared before the occurrence of main creep fracture in coal.

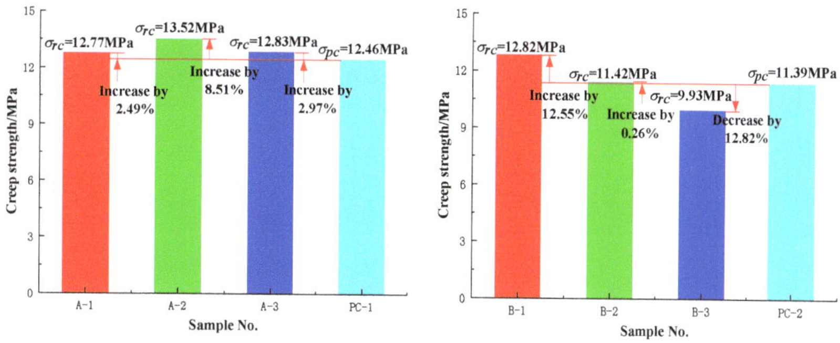

(a) (b)

Figure 9. Comparisons of σ_{pc} and σ_{rc} for composite samples and pure coal samples: (**a**) Comparisons of σ_{pc} and σ_{rc} of Group A composite samples and PC-1 pure coal sample; (**b**) Comparisons of σ_{pc} and σ_{rc} of Group B composite samples and PC-2 pure coal sample.

4. Creep Deformation Characteristics of Composite Samples

Data acquisition interval in these creep tests was relatively small, and the amount of data was large and dense, which were of great significances to the analysis of creep deformation characteristics of composite samples. Figure 10a,b show creep curves of A-1 and B-3 composite samples, respectively. The axial creep strains of Groups A and B composite samples at different stress levels are compared in Figure 11a,b, respectively.

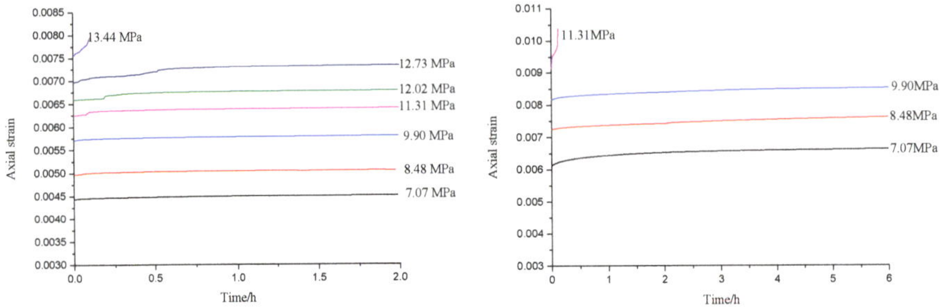

(a) (b)

Figure 10. Typical creep curves of sandstone-coal composite samples: (**a**) Creep curve of A-1 composite sample; (**b**) Creep curve of B-3 composite sample.

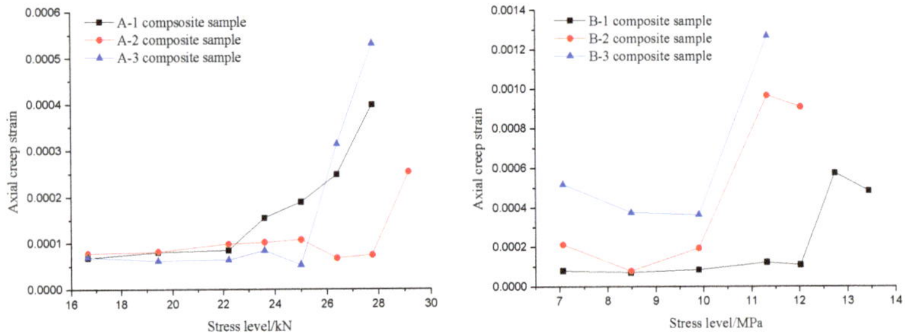

(a) (b)

Figure 11. Axial creep strains of sandstone-coal composite samples at different stress levels: (**a**) Axial creep strains of Group A composite samples; (**b**) Axial creep strains of Group B composite samples.

In Figure 10, at the beginning of each stress level, the composite sample occurred the instantaneous elastic response, producing the instantaneous elastic deformation. After the stress was stable, the composite sample entered the creep stage, and occurred the creep deformation. The creep deformation rate decreased with the loading time at the deceleration creeping stage, and it tended to be stable finally at the stable creeping stage. At each stress level, the creep curves of composite samples all appeared the deceleration creeping stage and stable creeping stage. Meanwhile, the creep curves of composite samples occurred the unstable creeping stage, and the corresponding creep deformation sharply increased at the last stress level. Then composite samples experienced the creep failure and instability.

In Figure 11, as a whole, the axial creep strain increased with the stress level, and under same stress level, the longer the loading time was, the higher the creep strain was. However, the phenomena of a low stress level with a large axial creep strain, a high stress level with a low axial strain, and a short stress level duration with a great axial creep strain under same stress level, were also found in Figure 11. For example, under the stress level of 11.31 MPa, the axial strain of A-1 composite sample was larger than that of B-1 composite sample. The main reason for these phenomena was that the creep process was a concurrent process of hardening and softening for composite sample at each stress level, which was the comprehensive embodiment of the competition between the hardening and softening. When creep softening was dominant, the corresponding creep strain was relatively large, while when creep hardening was dominant, the corresponding creep strain was relatively small. Meanwhile, before the unstable creeping stage, the axial creep strain-time curve did not increase steady, but occurred the slight or violent local increasing fluctuations, as shown in Figure 12. These also induced the abnormal axial creep strain variation phenomena in Figure 11.

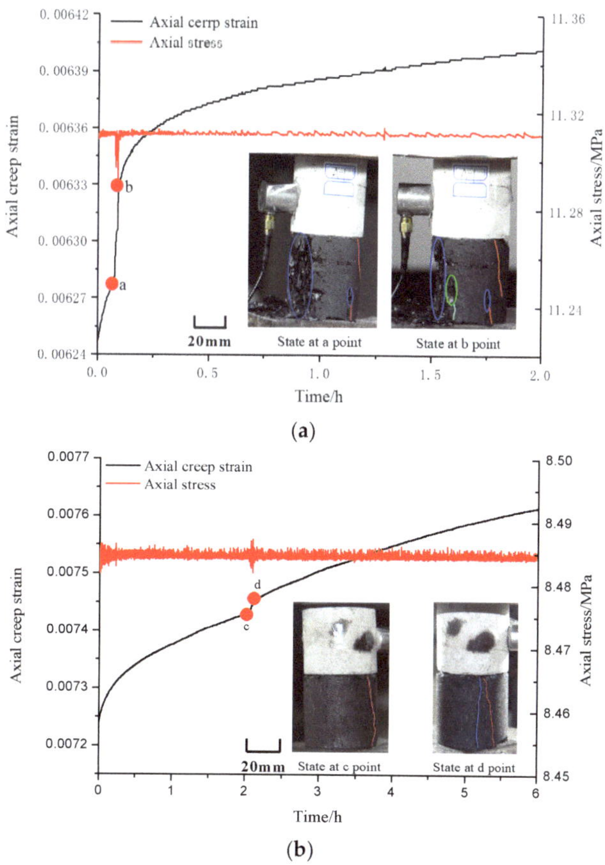

Figure 12. Axial creep strain fluctuations of sandstone-coal composite samples (the scale bar between the sample size in the figure and that of the real sample is 1:4): (**a**) Axial creep strain fluctuation of A-1 composite sample at the stress level of 11.31 MPa; (**b**) Axial creep strain fluctuation of B-3 composite sample at the stress level of 8.48 MPa.

In Figure 12a, the A-1 composite sample occurred the splitting ejection failure within coal before a point, and this failure was caused by the loading from the stress level of 8.48 MPa to the stress level of 9.90 MPa, which was analyzed comprehensively in Section 5. From a point to b point, the composite sample experienced the splitting ejection failure within coal around the initial splitting ejection failure zone again, as shown in the green elliptical area of Figure 12a. This process was accompanied with the macro-crack initiation and propagation near the new formed splitting ejection failure zone. These failures induced the linear sudden increase of axial creep strain, and the corresponding axial stress had a sudden drop. The axial creep strain was 0.000050 from a point to b point, which accounted for 32.26% of the whole axial creep strain at the stress level of 11.31 MPa. In Figure 12b, the B-3 composite sample occurred the splitting failure within coal before c point, and an obvious macro-tensile crack was found in coal. This failure was caused by the loading from the initial loading point to the first stress level of 7.07 MPa, as shown in the red line of Figure 12b, which was analyzed comprehensively in Section 5. From c point to d point, the composite sample experienced the splitting failure within coal again, and a blue macro-tensile crack was formed near initial red macro-tensile crack. These failures induced the sudden increase of axial creep strain, and the corresponding axial stress had serious fluctuations. The axial creep strain was 0.000021 from c point to d point, which accounted

for 5.59% of the whole axial creep strain at the stress level of 8.48 MPa. The fluctuation degree of creep axial strain caused by splitting ejection failure was stronger than that by tensile failure.

5. AE Characteristics of Composite Samples

In this investigation, the AE event was selected to analyze AE characteristics of composite samples under short-term uniaxial creep loading. Figures 13 and 14 give the variety curves of AE event and axial strain for A-1 and B-3 composite samples at creep stage for different stress levels, respectively. And a, b, c, d, e, f, g in Figure 13 refer to the stress levels of 7.07 MPa, 8.48 MPa, 9.90 MPa, 11.31 MPa, 12.02 MPa, 12.73 MPa, 13.44 MPa, respectively. And a, b, c, d in Figure 14 refer to the stress levels of 7.07 MPa, 8.48 MPa, 9.90 MPa, 11.31 MPa, respectively. Among them, the a, b, c, and d points in Figures 13 and 14 were same as those in Figure 12, respectively. The AE characteristics of composite samples under short-term uniaxial creep loading were as follows:

(1) At each stress level, the creep strain rate decreased gradually in the deceleration creeping stage, and the intensity and frequency of AE events were remarkably weakened. AE events experienced large fluctuations with peak values at the beginning of this stage. As the time increased, the composite sample entered the stable creeping stage with a constant creep strain rate. The corresponding intensity and frequency of AE events were relatively low and stable. Generally, the deceleration creeping stage and stable creeping stage occurred at a relatively low stress level, and the unstable creeping stage arose at a high stress level. As the composite sample entered the unstable creeping stage, the creep strain rate increased gradually. The composite sample occurred failure and instability. The intensity and frequency of AE events were strengthened again, which exhibited significant fluctuations with more peak values. At the end of this stage, the axial creep strain-time curve rose with a fast creep strain rate, and AE events displayed large fluctuations. The intensity and frequency of AE activities can reflect the creep damage evolution process of composite sample under short-term uniaxial creep loading. Therefore, under same stress level, the creep damage degree at the unstable creeping stage was the most serious, larger than that at the stages of deceleration creeping and stable creeping. The creep damage degree at the stable creeping stage was the lowest.

(2) Under same stress level duration, with the increase of the stress level, the intensity and frequency of AE events at the deceleration creeping stage and stable creeping stage were enhanced. These illustrated that the creep damage degree of the composite sample increased with the stress level. Until the occurrence of the unstable creeping stage at the last stress level, the accumulated creep damage exceeded the damage limit of the composite sample, and the composite sample experienced the failure and instability.

(3) Under same stress level, with the increase of the stress level duration, the AE activities were strengthened. The intensity and frequency of AE events at different stages were enhanced. These illustrated that creep damage degree of the composite sample increased with the stress level duration under same stress level. These also can verify the accuracy of experimental results, i.e., the creep strength at a large stress level duration was smaller than that at a small stress level duration.

In order to better describe the creep damage evolution of the composite sample, the creep damage coefficient δ was defined as

$$\delta = \frac{A}{D} \tag{2}$$

where, A was cumulative count of AE events at creep stage for a stress level; D was creep time for a stress level. The values of δ of A-1 and B-3 composite samples under different stress levels are compared in Figure 15.

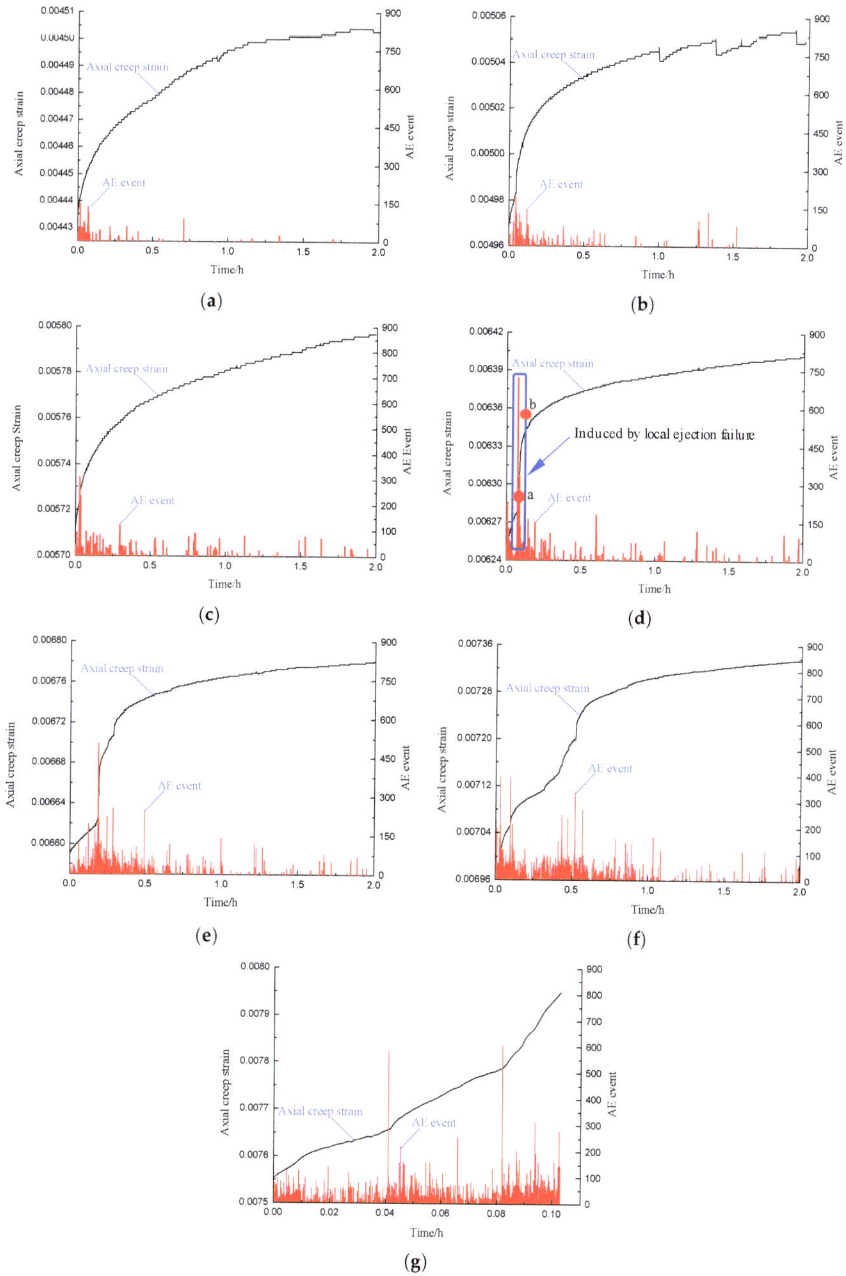

Figure 13. AE event variety curves of A-1 composite samples at creep stage for different stress levels: (**a**) AE event variety curve at the stress level of 7.07 MPa; (**b**) AE event variety curve at the stress level of 8.48 MPa; (**c**) AE event variety curve at the stress level of 9.90 MPa; (**d**) AE event variety curve at the stress level of 11.31 MPa; (**e**) AE event variety curve at the stress level of 12.02 MPa; (**f**) AE event variety curve at the stress level of 12.73 MPa; (**g**) AE event variety curve at the stress level of 13.44 MPa.

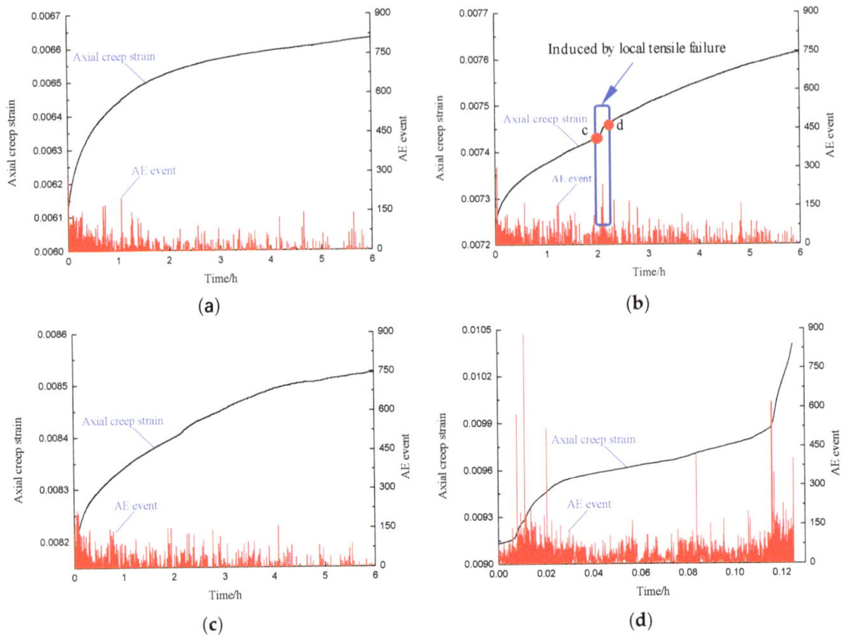

Figure 14. AE event variety curves of B-3 composite samples at creep stage for different stress levels: (**a**) AE event variety curve at the stress level of 7.07 MPa; (**b**) AE event variety curve at the stress level of 8.48 MPa; (**c**) AE event variety curve at the stress level of 9.90 MPa; (**d**) AE event variety curve at the stress level of 11.31 MPa.

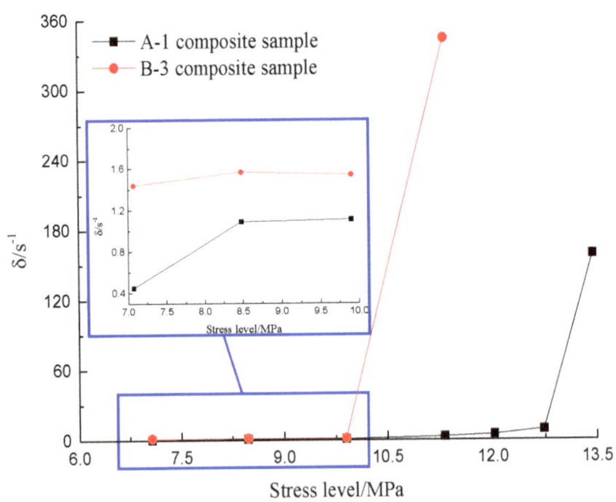

Figure 15. Comparison of δ values for A-1 and B-3 composite samples.

In Figure 15, under same stress level duration, δ increased with the stress level. These illustrated that the creep damage increased with the increase of the stress level. At the last stress level, δ showed a sharp increase with the most serious creep damage for the composite sample. Under same stress level, δ at a large stress level duration was greater than that at a small stress level duration. These illustrated that the creep damage of the composite sample at a large stress level duration was larger than that at a small stress level duration.

If the composite sample occurred the local failure or fracture at a stress level, the AE event would experience the peak value. For instance, the ejection failure of A-1 composite

sample caused the sudden increase of AE event at the stress level of 11.31 MPa, as shown in Figure 13d. And the AE event of B-3 composite sample experienced the relatively small peak value at the stress level of 8.48 MPa, as shown in Figure 14b.

6. Failure Characteristics of Composite Samples

6.1. Macro-Failure Patterns of Composite Samples

Figure 16a–f shows the macro-failure images of A-1, A-2, A-3, B-1, B-2, and B-3 composite samples under short-term uniaxial creep loading, respectively. In Figure 16, the failures of the composite samples mainly occurred within the coal, and no obvious failures were found in the sandstone. During creep loading, the coal experienced the ejection failure, ejecting small coal blocks. Further, the vertical tensile cracks were also found in the coal, as shown in red lines of Figure 16. Therefore, the coal mainly displayed the splitting ejection failure. With the increase of the stress level duration, the defects in the coal had sufficient time to propagate and coalesce, and the failure degrees of coal in Group B composite sample were greater than that in Group A composite samples.

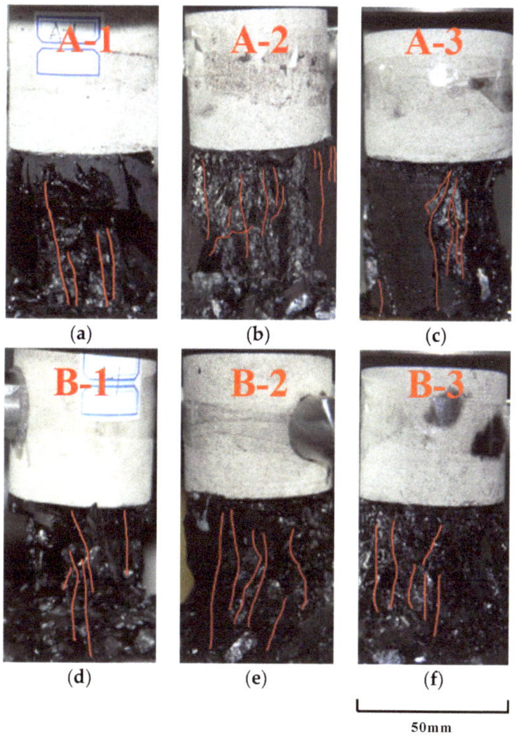

Figure 16. Macro-failure patterns of composite samples (the scale bar between the sample size in the figure and that of the real sample is 1:2): (**a**) Macro-failure pattern of A-1 composite sample; (**b**) Macro-failure pattern of A-2 composite sample; (**c**) Macro-failure pattern of A-3 composite sample; (**d**) Macro-failure pattern of B-1 composite sample; (**e**) Macro-failure pattern of B-2 composite sample; (**f**) Macro-failure pattern of B-3 composite sample.

6.2. Macro-Progressive Failure Characteristics

In the multi-step loading creep tests, the failure and instability of the composite sample were jointly caused by the creep damage due to the constant stress at each stress level, and by the loading damage between two adjacent stress levels as well as in the period from the initial loading point to the first stress level. These were attributed to the loading mechanism of our multi-step loading creep tests. The A-1 and B-3 composite samples were

selected to analyze the macro-progressive failure characteristics under different stress level durations, as shown in Figure 17I,II, respectively. Among them, the points of a~h represent the progressive failure points of the A-1 composite sample, and the points of i~o represent the progressive failure points of the B-3 composite sample.

In Figure 17I, there were two pre-existing vertical cracks on the surface of coal for A-1 composite sample. From the initial loading point to the second stress level (8.38 MPa), no obvious macro-failure was found on the surface of the composite sample, and the macro-failure occurred the deceleration creeping stage of the second stress level. Now, a small ejection failure (Blue ellipse in Figure 17I) occurred around the pre-existing vertical crack in coal accompanying by the tensile fracture. During the loading process from the second stress level to the third stress level (9.90 MPa), the composite sample experienced a relatively large failure. A tensile crack was first formed on the surface of coal, and then ejection failure occurred at the upper area of this crack, which was near the interface of sandstone and coal. Finally, the ejection failure zone could induce the chain destruction of the peripheral parts to form a large ejection failure zone along the tensile crack. At the deceleration creeping stage of the fourth stress level (11.31 MPa), the lower area of the large ejection failure zone experienced a small ejection failure again, and a small tensile fracture occurred in this area. When the axial stress reached 12.02 MPa (Fifth stress level), many macro-cracks were formed around the pre-existing vertical cracks and ejection failure zone, which were accompanied by small ejection failures. At the stable creeping stage of the stress level (12.73 MPa), more macro-cracks were generated on the surface of coal, and the ejection failure zones began to connect as a whole. At the last stress level (13.44 MPa), the propagation and coalescence of pre-existing vertical cracks and newborn macro-cracks made the coal become more broken, and a large ejection failure zone was formed on the surface of coal. Finally, these induced the failure of coal, and which further caused the whole instability and failure of the composite sample. Meanwhile, during this process, the rock rebound deformation enhanced the future failure of coal. Now, the composite sample lost the bearing capacity, and the axial stress decreased sharply with the rapid of axial strain.

In Figure 17II, the first macro-failure of B-3 composite sample occurred between the initial loading point and the first stress level (8.38 MPa). And a macro-tensile crack was formed on the surface of the coal. These indicated that the initial damage of coal was relatively large, and thus the creep strength of B-3 composite sample was comparatively small. There was no obvious macro-failure under the first stress level and between it and the second stress level (9.90 MPa). While, at the stable creeping stage of the second stress level, the coal displayed the tensile fracture, and a macro-tensile crack was found near the left area of the first macro-tensile crack. Between the second stress level and third stress level (11.31 MPa), more macro-tensile cracks were formed on the surface of the coal, and the lower part of the coal experienced a small ejection failure. At the stable creeping stage of the second stress level, the ejection failure zone induced the chain destruction of the peripheral parts to form a relatively large ejection failure zone in the coal, accompanied by the propagation and coalescence of macro-cracks. Then at the deceleration creeping stage of the fourth stress level (12.02 MPa), the composite sample occurred a large ejection failure. This caused the coal failure, and which then induced the whole instability and failure of the composite sample. Due to the large stress level duration, the rock rebound deformation was more serious, making the coal more broken than that of A-1 composite sample.

Meanwhile, when the composite experienced the macro-failure before final failure, a stress drop occurred on the stress-time curve. The axial stress decreased first and then increased, and this was the process from the failure to stability by adjusting the internal structures of the composite sample. The corresponding axial strain displayed fluctuations, and AE event occurred a peak value.

According to above analyses, the macro-failure of the composite sample occurred with the coal. The propagation and coalescence of pre-existing cracks and new-born cracks made the coal more broken, and the ejection failure zone induced the secondary ejection

failure of the peripheral parts to form a large ejection failure zone in the coal. These caused the final failure of coal, which resulted in the overall failure and instability of the composite sample. The rebound deformation of rock aggravates the fracture and movement of coal, promoting its failure degree, especially when the stress-level duration was large.

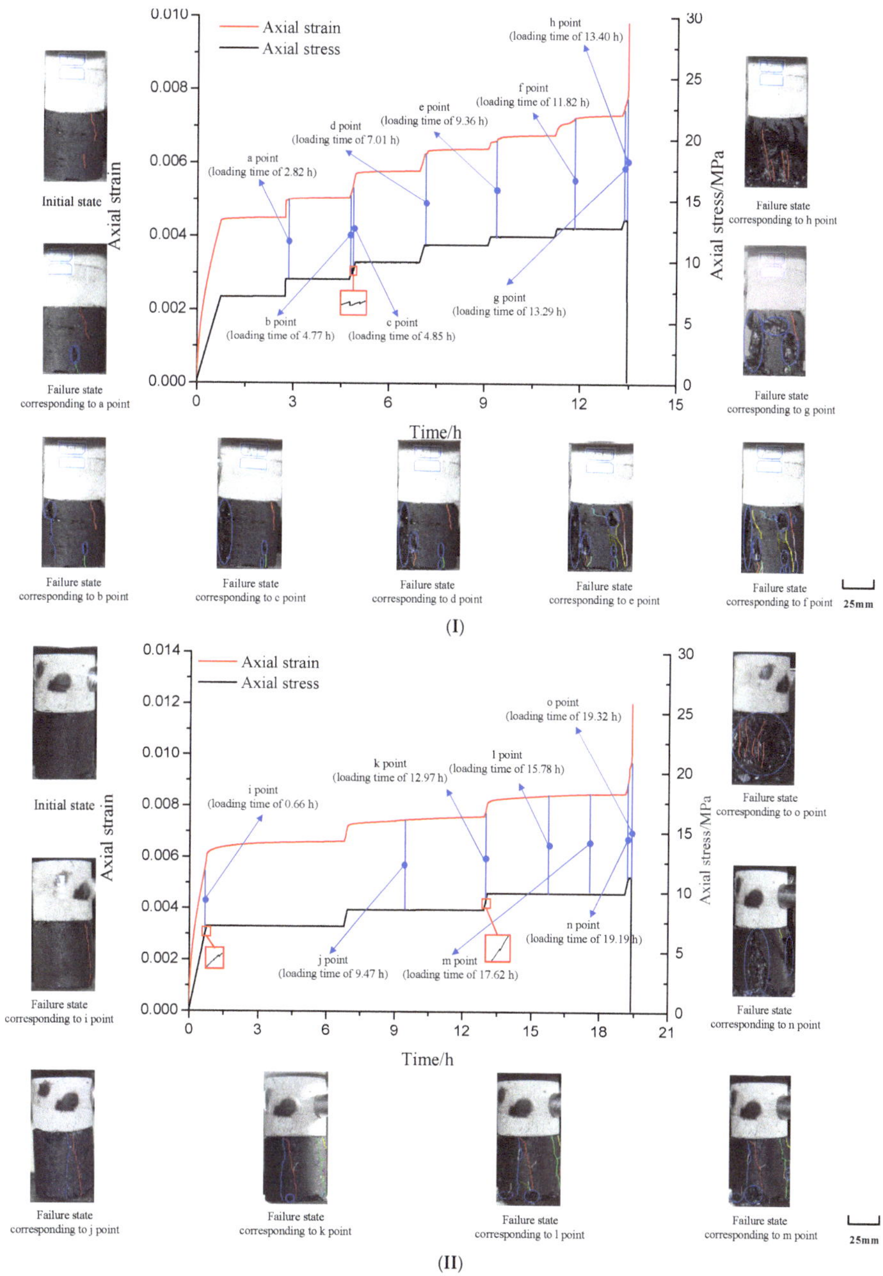

Figure 17. Macro-progressive failure processes of composite samples (the scale bar between the sample size in the figure and that of the real sample is 1:2.5): (**I**) Macro-progressive failure process of A-1 composite sample; (**II**) Macro-progressive failure process of B-3 composite sample.

7. Conclusions

In this paper, the uniaxial short-term multi-step loading creep tests with different stress-level durations were conducted on sandstone-coal composite samples to investigate their creep strengths, creep deformations, AE characteristics, and failure properties. The main conclusions are summarized as follows:

(1) The creep strength and creep coefficient of composite sample decreased with the stress-level duration. The average creep strength of Group A composite samples (stress level duration of 2 h) was 13.01 MPa with the average creep coefficient of 92.02%, and average creep strength of Group B composite samples (stress level duration of 6 h) was 11.39 MPa with the average creep coefficient of 80.55%. Compared with Group A composite samples, the average creep strength and creep coefficient of Group B composite samples both dropped by 12.45%. The creep strength of composite sample was mainly determined by the coal, which was influenced by the interactions with the sandstone. The creep deformation and damage within the sandstone induced by the stress levels reduced that within the coal, generally resulting in a higher creep strength for composite sample compared to pure coal sample.

(2) With the increase of the stress level, the axial creep strain of composite sample generally increased. Under same stress level, the larger the stress-level duration was, the greater the axial creep strain was. However, the micro failure or fracture and local macro failure or fracture within the composite sample caused the slight or sharp rise in axial creep strain, respectively.

(3) The AE characteristics of composite samples were related to the creep strain rate, stress level, stress level duration, and local failure or fracture within the composite sample, etc. Under same stress level, the frequency and intensity of AE signals were the most serious at the unstable creeping stage, and AE signals were the weakest at the stable creeping stage. While the frequency and intensity of AE signals at stages of deceleration creeping, stable creeping and unstable creeping increased with the increase of the stress level or stress level duration under same stress level duration or stress level. The local failure or fracture within the composite sample would enhance the AE activities, and the AE signals occurred peak values.

(4) The failures of the composite sample occurred with the coal, and no obvious failure was found in the sandstone. The propagation and coalescence of pre-existing cracks and new-born cracks made the coal more broken, and the ejection failure zone induced the secondary ejection failure of the peripheral parts to form a large ejection failure zone in the coal. These caused the final failure of coal, leading to the overall failure and instability of the composite sample. The rebound deformation of sandstone aggravates the fracture and movement of coal, promoting its failure degree, especially when the stress-level duration was large.

Author Contributions: Conceptualization, F.W.; methodology, F.W.; software, D.Y.; validation, J.Z., F.L. and C.Z.; data curation, J.Z.; writing—original draft preparation, D.Y.; writing—review and editing, D.Y. and F.F.; funding acquisition, F.W. All authors have read and agreed to the published version of the manuscript.

Funding: This work was supported by the National Science Foundation of China (51904167, 52074169), the Taishan Scholars Project, the Taishan Scholar Talent Team Support Plan for Advantaged & Unique Discipline Areas, the SDUST Research Fund, the Open Research Fund for the Key Laboratory of Safety and High-efficiency Coal Mining (JYBSYS2019201), the open funding of State Key Laboratory of Mining Disaster Prevention and Control Co-Founded by Shandong Province and The Ministry of Science and Technology (SICGM202104).

Data Availability Statement: The data are available and explained in this article; readers can access the data supporting the conclusions of this study.

Conflicts of Interest: The authors declare no conflict of interest.

References

1. Metallurgical Industry Energy Saving Committee of China Energy Saving Association. *China's Steel Industry Energy Saving and Low Carbon Development Report*; Metallurgical Industry Energy Saving Committee of China Energy Saving Association: Beijing, China, 2019. (In Chinese)
2. Gao, M.; Xie, J.; Gao, Y.; Wang, W.; Li, C.; Yang, B.; Liu, J.; Xie, H. Mechanical behavior of coal under different mining rates: A case study from laboratory experiments to field testing. *Int. J. Min. Sci. Technol.* **2021**, *31*, 825–841. [CrossRef]
3. Gao, M.; Xie, J.; Guo, J.; Lu, Y.; He, Z.; Li, C. Fractal evolution and connectivity characteristics of mining-induced crack networks in coal masses at different depths. *Géoméch. Geophys. Geo-Energy Geo-Resour.* **2021**, *7*, 1–15. [CrossRef]
4. Chen, S.; Yin, D.; Cao, F.; Liu, Y.; Ren, K. An overview of integrated surface subsidence-reducing technology in mining areas of China. *Nat. Hazards* **2016**, *81*, 1129–1145. [CrossRef]
5. Lama, R.; Bodziony, J. Management of outburst in underground coal mines. *Int. J. Coal Geol.* **1998**, *35*, 83–115. [CrossRef]
6. Cui, G.; Yang, L.; Fang, J.; Qiu, Z.; Wang, Y.; Ren, S. Geochemical reactions and their influence on petrophysical properties of ultra-low permeability oil reservoirs during water and CO_2 flooding. *J. Pet. Sci. Eng.* **2021**, *203*, 108672. [CrossRef]
7. Feng, G.R.; Bai, J.W.; Shi, D.X.; Qi, T.Y.; Wang, P.F.; Guo, J.; Wang, S.Y.; Kang, L.X. Key pillar theory in the chain failure of residual coal pillars and its application prospect. *J. China Coal Soc.* **2021**, *46*, 164–179. (In Chinese)
8. Bertuzzi, R.; Douglas, K.; Mostyn, G. An Approach to model the strength of coal pillars. *Int. J. Rock Mech. Min. Sci.* **2016**, *89*, 165–175. [CrossRef]
9. Chen, S.-J.; Guo, W.-J.; Zhou, H.; Shen, B.; Liu, J.-B. Field investigation of long-term bearing capacity of strip coal pillars. *Int. J. Rock Mech. Min. Sci.* **2014**, *70*, 109–114. [CrossRef]
10. Cui, G.; Wang, W.; Dou, B.; Liu, Y.; Tian, H.; Zheng, J.; Liu, Y. Geothermal Energy Exploitation and Power Generation via a Single Vertical Well Combined with Hydraulic Fracturing. *J. Energy Eng.* **2022**, *148*, 04021058. [CrossRef]
11. Salamon, M. Stability, instability and design of pillar workings. *Int. J. Rock Mech. Min. Sci. Géoméch. Abstr.* **1970**, *7*, 613–631. [CrossRef]
12. Salamon, M.D.G.; Ozbay, M.U.; Madden, B.J. Life and design of board-and-pillar workings affected by pillar scaling. *J. South. Afr. Inst. Min. Metall.* **1998**, *98*, 135–145.
13. Chen, S.J.; Yin, D.W.; Zhang, B.L.; Ma, H.F.; Liu, X.Q. Study on mechanical characteristics and progressive failure mechanism of roof-coal pillar structure body. *Chin. J. Rock Mech. Eng.* **2017**, *37*, 1588–1598. (In Chinese)
14. Yin, D.W.; Chen, S.J.; Xing, W.B.; Huang, D.M.; Liu, X.Q. Experimental study on mechanical behavior of roof-coal pillar structure body under different loading rates. *J. China Coal Soc.* **2018**, *43*, 1249–1257. (In Chinese)
15. Wang, T.; Jiang, Y.; Zhan, S.; Wang, C. Frictional sliding tests on combined coal-rock samples. *J. Rock Mech. Geotech. Eng.* **2014**, *6*, 280–286. [CrossRef]
16. Espinoza, D.N.; Pereira, J.-M.; Vandamme, M.; Dangla, P.; Vidal-Gilbert, S. Desorption-induced shear failure of coal bed seams during gas depletion. *Int. J. Coal Geol.* **2015**, *137*, 142–151. [CrossRef]
17. Zhao, T.-B.; Guo, W.-Y.; Lu, C.-P.; Zhao, G.-M. Failure characteristics of combined coal-rock with different interfacial angles. *Géoméch. Eng.* **2016**, *11*, 345–359. [CrossRef]
18. Li, W.; Bai, J.; Cheng, J.; Peng, S.; Liu, H. Determination of coal–rock interface strength by laboratory direct shear tests under constant normal load. *Int. J. Rock Mech. Min. Sci.* **2015**, *77*, 60–67. [CrossRef]
19. Yin, D.W.; Chen, S.J.; Liu, X.Q.; Ma, H.F. Effect of joint angle in coal on failure mechanical behavior of roof rock-coal combined body. *Q. J. Eng. Geol. Hydrogeol.* **2018**, *51*, 202–209. [CrossRef]
20. Yin, D.W.; Chen, S.J.; Ge, Y.; Liu, R. Mechanical properties of rockecoal bi-material samples with different lithologies under uniaxial loading. *J. Mater. Res. Technol.* **2021**, *10*, 322–338. [CrossRef]
21. Chen, S.J.; Yin, D.W.; Jiang, N.; Wang, F.; Guo, W.J. Simulation study on effects of loading rate on uniaxial compression failure of composite rock-coal layer. *Geomech. Eng.* **2019**, *17*, 333–342.
22. Yin, D.W.; Chen, S.J.; Sun, X.Z.; Jiang, N. Strength characteristics of roof rock-coal composite samples with different height ratios under uniaxial loading. *Arch. Min. Sci.* **2019**, *64*, 307–319.
23. Zhao, Z.; Lv, X.; Wang, W.; Tan, Y. Damage evolution of bi-body model composed of weakly cemented soft rock and coal considering different interface effect. *SpringerPlus* **2016**, *5*, 292. [CrossRef]
24. Zuo, J.P.; Chen, Y.; Zhang, J.W.; Wang, J.T.; Sun, Y.J. Failure behavior and strength characteristics of coal-rock combined body under different confining pressures. *J. China Coal. Soc.* **2016**, *41*, 2706–2713. (In Chinese)
25. Lu, J.; Huang, G.; Gao, H.; Li, X.; Zhang, D.; Yin, G. Mechanical Properties of Layered Composite Coal–Rock Subjected to True Triaxial Stress. *Rock Mech. Rock Eng.* **2020**, *53*, 4117–4138. [CrossRef]
26. Chen, S.J.; Yin, D.W.; Liu, H.M.; Chen, B.; Jiang, N. Effects of coal's initial macro-cracks on rockburst tendency of rock-coal bi-material samples. *R. Soc. Open Sci.* **2019**, *6*, 181795. [CrossRef]
27. Huang, B.X.; Liu, J.W. The effect of loading rate on the behavior of samples composed of coal and rock. *Int. J. Rock Mech. Min. Sci.* **2013**, *61*, 23–30. [CrossRef]
28. Nie, B.S.; He, X.Q.; Zhu, C.W. Study on mechanical property and electromagnetic emission during the fracture process of combined coal-rock. *Prog. Earth Planet. Sci.* **2009**, *1*, 281–287.
29. He, X.; Chen, W.; Nie, B.; Mitri, H. Electromagnetic emission theory and its application to dynamic phenomena in coal-rock. *Int. J. Rock Mech. Min. Sci.* **2011**, *48*, 1352–1358. [CrossRef]

30. Jin, P.; Wang, E.; Liu, X.; Huang, N.; Wang, S. Damage evolution law of coal-rock under uniaxial compression based on the electromagnetic radiation characteristics. *Int. J. Min. Sci. Technol.* **2013**, *23*, 213–219. [CrossRef]
31. Qu, X. *Experimental Study on Influence of Mechanical Properties of Roof and Floor on Stability of Strip Coal Pillar*; Shandong University of Science and Technology: Qingdao, China, 2018. (In Chinese)
32. Yin, D.W. *Experimental Study on Stabilities of Roof-Coal Pillar Structural Body*; Shandong University of Science and Technology: Qingdao, China, 2018. (In Chinese)
33. Guo, W.Y.; Tan, Y.L.; Yu, F.H.; Zhao, T.B.; Hu, S.C.; Huang, D.M.; Qin, Z. Mechanical behavior of rock-coal-rock specimens with different coal thicknesses. *Geomech. Eng.* **2108**, *15*, 1017–1027.
34. Zhang, H.; Wan, Z.; Zhang, Y.; Wu, D. Mechanical Properties and Failure Behavior of Composite Samples. *Adv. Mater. Sci. Eng.* **2018**, *2018*, 1–16. [CrossRef]
35. Zhang, H.; Elsworth, D.; Wan, Z. Failure response of composite rock-coal samples. *Géoméch. Geophys. Geo-Energy Geo-Resour.* **2018**, *4*, 175–192. [CrossRef]
36. Zhang, Z.T.; Liu, J.F.; Wang, L.; Yang, H.T.; Zuo, J.P. Effects of combination mode on mechanical properties and failure characteristics of the coal rock combinations. *J. China Coal Soc.* **2012**, *37*, 1677–1681. (In Chinese)
37. Fatemeh, S.R.; Mark, D.Z. Comparison of short-term and long-term creep experiments in shales and carbonates from unconventional gas reservoirs. *Rock Mech. Rock Eng.* **2018**, *51*, 1995–2014.
38. Cao, P.; Wen, Y.D.; Wang, Y.X.; Yuan, H.P.; Yuan, B. Study on nonlinear damage creep constitutive model for high-stress soft rock. *Environ. Earth Sci.* **2016**, *75*, 900. [CrossRef]
39. Maranini, E.; Brignoli, M. Creep behaviour of a weak rock: Experimental characterization. *Int. J. Rock Mech. Min. Sci.* **1999**, *36*, 127–138. [CrossRef]
40. Yang, S.-Q.; Jing, H.-W.; Cheng, L. Influences of pore pressure on short-term and creep mechanical behavior of red sandstone. *Eng. Geol.* **2014**, *179*, 10–23. [CrossRef]
41. Yang, D.; Chen, L.; Yang, S.; Chen, W.; Wu, G. Experimental investigation of the creep and damage behavior of Linyi red sandstone. *Int. J. Rock Mech. Min. Sci.* **2014**, *72*, 164–172. [CrossRef]
42. Zong, Y.; Han, L.; Jin, Y.; Zhao, W.; Meng, L. Experimental Investigation on the Post-Peak Short-Term and Creep Behavior of Fractured Sandstone. *Energies* **2020**, *13*, 598. [CrossRef]
43. Yang, Y.J.; Wang, D.C.; Zhao, N.N.; Chen, S.J. Acoustic emission characteristics of coal creep under step load. *J. Basic Sci. Eng.* **2013**, *21*, 159–166. (In Chinese)
44. Chen, S.J.; Guo, W.J.; Yang, Y.J. Experimental study of creep model and failure characteristics of coal. *Rock Soil Mech.* **2009**, *30*, 2595–2599. (In Chinese)
45. Cao, S.G.; Liu, Y.B.; Zhang, L.Q.; Jiang, Y.D. Experimental on acoustic emission of outburst-hazardous coal under uniaxial compression and creep. *J. China Coal Soc.* **2007**, *32*, 1264–1268. (In Chinese)
46. Wu, F.; Chen, J.; Zou, Q.L. A nonlinear creep damage model for salt rock. *Int. J. Damage Mech.* **2019**, *28*, 758–771. [CrossRef]
47. Wu, L.X.; Wang, J.Z. Preliminary exploration to rheology and micro-effect characteristics of coal. *Chin. J. Mech. Eng.* **1996**, *15*, 328–332. (In Chinese)

minerals

MDPI

Article

Optimal Support Solution for a Soft Rock Roadway Based on the Drucker–Prager Yield Criteria

Minghui Ma [1,2], Qifeng Guo [1,*], Jiliang Pan [1,*], Chi Ma [1] and Meifeng Cai [1]

[1] School of Civil and Resource Engineering, University of Science and Technology Beijing, Beijing 100083, China; b20190007@xs.ustb.edu.cn (M.M.); d202110033@xs.ustb.edu.cn (C.M.); caimeifeng@ustb.edu.cn (M.C.)
[2] Shandong Gold Mining (Laizhou) Limited Company Jiaojia Gold Deposit, Laizhou 261400, China
* Correspondence: guoqifeng@ustb.edu.cn (Q.G.); b20170008@xs.ustb.edu.cn (J.P.)

Abstract: Through theoretical calculation, the stress and deformation of surrounding rock can be analyzed, providing guidance for the support design and optimization of soft rock roadways. In this paper, theoretical solutions for both the optimal support pressure and the allowable maximum displacement of surrounding rock are derived from the Drucker–Prager (DP) yield criteria and the steady creep criterion expressed by the third invariant of deviator stress. The DP criterion with different parameters is compared and analyzed with an engineering example. Then, based on the calculation results the effects of long-term strength, cohesion, and internal friction angle of soft rock on the maximum plastic zone radius and allowable maximum displacement of roadway are discussed. The results show that the optimal support solution of soft rock roadways based on the DP criteria can not only reasonably reflect the intermediate principal stress but can also be used to compare and discuss the influence of different DP criteria on the calculation results. The higher the long-term strength of the rock surrounding a roadway is, the smaller the optimal support force is and the larger the allowable maximum displacement is. When the calculated long-term strength of soft rock can ensure that the deformation of the roadway does not exceed the allowable maximum displacement, the roadway can maintain long-term stability without support. With an increase in the cohesion or internal friction angle of soft rock, the radius of the plastic zone decreases gradually and the allowable maximum displacement is reduced by degrees. The use of grouting and other means to improve the strength of surrounding rock can effectively reduce the roadway deformation and save support costs. This new theoretical solution can consider different intermediate principal stress effects and different DP strength criteria, enabling the parameters to become easier to determine. It has a wider range of applications, and the calculation results better demonstrate the strength potential of the surrounding rock.

Keywords: roadway; soft rock; optimal support; Drucker–Prager yield criteria; elasto-plastic analysis

Citation: Ma, M.; Guo, Q.; Pan, J.; Ma, C.; Cai, M. Optimal Support Solution for a Soft Rock Roadway Based on the Drucker–Prager Yield Criteria. *Minerals* **2022**, *12*, 1. https://doi.org/10.3390/min12010001

Academic Editors: Diyuan Li, Zhenyu Han, Xin Cai, Shijie Xie and Yosoon Choi

Received: 9 November 2021
Accepted: 13 December 2021
Published: 21 December 2021

1. Introduction

Rock roadways exist widely in underground spaces, such as mining engineering, civil engineering, and tunnel engineering works. For deep geotechnical engineering works, some roadways will gradually transform from shallow hard rock roadways to deep soft rock roadways. Therefore, the problem of soft rock support has become a major safety issue that needs to be solved urgently [1–3]. Soft rock has obvious creep deformation characteristics. In engineering, serious extrusion deformation will occur in all directions of the roadway, which can lead to the instability and failure of the surrounding rock support structure [4–7]. In the process of the support design of soft rock roadways, it is generally necessary to optimize the support parameters reasonably by theoretical calculation, guide engineering design through theoretical calculation results, and reduce the uncertainty caused by the engineering analogy method used [8–10].

In the process of underground roadway excavation, the stress field will redistribute around the excavation area. The redistribution of the stress field leads to convergence

deformation of cavities produced by excavation. The size of deformation is related to rock mass properties, in situ stress, and support conditions [11–14]. It is important to analyze the distribution of the stress field and the displacement field of surrounding rock by theoretical calculation [15–17]. According to the creep mechanism of rock and the rheological control principle of soft rock, some scholars have established the optimal support calculation method for soft rock and solved the optimal support force and the allowable maximum displacement of surrounding rock [18–21]. Among them, the DP criteria not only eliminates the singularity of the yield surface of the hexagonal pyramid, but also reflects the effect of intermediate principal stress and hydrostatic pressure, which has been widely used in underground engineering [22–24]. However, the existing optimal support solutions for soft rock roadways have not discussed the differences in DP criteria [25–27].

The instability of excavation is usually caused by the excessive concentration of stress in the rock mass near the excavation, the excessive stress of supporting components, or the change of rock deformation and strength characteristics [28–32]. The deformation pressure of viscoelastic rock mass on underground roadway support depends on the properties of the surrounding rock and the rock–support interaction. Therefore, based on DP series criteria reflecting intermediate principal stress, this paper deduces the analytical solution of optimal support force and the allowable maximum displacement of surrounding rock in a circular roadway. In doing so, it compares different DP criteria by an engineering example and discusses the effects of long-term strength, cohesion, and internal friction angle of soft rock on the maximum plastic zone radius and allowable maximum displacement of a roadway. The research results can provide theoretical guidance for the rational design and optimization of soft rock roadway supports.

2. Drucker–Prager Yield Criteria

The yield surface of the Mohr–Coulomb (MC) yield criterion is an irregular hexagonal pyramid in three-dimensional stress space. To eliminate the singularity of the yield surface on the cone top and the ridgeline, Drucker and Prager proposed a smooth conical yield surface that is inscribed in the MC yield criterion hexagonal pyramid [33]. According to the relative positional relationship between the DP yield criterion and the MC yield criterion on the π plane, the DP yield criteria are derived [34]. The DP yield criteria can be given by:

$$\alpha I_1 + \sqrt{J_2} = k \tag{1}$$

where the parameters α and k are related to the cohesion c and the internal friction angle φ of the surrounding rock. According to the matching relationship with the MC criterion, the corresponding parameter expressions are shown in Table 1. I_1 is the first invariant of the stress tensor, $I_1 = \sigma_1 + \sigma_2 + \sigma_3$; J_2 is the second invariant of stress deviator, $J_2 = [(\sigma_1 - \sigma_2)^2 + (\sigma_2 - \sigma_3)^2 + (\sigma_3 - \sigma_1)^2]/6$; and σ_1, σ_2, and σ_3 are the large, medium, and small principal stresses of the surrounding rock, respectively.

Table 1. Parameter expressions of Drucker–Prager yield criteria.

Serial Number	Criterion Types	α	k
DP1	The MC criterion based on external corner circumscribed, a circle yield criterion	$\frac{2\sin\varphi}{\sqrt{3}(3-\sin\varphi)}$	$\frac{6c\cos\varphi}{\sqrt{3}(3-\sin\varphi)}$
DP2	The MC criterion based on inner corner circumscribed, a circle yield criterion	$\frac{2\sin\varphi}{\sqrt{3}(3+\sin\varphi)}$	$\frac{6c\cos\varphi}{\sqrt{3}(3+\sin\varphi)}$
DP3	The MC criterion based on matching circles, for plain strain problems with associated flow rules	$\frac{\sin\varphi}{\sqrt{3}\sqrt{3+\sin^2\varphi}}$	$\frac{3c\cos\varphi}{\sqrt{3}\sqrt{3+\sin^2\varphi}}$
DP4	The MC criterion based on equivalent area, a circle yield criterion	$\frac{2\sqrt{3}\sin\varphi}{\sqrt{2\sqrt{3}\pi(9-\sin^2\varphi)}}$	$\frac{2\sqrt{3}c\cos\varphi}{\sqrt{2\sqrt{3}\pi(9-\sin^2\varphi)}}$
DP5	The MC criterion based on matching circles, for plain strain problems with non-associated flow rules	$\frac{\sin\varphi}{3}$	$c\cos\varphi$

The expression of the DP criteria under plane strain condition is as follows:

$$\sigma_1 = M\sigma_3 + N \tag{2}$$

where $M = \frac{1+3\alpha}{1-3\alpha}$ and $N = \frac{2k}{1-3\alpha}$. Because the internal friction angle φ is always greater than 0, $\alpha \neq 0$ and $M \neq 1$.

3. Optimum Support Calculation of Roadway

3.1. Basic Assumptions

To carry out the elastic-plastic analysis of surrounding rock, the following assumptions are made:

(1) The cross section of the roadway is circular, and the length is infinite, so it can be simplified as a plane strain problem.
(2) The surrounding rock of the roadway is a continuous, homogeneous, and isotropic elastic-plastic material.
(3) Ignoring the effects of the surrounding rock weight on the yield, the original rock stress can be simplified as a uniform stress distribution. The roadway is under the conditions of uniform in situ stress and support force.

3.2. Elastic-Plastic Analysis

Taking the circular roadway under the combined action of uniform in situ stress p_0 and uniform support force p_i shown in Figure 1 as an example, the elastic-plastic stresses of the surrounding rock during excavation are solved in this section. In Figure 1, r_i is the excavation radius of the roadway; R is the radius of the plastic zone of the surrounding rock; u_0 is the displacement of the roadway wall; and σ_r and σ_θ are the radial and tangential stresses, respectively.

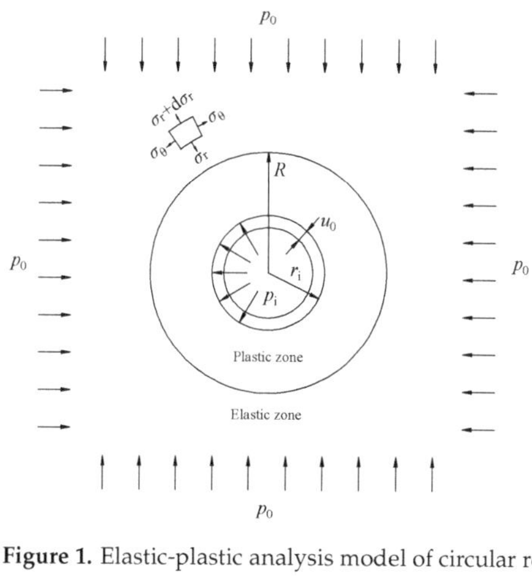

Figure 1. Elastic-plastic analysis model of circular roadway. p_0: uniform in situ stress, p_i: uniform support force, u_0: the displacement of the roadway wall, σ_r: radial stresses, σ_θ: tangential stresses, r_i: the excavation radius of the roadway, R: the radius of the plastic zone of the surrounding rock.

In the plastic zone of the surrounding rock, the tangential stress σ_θ and the radial stress σ_r are the maximum and minimum principal stresses, respectively. Equation (2) can be written as:

$$\sigma_\theta = M\sigma_r + N \tag{3}$$

The differential equation of equilibrium for the axisymmetric problem can be expressed as:

$$\frac{d\sigma_r}{dr} + \frac{\sigma_r - \sigma_\theta}{r} = 0 \tag{4}$$

where r is the radius of calculation area of the circular roadway.

By taking the stress at the inner wall of roadway as the boundary condition, the stress field distribution in the plastic zone is obtained as follows [12,13]:

$$\left.\begin{array}{l}\sigma_r^{\mathrm{p}}=(p_i+\frac{N}{M-1})(\frac{r}{r_i})^{M-1}-\frac{N}{M-1}\\ \sigma_\theta^{\mathrm{p}}=M(p_i+\frac{N}{M-1})(\frac{r}{r_i})^{M-1}-\frac{N}{M-1}\end{array}\right\} \quad (5)$$

where p_i is the support force; r_i is the radius of roadway; and σ_r^{p} and $\sigma_\theta^{\mathrm{p}}$ are the radial stress and tangential stress in the plastic zone, respectively.

Assuming that the radial stress at the interface between the elastic zone and the plastic zone of surrounding rock is $\sigma_r^{\mathrm{e-p}}$, based on thick-walled cylinder theory, the stress field distribution in the elastic zone can be obtained as follows:

$$\left.\begin{array}{l}\sigma_r^{\mathrm{e}}=p_0-(p_0-\sigma_r^{\mathrm{e-p}})\frac{R^2}{r^2}\\ \sigma_\theta^{\mathrm{e}}=p_0+(p_0-\sigma_r^{\mathrm{e-p}})\frac{R^2}{r^2}\end{array}\right\} \quad (6)$$

where p_0 is the uniform in situ stress; R is the radius of the plastic zone; σ_r^{e} and $\sigma_\theta^{\mathrm{e}}$ are the radial stress and tangential stress in the elastic zone. Respectively; and E and ν are the elastic modulus and Poisson's ratio of the surrounding rock, respectively.

Since the stress at the elastic–plastic interface of surrounding rock is continuous, the expressions of radial stress at the elastic–plastic interface and the radius of the plastic zone can be obtained as follows:

$$\sigma_r^{\mathrm{e-p}}=\frac{2p_0-N}{M+1} \quad (7)$$

$$R=r_i\left(\frac{\sigma_r^{\mathrm{e-p}}+\frac{N}{M-1}}{p_i+\frac{N}{M-1}}\right)^{\frac{1}{M-1}} \quad (8)$$

By substituting Equation (7) into Equation (6), the stress solutions in the elastic zone are obtained as follows:

$$\left.\begin{array}{l}\sigma_r^{\mathrm{e}}=p_0-\frac{(M-1)p_0+N}{M+1}\left(\frac{\frac{2p_0-N}{M+1}+\frac{N}{M-1}}{p_i+\frac{N}{M-1}}\right)^{\frac{2}{M-1}}(\frac{r_i}{r})^2\\ \sigma_\theta^{\mathrm{e}}=p_0+\frac{(M-1)p_0+N}{M+1}\left(\frac{\frac{2p_0-N}{M+1}+\frac{N}{M-1}}{p_i+\frac{N}{M-1}}\right)^{\frac{2}{M-1}}(\frac{r_i}{r})^2\end{array}\right\} \quad (9)$$

During the excavation of the roadway, the support force of surrounding rock is 0, and the initial plastic zone radius can be calculated by Equation (8). The initial plastic zone radius R_0 can be expressed as follows:

$$R_0=r_i\left[1+\frac{(M-1)(2p_0-N)}{(M+1)N}\right]^{\frac{1}{M-1}} \quad (10)$$

In the support stage, the elastic stress solution at the elastic–plastic interface can be obtained by substituting Equation (10) with Equation (9). The radial and tangential stresses at the elastic–plastic interface are as follows:

$$\left.\begin{array}{l}\sigma_r^{\mathrm{e}}\big|_{r=R_0}=p_0-\frac{(M-1)p_0+N}{M+1}\left[\frac{N}{(M-1)p_i+N}\right]^{\frac{2}{M-1}}\\ \sigma_\theta^{\mathrm{e}}\big|_{r=R_0}=p_0+\frac{(M-1)p_0+N}{M+1}\left[\frac{N}{(M-1)p_i+N}\right]^{\frac{2}{M-1}}\end{array}\right\} \quad (11)$$

For the plane strain problem, the Z-direction strain is 0, meaning that the intermediate principal stress σ_z can be derived by Hooke's law as follows:

$$\sigma_z=\nu(\sigma_\theta+\sigma_r) \quad (12)$$

The following equation can be obtained by ordering the principal stresses at the elastic–plastic interface:

$$\sigma_1 = \sigma_\theta^e|_{r=R_0},\ \sigma_2 = \sigma_z^e|_{r=R_0},\ \sigma_3 = \sigma_r^e|_{r=R_0} \tag{13}$$

Then, combined with Equations (11)–(13), the average principal stress σ_m can be obtained as follows:

$$\sigma_m = \frac{\sigma_1 + \sigma_2 + \sigma_3}{3} = \frac{2(1+\nu)}{3} p_0 \tag{14}$$

3.3. Optimal Support Solution

A creep is a special form of elastic-plastic deformation of rock, which is a phenomenon where the strain increases with time under the condition of keeping the stress constant. For example, soft rock or hard rock under a high level of stress usually have obvious creep characteristics and show serious extrusion deformation from all directions of underground engineering, which often leads to the instability and failure of supports [35,36]. Therefore, this kind of rock can be considered as an elastic-plastic material with rheological properties. The typical creep curve of rock mainly includes three stages: temporary creep (AB), stable creep (BC), and accelerated creep (CD), as shown in Figure 2.

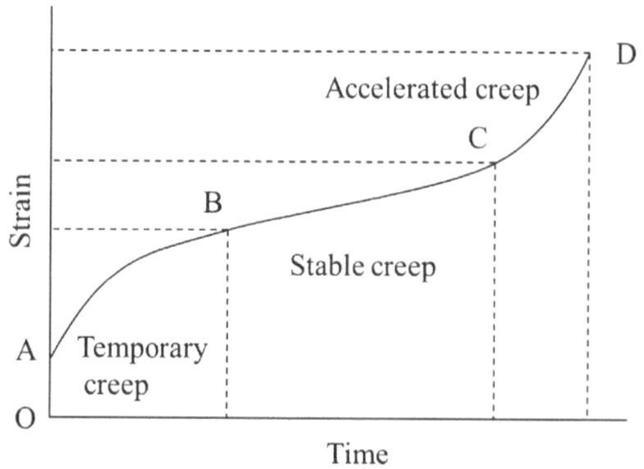

Figure 2. Typical creep curve of rock.

Many engineering practices show that excessive harmful deformation is the main reason for the instability of underground engineering. The establishment and application of the creep model are some of the core components in the study of the deformation characteristics of soft rock. The steady creep criterion expressed by the third invariant of deviator stress is as follows [30]:

$$\sqrt[3]{\frac{27}{2} J_3} \le \sigma_L \tag{15}$$

$$J_3 = (\sigma_1 - \sigma_m)(\sigma_2 - \sigma_m)(\sigma_3 - \sigma_m) \tag{16}$$

where σ_L is the long-term strength of rock and J_3 is the third invariant of the deviator stress.

By substituting Equations (13) and (14) into Equation (15) and making both sides of the equation equal, the optimum support force of roadway is obtained as follows:

$$p_{i-\min} = \frac{N}{M-1}\left[\frac{1}{3}\sqrt{\frac{\sigma_L^3}{(1-2\nu)p_0} + (1-2\nu)^2 p_0^2 [\frac{N}{M-1} + (\frac{M-1}{M+1}) p_0]^{-1}}\right]^{-\frac{M-1}{2}} - \frac{N}{M-1} \tag{17}$$

The optimal support state occurs when the roadway support force p_i is equal to the optimum support force $p_{i-\min}$. In this case, the radius of the plastic zone is:

$$R_{\max} = r_i \left[\left(\frac{2p_0 - N}{M+1} + \frac{N}{M-1} \right) \left(p_{i-\min} + \frac{N}{M-1} \right)^{-1} \right]^{\frac{1}{M-1}} \tag{18}$$

where $R_{\max}$ is the maximum radius of the plastic zone.

Based on thick-walled cylinder theory, the displacement in the elastic zone is:

$$u^{\mathrm{e}} = \frac{1+\nu}{E} (p_0 - \sigma_r^{\mathrm{e-p}}) \frac{R^2}{r} \tag{19}$$

At the elastic–plastic interface, $r = R$, and the displacement of surrounding rock at the elastic–plastic interface is:

$$u^{\mathrm{e-p}} = \frac{(1+\nu)R}{E} (p_0 - \sigma_r^{\mathrm{e-p}}) \tag{20}$$

In the optimum support condition, the initial elastic displacement at the elastic–plastic interface is as follows:

$$u_0^{\mathrm{e}} = \frac{(1+\nu)R_{\max}}{E} \frac{(M-1)p_0 + N}{M+1} \tag{21}$$

The creep modulus of rock can be obtained by laboratory rheological testing. By replacing the elastic modulus E of Equation (21) with the creep modulus E_{c}, the creep displacement at the elastic–plastic interface can be obtained as follows:

$$u^{\mathrm{c}} = \frac{(1+\nu)R_{\max}}{E_{\mathrm{c}}} \frac{(M-1)p_0 + N}{M+1} \tag{22}$$

The displacement of the plastic zone is composed of initial elastic displacement and stable creep displacement. According to Equation (21) and Equation (22), the stable displacement of roadway can be calculated as follows:

$$\triangle u = u^{\mathrm{c}} - u_0^{\mathrm{e}} = \frac{(1+\nu)R_{\max}[(M-1)p_0 + N]}{M+1} \left(\frac{E - E_c}{E \cdot E_c} \right) \tag{23}$$

Taking $u|_{r=R_{\max}} = \triangle u$ as the displacement boundary condition, the displacement of the roadway wall in the optimum support condition is obtained as:

$$u_{0-\max} = \frac{R_{\max}}{r_i} \triangle u = \frac{(1+\nu)R_{\max}^2[(M-1)p_0 + N]}{(M+1)r_i} \left(\frac{E - E_c}{E \cdot E_c} \right) \tag{24}$$

It can be seen that different DP yield criteria correspond to different parameters, M and N. By substituting Equation (17) into Equation (18) to obtain the maximum radius of plastic zone $R_{\max}$, and then substituting Equation (18) into Equation (24), the allowable maximum displacement at the roadway wall can be obtained.

4. Example Studies and Discussion

4.1. Roadway Parameters

The Beizao Coal Mine is located in Longkou City, Shandong Province, China, which is a typical soft rock mine with complex geological conditions; especially with continuous extensions to mining level, roadway deformation is a prominent problem. Taking the soft rock roadway in the Beizao Coal Mine as an engineering example, the effects of the DP criteria and the mechanical parameters of surrounding rock on roadway support can be analyzed. The geometric and mechanical parameters of the roadway are shown in Table 2. Yieldable U-shaped steel ribs are used to support the roadway and the in situ measured roadway wall displacement is 18.7 mm.

Table 2. Geometrical and mechanical parameters.

Symbol	Description	Value
r_i/m	Excavation radius	2.0
p_0/MPa	In situ stress	5.6
E/MPa	Elastic modulus	1500
E_c/MPa	Creep modulus	400
ν	Poisson's ratio	0.24
c/MPa	Cohesion	0.71
$\varphi/^\circ$	Internal friction angle	23.6
σ_c/MPa	Uniaxial compressive strength	29.0
σ_L/MPa	Long-term strength	6.38

4.2. The Effect of Yield Criteria

The calculation results for the optimal support force, maximum plastic zone radius, and allowable maximum displacement of surrounding rock under different DP criteria are shown in Table 3.

Table 3. Calculation results of different DP yield criteria.

Serial Number	$p_{i-\min}$/MPa	$R_{\max}$/m	$u_{0-\max}$/mm
DP1	0.76	2.33	23.69
DP2	0.20	3.71	46.03
DP3	0.16	4.06	52.72
DP4	0.25	3.43	41.10
DP5	0.18	3.85	48.66

It can be seen that the optimal support force, the maximum radius of the plastic zone, and allowable maximum displacement calculated by different DP criteria are different. The optimal supporting force obtained by the DP1 criterion is the largest, while that obtained by the DP3 criterion is the smallest. The optimum support force of the DP3 criterion is only 21% of the DP1 criterion. The effect of intermediate principal stress σ_2 on surrounding rock strength is equal to the minimum principal stress σ_3 according to the DP1 criterion, which will exaggerate the influence of intermediate principal stress. Therefore, the maximum plastic zone radius and allowable maximum displacement obtained by the DP1 criterion are the smallest of the DP series criteria. The maximum plastic zone radius calculated by the DP3 criterion is 1.74 times that of the DP1 criterion, and the allowable maximum displacement is 2.23 times that of the DP1 criterion. This means that the DP1 criterion and the DP3 criterion are the upper and lower limits of the DP series criteria respectively. When the DP criteria are used for the elastic-plastic analysis of roadway surrounding rock, the appropriate DP yield criterion should be selected according to the actual engineering background and mechanical parameters of surrounding rock.

4.3. The Effect of Long-Term Strength

Figure 3 shows the rule of variation of the optimal support force and allowable maximum displacement under different long-term strengths. It can be seen that the larger the long-term strength is, the smaller the optimal support force is and the larger the allowable maximum displacement is. In other words, the greater the long-term strength of surrounding rock is, the more stable the roadway is. From Figure 3a, it can be seen that the overall performance of the optimal support force is DP1 > DP4 > DP2 > DP5 > DP3. With the increase in the long-term strength of surrounding rock, the optimal support forces calculated by different DP criteria become closer and closer, and will eventually completely coincide and achieve the ideal state without support. The reason for this change is that the larger the long-term strength of rock mass is, the smaller the plastic zone of roadway is, which leads to the weakening of the effect of the yield criterion. As can be seen from Figure 3b, the allowable maximum displacement is DP3 > DP5 > DP2 > DP4 > DP1. The

results of the DP3 criterion are relatively conservative. Using the DP3 criterion in roadway support design can improve safety, but will increase the support cost.

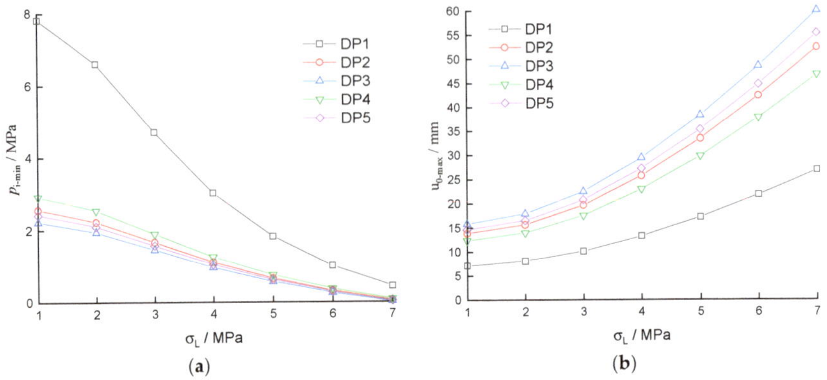

Figure 3. Effect of long-term strength on support: (**a**) optimal support force; (**b**) allowable maximum displacement.

The long-term strength of rock has a key impact on the optimal support parameters of the roadway. Under certain conditions, the surrounding rock can reach a stable creep stage by relying on its own strength without any support. The influence of surrounding rock cohesion and internal friction angle on the optimal support parameters of roadway is also very important, and the influence of cohesion becomes more prominent. In practical engineering, the real strength parameters of rock should be measured reasonably, and their variability should be considered.

4.4. The Effect of Strength Parameters

Figures 4 and 5 show the variation in the maximum plastic zone radii and allowable maximum displacement under different cohesion and internal friction angles, respectively. It can be seen that the influence trend of cohesion and internal friction angle on roadway deformation is roughly the same; that is, with the increase in cohesion or internal friction angle, the radius of the maximum plastic zone is smaller and smaller, and the allowable maximum displacement is also gradually reduced. The reason for this change is that the increase in cohesion or internal friction angle increases the bearing capacity of rock mass, thus reducing the failure range and restraining the plastic deformation of surrounding rock.

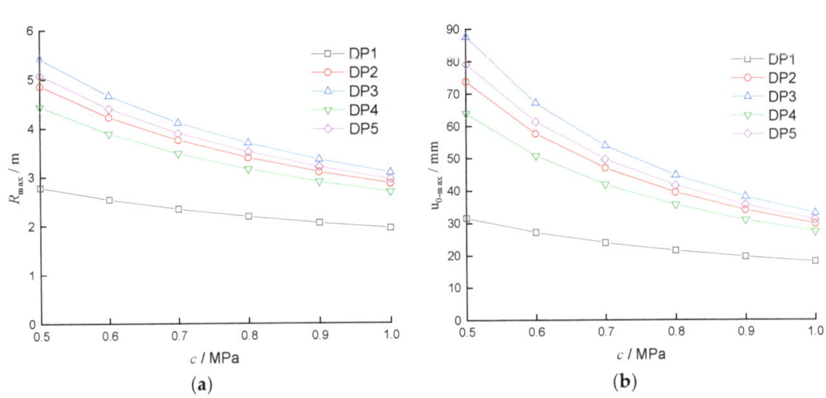

Figure 4. Effect of cohesion on radius and maximum displacement of plastic zone: (**a**) plastic zone radius; (**b**) allowable maximum displacement.

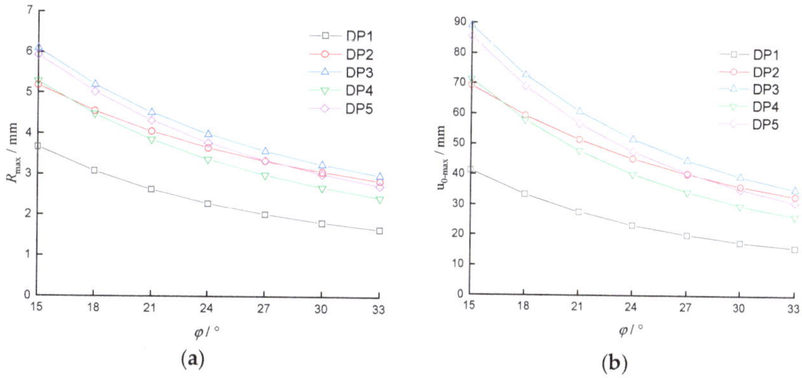

Figure 5. Effect of internal friction angle on radius and maximum displacement of plastic zone: (**a**) plastic zone radius; (**b**) allowable maximum displacement.

Taking the DP1 criterion as an example, when the cohesion increases from 0.5 MPa to 1.0 MPa, the radius of the plastic zone decreases from 2.78 m to 1.95 m, which is a reduction of nearly 30%. The allowable maximum displacement was reduced by 43% from 31.58 mm to 18.10 mm. When the internal friction angle increases from 15 to 33 degrees, the radius of plastic zone decreases from 3.67 m to 1.65 m, a reduction of 55%. The allowable maximum displacement is reduced from 41.30 mm to 15.86 mm, which is a decrease of 62%. It can be seen that the strength parameters of rock mass have great influence on the roadway support. The mechanical properties of engineering surrounding rock should be fully considered in support design. For example, the strength of surrounding rock can be improved by grouting, which can effectively reduce the deformation of the roadway.

4.5. Novelty and Comparability Analysis

Based on the DP yield criteria and the steady-state creep criterion expressed by the third invariant of deviatoric stress, the theoretical solutions of optimal support pressure and allowable maximum displacement of surrounding rock are derived. The theoretical solution can consider different degrees of intermediate principal stress effects and the selection of different DP strength criteria. The formula is concise and the parameters are easy to determine, which has important theoretical and practical engineering significance.

The intermediate principal stress significantly affects the optimal support parameters of a roadway. The greater the effect of intermediate principal stress is, the smaller the optimal support force of the corresponding roadway will be and the greater the allowable maximum displacement of the surrounding rock will be. Therefore, the calculation results of the Mohr–Coulomb criterion are too conservative without considering the influence of intermediate principal stress. The DP strength criterion fully considers the influence of intermediate principal stress, and the calculation results can better demonstrate the strength potential of the surrounding rock. In addition, the optimal roadway support solution of the J_2 stable creep criterion needs to add non-negative conditions under the root sign, meaning that the J_3 criterion solution has a wider application range. According to the J_3 criterion, the optimal support force of a roadway is large and the allowable maximum displacement of the surrounding rock is small.

5. Conclusions

(1) An analytical solution of optimal support force and allowable maximum displacement of surrounding rock for circular roadway based on the DP series criteria is proposed. The proposed optimal support solution can not only reflect the intermediate principal stress reasonably, but also can allow us to compare and discuss the influence of different DP criteria on the calculation results. The new theoretical solution can consider different intermediate principal stress effects and different DP

strength criteria, and the parameters are easier to determine. It has a wider range of applications, and the calculation results can better demonstrate the strength potential of the surrounding rock.

(2) The long-term strength of rock surrounding a roadway has a significant impact on the optimal support force and the allowable maximum displacement. The higher the long-term strength of rock surrounding a roadway is, the smaller the optimal support force will be and the larger the allowable maximum displacement will be. When the calculated long-term strength of soft rock can ensure that the deformation of the roadway does not exceed the allowable maximum displacement, the roadway can maintain long-term stability without support.

(3) The influence of the parameters of surrounding rock on roadway support and deformation is also significant. With the increase in cohesion or internal friction angle, the radius of the plastic zone of surrounding rock becomes smaller and smaller and the allowable maximum displacement also decreases gradually. The use of grouting and other means to improve the strength of surrounding rock can effectively reduce the deformation of a roadway. The analytical solution can also provide theoretical guidance for engineering practice.

Author Contributions: Conceptualization, M.M. and Q.G.; methodology, M.M. and Q.G.; validation, M.M. and Q.G.; formal analysis, J.P. and C.M.; investigation, J.P. and C.M.; resources, Q.G. and M.C.; writing—original draft, M.M. and Q.G.; writing—review and editing, Q.G. and J.P.; supervision, M.C.; project administration, M.C.; funding acquisition, Q.G. and M.C. All authors have read and agreed to the published version of the manuscript.

Funding: This research was funded by the Fundamental Research Funds for the Central Universities (Grant No. FRF-IDRY-20-032) and the National Natural Science Foundation of China (No. 51974014).

Data Availability Statement: Not applicable.

Conflicts of Interest: The authors declare that they have no conflict of interest.

References

1. Yang, X.; Pang, J.; Liu, D.; Liu, Y.; Tian, Y.; Ma, J.; Li, S. Deformation mechanism of roadways in deep soft rock at Hegang Xing'an Coal Mine. *Int. J. Min. Sci. Technol.* **2013**, *23*, 307–312. [CrossRef]
2. Shen, B. Coal mine roadway stability in soft rock: A case study. *Rock Mech. Rock Eng.* **2014**, *47*, 2225–2238. [CrossRef]
3. Yang, S.Q.; Chen, M.; Jing, H.W.; Chen, K.F.; Meng, B. A case study on large deformation failure mechanism of deep soft rock roadway in Xin'An coal mine, China. *Eng. Geol.* **2017**, *217*, 89–101. [CrossRef]
4. Cao, P.; Youdao, W.; Yixian, W.; Haiping, Y.; Bingxiang, Y. Study on nonlinear damage creep constitutive model for high-stress soft rock. *Environ. Earth Sci.* **2016**, *75*, 900. [CrossRef]
5. Ma, D.; Zhang, J.; Duan, H.; Huang, Y.; Li, M.; Sun, Q.; Zhou, N. Reutilization of gangue wastes in underground backfilling mining: Overburden aquifer protection. *Chemosphere* **2021**, *264*, 128400. [CrossRef]
6. Wang, Q.; Pan, R.; Jiang, B.; Li, S.C.; He, M.C.; Sun, H.B.; Wang, L.; Qin, Q.; Yu, H.C.; Luan, Y.C. Study on failure mechanism of roadway with soft rock in deep coal mine and confined concrete support system. *Eng. Fail. Anal.* **2017**, *81*, 155–177. [CrossRef]
7. Chen, Y.; Meng, Q.; Xu, G.; Wu, H.; Zhang, G. Bolt-grouting combined support technology in deep soft rock roadway. *Int. J. Min. Sci. Technol.* **2016**, *26*, 777–785. [CrossRef]
8. Wang, C.; Wang, Y.; Lu, S. Deformational behaviour of roadways in soft rocks in underground coal mines and principles for stability control. *Int. J. Rock Mech. Min. Sci.* **2000**, *37*, 937–946. [CrossRef]
9. Kang, H.P.; Lin, J.; Fan, M.J. Investigation on support pattern of a coal mine roadway within soft rocks—A case study. *Int. J. Coal Geol.* **2015**, *140*, 31–40. [CrossRef]
10. Yoshinaka, R.; Tran, T.V.; Osada, M. Non-linear, stress-and strain-dependent behavior of soft rocks under cyclic triaxial conditions. *Int. J. Rock Mech. Min. Sci.* **1998**, *35*, 941–955. [CrossRef]
11. Walton, G.; Delaloye, D.; Diederichs, M.S. Development of an elliptical fitting algorithm to improve change detection capabilities with applications for deformation monitoring in circular tunnels and shafts. *Tunn. Undergr. Space Technol.* **2014**, *43*, 336–349. [CrossRef]
12. Wu, K.; Shao, Z.; Qin, S. An analytical design method for ductile support structures in squeezing tunnels. *Arch. Civ. Mech. Eng.* **2020**, *20*, 1–13. [CrossRef]
13. Wu, K.; Shao, Z.; Qin, S.; Zhao, N.; Hu, H. Analytical-Based Assessment of Effect of Highly Deformable Elements on Tunnel Lining Within Viscoelastic Rocks. *Int. J. Appl. Mech.* **2020**, *12*, 2050030. [CrossRef]

14. Wu, K.; Shao, Z.; Qin, S.; Li, B. Determination of deformation mechanism and countermeasures in silty clay tunnel. *J. Perform. Constr. Facil.* **2020**, *34*, 04019095. [CrossRef]
15. Song, F.; Wang, H.; Jiang, M. Analytically-based simplified formulas for circular tunnels with two liners in viscoelastic rock under anisotropic initial stresses. *Constr. Build. Mater.* **2018**, *175*, 746–767. [CrossRef]
16. Song, F.; Wang, H.; Jiang, M. Analytical solutions for lined circular tunnels in viscoelastic rock considering various interface conditions. *Appl. Math. Model.* **2018**, *55*, 109–130. [CrossRef]
17. Pan, J.; Guo, Q.; Ren, F.; Cai, M. Comparative analysis of different strength criteria for deep-buried rock roadway under seepage. *J. China Coal Soc.* **2019**, *44*, 3369–3378.
18. Xu, S.Q.; Yu, M.H. The effect of the intermediate principal stress on the ground response of circular openings in rock mass. *Rock Mech. Rock Eng.* **2006**, *39*, 169–181. [CrossRef]
19. Ma, D.; Duan, H.Y.; Zhang, J.X.; Feng, X.J.; Huang, L. Experimental investigation of creep-erosion coupling mechanical properties of water inrush hazards in fault fracture rock masses. *Chin. J. Rock Mech. Eng.* **2021**, *40*, 1751–1763.
20. Park, K.H.; Kim, Y.J. Analytical solution for a circular opening in an elastic–brittle–plastic rock. *Int. J. Rock Mech. Min. Sci.* **2006**, *43*, 616–622. [CrossRef]
21. Sharan, S.K. Exact and approximate solutions for displacements around circular openings in elastic–brittle–plastic Hoek–Brown rock. *Int. J. Rock Mech. Min. Sci.* **2005**, *42*, 542–549. [CrossRef]
22. Guo, Q.; Pan, J.; Wu, X.; Xi, X.; Cai, M. A new unified solution for circular tunnels based on generalized SMP criterion considering the strain softening and dilatancy. *Adv. Civ. Eng.* **2019**, *2019*, 1684707. [CrossRef]
23. Pan, J.; Gao, Z.; Ren, F. Effect of strength criteria on surrounding rock of circular roadway considering strain softening and dilatancy. *J. China Coal Soc.* **2018**, *43*, 3293–3301.
24. Fan, H.; Wang, L.; Wang, K. Stability Analysis of Surrounding Rock in Circular Tunnels Based on Critical Support Pressure. *Adv. Civ. Eng.* **2020**, *2020*, 8870928. [CrossRef]
25. Qiuyan, F.; Weishen, Z. Method for calculating optimum support of soft rock. *Chin. J. Geotech. Eng.* **1997**, *19*, 77–83.
26. Cui, X.H.; Gao, Y.F.; Fan, Q.Z.; Niu, X.L. Optimum calculating method for constant resistant supporting of roadway surrounding rock. *Chin. J. Geotech. Eng.* **2006**, *28*, 674–678.
27. Zeng, K.; Yang, W.; Zhang, C.; Dai, H. Optimal Support in Soft Rock Tunnel Based on the Unified Strength Theory and Steady Creep Guideline. *Adv. Eng. Sci.* **2018**, *50*, 62–70.
28. Filcek, H.; Kwaśniewski, M.A. Fundamentals of Mine Roadway Support Design: Rock–Support Interaction Analysis. In *Analysis and Design Methods*; Pergamon: Bergama, Turkey, 1993; pp. 671–699.
29. Pan, J.; Wu, X.; Guo, Q.; Xi, X.; Cai, M. Uniaxial experimental study of the deformation behavior and energy evolution of conjugate jointed rock based on AE and DIC methods. *Adv. Civ. Eng.* **2020**, *2020*, 8850250. [CrossRef]
30. Guo, Q.; Pan, J.; Cai, M.; Zhang, Y. Analysis of Progressive Failure Mechanism of Rock Slope with Locked Section Based on Energy Theory. *Energies* **2020**, *13*, 1128. [CrossRef]
31. Guo, Q.; Pan, J.; Cai, M.; Zhang, Y. Investigating the effect of rock bridge on the stability of locked section slopes by the direct shear test and acoustic emission technique. *Sensors* **2020**, *20*, 638. [CrossRef]
32. Xi, X.; Wu, X.; Guo, Q.; Cai, M. Experimental investigation and numerical simulation on the crack initiation and propagation of rock with pre-existing cracks. *IEEE Access* **2020**, *8*, 129636–129644. [CrossRef]
33. Alejano, L.R.; Bobet, A. Drucker–Prager criterion. In *The ISRM Suggested Methods for Rock Characterization, Testing and Monitoring: 2007–2014*; Springer: Cham, Switzerland, 2012; pp. 247–252.
34. Jiang, H.; Xie, Y. A note on the Mohr–Coulomb and Drucker–Prager strength criteria. *Mech. Res. Commun.* **2011**, *38*, 309–314. [CrossRef]
35. Chu, Z.; Wu, Z.; Liu, Q.; Liu, B. Analytical Solutions for Deep-Buried Lined Tunnels Considering Longitudinal Discontinuous Excavation in Rheological Rock Mass. *J. Eng. Mech.* **2020**, *146*, 04020047. [CrossRef]
36. Chu, Z.; Wu, Z.; Liu, B.; Liu, Q. Coupled analytical solutions for deep-buried circular lined tunnels considering tunnel face advancement and soft rock rheology effects. *Tunn. Undergr. Space Technol.* **2019**, *94*, 103111. [CrossRef]

minerals

MDPI

Article

Research on the Method and Model for Calculating Impact Load in the Rockburst Tunnel

Zhiwei Yan [1,2], Dagang Liu [1,*], Zhilong Wang [1,2], Daming Zhao [1,2] and Hongtao Tian [1,2]

1 School of Civil Engineering, Southwest Jiaotong University, Chengdu 610031, China; 2019210073@my.swjtu.edu.cn (Z.Y.); zlwang@my.swjtu.edu.cn (Z.W.); daming_zhao@my.swjtu.edu.cn (D.Z.); tianhingtao18@126.com (H.T.)

2 Key Laboratory of Transportation Tunnel Engineering, Ministry of Education, Southwest Jiaotong University, Chengdu 610031, China

* Correspondence: ldg980272@163.com

Abstract: Among several design methods of tunnel supporting structure, the load-structure method is widely used in different countries, but the determination of load is essential in this design method. The problem of rockburst is becoming more prominent as tunnel engineering enters the deep underground space. However, the research on the impact load on the supporting structure is insufficient in relevant fields. Therefore, from the perspective of energy, this paper deduces the method and model for calculating the impact load of the rockburst tunnel acting on the supporting structure by using the method of structural mechanics first, after the location effect of impact load is determined under different section types and different section sizes. The results indicated that: dynamic load factor K is related to the stiffness EI and supporting size coefficient K_0 of the supporting structure, also the difference of impact load in different sections is proved. Tunnel rockburst-prone location is related to lateral pressure coefficient, thus when $\lambda = 1$, the probability of rockburst in the whole circular tunnel is the same, while side wall and vault are prone to rockburst in single-track horseshoe tunnel, and the side wall is prone to rockburst in double-track horseshoe tunnel; furthermore when $\lambda > 1$, the vault and the inverted arch are prone to rockburst; additionally, when $\lambda < 1$, the rockburst is most likely to occur in the arch waist of the circular tunnel and the side walls and the arch waist of the horseshoe tunnel. Finally, the rockburst tunnel's local load-structure calculation model and the calculation process based on the model are provided.

Keywords: rockburst tunnel; impact load; calculation model; energy method

Citation: Yan, Z.; Liu, D.; Wang, Z.; Zhao, D.; Tian, H. Research on the Method and Model for Calculating Impact Load in the Rockburst Tunnel. *Minerals* **2022**, *12*, 13. https://doi.org/10.3390/min12010013

Academic Editors: Diyuan Li, Zhenyu Han, Xin Cai, Shijie Xie and Abbas Taheri

Received: 27 October 2021
Accepted: 19 December 2021
Published: 22 December 2021

1. Introduction

Geotechnical engineering has long been concerned with deep underground space, and tunnel engineering's buried depth is increasing, resulting in higher and higher in-situ stress environments. Compared with the creep-mechanical behavior of soft rock, the hard rock under a high-stress level is prone to rockburst, posing considerable challenges [1,2]. For example, in Sichuan, a mountainous province in the west of China, the phenomenon of high in-situ stress exists in the deep and long tunnels that serve as the expressway's control engineering, and the majority of them encounter the problem of rockburst [1–4]. Among the railway tunnels, Sichuan-Tibet Railway has a high in-situ stress level which is extremely rare in the history of railway tunnel construction, which is expected to exceed 70 MPa. Rockburst also occurs frequently in underground projects in the field of hydropower, such as the diversion tunnels of Yuzixi level 1 and 2 hydropower stations, Baihetan Taiping Yi, Ertan, Tianshengqiao, and Jinping level 2 hydropower stations, etc. [5–7]. Rockburst not only damages construction equipment and slows construction progress, but also endangers the lives and safety of on-site workers [8].

Rockburst disasters in tunnels are extremely destructive and unpredictable, so enough attention must be paid to tunnels where rockburst may occur. Many types of research

on rockburst mainly focus on mechanism research, prediction and early warning, and instability control of rockburst [9–15]. Among them, appropriate supporting structures are an important part of rockburst-tunnel instability control. However, there are relatively few research results on the design method of rockburst tunnel supporting structures.

At present, more and more calculation methods of tunnel deformation have been proposed [16], the design methods of deep-buried hard rock tunnels for drilling, blasting, anchoring, and grouting in various countries are mainly divided into engineering analogy, load-structure, layer-structure, and theoretical calculation method. Among them, the stratiform-structure method and theoretical calculation method are affected by uncertain factors such as design parameters and supporting structure, and the support design process is more complex. To a large extent, it still depends on practical design. The design of rockburst tunnel supporting structures in China and other countries focuses mainly on engineering analogy. In practical engineering, the stratiform-structure method is used for verification and analysis.

The load-structure method has been widely adopted in some countries, such as China, Japan, Germany, etc. [17] because of its simple physical model, comprehensive analysis of various loads and rock mass states, and fast calculation speed. The determination of load is essential in this design method. However, through theauthor's investigation, there is no complete calculation method to estimate the impact load acting on the supporting structure of the rockburst tunnel. Therefore, to adopt a load-structure method in designing rockburst tunnel supporting structures, it is necessary to conduct an in-depth study on its impact load.

Therefore, based on the data research and theoretical analysis method, from the perspective of energy and using the method of structural mechanics, this paper deduces the calculation formula of the impact load acting on the supporting structure of the rockburst tunnel. In addition, the rockburst-prone parts of the circular tunnel and the horseshoe tunnel are studied utilizing theoretical analysis, numerical simulation, and data investigation. Finally, the local load-structure model of the rockburst tunnel and its calculation flow is derived.

2. Calculation Method of Impact Load in the Rockburst Tunnel

The calculation method of impact load acting on the supporting structure mainly includes three parts: the amount of kinetic energy released, the size and velocity of the blasting block, and the load acting on the supporting structure. Firstly, the strain energy stored in the rock mass and the kinetic energy of the explosive block transformed from the released energy are calculated employing energy analysis. Secondly, the size and velocity of the blasting block are determined. Finally, after obtaining the two key parameters, the size and velocity of the blasting block, the load acting on the supporting structure is estimated by the structural mechanic's method.

2.1. Surrounding Rock Strain Energy of Tunnel

According to the material mechanics and elasticity mechanics, it can be known that the strain energy of the surrounding rock in triaxial stress state can be calculated according to Equation (1), which is the calculation formula of elastic strain energy of surrounding rock per unit volume expressed by stress tensor.

$$\begin{cases} U = \frac{1}{2E}\left[(1+\mu)\sigma_{ij}\sigma_{ji} - \mu\Theta^2\right] \\ \Theta = \sigma_{11} + \sigma_{22} + \sigma_{33} \end{cases} \tag{1}$$

where U is the strain energy density of surrounding rock, E and u are the elastic modulus and Poisson's ratio of surrounding rock, respectively.

In the secondary stress state after tunnel excavation, the directions of the three stress spindles are, respectively, the tangent direction of the excavation contour plane, the axial

direction of the tunnel and the normal direction of the contour plane, and $\sigma_3 = 0$ near the contour plane.

It can be known that the strain energy formula of surrounding rock surrounding the tunnel after excavation is as follows:

$$U = \frac{1}{2E}\left[\sigma_1{}^2 + \sigma_2{}^2 - 2\mu\sigma_1\sigma_2\right] \tag{2}$$

In terms of vertical initial ground stress and lateral pressure coefficient,

$$U = \frac{1-\mu^2}{2E}\sigma_1{}^2 = \frac{1-\mu^2}{2E}\sigma_v{}^2(1+\lambda+2(1-\lambda)\cos 2\theta)^2 \tag{3}$$

With different θ values, the formula for calculating the strain energy density contained in the vault, shoulder, waist, side wall, and inverted supply of the tunnel under different initial vertical stresses and lateral pressure coefficients can be obtained:

Vault and invert: $\theta = 90°$

$$U = \frac{1-\mu^2}{2E}\sigma_v{}^2(3\lambda - 1)^2 \tag{4}$$

Arch shoulders and side walls: $\theta = 45°$

$$U = \frac{1-\mu^2}{2E}\sigma_v{}^2(1 + \lambda)^2 \tag{5}$$

Hence: $\theta = 0°$

$$U = \frac{1-\mu^2}{2E}\sigma_1{}^2 = \frac{1-\mu^2}{2E}\sigma_v{}^2(3 - \lambda)^2 \tag{6}$$

2.2. Determination of Released Kinetic Energy

In the process of rock failure, shown in Figure 1, energy stored in rock mass will not be completely released as kinetic energy, some will be dissipated in the form of thermal energy and surface energy. Different scholars have derived the calculation formula of dissipated energy, respectively. However, the calculation formula contains many parameters that are difficult to obtain in specific projects, and the calculation is complex. In this paper, the specific size of energy other than kinetic energy is not considered, and, instead, the focus is on how much of the strain energy stored in a rock mass can be converted into kinetic energy and released in the way of rockburst.

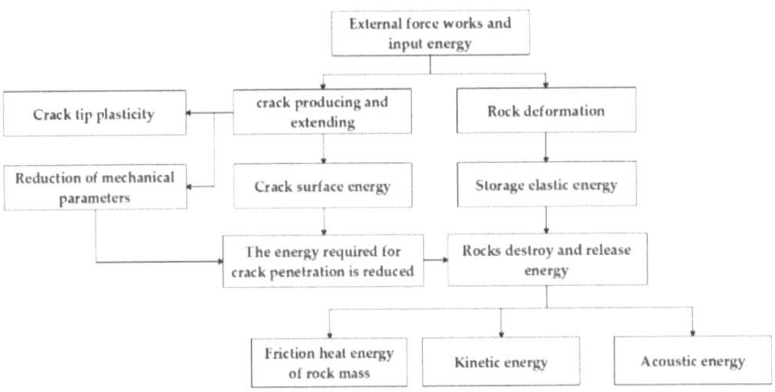

Figure 1. Transformation and transfer relations among various energies in the process of rock de-formation and failure.

The calculation of the released kinetic energy involves two processes. The first process is external force doing work, and energy being accumulated in the rock mass. Here, some

energy will be used to expand the cracks between degraded surrounding rocks. When the critical condition is reached, rockburst occurs and energy is released, this part of the energy will not be released. The strain energy contained in the surrounding rock includes elastic strain energy and plastic strain energy. The other process is that when rockburst occurs, a part of the energy is released in the form of acoustic energy and thermal energy, rather than all in the form of kinetic energy. The following analyzes the two processes, respectively, to determine the kinetic energy that can be released during rockburst.

The energy release rate of the first process can be determined by experimental methods, which refer to the acquisition method of the rockburst tendency index, customarily expressed as W_{et} (also known as elastic energy index in China) in the Polish national standard [17]. In the uniaxial compressive-strength test of rock, the specimen is loaded to 70–80% of its peak strength and then unloaded 0.05 times. In the process of loading and unloading tests, the stress–strain curves of loading and unloading of rocks under uniaxial stress state can be finally obtained, to calculate the ratio between the elastic strain energy released by unloading (Φ_{SP}) and the strain energy consumed (Φ_{ST}), which is known as the rockburst tendency index (W_{et}):

$$W_{et} = \Phi_{SP} / \Phi_{ST} \quad (7)$$

The test curve of the rockburst tendency index is shown in Figure 2:

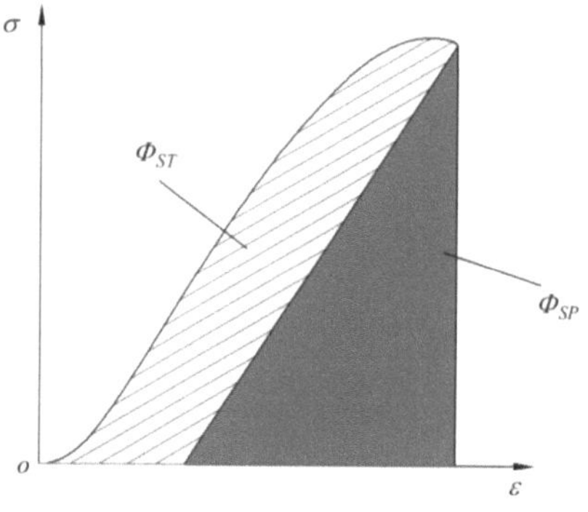

Figure 2. The test curve of W_{et}.

According to the definition of the rockburst tendency index, the value of W_{et} reflects the relationship between the energy stored in rock and released during rockburst, that is, the relationship between the sum of elastic strain energy and plastic strain energy. In the same state, the higher the W_{et} value, the greater the energy stored in the rock, the greater the energy it can release, and there is a higher possibility of higher-grade rockburst.

According to the Polish national standard, the values of W_{et} corresponding to different grades of rockburst are as follows [18]:

$W_{et} \geq 5.0$, Severe and extremely severe grade;

$2.0 \leq W_{et} < 5.0$, Mild and medium grade;

$W_{et} < 2.0$, No rockburst.

In different grades of rockburst, different W_{et} values are selected to calculate the proportion of elastic strain energy in all strain energy contained in the rock mass. When the test conditions are available, the surrounding rock of a specific tunnel engineering site should be used for rock uniaxial compressive strength test to obtain the values of W_{et}. Then, the proportion of elastic strain energy is determined under the surrounding rock condition.

In case of a lack of test conditions, the recommended value can be obtained according to Table 1 regarding Polish national standards.

Table 1. The proportion of elastic strain energy of surrounding rock under different grades of rockburst.

Rockburst Grade	W_{et}	The Proportion of Elastic Strain Energy	Recommended Values
mild	<2	<66.7%	65%
medium	2~3.5	66.7~77.8%	73%
severe	3.5~5	77.8~83.3%	80%
extremely severe	>5	>83.3%	85%

In the second process, the proportion of kinetic energy in all elastic strain energy released can refer to [17], which shows the relationship between the proportion of kinetic energy in various energy released during rock shear failure and the internal friction angle in Figure 3.

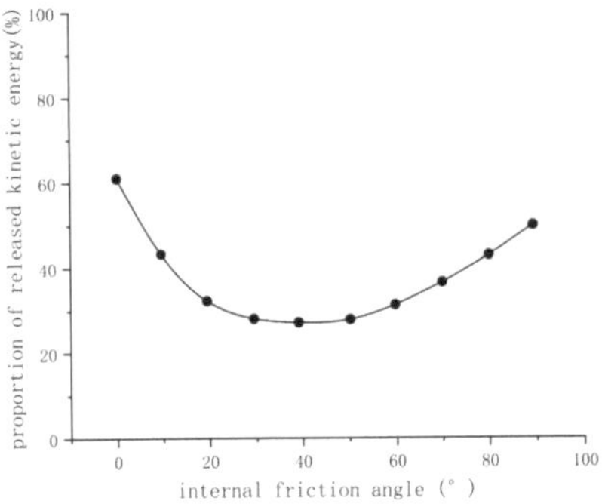

Figure 3. The relationship between released kinetic energy and the internal friction angle.

As shown in Figure 3, the proportion of kinetic energy in the released energy is the smallest when the internal friction angle is 45°, accounting for only 30%. As the internal friction angle approaches 0° and 90°, the proportion of kinetic energy gradually increases until it reaches the maximum value, 60%. The internal friction angle is given, the value can be taken according to Figure 3. In case of lack of data, it can be taken as 60% according to the most unfavorable situation.

2.3. Subsection

The characteristics of blasting blocks in different types of rockburst are given in the literature [19]. Generally speaking, it is a conch shell or flake in exfoliation rockburst, an elongated elliptic fragment in slight-ejection rockburst, and a block in burst rockburst. The main mode of rockburst is exfoliation in mild and medium grades and ejection in extremely severe grades. Exfoliation rockburst may occur several times, so it can not completely reflect the whole process if only the characteristics of a single rockburst are counted. Meanwhile, exfoliation can also be understood as ejection with a small velocity. In addition, the blasting blocks of ejection rockburst are not complete, more than one block can be ejected from the same blasting pit at a time, and some blocks will break into several pieces after landing. It is more reasonable to comprehensively analyze and calculate the size of the blasting block from the perspective of the blasting pit.

To study the characteristics of blasting pits, 14 tunnels in different countries were investigated [19,20], as shown in Table 2.

Based on Table 2, the shape of the blasting pit can be divided into three categories: V-shape (including ladder shape, right-angle shape, and wedge shape), dome shape (including arc long groove and dome shape), and nest shape (including nest shape, honeycomb shape, and pot bottom shape). Using the tunnel as the basic unit, each tunnel contains

one or two main blasting pit shapes, and the proportion of each blasting pit shape can be calculated, as illustrated in Figure 4.

Table 2. Statistics of craters of different rockburst tunnels.

Code	Tunnel Name	Mileage	Rockburst Grade	Characteristics of Blasting Pit
1	Futangba highway tunnel	K46 + 964~K46 + 944	severe	Arc long groove
2	Futangba highway tunnel	K46 + 613.5~K46 + 581	severe	Arc long groove
3	Futangba highway tunnel	K46 + 964~K46 + 581	medium	Large scale continuous distribution
4	Futangba highway tunnel	K46 + 964~K46 + 581	mild	Small scale sporadic distribution
5	Diversion tunnel of			5
6	Diversion tunnel of			6
27	Diversion Tunnel of SQirehatar Hydropower Station	-	-	Mainly V-shape

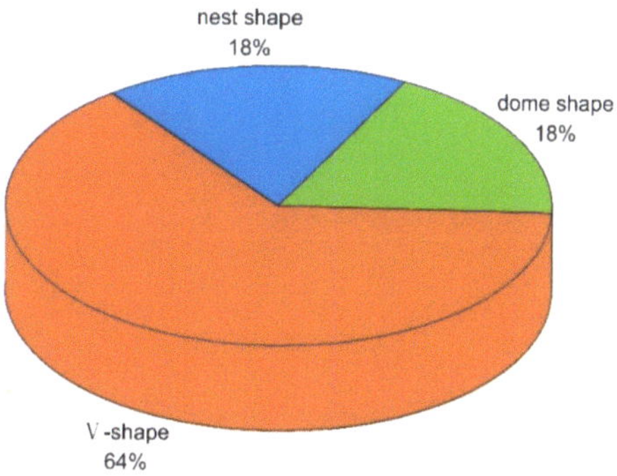

Figure 4. The proportion of different blasting pit shapes.

As shown in Figure 4, the V-shaped blasting pit is the most common, accounting for 64% of all blasting pits. Therefore, in the subsequent calculation of blasting block volume, the blasting pit is considered as a V-shaped blasting pit a whereas the blasting block is considered as the corresponding cone. Figure 5 is a photo of a V-shaped blasting pit at the rockburst site.

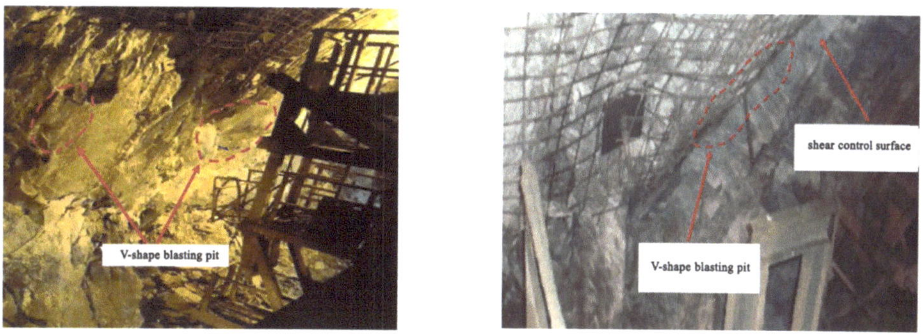

(a) (b)

Figure 5. V-shaped blasting pit at the rockburst site: (**a**) V-shape blasting pit without supporting structure; (**b**) V-shape blasting pit with supporting structure.

It is important to note that the blasting block is not necessarily the shape of the scattered blocks we see after the rockburst. The impact of the blasting block may also form the final shape after it falls to the ground.

To study the influence depth under different grades of rockburst, 166 rockburst sections of seven tunnels such as the Futangba highway tunnel are counted [20], and the statistical results are shown in Table 3.

Table 3. Statistical table of influence depth of rockburst of different grades.

Number	Tunnel Name	Mileage	Length/Times	Grade	Influence Depth
1	Futangba highway tunnel	K46 + 964~K46 + 944	20 m	severe	average: 1.0 m; maximum: 1.5 m
2		K46 + 613.5~K46 + 581	32.5 m	severe	average: 1.5 m; maximum: 2.0 m
3	Erlangshan highway tunnel	K260 + 080~K260 + 240	160 m	mild	0.5 m
4		K260 + 270~K260 + 330	60 m	mild	0.5 m
5		K260 + 996~K261 + 015	19 m	severe	2 m
6	Long exploratory tunnel of Jinping II Hydropower Station	4000 m	61 times	mild, medium	mainly: 10~30 cm; maximum: 55 cm
164	Diversion Tunnel of SqCirehatar Hydropower Station	10,521~10,530		mild	20~30 cm
165		10,521~10,530		mild	20 cm
166		11,170~11,190	20 m	mild	50 cm

The influence depth of blasting pit under different rockburst grades is shown in Table 4.

Table 4. Influence depth of rockburst.

Rockburst Grade	Mild	Medium	Severe	Extremely Severe
Influence depth	<0.5 m	0.5 m~1 m	1 m~3 m	>3 m

The influence depth of various rockburst grades is chosen, and the volume of the blasting block can be calculated using the shape of a square cone. Rockburst mainly occurs in hard rocks such as granite, limestone, marble, and gneiss. In specific projects, on-site rocks should be selected to test their density. When the test is not available on-site, it can be taken as 2.75 t/m^3 because of the slight difference in density of the above rock masses. The mass of the blasting block can be obtained finally by multiplying the density by the volume.

After the released kinetic energy and the mass of the blasting block are determined, the velocity of the blasting block can be obtained by the kinetic energy formula.

The kinetic energy formula:

$$E_k = \frac{mv^2}{2} \tag{8}$$

where E_k—kinetic energy of the blasting block (J); m—a mass of the blasting block (kg); V—velocity of blasting block (m/s).

$$v = \sqrt{\frac{2E_k}{m}} \tag{9}$$

The velocity of the blasting block obtained by calculating shall conform to the actual situation. Therefore, the research on the velocity of the blasting block is carried out. From the perspective of energy balance, Zuo et al. [21] studied the velocity of the blasting block by combining static loading and dynamic loading. McGarr [22] studied the velocity of the blasting block from the buckling instability of the beam. In addition, the maximum velocity of blasting blocks observed at different sites is 10 m/s [23,24]. According to the extensive research into various aspects, the maximum velocity of blasting blocks can exceed 10 m/s, but only in rare cases.

2.4. Dynamic Load Factor

The blasting block acts on the supporting structure at a certain velocity to cause deformation. In this process, the change of blasting block height is the vertical component of the displacement value of the supporting structure at its action point, and the displacement

value is minimal. Therefore, the change of gravitational potential energy of the blasting block is so small that it does not have to be considered. All the kinetic energy of the blasting block is transformed into the strain energy of the structure.

Kinetic energy: $E_k = \frac{mv^2}{2}$; strain energy: $V_d = \frac{F\Delta}{2}$. They are equal:

$$\frac{mv^2}{2} = \frac{F\Delta}{2} \tag{10}$$

where F—maximum impact force (kN); Δ—maximum deflection of impact site (mm).

According to structural mechanics, when the supporting structure is considered as a linear elastic system, it can be known that F and Δ should meet the relationship shown in Equation (11):

$$\Delta = \frac{K_0 F}{EI} \tag{11}$$

where E—Elastic modulus of supporting structure; I—Sectional bending coefficient; K_0—Support dimension coefficient.

In the formula, support dimension coefficient K_0 is a coefficient related to support section size and support section type, reflecting the difference of load-deformation relationship between different tunnel sections. When the tunnel is circular, it can be calculated according to Equation (12):

$$K_0 = \frac{\pi - 3}{2} R^2 \tag{12}$$

where R—Equivalent radius of tunnel (m).

The simultaneous Formulas (10)–(12) can be solved as follows:

$$\begin{cases} \Delta = \sqrt{\frac{K_0 m v^2}{EI}} \\ F = \sqrt{\frac{m v^2 EI}{K_0}} \end{cases} \tag{13}$$

where Δ—Maximum deflection of impact site (mm); F—Impact load expressed in the form of concentrated load (kN).

It can also be expressed by kinetic energy:

$$F = \sqrt{\frac{2E_k VEI}{K_0}} \tag{14}$$

where E_k—Released kinetic energy per unit volume of surrounding rock (J); V—Volume of surrounding rock releasing kinetic energy (m^3).

In structural dynamics, dynamic load factor K is used to represent the action of dynamic load. The dynamic load factor represents the ratio of the deformation of the structure under dynamic load to the deformation of the structure under self-weight. Moreover, it can also be expressed as the multiple of the impact load equivalent to the self-weight.

$$K = \frac{F}{mg} = \frac{\Delta}{\Delta_{st}} = \sqrt{\frac{v^2 EI}{K_0 m g^2}} \tag{15}$$

According to Equation (15), the dynamic load factor of impact load acting on the supporting structure is related to the stiffness of the supporting structure. Therefore, the impact load on the supporting structure is different depending on the supporting parameters.

It should be noted that when the supporting structure is considered as a linear elastic system, Equations (13)–(15) are universal. They are not only applicable to circular tunnels but also railway or highway structures with sectional shapes such as three-center circles and four-center circles. Different section sizes and different section types correspond to different support size coefficients K_0.

2.5. Impact Load

Increase the self-weight of blasting block to multiple of the dynamic load factor, and act on the structure in the form of uniform load, which can be equivalent to the impact load.

$$q = \frac{Kmg}{a^2} \tag{16}$$

Alternatively, expressed in the form of kinetic energy:

$$q = \sqrt{\frac{2E_k VEI}{K_0 a^4}} \tag{17}$$

where q—Impact load (Pa); a—length of a side of the blasting block (m).

The underlying principles of both expressions are the same. The significance of using different expressions is that the expression of Equation (16) is more concise, but it requires many assumptions. When the rockburst on site is different from the assumptions, the application of this equation will be limited. The expression of Equation (17) is more basic. In addition, when on-site parameters are different from the assumption, the relevant parameters can be directly adjusted to be more appropriate for the on-site situation, however the expression is complicated as a result.

3. The Acting Location of Impact Load in the Rockburst Tunnel

The impact load is the local load, applied to the location where rockburst is most likely to occur in the surrounding rock of the tunnel, or the location where the damage degree is the largest after rockburst occurs. Therefore, in this section, the authors research the acting location of impact load in circular tunnels and horseshoe tunnels.

3.1. Calculation Model and Calculation Parameters

The calculation model includes two section sizes, single-track tunnel, and double-track tunnel, and two-section types, circular section and horseshoe section, for each section. The section size and calculation model are shown in Figure 6.

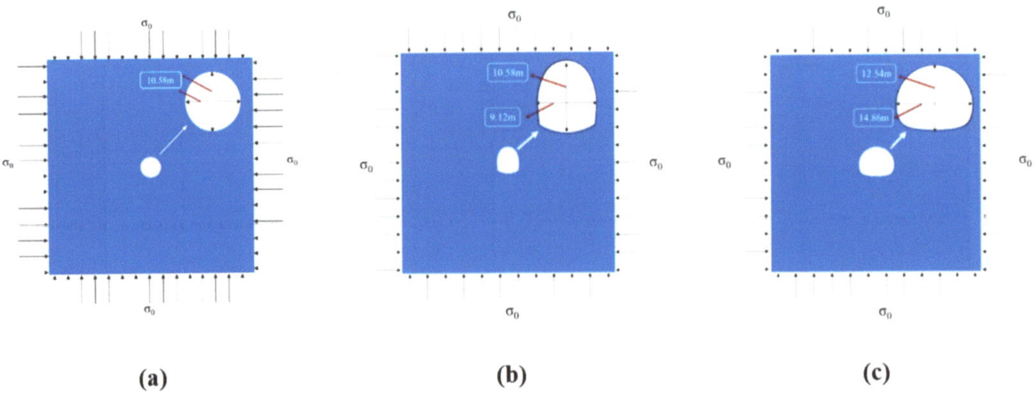

Figure 6. Numerical calculation model: (**a**) Circular; (**b**) Single-track; (**c**) Double-track.

Considering in-situ stress, section type, and lateral pressure coefficient, 18 working conditions are combined. The specific calculation conditions are listed in Table 5, and the numerical calculation parameters are shown in Table 6.

Table 5. Calculation condition table.

Section Type	In-Situ Stress (MPa)	Lateral Pressure Coefficient	Calculation Times
Circular tunnel	30 MPa	λ = 0.5	1
		λ = 1	1
		λ = 1.5	1
	40 MPa	λ = 0.5	1
		λ = 1	1
		λ = 1.5	1
Single-track	30 MPa	λ = 0.5	1
		λ = 1	1
		λ = 1.5	1
	40 MPa	λ = 0.5	1
		λ = 1	1
		λ = 1.5	1
Double-track	30 MPa	λ = 0.5	1
		λ = 1	1
		λ = 1.5	1
	40 MPa	λ = 0.5	1
		λ = 1	1
		λ = 1.5	1
Total			18

Table 6. Physical and mechanical parameters of surrounding rock.

Lithology	Average Unit Weight (kN/m^3)	Elasticity Modulus (GPa)	Poisson's Ratio	Cohesion (MPa)	Internal Friction Angle (°)	Uniaxial Compressive Strength (MPa)
Granite	26.9	33	0.23	2	54	90.48

The commonly used E. Hoek determination method [25] is selected to judge rockburst level by the ratio of the maximum tangential stress of tunnel section to the uniaxial compressive strength of rock, as shown in Formula (18).

$$\begin{cases} \sigma_\theta / R_c < 0.42 \text{ (Mild grade)} \\ 0.42 \le \sigma_\theta / R_c < 0.56 \text{ (Medium grade)} \\ 0.56 \le \sigma_\theta / R_c < 0.7 \text{ (Severe grade)} \\ \sigma_\theta / R_c \ge 0.7 \text{ (Extremely severe grade)} \end{cases} \quad (18)$$

where σ_θ—Maximum tangential stress of tunnel section; R_c—Uniaxial compressive strength of rock.

3.2. Analysis of the Acting Location of Impact Load

3.2.1. Circular Tunnel

The calculated maximum tangential stress of the tunnel section cannot directly reflect the occurrence range of rockburst. To have an intuitive display effect, the product of the determination coefficient in the limit state and the uniaxial compressive strength of rock are taken as the critical tangential stress of rockburst, and its contour line is drawn in the figure to obtain the occurrence range of rockburst of different grades. The occurrence range of extremely severe rockburst after treatment is shown in Figures 7 and 8.

Figures 7 and 8 reveal that, with the increase of in-situ stress, the influence depth in the surrounding rock where rockburst may occur gradually increases. However, the location where the influence depth is the deepest does not change with the increase of in-situ stress. No matter what the in-situ stress is, when λ = 1, the depth where rockburst may occur in the whole ring of the tunnel is the same; but when λ = 0.5, the depth where rockburst may occur at the tunnel arch waist is deepest; also, when λ = 1.5, the depth where rockburst may occur at the tunnel vault and invert is most resounding. The relevant conclusions are consistent with the theoretical analysis in the previous section.

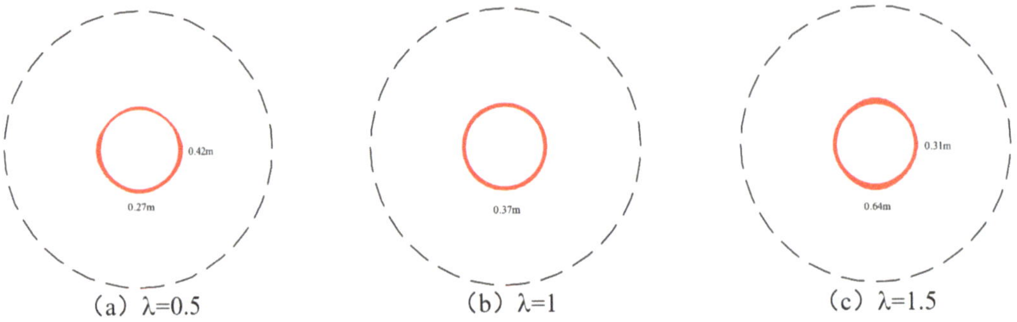

Figure 7. Range of extremely strong rockburst when σ_V = 30 Mpa: (**a**) σ_H/σ_V = 0.5; (**b**) σ_H/σ_V = 1.0; (**c**) σ_H/σ_V = 1.5.

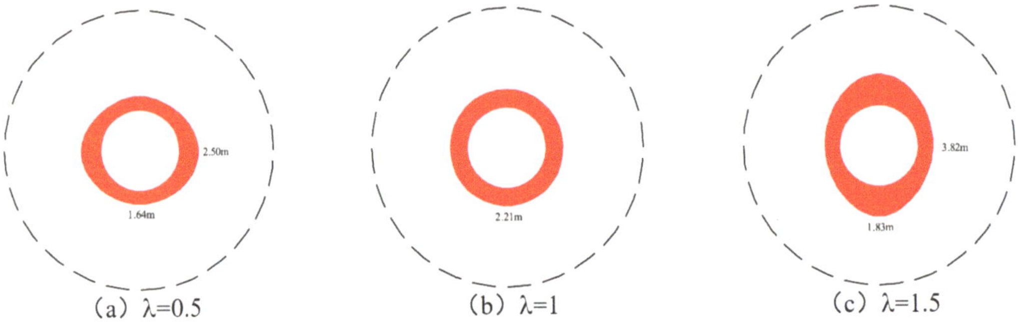

Figure 8. Range of extremely strong rockburst when σ_V = 40 Mpa: (**a**) σ_H/σ_V = 0.5; (**b**) σ_H/σ_V = 1.0; (**c**) σ_H/σ_V = 1.5.

The construction drainage tunnel of Jinping II hydropower station is a circular tunnel with complete surrounding rock and few structural planes. The buried depth of the drainage tunnel is generally significant. Except that gravity and structural stress act together within a certain range at the inlet and outlet, most of the tunnels are mainly gravity stress. Statistics were made on 137 parts where rockburst occurred in 115 sections ofdrainage tunnels [26], as shown in Table 7.

Table 7. Location statistics of rockburst in a construction drainage tunnel.

Number	Mileage	Rockburst Grade	Location
1	SK14 + 441.00~14 + 426.00	severe	right wall
2	SK14 + 415.00~14 + 390.00	severe	right wall
3	SK14 + 415.00~14 + 390.00	severe	left wall, vault
4	SK14 + 252.00~14 + 250.00	medium	right spandrel
5	SK14 + 251.00~14 + 245.00	severe	right spandrel
6	SK14 + 230.00~14 + 227.00	severe	right spandrel
7	SK14 + 226.00~14 + 224.00	medium	left spandrel
8	SK14 + 222.00~14 + 220.00	medium	right spandrel
9	SK13 + 039.00~13 + 043.00	mild	left spandrel
114	SK9 + 877.00~9 + 875.00	severe	Vault, spandrel
115	SK9 + 872.00~9 + 869.00	severe	spandrel, left wall

The location is divided into three parts: spandrel, arch waist, and vault, and the proportion of rockburst in each part is shown in Figure 9.

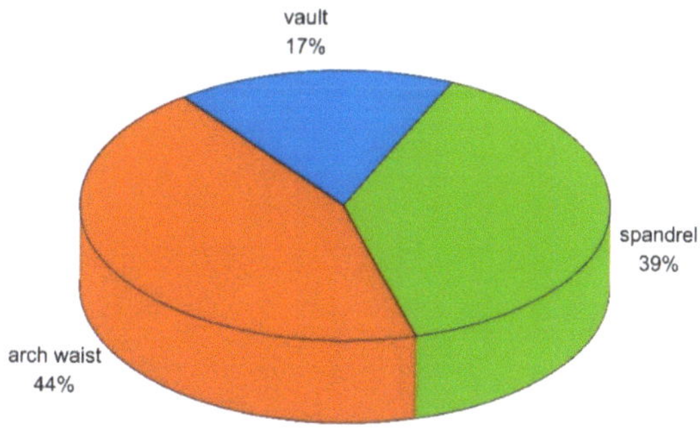

Figure 9. Statistics of rockburst locations.

As shown from Table 7 and Figure 9, the frequency of rockburst in different parts is arch waist > spandrel > vault. Considering that the in-situ stress is mainly gravity, the conclusion is consistent with the previous two sections.

3.2.2. Subsubsection

The same method can be used to study the location where rockburst most likely occurs in horseshoe tunnels under different in-situ stress and different lateral pressure coefficients. The rockburst occurrence range is treated in the same way as the circular tunnel. Severe rockburst is represented in blue, and extremely severe rockburst is represented in red, the occurrence range of severe and extremely severe rockburst in single-track and double-track tunnels can be obtained after treatment, as shown in Figures 10–13.

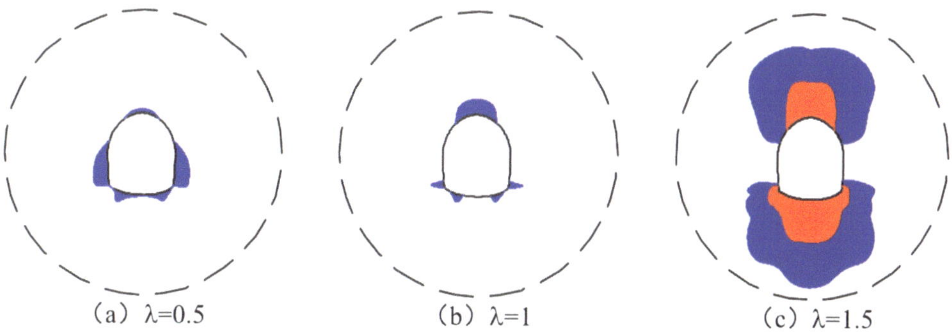

Figure 10. Range of strong and extremely strong rockburst in a single-track tunnel when σ_v = 30 MPa.

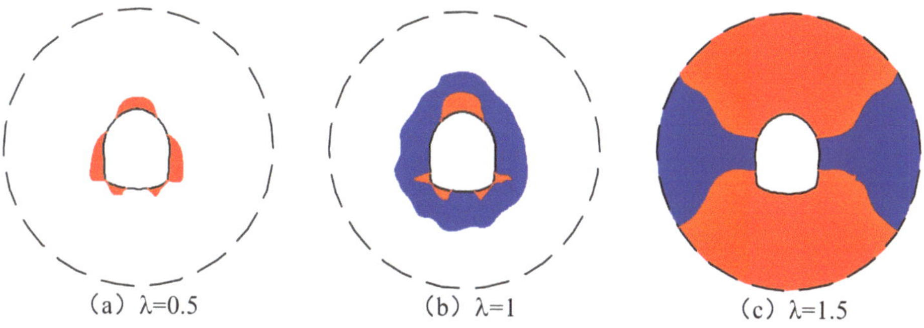

Figure 11. Range of strong and extremely strong rockburst in a single-track tunnel when σ_v = 40 MPa.

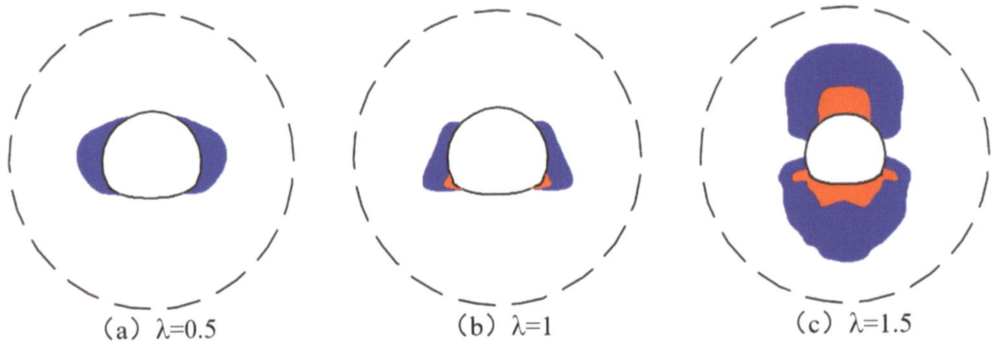

Figure 12. Range of strong and extremely strong rockburst in a double-track tunnel when σ_V = 30 MPa.

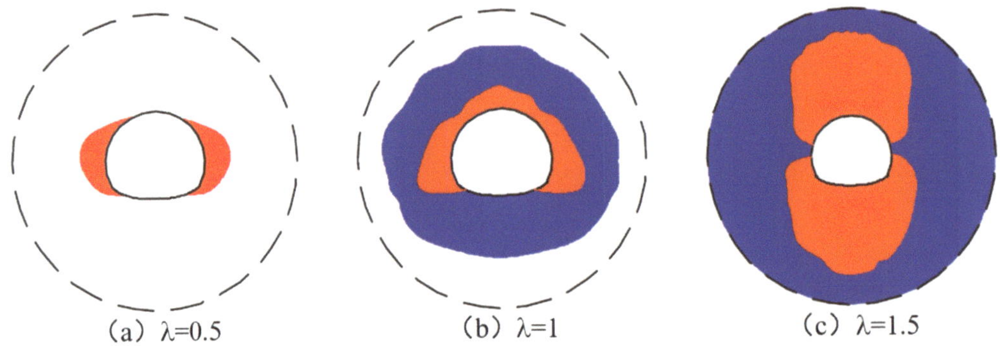

Figure 13. Range of strong and extremely strong rockburst in a double-track tunnel when σ_V = 40 MPa.

As can be seen from Figures 10–13, when the in-situ stress is mainly vertical, the most damaged parts are the side wall and arch waist; when the vertical stress and horizontal stress are equal, the most damaged parts are different between single-track tunnels and double-track tunnels: damage degree of a side wall and the vault is approximately equal in single-track tunnels, while the damage degree of side wall is the largest, followed by the arch waist in double-track tunnels; when the in-situ stress is mainly horizontal, the most damaged parts are vault and invert.

The 2# and 4# diversion tunnels of Jinping II hydropower station are horseshoe-shaped sections with a diameter of 13 m. In reference [27], the location of stress-type failure that occurred in the complete or relatively complete surrounding rock of tunnel 2# and tunnel 4# was recorded. In this process, special attention was paid to investigating actual geological conditions, to avoid counting the damage controlled by the structural plane. All the surrounding rocks of the tunnel in statistics are T2b marble.

The locations of 105 rockbursts that occurred in the 2# and 4# diversion tunnels with largely buried depths (1900–2400 m and 2160–2500 m) were statistically analyzed. Among 154 locations of 105 rockbursts, the proportion of each location is shown in Figure 14.

The super-buried depths are at 1900~2400 m and 2160~2500 m, and the in-situ stress in this section is mainly vertical stress. As can be seen from Figure 14, under this in-situ stress condition, the frequency of rockburst in different parts: side wall > arch waist > spandrel > vault. The horseshoe tunnel used at the site is a double-track tunnel. When the in-situ stress is mainly vertical, rockburst occurs most frequently on the side wall and arch waist.

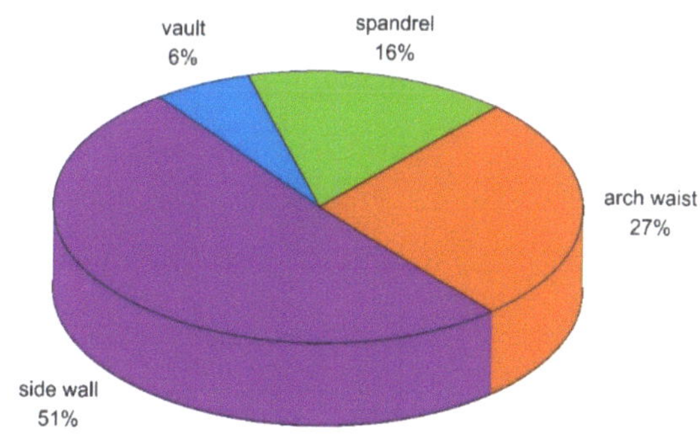

Figure 14. The proportion of each part of the rockburst.

3.2.3. Suggestions for the Application of Action Location

Table 8 summarizes the rockburst prone location under different section types based on the previous section's research.

Table 8. Summary of rockburst prone sites under different conditions.

Section Type	In-Situ Stress	Prone Sites
circular	$\lambda = 1$	the whole ring
	$\lambda > 1$	vault and invert
	$\lambda < 1$	arch waist
horseshoe	mainly vertical stress	side wall and arch waist
	vertical stress and horizontal stress are equal	single-track: side wall and vault
	mainly horizontal stress	double-track: side wall

4. The Rockburst Tunnel Calculation Model and Process

4.1. Calculation Model of Rockburst Tunnel

The previous section gives the calculation method of impact load and the determination method of action location and range of impact load. After both are determined, the local load-structure model can be used for numerical calculation. Similar to the conventional load-structure model, the radial spring which can only be pressed, and the tangential spring which can be both be pulled and pressed are used to simulate the interaction between surrounding rock and supporting structure. The difference is the scope of action, which changes from the whole structure to the local structure. It should be noted that the action range of impact load may not be continuous and may be divided into two parts in some cases. It is also possible that rockburst may occur in the whole ring when the in-situ stress is too large, and at this time, the action range of impact load is the whole supporting structure.

The schematic diagram of the final local load-structure model is shown in Figure 15.

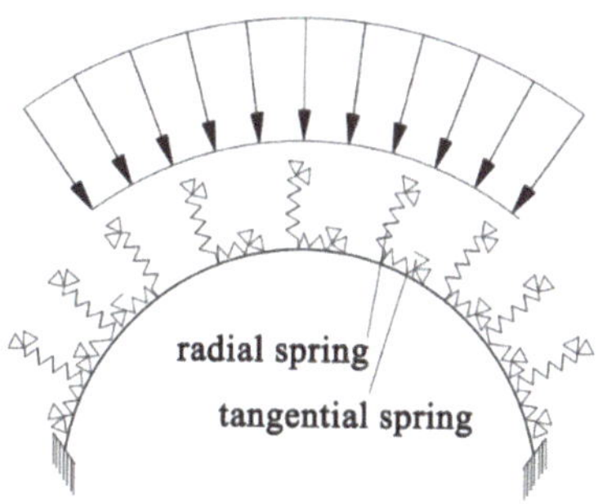

Figure 15. Schematic diagram of a local load-structure model of the rockburst tunnel.

4.2. Calculation Process of Rockburst Tunnel

The load acting on the supporting structure is related to the size and velocity of the blasting block and the stiffness of the supporting structure. Even if the same size block acts on the supporting structure at the same speed, the supporting structure will bear different loads and deformation if the stiffness of the supporting structure is different. When the stiffness of the supporting structure is large, the deformation under the impact of blasting block is small and the impact load is large. When the stiffness of the supporting structure is small, its deformation under the impact of blasting block is large and the impact load is small. If a rockburst occurs without a supporting structure, in that case, even if the blasting block is ejected at a certain speed, there is no load acting on the supporting structure because the supporting structure does not exist.

Through the above analysis, it can be seen that the impact load acting on the supporting structure is not determined. In practical engineering, the type and parameters of supporting structure can be determined by analogy with similar engineering first, and then the impact load acting on supporting structure can be calculated accordingly. The local load-structure method is used to determine the supporting structure's safety factor and make a quantitative judgment of its safety. For the supporting structure where the safety factor does not meet the code's requirements or meets the requirements but has a small safety reserve, the supporting parameters can be adjusted. The above calculation can be carried out according to the adjusted supporting parameters until the safety needs of the project are met as shown in Figure 16.

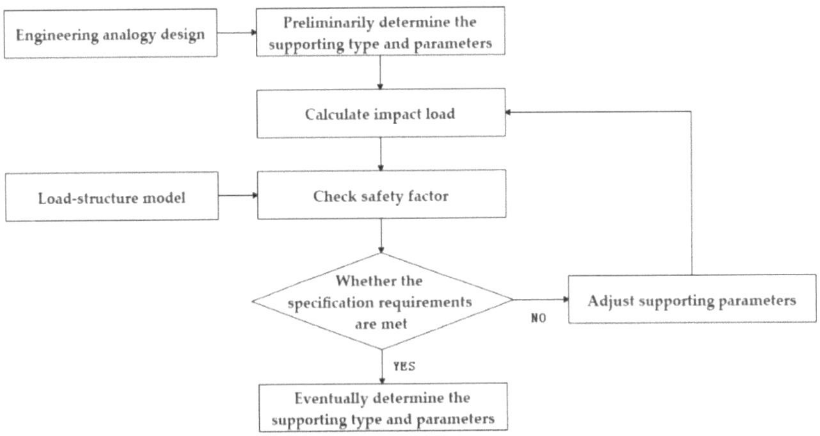

Figure 16. Design process.

5. Example Analysis

Assuming that the surrounding rock in the rockburst section is granite, the main surrounding rock parameters are selected as follows: elastic modulus E = 20 GPa, Poisson's ratio μ = 0.25, the in-situ stress is 40 MPa and the lateral pressure coefficient is 1, and the maximum strain energy density can be calculated as 150 kJ/m^3 according to Section 2.1, the corresponding released kinetic energy is 76.5 kJ/m^3 according to Section 2.2, and the corresponding supporting parameters are shown in Tables 9 and 10.

Table 9. Supporting parameters of the rockburst section in single-track by drilling and blasting method.

Rockburst Grade	Surrounding Rock Grade	Shotcrete (cm)	Steel Frame
mild	II III	5 8	
medium	II III	5 8	
severe	II, III	Initial spraying: 5 cm; Respraying: 15 cm	Set in arch and wall, HW125 type, Spacing 1.2 m
extremely severe	II, III	Initial spraying: 5 cm; Respraying: 18 cm	Set in arch and wall, HW150 type, Spacing 1 m

Table 10. Supporting parameters of the rockburst section in double-track by drilling and blasting method.

Rockburst Grade	Surrounding Rock Grade	Shotcrete (cm)	Steel Frame
Granite	II III	5 12	– –
medium	II III	5 12	– –
severe	II, III	Initial spraying: 5 cm; Respraying: 19 cm	Set in arch and wall, HW150 type, Spacing 1.2 m
extremely severe	II, III	Initial spraying: 5 cm; Respraying: 21 cm	Set in arch and wall, HW175 type, Spacing 1 m

6. Calculation of Impact Load

The location and range of rockburst were obtained by numerical simulation method, for the single-track and double-track types, and statistics were made for the results, as shown in Table 10. The results showed that rockburst might occur in the whole ring, but at different levels. However, according to the experience on-site, even if rockbursts repeatedly occur at the same section, they do not necessarily occur simultaneously, but often in order. When different grades of rockburst act separately, the impact on the structure is the most adverse. Therefore, when applying load, only one level of load shall be applied at a time. Taking extremely severe rockburst as an example, the load is applied to the vault with a range of 4.45 m in the single-line tunnel and the load is applied to the side wall with a range of 7.45 m in the double-track tunnel. Other levels of rockbursts refer to extremely severe rockbursts, which are used to determine the location and scope of impact load, as shown in Table 11.

Table 11. Parts and range of impact load.

Section Type	Rockburst Grade	Acting Parts	Acting Range
single-track	mild	vault	0.5 m
	medium	vault	1 m
	severe	vault	3 m
	extremely severe	vault	4.45 m
double-track	mild	side wall	0.5 m
	medium	side wall	1 m
	severe	side wall	3 m
	extremely severe	side wall	7.45 m

The final kinetic energy release of surrounding rock per unit volume and the calculated velocity of blasting block are shown in Table 12.

Table 12. Calculation table of releasing kinetic energy and bursting block velocity of different levels of rockburst.

Rockburst Grade	The Kinetic Energy Released Per Unit Volume of Surrounding Rock (kJ/m^3)	The Velocity of Bursting Block (m/s)
mild	5	1.91
medium	10	2.70
severe	35	5.05
extremely severe	76.5	7.46

According to the determined supporting parameters and release kinetic energy, the impact loads under different rockburst grades are shown in Table 13.

Table 13. Impact load calculation table.

Section Type	Rockburst Grade	Impact Load (kPa)
single-track	mild	51.29
	medium	103.80
	severe	443.18
	extremely severe	699.77
double-track	mild	36.50
	medium	135.70
	severe	414.57
	extremely severe	598.50

It can be seen from Table 13 that the impact loads of single-track tunnels and double-track tunnels are not the same. On the one hand, the shape and size of the two sections are different, which leads to the difference of supporting dimension coefficient K_0. On the other hand, the difference in thickness directly affects the flexural stiffness *EI* value of the supporting structure. However, in terms of results, there was not much difference overall.

Safety Factor Checking

The factor of safety is checked by ANSYS numerical simulation software, and the calculation model by the local load-structure model. The established structure model of the single-track tunnel is shown in Figure 17.

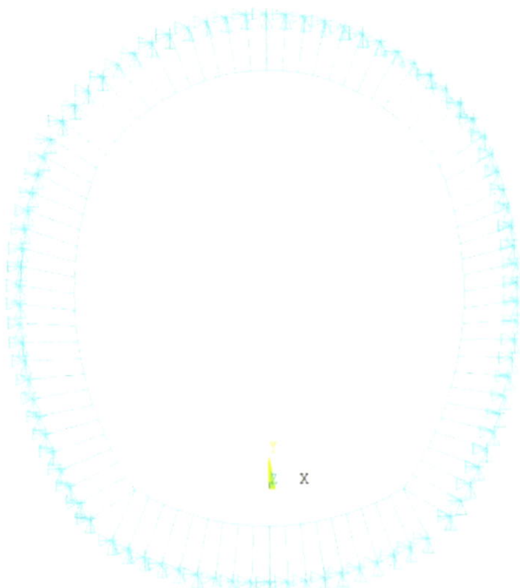

Figure 17. Schematic diagram of the geometric model.

Taking the single-track tunnel as an example, the supporting structure's bending moment and axial diagram are calculated, as shown in Figures 18 and 19.

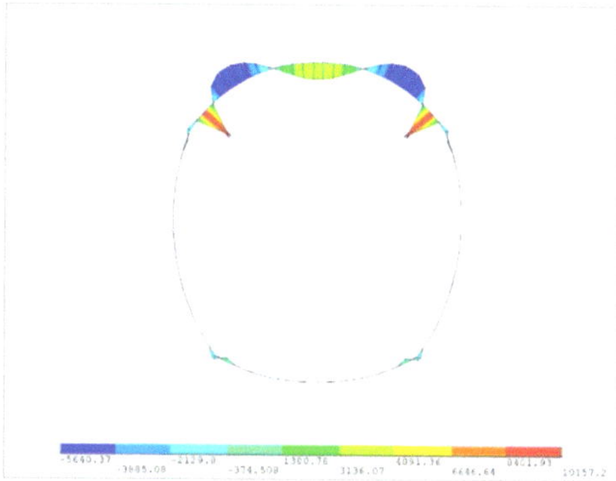

Figure 18. Bending moment diagram of supporting structure.

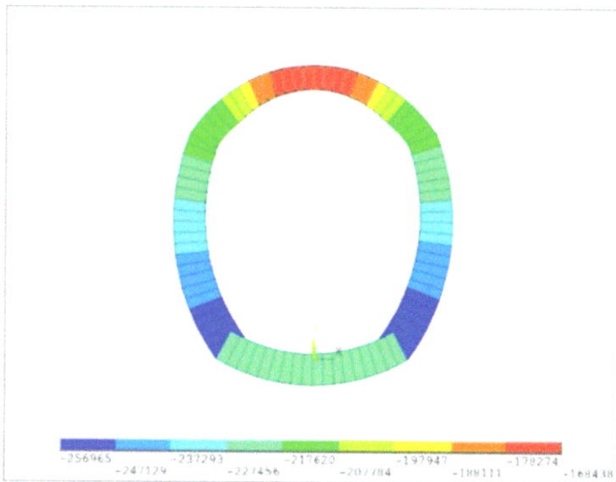

Figure 19. Axial diagram of supporting structure.

The final calculation results of the rockburst safety factor are summarized in Table 14.

Table 14. Safety factor check table of rockburst tunnel.

Section Type	Rockburst Grade	Minimum Safety Factor	Pull or Press	Part	Control Standards
single-track	mild	3.24	pull	vault	2.7 (pull)
	medium	2.88	pull	vault	2.7 (pull)
	severe	1.83	pull	vault	1.8 (pull)
	extremely severe	1.35	pull	vault	1.8 (pull)
double-track	mild	3.95	press	side wall	1.8 (press)
	medium	2.94	press	side wall	1.8 (press)
	severe	2.12	pull	side wall	1.8 (pull)
	extremely severe	1.47	pull	side wall	1.8 (pull)

According to the calculation results in Table 14, except for extremely severe rockburst, rockburst of other levels can meet the requirements of minimum factor safety, but the safety reserve is small. In practical engineering, parameters can be adjusted in real-time according to relevant calculation results to ensure the safety of the supporting structure. The safety of the supporting structure can be improved by adopting other auxiliary construction measures, such as stress release holes and high-pressure water injection.

7. Conclusions

From the perspective of energy with the method of structural mechanics, this paper deduces the calculation method of impact load of rockburst tunnel acting on the supporting structure, determines the action location of impact load under different section types and different section sizes, and derives the local load-structure calculation model and the calculation process designed by this model. The main conclusions are as follows:

(1) Using the method of structural mechanics, the calculation method of impact load of rockburst tunnel acting on the supporting structure is deduced. The expression form of impact load is uniform load. The dynamic load factor K in the formula is related to the stiffness EI of the supporting structure and the supporting dimension coefficient K_0. The support dimension coefficient K_0 is related to the size and type of support section, which reflects the difference in load-deformation relations between tunnel sections.

(2) Rockburst-prone parts of circular tunnel are related to lateral pressure coefficient: when $\lambda = 1$, the probability of rockburst occurring in the whole ring is the same; when $\lambda > 1$, the vault and inverted roof are most prone to rockburst; when $\lambda < 1$, the arch waist is most prone to rockburst. In the horseshoe tunnel, when the vertical stress is dominant, the parts prone to rockburst are the side walls on both sides and the arch waist. When the vertical and horizontal stresses are equivalent, the rockburst-prone parts of the single-track tunnel are side wall and vault. The double-track tunnel is a side wall; when the horizontal stress is dominant, the prone parts to rockburst are vault and invert.

(3) Finally, the local load-structure model based on the impact load calculation method can be used for quantitative estimation of the rockburst tunnel. The type and parameters can be determined by analogy with similar projects during calculation. After calculating the impact load, the parameters of the supporting structure whose factor of safety does not meet the requirements can be adjusted until the project's safety requirements are met.

Author Contributions: Conceptualization, Z.Y. and H.T.; Methodology, Z.Y.; Software, Z.Y. and D.Z.; Validation, Z.Y., D.L., Z.W., D.Z. and H.T.; Formal analysis, Z.Y., D.L. and D.Z.; Investigation, D.L. and H.T.; Resources, Z.W. and H.T.; Data curation, D.Z.; Writing—original draft preparation, Z.Y.; Writing—review and editing, D.L.; Visualization, Z.Y. and D.Z.; Supervision and project administration: Z.Y. and D.L. All authors have read and agreed to the published version of the manuscript.

Funding: This research was funded by Science and technology research and development project of China National Railway Group (P2019G038-5).

Data Availability Statement: The data presented in this study are available on request from the corresponding author.

Acknowledgments: We acknowledge SWJTU (Southwest Jiaotong University) and the staff for their technical support and assistance.

Conflicts of Interest: The authors declare no conflict of interest.

References

1. Cheng, L.; Jianfeng, L.; Yi, R.; Chao, L.; Yilin, L. Study on Very Long-Term Creep Tests and Nonlinear Creep-Damage Constitutive Model of Salt Rock. *Int. J. Rock Mech. Min. Sci.* **2021**, 146–156.
2. Lyu, C.; Liu, J.; Ren, Y.; Liang, C.; Zhang, Q. Study on Long-Term Uniaxial Compression Creep Mechanical Behavior of Rocksalt-Mudstone Combined Body. *Int. J. Damage Mech.* **2021**, 1–19. [CrossRef]
3. Xie, H.; Pei, J.; Zuo, J.; Zhang, R. Investigation of Mechanical Properties of Fractured Marbles by Uniaxial Compression Tests. *J. Rock Mech. Geotech. Eng.* **2011**, *3*, 302–313. [CrossRef]
4. Zhou, H.; Chen, S.; Li, H.; Liu, T.; Wang, H. Rockburst Prediction for Hard Rock and Deep-Lying Long Tunnels Based on the Entropy Weight Ideal Point Method and Geostress Field Inversion: A Case Study of the Sangzhuling Tunnel. *Bull. Eng. Geol. Environ.* **2021**, *80*, 3885–3902. [CrossRef]

5. Gong, Q.M.; Yin, L.J.; Wu, S.Y.; Zhao, J.; Ting, Y. Rock Burst and Slabbing Failure and Its Influence on Tbm Excavation at Headrace Tunnels in JINPING II Hydropower Station. *Eng. Geol.* **2012**, *124*, 98–108. [CrossRef]
6. Naji, A.M.; Emad, M.Z.; Rehman, H.; Yoo, H. Geological and Geomechanical Heterogeneity in Deep Hydropower Tunnels: A Rock Burst Failure Case Study. *Tunn. Undergr. Sp. Technol.* **2019**, *84*, 507–521. [CrossRef]
7. Xue, R.; Liang, Z.; Xu, N.; Dong, L. Rockburst Prediction and Stability Analysis of the Access Tunnel in the Main Powerhouse of a Hydropower Station Based on Microseismic Monitoring. *Int. J. Rock Mech. Min. Sci.* **2020**, *126*, 1365–1609. [CrossRef]
8. He, S.; Lai, J.; Zhong, Y.; Wang, K.; Xu, W.; Wang, L.; Liu, T.; Zhang, C. Damage Behaviors, Prediction Methods and Prevention Methods of Rockburst in 13 Deep Traffic Tunnels in China. *Eng. Fail. Anal.* **2021**, *121*, 105178. [CrossRef]
9. Zhou, Y.; Zhao, D.; Li, B.; Wang, H.; Zhang, Z. Fatigue Damage Mechanism and Deformation Behaviour of Granite under Ultrahigh-Frequency Cyclic Loading Conditions. *Rock Mech. Rock Eng.* **2021**, *54*, 4723–4739. [CrossRef]
10. Han, Z.; Li, D.; Zhou, T.; Zhu, Q.; Ranjith, P.G. Experimental Study of Stress Wave Propagation and Energy Characteristics across Rock Specimens Containing Cemented Mortar Joint with Various Thicknesses. *Int. J. Rock Mech. Min. Sci.* **2020**, *131*, 104352. [CrossRef]
11. Li, D.; Han, Z.; Sun, X.; Zhou, T.; Li, X. Dynamic Mechanical Properties and Fracturing Behavior of Marble Specimens Containing Single and Double Flaws in SHPB Tests. *Rock Mech. Rock Eng.* **2019**, *52*, 1623–1643. [CrossRef]
12. Müller, W. Numerical Simulation of Rock Bursts. *Min. Sci. Technol.* **1991**, *12*, 27–42. [CrossRef]
13. Cook, N.; Hoek, E.P.; Pretorius, J.; Ortlepp, W.D.; Salamon, M. Rock Mechanics Applied to the Study of Rockbursts. *J. S. Afr. Inst. Min. Metall.* **1966**, *66*, 435–528.
14. Wang, J.; Zuo, T.; Li, X.; Tao, Z.; Ma, J. Study on the Fractal Characteristics of the Pomegranate Biotite Schist under Impact Loading. *Geofluids* **2021**, *2021*, 1570160. [CrossRef]
15. Ma, J.; Li, X.; Wang, J.; Tao, Z.; Zuo, T.; Li, Q.; Zhang, X. Experimental Study on Vibration Reduction Technology of Hole-by-Hole Presplitting Blasting. *Geofluids* **2021**, *2021*, 5403969. [CrossRef]
16. Fang, Q.; Wang, G.; Yu, F.; Du, J. Analytical Algorithm for Longitudinal Deformation Profile of a Deep Tunnel. *J. Rock Mech. Geotech. Eng.* **2021**, *13*, 845–854. [CrossRef]
17. Mingqing, X.; Shaofeng, W.; Libao, C. Research on the Design Method of Primary Support of Tunnel Based on the Load-Structure Method. *J. Railw. Eng. Soc.* **2018**, *35*, 60.
18. Li, T.; Meng, L.; Wang, L. *Stability of High-Stress Tunnels and Prevention of Rockburst and Large Deformation Disasters*; Science Press: Beijing, China, 2016.
19. Cai, M.; Kaiser, P. Assessment of Excavation Damaged Zone Using a Micromechanics Model. *Tunn. Undergr. Sp. Technol.* **2005**, *20*, 301–310. [CrossRef]
20. Gao, Y.; Zhang, H.; Zhao, G.; Cheng, X. *Study on Rockburst Mechanism and Prevention Practice of Deep Buried Long Tunnel*; China Water & Power Press: Beijing, China, 2014.
21. Zuo, Y.J.; Li, X.B.; Zhou, Z.L. Determination of Ejection Velocity of Rock Fragments during Rock Burst in Consideration of Damage. *J. Cent. South Univ. Technol.* **2005**, *12*, 618. [CrossRef]
22. Mcgarr, A. A Mechanism for High Wall-Rock Velocities in Rockbursts. *Pure Appl. Geophys.* **1997**, *150*, 381–391. [CrossRef]
23. Linkov, A.M. Rockbursts and the Instability of Rock Masses. *Int. J. Rock Mech. Min. Sci. GeoMech. Abstr.* **1996**, *33*, 727–732. [CrossRef]
24. Ortlepp, W. High Ground Displacement Velocities Associated with Rockburst Damage. *Rockbursts Seism. Min.* **1993**, *93*, 101–106.
25. Hoek, E.; Brown, E.T. Practical Estimates of Rock Mass Strength. *Int. J. Rock Mech. Min. Sci.* **1997**, *34*, 1165–1186. [CrossRef]
26. Liu, L.; Ling, Y.; Li, N.; Pang, T. *Research and Case Analysis of Rock Burst Disaster in Deep Hydraulic Tunnel*; China Water & Power Press: Beijing, China, 2019.
27. Yamashi, F.; Zhang, C.; Shao-Jun, L.; Qiu, S.-L.; Zhang, C. *Dynamic Design Method for Deep Tunnels in Hard Rock*; Science Press: Beijing, China, 2013.

minerals

MDPI

Article

The Meso-Structural Characteristics of Crack Generation and Propagation during Rock Fracturing

Yanbo Zhang [1], Guangyu Gao [1], Shaohong Yan [2], Xulong Yao [1,*], Xiangxin Liu [1], Peng Liang [1] and Yuedong Xu [1]

1 College of Mining Engineering, North China University of Science and Technology, Tangshan 063210, China; fzdn44444@163.com (Y.Z.); hblgggy@163.com (G.G.); liuxiangxin9@163.com (X.L.); hnlp87@163.com (P.L.); xuyuedong1996@163.com (Y.X.)

2 College of Science, North China University of Science and Technology, Tangshan 063210, China; shaohong@ncst.edu.cn

* Correspondence: yaoxulonglg@126.com

Abstract: Meso-structure is an important factor affecting the characteristics of rock fracture. To determine the factors influencing the internal meso-structural characteristics upon the crack generation and extension, rock samples were tested under uniaxial cyclic loading and unloading and examined using computed tomography (CT) scanning. CT scanning was used to visualize and investigate the entire process of fracture source generation and its development in three dimensions, and finally the location information of the fracture source was determined. The mineral composition and structure along the fracture path inside the specimen were studied by using a polarizing microscope, and the evolution of fracture propagation around mineral particles was revealed based on its mineralogical characteristics. Results indicate that based on the fracture source around different rock meso-structure types, the initial fracture source can also be divided into different types, namely, the primary porosity type, the micro-crack type, and the mineral grain type. The strength characteristics of mineral grains can determine whether the crack extends around the gravel or through it. The hard grains at the crack-tip promote the transformation of tensile stress to shear stress, which lead to the change in the direction of crack extension and bifurcation. The spatial shape of the cracks after rock fracture is related to the initial distribution of minerals and is more complicated in areas where minerals are concentrated. The crack extension around gravel particles also generates a mode of failure, affecting large grains with gravel spalling from the matrix. The findings provide a study basis for identifying the potentially dangerous areas and provide early warning for the safety of underground engineering construction operations.

Keywords: meso-structure; initial fracture source; crack propagation path; mineral particles

Citation: Zhang, Y.; Gao, G.; Yan, S.; Yao, X.; Liu, X.; Liang, P.; Xu, Y. The Meso-Structural Characteristics of Crack Generation and Propagation during Rock Fracturing. *Minerals* **2022**, *12*, 94. https://doi.org/10.3390/min12010094

Academic Editors: Diyuan Li, Zhenyu Han, Xin Cai, Shijie Xie and Stephen Laubach

Received: 11 December 2021
Accepted: 10 January 2022
Published: 14 January 2022

1. Introduction

Rock is a natural geological body formed from various minerals and interstitials. Many irregular, randomly distributed pores and cracks, or defects, are present within rocks, and the meso-structure of rock is complicated [1]. Natural defects represent potential hazards in engineering rock masses, such as rock slopes and underground roadways [2]. The macroscopic (macro-scale (>10^{-1} m), meso-scale (10^{-4}~10^{-1} m), micro-scale (10^{-8}~10^{-4} m) [3,4]) failure is the result of the cumulative meso-structure evolutions, and the source of fracture is often closely related to the meso-structure [5–7].

The study of meso-structural effects on the generation, expansion, and evolution of fracture sources, exploring the generation mechanism and evolution rules of fracture sources, will reveal potentially dangerous areas prone to instability. Some scholars studied the internal meso-structure, e.g., initial defects, mineral types, and particle size, to reveal the relationship between meso-structure and macro-fracture [8–10]. Li et al. [11] studied the rheological properties of tailings and cement with various structural characteristics.

Cao et al. [12,13] added external fibrous materials with various meso-structural characteristics to concrete and backfill. The effects of this material on the strength, crack formation, and expansion of concrete and backfill materials were studied by using CT scanning and scanning electron microscopy. Ji [14] and Song [15] studied the fracture process of rocks by uniaxial loading and SEM, and found that the meso-structural destruction of weakly cemented rocks mainly occurred between the particles of the cemented material. Hasanipanah et al. [16] and Huang et al. [17] used different algorithms to predict the shear strength of the rocks with good prediction results. Ge et al. [18] studied the variation law of rock surface roughness under load by means of numerical simulation. Han et al. [19] established an evaluation index of crack extension capacity based on the deformation characteristics of rocks under triaxial compression, and compared this for rocks with different lithologies. Computed tomography (CT) has been used to extract the information about pore or crack morphology from rock specimens or rock-like materials, to compare and evaluate the changes in the meso-scale parameters, and quantitatively characterize the fractures therein [20–22]. To study the process of the fracture initiation, a single fissure, double fissures, and multiple fissures were prefabricated on rock samples or rock-like materials [23–25], and a single hole, double holes, and a combination of holes and cracks were prefabricated on rock specimens [26–28]. Chemical processes have a much greater impact on fracture extension than we thought. Many scholars have studied the fracture behavior of rocks under load after immersion in chemical solutions from the perspective of chemical reactions and revealed the damage mechanism of rocks after chemical reactions [29–31].

In terms of numerical simulation experiments involving rock fracturing, many scholars [32–35] found that the peak strength and Poisson's ratio of rock are closely related to the mineral particle sizes. The effects of various particle gradations on tensile and compressive strengths were explored by adjusting the proportion of particles [36]. Bai et al. [37] established a pore distribution model, and analyzed the effect of the pore structure, such as the number, size, and location of pores, on the tensile strength of rocks. Yu et al. [38] established a three-dimensional (3D) model to characterize the meso-structure of minerals by using a digital image-processing technique, elucidating the thermal cracking mechanisms in rock.

In summary, most scholars studied the fracture initiation process and mechanical mechanism of fracture by prefabricating holes and fractures, but the reality is that there are multiple meso-structures within the rock, and the combination of these structures leads to fracture source generation and fracture extension. The association between fracture sources and meso-structures, and the types of fracture sources, are not well-defined, and there is a lack of in-depth research. The deep rocks are often in a state of stress redistribution and cyclic loading, and few studies have focused on the types of fracture in this state. Scholars have studied a lot about the influence of meso-structure on fracture extension, and very little about the source of fracture. Rocks are "black-box" materials, which cannot be directly observed inside, while CT scanning technology can reveal the internal structural features of rocks, providing technical support to explore the source of the fracture. To study the crack generation and propagation during rock fracturing on a meso-structural scale, a multi-stage loading experiment and CT scanning were performed. The distribution of internal pores or cracks was monitored, and the 3D reconstruction of rock specimens was undertaken using CT scanning and polarized-light microscopy. Finally, the influence of initial pores and micro-cracks and mineral particle distribution and strength were studied and discussed on the generation and propagation path of fracture sources.

2. Experimental Design

2.1. Specimen Selection

Deep mining has become the main battlefield for resource development. With the increase in mining depth, the change of mechanical properties of the rock mass is mainly manifested as the deterioration of rock strength and the accumulated deformation of the meso-structure under cyclic loading. This phenomenon is more common in the construction activities of geotechnical projects such as roadway boring and chamber excavation. The

sample comes from an open quarry and is a sandstone layer in the Upper Cretaceous strata. The rock mass is produced in layers, and the thickness of a single layer is generally greater than 2 m, which is a sedimentary layered deposit, and the type of rock is mainly neritic sedimentary rocks. Red sandstone has a diversity of mineral particle sizes, and was selected as the specimen. Five rock specimens were prepared, each measuring 100 × 100 × 100 mm (Figure 1).

Figure 1. The sandstone blocks prepared for strength testing and CT scanning. "CGRS-CT" indicates: Coarse Grain Red Sandstone-CT. The length, width, and height of the specimens are all 100 mm. The type of specimen is a sedimentary rock, and the mineral grains within it vary in size, with good CT imaging results. To ensure the consistency of the experiment, the laminar direction and the loading direction are parallel.

To determine the types and distribution of minerals, the rock slices were identified by polarizing microscopy. The rock we tested is mainly composed of clastic particles and interstitial substances: clastic particles include mainly quartz, feldspar, and cuttings. There are many methods of rock mineral grain size analysis, such as the laser method, screening method, light penetration method, and imaging methods. The particle size of mineral particles in rocks can be determined through these methods, and the imaging method is the most widely used to calculate the particle size distribution in the present work. It is noteworthy that the particle size distribution of mineral particles was roughly determined by polarized light microscopy and software, and then calculated by sieve analysis for purposes of correction of the initial data. The particle size was generally 2–8 mm, some particles were of 8–17 mm, and a few were between 0.5 and 2 mm (Figure 2). Most of them were subangular to sub-round, some were spherical, and the grains were poorly sorted. The interstitial materials (Figure 3) were mainly argillaceous, sand, sericite, carbonate minerals, and ferric oxide, which are mainly clay minerals, and show the pore type cementation between clastic particles.

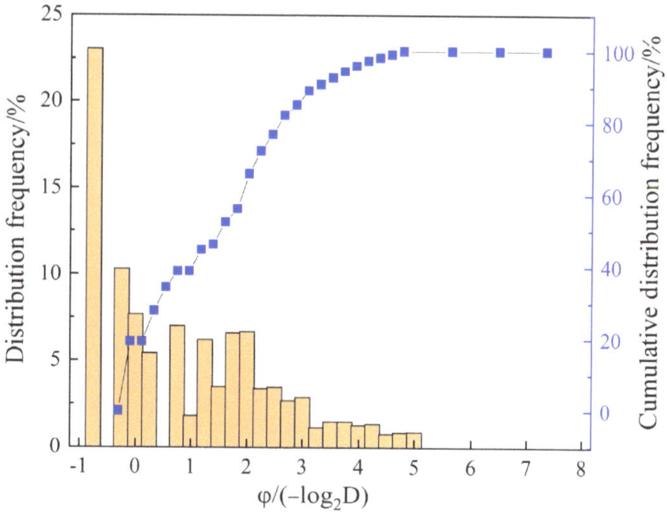

Figure 2. Particle size distribution. D indicates the particle size, φ is the common particle size unit for particle size analysis, and the relationship between the two can be expressed by the equation: $\varphi = -\log_2 D$. The rock interior has the highest percentage of coarse sand content, followed by medium sand, and the lowest content of silt sand.

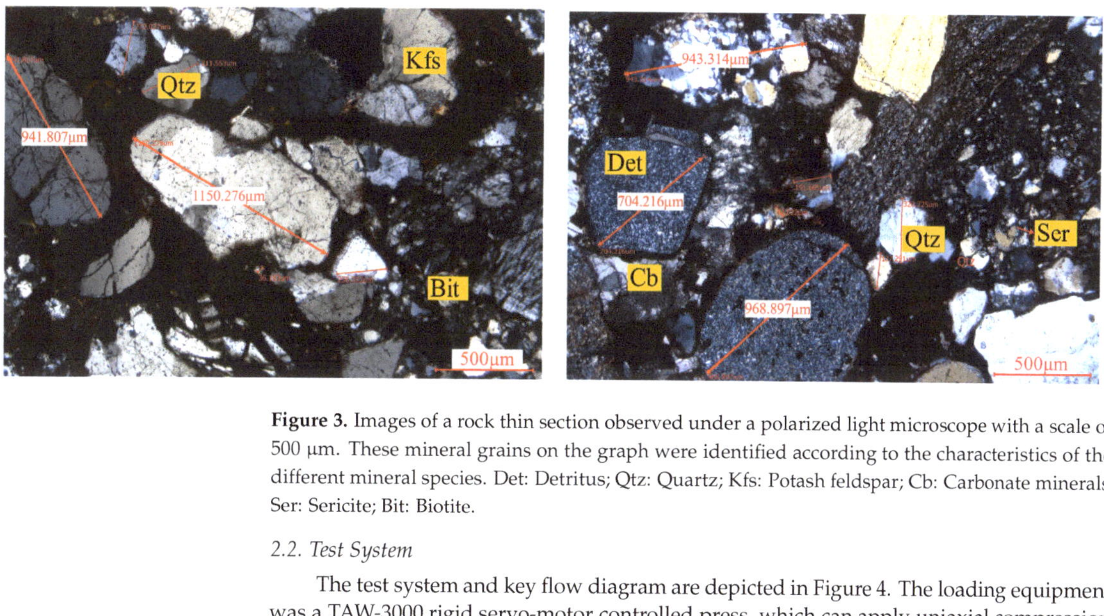

Figure 3. Images of a rock thin section observed under a polarized light microscope with a scale of 500 μm. These mineral grains on the graph were identified according to the characteristics of the different mineral species. Det: Detritus; Qtz: Quartz; Kfs: Potash feldspar; Cb: Carbonate minerals; Ser: Sericite; Bit: Biotite.

2.2. Test System

The test system and key flow diagram are depicted in Figure 4. The loading equipment was a TAW-3000 rigid servo-motor controlled press, which can apply uniaxial compression and step loading. The maximum axial load was 3000 kN, and the measurement accuracy was ±1%. The Phoenix v | tome | xm Industrial CT scanner produced by German General Electric Company was used as the CT scanning equipment. The instrument is capable of scanning, high-power microfocus, and high-resolution nano-focus. The resolution of the micro-focus detail was less than 1 μm.

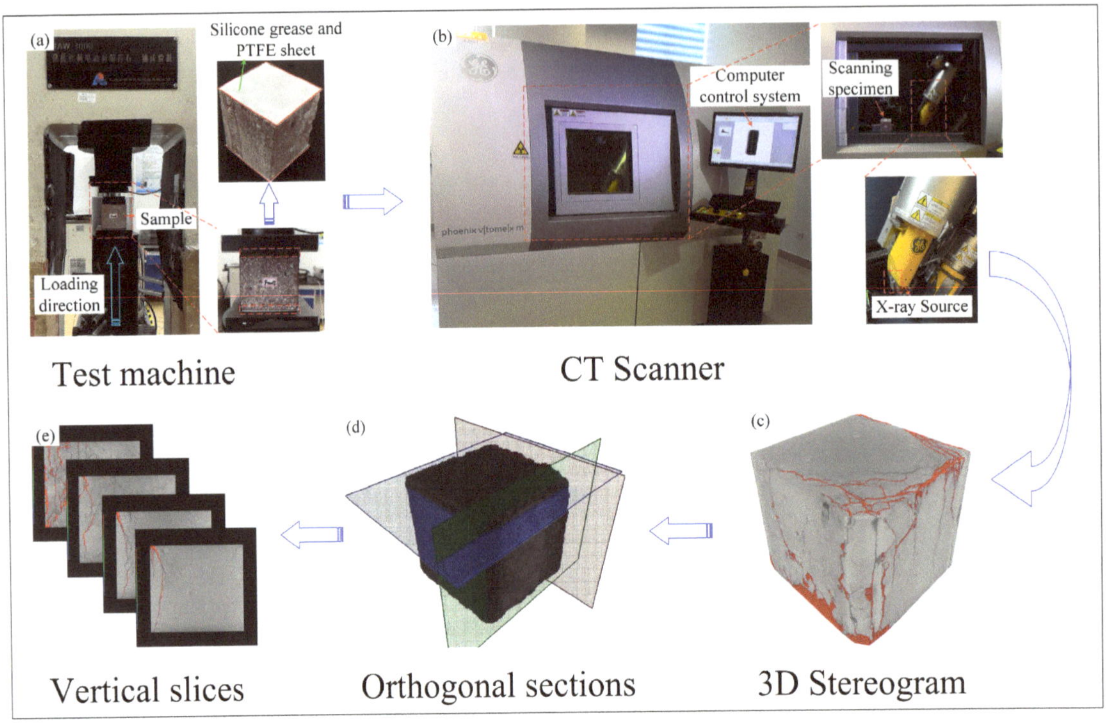

Figure 4. Experimental equipment and key flow diagram. (**a**) The specimens were tested with a uniaxial press after they were pre-treated. (**b**) The specimen was taken into the CT machine for scanning after it had been loaded and unloaded. (**c**) The 3D reconstructed model of the specimen was obtained after a series of processing of the rock images from the CT scan. (**d**) Slices in different directions can be selected by using the intercepted plane work of the software. (**e**) Intercepted sections of the specimen in the vertical direction for analysis.

2.3. Experimental Process

To specifically determine the time points of loading, unloading, and CT scanning, the average peak stress, σ_c, on the lithology was 41.93 MPa after testing (this was taken as the standard value). According to the stage division of the rock stress–strain curve [39], the stress and the corresponding load at each CT scanning point during the experiment were determined (Table 1).

Table 1. Scanning times and states of samples.

Scan Times	Stress Level	Corresponding Load Value/kN	Deformation Stage
First	0	0.00	Initial stage
Second	$0.3\sigma_c$	125.80	Compaction stage
Third	$0.7\sigma_c$	293.50	Stable fracture stage
Fourth	$0.9\sigma_c$	377.4.0	Unstable fracture stage
Fifth	–	–	After rupture

The sizes of most of the specimens studied by scholars have been small, however, which means that many meso-structural features are not included, so larger specimens were chosen for the present study. However, due to the limitations of the experimental conditions and CT equipment, the mechanical loading and unloading were performed separately from the CT scanning.

To maximize the information gleaned from CT scanning as to fracture source formation and crack expansion, rock specimens were tested under uniaxial cyclic loading and unloading and CT scanning: the circulating experimentation of "loading → unloading → CT scanning" was adopted.

(1) Before the start of the experiment, the surfaces of the rock samples were polished, and the standard position was marked on the upper and lower loading ends of the press. Lubricating silicone grease was applied to the upper and lower loading surfaces of the rock samples, and pads made from polytetrafluoroethylene (PTFE) sheets were used to avoid regional stress concentration at the loaded end.

(2) After the rock was loaded to the rated load for a period of time, then unloaded at the same rate, we removed the rock specimen and wrapped them in plastic film to avoid secondary damage during CT scanning.

The experimental scheme was designed according to the part of the division of the uniaxial compression stress–strain curve in the textbook of rock mechanics. The test is a cyclic loading and unloading test: each time, the specimen was loaded to a predetermined value and then unloaded to 0, then the specimen was removed and loaded again, and so on. To show the loading point and CT scanning point, the cyclic of loading and unloading and the phase of CT scanning were explained (Figure 5). The specific position of the CT scanning times was set as: (1) First (first scan in Figure 5): the specimen was subjected to the first time of the CT scan to obtain the initial microstructural information, (2) Second (second scan in Figure 5): the specimen was subjected to a load of up to 125.8 kN at a cross-head displacement rate of 0.15 mm/min. The stress level was $0.3\sigma_c$, and the deformation stage was one involving compaction of the specimen. CT scanning for the second time after unloading followed. (3) Third (third scan in Figure 5): the force was increased to 293.5 kN, then unloaded before undertaking to the third CT scan. The stress was $0.7\sigma_c$, and the deformation was such that stable fracturing occurred. (4) Fourth (fourth scan in Figure 5): the force was increased to 377.4 kN, then unloaded before undertaking another CT scan. The stress was $0.9\sigma_c$, and the deformation stage led to unstable fracturing. (5) Finally (fifth scan in Figure 5): the fracturing specimen were scanned by CT for the fifth time.

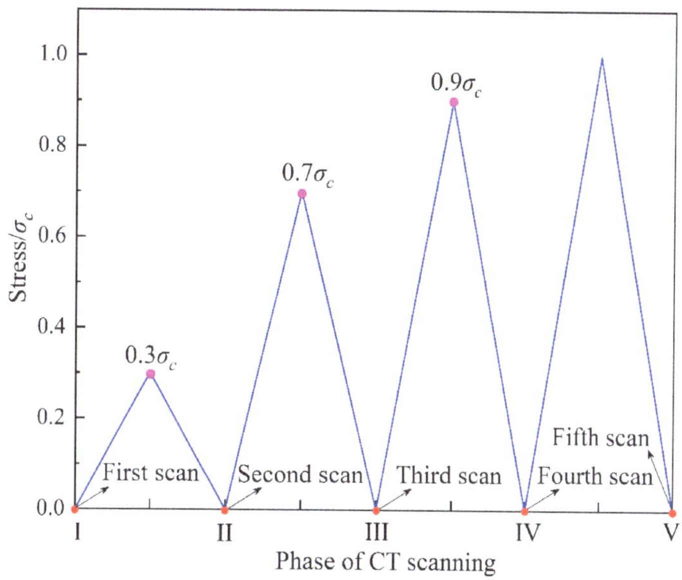

Figure 5. Schematic diagram of the cyclic loading and unloading process and CT scanning points. Each time, the specimen was loaded to a predetermined intensity value and then unloaded to 0, the CT scan was then performed, and so on. The red dots show the CT scan points.

3. Experimental Results

3.1. Location of Fracture Sources

To determine the areas representing fracture sources, VG software was used to reconstruct specimens at different stages after CT scanning (Figure 6). The scanned slices were input to the 3D reconstruction software VG. After threshold segmentation, cutting pages, defining the sample surface, and calculation, the digital models of the 3D fracture of the rock were obtained, and the internal holes and fractures of the models were marked with different colors. To better show the distribution of fractures, the transparency of the model was adjusted to 90%. At the beginning of the loading stage, the initial fracture sources were randomly distributed, and small fractures were difficult to distinguish from the original defects. In the present study, the focus was on the fracture sources related to the main crack. Crack expansion may also lead to the development of its relationship to the applied load [40].

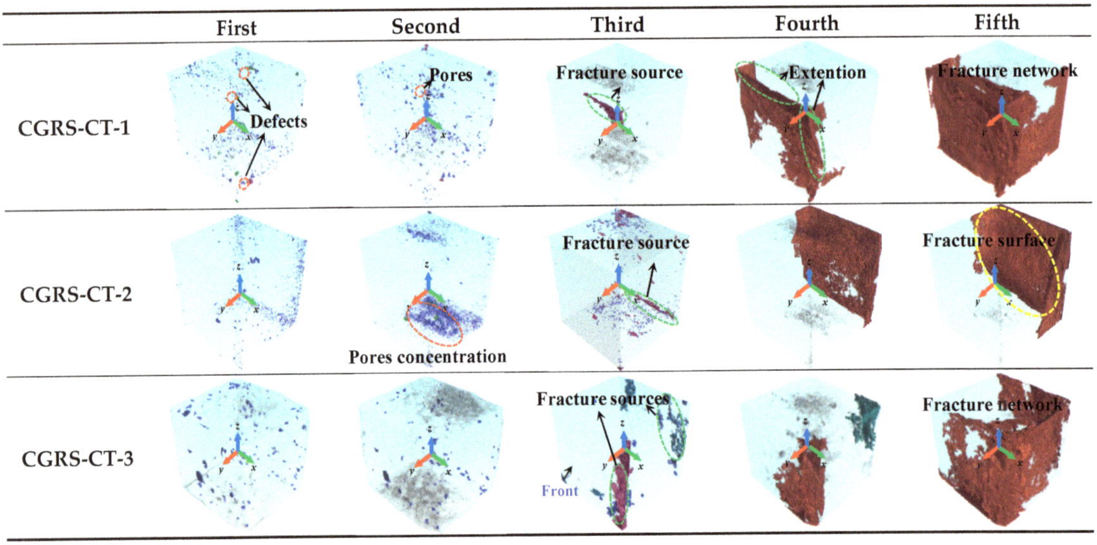

Figure 6. Three-dimensional crack spatial distribution of samples in different stages of the fracturing process. Horizontal graphs represent the same specimen in 3D at different scanning stages. The morphological distribution of cracks from sprouting to final rupture after loading of the specimens was recorded in detail.

Many kinds of main-fracture sources were found in specimens (Figure 6), e.g., where the fracture sources for both CGRS-CT-1 and CGRS-CT-2 were generated in the region of the specimen near the lateral edge, and there were two sources for CGRS-CT-3, one at the edge of the specimen, and others in a flake shape, parallel to the side surface. During sustained loading to the unstable fracturing stage, specimen CGRS-CT-1 started to expand horizontally and longitudinally based on the fracture sources, CGRS-CT-2 continued to expand, and the fracture sources at both ends of CGRS-CT-3 continued to develop and extend and were likely to converge. After final fracturing, the macroscopic cracks fully developed and formed a crack network of interconnected and interwoven fractures. In summary, it can be preliminarily determined that the fracture source of specimens CGRS-CT-1 and CGRS-CT-2 is the cavity. Under load, the cavity gradually evolved, leading to the generation of cracks. After loading and unloading again, the bearing capacity of the sample decreased, and finally, the crack expanded further and led to the formation of the final fracture surface. However, the fracture source of sample CGRS-CT-3 warrants further analysis.

3.2. Crack Propagation during Rock Fracturing Process

To further distinguish the fracture sources of specimens, and clarify the trend of crack expansion, specimen CGRS-CT-3 was taken as an example of these aspects in both the spatial and temporal domains. Figure 7 shows the longitudinal CT slices of specimen CGRS-CT-3 at different loading stages (the yellow and red curves indicate cracks).

- First (Figure 7b) and second (Figure 7c) scans: the specimens showed no cracks in the initial state and the compaction stage.
- Third scan (Figure 7d): In the corresponding elastic phase, fractures could be found in all slices. On the slice at 40 mm, the fracture occurred at the edge of the specimen. On the slices at 20 and 0 mm, the number of microfractures gradually increased. On the slice at −20 mm, evidence of secondary fracture production can be found in a vertical distribution parallel to the end of the specimen. On the slice at −40 mm, the fracture on the right-hand side of the sample penetrated both end surfaces and extended to the middle area.
- Fourth scan (Figure 7e): After the unstable fracture stage, the generation and expansion of microfractures could be found on the slices at 40 and 20 mm, and compared with the previous stage, the distribution of fractures did not change much. On the slice at 0 mm, the number of fractures on the right-hand side of the sample gradually increased with a tendency to penetrate the upper and lower end surfaces. On the slice at −20 mm, the fractures on the right-hand side were approximately vertically parallel, with the upper and lower end surfaces connected. On the slice at −40 mm, the transverse fracture on the left-hand side and the vertical fracture on the right were connected to each other.
- Fifth scan (Figure 7f): After the final fracture, compared with the previous stage, on the slices at 40 and 20 mm, both the length of the fracture and the degree of fracture propagation were increased. On the slice at −20 mm, there were many more branching fractures, which were distributed on both sides of the sample. On the slice at −40 mm, the branching fractures were inter-connected, and a more complex fracture network was formed.

(a)

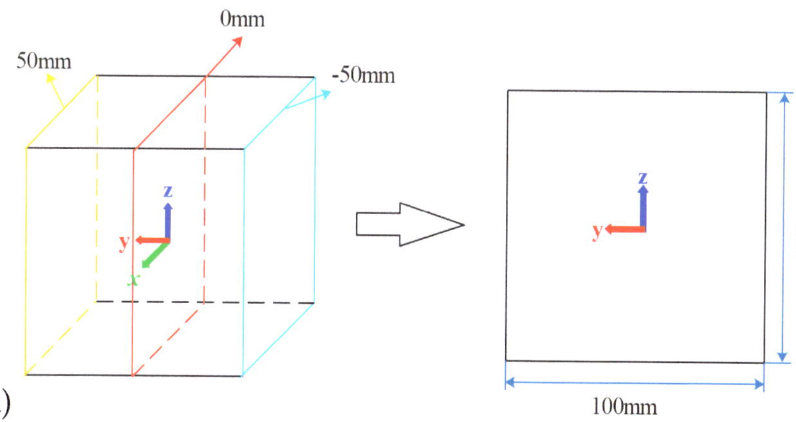

Figure 7. *Cont.*

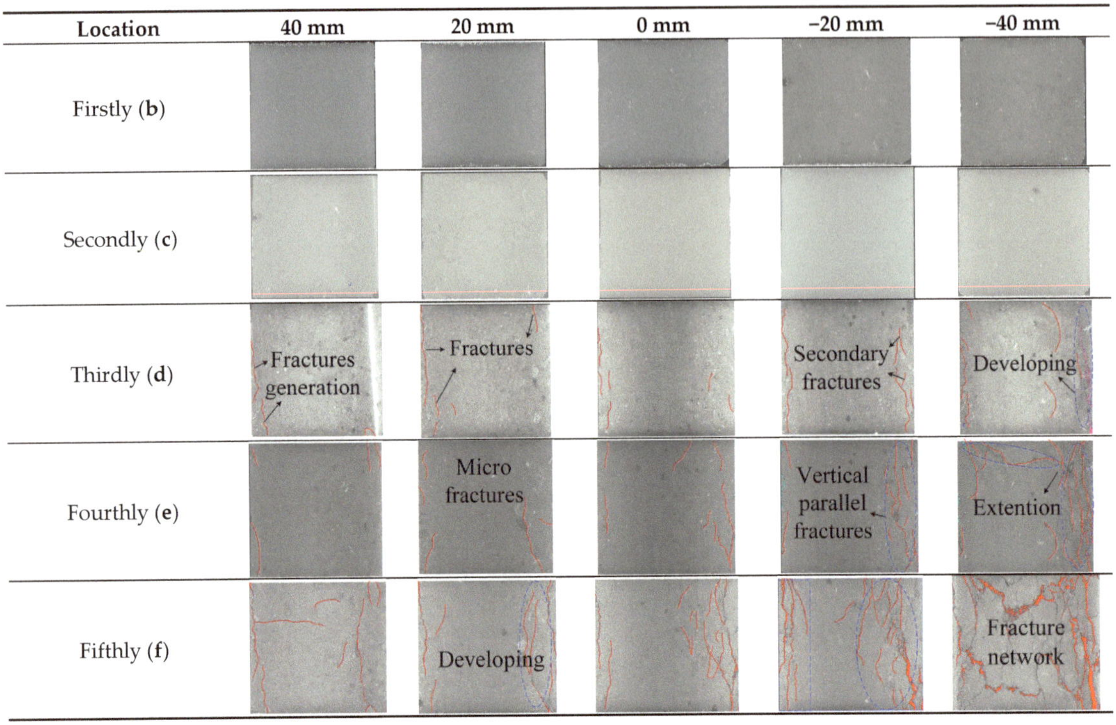

Figure 7. Slices of specimens in different positions. (**a**) Schematic diagram of section location, the specimen has a side length of 100 mm, the center of the specimen is the origin, the middle position is at 0 mm, and the sides are 50 and −50 mm each. The distribution of the fracture extension after the specimen has been loaded and unloaded several times. (**b**) Slices of the first CT scan. (**c**) Slices of the second CT scan. (**d**) Slices of the third CT scan. (**e**) Slices of the fourth CT scan. (**f**) Slices of the fifth CT scan.

3.3. *Types of Fracture Sources Based on Meso-Structure Scale*

3.3.1. Fracture Sources of Pore Structure

There is a positive relationship between pore structure and crack initiation. The pore structure reduced the strength of specimens, readily causing local stress concentration, and rapidly reaching the stress required for crack initiation. Then, the cracks were mostly generated at the pore tip [41]. The meso-scale pores were one fracture source, and pore concentration and pore volume can be used to depict the proportion of pores. To ascertain that the fracture source of samples CGRS-CT-1 and CGRS-CT-2 was caused by pore concentration, the area of pore concentration was determined by coordinates, and this area was selected as the object of study and the percentage of pore volume in this area to the total pore volume was calculated. The proportion of pore volume of specimen CGRS-CT-1 was 18.75%, in specimen CGRS-CT-2 it reached to 23.56%, and in specimen CGRS-CT-3 it was 34.09% (Table 2).

Table 2. Pore volume and proportion.

Sample Number	Pore Volume at Fracture Source/mm^3	Proportion of Total Pore Volume /%
CGRS-CT-1	906.1	18.75
CGRS-CT-2	386.8	23.56
CGRS-CT-3	362.75	34.09

3.3.2. Fracture Sources of Micro-Cracks

The presence of micro-cracks also affected the crack initiation. Specimen CGRS-CT-3 had a micro-crack in its initial state, so the transverse and longitudinal CT slices were selected for micro-crack analysis (Figure 8).

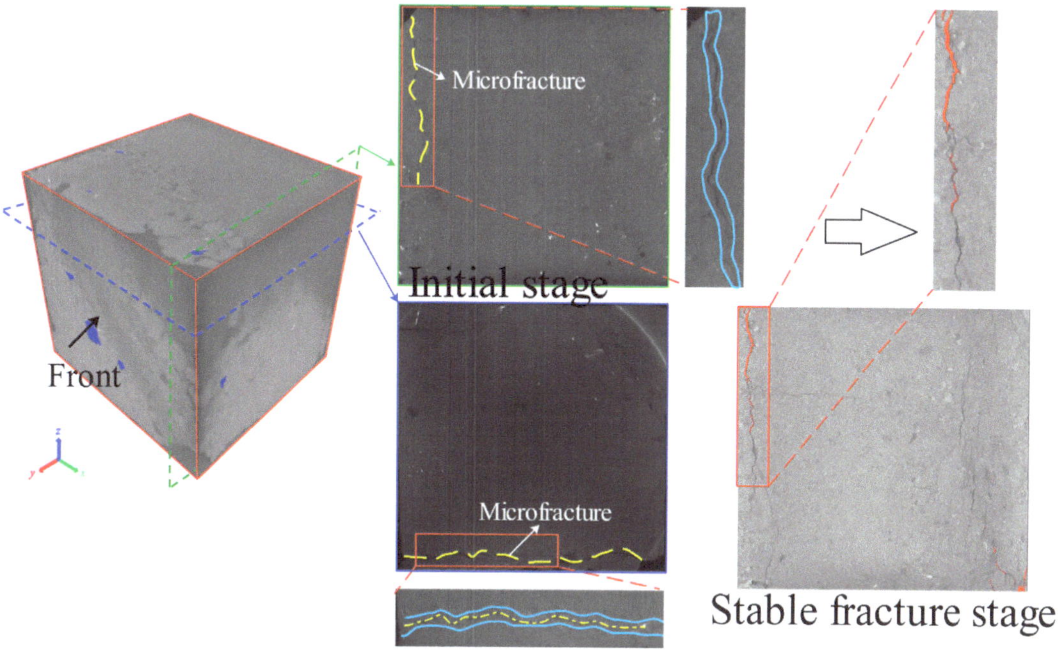

Figure 8. Schematic diagram of CT section. Extension and matching of microfractures in the initial and stable rupture stages.

The position of the crack and the initial micro-crack overlapped, indicating that the crack was closely related to the micro-crack. When the crack angle ran parallel to the direction of the applied load, most of the cracks were generated from the crack-tip [42]. The micro-cracks expanded and developed under continuous loading, forming a flake-like fracture network in the elastic stage.

3.3.3. Fracture Sources of Mineral Particles

Mineral particles affect crack initiations, e.g., specimen CGRS-CT-3 was taken as an example to explore the characteristics of the mineral particle distributions. In Figure 9, the analysis of the longitudinal slice at this position indicates a mineral particle (particle R) at the middle and lower position of the fracture path. A large quartz particle was present near the side surface, which was near particle R. The crack expanded around it and was likely to extend in other directions.

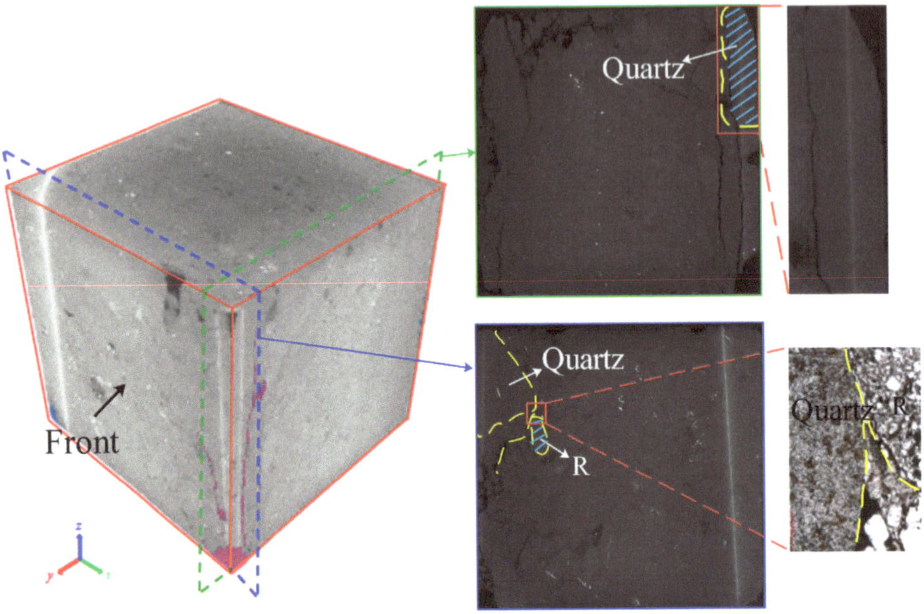

Figure 9. Results of fracture surface scanning. The meso-structural features at the source of fracture in rock sections of different orientations are shown, and it can be seen that the source of fracture there is the surrounding structure of quartz particles and R particles.

Point-to-point contact mostly arose between the large pieces of quartz, cuttings, and particle R (Figure 9). Particles formed the framework in this area, and cement and interstitial matter filled the contact surface. The particles were prone to deformation under the external force, leading to cracking.

4. Discussion

4.1. Physical Mechanisms of Fracture Source Classification

Mineral particles and microfractures are encountered along the tip path, which change the state of extension of the cracks. The pore distribution and the existence of initial micro-cracks provide necessary conditions for their generation. The cementation and contact methods between mineral particles have a positive effect on subsequent crack generation. The location of fracture sources is likely to affect the strength of the mineral particles [43].

Compared with other areas, there are no large particles (>2–3 mm) in which there is mainly sandstone matrix. The rock cuttings are distributed in aggregated form, and the mechanical properties are weak. Thus, it is prone to generation of fracture sources. The fracture sources on the right-hand side of specimen CGRS-CT-3 generated by the initial micro-cracks continue to develop into a fracture surface. The fracture sources on the left are generated at the boundary of the mineral grains. The point-contacted particles are prone to dislocation and cementation under loading. The cementing material ruptures and produces fracture sources.

According to the different microscopic characteristics around the fracture sources, the fracture sources can be divided into three different types: the pore-based, the micro-crack-based, and the mineral particle-based.

4.2. Influence of Rock Physical Properties on Crack Propagation

4.2.1. Mineral Particle Strength

The crack propagation path is affected by the strength of the particles. The cracking path has places where mineral particles and debris aggregates are cemented as well as low-strength carbonate minerals and iron oxide debris aggregates, through which fractures will propagate (Table 3). The physical and mechanical properties of minerals, such as quartz and feldspar, and carbonate minerals are different, including the strength, and with continuous external loading, the deformation of the particles also varies [44]; thus, cracks easily expand and extend between them. Cracks mainly follow the path with the least energy required for expansion [45].

Table 3. Distribution of mineral characteristics in key areas of the fracture path. Area I is a region containing a complex fracture network, and there are many quartz particles, feldspar, and rock debris. Area II is a region wherein the fracture extends from the end to the middle, and the particles here are very fine and diverse. Area III contains the intersections of secondary fractures and main fractures. The mineral particles are mostly quartz sand and gravel with complex composition. Area IV is an area wherein the secondary fractures become concentrated.

Critical Area of Fracture Path	Single Polarized Slice	Orthogonal Polarizing Sheet
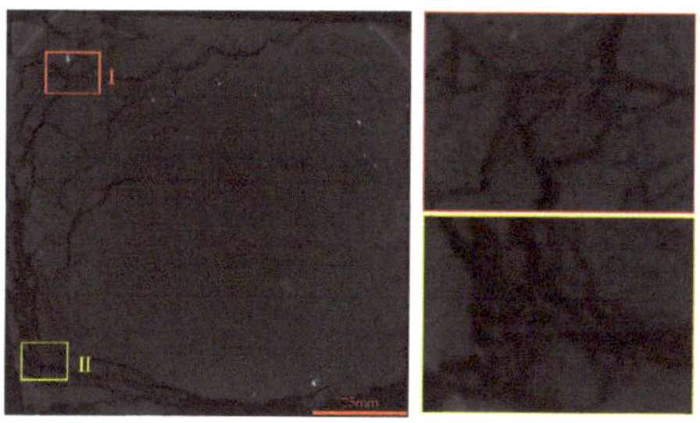	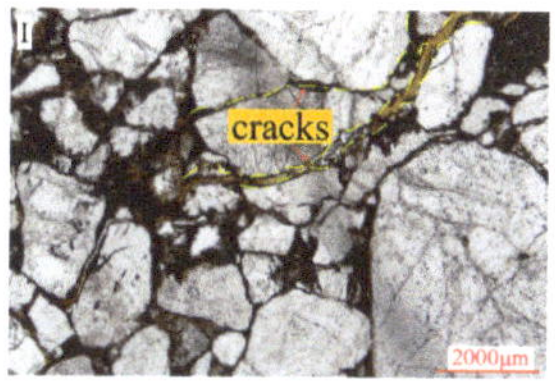	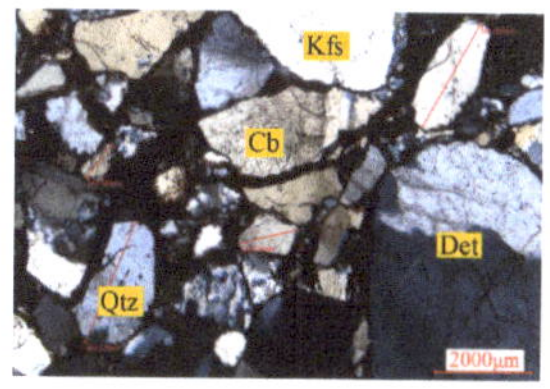
		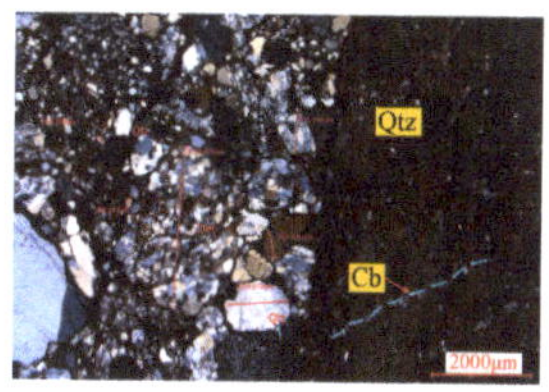
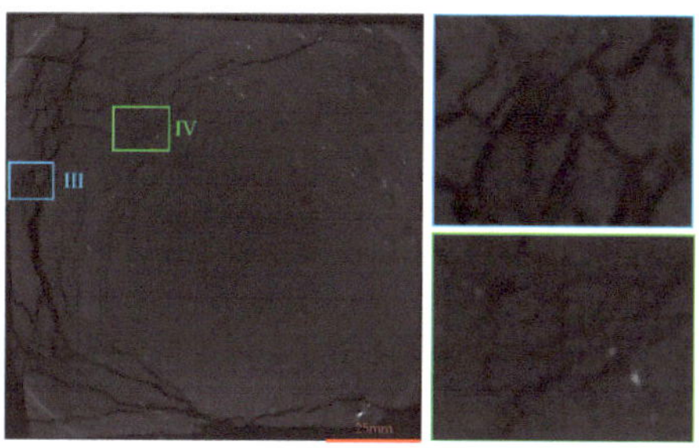	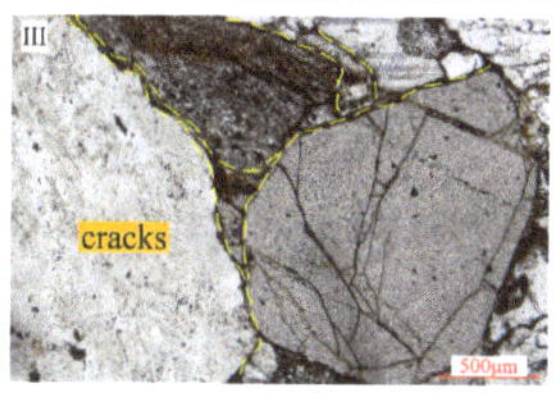	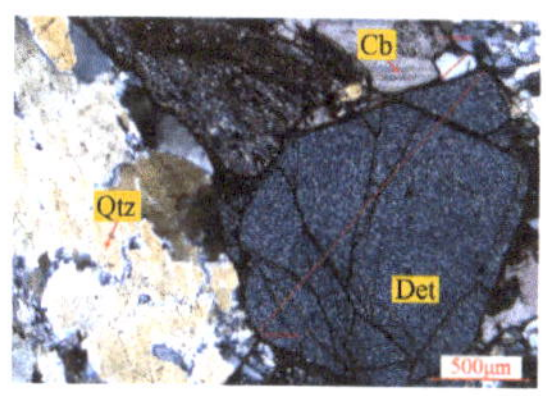
	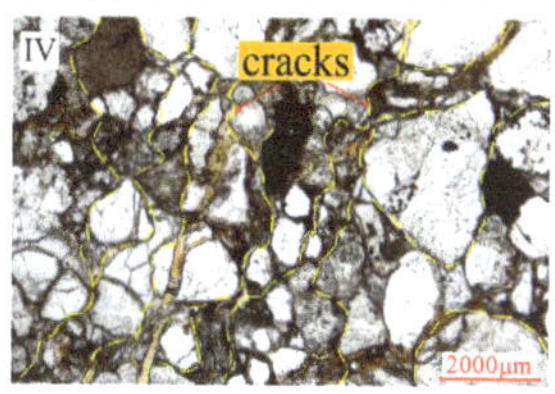	

Cracks mostly expand along areas with weak mechanical properties, while the extension of tensile cracks is small, and the cracks are likely to change in direction due to the influence of the microstructure on the path [46]. A hard mineral particle could be observed

on the crack propagation path at the upper end of specimen CGRS-CT-3, that is, point *e* (Figure 10), which serves as the fracture source for the next period, and the crack expands from here. After being blocked, multiple branching cracks appear.

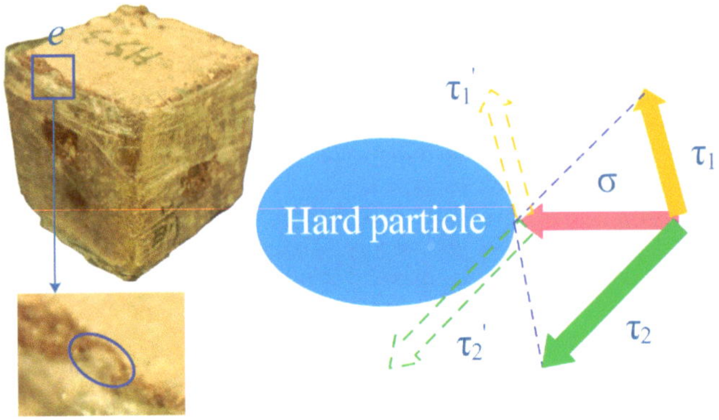

Figure 10. Bifurcation point and stress diagram. Point *e* is a hard particle, the fracture was subjected to shear stress when passing through the particle, and then a bifurcation occurs.

4.2.2. Mineral Spatial Distributions

To evaluate the influence of the spatial distribution of minerals on crack expansion, the initial state of the mineral distribution inside the specimen after CT scanning was extracted. The result shows that the overall mineral distribution is similar to the fracture network distribution after failure. The specific distribution can be divided into two parts (Figure 11a). Many complex particles are observed in the middle and upper parts, and the lower part, distributed in bands. The overall spatial distribution of the fracture network after destruction is closely related to the particle distribution in the initial state (Figure 11b,c). Most of the particles at the upper end of the specimen are already discrete, and the particles and gravel show signs of collapsing and spalling (Figure 11d). This special failure mode is consistent with "quasi-phase transition" theory [47]. The cracks continue to extend between the particles, and the nature of the specimen itself and the boundary conditions limit the internal coarse particles or gravel will collapse or spall.

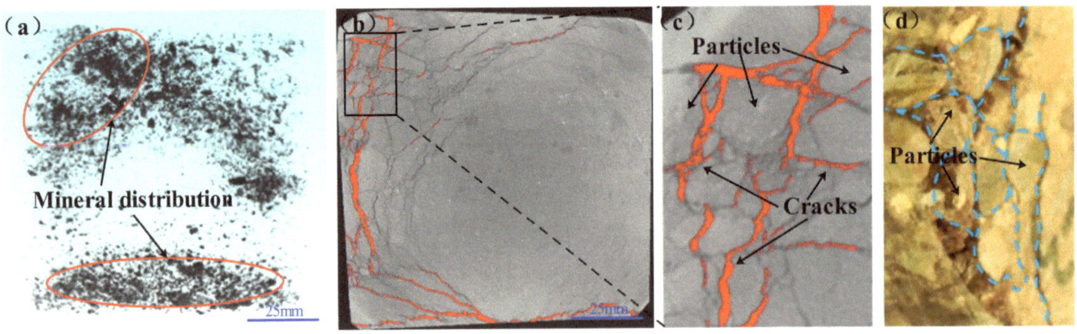

Figure 11. Top view of mineral distribution and fracture. (**a**) Mineral distribution area. (**b**) Part of the fracture network. (**c**) Particles and cracks. (**d**) Particles of real rock. Comparing (**a**) and (**b**), it is obvious to see similarities between the two, the fractures are more complex in the area where mineral particles are concentrated, and the mineral distribution area is consistent with the distribution of fractures.

There exist many kinds of point-to-point or point-to-surface areas, and high local stress concentration will appear in these areas. The cracks expand between the contact

surfaces, which are distributed in a net-like shape. The cementation materials between the particles affect the mechanical properties of the rock [48]. The spatial distribution of mineral particles is of great importance and influences the distribution morphology of the fracture network.

5. Conclusions

(1) The generation of rock fracture sources was visualized by CT scanning: There are three types of fracture source (pore-based, microfracture-based, and mineral particle-based). The pores and the initial micro-cracks were the initial weak surfaces. The fracture sources of mineral particles' type were formed by destruction of the cemented materials under load.

(2) The expansion of the fracture path and mineral grains was revealed with the help of a polarizing microscope. According to the analysis of mineralogical characteristics, the mechanical properties of different mineral particles along the fracture propagation path can be obtained. Finally, the modes of propagation of cracks between particles with different intensities were obtained. The crack expansion around mineral particles can be determined by the strength of the mineral particle. The cracks between high-strength mineral particles, such as quartz, expanded around it, and the low-strength carbonates and rock cuttings and other particles expanded through it.

(3) The spatial distribution of cracks in the sample and the mineral distribution state in the original state were compared and studied by 3D reconstruction of the model. It can be concluded that there is a close relationship between the fracture evolution path and the fracture morphology: their distributions maintain a high degree of consistency, and the fracture morphology is more complicated in areas where minerals are concentrated. The expansion of cracks at the boundary between large particles and gravel also causes gravel collapse and shedding.

Author Contributions: Writing—original draft preparation, G.G.; software, S.Y.; data curation, X.L., X.Y. and P.L.; supervision, Y.X.; funding acquisition, Y.Z. All authors have read and agreed to the published version of the manuscript.

Funding: This research was funded by the National Natural Science Foundation of China (grant number 51774138, 51804122, 52074123), the Natural Science Foundation of Hebei Province(grant number E2021209052, E2021209148).

Institutional Review Board Statement: Not applicable.

Informed Consent Statement: Not applicable.

Data Availability Statement: Data is contained within the article.

Acknowledgments: The authors would like to thank all the technicians who worked on the experimental process. The authors also would like to express their sincere appreciation to the reviewers for their constructive comments on the revision and improvement of the manuscript. All the support is gratefully acknowledged.

Conflicts of Interest: The authors declare no conflict of interest.

References

1. Xu, S.J.; Yin, X.T.; Dang, F.N. Mechanical characteristics of rock and soil affected by particle size of crystal and mineral. *Rock Soil Mech.* **2009**, *30*, 2581–2587. [CrossRef]
2. Qin, Y.D.; Wang, H.; Wang, Z.X. Failure mechanism and control analysis of fractured and easily weathered roadway in deep mining. *China Min Mag.* **2020**, *29*, 388–393. [CrossRef]
3. Anders, M.H.; Laubach, S.E.; Scholz, C.H. Microfractures: A review. *J. Struct. Geol.* **2014**, *69*, 377–394. [CrossRef]
4. Wittmann, F.H.; RoelfstraP, E.; Sadouki, H. Simulation and analysis of composite structures. *Mater. Sci. Eng.* **1985**, *68*, 239–248. [CrossRef]
5. Yu, Q.; Yang, T.; Zheng, C.; Tang, C.A.; Wang, P.T. Numerical analysis of influence of rock mesostructure on its deformation and strength. *Rock Soil Mech.* **2011**, *32*, 3468–3472. [CrossRef]
6. Zhao, B.; Wang, Z.Y.; Wu, J.P. Relation between mineralogical composition and microstructure to the mechanical properties of rock materials. *Coal Geol. Explor.* **2013**, *41*, 59–63, 67. [CrossRef]

7. Zhang, G.K.; Li, H.B.; Xia, X.; Li, J.R.; Li, X.F.; Song, T. Effects of microstructure and micro parameters on macro mechanical properties and failure of rock. *Chin. J. Rock Mech. Eng.* **2016**, *35*, 1341–1352. [CrossRef]
8. Deng, C.F.; Liu, J.F.; Chen, L.; Li, Y.; Xiang, G. Fracture mechanical behaviors and acoustic emission characteristics of Beishan granites with different particle sizes. *Rock Soil Mech.* **2016**, *37*, 2313–2320. [CrossRef]
9. Zhao, Y.C.; Liu, H.L.; Yang, T.H.; Xu, T.; Li, Y.; Hou, X.G.; Qin, Y. Effect of mesozoic sandstone meso structure on strength and energy dissipation characteristic. *J. China Coal Soc.* **2017**, *42*, 452–459. [CrossRef]
10. Zhao, K.; Zhou, Y.T.; Zeng, P.; Yang, Z.Y.; Liu, C.F. Experimental study on acoustic emission characteristics of rock-like materials with different particle sizes under three-point bending. *J. China Coal Soc.* **2018**, *43*, 3107–3114. [CrossRef]
11. Li, J.; Yilmaz, E.; Cao, S. Minerals influence of solid content, cement/tailings ratio, and curing time on rheology and strength of cemented tailings backfill. *Minerals* **2020**, *10*, 922. [CrossRef]
12. Cao, S.; Xue, G.L.; Yilmaz, E.; Yin, Z.Y.; Yang, F.D. Utilizing concrete pillars asan environmental mining practice in underground mines. *J. Clean. Prod.* **2021**, *278*, 123433. [CrossRef]
13. Cao, S.; Yilmaz, E.; Yin, Z.Y.; Xue, G.L.; Song, W.D.; Sun, L.J. CT scanning of internal crack mechanism and strength behavior of cement-fibertailings matrix composites. *Cem. Concr. Compos.* **2020**, *116*, 103865. [CrossRef]
14. Ji, H.G.; Jiang, H.; Song, C.Y.; Liu, Z.Q.; Tan, J.; Liu, Y.J.; Wu, Y.F. Analysis on the microstructure evolution and fracture morphology during the softning process of weakly cemeted sandstone. *J. China Coal Soc.* **2018**, *43*, 993–999. [CrossRef]
15. Song, C.Y.; Ning, F.B. Progress on the Association between Mesostructural Parameters and Macromechanical Behaviors of Weakly Cemented Rocks. *Met. Mine.* **2018**, *12*, 1–9. [CrossRef]
16. Hasanipanah, M.; Meng, D.; Keshtegar, B.; Behrooz, K.; Nguyen, T.T.; Duc, K.T. Nonlinear models based on enhanced Kriging interpolation for prediction of rock joint shear strength. *Neural Comput. Appl.* **2021**, *33*, 4205–4215. [CrossRef]
17. Huang, J.D.; Zhang, J.; Gao, Y. Intelligently predict the rock joint shear strength using the support vector regression and Firefly Algorithm. *Lithosphere* **2021**, *2021*, 2467126. [CrossRef]
18. Ge, Y.F.; Tang, H.M.; Eldin, M.A.; Wang, L.Q.; Wu, Q.; Xiong, C.R. Evolution process of natural rock joint roughness during direct shear tests. *Int. J. Geo.* **2016**, *17*, E4016013. [CrossRef]
19. Han, Q.; Feng, X.T.; Yang, C.X.; Kong, R.; Zhao, J.; Zhang, Y. Evaluation of the crack propagation capacity of hard rock based on stress-induced deformation anisotropy and the propagation angle of volumetric strain. *Rock. Mech. Rock. Eng.* **2021**, *54*, 6585–6603. [CrossRef]
20. Sun, W.; Wu, A.; Hou, K.; Yang, Y.; Liu, L.; Wen, Y. Real-time observation of meso-fracture process in backfill body during mine subsidence using X-ray CT under uniaxial compressive conditions. *Constr. Build. Mater.* **2016**, *113*, 153–162. [CrossRef]
21. Liu, S.; Wang, Z.; Zhang, L. Experimental study on the cracking process of layered shale using X-ray microCT. *Energy Explor. Exploit.* **2018**, *36*, 297–313. [CrossRef]
22. Duan, Y.; Li, X.; Zheng, B.; He, J.; Hao, J. Cracking Evolution and Failure Characteristics of Longmaxi Shale Under Uniaxial Compression Using Real-Time Computed Tomography Scanning. *Rock. Mech. Rock. Eng.* **2019**, *52*, 3003–3015. [CrossRef]
23. Wang, M.; Wan, W.; Zhao, Y. Experimental study on crack propagation and the coalescence of rock-like materials with two preexisting fissures under biaxial compression. *Bull. Eng. Geol. Environ.* **2020**, *79*, 3121–3144. [CrossRef]
24. Pu, C.Z.; Cao, P. Failure characteristics and its influencing factors of rock-like material with multi-fissures under uniaxial compression. *Trans. Nonferrous Met. Soc. China* **2012**, *22*, 185–191. [CrossRef]
25. Wang, D.J.; Tang, H.M.; Elsworth, D.; Wang, C.Y. Fracture evolution in artificial bedded rocks containing a structural flaw under uniaxial compression. *Eng. Geol.* **2019**, *250*, 130–141. [CrossRef]
26. Du, M.R.; Jing, H.W.; Su, H.J.; Zhu, T.T. Experimental study of strength and failure characteristics of sandstone containing prefabricated elliptical hole. *J. China Univ. Min. Technol.* **2016**, *45*, 1164–1171. [CrossRef]
27. Zhu, T.T.; Jing, H.W.; Su, H.J.; Yin, Q. Experimental investigation on mechanical behavior of sandstone with coupling effects under uniaxial compression. *J. China Coal Soc.* **2015**, *40*, 1518–1525. [CrossRef]
28. Han, G.S.; Jing, H.W.; Su, H.J.; Zhu, T.T.; Du, M.R.; Xiong, F.; Wu, J.Y. Experimental study on compressive strength and fracture characteristics of sandstone containing double circular cavities. *J. China Coal Soc.* **2017**, *42*, 871–878. [CrossRef]
29. Laubach, S.E.; Lander, R.H.; Criscenti, L.J.; Anovitz, L.M.; Urai, J.L.; Pollyea, R.M.; Hooker, J.N.; Narr, W.; Evans, M.A.; Kerisit, S.N.; et al. The role of chemistry in fracture pattern development and opportunities to advance interpretations of geological materials. *Rev. Geophys.* **2019**, *57*, 1065–1111. [CrossRef]
30. Karfakis, M.G.; Akram, M. Effects of chemical solutions on rock fracturing. *Int. J. Rock Mech. Min. Sci. Geomech. Abstr.* **1993**, *30*, 1253–1259. [CrossRef]
31. Miao, S.J.; Cai, M.F.; Guo, Q.F.; Wang, P.T.; Liang, M.C. Damage effects and mechanisms in granite treated with acidic chemical solutions. *Int. J. Rock Mech. Min. Sci.* **2016**, *88*, 77–86. [CrossRef]
32. Přikryl, R. Some microstructural aspects of strength variation in rocks. *Int. J. Rock Mech. Rock. Eng.* **2001**, *38*, 671–682. [CrossRef]
33. Han, Z.H.; Zhang, L.Q.; Zhou, J. Effect of mineral particle size heterogeneity on mechanical properties in PFC2D simulation. *J. Eng. Geol.* **2019**, *27*, 706–716. [CrossRef]
34. Han, Z.H.; Zhang, L.Q.; Zhou, J.; Yuan, G.X.; Wang, P.J. Uniaxial compression test and numerical studies of grain size effect on mechanical properties of granite. *J. Eng. Geol.* **2019**, *27*, 497–504. [CrossRef]
35. Sun, C.; Ao, Y.H.; Zhang, J.M.; Wang, S. Particle flow of meso-fracture characteristics and macro-scale effect of granites. *Chin. J. Geotech. Eng.* **2020**, *42*, 1687–1695. [CrossRef]

36. Ye, G.; Cao, H.; Gao, Q.; Zhang, Z.; Wang, T.Y. Numerical simulation study on the influence of particle proportion on rock mechanics characteristics. *J. Mech.* **2019**, *25*, 1129–1137. [CrossRef]
37. Bai, Q.; Tu, S.; Zhang, C. DEM investigation of the fracture mechanism of rock disc containing hole(s) and its influence on tensile strength. *Theor. Appl. Fract. Mech.* **2016**, *86*, 197–216. [CrossRef]
38. Yu, Q.L.; Ranjith, P.G.; Liu, H.Y.; Yang, T.H.; Tang, C.A.; Yang, S.Q. A Mesostructure-based Damage Model for Thermal Cracking Analysis and Application in Granite at Elevated Temperatures. *Rock. Mech. Rock. Eng.* **2015**, *48*, 2263–2282. [CrossRef]
39. Cai, M. *Rock Mechanics and Engineering*, 2nd ed.; Science Press: Beijing, China, 2018; pp. 44–46.
40. Liu, X.X.; Wu, L.X.; Zhang, Y.B.; Wang, S.Z.; Yao, X.L.; Wu, X.Z. The characteristics of crack existence and development during rock shear fracturing evolution. *Bull. Eng. Geol. Environ.* **2020**, *80*, 1–12. [CrossRef]
41. Li, J.; Kong, X.C.; Song, M.S.; Wang, Y.; Wang, H.; Liu, X.L. Study on the influence of reservoir rock micro-pore structure on rock mechanical properties and crack propagation. *Rock Soil Mech.* **2019**, *40*, 4149–4156, 4164. [CrossRef]
42. Wang, H.; Li, Y.; Cao, S.G.; Pan, R.K.; Yang, H.Y.; Zhang, K.W.; Liu, Y.B. Brazilian splitting test study on crack propagation process and macroscopic failure mode of pre-cracked black shale. *Chin. J. Rock Mech. Eng.* **2020**, *39*, 912–926. [CrossRef]
43. Li, M.; Guo, Y.; Wang, H.; Li, Z.; Hu, Y. Effects of mineral composition on the fracture propagation of tight sandstones in the Zizhou area, east Ordos Basin, China. *J. Nat. Gas. Sci. Eng.* **2020**, *78*, 103334. [CrossRef]
44. Li, B.F.; Zuo, Y.J.; Li, W.; Cao, J. Analysis on fracture mechanics of granite containing flaws based on digital image processing. *Mech Pract.* **2016**, *38*, 262–268. [CrossRef]
45. Zhang, C.; Yang, Q.J.; Cao, W.G. Simulation method of rock deformation and failure process considering initial imperfection and elastic modulus. Chin. *J. Appl. Mech.* **2020**, *37*, 286–292. [CrossRef]
46. David, C.; Menéndez, B.; Bernabé, Y. The mechanical behaviour of synthetic sandstone with varying brittle cement content. *Int. J. Rock Mech. Min. Sci.* **1998**, *35*, 759–770. [CrossRef]
47. Song, C.Y.; Ji, H.G.; Zeng, P.; Sun, L.H.; Tan, J. Phase-like transition characteristics of uniaxial compression failure of weakly cemented coarse-grained sandstone in western china. *J. Min. Saf. Eng.* **2020**, *37*, 1027–1036. [CrossRef]
48. Song, C.Y. The Analysis and Application of mesoscopic Structure Characteristic and Deformation and Failure Mechanism of Weak Cemeted Sandstone. Ph.D. Thesis, Beijing University of Science and Technology, Beijing, China, 2017.

minerals

Article

Experimental and Analytical Investigation into the Synergistic Mechanism and Failure Characteristics of the Backfill-Red Sandstone Combination

Wen Zhang [1,*], Chengyuan Yan [1], Guoyue Zhou [1], Jinping Guo [1], Yanyu Chen [1], Baohua Zhang [1] and Saisai Wu [1,2,*]

1 School of Resources Engineering, Xi'an University of Architecture and Technology, Xi'an 710055, China; ycy@xauat.edu.cn (C.Y.); guojinping@xauat.edu.cn (J.G.); zhouguoyue@xauat.edu.cn (G.Z.); chenyanyu@xauat.edu.cn (Y.C.); zhangbaohua@xauat.edu.cn (B.Z.)

2 International Joint Research Laboratory of Henan Province for Underground Space Development and Disaster Prevention, Henan Polytechnic University, Jiaozuo 454003, China

* Correspondence: zhangwen@xauat.edu.cn (W.Z.); saisai.wu@xauat.edu.cn (S.W.)

Abstract: The stability of underground goaf in filling mining is dominated by the interaction mechanism of the backfill-surrounding rock combination. In order to investigate the interaction mechanism and failure characteristics of the backfill-surrounding rock combination, backfill-red sandstone combinations with three different cement–sand ratios were prepared for uniaxial compression tests. The deformation and failure characteristics of the specimens were analyzed. It was found that at the cement–sand ratio of 1:4, the backfill and red sandstone interacted with and restricted each other, and the through cracks appeared in the whole specimens, which indicated that the backfill and red sandstone can jointly resist external loads and play a role in common bearing. However, with the decrease of the cement–sand ratio, the stress mainly acts on the backfill, and the deformation observed in the backfill is large while there is no obvious rupture in the rock. Based on the failure characteristics and the stress–strain curves of the specimens, the damage constitutive relationship that can describe the failure process and deformation characteristics is proposed. Correlated with the experiment results, the damage constitutive equation is established in three stages including compaction pre-synergy stage, quasi-elastic synergy deformation stage and rupture deformation stage. The failure characteristics observed in each stage are analyzed. The research results are of great significance to accurately understanding the interaction between backfill and surrounding rock, which can be used to design and select the mixture ratio of the filling materials.

Keywords: backfill-red sandstone combination; uniaxial compression test; stress–strain curve; synergistic deformation and failure characteristics; damage constitutive model

Citation: Zhang, W.; Yan, C.; Zhou, G.; Guo, J.; Chen, Y.; Zhang, B.; Wu, S. Experimental and Analytical Investigation into the Synergistic Mechanism and Failure Characteristics of the Backfill-Red Sandstone Combination. *Minerals* **2022**, *12*, 202. https://doi.org/10.3390/min12020202

Academic Editors: Diyuan Li, Zhenyu Han, Xin Cai and Shijie Xie

Received: 13 December 2021
Accepted: 3 February 2022
Published: 4 February 2022

1. Introduction

Deep mining is one of the major technical issues faced by the mining industry. With the increase of mining depth, the ground pressure and roadway offset increase resulting in frequent occurrence of rock burst, which significantly increases the difficulties in maintaining the stability of roadway and stope. The filling mining method is more and more widely used for its high resource recovery rate, which also effectively prevents roof caving and stope collapse, as well as meets the requirements of green mine construction. In the process of underground filling, the surrounding rock and filling form a kind of heterogeneous, discontinuous and nonlinear special support. The surrounding rock and filling work together to bear the load of overburden, which has complex mechanical properties and deformation and failure mechanisms [1–3]. Therefore, it is necessary to establish the common bearing mechanical model of backfill-surrounding rock combination (BSRC) and study the damage evolution characteristics of surrounding rock and backfill after reasonable matching, which

has an important reference value for exploring the damage and failure mechanism of the BSRC model in deep mining.

In order to improve the safety of underground filling mining areas, a lot of research on the mechanical properties, loss and failure characteristics of rock and backfill have been carried out world widely [4–7]. In terms of rock damage and failure, many scholars have analyzed the damage evolution in the process of rock deformation, and failure through experimental means and established constitutive models [8–11]. Bustamante et al. [12] proposed a nonlinear constitutive relation to simulate the properties of sandstone and studied the compressional and shear wave properties of rock under compression, finding that the compressional wave velocity is related to the compressive load. Zhao et al. [13] discussed the fracture toughness and subcritical crack propagation of rock under different environments, established a simple and practical rheological fracture model of rock crack and proposed a modified shear strength model of filled joint. Shahin et al. [14], based on viscoelastic and nonlinear dynamics, gave a mechanical interpretation of compaction creep and defined a constitutive operator to describe the evolution of compaction creep, so as to evaluate the spontaneous accumulation of pore collapse in the active compaction zone. In addition, by comprehensively considering the effects of damage threshold and residual strength, the Drucker Prager criterion, Harris function, Nishihara shear creep model and Burgers shear creep model of rock micro unit strength were improved; the deformation characteristics and long-term shear strength of rock during creep were analyzed, and the statistical constitutive models of rock damage and strain softening were established, respectively [15–19]. In terms of backfill, many scholars have conducted a lot of research on the properties and failure modes of filling materials [20,21]. Shahsavari et al. [22] considered the different behaviors and irreversible thermodynamics of concrete in tension and compression, decomposed the Gibbs potential energy of concrete into three parts, elasticity, damage and healing, and obtained the evolution equation of damage and healing variables by using Clausius Duhem inequality. In addition, lots of laboratory tests on filling materials with different lime sand ratio have been conducted. Based on the experimental results and damage mechanics theory and combined with the research results of fracture characteristics of different filling materials, the defect coupling constitutive models based on continuous damage mechanics were established [23–25]. With the wide application of filling mining method, the existing research on the deformation and failure of filling material and backfill under load is also more in-depth.

The above studies only focus on the damage evolution process of rock or backfill under different conditions. Considering that in the process of underground filling, the backfill is greatly affected by confining pressure from the surrounding rock. Koupouli et al. [26] showed that the interface of backfill with low cement content showed strain hardening behavior, and the shear strength between backfill and surrounding rock was affected by cohesion and internal friction angle. Fang et al. [27] experimentally studied the influence of different curing temperatures on the shear behavior and strength of the CPB-rock interface and found that higher curing temperature could improve cement hydration and self-drying rate, thus increasing the peak shear stress at the interface between CPB and rock in the early stage. Falaknaz et al. [28] used the numerical simulation method to create the response of two adjacent filling stopes. Based on the explicit relationship between Poisson's ratio and the friction angle in the backfill, the stress state of the filling stope was changed. The simulation results expressed in stress, displacement and strain explained the influence of different parameters and evaluated the interaction between the filling material and surrounding rock to ensure the safe application of the filling material. The above research on the combined action of surrounding rock and backfill mainly involved their mechanical action mechanism and only studied the failure law, without considering the synergistic effect, and thus the damage constitutive model was established [29–32]. The surrounding rock and backfill were regarded as independent individuals and studied their roles under different combination structure modes, respectively. However, under the action of load, the surrounding rock and backfill form a composite by compression contact coupling. The

properties and strength of the two materials are different, and the deformation and failure mode and law will inevitably change.

In this study, the backfill-red sandstone combination is prepared by red sandstone and full tailings cemented backfill with different cement–sand ratios. The uniaxial compression tests are carried out, and the deformation and failure characteristics of the composites in the uniaxial compression process are analyzed. The appropriate cement–sand ratio to realize the synergy between red sandstone and backfill is obtained. According to the stress–strain curve of the specimens during the compressing process, the damage constitutive relationship of the backfill-red sandstone combination under uniaxial compression is analyzed. Combined with the damage constitutive and elastic–plastic theory, the damage process is divided into three stages including compaction pre-synergy stage, quasi-elastic synergy deformation stage and rupture deformation stage. The damage constitutive model of the backfill-red sandstone combination is established which is able to describe the damage evolution law of the synergistic action between the surrounding rock and backfill. The obtained results provide an important reference value for improving the damage theory and failure mechanism of interaction between surrounding rock and backfill.

2. Experiment

2.1. Experimental Materials

2.1.1. Tailings

The tailings selected as aggregates in this research were collected from an iron mine in Shaanxi, China. The main chemical compositions of tailings obtained through analysis are shown in Table 1. The main chemical compositions are Fe_2O_3 and SiO_2, accounting for 88% of the total solid weight. The content of SiO_2 is 67.72%, indicating that the tailings are high-silicon iron tailings.

Table 1. Composition of iron tailings.

Composition	**SiO_2**	**Al_2O_3**	**Fe_2O_3**	**CaO**	**MgO**	**Na_2O**	**Other**
Content (%)	67.72	2.05	20.65	4.66	3.16	0.5	1.26

The particle size distribution is illustrated in Figure 1. The average particle size of the tailings d_{50} is 86.40 μm, the non-uniformity coefficient C_u is 5.70, and the curvature coefficient C_c is 1.39, indicating that the iron tailings have good gradation, compaction and mechanical properties.

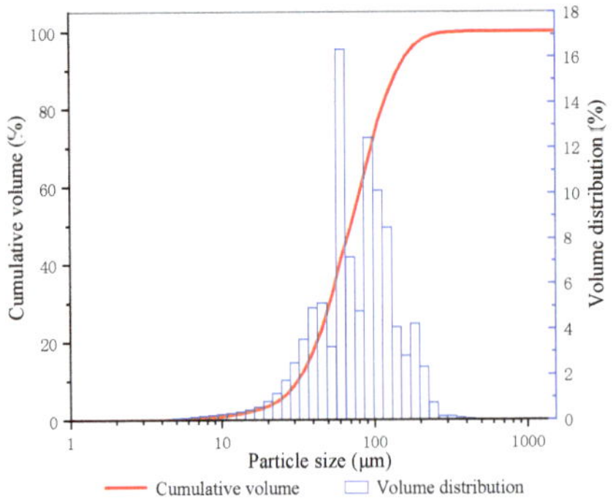

Figure 1. Particle size distribution of the tailings.

2.1.2. Backfill Material

The backfill materials are composed of iron tailings, P.O42.5 cement and city tap water. P.O42.5 cement is used as gelling agents in the experiment. By adjusting the quality of cement, iron tailings and water, the filling slurry with different cement–sand ratio can

be prepared. The backfill samples were selected with different cement–sand ratios (1:4, 1:6, and 1:8, respectively), and material concentration of 80%. According to the designed material ratio, three specimens with the same cement–sand ratio were in each group, and three groups of nine standard cylindrical specimens with dimensions of 50 and 100 mm in diameter and height were prepared. Nine specimens were marked and placed in a curing box for curing for 28 days with the curing temperature of 20 ± 5 °C and relative humidity of 95% ± 5%. Then, the backfill specimens were taken out for UCS test. The strength characteristics of specimens are shown in Table 2. With the change of the cement–sand ratio, there is an obvious gradient change in the compressive strength of the backfill. The strength of the backfill increase with the increase of the cement–sand ratio. The larger the cement–sand ratio, the higher the strength of the backfill.

Table 2. Strength characteristics of backfill.

No.	Solid Content /wt%	Cement–Sand Ratio	UCS (MPa)	Elastic Modulus (GPa)
B4-1	80	1:4	6.41	1.09
B4-2		1:4	6.47	1.12
B4-3		1:4	6.93	1.14
B6-1		1:6	4.21	0.46
B6-2		1:6	4.52	0.62
B6-3		1:6	4.46	0.57
B8-1		1:8	3.34	0.25
B8-2		1:8	3.36	0.27
B8-3		1:8	3.07	0.19

2.1.3. Red Sandstone

The surrounding rock in this research was red sandstone with good uniformity. The average density of red sandstone is 2132 kg/m^3; the uniaxial compressive strength (UCS) of red sandstone cylinder specimen with 50 mm in diameter and 100 mm in height is 22.39 MPa (average value).

By comparison, it is observed that there is a large difference in uniaxial compressive strength between the red sandstone and backfill. The backfill-red sandstone combination specimen is composed of two materials with different strength. How to realize a synergistic effect between two different materials with large strength difference is the focus of this paper.

2.2. Specimen Preparation

In strict accordance with the standard for test methods of engineering rock mass, code for rock test of water conservancy and Hydropower Engineering (DL/t5368-2007) and the standards recommended by the international society of rock mechanics, the red sandstone is prepared into a standard cylindrical sample with a diameter of 50 mm and a height of 50 mm, as shown in Figure 2a. The full tailings from iron ore, cement and water were used to make the slurry with a cement–sand ratio of 1:4, 1:6, 1:8 and mass fraction of 80%. The standard red sandstone sample (Φ 50 mm × H 50 mm) was placed in cylinder mold with size Φ 50 mm × H 100 mm, and then, the evenly stirred filling slurry with different proportions was slowly injected into the cylinder mold until the liquid level was flat with the mold, as shown in Figure 2b. After the filling slurry was fully cemented and solidified, the surface was troweled. Then, the prepared composite specimens were placed in the standard curing box for curing for 28 days. The curing environment was the same as in the backfill experiment. In total, 3 groups of backfill-red sandstone combination specimens with different cement–sand ratio were finally prepared, as shown in Figure 2c.

2.3. Testing Procedures

The microcomputer controlled electro-hydraulic servo universal testing machine after improvement was used for the uniaxial compression test, and the loading adopted displacement control mode, as shown in Figure 3. The composite specimen was placed on the pressure plate. Due to the different material properties of the backfill body and red

sandstone, the strain gauges were pasted on the backfill and red sandstone, respectively, to collect the strain data of the two parts. The attached positions are shown in Figure 3. (Strain gauge BFH120-10AA-D150, sensitivity coefficient 2.0 ± 1%, resistance 120 Ω). The strain gauges were connected to the strain measuring instrument to obtain the strain data during the compression process of the specimen. The strain instrument transmitted data to the computer to obtain the actual stress–strain curve of the composite specimen. During the loading process, especially near the peak failure, the failure process was carefully observed and photographed. When the specimen entered the residual strength stage, it could be seen that with the increase of strain, the stress remained basically unchanged, and the test stopped.

Figure 2. Preparation process of backfill-red sandstone combination specimens: (**a**) red sandstone samples (Φ 50 mm × H 50 mm); (**b**) backfill with different cement–sand ratios; (**c**) backfill-red sandstone combination specimens.

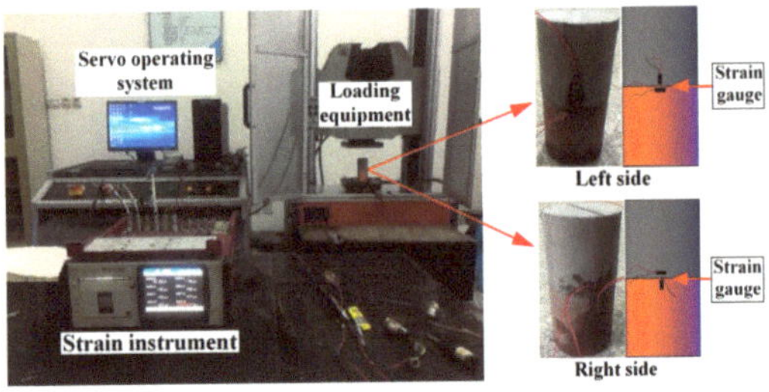

Figure 3. Experimental equipment.

The mechanical parameters obtained from the uniaxial compression test are shown in Table 3. σ_p and σ_r represent the peak stress and residual stress, respectively. ε_p and E represent the peak strain and elastic modulus. The variables of σ_p, σ_r, ε_p and E were obtained from the stress–strain curves. The elastic modulus E is defined as the tangent of the secants corresponding to 30%–60% of the peak stress [33].

Table 3. Mechanical parameters of backfill-red sandstone combination specimens.

Composite Specimen	Solid Content /wt%	Cement–Sand Ratio	σ_p (MPa)	σ_r (MPa)	ε_p	E (GPa)
HB4-1			9.32	4.08	1.09×10^{-3}	8.95
HB4-2		1:4	9.88	4.13	1.16×10^{-3}	9.45
HB4-3			10.29	4.08	1.39×10^{-3}	10.77
HB6-1			7.44	3.57	2.03×10^{-3}	4.58
HB6-2	80	1:6	5.40	3.46	2.06×10^{-3}	2.43
HB6-3			6.32	3.46	2.03×10^{-3}	3.34
HB8-1			5.55	/	2.26×10^{-3}	2.57
HB8-2		1:8	5.40	/	2.18×10^{-3}	2.48
HB8-3			4.48	/	2.11×10^{-3}	2.28

3. Results and Analysis

3.1. Stress–Strain Curves

Due to the different material properties of the backfill and red sandstone, the strain gauges were pasted on the backfill and red sandstone, respectively, to collect the strain data from the two parts. The stress–strain curves of the backfill and red sandstones were obtained in the uniaxial compression tests, as shown in Figures 4–6. The different lines in the stress–strain curves represent different parts of the specimens rather than different materials. The red line represents the stress–strain curve drawn by the data collected from the strain gauges affixed to the red sandstone during loading, while the blue line is the stress–strain curve drawn by the data from the strain gauges on the upper backfill. The gray line represents the synthesis of the two parts, namely, the stress–strain curve of the backfill-red sandstone composite specimen from the load equipment.

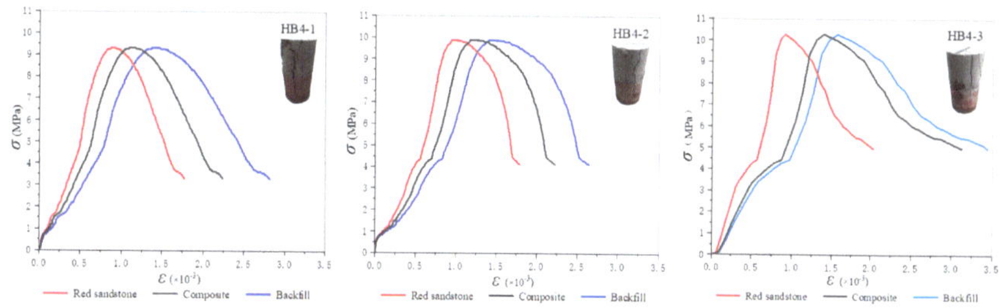

Figure 4. Stress–strain curves of the specimens at the cement–sand ratio of 1:4.

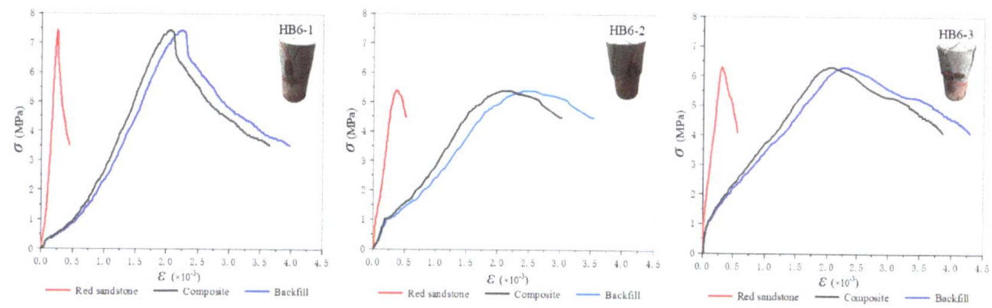

Figure 5. Stress–strain curves of the specimens at the cement–sand ratio of 1:6.

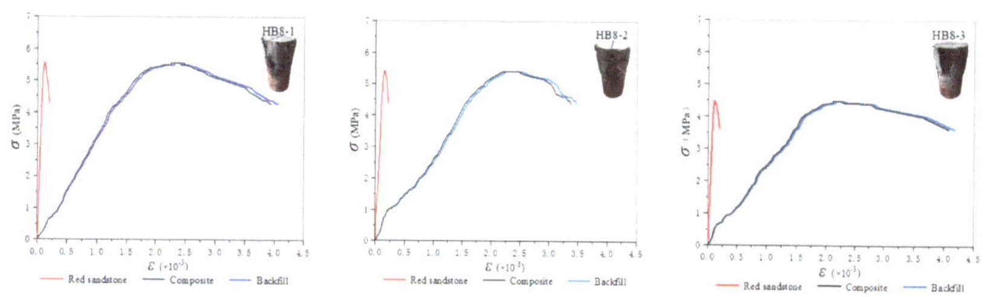

Figure 6. Stress–strain curves of the specimens at the cement–sand ratio of 1:8.

The stress–strain curves of the specimens with the cement–sand ratio of 1:4 (HB4) in the uniaxial compression tests are displayed in Figure 4. According to the stress–strain curves, the failure process could be separated into several stages. The first stage is compaction pre-synergy stage. During this period, the microcracks or gaps in the composite are continuously compacted with the strain increased rapidly for both the backfill and red sandstone. With the continuous increases of axial load, there is an obvious inflection point in the stress–strain curve of the specimen; the value of this inflection point is about three quarters of the compressive strength of the backfill with cement–sand ratio of 1:4.

The reason for this phenomenon is that there is an interface between the backfill and red sandstone, which is different from the complete rock sample, resulting in the plastic yield of the upper backfill under load. This stage is also a key preparation stage for the synergy between the upper backfill and the lower red sandstone. The second stage is the quasi-elastic synergy deformation stage. It occurs between the inflection point in the previous stage and the peak stress. At this stage, the stress–strain curve of the specimen is approximately linear. The reason for this phenomenon is that the upper backfill and the lower red sandstone basically achieve synergy. The upper backfill and the lower red sandstone bear the load together. The third stage is the rupture deformation stage. After reaching the peak stress, the stress decreases rapidly while the strain increases rapidly. The curve shows an obvious parabolic shape. At this stage, the upper backfill and the lower red sandstone fully realize the synergistic support. The micro ruptures in the composite continue to expand, producing new ruptures and large through ruptures in the specimens. During this process, the strength of the composite specimens did not completely lose. The bearing capacity gradually decreased with the increase of strain, showing an obvious softening phenomenon.

The deformation trends of the curves in Figure 4 are consistent, the failure modes of the upper and lower components of the backfill-red sandstone combination specimens are basically the same, and the shear failure occurred in both the red sandstone part and the backfill part. This observation indicated that the upper backfill and the lower red sandstone can achieve coordinated deformation and jointly resist damage. In the compaction pre-synergy stage, due to the interface between the upper backfill and the lower sandstone, there is an obvious inflection point in the stress–strain curve. At this stage, the upper backfill first reaches plastic yield, but there is no crack on the surface. Therefore, in the next stage, the synergistic effect between the upper backfill and the lower red sandstone is obvious, the stress–strain curve is approximately linear, and the stress increases continuously with the increase of strain. After reaching the peak value, the composite specimen produces through cracks as a whole, and the stress does not decrease rapidly, but there is a small section of basic stability maintenance state, at which time the strain continues to increase. With the continuous increase of load, the strain continues to increase, while the stress begins to decrease rapidly, and finally reaches the residual strength of the specimen.

The stress–strain curves of the specimens with the cement–sand ratio of 1:6 (HB6) are shown in Figure 5. Due to the huge difference in the deformation between the backfill and red stone, the upper backfill and the lower red sandstone failed to realize the overall synergy resistance during the compression tests. Therefore, compared with the 1:4 composite specimens, there is no obvious inflection point in the stress–strain curve. Additionally, after reaching the peak stress, only the upper backfill breaks and produces plastic deformation. During the loading process, due to the large deformation of the upper backfill, there was no obvious through rupture in the lower red sandstone, which indicated the failure mostly occurred in the backfill parts.

Figure 6 shows the stress–strain curve of the specimens with the cement–sand ratio of 1:8 (HB8) under uniaxial compression. The deformation process of the composite specimen is similar to that of HB6. With the continuous application of load, the strain increase rate is significantly greater than the stress increase rate, and the stress–strain curve of the backfill part basically coincides with the stress–strain curve of the composite specimen. The stress–strain curve of the lower red sandstone is obviously different from that of the backfill. During the loading process, the deformation of the backfill is large. After the stress reaches the peak value, there is a step decline and obvious plastic deformation, which further shows that the load mainly acts on the backfill. The experiment ended with the destruction of the backfill without obtaining the residual strength of combination.

Comparing the variation characteristics of stress–strain curves of the specimens with three different cement–sand ratios, it can be found that the strain of red sandstone gradually decreases with the decrease of the cement–sand ratio. When the cement–sand ratio is 1:8, the failure of red sandstone at the bottom of the composite is not obvious. The strain of the

backfill gradually increases with the decrease of the cement sand ratio. The final stress and strain mainly act on the backfill, and the main damage of the composite is the backfill.

3.2. Macroscopic Failure Patterns

Three composite specimens with the most representative failure characteristics are selected from nine specimens, as shown in Figure 7. Figure 7a shows the macroscopic failure phenomenon of composite specimen HB4-3. It is obvious that the cracks occur in both the backfill and red sandstone. The backfill first has cracks and large deformation. With the continuous increase of stress, the red sandstone splits after reaching the peak, indicating that the backfill limits the evolution of rock displacement and damage to a certain extent. Figure 7b shows the macroscopic failure phenomenon of the composite specimen HB6-1. Cracks first appear in the backfill and extend downward to the contact surface. The edges and corners of red sandstone specimens are slightly damaged, and the cracks in the upper backfill are further expanded and damaged, finally showing a "V" rupture. Figure 7c shows the macroscopic failure phenomenon of composite specimen HB8-1. With the continuous increase of load, the red sandstone was not damaged, but the backfill was seriously damaged, and finally the occurrence of failure was "flaking".

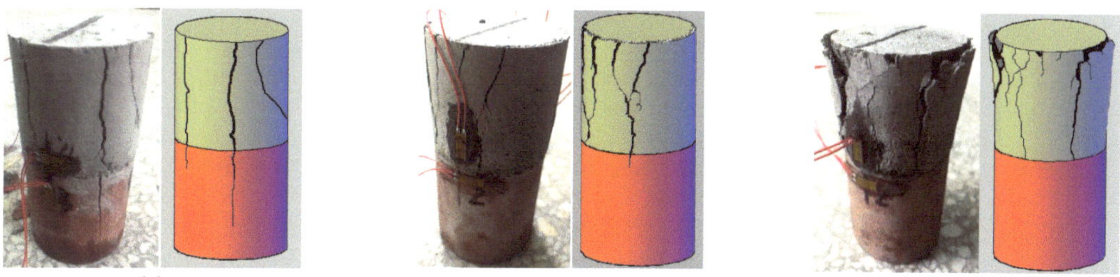

(a) (b) (c)

Figure 7. Macroscopic failure phenomenon of composite specimens: (**a**) HB4-3, (**b**) HB6-1, and (**c**) HB8-1.

3.3. Deformation and Failure Characteristics

The backfill-red sandstone combination is composed of two different materials, which is different from the intact rock or backfill, and its bearing mechanism is different. When the composite specimen is subjected to load, the two materials interact and restrict each other. Nine stress–strain curves are summarized and analyzed, and three groups of typical stress–strain curves of backfill-red sandstone combinations with different ratios are obtained, as shown in Figure 8. It can be seen that the peak strength of backfill-red sandstone combination specimens decreases with the decrease of the cement–sand ratio. The residual strength of HB4 is large. At the cement–sand ratio of 1:4, the red sandstone and the backfill jointly resist the external load, and the stress continues to increase with the increase of strain. For the composites HB6, the residual strength is smaller. With the increase of load, the strain increases rapidly, but the strain of the lower part of the red sandstone increases slowly and remains constant after reaching the peak. Therefore, the red sandstone part has not undergone large deformation, which shows that the strength of the backfill under this cement–sand ratio fails to match well with the red sandstone, and the synergy cannot be realized. At the cement–sand ratio of 1:8, the load completely acts on the backfill, and the strain of the red sandstone almost tends to zero. The stress–strain curve of composite specimen with the cement–sand ratio of 1:8 can be approximately regarded as the stress–strain of the backfill. The difference in the mechanical properties and failure patterns of the specimens suggest that it is necessary to reasonably optimize the cement–sand ratio of the backfill to realize synergy support of the red sandstone and the backfill.

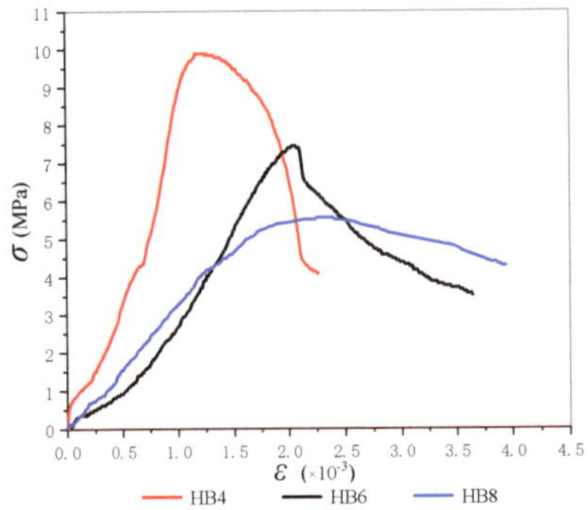

Figure 8. Comprehensive stress–strain curves of backfill-red sandstone combination specimens with different cement–sand ratios.

4. Synergy Support Damage Constitutive Model

According to the stress–strain curves as well as failure characteristics of the specimens, it is found that at the cement–sand ratio of 1:6, the upper backfill is completely destroyed, while the lower red sandstone corner is slightly damaged. The backfill and red sandstone cannot achieve the role of synergy support. When the cement–sand ratio is 1:8, the stress–strain curve of the upper backfill part basically coincides with the composite specimen. The main damage of the composite is the backfill while the lower red sandstone is almost not damaged. The overall compressive strength and elastic modulus are small, and the backfill and red sandstone fail to achieve synergy support. Only when the cement–sand ratio is 1:4, the cracks occur in both the backfill and red sandstone. The composite specimens composed of red sandstone and backfill realize the synergy support and common bearing.

For the backfill-red sandstone combinations with the cement–sand ratio of 1:4, the loading damage can be equivalent to the coupling of two damage states of backfill and red sandstone. The first part is the damage state of the initial backfill, that is, the initial damage caused by the initial defect on the upper part of the delamination plane. The second part is the common damage state of the backfill and the loading of red sandstone part, that is, the damage caused by the continuous loading of the backfill-red sandstone combination. The third part is that the backfill and red sandstone bear together until damage and rupture expansion occur. Layered structure and load weaken the cohesion of materials with different mechanical mechanisms. Combined with the damage and failure characteristics of backfill-red sandstone combination specimens, the uniaxial compression failure process can be divided into three stages: compaction pre-synergy stage (OA), quasi-elastic synergy deformation stage (AB) and rupture deformation stage (BC) (Figure 9).

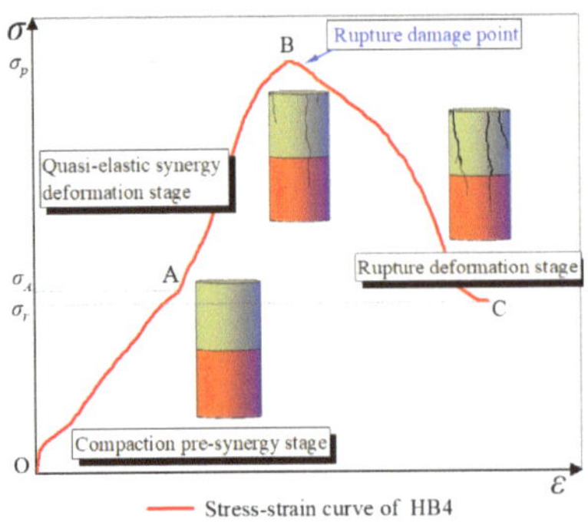

Figure 9. Stress–strain curve of backfill-red sandstone combination specimen.

4.1. Constitutive Model in Compaction Pre-Synergy Stage (OA)

From a macroscopic point of view, the backfill-red sandstone combinations did not rupture at the beginning of the loading process. However, it can be found that due to the mutual bonding between the upper backfill and red sandstone, the microcracks at the contact surface between the backfill and the rock compacted rapidly. With the increase of strain, the stress obviously shows a non-linear elastic rise. At this time, the microcracks in the upper backfill are slowly compacted with the increase of load, and the micro elements in the internal structure of the backfill are damaged. With the continuous increase of stress, the composite specimen enters the non-linear elastic failure stage, and the stress–strain curve shows the OA segment, which is called the compaction pre-synergy stage. According to experimental data and stress–strain curve, it is identified that there is a certain relationship among the elastic modulus of the backfills, red sandstone and the combinations. In practical engineering, the elastic modulus of the backfill after curing and consolidation is lower than red sandstone. At this time, the main function of the backfill is to transfer load. Therefore, the concept of relative contribution degree is introduced to preprocess the initial weight of elastic modulus of red sandstone and backfill, that is, the contribution degree η is set. According to the bearing capacity characteristics of red sandstone and backfill during loading, the response model of their elastic modulus during initial loading is obtained:

$$E_0 = \eta_1 E_c + \eta_2 E_h \tag{1}$$

where E_0 is the initial elastic modulus of the combination; η_1 is the relative contribution of backfill; η_2 is the relative contribution of red sandstone; E_c is the elastic modulus of the backfill; E_h is the elastic modulus of red sandstone.

Under load, the deformation mechanism of the internal meso element of the combination specimen is complex. The damage theory is introduced to characterize the change law of the meso element. At this stage, the meso element strength of backfill-red sandstone combination specimen tends to follow Loland damage model [34]. Under the action of continuous load, the relationship between stress and strain of the OA segment can be obtained:

$$\widetilde{\sigma} = \frac{E_0 \varepsilon}{(1 - D_0)} = \frac{(\eta_1 E_c + \eta_2 E_h)\varepsilon}{(1 - D_0)} \tag{2}$$

where $\widetilde{\sigma}$ is the effective stress; ε is strain; D_0 is the initial damage, and the specimen has certain damage before loading. Because the current technical conditions cannot well measure the damage in the initial stage, it is considered that $D_0 = 0$.

Therefore, the damage constitutive model of the OA segment can be obtained as follows:

$$\sigma = \widetilde{\sigma}(1 - D_e) \tag{3}$$

where D_e is the damage variable of the OA segment.

According to the stress–strain curve of the OA segment, the microcracks in the backfill are closed under the action of load, resulting in micro element structure damage. The damage value D_e is expressed as:

$$D_e = D_0 + C\varepsilon^{\beta} \tag{4}$$

where C and β are the backfill material parameters.

Based on the stress–strain curve and the boundary conditions:

$$\text{(i) } \varepsilon = \varepsilon_p, \sigma = \sigma_p; \text{ (ii) } \varepsilon = \varepsilon_p, \frac{d\sigma}{d\varepsilon} = 0 \tag{5}$$

The constant C, β is

$$C = \frac{1 - D_0}{1 + \beta}\varepsilon_p \tag{6}$$

$$\beta = \frac{\sigma_p}{(\eta_1 E_c + \eta_2 E_h)\varepsilon_p - \sigma_p} \tag{7}$$

4.2. Constitutive Model in Quasi-Elastic Synergy Deformation Stage (AB)

Under the action of continuous load, the backfill-red sandstone combination has no macro rupture. At the micro level, the internal deformation and failure of the upper backfill occurred slowly. The internal structure of the lower red sandstone is also damaged, resulting in micro element damage. The backfill-red sandstone combination is like linear elastic deformation, and its internal original damage begins to intensify and evolve until it exceeds the peak stress failure. The stress–strain curve shows the AB segment, which is called the quasi-elastic synergy deformation stage of backfill and red sandstone. Considering the contribution of backfill and red sandstone to the composite, the response model of elastic modulus at this stage is obtained:

$$E_p = \eta_1' E_c + \eta_2' E_h \tag{8}$$

where E_p is the peak elastic modulus of composite; η_1' is the relative contribution of backfill in the AB segment; η_2' is the relative contribution of red sandstone in the AB segment.

According to the stress–strain curve of the AB segment, the damage constitutive model of stress and strain can be obtained:

$$\sigma = E_p \varepsilon (1 - D_p) = (\eta_1' E_c + \eta_2' E_h)\varepsilon(1 - D_p) \tag{9}$$

where ε is the strain of the composite specimen in section AB; D_p is the damage variable of the AB segment.

At this stage, the micro element strength of backfill-red sandstone combination specimen obeys Weibull distribution [35,36], the probability density function is

$$P(F) = \begin{cases} 0, & F < 0 \\ \frac{b}{F_0}\left(\frac{F}{F_0}\right)^{b-1} \exp\left[-\left(\frac{F}{F_0}\right)^b\right], & F \geq 0 \end{cases} \tag{10}$$

where F is the stress level of the composite specimen in the AB segment; F_0 is the characteristic stress; b is the distribution parameter.

$$F = (\eta_1' E_c + \eta_2' E_h)\varepsilon \tag{11}$$

$$F_0 = (\eta_1' E_c + \eta_2' E_h)\varepsilon_0 \tag{12}$$

where ε_0 is the strain parameter of the AB segment.

Thus, substituting Equations (11) and (12) into Equation (10) can obtain:

$$P(\varepsilon) = \frac{b}{\varepsilon_0}\left(\frac{\varepsilon}{\varepsilon_0}\right)^{b-1} \exp\left[-\left(\frac{\varepsilon}{\varepsilon_0}\right)^b\right] \tag{13}$$

By integrating Equation (13), the damage variable expression can be obtained as follows:

$$D_p = \int P(\varepsilon) d\varepsilon = \exp\left(-\left(\frac{\varepsilon}{\varepsilon_0}\right)^b\right) \tag{14}$$

Substituting Equation (14) into Equation (9) can be obtained:

$$\sigma = (\eta_1' E_c + \eta_2' E_h)\varepsilon\left(1 - \exp\left(-\left(\frac{\varepsilon}{\varepsilon_0}\right)^b\right)\right) \tag{15}$$

In the stress–strain relationship curve of the composite, the peak stress σ_p and the corresponding strain ε_p also meet the boundary conditions of Equation (5). Thus:

$$\sigma_p = \left(\eta_1' E_c + \eta_2' E_h\right)\varepsilon_p\left(1 - \exp\left(-\left(\frac{\varepsilon_p}{\varepsilon_0}\right)\right)^b\right) \tag{16}$$

The partial differential equation obtained from Equation (15) is

$$\frac{d\sigma}{d\varepsilon} = \left(\eta_1' E_c + \eta_2' E_h\right)\left(1 + \exp\left(-\left(\frac{\varepsilon}{\varepsilon_0}\right)^b\right)\left(b\left(\frac{\varepsilon}{\varepsilon_0}\right)^b - 1\right)\right) \tag{17}$$

Substitute Equation (17) into Equation (5) to obtain:

$$b\left(\frac{\varepsilon_p}{\varepsilon_0}\right)^b = 1 \tag{18}$$

According to Formulas (16) and (18), it can be obtained:

$$\varepsilon_0 = \frac{\varepsilon_p}{\left(\frac{1}{b}\right)^{\frac{1}{b}}} \tag{19}$$

$$b = \frac{1}{\ln\left(\frac{\left(\eta_1' E_c + \eta_2' E_h\right)\varepsilon_p}{\left(\eta_1' E_c + \eta_2' E_h\right)\varepsilon_p - \sigma_p}\right)} \tag{20}$$

4.3. Constitutive Model in Rupture Deformation Stage (BC)

After reaching the peak value, both the upper backfill and the lower red sandstone are damaged, and the backfill appears complex damage phenomena such as cracks propagation, bifurcation and detour. The cracks of the red sandstone develop rapidly and show splitting cracks in varying degrees. At this time, the stress of the composite specimen gradually weakens, and its bearing capacity decreases rapidly with the increase of axial strain. The stress–strain curve shows the BC segment, which is called the rupture deformation stage of backfill-red sandstone combination specimen. At this stage, according to D. Krajcinovic [37], it is assumed that the damage and strain principal axis coincide with the material principal axis in the process of the uniaxial compression. Assume the equivalent stress is $\widetilde{\sigma} = \sigma(I - D)^{-1}$, and the equivalent strain is $\widetilde{\varepsilon} = \varepsilon(I - D)$, the equation $\widetilde{E} = E(I - D)^2$ can be obtained. Thus, the constitutive equation is

$$\{\sigma\} = \left[\widetilde{E}\right]^{-1}\{\varepsilon\} \tag{21}$$

According to the stress–strain curve of the BC segment, the constitutive relationship of the backfill at this time is as follows:

$$\sigma = \left(\eta_1' E_c + \eta_2' E_h\right)\varepsilon(1 - D_r)^2 \tag{22}$$

where σ is the stress of the BC segment; ε is the strain of the BC segment; D_r is the damage variable of the BC segment.

Considering the residual strength of composite specimens and the damage characteristics of actual stress–strain curves, the above constitutive equation was improved:

$$\sigma = \left(\eta_1' E_c + \eta_2' E_h\right)\varepsilon(1 - D_r)^2 + \sigma_r {D_r}^2 \tag{23}$$

where σ_r is residual strength of the backfill-red sandstone combination.

For the whole process of the BC segment, severe damage will occur in the backfill-red sandstone combination after reaching the stress peak. The damage variable can be expressed as

$$D_r = A|\varepsilon|^N \tag{24}$$

where A and N are the material parameters of the composite.

Considering the boundary conditions, the following is obtained from Equation (5):

$$N = \frac{\sqrt{(\eta_1' E_c + \eta_2' E_h)}}{2\sqrt{(\eta_1 E_c + \eta_2 E_h) - \sqrt{(\eta_1' E_c + \eta_2' E_h)}}} \tag{25}$$

$$A = \frac{1}{(2N+1)|\varepsilon_P|^N} \tag{26}$$

5. Model Validation

5.1. Relative Contribution Value

For determining the relative contribution value, the mechanical interaction model between backfill and red sandstone is analyzed. In order to simplify its mechanical model, lateral constraints and lateral deformation of backfill are temporarily ignored in the laboratory test, and the model shown in Figure 10 is obtained. E_c and ε_c are the elastic modulus and strain of backfill, E_h and ε_h are the elastic modulus and strain of red sandstone, respectively. E_h is significantly higher than E_c.

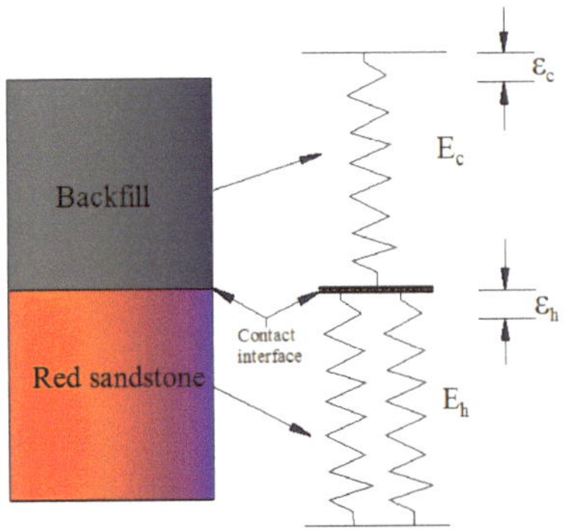

Figure 10. Mechanical analysis model of backfill-red sandstone combination specimen.

According to the failure characteristics of the specimens, the load in the initial stage mainly acts on the microcracks compaction process of the backfill and the surface contact between the backfill and red sandstone. Based on the characteristics of stress–strain curve and engineering experience, the relative contribution of the elastic modulus of red sandstone at this stage to the elastic modulus of the initial composite can be calculated and distributed as follows:

$$\eta_2 = \frac{|E_c - E_h|}{E_c \cdot E_h} \tag{27}$$

At this time, the relative contribution of the elastic modulus of the backfill to the elastic modulus of the initial composite is

$$\eta_1 = 1 - \eta_2 \tag{28}$$

In the next stage, with the continuous loading, the upper backfill and the lower red sandstone bear the load together. At this time, the relative contribution of the elastic modulus of the red sandstone to the elastic modulus of the composite is

$$\eta_2' = \frac{E_c \cdot E_h}{E_c + E_h} \tag{29}$$

At this time, the relative contribution of the elastic modulus of the backfill to the elastic modulus of the composite is

$$\eta_1' = 2 - \eta_2' \tag{30}$$

In the initial stage, the microcracks in the upper backfill and the microcracks in the stratified contact surface are compacted under the action of load. This process is the full combination process of the upper backfill and the lower red sandstone, so as to achieve the purpose of synergy. The lower red sandstone bears the load and the gravity of the upper backfill, so its relative contribution is high. In the next stage, with the continuous increase of load, the upper backfill and the lower red sandstone are fully combined to achieve the synergistic support effect. In this stage, the relative contribution of the upper backfill and the lower red sandstone is basically the same. The calculation results of relative contribution are shown in Table 4.

Table 4. Values based on relative elastic modulus contribution.

No.	E_c (GPa)	E_h (GPa)	η_1	η_2	E_0 (GPa)	η_1'	η_2'	E_p (GPa)
HB4-1	1.09	9.32	0.20	0.80	7.67	1.02	0.98	10.24
HB4-2	1.12	9.76	0.21	0.79	7.94	0.99	1.01	10.97
HB4-3	1.14	11.15	0.21	0.79	9.05	0.98	1.02	12.49

5.2. Verification of Damage Parameters and Constitutive Equations

In order to verify the deduced constitutive model, three backfill-red sandstone combination specimens with the cement–sand ratio of 1:4 are compared and analyzed. The mechanical parameters of three backfill-red sandstone combination specimens are shown in Table 5.

Table 5. Mechanical parameters of composite specimen.

No.	σ_p (MPa)	E_0 (GPa)	ε_p (10^{-3})	σ_r (MPa)	E_p (GPa)	β	C	b	ε_0	N	A
HB4-1	9.32	7.67	1.09	4.08	10.24	4.15	0.21	0.61	0.48	3.19	0.10
HB4-2	9.88	7.94	1.16	4.12	10.97	1.93	0.39	0.94	1.07	2.94	0.09
HB4-3	10.29	9.05	1.39	4.08	12.49	1.46	0.56	1.11	1.52	2.84	0.06

Based on the data in Table 5, the stress value of intersections of the three-stages are calculated (Table 6). The error between theoretical calculation and experiment value is slight. The theoretical curves of the constitutive model are constructed and compared with the test curves. As shown in Figure 11, the theoretical curves of the model are consistent with the test curves, which confirmed the validities of the proposed constitutive model to described characteristics of the whole damage deformation stage of the specimens.

Table 6. Data table of intersections of composite specimen at each stage.

No.	Actual σ_A (MPa)	OA Section σ_A (MPa)	Actual σ_p (MPa)	AB Section σ_p (MPa)	Actual σ_r (MPa)	BC Section σ_r (MPa)
HB4-1	4.73	4.52	9.32	9.49	4.08	3.45
HB4-2	4.33	4.29	9.88	9.93	4.08	3.50
HB4-3	4.43	4.39	10.29	10.18	4.12	3.54

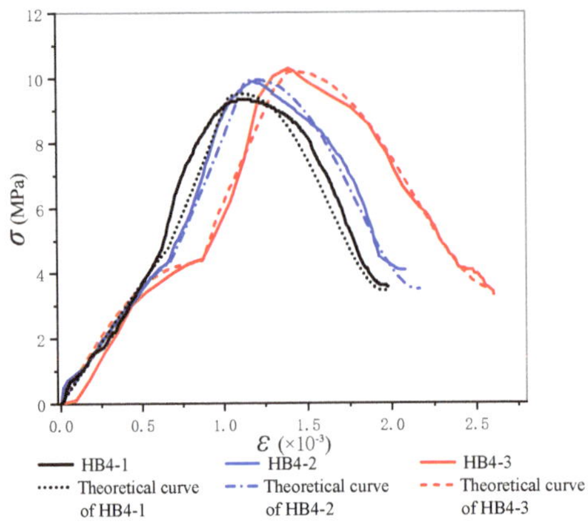

Figure 11. Damage constitutive model compared with the test.

It can be found that the experimental curves basically have three stages: compaction pre-synergy stage, quasi-elastic synergy deformation stage and rupture deformation stage (Figure 12). The three types of damage constitutive relations established according to these stages are in good agreement with the experimental curves. However, due to the influence of experimental environment, experimental errors and other factors, there are still some differences in each group of data. According to the damage deformation characteristics and test results in each stage, the damage constitutive equation and damage deformation characteristic diagram of each stage are obtained (Table 7).

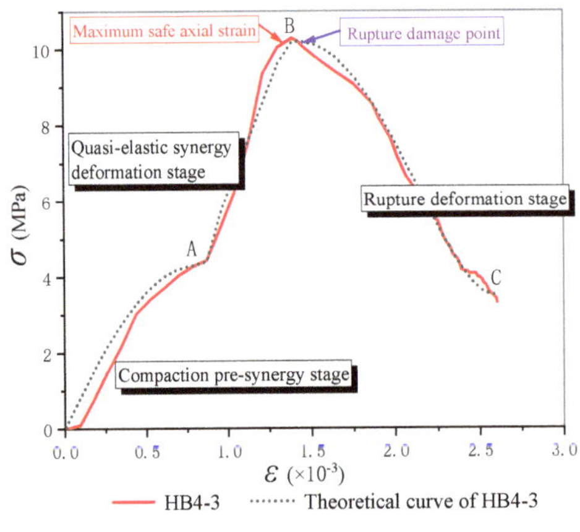

Figure 12. Theoretical curve with phased characteristic diagram.

5.3. Damage Variable and Evolution Law

According to its load damage, the synergy deformation of the specimens can also be divided into three stages. The first stage is the microcracks compaction process of the upper backfill and layered contact surface. The second stage is the common bearing process of the upper backfill and the lower red sandstone. The third stage is the rupture process of backfill-red sandstone combination. The damage patterns in each stage are different. According to the established damage constitutive model, the damage evolution process of backfill-red sandstone combination can be characterized, and the relationship between damage variables and strain can be obtained, as shown in Figure 13.

Table 7. Constitutive equation and damage of each stage.

Stage Division	Damage Constitutive Model	Damage Deformation Diagram	Description of Deformation Characteristics
Section OA: compaction pre-synergy stage	Constitutive equation: $\sigma = (\eta_1 E_c + \eta_2 E_h) \cdot \varepsilon\left(1 - C \cdot \varepsilon^{\beta}\right)$ Actual parameter equation: $\sigma = 9.05 \cdot \varepsilon\left(1 - 0.56 \cdot \varepsilon^{1.46}\right)$		At this stage, the internal microcracks of the upper backfill and its stratified contact surface with the lower red sandstone are compacted and damaged.
Section AB: quasi-elastic synergy deformation stage	Constitutive equation: $\sigma = (\eta_1' E_c + \eta_2' E_h) \cdot \varepsilon\left(1 - \exp\left(-\frac{\varepsilon}{\varepsilon_0}\right)^b\right)$ Actual parameter equation: $\sigma = 12.49 \cdot \varepsilon\left(1 - \exp\left(-\frac{\varepsilon}{1.52}\right)^{1.11}\right)$		At this stage, there is no crack in the backfill-red sandstone combination as a whole, but the edges and corners of the upper backfill are damaged.
Section BC: rupture deformation stage	Constitutive equation: $\sigma = (\eta_1' E_c + \eta_2' E_h) \cdot \varepsilon\left(1 - A \cdot \varepsilon^N\right)^2 + \sigma_r \cdot \left(1 - A \cdot \varepsilon^N\right)^2$ Actual parameter equation: $\sigma = 12.49 \cdot \varepsilon\left(1 - 0.06 \cdot \varepsilon^{2.84}\right)^2 + 4.08 \cdot \left(1 - 0.06 \cdot \varepsilon^{2.84}\right)^2$	 	At this stage, when the stress exceeds the peak value for a period of time, the through crack is rapidly generated and expanded. With the continuous application of load, the fissure of the backfill presents V-shaped expansion, and the phenomenon of "rib spalling" occurs locally.

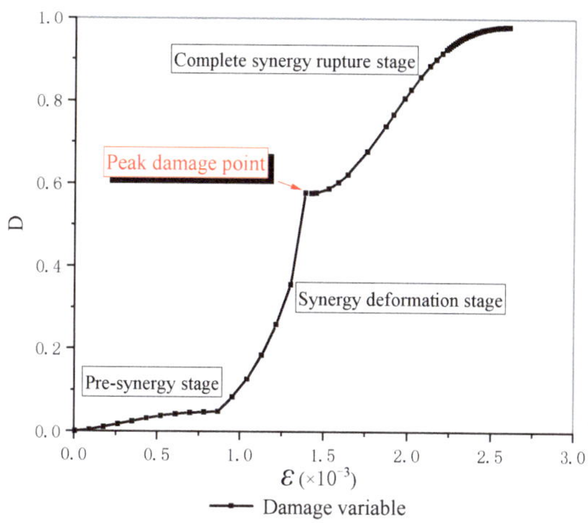

Figure 13. Damage variable change of HB4-3.

According to Figure 13, it can be found that in the process of the uniaxial compression, the initial damage in the upper backfill is less than 0.1 due to the influence of microcracks. With the continuous increase of stress, the lower red sandstone and the upper backfill bear the load together. The damage increases sharply, and cracks appear after exceeding the

peak stress. With the continuous action of load, the damage of backfill-red sandstone combination continued, and the specimen was completely destroyed. The macro failure patterns, micro damage variables, overall deformation and failure characteristics all prove that the backfill-red sandstone combination realize the synergistic support at the cement–sand ratio of 1:4. This provides a reference for the study of synergistic support between backfill and surrounding rock in the process of underground filling. The results also provide a basis for the damage evolution law and failure process of the synergy support between backfill and surrounding rock.

5.4. Synergy Analysis between Backfill and Red Sandstone

In the process of actual backfill mining, the backfill and surrounding rock form a combination to bear ground pressure together. This leads to an interaction system between rock mass and backfill. From the compression tests, it was found that the interaction and matching effect between the backfill and red sandstone were affected by the cement–sand ratio. The larger the cement–sand ratio, the higher the strength of the backfill and the greater load-carrying capacity of backfill-red sandstone composite. When the cement–sand ratio was 1:4, the part of backfill had the highest strength, which can fully cooperate with red sandstone to resist the load. The peak stress reached 10.29 MPa. The cracks occurred in both the backfill and red sandstone and propagated to form split failure. When the cement–sand ratio was 1:8, the backfill failed to cooperate with red sandstone. Only the backfill part of composite broke completely. Therefore, the strength of backfill is the key to realize the synergistic support for backfill mining. In the actual mining engineering, in order to save the cost, mining industry is looking always to lower cement ratio in their mixtures. To solve this problem, the new cementing material with low cost can be selected in the future to obtain the backfill with low ratio and high strength.

In this research, the backfill with cement–sand ratio 1:4 had a good matching effect with the red sandstone. During the load-carrying process, the backfill and red sandstone complemented each other and each played their own role. The two achieved synergistic support. According to stress–strain curves and failure characteristics, their functions were analyzed. The relative contribution degrees of backfill and red sandstone were proposed and calculated. The damage constitutive models were established. From the viewpoint of damage, the synergistic mechanism between the backfill and red sandstone was analyzed. In the initial stage, the backfill and red sandstone bore the load together. The interaction of two parts produced the resistance force inside them, and the internal structure was damaged. At this stage, the damage changed slowly, and it provided the foundation for the matching of the two. In the second stage, the strength of backfill matched with the red sandstone, and the first inflection point of the damage variable curve appeared. Under the stress at this point, the backfill should be broken, but in fact, there were no macroscopic cracks in it. For damage variable, the change rate accelerated with the increase of load from this point on. At the end of this stage, the second inflection point of variable damage appeared, and there was peak stress at this point. Then, in the third stage, the synergistic effect between backfill and red sandstone continued. With the continuous increase of load, the rate of damage change decreased compared with that in the previous stage. The through crack was rapidly generated and expanded, and the composite was finally damaged. The damage process reflected the synergistic support between the backfill and red sandstone; the bearing and support effect of composite was no longer dependent on one of them because their support depended on both.

6. Conclusions

Uniaxial compression tests on the backfill-red sandstone combination with different cement–sand ratios were carried out. It was found that at the cement–sand ratio of 1:4, backfill and red sandstone realized synergistic support; the two parts interacted and restricted each other, jointly resisted external loads and played a role in common bearing. However, with the decrease of the cement–sand ratio, the stress mainly acted on the filling

materials, and the deformation observed in the backfill was large while there was no obvious rupture in the red sandstone. The backfill-red sandstone combination with the cement–sand ratio of 1:4 had a splitting failure and obvious ruptures. With the cement–sand ratio of 1:6, the upper backfill of combination specimen had a V-shaped rupture, and the lower red sandstone had some small cracks. With the cement–sand ratio of 1:8, the upper backfill of the backfill-red sandstone combination had a "slope" rupture, and there were no cracks in the lower red sandstone.

According to the stress–strain curve of the specimens, the damage failure process is divided into three stages: compaction pre-synergy stage, quasi-elastic synergy deformation stage and rupture deformation stage. The damage constitutive models in each stage are established, respectively. The damage of backfill-red sandstone combination under load is equivalent to the coupling of two damage states of backfill and red sandstone. The theoretical curves obtained from the damage constitutive model are highly consistent with the experimental curves which confirmed the validities of the proposed constitutive model to describe the coupling relationship between the red sandstone and the backfill. The results obtained from this study are of great significance for accurately understanding the synergy relationship between backfill and surrounding rock and for selecting reasonable backfill strength and material ratio in goaf.

Author Contributions: Conceptualization, C.Y. and W.Z.; methodology, C.Y.; investigation, G.Z., B.Z. and Y.C.; writing—original draft preparation, C.Y. and W.Z.; writing—review and editing and supervision, W.Z., S.W. and J.G.; funding acquisition, W.Z. All authors have read and agreed to the published version of the manuscript.

Funding: This research was funded by the National Natural Science Foundation of China (Nos. 51904220, 52004196) and Science and Technology Bureau of the Beilin, Xian (GX2016).

Data Availability Statement: Not applicable.

Conflicts of Interest: The authors declare no conflict of interest.

References

1. Li, X.B.; Gong, F.Q.; Tao, M.; Dong, L.J.; Du, K.; Ma, C.D.; Zhou, Z.L.; Yin, T.B. Failure mechanism and coupled static-dynamic loading theory in deep hard rock mining: A review. *J. Rock Mech. Geotech. Eng.* **2017**, *9*, 767–782. [CrossRef]
2. Huang, Z.; Yilmaz, E.; Cao, S. Analysis of strength and microstructural characteristics of mine backfills containing fly ash and desulfurized gypsum. *Minerals* **2021**, *11*, 409. [CrossRef]
3. Fang, K.; Fall, M. Shear behavior of the interface between rock and cemented backfill: Effect of curing stress, drainage condition and backfilling rate. *Rock Mech. Rock Eng.* **2020**, *53*, 325–336. [CrossRef]
4. Razyq, N.; Luo, G.H.; Neil, R.; Andy, F.; Johns, M.L.; Fridjonsson, E.O. Understanding the microstructural evolution of hypersaline cemented paste backfill with low-field NMR relaxation. *Cem. Concr. Res.* **2021**, *147*, 106516.
5. Hefni, M.; Hassani, F. Experimental development of a novel mine backfill material: Foam mine fill. *Minerals* **2020**, *10*, 564. [CrossRef]
6. Yang, W.D.; Li, G.Z.; Ranjith, P.G.; Fang, L.D. An experimental study of mechanical behavior of brittle rock-like specimens with multi-non-persistent joints under uniaxial compression and damage analysis. *Int. J. Damage Mech.* **2019**, *28*, 1–33. [CrossRef]
7. Sirdesai, N.N.; Singh, T.N.; Ranjith, P.G.; Singh, R. Effect of varied durations of thermal treatment on the tensile strength of red sandstone. *Rock Mech. Rock Eng.* **2017**, *50*, 205–213. [CrossRef]
8. Kim, E.; Changani, H. Effect of water saturation and loading rate on the mechanical properties of red and buff sandstones. *Int. J. Rock Mech. Min. Sci.* **2016**, *88*, 23–28. [CrossRef]
9. Khanlari, G.; Sahamieh, R.Z.; Abdilor, Y. The effect of freeze-thaw cycles on physical and mechanical properties of upper red formation sandstones, central part of Iran. *Arab. J. Geosci.* **2015**, *8*, 5991–6001. [CrossRef]
10. Pan, X.K.; Berto, F.; Zhou, X.P. Creep damage behaviors of red sandstone subjected to uniaxial compression after high-temperature heat treatment using acoustic emission technology. *Fatigue Fract. Eng. Mater. Struct.* **2021**, *45*, 302–322. [CrossRef]
11. Chen, D.H.; Chen, H.E.; Zhang, W.; Lou, J.Q.; Shan, B. An analytical solution of equivalent elastic modulus considering confining stress and its variables sensitivity analysis for fractured rock masses. *J. Rock Mech. Geotech. Eng.* 2021, *in press*. [CrossRef]
12. Bustamante, R.; Rajagopal, K.R. A nonlinear model for describing the mechanical behaviour of rock. *Acta Mech.* **2018**, *229*, 251–272. [CrossRef]
13. Zhao, Y.L.; Zhang, L.Y.; Wang, W.J.; Liu, Q.; Tang, L.M.; Cheng, G.M. Experimental study on shear behavior and a revised shear strength model for infilled rock joints. *Int. J. Geomech.* **2020**, *20*, 04020141. [CrossRef]

14. Shahin, G.; Marinelli, S.; Buscarnera, G. Viscoplastic interpretation of localized compaction creep in porous rock. *J. Geophys. Res. Solid Earth* **2019**, *124*, 017498. [CrossRef]
15. Kolo, I.; Rashid, K.A.; Sousa, R.L.; Lonetti, P. Computational modelling of fracture propagation in rocks using a coupled elastic-plasticity-damage model. *Math. Probl. Eng.* **2016**, *5*, 1–15. [CrossRef]
16. Zhu, C.; He, M.C.; Jiang, B.; Qin, X.Z.; Yin, Q.; Zhou, Y. Numerical investigation on the fatigue failure characteristics of water-bearing sandstone under cyclic loading. *J. Mt. Sci.* **2021**, *18*, 3348–3365. [CrossRef]
17. Khalij, L.; Pagnacco, E.; Troian, R. Fatigue criterion improvement of Gough and Nishihara & Kawamoto to predict the fatigue damage of a wide range of metallic materials. *Int. J. Fatigue* **2017**, *99*, 137–150.
18. Lee, S.H.; Han, J.T. A study on comparison of strength parameters of hydrostatic pressure-dependent yield criteria. *J. Korea Acad. Ind. Coop. Soc.* **2016**, *17*, 529–535.
19. Lin, H.; Zhang, X.; Cao, R.H.; Wen, Z.J. Improved nonlinear Burgers shear creep model based on the time-dependent shear strength for rock. *Environ. Earth Sci.* **2020**, *79*, 1–9. [CrossRef]
20. Zhu, C.; He, M.C.; Zhang, X.H.; Tao, Z.G.; Yin, Q.; Li, L.F. Nonlinear mechanical model of constant resistance and large deformation bolt and influence parameters analysis of constant resistance behavior. *Rock Soil Mech.* **2021**, *42*, 1911–1924.
21. Kaimonov, M.V.; Khokholov, Y.A. Selection of frozen backfill mixture composition. *J. Min. Sci.* **2019**, *55*, 857–864. [CrossRef]
22. Shahsavari, H.; Baghani, M.; Sohrabpour, S.; Naghdabadi, R. Continuum damage-healing constitutive modeling for concrete materials through stress spectral decomposition. *Int. J. Damage Mech.* **2016**, *25*, 900–918. [CrossRef]
23. Sainoki, A.; Mitri, H.S. Quantitative analysis with plastic strain indicators to estimate damage induced by fault-slip. *J. Rock Mech. Geotech. Eng.* **2018**, *10*, 1–10. [CrossRef]
24. Koohestani, B.; Koubaa, A.; Belem, T.; Bussière, B.; Bouzahzah, H. Experimental investigation of mechanical and microstructural properties of cemented paste backfill containing maple-wood filler. *Constr. Build. Mater.* **2016**, *121*, 222–228. [CrossRef]
25. Popczyk, M.K.; Jendrus, R. Impact of ash and water mixture density on the process of gob grouting in view of laboratory tests. *Arch. Min. Sci.* **2019**, *64*, 625–634.
26. Koupouli, N.J.F.; Belem, T.; Rivard, P.; Effenguet, H. Direct shear tests on cemented paste backfill-rock wall and cemented paste backfill-backfill interfaces. *J. Rock Mech. Geotech. Eng.* **2016**, *8*, 472–479. [CrossRef]
27. Fang, K.; Fall, M. Effects of curing temperature on shear behaviour of cemented paste backfill-rock interface. *Int. J. Rock Mech. Min. Sci.* **2018**, *112*, 184–192. [CrossRef]
28. Falaknaz, N.; Aubertin, M.; Li, L. Numerical investigation of the geomechanical response of adjacent backfilled stopes. *Can. Geotech. J.* **2015**, *52*, 1507–1525. [CrossRef]
29. Jafari, M.; Shahsavari, M.; Grabinsky, M. Experimental study of the behavior of cemented paste backfill under high isotropic compression. *J. Geotech. Geoenviron. Eng.* **2020**, *146*, 06020019. [CrossRef]
30. Siddiquee, M.S.A.; Noguchi, T.; Hirakawa, D. Computational simulation of time-dependent behavior of soil–structure interaction by using a novel creep model: Application to a geosynthetic-reinforced soil physical model. *Comput. Geotech.* **2015**, *66*, 180–188. [CrossRef]
31. Jafari, M.; Shahsavari, M.; Grabinsky, M. Drained triaxial compressive shear response of cemented paste backfill (CPB). *Rock Mech. Rock Eng.* **2021**, *54*, 3309–3325. [CrossRef]
32. Pachoud, A.J.; Schleiss, A.J. Stresses and displacements in steel-lined pressure tunnels and shafts in anisotropic rock under quasi-static internal water pressure. *Rock Mech. Rock Eng.* **2016**, *49*, 1263–1287. [CrossRef]
33. Tu, B.B.; Liu, L.; Cheng, K.L.; Zhang, B.; Zhao, Y.Z.; Yang, Q.X.; Song, K. A Constitutive Model for Cemented Tailings Backfill Under Uniaxial Compression. *Front. Phys.* **2020**, *8*, 1–9. [CrossRef]
34. Yang, H.F.; Liu, C.L.; Jiang, J.S. Damage constitutive model of stirrup-confined recycled aggregate concrete after freezing and thawing cycles. *Constr. Build. Mater.* **2020**, *253*, 119100. [CrossRef]
35. Klakattawi, H.S. The Weibull-Gamma distribution: Properties and applications. *Entropy* **2019**, *21*, 438. [CrossRef] [PubMed]
36. Ahmad, A.B.A.; Ghazal, M.G.M. Exponentiated additive Weibull distribution. *Reliab. Eng. Syst. Saf.* **2020**, *193*, 106663. [CrossRef]
37. Krajcinovic, D.; Trafimow, J.; Sumarac, D. Simple constitutive model for a cortical bone. *J. Biomech.* **1987**, *20*, 779–784. [CrossRef]

minerals

MDPI

Article

Creep Constitutive Model and Numerical Realization of Coal-Rock Combination Deteriorated by Immersion

Xuebin Li [1], Xuesheng Liu [1,2,*], Yunliang Tan [1], Qing Ma [1], Baoyang Wu [2] and Honglei Wang [1]

1 College of Energy and Mining Engineering, Shandong University of Science and Technology, Qingdao 266590, China; xuebin9311@163.com (X.L.); yunliangtan@163.com (Y.T.); 15764250774@163.com (Q.M.); 18340017056@163.com (B.W.); wanghonglei1021@163.com (H.W.)

2 State Key Laboratory of Water Resources Protection and Utilization in Coal Mining, National Energy Investment Group Co., Ltd., Beijing 100013, China

* Correspondence: xuesheng1134@163.com

Abstract: Coal-rock combination refers to the coal and rock as a whole, and the failure of the whole structure of the combination is the main cause for the instability of the deep underground engineering. In deep underground engineering, the coal-rock combination is usually under certain hydrogeological conditions, and it is prone to seepage and rheological failure instability accidents due to the long-term action of water and stress. In this study, the creep constitutive model of coal-rock combination considering the influence of moisture content was established based on the Burgers creep model. According to the experimental results of triaxial creep of rock, the relationship between the moisture content and the parameter of the Burgers creep model was derived, and the correctness of the constitutive model in this study was verified. Then, through the C++ language, the core equation of the model was modified, and the numerical calculation of the model was realized by introducing the coal-rock combination creep model considering the influence of moisture content into FLAC3D numerical simulation software. Finally, the model was used to simulate and study the creep characteristics of coal-rock combination with different moisture contents under triaxial loading. The results showed that the stress environment and moisture content have significant effects on the creep characteristics of the coal-rock combination. Under the same stress state, with the increased of moisture content, the strain rate of the coal-rock combination exhibited a non-linear rapid increase in the constant-velocity creep stage, the limit creep deformation and the instantaneous elastic deformation increased, and the viscosity coefficient was significantly decreased. For example, when the axial stress was 5 MPa and the moisture content increased from 0% to 1.5%, the strain rate increased by 44.06%, the limit creep deformation increased by 20%, the instantaneous elastic deformation increased 10.53%, and the viscosity coefficient decreased by about 50%. When the moisture content is 0%, the axial stress increased from 5 to 14 MPa, and the limit creep deformation increased nearly four times. With the increase of moisture content, this value will further expand. The research conclusions can provide a certain reference basis for the long-term stability control of surrounding rock in underground engineering affected by the water.

Keywords: coal-rock combination; creep; constitutive model; moisture content; numerical simulation

Citation: Li, X.; Liu, X.; Tan, Y.; Ma, Q.; Wu, B.; Wang, H. Creep Constitutive Model and Numerical Realization of Coal-Rock Combination Deteriorated by Immersion. *Minerals* **2022**, *12*, 292. https://doi.org/10.3390/min12030292

Academic Editors: Diyuan Li, Zhenyu Han, Xin Cai, Shijie Xie and Gianvito Scaringi

Received: 29 December 2021
Accepted: 21 February 2022
Published: 25 February 2022

1. Introduction

Creep behavior and the long-term strength of coal and rock is an important factor that influences the instability and destruction of the deep underground engineering, which is closely related to the long-term action of deep high stress. In addition, the creep characteristic of coal and rock in deep underground engineering such as deep-buried diversion tunnels and high dam foundations is also affected by the hydrological environment [1,2]. With the immersion time increasing, the internal damage of coal-rock intensifies. As a result, the mechanical properties changed greatly, and the creep deformation and deformation rate

increased significantly [3], which will inevitably have a serious impact on the long-term stability of the deep underground engineering. The coal-rock combination is a multi-layer combination structure that is usually seen as a whole composed of the coal seam and its roof and floor strata. In the deep underground space, deformation and failure of the coal-rock combination are often the main causes of coal mining disasters [4–7]. Therefore, the investigation and better understanding of the creep characteristics of coal-rock combination immersed in water under deep high stress conditions is of key importance in the scientific design and safe operation of deep underground engineering, especially for deep large deformation and long-term deformation.

In recent years, many studies on the mechanical properties of coal and rock with different moisture content have been carried out, which provides a theoretical basis for the study of the long-term stability of underground engineering under the water environment. For the rock under the influence of different moisture content, Li et al. [8,9] investigated the wetting-induced deformation behavior of rock mass in fault fracture zones, and established a new time-varying effect model to describe the wetting deformation of the rock mass. Yao et al. [10] obtained the evolution law of long-term strength and internal crack propagation of rock under different moisture contents by triaxial creep test. Hashiba et al. [11] and Mirosława [12] studied the mechanical properties of rock under different moisture contents through the uniaxial compression test; Li et al. [13] conducted experiments on the rheological properties of the rock in the saturated state and found that with the increase of moisture content, the strength of the rock specimen gradually decreased, and the strain rate and deformation increased significantly. Shtumpf [14] proposed a method to measure the moisture content of rock mass and revealed that the increase of moisture content makes the strength and deformation characteristics of rock mass change significantly; Ojo et al. [15], Vásárhelyi [16] and Ván [17] conducted relevant mechanical tests on rocks under saturated conditions, and obtained the stress–strain law of rocks under saturated conditions. In addition, they proposed the wet strength and ideal water sensitivity of rock must be established to assess the potential changes of strength and deformation for practical engineering design. Okubo et al. [18] showed that moisture content and time were the main reasons for the change of rock mechanical properties. In addition, many scholars have carried out a lot of correlation studies on the influence of rock creep failure under different moisture contents through laboratory tests and numerical simulation [19–28]. For the coal under the influence of different moisture content, Poulsen et al. [29] obtained that the strength of coal and coal-bearing rock decreases with the increase of moisture content by the physical experiment; Wang et al. [30] found that with the increase of moisture content, the strength, elastic modulus, and cohesion of soft coal increased first and then decreased, and the Poisson's ratio showed an increasing trend by the uniaxial and triaxial compression tests; Talapatra and Karim [31] studied the influence of moisture content on the deformation and permeability of coal, and found that the deformation and permeability of coal increased with the increase of moisture content. In addition, some scholars have studied the creep failure characteristics of coal with different moisture content by the acoustic emission test [32–34].

In terms of the creep model, the characteristics of various rock materials in the attenuation and steady creep stages can be accurately described by the traditional Burgers model, which is widely used in rock engineering. Therefore, many scholars often use the burgers model as the basis for establishing the creep model and improving it. For example, based on the classical Burgers model, Tang et al. [35] introduced the creep empirical element of moisture content to improve the study of rock creep properties under the influence of moisture content. Yang et al. [36] improved the Burgers creep model considering the influence of parameter degradation in the rock mass. Based on the Burgers model, Xue et al. [37] established the creep damage model considering the temperature and volume stress in the deep salt rock formation. In addition, Zhao [38] established a rock creep mechanical model considering moisture content based on the unified theological mechanical model theory. Sun et al. [39] proposed an improved Nishihara model considering moisture content by

using the damage theory, and many scholars studied the creep constitutive models of rocks with different moisture contents after freezing and thawing [40–42].

In the aspect of secondary development of creep model, Li [43] and Chen [44] carried out secondary development of a coal-rock creep model, and introduced an example to verify, provide solutions and ideas for the customization of the creep model and the research of coal-rock with different moisture content. Kong et al. [45] developed the modified Naylor constitutive model with VC++(7.0) environment, and a series of simulations of triaxial tests are performed to verify the correctness of the compiled model; He and Chen [46] developed a simplified concrete damage constitutive model in FLAC3D environment, and the model has a certain accuracy. In addition, many scholars carried out the secondary development through numerical simulation software such as ABAQUS and PFC to realize the numerical calculation of the self-defined creep constitutive model of coal and rock [47–50].

In summary, a lot of research work has been carried out on the creep characteristics of coal and rock under different moisture contents, and a lot of progress has been made. However, since the coal in the deep underground space does not exist alone, most of them appear in the combinatorial structure of "coal-rock". If the coal or rock mass is taken as the research object alone, the result is one-sided and the solution to the deep underground engineering is lacking. In the meanwhile, considering the long-term effect of deep high stress of coal-rock combination in the deep underground space, it is of practical engineering significance to study the creep characteristics of the coal-rock combination with different moisture contents under triaxial loading.

Therefore, the creep constitutive model of the coal-rock combination considering the influence of moisture content was established in this paper based on the Burgers creep model. The relationship between moisture content and the correlation coefficients of the model was obtained combined with the experimental results of triaxial creep of rock. Then the experimental data were inverted, and the fitting test was carried out with the original experimental data. The creep constitutive model of coal-rock combination considering the influence of moisture content was introduced into FLAC3D software through the Visual Studio 2010 software, and the creep characteristics of coal-rock combination with different moisture content under triaxial loading conditions were obtained by using this model, and the correctness of the constitutive model was verified.

2. Creep Constitutive Model and Numerical Realization

2.1. Creep Constitutive Model of Coal-Rock Combination Considering the Influence of Moisture Content

Creep characteristics of the coal-rock combination are quite different compared with coal or rock, especially when they are immersed in water for a long time, their creep characteristics are more complicated. The influence of moisture content on the coal-rock combination cannot be described in the existing creep constitutive model. One of the reasons is that the key mechanical parameters in the constitutive model are generally constant, and the rock parameters do not change with the moisture content, which is linear creep. However, in fact, such a constitutive model does not reflect the true change law of coal-rock creep. Among the traditional creep models, the Burgers model has creep characteristics such as instantaneous elasticity, creep, and stress relaxation, which can better reflect the creep characteristics of the coal-rock combination. By adjusting the four parameters of the shear modulus and viscosity coefficient of Kelvin body and the shear modulus and viscosity coefficient of Maxwell body through the model, the creep characteristics of all rock materials can be obtained. Based on this, if the change in the above-mentioned four parameters with moisture content can be obtained, the relationship equation can be established, and the Burgers model can be improved to study the creep characteristics of coal-rock under different moisture contents. That is the creep constitutive model of coal-rock combination considering the influence of moisture content. The creep model is shown in Figure 1. In Figure 1, σ is the stress, $E_1(\omega)$ and $E_2(\omega)$ are the elastic moduli of Kelvin's body and Maxwell's body affected by moisture content, $\eta_1(\omega)$ and $\eta_2(\omega)$

are the viscosity coefficient of Kelvin's body and Maxwell's body affected by moisture content, and ω is the moisture content.

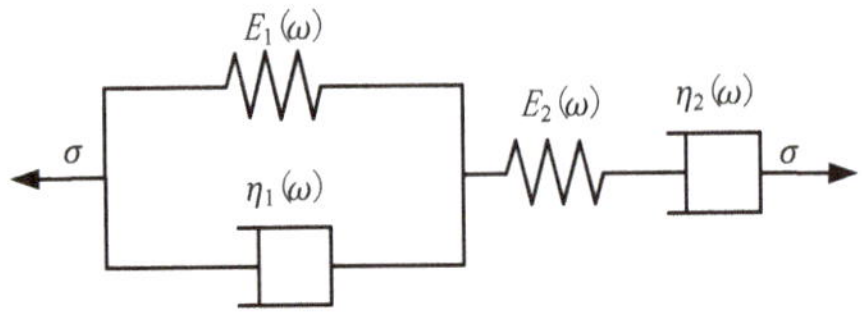

Figure 1. The creep constitutive model of coal-rock combination considering the influence of moisture content.

Burgers creep equation is known as:

$$\varepsilon = \left[1 - \exp\left(-\frac{E_1}{\eta_1}t\right)\right] + \frac{\sigma}{E_2} + \frac{\sigma}{\eta_2}t \tag{1}$$

Then, through the experimental data of triaxial compression creep tests of oil shale under different moisture content (obtained from Wang et al. [51]), and the creep experiment data of the oil shale in different moisture content (the axial stress is 3 MPa and the confining pressure is 0.5 MPa) are substituted into Equation (1). The inline () and beta () functions in MATLAB software are used to invert and fit the model parameters. MATLAB software has powerful functions of numerical calculation and nonlinear fitting, and the built-in multiple optimization algorithms make it more adaptable. The values of E_1, E_2, η_1, η_2 at each moisture content are obtained by calculation, and the corresponding relationship between moisture content and model parameters is shown in Table 1.

Table 1. Correspondence between moisture content and model parameters.

Moisture Content ω (%)	E_1 (GPa)	E_2 (GPa)	η_1 (GPa·h)	η_2 (GPa·h)	R^2
0	1.170	0.417	5.824	268.99	0.999
0.67	1.225	0.327	6.684	103.24	0.998
1.33	1.410	0.313	3.83	51.64	0.998
1.75	1.426	0.312	0.909	29.20	0.993

To verify the correctness of the constitutive model, when the moisture content is 0.67%, the model parameters are used to invert the test data. The inversion test data are compared with the original test data, as shown in Figure 2. The comparison results showed that the experimental data after inversion using the above model parameters were not much different from the original experimental data, and the correlation coefficient R^2 is 0.998, which can verify the correctness of the constitutive model.

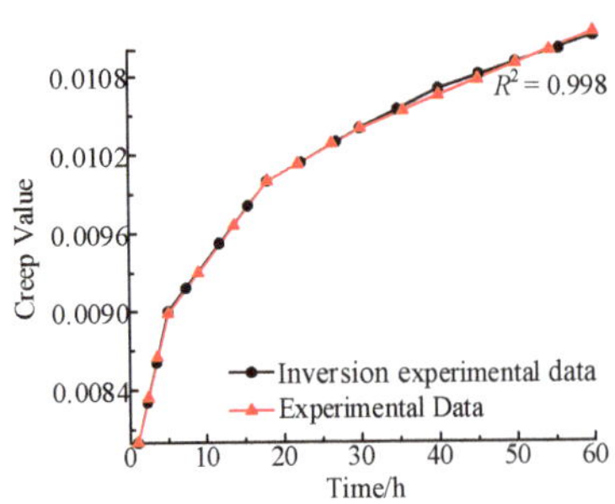

Figure 2. The fitting curve of experimental data and inversion experimental data.

The data in Table 1 were imported into Origin software for fitting analysis, the moisture content was taken as the X-axis, and each parameter in the model was the Y-axis. The parameters were fitted by the Exp3P2 model. The fitting results were shown in Figure 3.

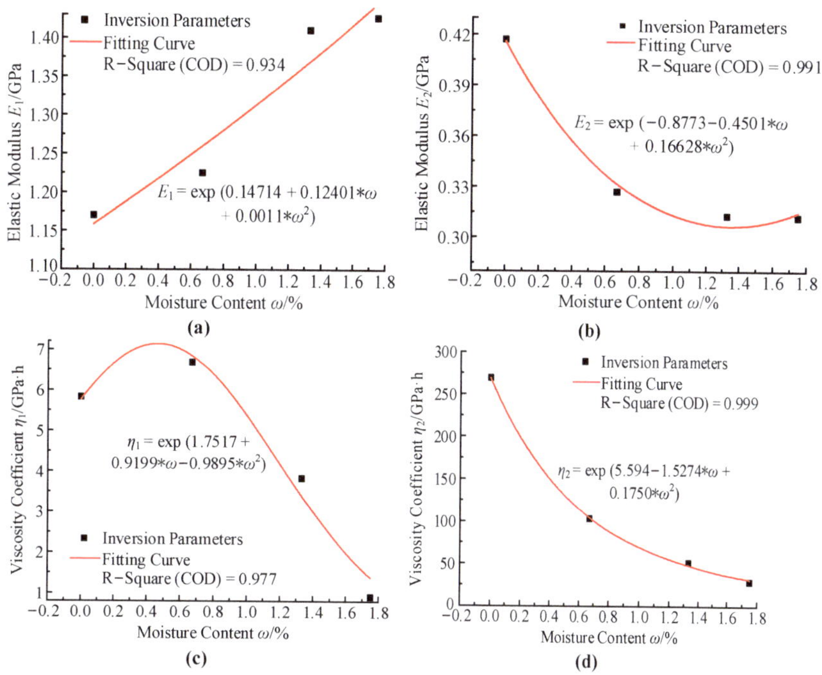

Figure 3. The fitting relationship curve of each parameter in the model and moisture content: (**a**) Fitting curve between elastic modulus E_1 and moisture content; (**b**) fitting curve between elastic modulus E_2 and moisture content; (**c**) fitting curve between viscosity coefficient η_1 and moisture content; (**d**) fitting curve between viscosity coefficient η_2 and moisture content.

It could be seen that the relationship between the moisture content and the parameters in the model was:

$$\begin{aligned} E_1 &= \exp(0.14714) * \exp(0.12401 * w) * \exp\left(0.0011 * w^2\right) \\ E_2 &= \exp\left(-0.8773 - 0.4501 * w * \exp\left(0.16628 * w^2\right)\right. \\ \eta_1 &= \exp(1.7517) * \exp\left(0.9199 * w - 0.9895 * w^2\right) \\ \eta_2 &= \exp(5.594 - 1.5274 * w) * \exp\left(0.1750 * w^2\right) \end{aligned} \tag{2}$$

2.2. Numerical Calculation Realization of the Creep Model

To facilitate the programming of the constitutive model, and through the interface between Visual Studio 2010 and FLAC3D simulation software, the core equation was modified by C++ language, and creep constitutive model of coal-rock combination considering the influence of moisture content was introduced into FLAC3D software.

In order to realize the numerical calculation of the model, it was necessary to write a one-dimensional constitutive equation into a three-dimensional differential form. By using the deviator strain rate relationship between Kelvin body and Maxwell body in Burgers model [43] in a three-dimensional state,

$$\Delta e_{ij} = \left(\frac{b}{a} - 1\right) e_{ij}{}^{K,O} + X S_{ij}{}^{N} + Y S_{ij}{}^{O} \tag{3}$$

$$S_{ij}{}^N = \frac{1}{X}\left[\Delta e_{ij} - (\frac{b}{a} - 1)e_{ij}{}^{K,O} - YS_{ij}{}^O\right] \tag{4}$$

where Δe_{ij} is the total deviator strain rate of the model, $S_{ij}{}^N$ and $S_{ij}{}^O$ are the first and second deviator stress tensor in a time step, respectively, and $e_{ij}{}^{K,O}$ is the second deviator strain tensor in a time step [44].

In Equations (4) and (5), a, b, X, and Y have the following relationships:

$$\begin{aligned} a &= 1+\frac{G_1\Delta t}{2\eta_1} \\ b &= 1-\frac{G_1\Delta t}{2\eta_1} \\ X &= \frac{1}{2G_2}+\frac{\Delta t}{4\eta_2}+\frac{\Delta t}{4a\eta_1} \\ Y &= \frac{\Delta t}{4\eta_2}+\frac{\Delta t}{4a\eta_1}-\frac{1}{2G_2} \end{aligned} \tag{5}$$

where G_1 is the shear modulus of Kelvin body, GPa; G_2 is the shear modulus of Maxwell body, GPa.

In the aspect of secondary development of creep model, the total stress should be decomposed into spherical stress tensor σ_m and deviator stress tensor S_{ij}:

$$\sigma_{ij} = \sigma_m\delta_{ij} + S_{ij} \tag{6}$$

The increment relation of the spherical stress tensor is:

$$\sigma_m{}^N = \sigma_m{}^O + K\Delta\varepsilon_{kk} \tag{7}$$

where, $\sigma_m{}^N$, $\sigma_m{}^O$ are the spherical stress tensor in a time step, K is the bulk modulus, $\sigma_m{}^O = (\sigma_{11}+\sigma_{22}+\sigma_{33})/3$, $\Delta\varepsilon_{kk} = \Delta\varepsilon_{11}+\Delta\varepsilon_{22}+\Delta\varepsilon_{33}$

Substituting Equations (4) and (7) into Equation (6), the total stress tensor of the model was

$$\sigma_{ij}{}^N = \sigma_m{}^N\delta_{ij} + S_{ij}{}^N \tag{8}$$

The numerical calculation of the model was realized based on the visual studio 2010 software, and the core equation of the model was realized by the programming language. The key work included the definition of model variables, the modification of the functions such as Properties(), Initialize(), Run(), etc. The code editing window is shown in Figure 4.

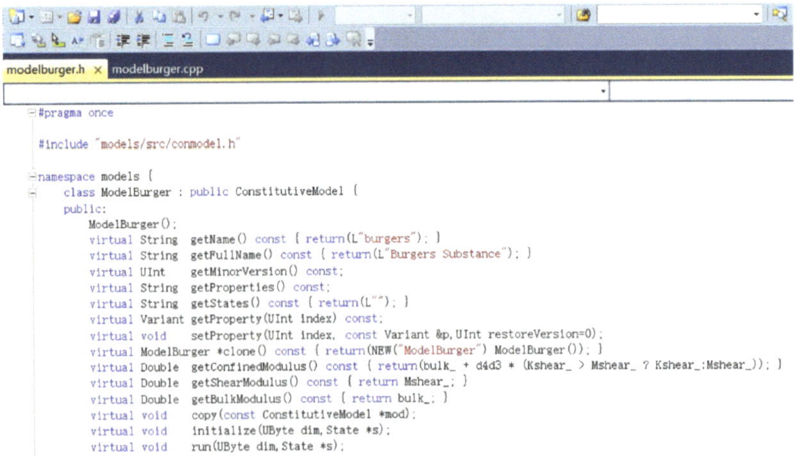

Figure 4. Code editing window.

3. Numerical Simulation Research

3.1. Simulation Scheme

In this study, the numerical simulation model established in this paper was cylindrical to facilitate the simulation calculation. The upper part of the model is coal and the lower part is siltstone. The size of the numerical model was Φ 50 mm $\times$ L 100 mm, and the displacement constraint is applied at the bottom. The confining pressure was set to 3 MPa, and the axial stress was loaded in order of 5, 8, 11, 14 MPa. Based on the above experimental results, we set up four groups of moisture contents to study the creep characteristics of coal-rock combination, and the creep time was 12 h. The specific simulation scheme is shown in Table 2. Three monitoring points were set at the top of the model to record the creep value of the model. The creep constitutive model of coal-rock combination considering the influence of moisture content established in this paper was used for simulation. The model parameters were obtained from the conversion Equation (2) of moisture content in a C++ environment. The physical and mechanical parameters of coal and siltstone were shown in Table 3, and the numerical model was shown in Figure 5.

Table 2. Simulation scheme.

Moisture Content ω (%)	Confining Pressure (MPa)	Axial Stress (MPa)	Creep Time (h)
0	3	5	12
0.5	3	8	12
1.0	3	11	12
1.5	3	14	12

Table 3. Physical and mechanical parameters.

Lithology	Density (kg/m^3)	Cohesion (MPa)	Friction Angle (°)	Poisson's Ratio	Tensile Strength (MPa)
Coal	1628	3.05×10^6	32	0.32	1.05×10^6
Siltstone	2637	6.51×10^6	38	0.2	5.29×10^6

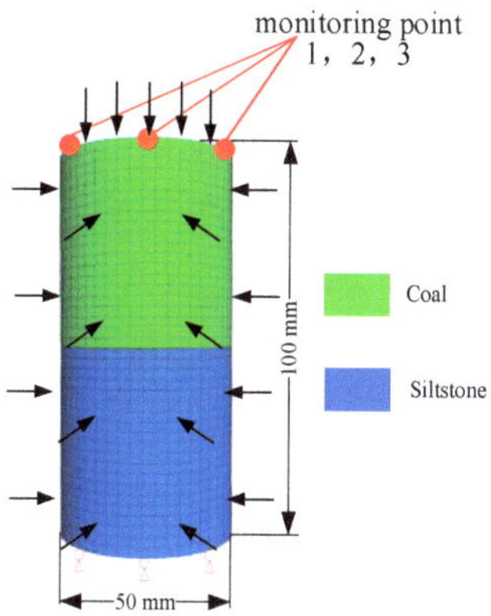

Figure 5. Numerical calculation model.

3.2. Simulation Result Analysis

3.2.1. Deformation Characteristics

According to the data collected from the simulation records, the strain–time curves under different axial stress were obtained by using Origin function drawing software, as shown in Figure 6. It could be seen from the figure that the strain-time curve of coal-rock combination under triaxial stress accorded with the first and second stages of rock typical

creep curve, which is the initial creep and constant creep stage. It failed to reflect its plastic characteristics in the simulation due to the Burgers model belonging to the viscoelastic model, so the accelerated creep stage could not be reflected. With the application of loading, the coal-rock combination first underwent instantaneous elastic deformation in the initial creep stage, which was caused by the closure of microcracks in the rock. At the same time, it could be seen from Figure 6 that the mechanical properties of the coal-rock combination changed due to the influence of the moisture content (ω) and stress under the same confining pressure. After entering the creeping stage, the creep deformation was more obvious, which was reflected in the limit creep deformation of the combination increases with the increase of moisture content and the stress. For example, when the axial stress was 5 MPa, the moisture content increased from 0% to 1.5%, and the limit creep deformation increased by about 20%. When the moisture content is 0%, the axial stress increased from 5 to 14 MPa, and the limit creep deformation increased nearly four times. The limit creep deformation was controlled by the limit creep deformation modulus, so the limit creep deformation modulus decreased with the increase of moisture content, which was also in line with the characteristics of the influence of water on the timeliness of rock. According to the strain–time curve under the influence of different stress states and different moisture content, the magnitude of the strain was positively correlated with the moisture content and axial stress.

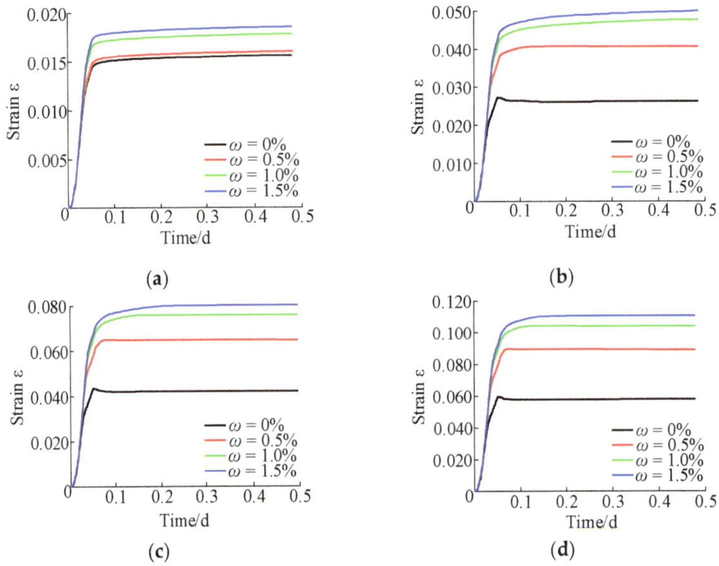

Figure 6. Strain-time curves of coal-rock combination under different moisture content: (**a**) Axial stress 5 MPa; (**b**) Axial stress 8 MPa; (**c**) Axial stress 11 MPa; (**d**) Axial stress 14 MPa.

Instantaneous elastic deformation is also a very important stage in the creep process of coal-rock combination. Take the axial stress of 5 MPa as an example, as shown in Figure 7. It could be seen from the figure that the greater the moisture content, the greater the instantaneous elastic deformation under the same stress state. When the moisture content increased from 0% to 1.5%, the instantaneous elastic deformation increased by approximately 10.53%. The magnitude of the instantaneous elastic deformation was controlled by the instantaneous elastic deformation modulus. Therefore, it could be known from Hooke's law that the instantaneous elastic deformation modulus decreased as the moisture content increased.

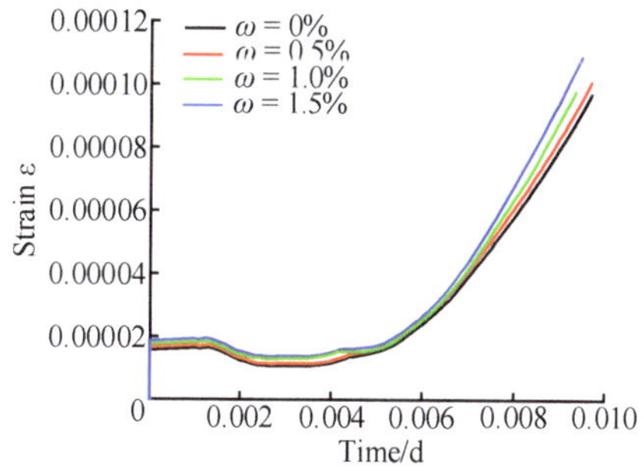

Figure 7. Instantaneous elastic deformation stage when the axial stress is 5 MPa.

3.2.2. Strain Rate Characteristics

The important indicator to show the creep speed of the coal-rock combination was strain rate. The different moisture contents had an important influence on the strain rate. The greater the moisture content of the coal-rock combination was, the greater the strain rate was. It would eventually become stable, which is entering the constant creep stage. At this time the strain rate was approximately a constant value. Strictly speaking, the constant creep stage did not exist, but the stage was divided according to the experimental phenomenon [52]. Then, the strain rate–moisture content curves under different axial stress were as shown in Figure 8. It could be seen from the figure that under the same stress state, the strain rate increased with the increase of moisture content. For example, when the axial stress was 5 MPa, the moisture content increased from 0% to 1.5%, and the strain rate increased by 44.06%. At the same time, when the moisture content was the same, the strain rate increased with the increase of axial stress. For example, when the moisture content was 0.5%, the axial stress increased from 5 to 14 MPa, and the strain rate increased by 40.38%.

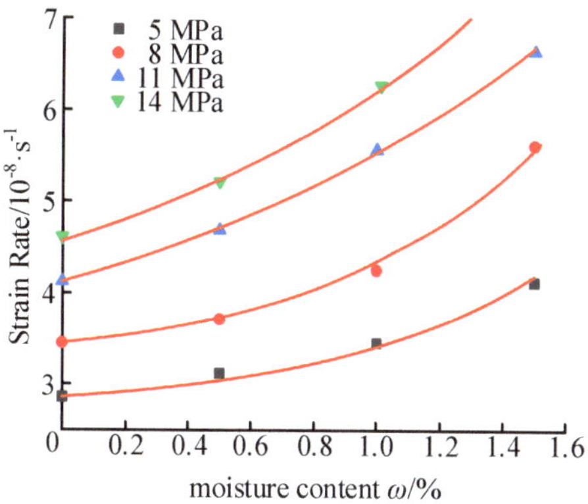

Figure 8. Strain rate–moisture content curve of coal-rock combination.

3.2.3. Viscosity Coefficient Characteristics

A characteristic index reflecting rock creep and flow was the viscosity coefficient, which reflected the rheological properties of materials, and could be expressed by the slope of the straight line segment of the stress–strain rate curve. As can be seen from Figure 9, when the moisture content was 0%, 0.5%, 1.0%, and 1.5%, the corresponding viscosity coefficients were 4.76×10^{14}, 3.85×10^{14}, 3.61×10^{14}, and 2.38×10^{14} Pa·s, respectively. It could be seen that with the increase of moisture content, the viscosity coefficient of the coal-rock combination gradually decreased. The reason may be that water changes the internal structure and properties of the combination, resulting in a decrease in the

viscosity coefficient, so the strain rate increased. This further explained the rise of the above strain rate.

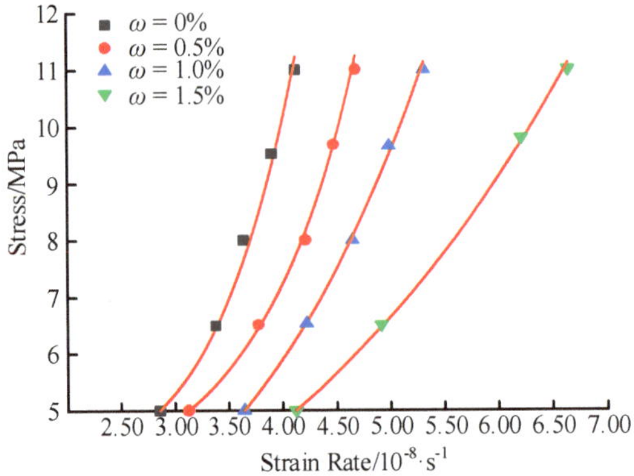

Figure 9. Stress–strain rate curve of coal and rock mass.

4. Discussion

Creep instability failure is a common failure form in underground engineering, which is closely related to stress and the geological environment, so the groundwater will inevitably affect the creep performance of coal and rock. In order to reveal the influence of water on the creep characteristics of underground engineering, scholars and engineers have done a lot of research work on pure coal or pure rock samples and obtained the influence of water on the creep characteristics of the samples. However, in underground engineering, coal and rock do not exist in the form of a monomer, but in a multi-layer structure in which the coal and rock coexist, that is, in the form of combination. Therefore, it is of practical engineering significance to study the creep characteristics of the coal-rock combination after water immersion deterioration.

Compared with previous studies, this study improved the traditional Burgers model, established the relationship between model parameters and moisture content, and then put forward a creep constitutive model of coal-rock combination considering the influence of moisture content. The model can well describe the creep characteristics of combination samples after immersion, and the advantages of high fitting accuracy and simple parameter acquisition are obvious. At the same time, the numerical calculation of the user-defined constitutive model is realized by using the secondary development of FLAC3D simulation software.

In the existing studies, Okubo et al. [16] showed that moisture content and time are the main reasons for the change of rock mechanical properties; in this study, through the establishment of creep constitutive model of coal-rock combination considering the influence of moisture content, and by the parameter inversion and fitting verification, the correlation between moisture content and various mechanical parameters in the model was finally obtained. The deformation characteristics, strain rate characteristics, and viscosity coefficient characteristics of coal-rock combination are analyzed qualitatively and quantitatively through numerical simulation. Vásárhelyi [14], through the comparative test of limestone in the dry and saturated state, shows that: the physical and mechanical parameters of rock in the saturated state are about 66% of that in the dry state. In this study, the numerical simulation showed that when the moisture content increased from 0% to 1.5%, the strain rate increased by 44.06%, the limit creep deformation increased by 20%, and the instantaneous elastic deformation increased by about 10.53%, and the viscosity coefficient decreased about 50%. The difference between the two test results is due to the differences in the test pieces used, the former uses limestone, and the latter uses coal-rock combinations. Another reason may be the difference in loading method, the former uses the

uniaxial loading, and this paper is the triaxial loading. Compared with previous studies, the creep characteristics of coal-rock combination under different water content can provide a more accurate reference for the design and construction of underground engineering under water immersion conditions.

As for the creep mechanical properties of coal-rock combination, this paper only analyzed the initial and constant creep stages due to the limitations of Burgers model itself, and the accelerated creep stage of coal-rock combination is not involved. In the future, it will be necessary to carry out in-depth research on triaxial compression creep experiments and nonlinear creep constitutive models of coal-rock combination.

5. Conclusions

On the basis of the present analysis, the following conclusions are drawn:

The Burgers model was modified by creep experimental data, and the creep constitutive model of coal-rock combination considering the influence of moisture content was established, and the parameters of the model were obtained. The experimental data are inversed by using the model parameters, and the inversed experimental data are compared with the original experimental data. The correlation coefficient R^2 is 0.998, which showed that the model had the advantages of high fitting precision and simple parameter acquisition. Then, in order to use the model to study the creep characteristics of coal-rock combination under different moisture content, the numerical calculation of the model is realized by the VS platform;

Creep deformation characteristics of coal-rock combination are closely related to moisture content and stress environment. Under the same stress state, with the increase of moisture content, the limit creep deformation modulus and instantaneous elastic deformation modulus of coal-rock combination decreased, and the corresponding deformation also increased significantly. When the axial stress is 5 MPa, the moisture content increased from 0% to 1.5%, the limit creep deformation increased by about 20%, and the instantaneous elastic deformation increased by about 10.53%. Under the same moisture content, with the increase of axial stress, the limit creep deformation modulus decreased, and the corresponding deformation modulus increased significantly. When the moisture content is 0%, the axial stress increased from 5 to 14 MPa, the limit creep deformation increased nearly four times.

The characteristics of strain rate and viscosity coefficient of coal-rock combination are closely related to moisture content and stress environment. Under the same stress state, the strain rate of the coal-rock combination increased nonlinearly and rapidly with the increase of moisture content, while the viscosity coefficient decreased significantly. When the axial stress was 5 MPa and the moisture content increased from 0% to 1.5%, the strain rate increased by about 44.06% and the viscosity coefficient decreased by about 50%. When the moisture content was the same, the axial stress increased from 5 to 14 MPa, the strain rate increased by 67.74%.

Author Contributions: Conceptualization, X.L. (Xuebin Li); Investigation, B.W. and H.W.; Methodology, X.L. (Xuebin Li); Software, Q.M.; Supervision, X.L. (Xuesheng Liu) and Y.T.; Writing–original draft, X.L. (Xuebin Li); Writing–review & editing, X.L. (Xuesheng Liu) and Y.T. All authors have read and agreed to the published version of the manuscript.

Funding: The research described in this paper was financially supported by the National Natural Science Foundation of China (No. 52174122, No. 52074168, No. 51874190), Open Fund of State Key Laboratory of Water Resource Protection and Utilization in Coal Mining (Grant No. GJNY-18-73.5), The Climbing Project of Taishan Scholar in Shandong Province (No. tspd20210313).

Data Availability Statement: The data used to support the findings of this study are available from the first author upon request.

Acknowledgments: The author would like to thank reviewers and editors for their kind work.

Conflicts of Interest: The authors declare no conflict of interest.

References

1. He, M.C.; Xie, H.P.; Peng, S.P.; Jiang, Y.D. Study on rock mechanics in deep mining engineering. *Chin. J. Rock. Mech. Eng.* **2005**, *24*, 2803–2813.
2. Cao, P.; Wan, L.H.; Wang, Y.X.; Huang, Y.H.; Zhang, X.Y. Viscoelasto-plastic properties of deep hard rocks underwater environment. *T. Nonferr. Metal. Soc.* **2011**, *21*, 2711–2718. [CrossRef]
3. Liu, J.S.; Jing, H.W.; Meng, B.; Wang, L.G.; Zhang, X.D.; Yang, J.J. Research on the effect of moisture content on the creep behavior of weakly cemented soft rock and its fractional-order model. *Rock. Soil Mech.* **2020**, *41*, 2609–2618.
4. Fan, D.Y.; Liu, X.S.; Tan, Y.L.; Song, S.L.; Ning, J.G.; Ma, Q. Numerical simulation research on response characteristics of surrounding rock for deep super-large section chamber under dynamic and static combined loading condition. *J. Cent. South Univ.* **2020**, *27*, 3544–3566. [CrossRef]
5. Sarfarazi, V.; Haeri, H.; Shemirani, A.B.; Hedayat, A.; Hosseini, S.S. Investigation of ratio of TBM disc spacing to penetration depth in rocks with different tensile strengths using PFC2D. *Comput. Concr.* **2017**, *20*, 429–437.
6. Song, S.L.; Liu, X.S.; Tan, Y.L.; Fan, D.Y.; Ma, Q.; Wang, H.L. Study on failure modes and energy evolution of coal-rock combination under cyclic loading. *Shock. Vib.* **2020**, *2020*, 1–16. [CrossRef]
7. Liu, X.S.; Tan, Y.L.; Ning, J.G.; Lu, Y.W.; Gu, Q.H. Mechanical properties and damage constitutive model of coal in coal-rock combined body. *Int. J. Rock. Mech. Min.* **2018**, *110*, 140–150. [CrossRef]
8. Li, D.Y.; Du, S.H.; Zhang, C.S.; Mao, D.W.; Ruan, B. Time-dependent deformation behavior of completely weathered granite subjected to wetting immersion. *Rock. Mech. Rock. Eng.* **2021**, *54*, 6373–6391. [CrossRef]
9. Li, D.Y.; Wang, W.J. Quantitative analysis of the influence of saturation on rock strength reduction considering the distribution of water. *Geomech. Geophys. Geo-Energ. Geo-Resour.* **2019**, *5*, 197–207. [CrossRef]
10. Yao, Q.L.; Zhu, L.; Huang, Q.H.; Yang, P.; Xu, Z. Experimental study on the effect of moisture content on creep characteristics of fine-grained feldspar lithic sandstone. *J. Min. Saf. Eng.* **2019**, *36*, 1034–1042+1051.
11. Hashiba, K.; Fukui, K. Effect of water on the deformation and failure of rock in uniaxial tension. *Rock. Mech. Rock. Eng.* **2015**, *48*, 1751–1761. [CrossRef]
12. Bukowska, M. Influence of height-to-diameter ratio samples and humidity on the geomechanical properties of sandstones. In Proceedings of the 15th International Multidisciplinary Scientific GeoConference 2015 (SGEM 2015), Albena, Bulgaria, 18–24 June 2015; Volume 3, pp. 391–398.
13. Li, Y.; Zhu, W.S.; Bai, S.W.; Yang, C.H. Uniaxial experimental study on rheological properties of granite in air-dried and saturated states. *Chin. J. Rock. Mech. Eng.* **2003**, *22*, 1673–1677.
14. Shtumpf, G.G. Water-physical properties of coal-bearing rocks. *J. Min. Sci.* **1993**, *29*, 530–536. [CrossRef]
15. Ojo, O.; Brook, N. The effect of moisture on some mechanical properties of rock. *Int. J. Min. Sci. Technol.* **1990**, *10*, 145–156. [CrossRef]
16. Vásárhelyi, B. Statistical analysis of the influence of moisture content on the strength of the miocene limestone. *Rock. Mech. Rock. Eng.* **2005**, *38*, 69–76. [CrossRef]
17. Vásárhelyi, B.; Ván, P. Influence of water content on the strength of rock. *Eng. Geol.* **2006**, *84*, 70–74. [CrossRef]
18. Okubo, S.; Nishimatsu, Y.; He, C.; Chun, S.Y. Loading rate dependency of uniaxial compressive strength of rock under water-saturated condition. *J. Soc. Mater. Sci. Jpn.* **1992**, *41*, 403–409. [CrossRef]
19. Varun, M.; Eunhye, K. Effects of water on rock fracture properties: Studies of mode I fracture toughness, crack propagation velocity, and consumed energy in calcite-cemented sandstone. *Geomech. Eng.* **2019**, *17*, 57–67.
20. Eunhye, K.; Adriana, G.; Hossein, C. Fragmentation and energy absorption characteristics of Red, Berea and Buff sandstones based on different loading rates and moisture contents. *Geomech. Eng.* **2018**, *14*, 151–159.
21. Edward, M.; Van, E. Mechanisms of strength reduction due to moisture in coal mine shales. *Int. J. Rock. Mech. Min.* **1976**, *13*, 61–67.
22. Liu, X.S.; Tan, Y.L.; Ning, J.G.; Tian, C.L.; Tian, Z.W. Energy criterion of abutment pressure induced strain-mode rockburst. *Rock. Soil Mech.* **2016**, *37*, 2929–2936.
23. Erguler, Z.A.; Ulusay, R. Water-induced variations in mechanical properties of clay-bearing rocks. *Int. J. Rock. Mech. Min.* **2009**, *46*, 355–370. [CrossRef]
24. Hawkins, A.B.; Mcconnell, B.J. Sensitivity of sandstone strength and deformability to changes in moisture content. *Q. J. Eng. Geol. Hydrog.* **1992**, *25*, 115–130. [CrossRef]
25. Gu, H.L.; Tao, M.; Wang, J.X.; Jiang, H.B.; Li, Q.Y.; Wang, W. Influence of water content on dynamic mechanical properties of coal. *Geomech. Eng.* **2018**, *16*, 85–95.
26. Haeri, H.; Shahriar, K.; Marji, M.F.; Vand, P.M. Simulating the bluntness of TBM disc cutters in rocks using displacement discontinuity method. In Proceedings of the 13th International Conference on Fracture, Beijing, China, 16–21 June 2013; Volume 2, pp. 1414–1423.
27. Kirby, S.H. Introduction and digest to the special issue on chemical effects of water on the deformation and strengths of rocks. *J. Geophys. Res.-Sol. Earth* **1984**, *89*, 3991–3995. [CrossRef]
28. Afrouz, A.; Harvey, J.M. Rheology of rocks within the soft to medium strength range. *Int. J. Rock. Mech. Min. Geo. Abstr.* **1974**, *11*, 281–290. [CrossRef]

29. Poulsen, B.A.; Shen, B.; Williams, D.J.; Huddlestone-Holmes, C.; Erarslan, N.; Qin, J. Strength reduction on saturation of coal and coal measures rocks with implications for coal pillar strength. *Int. J. Rock. Mech. Min.* **2014**, *71*, 41 52. [CrossRef]
30. Wang, L.; Zhu, C.Q.; Yin, Z.Q.; Hou, J.L. Research on soft coal mechanics characteristic test for moisture content effect. *J. Min. Saf. Eng.* **2016**, *33*, 1145–1151.
31. Talapatra, A.; Karim, M.M. The influence of moisture content on coal deformation and coal permeability during coalbed methane (CBM) production in wet reservoirs. *J. Petrol. Explor. Prod. Technol.* **2020**, *10*, 1907–1920. [CrossRef]
32. Zhang, T.J.; Chen, Z.Y.; Pan, H.Y. Experimental study on the effect of water on the creep acoustic emission characteristics of coal containing pores. *Coal. Sci. Technol.* **2020**, 1–9. Available online: https://kns.cnki.net/kcms/detail/11.2402.TD.20200207.1449.019.html (accessed on 20 October 2021).
33. Liu, X.S.; Fan, D.Y.; Tan, Y.L.; Ning, J.G.; Song, S.L.; Wang, H.L.; Li, X.B. New detecting method on the connecting fractured zone above the coal face and a case study. *Rock. Mech. Rock. Eng.* **2021**, *54*, 4379–4391. [CrossRef]
34. Liu, X.S.; Song, S.L.; Tan, Y.L.; Fan, D.Y.; Ning, J.G.; Li, X.B.; Yin, Y.C. Similar simulation study on the deformation and failure of surrounding rock of a large section chamber group under dynamic loading. *Int. J. Min. Sci. Technol.* **2021**, *31*, 495–505. [CrossRef]
35. Tang, Z.H.; Wang, H.J.; Deng, H.; Li, W.C.; He, Y.F.; Li, P.W.; Yang, F. Creep model considering water content and its application in tunnel deformation. *Pearl. River.* **2020**, *41*, 73–80.
36. Yang, W.D.; Zhang, Q.Y.; Zhang, J.G.; He, R.P.; Zeng, J.Q. Second development of improved Burgers creep damage constitutive model of rock based on FLAC3D. *Rock. Soil Mech.* **2010**, *31*, 1956–1964.
37. Xue, D.J.; Lu, L.L.; Yi, H.Y.; Wu, Z.D.; Zhang, Q.S.; Zhang, Z.P. A fractional Burgers model for uniaxial and triaxial creep of damaged salt-rock considering temperature and volume-stress. *Chin. J. Rock. Mech. Eng.* **2021**, *40*, 315–329.
38. Zhao, L.C. Secondary development and engineering application of creep damage model of water bearing sandstone. *J. Yangtze River Sci. Res. Inst.* **2021**, *38*, 105–112, 120.
39. Sun, X.M.; Miao, C.Y.; Jiang, M.; Zhang, Y.; Yang, L.; Guo, B. Experimental and theoretical study on creep behaviors of sandstone with different moisture contents based on modified Nishihara model. *Chin. J. Rock. Mech. Eng.* **2021**, *40*, 2411–2420.
40. Gao, W. Influence of interaction between coal and rock on the stability of strip coal pillar. *Geomech. Eng.* **2018**, *16*, 151–157.
41. Rabat, Á.; Cano, M.; Tomás, R. Effect of water saturation on strength and deformability of building calcarenite stones: Correlations with their physical properties-sciencedirect. *Constr. Build. Mater.* **2020**, *232*, 117–259. [CrossRef]
42. Liu, X.S.; Fan, D.Y.; Tan, Y.L.; Song, S.L.; Li, X.F.; Ning, J.G.; Gu, Q.H.; Ma, Q. Failure evolution and instability mechanism of surrounding rock for close-distance chambers with super-large section in deep coal mines. *Int. J. Geomech.* **2021**, *21*, 04021049. [CrossRef]
43. Li, S.J. Secondary Development and Application of Rock Water-Damage Creep Model Based on FLAC3D. Master's Thesis, Hefei University of Technology, Hefei, China, March 2018.
44. Chen, L.W.; Li, S.J.; Chen, Y.F.; Zhang, K.X.; Liu, Y.X. Further development and application of a creep damage model for water-bearing rocks. *Chin. J. Solid. Mech.* **2018**, *39*, 642–651.
45. Kong, J.; Yuan, J.; Pan, X. Secondary development of flac3d and application of naylor K-G constitutive model. *Comput. Struct. Eng.* **2009**, 1267–1273. [CrossRef]
46. He, J.P.; Chen, W.Z. A simplified concrete damage constitutive model. *Adv. Mater. Res.* **2013**, *671–674*, 1663–1671. [CrossRef]
47. Shi, Y.L.; Li, H.W.; Wang, W.D.; Hou, C. A fiber model based on secondary development of abaqus for elastic–plastic analysis. *Int. J. Steel. Struct.* **2018**, *18*, 1560–1576. [CrossRef]
48. Haeri, H.; Hedayat, A.; Sarfarazi, V.; Shemirani, A.B. A review paper about experimental investigations on failure behaviour of non-persistent joint. *Geomech. Eng.* **2017**, *13*, 535–570.
49. Dai, Z.P.; Zhao, C.; Zhao, C.F. The development and implementation of improved duncan-chang constitutive model in ABAQUS. *Appl. Mech. Mater.* **2011**, *99–100*, 965–971. [CrossRef]
50. Haeri, H.; Sarfarazi, V.; Zhu, Z. Effect of normal load on the crack propagation from pre-existing joints using particle flow code (PFC). *Comput. Concr.* **2017**, *19*, 99–110. [CrossRef]
51. Wang, J.G.; Liang, B.; Tian, M. Study of creep characteristics produced by nonlinear damage of oil shale in hydrous state. *J. Exper. Mech.* **2014**, *29*, 112–118.
52. Yuan, Q.; Tan, C.; Li, L.L. Experimental investigation creep anisotropy characteristics of phyllite on uniaxial compression. *Water. Resour. Power.* **2019**, *37*, 148–151+78.

minerals

MDPI

Article

Experimental Study on the Characteristics and Formation Mechanism of Dynamic Filter Cake for Slurry Shield Tunneling

Yang Bai [1,2], Binsong Jiang [1,*], Le Yang [1], Yunqiu Liu [1], He Zheng [1] and Yapeng Li [1]

1 State Key Laboratory for Geomechanics and Deep Underground Engineering, China University of Mining and Technology, Xuzhou 221116, China; 10690@jsjzi.edu.cn (Y.B.); yangle55yonghuhao@163.com (L.Y.); liuyunqiu@cumt.edu.cn (Y.L.); zhenghehhu@163.com (H.Z.); liyapg@163.com (Y.L.)
2 College of Transportation Engineering, Jiangsu Vocational Institute of Architectural Technology, Xuzhou 221116, China
* Correspondence: db17220010b0@cumt.edu.cn

Abstract: The key to guaranteeing excavation face stability in slurry shield tunneling is the formation of an impermeable dynamic filter cake. At the same time, the effect of the cutter head and rotation speed should be taken into account. We studied the characteristics and formation mechanism of the dynamic filter cake using a newly developed experimental apparatus. The experiment results show that the hysteretic infiltration zone appeared in the curves of stepped loading filtration while the cutter head was rotating, and the volume of water filtration increased by 11.2% compared to when the cutter head stopped. The higher the rotation speed was, the lower the conversion rate of the effective stress was. Under the same rotation speed, the formation time of the 6-cutter arm was almost 5 s slower than that of the 5-cutter arm. As the cutter arms and the rotation speed increased, the stratum's electrical conductivity increased and stabilized at a distance of 20 cm from the cutter head. The filter cake transited from 'filter cake plus an infiltration zone' to 'an infiltration zone without a filter cake' with the increase of the rotation speed. The thickness of the dynamic filter cake was smaller than that of the static filter cake, the thickness of 10 groups decreased significantly, and the average thickness decreased by 76.15% at 1.0 rpm. The mesoscopic formation process of the dynamic filter cake can be divided into six stages. This study revealed the slurry penetration mechanism and filter cake characteristics present under cyclic damage by the shield cutter head to the filter cake and soil and provided theoretical support on how to maintain the stability of the excavation face during slurry shield tunneling.

Keywords: filter cake; shield tunneling; excavation face stability; cutter head; slurry permeation

Citation: Bai, Y.; Jiang, B.; Yang, L.; Liu, Y.; Zheng, H.; Li, Y. Experimental Study on the Characteristics and Formation Mechanism of Dynamic Filter Cake for Slurry Shield Tunneling. *Minerals* **2022**, *12*, 331. https://doi.org/10.3390/min12030331

Academic Editors: Diyuan Li, Zhenyu Han, Xin Cai, Shijie Xie and Michael Hitch

Received: 25 January 2022
Accepted: 3 March 2022
Published: 7 March 2022

1. Introduction

In the past few years, due to their extensive range of geological applications, low disturbance to surrounding rock, and high excavation face stability, slurry pressure balance shield (SPB) technology has been widely used in shield tunnels [1–4], such as the fourth tunnel of the Elbe River in Germany, the Westerschelde tunnel in The Netherlands, the Eurasia tunnel in Turkey, the Tokyo Bay submerged highway tunnel in Japan, and the Nanjing Yangtze River and Hangzhou Qianjiang tunnels in China [5–8]. However, when slurry shield tunnels are constructed in a stratum in which the water pressure is more than 0.5 Mpa and the permeability coefficient is more than 10^{-3} cm/s, the excavation face will become unstable, causing seawater backflow and slurry spillover if special measures are not taken. This can lead to major accidents, such as tunnel collapse [9,10].

During slurry shield tunneling, the balance between the stratum's initial stress field and the underwater seepage field is broken, so the tunnel's excavation surface is prone to large deformation or instability failure [11–15]. Therefore, it is especially crucial to guarantee the stability of the slurry shield's excavation face. Based on Horn's theory [16],

Anagnostou et al. [17] proposed a wedge failure model for slurry shield tunnels and studied the complicated relationships among the slurry shear strength, permeability, suspension parameters, slurry pressure, tunnel alignment, and safety factors. Regarding slurry permeation, Broere et al. [18] established a mechanical model for excavation face stabilization. He also compared the calculated pore-pressure distribution with the actual results of the slurry shield in the sand stratum, indicating that the excess pore pressure had an important impact on the tunnel face stability. Zizka et al. [19] applied a method for predicting an increase in pore water pressure in front of the tunnel face and concluded that the permeation of slurry is related to time and the pore water pressure transfer follows Darcy's law.

Zumsteg et al. [20] found that increasing the slurry strength can guarantee the excavation face's stability, while the cutter head can be easily clogged. Considering the permeable filter cake, Chen et al. [21] established a two-dimensional numerical model and obtained the pore water pressure's maximum value through a transient seepage analysis. It was found that the efficiency of tunnel face stabilization deteriorated when the slurry pressure increased. Liu et al. [22] studied the failure mechanism of slurry shields through a series of model tests and the vertical displacement of the ground surface and the earth pressure distribution in strata with different covering depths. It was found that the vertical displacement was mainly distributed in a triangular region bounded by the fracture surface above the cutter head, and the horizontal displacement was mainly distributed in the excavation face. The above studies did not consider the influence of the cutter head on the excavation face, and this effect needs to be discussed.

In highly permeable stratum, the underwater environment is complex [23–25]. The timely formation of a safe and effective filter cake is the key to guaranteeing the stability of the excavation face in slurry balance shield tunnels and reducing the project risks [26–28]. Through designing the slurry filtration column test, Fritz [29] confirmed that the addition of certain proportions of bentonite, polymer, sand, and vermiculite could improve the stability of the excavation face for the high-permeability stratum (the maximum permeability value is 10^{-3} m/s). Zdenek et al. [30] investigated the pressure transfer mechanism interaction with cutting tools for the transfer of pressure in areas of deep penetration exceeding the cutting tool depth. They proposed that a linear pore pressure distribution can deliver more efficient support pressure transfer due to the lower slurry penetration depth with increased stagnation. Min et al. [10] used a self-made slurry filtration device to investigate filter cake formation under conditions with different slurry characteristics and stratum combinations and categorized the slurry particle accumulating form into three types, as shown in Figure 1: an intact filter cake, filter cake plus a permeation, and an infiltration zone without a filter cake.

Figure 1. Three types of filter cake formation. (**A**) is an intact filter cake after the slurry particle accumulating; (**B**) is a filter cake plus permeation zone after the slurry particle accumulating; (**C**) is an infiltration zone without a filter cake after the slurry particle accumulating.

Through the slurry filtration model test, Liu et al. [31] studied the influence of the pore parameters of the stratum on the characteristics of the filter cake and established a two-parameter model for formation fluid loss. They proposed that the average pore volume per particle (v) of the formation soil significantly affected the filter cake formation. Xu et al. [32] conducted an experimental study on the penetration process of bentonite slurry in saturated sand and found that the filter cake was only formed in front of the cutter head. The slurry proportion was shown to have a remarkable effect on the stability of the excavation face—the greater the bentonite concentration was, the lower the permeability of the sand and filter cake were. Lin et al. [33] considered the clogging behavior of the slurry particles and the filtering ability of the ground, and the particle retention rate was proposed to evaluate the degree of matching between the slurry and the ground. Mao et al. [34] established a constant pressure slurry permeation cake model of a slurry shield excavation face in sandy stratum and analyzed the effects of the permeation time, slurry concentration, slurry pressure, and initial stratum porosity on filter cake formation. He summarized the variation law of soil porosity at the excavation face during periodic slurry permeation. Obviously, the cutting effect of the cutter head has a negative effect on the formation of the filter cake. Few studies have investigated the formation mechanism and characteristics of the filter cake constructed under the influence of the 'machine–soil' relationship using experimental methods. Therefore, it is necessary to carry out an experiment of dynamic filter cake formation.

The filter cake formation process is complicated. The slurry proportion, slurry pressure, stratum conditions, cutter head structure, rotation speed, and cutter layout all affect the penetration process and the filter cake quality. However, previous studies all focused on the static filter cake, ignoring the effects of periodic cutting and slurry pressure fluctuation on the surrounding soil, as well as the dynamic filter cake formation process [14,17,18,35]. Therefore, a self-made test apparatus was developed to establish a dynamic infiltration model in the saturated sand stratum. The features of this apparatus are as follows: Dynamic filter cake experiments of the slurry shield can be carried out. A replaceable cutter head type experiment apparatus can be designed. As the cutter head of the shield machine used in each shield tunnel is designed uniquely, the apparatus can be used for relevant experiments for each shield machine by making and replacing the experiment cutter head. By applying TDR equipment (TDR is a time-domain reflection technique that can measure the conductivity of soil), the slurry penetration distance can be tracked more accurately through the change in soil conductivity. It is an experimental apparatus system that can be used for future shield machine selection. The apparatus can also be used for filter cake-formation experiments in the cutter head-selection stage, and the characteristics of the cutter head can be selected from the perspective of filter cake formation, including the number of cutter arms, opening rate, and other key parameters.

In this study, a new dynamic filter cake test device was designed to investigate the dynamic filter cake characteristics and formation mechanism of the slurry shield. Through experimental research, the characteristics of the dynamic filter cake (water filtration, maximum penetration distance, formation time, and final state) were studied, as shown in Figure 2. The dynamic filter cake formation process is divided into six stages. The rotating speed of the cutter head is reduced in the process of shield tunneling, and the number of cutter arms and opening rate of the cutter head, are decreased in the shield selection stage to form a high-quality filter cake.

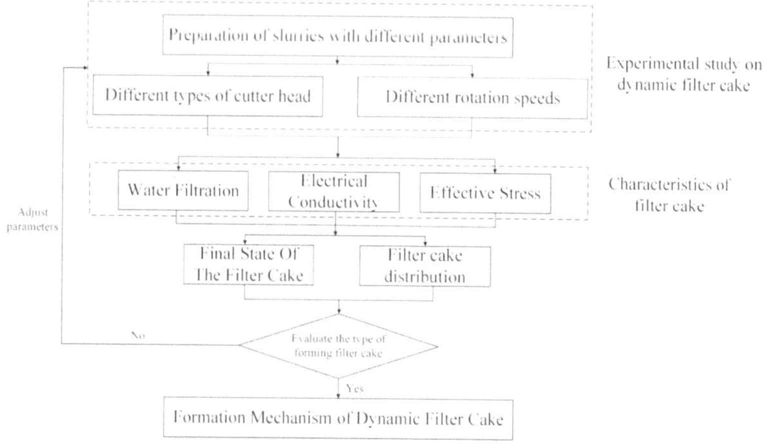

Figure 2. The experimental design.

2. Experimental Apparatus and Methods

2.1. Test Slurry and Stratum

In this test, clay from the middle and lower reaches of the Yangtze River was selected and immersed to make the base slurry. Referring to the slurry parameters commonly used in slurry shields, it was made with a bentonite to water ratio of 1:12 [36,37]. The mixture included bentonite, CMC (carboxymethyl cellulose), and Na_2CO_3 (sodium carbonate) in certain proportions and had a hydration time of about 24 h. As the main pulping material, bentonite has good water absorption, expansibility, and cohesiveness. CMC was used as the tackifier of the slurry, and Na_2CO_3 played roles in dispersion and uniformity in the slurry. To reduce the test error and ensure test repeatability and comparability among data, 10 groups of slurry (SL1–SL10) were tested, as shown in Figure 3. In physical shield tunneling, the values of slurry density and viscosity are strictly controlled within certain ranges. The density of the slurry shield should be 1.05–1.30 g/cm^3, and the viscosity should be 20–35 s. Thus, as shown in Table 1, the physical–mechanical parameters of the test slurry were selected. The properties of the slurry samples were measured in the laboratory. Specifically, a 1002 type slurry-specific gravity scale and a Russian funnel viscometer were used to measure the slurry's density and viscosity. The relative density error was within 0.01 g/cm^3, the error in the slurry viscosity was within ± 1 s, the 10 groups of slurry samples had good physical stability, and the 2 h bleeding rate was less than 2%. The grain size distribution of the 10 groups of slurry measured by the MS2000 laser particle size analyzer is shown in Figure 4.

Silty–fine and medium–coarse sands, which have difficulty forming a filter cake in the middle and lower reaches of the Yangtze River, were collected as the main standard sample and used to prepare the test sand. To control the grain size and gradation, the stratum was divided into two types of sand with different particle dimensions. As determined by the geological investigation, the basic parameters of the test soil are shown in Table 2. The dry density of the stratum and the parameters of the saturated sand sample were strictly controlled using the layered compaction method. The test stratum was divided into layers by the rectangular open funnel. The initial dry density was 1.61 g/cm^3, the effective weight was 9.98 KN/m^3, and the stratum porosity was 0.41. After a layer of the sand sample had been prepared, the water was slowly immersed until all layers were saturated, and the permeation column was sealed with a top cover. According to ASTM (American Society of Testing Materials) D2432–06 (ASTM,2006) [38], the permeability coefficient of the stratum was measured as 6.5×10^{-3} m/s through the long-head permeation test.

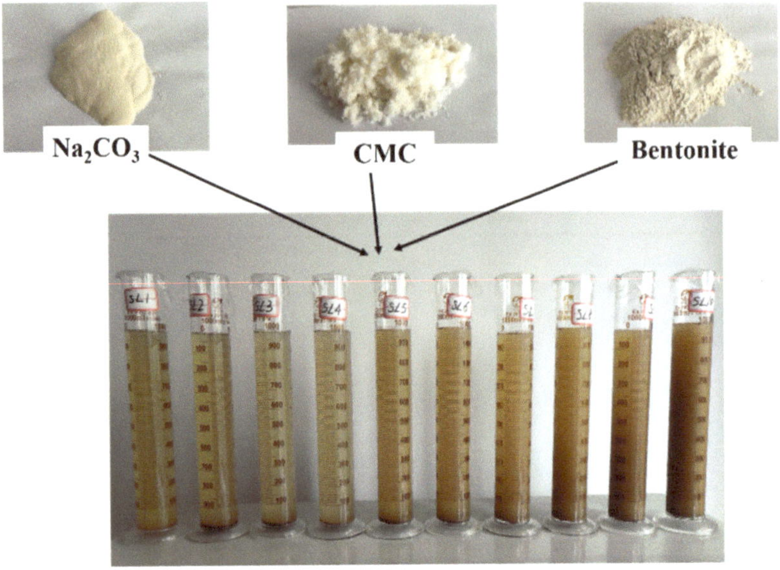

Figure 3. The ten groups of slurry samples used for testing.

Table 1. Physical parameters of the testing slurry.

Test No.	Bentonite Content	Slurry Relative Density	Slurry Viscosity	Clay Content	CMC Content	$d_{85}/\mu m$
SL1	62 g	1.05	28 s	60 g	5 g	170
SL2	62 g	1.10	30 s	80 g	6 g	168
SL3	62 g	1.15	30 s	100 g	6 g	173
SL4	83 g	1.20	32 s	120 g	8 g	212
SL5	83 g	1.25	32 s	140 g	8 g	206
SL6	62 g	1.12	22 s	50 g	5 g	140
SL7	62 g	1.15	25 s	100 g	5 g	138
SL8	83 g	1.20	30 s	120 g	5 g	136
SL9	100 g	1.25	30 s	130 g	5 g	135
SL10	100 g	1.28	32 s	150 g	5 g	128

Table 2. Basic properties of the strata.

Sand No.	Name of Sand	Density	Permeability Coefficient	Internal Friction Angle	Void Ratio
S1	Silty–fine sand	$1.45\ g/cm^3$	5.7×10^{-3}	31.8°	0.54
S2	Medium–coarse sand	$1.86\ g/cm^3$	1.08×10^{-2}	33.5°	0.62

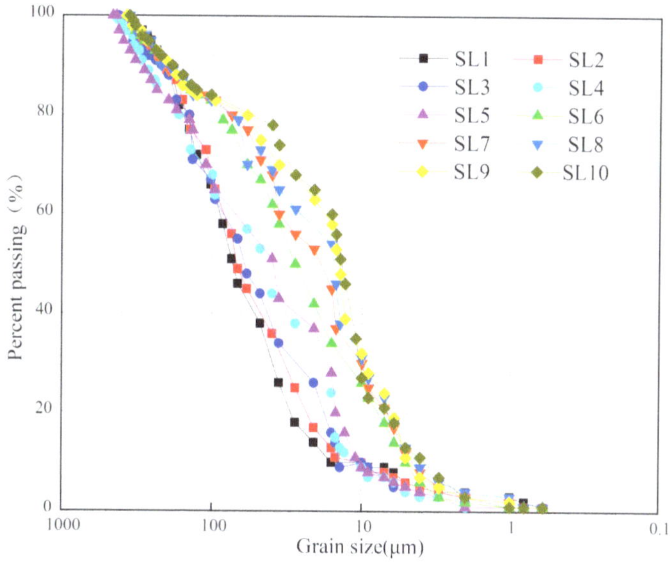

Figure 4. Grain size distribution of the slurry.

2.2. Experiment System

The previously developed device can only simulate the static filter cake [29,32,36,39]. This was based on the slurry infiltration principle of the slurry shield machine, the balance between compressed air, slurry in the slurry chamber, and the water and earth pressure, as shown in Figure 5. A dynamic filter cake test system, as shown in Figure 6, was independently developed. The slurry chamber and cutter head of the slurry shield machine are key components in the simulation of dynamic filter cake formation. This apparatus reflects the interaction between the slurry chamber, cutter head, and stratum of the physical shield machine well. The test system consists of a slurry filtration column, slurry mixing tank, air pressure source, connecting shaft, motor and cutter head, controller, and measuring devices. The top of the filtration column is fixed with a specific low-speed motor. The motor is equipped with a cutter head governor to adjust the rotation speed and direction, and the speed range is 0–9 rpm with an accuracy level of 0.1 rpm. The dimensions of the test apparatus and measurement devices are shown in Figure 7.

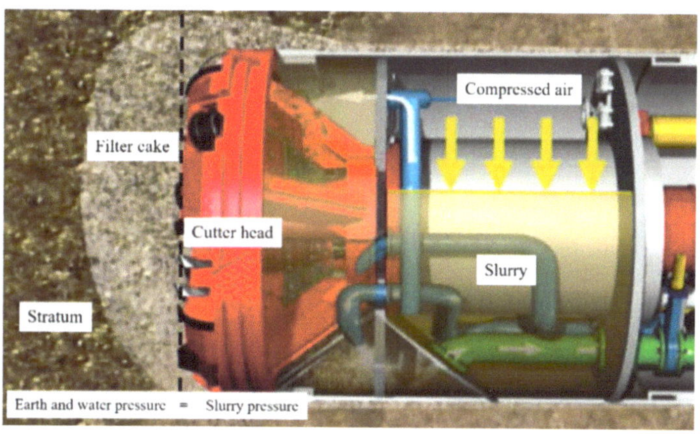

Figure 5. Principles of the slurry pressure balance shield.

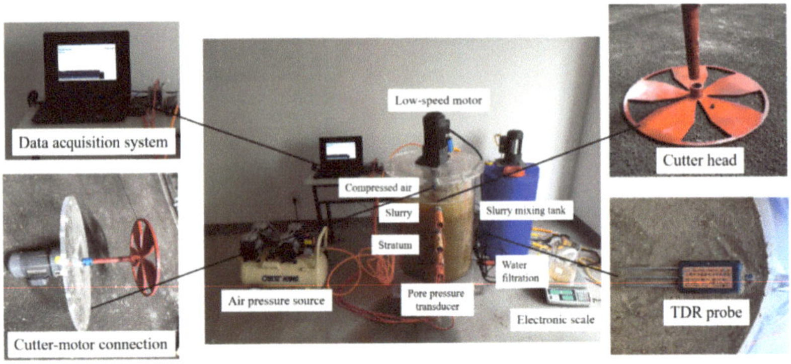

Figure 6. Slurry permeation test system for slurry shield tunneling.

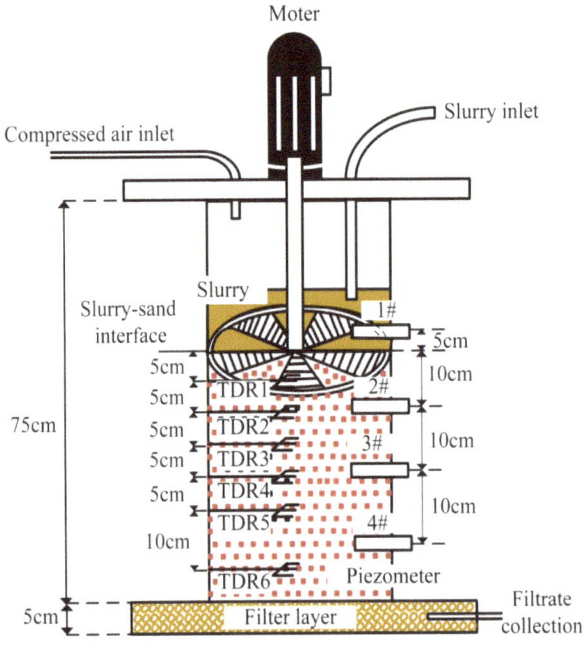

Figure 7. The dimensions of the test apparatus and measurement devices.

The filtration column has a height of 80 cm and an outer diameter of 42 cm. Time-domain reflectometries (TDR) were placed in 6 holes on the sidewall to measure the electrical conductivity after the penetration of the stratum by slurry so as to determine the penetration distance and diffusion scope of the slurry under the influence of the cutter head's rotation. The conductivity range of the TDR is 0–20,000 µs/cm, and the accuracy level is ±3%. Simultaneously, 4 vibrating wire-type pore pressure transducers were installed on the other sidewall to measure the slurry pressure and pore pressure in the filtration column. The measurable range is 0~600 Kpa, and the accuracy is 1.0 Kpa. Cutter heads with a large opening rate are widely selected during slurry shield tunneling in underwater sandy soil [36,37]. Two types of cutter head were designed in this test, as shown in Figure 8. The CH1 type is characterized by 5 cutter arms with an opening rate of 36%, and the CH2 type is characterized by 6 cutter arms with an opening rate of 28%. The outer diameter of the cutter head (400 mm) is made of alloy steel and has a thickness

of 15 mm. In addition, the upper part of the cutter is covered with a large number of grid-like alloys.

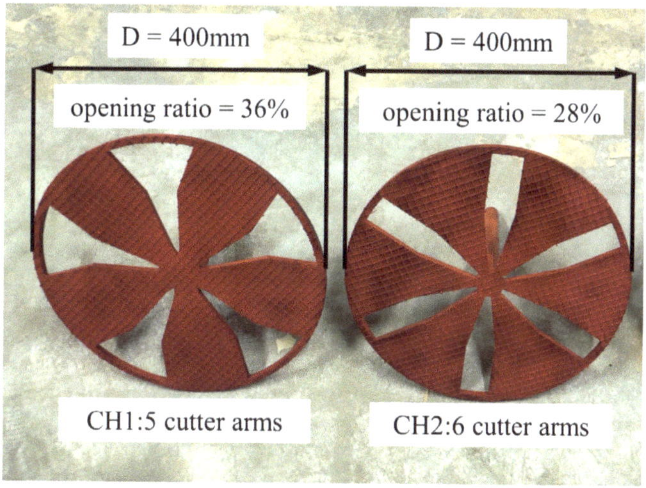

Figure 8. Two types of cutter head.

2.3. Experiment Conditions and Production Procedure

Once the test instrument had been assembled and debugged, the prepared slurry was injected into the mixing tank and continuously stirred for more than 1 h. The basic parameters of the TDR were calibrated, and ECp (the conductivity of the initial sample) was calculated. After turning the motor on, the rotation speed was adjusted to 1.0 rpm to observe the rotation of the cutter head. Simultaneously, the motor rotation was measured for more than 3 min, and the number of rotations was recorded per minute until the frequency converter's indication was consistent with the rotation number of the cutter head.

After setting the air pressure, the valve was opened to slowly infiltrate the stratum with slurry, and the starting time was recorded. Considering the unfavorable situation of slurry shield tunneling [1–4,37], the pressure was loaded to 0.5 MPa, with each step being 0.1 Mpa to facilitate the observation of the slurry infiltration process. At the same time, the water filtration volume under each level of pressure was recorded at 10-s intervals. It should be noted that the parameters used in the experiment (rotating speed, slurry pressure, and experiment duration) were chosen according to physical engineering records. With the water filtration becoming stable, the next level of air pressure was applied under the same rotation speed and cutter head direction. Statistics for the rotation speed of the cutter head [37] show that, for the medium–coarse sand stratum, the rotation speed is generally controlled within the range of 0.5 to 1.0 rpm. Therefore, three typical rotation speeds were chosen to investigate the formation of the dynamic filter cake. The typical rotation speeds were 0 rpm for the segment assembly state, 0.5 rpm, and 1.0 rpm for the slurry shield tunneling state.

3. Results and Discussion

3.1. Water Filtration

In the slurry permeation test, the water filtration volume at a certain time can reflect the quality of the filter cake: the less water filtration, the better the quality of the filter cake and the greater the stability of the excavation face [32,36]. The water filtration time curves considering the effects of the cutter head structure and rotation are shown in Figure 9. We found that the SL1–SL10 slurry sample curves had a stepped frame distribution with the two cutter head types and three different speeds; that is, the permeation speed increased rapidly and then gradually decreased within 10 s for each loading pressure when the

cutter head stopped. As shown in Figure 9A,B, as the slurry pressure increased, the slurry particles penetrated into the soil to form a filter cake, and the stress was transmitted to the soil particles through the filter cake.

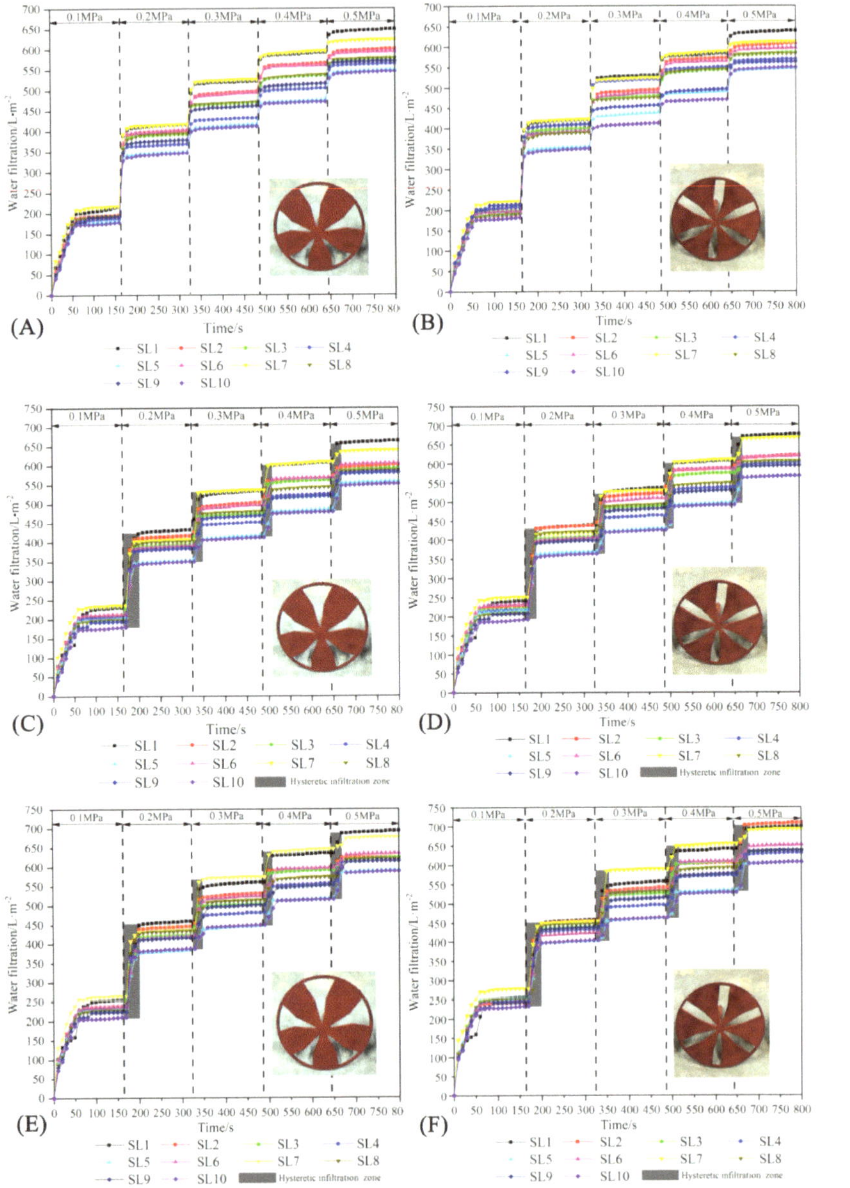

Figure 9. Water filtration time curves: (**A**) cutter head CH1, rotation speed of the cutter head = 0 rpm; (**B**) cutter head CH2, rotation speed of the cutter head = 0 rpm; (**C**) cutter head CH1, rotation speed of the cutter head = 0.5 rpm; (**D**) cutter head CH2, rotation speed of cutter head = 0.5 rpm, (**E**) cutter head CH1, rotation speed of the cutter head = 1.0 rpm, (**F**) cutter head CH2, rotation speed of the cutter head = 1.0 rpm.

The water in the slurry penetrated through the pores, and the pressure increased until reaching stability. Through the continuous accumulation of slurry particles, the water filtration volume increased and then gradually became stable. Therefore, the curve has a stepped distribution. As shown in Figure 9C–F, as the rotation speed increased, the final water filtration volume per unit of area slowly increased. For the shaded area, the slurry permeation time was longer than the working duration of the cutter head, showing a hysteretic infiltration zone and indicating that there was a delay of about 10–20 s while the cutter head stopped. The cutter head rotation decreased the water permeation speed during cutting, and there were slurry particles, and water stuck on the cutter head when the slurry passed through the opening of the rotating cutter head. As the rotation speed increased, the hysteretic infiltration zone gradually shortened, suggesting that the delayed effect of water filtration became weaker over a longer time scale.

The cutter head structure (the number of cutter arms and opening ratio) and rotation speed were found to have essential impacts on the volume of water filtration. The final water filtration volume produced with different cutter heads and rotation speeds for the SL5 slurry sample (SL5 slurry is also selected as the final state) is shown in Figure 10A. When the cutter head stopped, the differences in the water filtration volume were minute. As the cyclic damage to soil and filter cake increased with the cutter head speed, the final water filtration volume continued to increase. When the cutter head rotated, the water filtration volume of the 6-cutter arm was more than that of the 5-cutter arm. The more cutter arms there were, the greater the soil destruction was, increasing water filtration and reducing the quality of the filter cake.

Taking SL4 and SL10, for example, the viscosity of the two slurries was 32 s, and the density of SL4 was less than that of SL10. The experiment's results revealed that the greater the relative density and viscosity of the slurry were, the lower the water filtration was and the better the quality of the filter cake was, as shown in Figure 10B. When the cutter head speed was 0, 0.5, or 1.0 rpm, the infiltration volume of SL4 was greater than that of SL10. The main reason for this is that the slurry density represents the content of solid particles in the slurry. When the slurry density is high, there are many solid particles in the slurry, which makes it easier to accumulate and block the soil and form a high-quality filter cake. Therefore, the volume of water filtration in high-density soil is relatively low. Taking SL3 and SL7, for example, the density of the two slurries was found to be 1.15 g/cm^3. The viscosity of SL3 was greater than that of SL7. As is shown in Figure 10C, when the cutter head speed was 0, 0.5, or 1.0 rpm, the water filtration volume of SL3 was less than that of SL7. This can be attributed to the viscosity, which represents the physical stability of the slurry: the higher the viscosity of the slurry is, the stronger the binding forces between particles and water are. Once the particles in the slurry clog the pores during filter cake formation, it is difficult for the water in the slurry to penetrate the soil and reduce the water filtration volume.

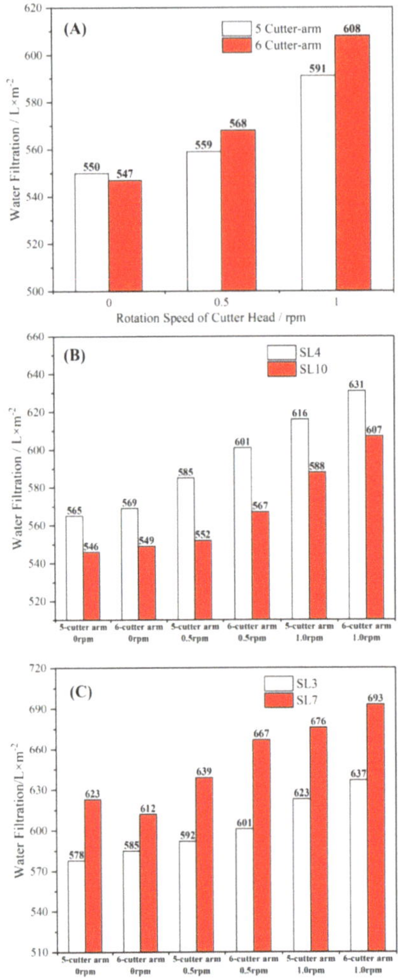

Figure 10. Influences of the cutter head and slurry parameters on water filtration. (**A**) is the final water filtration volume produced with different cutter heads and rotation speeds for the SL5 slurry sample. (**B**) is the final water filtration volume produced with different cutter heads and rotation speeds for the SL4 and SL5 slurry samples. (**C**) is the final water filtration volume produced with different cutter heads and rotation speeds for the SL3 and SL7 slurry samples.

3.2. Effective Stress

Referring to the actual slurry shield construction experience, a slurry pressure of 0.2 Mpa was selected. According to research by Wei et al. [37], the conversion rate of effective stress ω can be used to represent the formation time and the characteristics of the filter cake. The time that the filter cake takes to reach a stable state from the beginning of the experiment is defined as the formation time: the shorter the formation time, the more compact the filter cake. ω can be calculated as follows:

$$\omega = \frac{P_s - P_w - P_e}{P_s - P_w} \times 100\% \quad (1)$$

where P_s is the slurry pressure, P_w is the stratum hydrostatic pressure, and P_e is the excess pore water pressure. P_w can be calculated according to the depth that the pressure

transducers are buried. P_e represents the data measured by the pore transducer minus the corresponding P_w. As shown in Figure 6, pressure transducer 2# was the closest to the filter cake. Thus, the effective stress conversion rate in this region was selected to determine the filter cake formation time.

Under a pressure of 0.2 Mpa, the value of ω at pressure transducer 1# increased rapidly and then achieved stability when the maximum effective stress conversion rate $\omega_{\max}$ was reached. As shown in Figure 11A,B, with the same slurry parameters and rotation speed, $\omega_{\max}$ and the formation time of different cutter heads were similar when the cutter head stopped. However, as the rotation speed increased, $\omega_{\max}$ and the formation time varied for different cutter arms. As shown in Figure 11C–F, when the rotation speed was 0.5 and 1.0 rpm, the average $\omega_{\max}$ values of the 5-cutter arm were 79.1% and 77%, respectively, higher than the values of 76.7% and 73.2% achieved by the 6-cutter arm. The average formation time occurred about 5 s earlier than that of the 6-cutter arm.

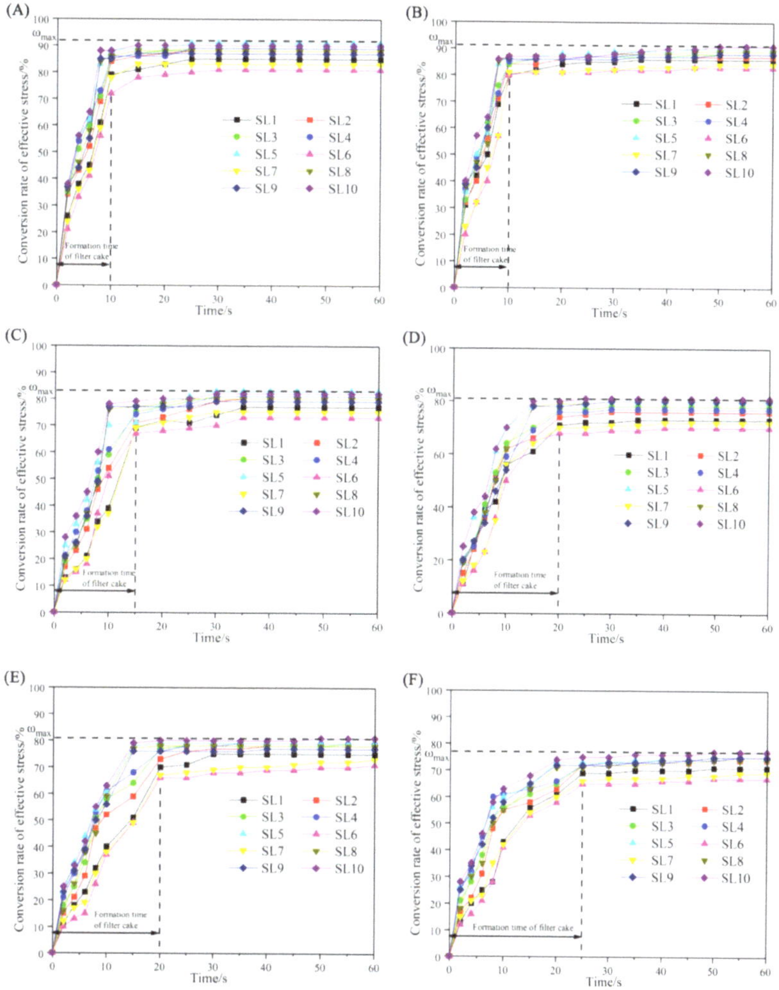

Figure 11. Variation curves of effect stress conversion: (**A**) cutter head CH1, rotation speed of cutter head = 0 rpm, (**B**) cutter head CH2, rotation speed of cutter head = 0 rpm, (**C**) cutter head CH1, rotation speed of cutter head = 0.5 rpm, (**D**) cutter head CH2, rotation speed of cutter head = 0.5 rpm, (**E**) cutter head CH1, rotation speed of cutter head = 1.0 rpm, (**F**) cutter head CH2, rotation speed of cutter head = 1.0 rpm.

For the same cutter head, the ω_{max} decreased continuously as the rotation speed increased. As shown in Figure 11, when the rotation speed was 0, 0.5, and 1.0 rpm, the average ω_{max} of the 5-cutter arm was 91%, 82%, and 80%, respectively, while the average ω_{max} of the 6-cutter arm was 90%, 81%, and 77%, respectively. The formation time increased as the rotation speed increased. For the 6-cutter arm cutter head, the filter cake formation time was about 25 s at a rotation speed of 1.0 rpm, about 5 s slower than with a rotation speed of 0.5 rpm.

3.3. Electrical Conductivity

As the main component used in the slurry shield tunneling, the pulping agent bentonite is a layered aluminosilicate clay mineral with montmorillonite. The unsaturated interlayer electrovalence makes the bentonite slurry particles negatively charged [21]. The conductivity of slurry particles infiltrating the sand with pressure was greater than that of the original stratum. Based on this principle, the slurry infiltrating range and distance were estimated to evaluate the quality of the filter cake. At the beginning of the test, the conductivity of the stratum (ECp) in the standard state was measured. To avoid the influence of the boundary effect, the six TDR probes were placed under the center of the cutter head. Setting the interface between the cutter head and the stratum as the origin and the vertical direction as the infiltration direction, the TDR1–TDR6 probes were buried at 5, 10, 15, 20, 25, and 35 cm below the slurry–sand interface, as shown in Figure 7. Based on the data acquisition system, the variation law of the stratum conductivity of SL1–SL10 slurries under different types of cutter heads and speeds was studied.

It was found that when the cutter head rotated and stopped, the variation trend for the electrical conductivity of the stratum was basically unchanged. As shown in Figure 12, the variation curves of the electrical conductivity can be divided into two parts: the descent stage and the stable stage. The first part is the descent stage, which includes the zone ranging from 0 to 20 cm below the cutter head–stratum interface. The negative charge of the slurry particles was shown to penetrate the stratum through the cutter opening, and the electrical conductivity of sand near the cutter head–stratum interface decreased obviously, mainly due to the cutter head's rotation. The electrical conductivity at 5 cm below the cutter head–stratum interface was three to four times the initial value. As the penetration distance increased, under the function of the permeation force, the slurry particles penetrated the stratum pores, making the electrical conductivity in this area steadily decrease, indicating that the slurry particles continued to decrease as the penetration distance increased. The average electrical conductivity dropped to two to three times that of the original stratum. The second part was the stable stage, and this included zones ranging from 20 to 35 cm below the cutter head–stratum interface, where the stratum was minimally affected by the cutter head rotation. Under the function of the permeation force, most of the particles penetrated into the front stratum pores, and only the water filtration passed through this stratum. Due to the small slurry particles in this area, the electrical conductivity was basically the same as that of the original stratum.

For the same type of cutter head, the rotation speed affected the electrical conductivity remarkably. As the rotation speed increased, the electrical conductivity in the descent stage of the curves increased obviously. As shown in Figure 12A–F, when the speed was 1.0 rpm, the electrical conductivity increased by an average of about 21% compared to that of the stopped cutter head, and the greater the rotation speed was, the more quickly the electrical conductivity. When the rotation speed was 1.0 rpm, the ratio of the electrical conductivity to the initial value approached 4–5, probably due because the cutter head's cutting enlarged the size and number of stratum pores, leading to more slurry particles becoming clogged and adsorbed.

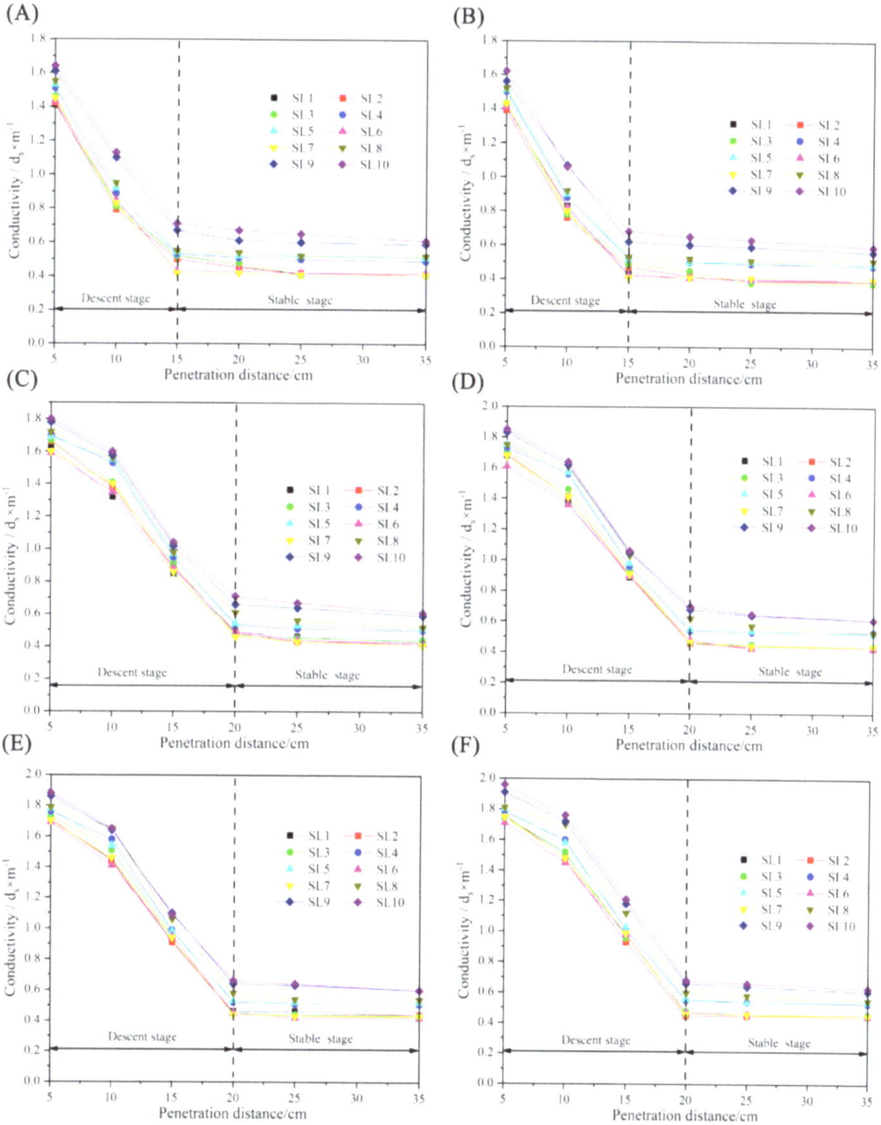

Figure 12. Variation curves of electric conductivity: (**A**) cutter head CH1, rotation speed of cutter head = 0 rpm, (**B**) cutter head CH2, rotation speed of cutter head = 0 rpm, (**C**) cutter head CH1, rotation speed of cutter head = 0.5 rpm, (**D**) cutter head CH2, rotation speed of cutter head = 0.5 rpm, (**E**) cutter head CH1, rotation speed of cutter head = 1.0 rpm, (**F**) cutter head CH2, rotation speed of cutter head = 1.0 rpm.

As shown in Figure 12A,B, when the cutter head stopped, the variation laws of electrical conductivity for the two cutter heads were consistent, and the electrical conductivity at the same measuring site was similar. When the cutter head rotated for a given slurry type and rotation speed, the variation law of electrical conductivity of different cutter heads appeared different. As shown in Figure 12C–F, in the descent stage, at a given measurement point, the conductivity of the 6-cutter arm was about 4% higher than that of the 5-cutter arm. This trend became more obvious as the rotation speed increased. The more cutter arms there were, the more frequently the stratum was cut within a certain period, and the more severely the internal structure of the stratum was damaged, leading to an increase in pores and penetration of the stratum by more slurry particles so that the electrical conductivity was raised.

As shown in Figure 12, the bentonite in the slurry sample appeared to have no effect on the slurry's maximum permeation distance. On the other hand, at the measurement site closer to the cutter head, the electrical conductivity of slurry sample SL10, which had more bentonite, had a maximum permeation distance of 16% greater than that of slurry sample SL1, which had less bentonite. This is because bentonite, with its negative charge, enhanced the electrical conductivity of the stratum.

3.4. The Final State of the Filter Cake

The filter cake formation penetration ×tests were carried out on 10 groups of slurry samples with different cutter head structures and rotation speeds. By analyzing the final slurry cake morphologies, it was found that the 10 groups formed similar slurry cake morphologies. In front of the center of the cutter head, the filter cake produced by the representative SL5 slurry was selected as the representative case.

As shown in Figure 13, when the cutter head stopped or rotated, a 'filter cake plus an infiltration zone' was formed with a clear boundary between the accumulation layer of the slurry particles and the stratum. As shown in Figure 13A,B, it is evident that when the CH1 and CH2 cutter heads stopped, the thickness of the filter cake was consistent—about 8 mm. As shown in Figure 13C–F, as the rotation speed increased continuously, the shapes of filter cakes constructed with two types of cutter head gradually transited from 'filter cake plus a permeation zone' to 'a permeation zone without filter cake', and the thickness of the filter cakes decreased gradually. Specifically, when the rotation speed increased to 0.5 rpm, the surface of the filter cake formed by the 5-cutter arm was smoother with a thickness of about 4 mm, while the 6-cutter arm cutter head formed a rougher filter cake surface with a thickness of about 3 mm, because the more cutter arms there were, the more times the stratum was cut within a certain time period and the thinner the filter cake became. As shown in Figure 13E,F, as the rotation speed increased, the thickness of the filter cake continued to decrease. The thicknesses of the filter cakes generated by the 5-cutter and 6-cutter arm were about 3 and 1 mm, respectively, because, under the combined functions of the cutter head cutting, slurry permeation, and air pressure, sand particles on the stratum surface were redistributed. The rotation of the cutter head destroyed the structure of the sand at the surface of the stratum, the thickness of the filter cake was thin, and the slurry particles completely penetrated the stratum pores. Based on the characteristics of the formed filter cake, it appears that the fewer cutter arms there were, the lower the permeability of the slurry film was, and the slower the speed of the cutter head was, making the set up more conducive to maintaining the stability of the shield excavation face.

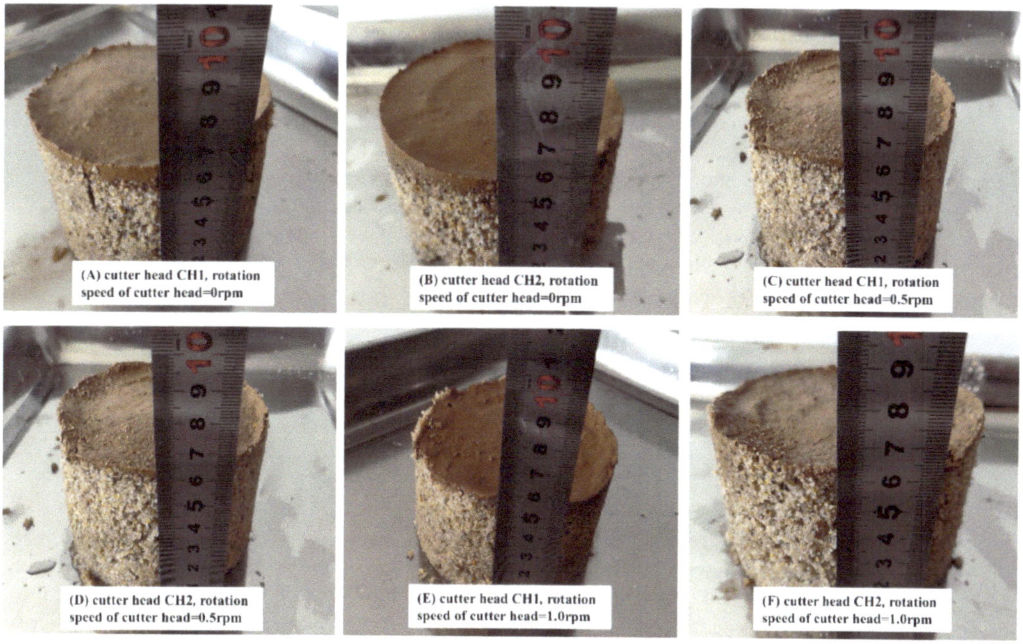

Figure 13. The final filter cake states: (**A**) cutter head CH1, rotation speed of cutter head = 0 rpm, (**B**) cutter head CH2, rotation speed of cutter head = 0 rpm, (**C**) cutter head CH1, rotation speed of cutter head = 0.5 rpm, (**D**) cutter head CH2, rotation speed of cutter head = 0.5 rpm, (**E**) cutter head CH1, rotation speed of cutter head = 1.0 rpm, (**F**) cutter head CH2, rotation speed of cutter head = 1.0 rpm.

As shown in Table 3, as the rotating speed of the cutter head increased, the thickness of the filter cake decreased. The cutter arms of the cutter head increased, and the damage frequency of the filter cake increased, leading to a decrease in the filter cake thickness. After rotation of the cutter head, the average thickness of the static filter cake among the 10 groups of samples was 6.5 mm. The average thickness at 0.5 rpm was 2.85 mm, and the average thickness at 1.0 rpm was 1.55 mm. The filter cake thickness at 0.5 rpm was 56.15% lower than that at static speed. Moreover, the thickness at 1.0 rpm was 76.15% lower than that at static speed; that is, the thickness of the static filter cake was significantly lower than that of the static filter cake.

As shown in Figure 14, the cutter head was removed after the test. By observing the filter cake on the cutter head, a thin filter cake attached to the surface of the cutter head was identified. This occurred because the mixture of slurry and bentonite particles has a certain viscosity. This filter cake can be considered to be firmly adhered to the cutter head and is not affected by the cutting of the cutter head. The filter cake attached to the center of the cutter head was found to be thicker than that at other sites, as the center of the cutter head is relatively static relative to the soil, meaning there is less damage to the formed filter cake, and the filter cake formed at the center of the cutter head is more compact. While the filter cake at the edge of the cutter head is relatively thin due to the high linear speed, when the edge rotates, this greatly disturbs the formed filter cake.

Table 3. Thickness of the Filter Cake (mm).

Slurry No.	Type of Cutter Head	Rotation Speed of the Cutter Head		
		0 rpm	**0.5 rpm**	**1.0 rpm**
SL1	5 cutter arms	6	2	1
	6 cutter arms	6	2	1
SL2	5 cutter arms	6	3	2
	6 cutter arms	6	2	1
SL3	5 cutter arms	6	3	2
	6 cutter arms	6	2	1
SL4	5 cutter arms	7	3	3
	6 cutter arms	7	3	1
SL5	5 cutter arms	8	4	3
	6 cutter arms	8	3	1
SL6	5 cutter arms	5	2	1
	6 cutter arms	5	1	1
SL7	5 cutter arms	6	2	1
	6 cutter arms	6	2	1
SL8	5 cutter arms	6	3	2
	6 cutter arms	6	3	1
SL9	5 cutter arms	6	4	2
	6 cutter arms	6	3	1
SL10	5 cutter arms	9	5	3
	6 cutter arms	9	5	2

Figure 14. Filter cake attached to the cutter head.

The filter cake formed from the test was dried and scanned by an electron microscope. The filter cake was observed at the same position at different magnifications. It can be seen from Figure 15A (magnified 500 times) that the particles in the filter cake accumulated in a disorderly manner, the particles of the filter cake differed in size, the particles were neither spherical nor smooth, and the maximum particle size reached 35.54 μm. Various contact modes were observed, mostly edge to surface and surface to surface, with good connectivity between particles. Slurry particles filled the formation pores under water pressure, and the pores between large particles were gradually filled and compacted, with the maximum pore size being 12.51–16.35 μm. The fine particles in the slurry gradually accumulated into a thin, small particle layer. Although there were pores between the particles, most formed a dense filter cake structure, achieving the effect of impermeability. At high magnification, as shown in Figure 15B (magnified 10,000 times), the filter cake particles were established in a squamous structure, and the maximum pore size of the adjacent squamous structure was 1.38 μm. This was an open pore with discontinuity and poor connectivity.

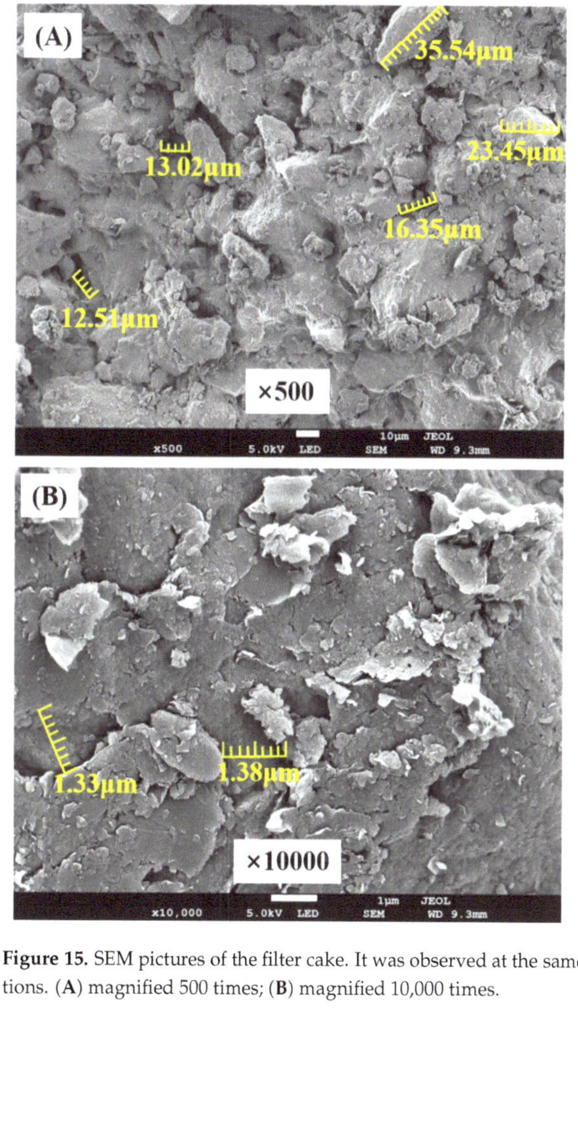

Figure 15. SEM pictures of the filter cake. It was observed at the same position at different magnifications. (**A**) magnified 500 times; (**B**) magnified 10,000 times.

4. Discussion

The experiment's results show that the dynamic filter cake formation time is longer than that of the static filter cake (when the rotating speed of the cutter head is 0 rpm), which indicates the process of regeneration after the destruction of the filter cake. With an increase in the rotating speed of the cutter head, the conductivity of the soil increases, which indicates that the cutter head destroys the newly formed filter cake and soil, and the slurry particles penetrate a larger area. The filter cake is thinner than the static filter cake, which means that the newly formed filter cake is destroyed repeatedly, resulting in a reduction in thickness. Above all, the mesoscopic formation process of the dynamic filter cake can be divided into six stages, as shown in Figure 16: initial slurry permeation, large-scale permeation of slurry, the initial formation of the filter cake, the failure of the filter cake, deep permeation of slurry, and final formation of the filter cake.

Stage I: The initial porosity of the sand was relatively high when the cutter head stopped. Under the function of the effective slurry pressure, the slurry particles passed through the cutter head opening and continuously penetrated the stratum pores, as shown in Figure 16A.

Stage II: The cutter head started to rotate and cut sand particles from the stratum surface. The original sand structure was destroyed, and the pores increased and enlarged, resulting in a large number of slurry particles infiltrating and filling in the stratum, as shown in Figure 16B.

Stage III: With the continuous permeation of slurry particles, a large number were accumulated in the pores of the sand. Then, they aggregated and settled down gradually on the surface of the stratum to form a micro-permeable cake (filter cake I), as shown in Figure 16C.

Stage IV: The cutter head rotated, and the newly formed filter cake structure was destroyed. Meanwhile, as the cutter head cut the sand particles, the structure of the sand was further destroyed, and the slurry particles in the stratum surface were mixed with sand particles, as shown in Figure 16D.

Stage V: The cutting of sand particles by the cutter head increased and enlarged the particle pores. Under the function of the effective slurry pressure, most of the slurry particles deeply penetrated the sand stratum in a larger area, and a small number of slurry particles stuck to the surface of the cutter head to form the inner filter cake (Figure 13). It should be noted that the new filter cake structure was not completely formed, as shown in Figure 16E.

Stage VI: As the slurry particles continuously penetrated the sand pores, when a large number of slurry particles accumulated in the pores of the damaged sand structure, slurry particles continued to settle on the surface, and the thickness of the slurry particles increased continuously, resulting in the formation of the outside of the filter cake (external filter cake). Then, the final filter cake (filter cake II) was formed together with the slurry particles, which initially attached to the cutter head panel (inner filter cake), as shown in Figure 16F.

With the continuous rotation of the cutter head, the 'destruction–formation' of the filter cake was repeated cyclically. From the perspective of the formation quality of the filter cake and the stability of the excavation face, a panel-type cutter head should be used to increase the area of a single cutter arm to reduce the damage done to the filter cake.

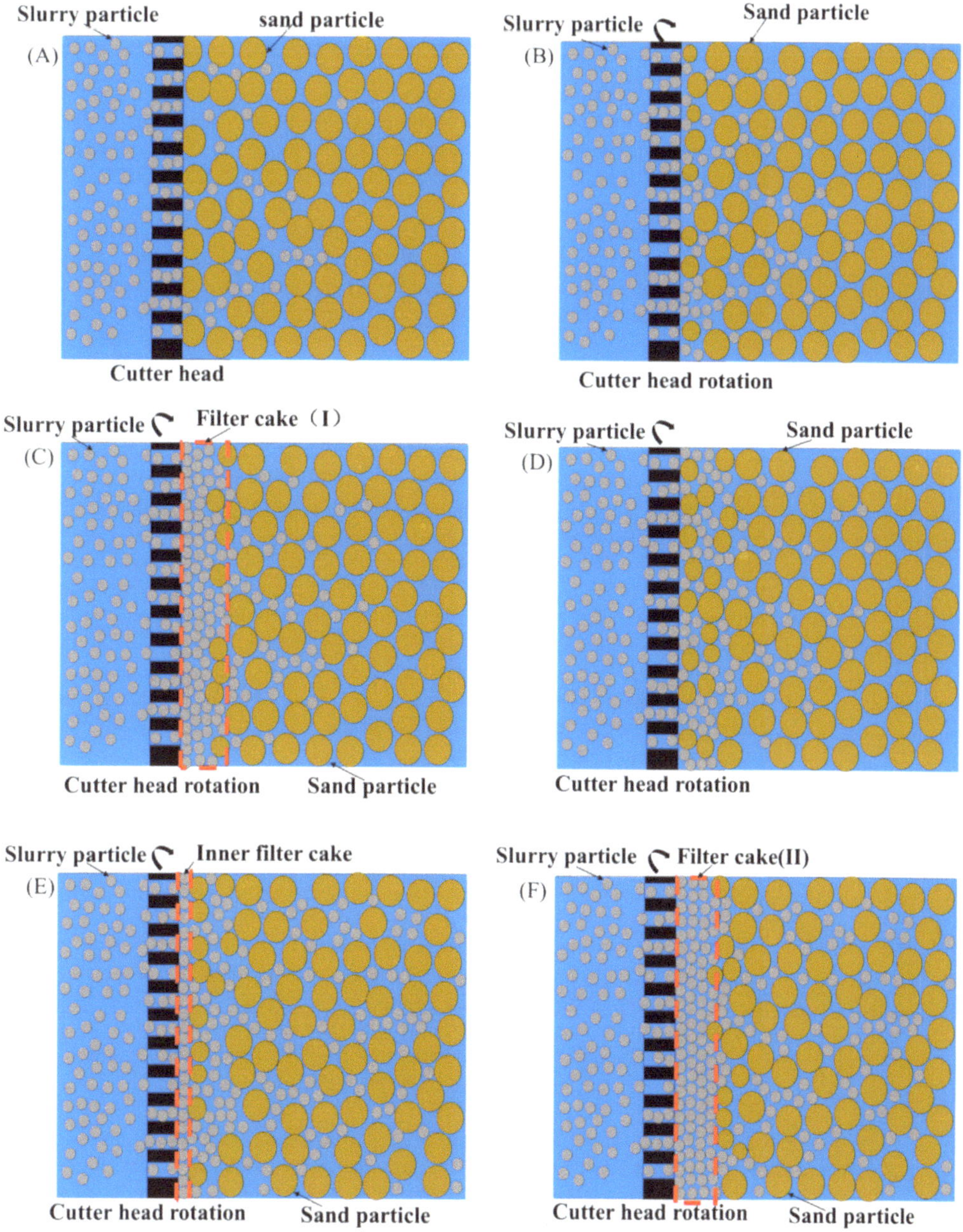

Figure 16. Stages of filter cake formation: (**A**) initial slurry permeation, (**B**) large scale slurry permeation, (**C**) initial filter cake formation, (**D**) the failure of filter cake, (**E**) deep permeation of slurry, (**F**) final filter cake formation.

5. Conclusions

In this study, a series of pressure filtration tests were carried out with 10 different slurries and 2 types of cutter head, using a self-made dynamic filter cake permeation test device. The characteristics of the filter cakes of the slurry shield tunnels were studied. Previous studies ignored the impact of cutter head cutting on the filter cake and soil [40,41]. Through the test, it was found that this impact will increase the permeability and forming time of filter cake expand the slurry penetration distance. The rotation speed and structure of the cutter head significantly impact the characteristics and quality of the filter cake and can easily cause instability of the tunnel excavation surface. Therefore, in practical engineering, the cutter head speed should be adjusted reasonably, and the cutter head selection should be carried out in combination with the influence of filter cake formation. The following conclusions were obtained by conducting the dynamic slurry penetration test:

(1) Considering the effect of the cutter head structure and rotation, the time curve of water filtration showed a stepped frame distribution, and the filtration volume increased rapidly in the early stage and gradually stabilized in the later stage of the test. The more cutter arms there were and the higher the rotation speed was, the greater the water filtration volume was, decreasing the impermeability of the filter cake.

(2) The variation curve of electric conductivity can be divided into two parts: the descent stage and the stable stage. When the cutter head rotated at 1.0 rpm, the penetration distance of the slurry was 1.5 times that when the cutter head stopped. Moreover, the formation of filter cake with a 6-cutter arm occurred about 5 s later than with the 5-cutter arm.

(3) As the rotation speed of the cutter head increased in saturated sand, the final form of the filter cake transited from a filter cake plus an infiltration zone to an infiltration zone without a filter cake, the thickness of the filter cake decreased continuously, and the filter cake's permeability increased, which decreased the stability of the excavation face.

(4) At the mesoscopic level, the dynamic formation process of the filter cake can be divided into six stages: the initial permeation of slurry particles, large-scale permeation after rotation of the cutter head, the initial formation of the filter cake, the failure of the filter cake, deep permeation of slurry, and the final formation of the filter cake.

Author Contributions: Methodology, Y.B. and L.Y.; Writing—original draft, Y.B.; Funding acquisition, B.J.; Investigation, L.Y.; Writing—review and editing. B.J. and Y.L. (Yunqiu Liu); Conceptualization, Y.L. (Yunqiu Liu) and Y.L. (Yapeng Li); Data curation, H.Z. and Y.L. (Yapeng Li); Formal analysis, Y.B. and Y.L. (Yangpeng Li); Validation, H.Z.; Visualization, Y.B. and L.Y. All authors have read and agreed to the published version of the manuscript.

Funding: This work was supported by the National Basic Research Program of China (2014CB046301), the National Nature Science Foundation of China (No. 51174196), Jiangsu Construction System Science Technology Project (2018ZD030), Research Projects of Jiangsu Vocational Institute of Architectural Technology (JYA320-19), and a high-end research project of the professional leader of teachers in Higher Vocational Colleges in Jiangsu Province (2021GRGDYX042).

Data Availability Statement: The data used to support the findings of this study are available from the first author upon request.

Acknowledgments: The author would like to thank reviewers and editors for their kind work.

Conflicts of Interest: The authors declare no conflict of interest.

References

1. Broere, W. Urban underground space: Solving the problems of today's cities. *Tunn. Undergr. Space Technol.* **2016**, *55*, 245–248. [CrossRef]
2. Yuan, D.J.; Shen, X.; Liu, X.Y.; Wu, J. Research on excavation face stability of slurry shield tunneling. *China J. Highw. Transp.* **2016**, *30*, 24–37.

3. Wang, Q. Study on reasonable slurry pressure of excavation face of slurry shield in water rich environment. *Highw. Eng.* **2019**, *44*, 151–155.
4. Zhu, W.; Qian, Y.J.; Min, M.L.; Wang, L.; Wang, C.; Xu, C.; Hu, J.N. The current status and some problems of slurry shield in China. *Tunn. Constr.* **2019**, *39*, 724–735.
5. Liu, G.Y.; Xu, C.; Zhang, Z.Y. Experimental study on slurry performance of slurry balance shield in sandy pebble stratum. *Subgrade Eng.* **2020**, *208*, 95–102.
6. Mo, Z.Z.; Wang, M.S.; Wang, H.B.; Li, H.B.; Qian, Y.J.; Luo, G.D.; Wang, H. Laboratory investigation on pore water pressure variation caused by filter cake effect during slurry-EPB shield tunneling in silty sand layer. *Rock Soil Mech.* **2019**, *40*, 2257–2263.
7. Liu, X.Y.; Wang, F.M.; Yuan, D.J.; Fang, H.Y.; Zhong, S.L. Range of support pressure for slurry shield and its factor analysis. *Chin. J. Geotech. Eng.* **2018**, *41*, 908–917.
8. Qi, W.Q.; Yang, Z.Y.; Jiang, Y.S.; Shao, X.K.; Yang, X.; He, Q. Structural Deformation of Existing Horseshoe-Shaped Tunnels by Shield Overcrossing. *KSCE J. Civ. Eng.* **2020**, *25*, 735–749. [CrossRef]
9. Hong, K.R. Typical underwater tunnels in the mainland of China and related tunneling technologies. *Engineering* **2017**, *3*, 871–879. [CrossRef]
10. Min, F.; Zhu, W.; Han, X. Filter cake formation for slurry shield tunneling in highly permeable sand. *Tunn. Undergr. Space Technol.* **2013**, *38*, 423–430. [CrossRef]
11. Liu, C.; Lu, Y.; Liu, L.; Lv, W.H. Analysis of clogging mechanism and filter-cake structure of slurry containing coarse-particle materials in sandy strata. *Mod. Tunn. Technol.* **2018**, *55*, 245–253.
12. Bai, Y.; Jiang, B.S.; Yang, L.; Ji, X.; Chen, X.X. Experimental study on filter cake formation of large diamater slurry GIL utility tunnel at Sutong. *Water Resour. Power* **2020**, *38*, 184–188.
13. Hernández, Y.Z.; Farfán, A.D.; Assis, A.P. A Three-dimensional analysis of excavation face stability of shallow tunnels. *Tunn. Undergr. Space Technol.* **2019**, *92*, 103062. [CrossRef]
14. Li, L.P.; Shang, C.S.; Chu, K.W.; Zhou, Z.Q.; Song, S.G.; Liu, Z.H.; Chen, Y.H. Large-scale geo-mechanical model tests for stability assessment of super-large cross-section tunnel. *Tunn. Undergr. Space Technol.* **2021**, *109*, 103756. [CrossRef]
15. Chung, J.H.; Daniel, D.E. Modified fluid loss test as an improved measure of hydraulic conductivity for bentonite. *Geotech. Test. J.* **2019**, *31*, 243–251.
16. Horn, N. Horizontal earth pressure on the vertical surfaces of the tunnel tubes. In Proceedings of the National Conference of the Hungarian Civil Engineering Industry, Budapest, Hungary, 18–21 June 1961; pp. 7–16.
17. Anagnostou, G.; Kovári, K. Face stability condition with earth pressure balanced shields. *Tunn. Undergr. Space Technol.* **1996**, *11*, 165–173. [CrossRef]
18. Broere, W. Influence of excess pore pressures on the stability of the tunnel face. In *Claiming the Underground Space*; ITA: Amsterdam, The Netherlands, 2003; pp. 759–765.
19. Zizka, Z.; Schoesser, B.; Thewes, M. Slurry Shield Tunneling: New Methodology for Simplified Prediction of Increased Pore Pressures Resulting from Slurry Infiltration at the Tunnel Face Under Cyclic Excavation Processes. *Int. J. Civ. Eng.* **2018**, *17*, 113–130. [CrossRef]
20. Zumsteg, R.; Puzrin, A.M.; Anagnostou, G. Effects of slurry on stickiness of excavated clays and clogging of equipment in fluid supported excavations. *Tunn. Undergr. Space Technol.* **2016**, *58*, 197–208. [CrossRef]
21. Chen, R.P.; Yin, X.C.; Li, Y.C.; Meng, F.Y.; Sa, C. Permeability of filter cake and its fluence on face stability of slurry shield-driven tunnels. *Chin. J. Geotech. Eng.* **2017**, *39*, 2102–2108.
22. Liu, J.J.; Chen, T.L.; Yao, M.H.; Wei, Y.X.; Zhou, Z.J. Experimental and numerical study on slurry fracturing of shield tunnels in sand stratum. *J. Zhejiang Univ.* **2020**, *54*, 1–12.
23. Jeblli, J.; Meguid, M.A.; Sedghinead, M.K. Excavation failure during micro-tunneling in fine sands. *Tunn. Undergr. Space Technol.* **2010**, *25*, 811–818. [CrossRef]
24. Kim, S.H.; Tonon, F. Face stability and required support pressure for TBM driven tunnels with ideal face membrane-drained case. *Tunn. Undergr. Space Technol.* **2010**, *25*, 526–542. [CrossRef]
25. Zumsteg, R.; Puzrin, A.M. Stickiness and adhesion of conditioned clay pastes. *Tunn. Undergr. Space Technol.* **2012**, *31*, 86–96. [CrossRef]
26. Zhou, Q.H. Analysis of ground disturbance induced by large diameter shield tunneling in silty clay soil. *Railw. Stand. Des.* **2020**, *64*, 158–163.
27. Li, D.; Han, Z.; Sun, X.; Zhou, T.; Li, X. Dynamic Mech anical Properties and Fracturing Behavior of Marble Specimens Containing Single and Double Flaws in SHPB Tests. *Rock Mech. Rock Eng.* **2019**, *52*, 1623–1643. [CrossRef]
28. Li, Y.H.; Liu, D.Z.; Yang, S. Development and application of physical simulation test system for small and medium-sized tunnels based on biaxial motor loading. *Chin. J. Geotech. Eng.* **2020**, *42*, 1556–1563.
29. Fritz, P. Additives for slurry shields in highly permeable ground. *Rock Mech. Rock Eng.* **2006**, *40*, 81–95. [CrossRef]
30. Zdenek, Z.; Britta, S.; Markus, T. Investigations on the transient support pressure transfer at the tunnel face during slurry shield drive Part 2: Case B– Deep slurry penetration exceeds tool cutting depth. *Tunn. Undergr. Space Technol.* **2021**, *118*, 104169.
31. Liu, C.; Tang, X.Y.; Gao, Y.F. Influence of pore characteristics of sand strata on filter cake formation under slurry shield. *Chin. J. Geotech. Eng.* **2017**, *39*, 2003–2008.

32. Xu, T.; Bezuijen, A. Experimental study on the mechanisms of bentonite slurry penetration in front of a slurry TBM. *Tunn. Undergr. Space Technol.* **2019**, *93*, 103052. [CrossRef]
33. Lin, Y.F.; Fang, Y.; He, C.; Wang, W. Experimental study on degree of match between slurry and ground based on particle retention rate. *Tunn. Undergr. Space Technol.* **2021**, *116*, 104105. [CrossRef]
34. Mao, J.H.; Yuan, D.J.; Yang, J.X.; Zhang, B. Theoretical study of porosity characteristic on excavation face of slurry shield in sand stratum. *Rock Soil Mech.* **2020**, *41*, 2283.
35. Zhang, Z.; Yin, T.; Huang, X.; Dias, D. Slurry filtration process and filter cake formation during shield tunneling: Insight from coupled CFD-DEM simulations of slurry filtration column test. *Tunn. Undergr. Space Technol.* **2019**, *87*, 64–77. [CrossRef]
36. Min, F.L.; Du, J.R.; Zhang, N.; Chen, X.G.; Lv, H.J.; Liu, L.C.; Yu, C.J. Experimental study on property change of slurry and filter cake of slurry shield under seawater intrusion. *Tunn. Undergr. Space Technol.* **2019**, *88*, 290–299. [CrossRef]
37. Wei, D.W.; Zhu, W.; Min, F.L. Experimental study of forming time of filter cake and conversion rate of slurry pressure in slurry shield in sand stratum. *Rock Soil Mech.* **2014**, *35*, 423–428.
38. *ASTM D2434-06*; Standard Test Method for Permeability of Granular Soils. American Society for Testing and Materials: West Conshohocken, PA, USA, 2006.
39. Liu, D.S.; Liu, X.R.; Lin, C.Q.; Xiong, F.; Han, Y.F.; Meng, Q.J.; Zhong, Z.L.; Chen, Q.; Weng, C.X. Experimental study and engineering application of slurry permeability mechanism of slurry shield in circular-gravel stratum. *Arab. J. Geosci.* **2020**, *13*, 1001–1012. [CrossRef]
40. Fan, W.C.; Sun, Z.C.; Li, F.Y.; Zhang, B.; Chen, Q.; Wang, F.M.; Wang, K. Research on prediction of tunneling parameters of super-large diameter slurry shield in composite strata of Shantou Bay tunnel. *Tunn. Constr.* **2020**, *40*, 1160–1168.
41. Li, D.Y.; Feng, F.; Li, X. Numerical simulation of rock failure under static and dynamic loading by splitting test of circular ring. *Eng. Fract. Mech.* **2018**, *188*, 184–201. [CrossRef]

minerals

MDPI

Article

Research on the Mechanical Properties and Damage Constitutive Model of Multi-Shape Fractured Sandstone under Hydro-Mechanical Coupling

Ying Zhang [1,2], Xu Wu [3], Qifeng Guo [1,2,*], Zhaohong Zhang [4], Meifeng Cai [1,2] and Limei Tian [1,2,5]

1 School of Civil and Resource Engineering, University of Science and Technology Beijing, Beijing 100083, China; zhangyingmail@ustb.edu.cn (Y.Z.); caimeifeng@ustb.edu.cn (M.C.); b20170009@xs.ustb.edu.cn (L.T.)
2 Beijing Key Laboratory of Urban Underground Space Engineering, University of Science and Technology Beijing, Beijing 100083, China
3 Beijing Municipal Engineering Research Institute, Beijing 100037, China; 13051510807@163.com
4 Jining Energy Development Group Co., Ltd., Jining 272000, China; jackzhangzhaohong@163.com
5 Institute of Architectural Civil Engineering, Langfang Teachers University, Langfang 065000, China
* Correspondence: guoqifeng@ustb.edu.cn; Tel.: +86-010-62333700

Abstract: In this paper, mechanical property tests of sandstone with multiple shapes of prefabricated fractures (single, T-shaped, and Y-shaped fractures) are carried out through the MTS815 rock mechanics testing machine and the Teledyne ISCO D-Series Pumps system. Considering the hydro-mechanical coupling effects, the experiments reveal the key thresholds, strength characteristics and deformation laws of multi-shape fractured sandstones during the progressive failure process. According to the elastic-plastic theory, the continuous damage theory and the statistical damage theory, a new damage model is constructed, which fully reflects the coupled effects among water, micro flaws and macroscopic prefabricated fractures. The crack closure stress σ_{cc}, crack initiation stress σ_{ci} and damage stress σ_{cd} of multi-shape fractured sandstone samples are determined by the proposed volumetric strain response method. In the range of 0–90°, the σ_{cc} and σ_{ci} of the multi-shape fractured sandstone samples are different, as well as the angles when the σ_{cd} and peak strength (σ_c) reach their peak values. The stress ratios (the σ_{cc}/σ_c, σ_{ci}/σ_c, and σ_{cd}/σ_c are collectively referred to as stress ratios) are hardly affected by the shape and inclination of the fractures inside the rock. According to strength analysis and deformation characteristics, the weakening effect of water has less of an influence on the strength than prefabricated fractures. The stress–strain curve obtained, based on the hydro-mechanical coupling test, is in good agreement with the theoretical curve generated by the damage constitutive model, verifying the rationality of the damage constitutive model. In addition, the fracture inclination only affects the numerical value of the total damage variable of multi-shape fractured sandstone samples, and has minor effects on its variation trend.

Keywords: multiple shapes of prefabricated fractures; hydro-mechanical coupling; mechanical properties; stress threshold; damage constitutive model

Citation: Zhang, Y.; Wu, X.; Guo, Q.; Zhang, Z.; Cai, M.; Tian, L. Research on the Mechanical Properties and Damage Constitutive Model of Multi-Shape Fractured Sandstone under Hydro-Mechanical Coupling. *Minerals* **2022**, *12*, 436. https://doi.org/10.3390/min12040436

Academic Editors: Diyuan Li, Zhenyu Han, Xin Cai, Shijie Xie and Gianvito Scaringi

Received: 24 January 2022
Accepted: 25 March 2022
Published: 31 March 2022

1. Introduction

In engineering activities, such as mining development, water conservancy and hydropower construction, and geothermal mining, underground fractured rock mass often exists in an environment in which the stress field and seepage field interact with each other, and there are many pores and fractures of different scales and shapes inside the fractured rock mass. The mechanical properties of fractured rocks become more complex, and the damage evolution law is also affected to a greater extent under hydro-mechanical coupling [1–3]. Therefore, studying the mechanical properties and damage constitutive model of the fractured rock mass during the damage and failure process under hydro-mechanical

coupling helps to reveal the evolution mechanism of the progressive failure of the fractured rock mass, and provide theoretical support and a scientific basis for projects in the field of underground geotechnical engineering.

Some achievements have been made in the study of rock mechanical properties under hydro-mechanical coupling. Yilmaz, Vasarhely, and Van analyzed the effect of water content on rock peak strength and elastic modulus through hydro-mechanical coupling experiments [4,5]. Kou et al. conducted a hydro-mechanical coupling test on prefabricated 45° single-fracture rocks. With the increase in water pressure, the peak strength of the samples decreased continuously, and the degree of deviatoric stress decrease in the post-peak deformation increased with the increase in the water pressure. The peak strength increases with the increasing confining pressure, and the degree of strain softening in the post-peak deformation also gradually increases with the increasing confining pressure [6]. Wang et al. studied the stress–strain law of rock samples with different water pressures and confining pressures under the action of coupling, including the characteristics of lateral strain and volume strain, and discussed the failure characteristic of fluid flow under hydro-mechanical coupling [7]. Wang et al. studied the mechanical and permeability properties of sandstone and limestone in a coal mine, before and after rock failure under hydro-mechanical coupling through laboratory experiments revealed the correlation between mechanical properties and permeability, and obtained the strength relationship between deformation and water pressure difference [8]. Song et al. studied the deformation, mechanics, fracture and strain energy characteristics of tunnel limestone through hydro-mechanical coupling experiments, which provided theoretical support for the safe excavation of water-rich tunnels [9]. Wang et al. combined laboratory tests and numerical simulation methods to study the effects of confining pressure and seepage pressure on the mechanical properties and permeability of red sandstone. In addition, new explorations can be carried out with the help of other advanced observation instruments to continuously enrich the research on the deformation and strength characteristics of rocks or rock samples with fractures under hydro-mechanical coupling [10].

In addition, many scholars established corresponding damage constitutive models for the damage evolution process of fractured rocks. Kawamoto et al. used the damage tensor to represent the discontinuous state distributed in the rock mass, and then proposed a damage constitutive model of the fractured rock mass, and studied the effect of joints on the damage, strength and deformation of the rock mass [11]. Swoboda et al. introduced the active damage tensor to reflect the phenomenon that cracks in the rock mass may close in the compression process [12]. The above studies did not consider the effect of micro-flaws in the rock materials. Some scholars fully combine the continuous damage theory, fracture damage theory and statistical strength theory, and consider randomly distributed micro-flaws, assuming that the rock micro-element strength obeys the normal distribution or Weibull distribution, further enriching and improving the development of rock damage constitutive models. Among them, Zhang et al. and Chen et al. established a macroscopic flaws damage evolution model from the perspective of fracture strain energy [13,14]. Xu and Yang proposed a fractured rock constitutive model that can reflect the initial fracture closure phase in compression. The proposed model can predict energy and damage behavior, and the model is verified by conventional triaxial compression tests of siltstone. The theoretical constitutive model is in reasonably agreement with the stress–strain curve [15]. Based on the Drucker–Prager criterion, and combined with the energy principle and fracture damage theory, Chen and Qiao established a macro–meso composite damage constitutive model for non-persistent jointed rock mass [16]. In summary, the research on the constitutive model has made abundant progress, but there are still some shortcomings. There are many studies on rock damage constitutive models using uniaxial compression tests, and rarely using triaxial tests or hydro-mechanical coupling tests.

The purpose of this study is to explore the mechanical properties and damage evolution laws of multi-shape fractured sandstone samples under hydro-mechanical coupling. First, the stress–strain curves of the multi-shape fractured sandstone samples under hydro-

mechanical coupling are obtained, and then the changing laws of mechanical parameters, such as strength, elastic modulus and Poisson's ratio, are analyzed. Finally, these values are compared to and verify the validity of the constitutive model by a mechanical test and numerical simulation test, and the evolution law of the total damage variable of multi-shape fractured sandstone are analyzed. The research results provide a useful reference for revealing the hydro-mechanical coupling mechanism in deep water-rich rock mass engineering.

2. Experimental Methodology

2.1. Rock Sample Preparation

In Figure 1, waterjet cutting and wire cutting equipment were combined to prefabricate the sandstone samples with different inclinations (α = 0°, 15°, 30°, 45°, 60°, 75°, 90°), single, T-shaped and Y-shaped fractures on standard samples (ϕ50 × 100 mm). The mineral composition of the sandstone mainly includes quartz (72.5%), feldspar (5%), debris (12.5%), and others (10%) by polarizing microscope and X-ray diffraction analysis. In Table 1, in order to study the mechanical characteristics of multi-shape fractured sandstone samples under hydro-mechanical coupling, the tests were carried out on multi-shape fractured sandstone samples with different angles, with a 10 MPa confining pressure and 3 MPa water pressure. In Table 2, two tests of intact sandstone samples were added, which were an intact sample without water pressure (10 MPa confining pressure) and an intact sample with water pressure (10 MPa confining pressure and 3 MPa water pressure) for comparative analysis. In this paper, the sample numbers SF, ST and SY represent single-fracture, T-shaped and Y-shaped fractured sandstone samples, and the following numbers represent the inclinations.

Table 1. Test scheme of fractured sandstone samples.

Fracture Inclination α	Sample No.			Confining Pressure/MPa	Water Pressure/MPa
	Single Fracture	T-Shaped Fracture	Y-Shaped Fracture		
0°	SF0	ST0	SY0	10	3
15°	SF15	ST15	SY15	10	3
30°	SF30	ST30	SY30	10	3
45°	SF45	ST45	SY45	10	3
60°	SF60	ST60	SY60	10	3
75°	SF75	ST75	SY75	10	3
90°	SF90	ST90	SY90	10	3

Table 2. Test scheme of intact sandstone samples.

Name	Sample No.	Confining Pressure/MPa	Water Pressure/MPa
Intact sample without water pressure	W1	10	-
Intact sample with water pressure	W2	10	3

2.2. Testing Procedure

In order to study the mechanical characteristics of sandstone with multiple shapes of fractures under hydro-mechanical coupling, the test equipment adopted the American MTS815 rock mechanics testing machine and Teledyne ISCO D-Series Pumps system, as shown in Figure 2. During the test, both the axial pressure and the confining pressure were controlled by the hydraulic servo system matched with MTS815, and the water pressure was controlled by the D-Series Pumps system. The deformation of the sandstone samples was measured using axial and hoop extensometers. During the installation of the sample, the intact and fractured sandstone samples were wrapped with heat-shrinkable tubes to prevent rock breakage debris from blocking the oil-return hole of the equipment. The water pressure was always kept lower than the confining pressure during the test. During the

axial loading process, the entire loading process was controlled by a combination of load and deformation (loop or axial). In the initial loading stage, the axial load control method was used for loading, and the loading rate was 300 N/s. When the load reached about 80% of the peak strength (55 MPa), the loading method was converted to deformation control, and the loading rate was 0.02 mm/min until the rock failed.

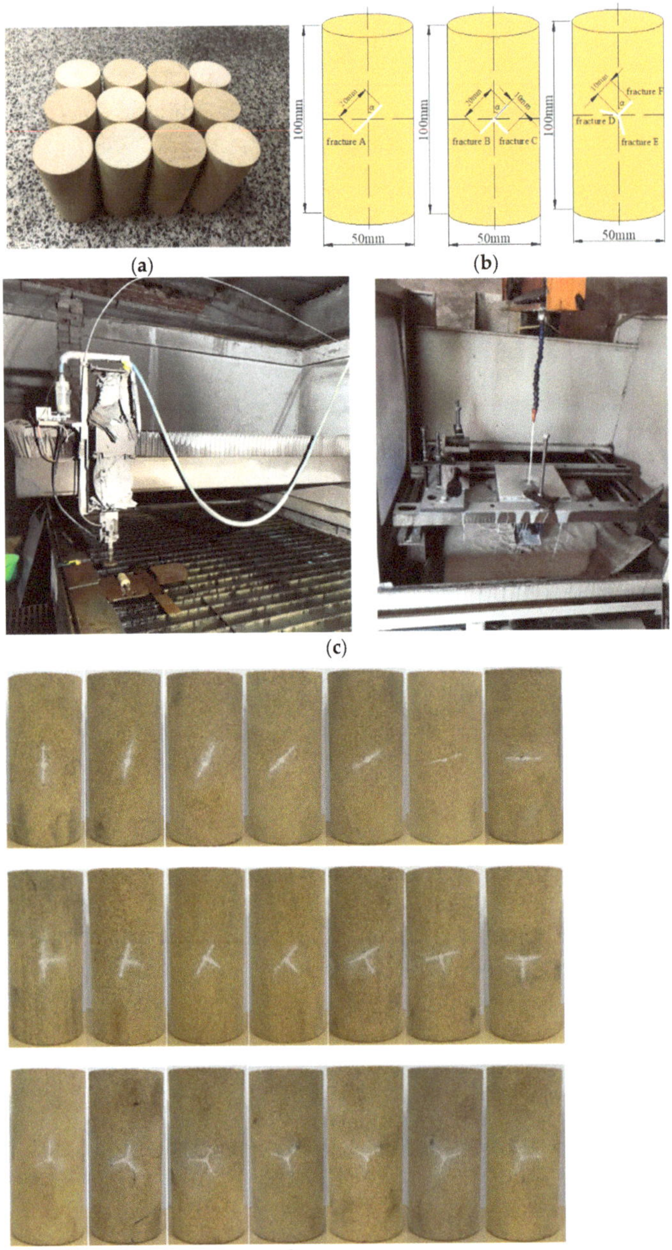

Figure 1. Preparation process of multi-shape fractured sandstone samples: (**a**) standard samples; (**b**) prefabricated fracture-size models; (**c**) fractures cutting equipment; and (**d**) prefabricated fracture samples.

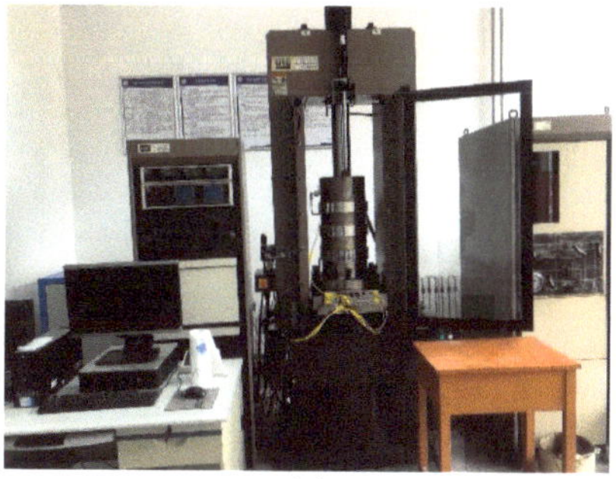

(**a**)

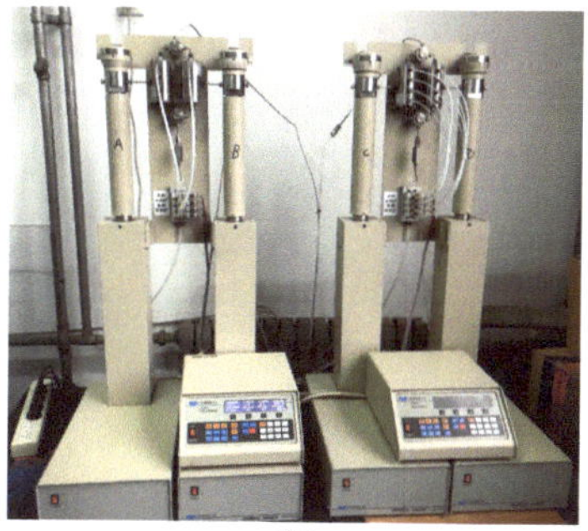

(**b**)

Figure 2. Test equipment: (**a**) MTS815 rock mechanics testing machine and (**b**) Teledyne ISCO D-Series Pumps.

3. Mechanical Properties of the Fractured Sandstone under Hydro-Mechanical Coupling

3.1. The Stress–Strain Law of the Fractured Sandstone during Progressive Failure

In Figure 3, the sample SF15 is taken as an example, and the progressive failure stage of the fractured sandstone sample is divided in detail. Martin and Chandler, Brace et al., Hoek and Bieniawski, and Hallbauer et al. conducted a detailed study on the failure of intact rock through experiments, and divided the stress–strain curve of the rock into five stages. This division method is widely accepted as: I. Crack closure, II. Elastic region, III. Stable crack growth, IV. Accelerated crack growth, and V. Post-peak region [17–21].

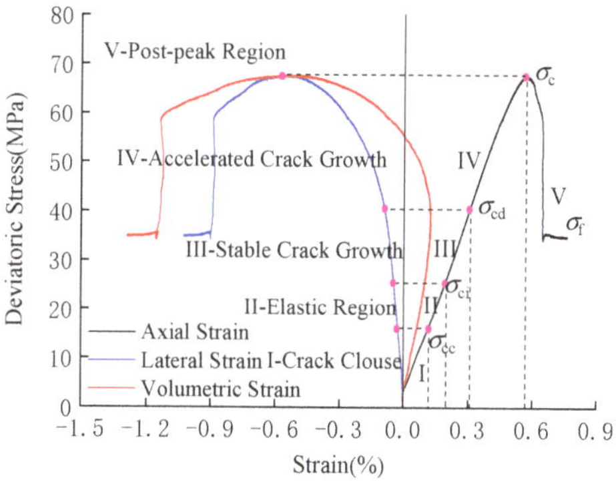

Figure 3. Stage division of stress–strain curve of the fractured sandstone sample under hydro-mechanical coupling (SF15-2).

3.2. Determination of the Threshold of the Fractured Sandstone during Progressive Failure

Crack closure stress (σ_{cc}), crack initiation stress (σ_{ci}), damage stress (σ_{cd}) and peak strength (σ_c) are the key indicators for dividing the progressive failure stage of rock samples and the key thresholds for measuring the macro-mechanical properties of fractured rock mass (Figure 3). To date, a lot of research has been conducted on σ_{ci}, and the commonly used methods to determine σ_{ci} are the volumetric strain method (VS) [18], lateral strain method (LS) [22], crack volumetric strain method (CVS) [17], and lateral strain response method (LSR) [23]. In addition, the acoustic emission method (AE) [24] and numerical simulation method (NS) [25] are also utilized.

However, the main methods used to determine σ_{cc} and σ_{ci} at the same time are VS, LS, and CVS. In the VS method, σ_{cc} is determined by the starting point of the linear segment in the axial stress–volume strain curve, and σ_{ci} is determined at the end point. The advantage is that the determination of σ_{cc} and σ_{ci} of the rock is intuitive, and the method is simple and easy to operate. The disadvantage is that it relies on artificial tangents to determine

σ_{cc} and σ_{ci}, and there are unavoidable subjectivity and value errors. The LS method uses the deviation from the linear segment in the axial stress–lateral strain curve to determine σ_{cc} at the start and end points to determine σ_{ci}. This method has the advantages of VS, removes the influence of the axial strain, and has less interference factors to determine σ_{cc} and σ_{ci}. The disadvantages are the same as VS. In the CVS method, the rock crack volume strain is obtained by calculating the difference between the total volume strain and the elastic volume strain during rock compression, and σ_{cc} and σ_{ci} are determined from the crack volume strain–axial strain curve. The advantage is that the artificial tangent is no longer used to determine σ_{cc} and σ_{ci}, and the process of determining σ_{cc} and σ_{ci} is not easy to confuse, and the obtained σ_{cc} and σ_{ci} values are relatively objective. The disadvantage is that it depends on the calculation of the elastic modulus and Poisson's ratio, and is sensitive to the change of Poisson's ratio, and it is easy to produce errors in judging the crack closure and crack initiation points.

In the LSR method, σ_{cd} is determined by the axial stress–volume strain curve. First, an reference line is formed by connecting the origin and σ_{cd}, and then the corresponding lateral strain value on the reference line is determined, the lateral strain difference is obtained by subtracting the reference lateral strain from the measured lateral strain, and, finally, the relationship between the lateral strain difference and the axial stress is drawn, and the peak point is σ_{ci} in the figure. The advantage is that the determination of the extreme value is unique, which avoids human error and ensures the objectivity of the result, and the disadvantage is that it is only applicable to hard rock. The drawing operations of the above four methods are shown in Figure 4. In addition, in the AE method, the stress value corresponding to the apparent acoustic emission behavior inside the rock is monitored by the acoustic emission system, which is σ_{ci}. The advantage is that the acoustic emission method is an important supplement to the conventional strain method, and the disadvantage is that the acoustic emission signal is easily disturbed by noise, and the rock may also have a strongly fluctuating acoustic emission signal during the fracture closure and linear elastic stages, which interferes with the accurate identification of σ_{ci}. In the NS method, the crack numbers–axial strain curve is drawn through simulation. In this curve, as the strain increases, the strain value corresponding to the first inflection point of the curve is obtained, and σ_{ci} is determined in the stress–strain curve according to this strain value. The advantage is that human error is avoided, and σ_{ci} can be easily and quickly determined through simulation; the disadvantage is that the simulation results need to be verified by experiments.

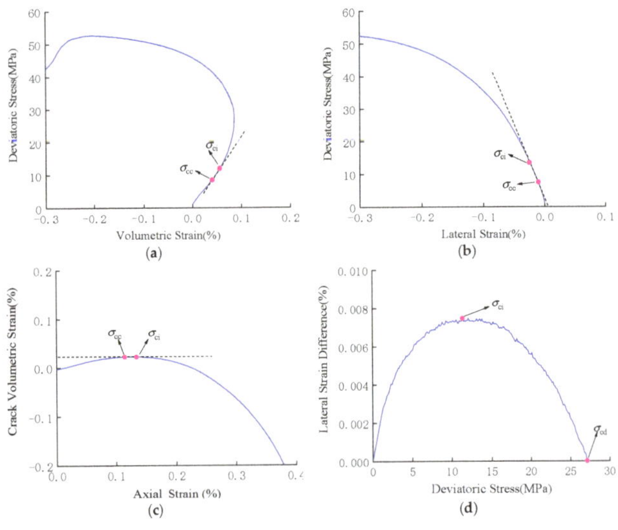

Figure 4. Different methods to determine the σ_{cc} or σ_{ci} values: (**a**) VS; (**b**) LS; (**c**) CVS; and (**d**) LSR.

In summary, according to the advantages of each method, this section adopts a new method to determine the σ_{cc}, σ_{ci} and σ_{cd} values of the fractured sandstone samples, which are called the volumetric strain response method (VSR). Due to the inhomogeneity of the particle size and structure in the rock, the microcracks and pores are compressed under stress until they are completely closed, and as the stress continues to increase, local tensile stress is generated inside the rock. Under the action of local tensile stress, microcracks occur between the particles with weak cohesion. Therefore, the physical meaning of the maximum relative volume strain difference is elucidated as the microcracks and pores are compressed, or via the initiation of new microcracks. The specific steps are as follows:

(1) In the deviatoric stress–volumetric strain curve, the stress value corresponding to the maximum volumetric strain is selected as σ_{cd}, as shown in Figure 5.

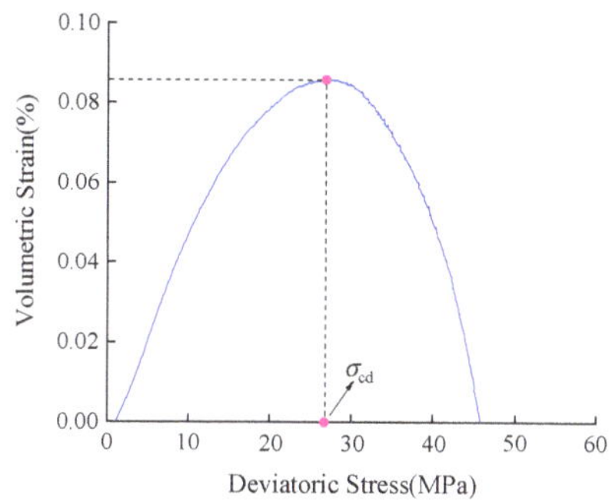

Figure 5. The σ_{cd} is determined based on volume strain–deviatoric stress curve.

(2) In the deviatoric stress–volumetric strain curve, a reference line is formed by connecting the starting point and σ_{cd}, and the corresponding volumetric strain value on the reference line is determined. The volumetric strain difference is obtained by subtracting the reference volumetric strain from the measured volumetric strain. Figure 6 shows the relationship between the volumetric strain difference and deviatoric stress, with the deviatoric stress value corresponding to the peak point σ_{ci}.

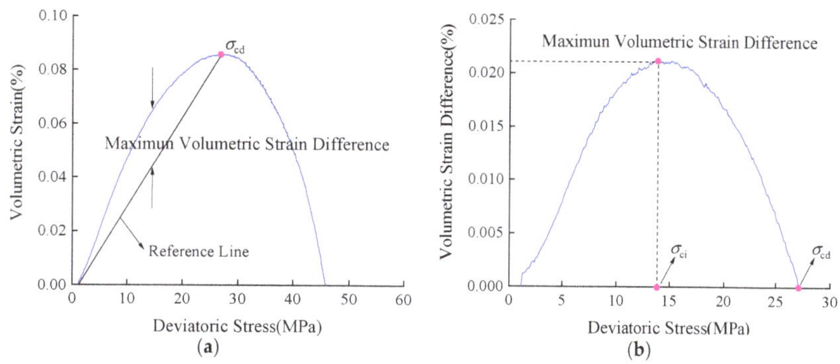

Figure 6. The value of σ_{ci} is determined based on the volume strain difference–deviatoric stress curve and reference line: (**a**) creating the reference line and (**b**) determining the σ_{ci} value.

(3) In the deviatoric stress–volume strain curve, based on the σ_{ci} value determined in (2), a reference line is formed by connecting the starting point with σ_{ci}, and the corresponding volumetric strain value on the reference line is determined. The volumetric strain difference is obtained by subtracting the reference volumetric strain from the measured volumetric strain. Figure 7 presents the relationship between the volumetric strain difference and deviatoric stress, with the deviatoric stress value corresponding to the peak point σ_{cc}.

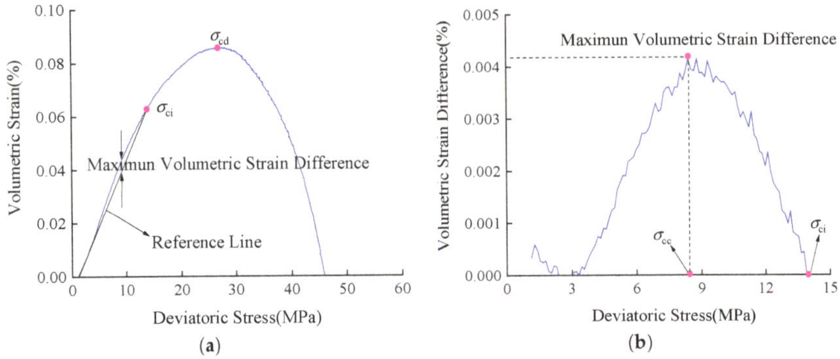

Figure 7. The value of σ_{cc} is determined based on the volume strain difference–deviatoric stress curve and reference line: (**a**) creating the reference line and (**b**) determining the σ_{cc} value.

In Tables 3 and 4, the σ_{cc} and σ_{ci} values of the intact samples and multi-shapes fractured samples are determined by various methods, such as VS, LS, CVS, LSR, and VSR. In order to thoroughly analyze the dispersion degree of σ_{cc} and σ_{ci} obtained by various methods, the standard deviation (SD) and coefficient of variation (CoV) were introduced. It can be observed from Table 3 that the mean SD of the σ_{cc} value obtained by the five methods are 0.87 MPa, 1.14 MPa, and 0.58 MPa, and the mean CoV are 6.57%, 7.29%, and 3.83%, respectively. It can be observed from Table 4 that the mean SD values of the σ_{ci} values obtained by the five methods are 0.75 MPa, 1.19 MPa, and 1.07 MPa, and the mean CoV are 4.38%, 6.71%, and 5.12%, respectively. However, the SD and CoV values of σ_{cc} and σ_{ci} of the intact sample with water pressure and without water pressure are higher than those of the samples with prefabricated fractures, and the SD and CoV values of most of the samples do not exceed 10MPa and 10% [26]. It is shown that each method can reasonably determine σ_{cc} and σ_{ci}, and it also shows the rationality of the method proposed in this paper.

Table 3. Summary of the sandstone progressive failure thresholds σ_{cc} and σ_{cd} under hydro-mechanical coupling.

Sample No.	σ_{cc}/MPa							σ_{cd}/MPa
	VS	LS	CVS	VSR	Mean	SD	CoV/%	
W1	22.59	22.75	21.68	17.30	21.08	2.56	12.14%	68.48
W2	18.17	24.02	22.23	16.54	20.24	3.47	17.14%	53.93
SF0	10.76	10.71	10.70	13.62	11.45	1.45	12.66%	29.07
SF15	16.46	19.67	20.26	19.42	18.95	1.70	8.97%	38.46
SF30	10.92	9.89	9.65	10.33	10.20	0.56	5.49%	29.14
SF45	12.89	13.27	13.38	13.08	13.16	0.22	1.67%	33.13
SF60	12.23	12.74	14.52	12.31	12.95	1.07	8.26%	27.82
SF75	11.55	12.13	12.93	12.78	12.35	0.63	5.10%	26.12
SF90	12.43	11.76	12.85	12.10	12.29	0.47	3.82%	24.96
ST0	15.86	16.71	17.11	13.56	15.81	1.59	10.06%	38.97
ST15	12.44	13.87	13.98	14.11	13.60	0.78	5.74%	33.77
ST30	8.86	8.38	10.50	8.81	9.14	2.42	10.18%	27.33
ST45	7.89	9.91	8.69	7.59	8.52	1.04	12.21%	21.66
ST60	14.84	14.69	15.14	15.07	14.94	0.21	1.41%	31.10
ST75	16.21	15.58	15.11	15.32	15.56	0.48	3.08%	30.13
ST90	18.64	17.34	15.77	19.05	17.70	1.48	8.36%	43.62
SY0	16.14	14.62	15.66	15.99	15.60	0.69	4.42%	29.73
SY15	15.77	16.13	16.31	14.91	15.78	0.62	3.93%	40.46
SY30	14.77	15.48	15.53	14.87	15.16	0.40	2.64%	34.63
SY45	15.57	16.46	17.32	15.27	16.16	0.93	5.75%	38.02
SY60	10.43	10.91	11.51	11.14	11.00	0.45	4.09%	27.89
SY75	15.50	15.58	15.31	15.02	15.35	0.25	1.63%	37.40
SY90	16.63	15.82	17.54	16.36	16.59	0.72	4.34%	43.00

Table 4. Summary of the sandstone progressive failure thresholds σ_{ci} and σ_c under hydro-mechanical coupling.

Sample No.	σ_{ci}/MPa								σ_c/MPa
	VS	LS	CVS	LSR	VSR	Mean	SD	Cov/%	
W1	33.47	33.69	26.33	39.60	32.63	33.14	4.71	14.21%	97.54
W2	25.76	31.17	29.12	31.39	28.22	29.13	2.32	7.96%	92.42
SF0	15.35	14.03	16.98	13.68	17.83	15.57	1.81	11.62%	78.73
SF15	24.38	25.31	24.16	22.07	24.18	24.02	1.19	4.95%	67.70
SF30	14.72	15.34	14.18	14.86	15.17	14.85	0.45	3.03%	57.04
SF45	17.26	17.62	17.04	17.71	17.66	17.46	0.29	1.66%	60.92
SF60	15.59	16.72	17.16	16.76	16.37	16.52	0.59	3.57%	58.15
SF75	15.17	15.23	15.46	14.66	15.86	15.28	0.44	2.88%	54.32
SF90	15.88	16.03	16.97	16.48	15.82	16.24	0.48	2.96%	61.99
ST0	20.43	21.16	19.58	20.41	20.90	20.50	0.60	2.93%	63.51
ST15	19.02	19.46	20.32	14.16	19.22	18.44	2.44	13.23%	64.58
ST30	12.66	13.44	16.63	11.36	13.93	13.60	1.95	14.34%	52.58
ST45	13.08	13.69	14.21	13.59	14.25	13.76	0.48	3.49%	59.19
ST60	20.57	21.21	21.15	22.01	20.64	21.12	0.58	2.75%	51.47
ST75	22.03	19.93	21.28	22.27	21.52	21.41	0.91	4.25%	55.83
ST90	23.01	23.09	20.70	24.17	24.00	22.99	1.38	6.00%	62.41
SY0	20.16	19.96	20.52	20.09	20.03	20.15	0.22	1.09%	54.77
SY15	22.27	23.75	21.89	25.03	22.40	23.07	1.30	5.64%	60.56
SY30	22.40	21.13	22.13	22.43	22.32	22.08	0.54	2.45%	55.53
SY45	21.44	21.87	23.82	22.73	23.03	22.58	0.94	4.16%	55.29
SY60	15.22	15.10	17.98	15.01	17.55	16.17	1.46	9.03%	54.33
SY75	22.18	23.75	20.55	23.91	20.07	22.09	1.77	8.01%	61.18
SY90	21.92	21.20	22.32	24.48	22.12	22.41	1.23	5.49%	54.70

3.3. An Analysis of the Strength Characteristics

As shown in Figures 8–10, in order to facilitate the comparison, the intact without water pressure (confining pressure: 10 MPa) and the intact with water pressure (confining pressure: 10 MPa; water pressure: 3 MPa) sample and multi-shape fractured samples are analyzed together. As can be observed from Figure 8a, the change law of the key threshold of the samples with different single fracture inclinations varies with the change of prefabricated fracture inclinations under hydro-mechanical coupling. The peak strength of the single fracture samples is the lowest at 75° and the highest at 0°; the σ_{cd} value is the lowest at 90° and the highest at 15°; the σ_{ci} value is the lowest at 30° and the highest at 15°; and the σ_{cc} value is the lowest at 30° and the highest at 15°. Among them, the variation law of the peak strength is evidently consistent with that of the prefabricated fracture rocks studied by Zhao Cheng et al. [27]. The peak strength is the lowest at 30° (60° in the present study) and the highest at 90° (0° in the present study). As can be observed from Figure 8b, the average σ_{cd}/σ_c value, the average σ_{ci}/σ_c value, and the average σ_{cc}/σ_c value of the single fracture samples under different angles are 0.48, 0.28, and 0.21, respectively.

As can be observed from Figure 9a, the variation law of the key threshold of samples with different T-shaped fracture samples with the prefabricated fracture inclination under the hydro-mechanical coupling is as follows: the peak strength is the lowest at 60° and the highest at 15°; the σ_{cd} value is the lowest at 45° and the highest at 90°; the σ_{ci} value is the lowest at 30° and the highest at 90°; the σ_{cc} value is the lowest at 45° and the highest at 90°. As can be observed from Figure 9b, the average σ_{cd}/σ_c value, average σ_{ci}/σ_c value, and average σ_{cc}/σ_c value of T-shaped fracture samples under different dip angles are 0.55, 0.32, and 0.23, respectively.

As can be observed from Figure 10a, the variation law of the key threshold of the samples with different Y-shaped fracture samples with the prefabricated fracture inclination under the hydro-mechanical coupling is as follows: the peak strength is the lowest at 60° and the highest at 75°; the σ_{cd} value is the lowest at 60° and the highest at 90°; the σ_{ci} value

is the lowest at 60° and the highest at 15°; and the σ_{cc} value is the lowest at 60° and the highest at 90°. As can be observed from Figure 10b, the average σ_{cd}/σ_c value, average σ_{ci}/σ_c value, and average σ_{cc}/σ_c value of the Y-shaped fracture samples under different dip angles are 0.63, 0.37, and 0.27, respectively.

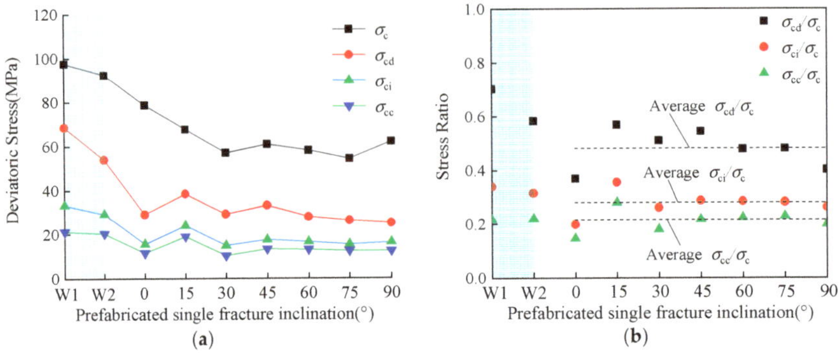

Figure 8. Relationship between fracture inclination and different thresholds, stress ratio of intact and single fracture samples: (**a**) relationship between the different thresholds and fracture inclination and (**b**) relationship between the stress ratio and fracture inclination.

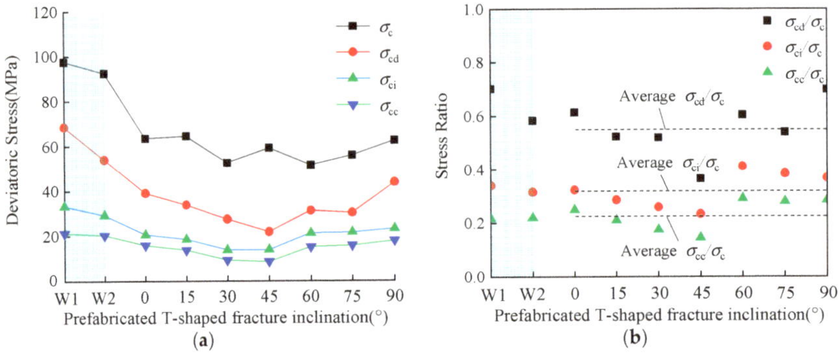

Figure 9. Relationship between the fracture inclination and different thresholds, and the stress ratios of the intact and T-shaped fracture samples: (**a**) relationship between the different thresholds and fracture inclination and (**b**) relationship between the stress ratio and fracture inclination.

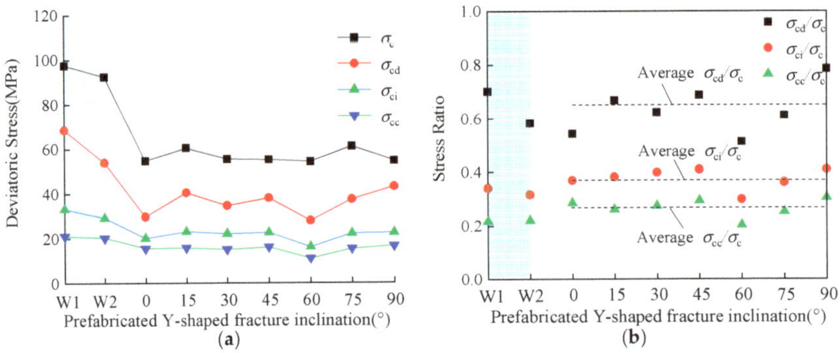

Figure 10. Relationship between the fracture inclination and different thresholds, and the stress ratios of the intact and Y-shaped fracture samples: (**a**) relationship between the different thresholds and fracture inclination and (**b**) relationship between the stress ratio and fracture inclination.

In summary, the samples, in descending order of peak strength, are the intact sample without water pressure, intact sample with water pressure, and all the samples with prefabricated fractures. The peak strength of the intact sample with water pressure is 5.25% lower than that of the intact sample without water pressure. The peak strength of the single fracture samples with different inclinations (0°, 15°, 30°, 45°, 60°, 75°, and 90°) decreased by 19.29%, 30.60%, 41.53%, 37.54%, 40.39%, 44.31%, and 36.45%, respectively, compared to the intact sample without water pressure, and decreased by 14.82%, 26.75%, 38.28%, 34.08%, 37.08%, 41.23%, and 32.93%, respectively, compared to the intact sample with water pressure. The peak strength of the T-shaped fracture samples with different inclinations (0°, 15°, 30°, 45°, 60°, 75°, and 90°) decreased by 34.89%, 33.79%, 46.09%, 39.32%, 47.23%, 42.76%, and 36.02%, respectively, compared to the intact sample without water pressure, and decreased by 31.28%, 30.12%, 43.11%, 35.95%, 44.31%, 39.59%, and 32.47%, respectively, compared to the intact sample with water pressure. The peak strength of the Y-shaped fracture samples with different inclinations (0°, 15°, 30°, 45°, 60°, 75°, and 90°) decreased by 43.85%, 37.91%, 43.07%, 43.32%, 44.30%, 37.28%, and 43.92%, respectively, compared to the intact sample without water pressure, and decreased by 40.74%, 34.47%, 39.91%, 40.17%, 41.21%, 33.80%, and 40.81%, respectively, compared to the intact sample with water pressure. It fully shows that the weakening effect of the water has less of an influence on the strength than the prefabricated fractures. In addition, there are no significant differences in the stress ratios for the intact sample without water pressure, the intact sample with water pressure, and the prefabricated fracture samples.

3.4. Deformation Characteristic Analysis

The variation trend of the elastic modulus and Poisson's ratio of the intact, single, T-shaped and Y-shaped fracture samples with fracture inclinations are shown in Figures 11–13. In Figure 11, the elastic modulus of the single fracture samples are the smallest at 30° and the largest at 0°, which are similar to the variation law of the peak strength. On the whole, the elastic modulus experiences a trend of first decreasing and then increasing with the increase in the fracture inclination, while the Poisson's ratio of the single fracture samples fluctuates to a certain extent with the increase in the fracture inclination. In Figure 12, the elastic modulus of the T-shaped fracture samples are the smallest at 60° and the largest at 75°. The elastic modulus generally experiences a decreasing–increasing–decreasing trend with the increase in the fracture inclination, while the Poisson's ratio of the T-shaped fracture samples fluctuates significantly with the increase in the fracture inclination. In Figure 13, the elastic modulus of the Y-shaped fracture samples are the smallest at 60° and the largest at 45°. The elastic modulus generally experiences a trend of increase–decrease–increase, with the increase in the fracture inclination, while the Poisson's ratio of the Y-shaped fracture samples fluctuates significantly with the increase in the fracture inclination. By comparing the intact samples without water pressure with the intact samples with water pressure and with the prefabricated fracture samples, it can be observed that the elastic modulus of the intact samples without water pressure is the largest, other samples are affected by the water pressure, and the single fracture samples are more affected by the water pressure, indicating that the fractures weaken the stiffness of the samples, which are consistent with the conclusion presented in reference [19].

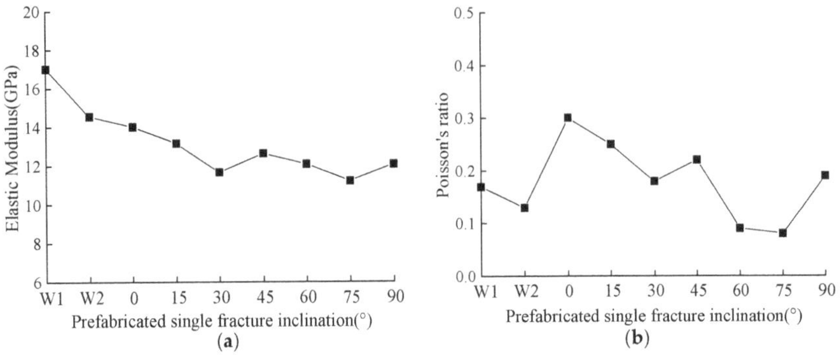

Figure 11. Relationship between the elastic modulus, Poisson's ratio, and fracture inclination of intact and single fracture samples: (**a**) relationship between the elastic modulus and fracture inclination and (**b**) relationship between the Poisson's ratio and fracture inclination.

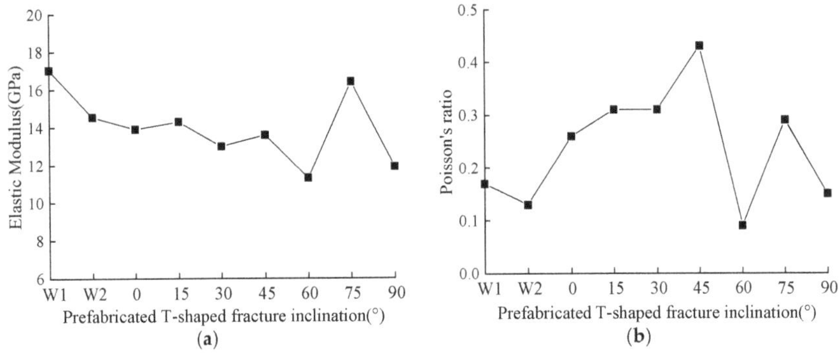

Figure 12. The relationship between the elastic modulus and Poisson's ratio of the intact and T-shaped fracture samples: (**a**) relationship between the elastic modulus and fracture inclination and (**b**) relationship between the Poisson's ratio and fracture inclination.

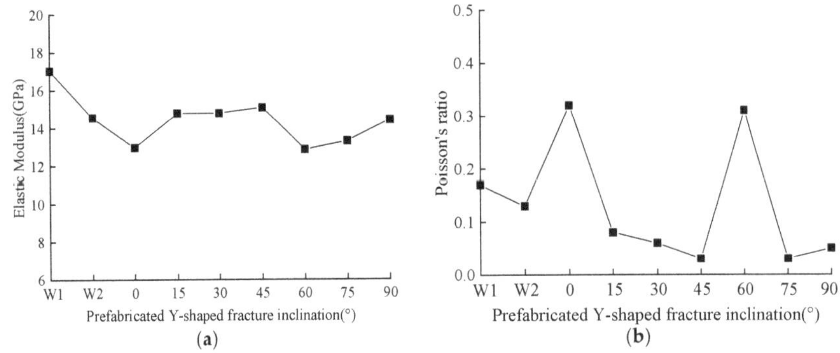

Figure 13. The relationship between the elastic modulus and the Poisson's ratio of the intact and Y-shaped fracture samples: (**a**) relationship between the elastic modulus and the fracture inclination and (**b**) relationship between the Poisson's ratio and the fracture inclination.

4. Damage Constitutive Model and Verification of the Multi-Shape Fractured Sandstone under Hydro-Mechanical Coupling

4.1. Construction of the Damage Variable for Multi-Shape Fractured Rock

According to the strain equivalence principle proposed by Lemaitre [28–30], it can be determined that the strain caused by the macroscopic nominal stress σ on the damaged

material is equivalent to the strain caused by the effective stress σ' on the non-damaged material, namely

$$\varepsilon = \frac{\sigma}{E'} = \frac{\sigma'}{E} = \frac{\sigma}{E(1-D)} \tag{1}$$

where E and E' are the elastic modulus of the non-damaged material and damaged material, respectively, D is the damage variable.

In this paper, the damage and failure of the multi-shape fractured sandstone samples during hydro-mechanical coupling loading are caused by the superposition or coupling of three types of damage: (1) the damage caused by water immersion before loading, (2) damage caused by random microcracks in the sample during loading, and (3) damage caused by the macroscopic cracks during loading.

It is very difficult to analyze the damage of the sandstone material under water action from the microscopic point of view. Therefore, the elastic modulus of the sandstone material before and after water action can be measured from the macroscopic point of view, and the expression of the damage variable under water action can be determined:

$$D_1 = 1 - \frac{E_1}{E_0} \tag{2}$$

where D_1 is the damage variable caused by water; E_0 and E_1 correspond to the elastic modulus of the sandstone samples before and after the damage under water.

In addition, the damage state of the prefabricated fractured sandstone samples after water action is regarded as the initial state, and the damage state caused by the load under the hydro-mechanical coupling after water action is regarded as the second damage state. Based on the Lemaitre strain equivalence hypothesis, the total damage variable of the fractured sandstone under the combined action of water action damage and hydro-mechanical coupling loading damage is

$$E_0(1-D_1)(1-D_2)\varepsilon_1 = E_0(1-D_0)\varepsilon_1 \tag{3}$$

Simplified as

$$D_0 = D_1 + D_2 - D_1 D_2 \tag{4}$$

where D_0 is the total damage variable of the fractured sandstone under the combined action of water action damage and hydro-mechanical coupling loading damage, D_2 is the coupling damage variable.

In the loading process under hydro-mechanical coupling, the loading damage variable is the coupling of micro-flaw damage and prefabricated fracture damage. Based on Lemaitre strain equivalence hypothesis, there is

$$\varepsilon_f = \varepsilon_a + \varepsilon_b - \varepsilon_c \tag{5}$$

where ε is the uniaxial strain of the sandstone materials, the footnotes a, b, and c represent the sandstone samples with prefabricated fractures and macrocracks, the sandstone samples containing only micro-flaws, the sandstone samples containing only macrocracks, and the initial damage sandstone samples, respectively.

Therefore, there is

$$\varepsilon_f = \frac{\sigma}{E_f} = \frac{\sigma}{E_a} + \frac{\sigma}{E_b} - \frac{\sigma}{E_0} \tag{6}$$

The damage variables caused by the micro-flaws and macrocracks are expressed as D_a and D_b, respectively, and their coupling damage variable is D_2. The following formula can be obtained based on continuous damage mechanics:

$$E_f = E_0(1-D_2) \tag{7}$$

$$E_a = E_0(1-D_a) \tag{8}$$

$$E_b = E_0(1 - D_b) \tag{9}$$

Substituting Equations (7)–(9) into Equation (6), the coupling damage variable D_2 in the loading process can be obtained as:

$$D_2 = 1 - \frac{(1 - D_a)(1 - D_b)}{(1 - D_a D_b)} \tag{10}$$

Substituting Equation (10) into Equation (4) and obtain:

$$D_0 = 1 - (1 - D_1)\frac{(1 - D_a)(1 - D_b)}{1 - D_a D_b} \tag{11}$$

4.2. Constitutive Model Construction of Multi-Shapes Fractured Rock

The distribution of micro-flaws in rock materials is random, which leads to great differences in the mechanical properties of the rock materials. Therefore, the rock materials are divided into micro-elements containing several flaws. It is assumed that the relationship between the statistical distribution density of the micro-element damage of rock materials and the damage variable is [31]:

$$dD = P(F)dF \tag{12}$$

where $P(F)$ is the probability density function of the rock-loading process, F is the strength parameter of the rock micro-element.

Assuming that the micro-element strength during rock loading obeys the Weibull distribution [32], the damage evolution equation of the sandstone samples under hydro-mechanical coupling is:

$$D_a = \int_0^F P(F)dF \tag{13}$$

$$D_a = \int_0^F \frac{n}{F_0}\left(\frac{F}{F_0}\right)^{n-1} \exp\left[-\left(\frac{F}{F_0}\right)^n\right] dF = 1 - \exp\left[-\left(\frac{F}{F_0}\right)^n\right] \tag{14}$$

Therefore, the constitutive model of the rock with only micro-flaws under hydro-mechanical coupling can be obtained:

$$\sigma_1 = E_1\varepsilon_1(1 - D_a) + 2\mu\sigma_3 = E_1\varepsilon_1 \exp\left[-\left(\frac{F}{F_0}\right)^n\right] + 2\mu\sigma_3 \tag{15}$$

where σ_1 and ε_1 are the axial stress and strain under hydro-mechanical coupling, respectively, E_1 is the elastic modulus.

From the above analysis, it can be observed that the key parameters to be determined in the damage constitutive equation in Equation (15) are n and F_0. Therefore, the micro-element strength of the rock is established by introducing the Drucker–Prager failure criterion [33]:

$$F = \alpha_0 I_1 + \sqrt{J_2} \tag{16}$$

$$\alpha_0 = \frac{\sqrt{3}\sin\varphi}{3\sqrt{3 + \sin^2\varphi}} \tag{17}$$

$$I_1 = \sigma_1^* + \sigma_2^* + \sigma_3^* \tag{18}$$

$$J_2 = \frac{1}{6}\left[(\sigma_1^* - \sigma_2^*)^2 + (\sigma_2^* - \sigma_3^*)^2 + (\sigma_3^* - \sigma_1^*)^2\right] \tag{19}$$

where φ is the internal friction angle of the rock; I_1 is the first invariant of the stress tensor; J_2 is the second invariant of the deviatoric stress tensor; σ_1, σ_2, and σ_3 are the principal normal stresses in the three principal stress directions; and the corresponding effective

stresses are σ_1^*, σ_2^*, and σ_3^*, respectively. In hydro-mechanical coupling tests (triaxial tests), $\sigma_2 = \sigma_3$ and $\sigma_2^* = \sigma_3^*$, according to Hooke's law:

$$\left.\begin{array}{l}\varepsilon_1 = (\sigma_1^* - 2\mu\sigma_3^*)^2/E_1 \\ \sigma_2^* = \sigma_3^* = \sigma_3/(1-D_a) \\ \sigma_1^* = \sigma_1/(1-D_a)\end{array}\right\} \tag{20}$$

$$F_1 = \alpha_0 I_1 + \sqrt{J_2} \tag{21}$$

$$I_1 = \frac{(\sigma_1+2\sigma_3)E_1\varepsilon_1}{\sigma_1-2\mu\sigma_3} \tag{22}$$

$$\sqrt{J_2} = \frac{(\sigma_1-\sigma_3)E_1\varepsilon_1}{\left[\sqrt{3}(\sigma_1-2\mu\sigma_3)\right]} \tag{23}$$

$$F = \frac{\sqrt{3}\sin\varphi}{3\sqrt{3+\sin^2\varphi}} \cdot \frac{(\sigma_1+2\sigma_3)E_1\varepsilon_1}{\sigma_1-2\mu\sigma_3} + \frac{(\sigma_1-\sigma_3)E_1\varepsilon_1}{\left[\sqrt{3}(\sigma_1-2\mu\sigma_3)\right]} \tag{24}$$

$$\begin{array}{l}\frac{d\sigma_1}{d\varepsilon_1} = E_1\exp\left[-\left(\frac{F}{F_0}\right)^n\right] + E_1\varepsilon_1\exp\left[-\left(\frac{F}{F_0}\right)^n\right]\left[-\frac{nF^{n-1}}{F_0^n}\right]\cdot\left\{\frac{\alpha_0 E_1\left[(\sigma_1+2\sigma_3)+\alpha_0 E_1\varepsilon_1\frac{d\sigma_1}{d\varepsilon_1}\right]}{\sigma_1-2\mu\sigma_3}\right. \\ \left. -\frac{E_1\varepsilon_1(\sigma_1+2\sigma_3)}{(\sigma_1-2\mu\sigma_3)^2}\frac{d\sigma_1}{d\varepsilon_1} + \frac{E_1(\sigma_1-\sigma_3)+E_1\varepsilon_1\frac{d\sigma_1}{d\varepsilon_1}}{\sqrt{3}(\sigma_1-2\mu\sigma_3)} - \frac{E_1\varepsilon_1(\sigma_1-\sigma_3)}{\sqrt{3}(\sigma_1-2\mu\sigma_3)^2}\frac{d\sigma_1}{d\varepsilon_1}\right\}\end{array} \tag{25}$$

From the stress–strain curve of the hydro-mechanical coupling test, it can be observed that the stress–strain curve meets the following boundary conditions:

$$\varepsilon_1 = \varepsilon_{1p},\ \sigma_1 = \sigma_{1p},\ d\sigma_1/d\varepsilon_1 = 0 \tag{26}$$

Substitute Equation (26) into Equation (25) to obtain the following:

$$\frac{nF_p^{n-1}}{F_0^n} = \frac{(\sigma_{1p} - 2\mu\sigma_3)}{E_1\varepsilon_{1p}\left[\alpha_0(\sigma_{1p}+2\sigma_3) + (\sigma_{1p}-\sigma_3)/\sqrt{3}\right]} \tag{27}$$

Substitute Equation (26) into Equation (15) to obtain the following:

$$\left(\frac{F_p}{F_0}\right)^n = -\ln\frac{\sigma_{1p}-2\mu\sigma_3}{E_1\varepsilon_{1p}} \tag{28}$$

Combine Equations (27) and (28) to obtain

$$n = \frac{-F_p(\sigma_{1p}-2\mu\sigma_3)}{E_1\varepsilon_{1p}\left[\alpha_0(\sigma_{1p}+2\sigma_3)+(\sigma_{1p}-\sigma_3)/\sqrt{3}\right]\ln\left[\frac{\sigma_{1p}-2\mu\sigma_3}{E_1\varepsilon_{1p}}\right]} \tag{29}$$

$$F_0 = \left[\frac{F_p^n}{-\ln\left[(\sigma_{1p}-2\mu\sigma_3)/E_1\varepsilon_{1p}\right]}\right]^{\frac{1}{n}} \tag{30}$$

In summary, the stress–strain relationship of the multi-shape fractured sandstone under hydro-mechanical coupling is:

$$\sigma = E_0(1-D_1)(1-D_2)\varepsilon_1 + 2\mu\sigma_3 = E_0(1-D_0)\varepsilon_1 + 2\mu\sigma_3 \tag{31}$$

4.3. Validation of the Constitutive Model under Hydro-Mechanical Coupling

In Table 5, in order to verify the constitutive model of the multi-shape fractured sandstone sample under hydro-mechanical coupling, according to the test results, the Weibull distribution parameters n and F_0 are solved by Equations (29) and (30). In order to avoid redundancy, single, T-shaped and Y-shaped fracture sandstone samples with fracture

inclinations of 0°, 45°, and 90° were selected as examples to compare and analyze the results of theoretical curves and experimental curves. As shown in Figure 14b,c,f, it can be observed that some differences occur in the coincidence degree of the curves before the peak stress, but the overall development law is consistent. The constitutive calculation curves of the remaining samples are highly coincidental with the experimental stress–strain curves. Table 6 compares and analyzes the error values of theoretical calculations and test results under different constitutive models, and it is found that the error of the constitutive model in this paper is the smallest. The constitutive model proposed in this paper is further verified by RFPA2D-FLOW software, through comprehensively comparing the experimental results and theoretical calculation results. A constant 10 MPa confining pressure was applied to the left and right boundaries of the numerical model during the simulation process, and the bottom boundary was fixed. An axial load speed of 3×10^{-5} m/s was applied to the top boundary to control the load until the sample completely lost its bearing capacity. In addition, the seepage behavior in the hydro-mechanical coupling test was simulated by the steady-state seepage model, the left and right boundaries of the sample were impermeable boundaries, and a water pressure of 3 MPa was applied to the lower part of the sample. The main key simulation parameters were the residual strength percentage 0.3%, friction angle 43°, permeability coefficient 0.08 m/d, effective stress coefficient 0.49, coupling coefficient 0.0979, and the assignment value of elastic modulus; the Poisson's ratio and compressive strength varied with fracture inclination. The simulation results are shown in Figure 14, and the simulation curves and the constitutive calculation curves are basically highly coincident with the measured stress–strain curves, which further verifies the rationality of the constitutive model. In general, the damage constitutive model established in this paper can reflect the stress–strain relationship of the multi-shape fractured sandstone samples better. However, there are also some differences between the damage constitutive model proposed in this paper and the experimental and numerical results, which are mainly caused by the uneven distribution of the mineral composition of the rock. In addition, in the process of simulation analysis, numerical simulation methods often need to simplify the boundary conditions and material properties, which more or less affects the analysis results, and the results obtained with different meshing accuracies are also different, which eventually leads to certain differences.

Table 5. Physical and mechanical properties of the samples with prefabricated fractures.

Sample No.	Peak Stress/MPa	Peak Strain/%	Total Damage Variable	Weibull Parameters	
				n	F_0
SF0	78.73	0.639	0.276	6.011	98.363
SF15	67.70	0.568	0.300	5.027	89.403
SF30	57.04	0.490	0.317	4.487	78.190
SF45	60.92	0.572	0.375	3.204	93.988
SF60	58.15	0.525	0.348	3.705	85.349
SF75	54.32	0.552	0.422	2.562	90.946
SF90	61.99	0.637	0.429	2.482	105.014
ST0	63.51	0.531	0.303	5.102	83.475
ST15	64.58	0.499	0.240	8.543	73.265
ST30	52.58	0.426	0.275	6.077	65.484
ST45	59.19	0.454	0.237	8.869	66.312
ST60	51.47	0.553	0.453	2.240	90.513
ST75	55.83	0.412	0.208	13.289	57.208
ST90	62.41	0.654	0.440	2.371	107.467
SY0	54.77	0.580	0.446	2.313	95.205
SY15	60.56	0.469	0.241	8.445	68.898
SY30	55.53	0.423	0.230	9.653	61.149
SY45	55.29	0.407	0.202	14.633	55.831
SY60	54.33	0.403	0.208	13.207	55.929
SY75	61.18	0.452	0.205	13.821	62.438
SY90	54.70	0.385	0.169	41.941	48.072

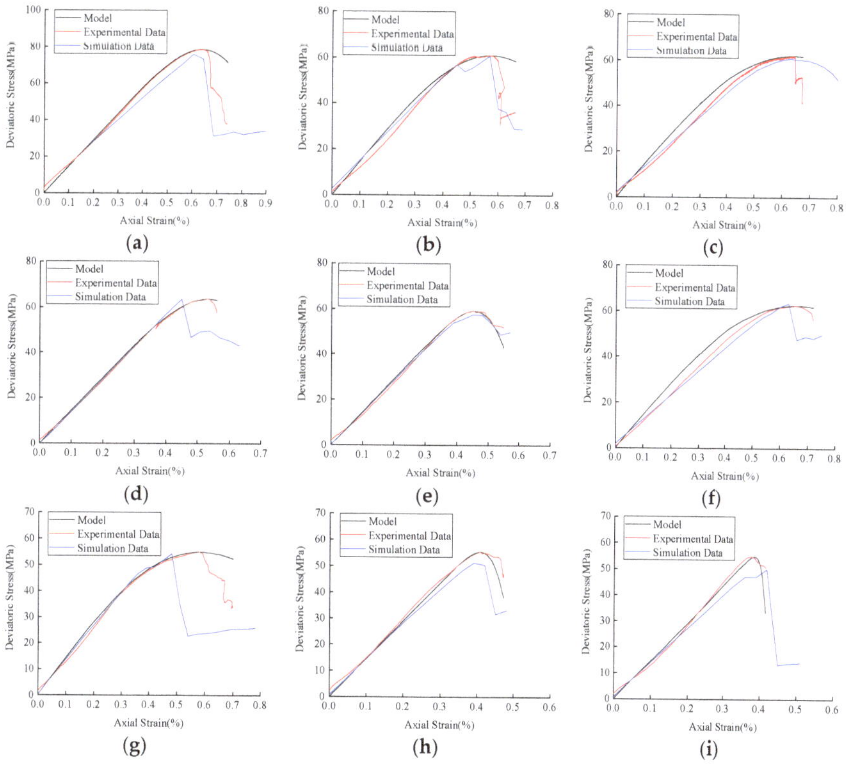

Figure 14. Comparison of the theoretical curves, simulation curves, and experimental curves of the multi-shape fractured samples: (**a**) SF0; (**b**) SF45; (**c**) SF90; (**d**) ST0; (**e**) ST45; (**f**) ST90; (**g**) SY0; (**h**) SY45; and (**i**) SY90.

Table 6. Typical constitutive model comparison table.

Source	Constitutive Model	Elastic Modulus Relative Error	Peak Strength Relative Error	Applicable Conditions
Reference [34]	$\sigma = E(1-D)\varepsilon = \frac{E(\varepsilon - m\varepsilon^{k+1})}{1-n_0}$	13%	2%	Chemical damage, intact rock, uniaxial
Reference [16]	$[\sigma] = [E][\varepsilon](1-\Omega_{12}) = [E][\varepsilon]\frac{(1-\Omega)(1-D_2)}{1-\Omega D_2}$	41%	6%	Fractured rock, uniaxial
Reference [35]	$\sigma = E_0(1-D)\varepsilon_1$ $D = D_c + D_{load} - D_c D_{load}$	2%	1%	Chemical damage, fractured rock, uniaxial
This article	$\sigma = E_0(1-D_0)\varepsilon_1 + 2\mu\sigma_3$ $D_0 = 1-(1-D_1)\frac{(1-D_a)(1-D_b)}{1-D_a D_b}$	1.8%	0.8%	Water damage, fractured rock, triaxial

In Figure 15, for the multi-shape fractured sandstone samples with different angles, the damage evolution curve increases slowly at first, then the curve rises sharply, and finally rises slowly and tends to a constant value 1. With the increase in the strain, the curve development and variation of multi-shape fractured sandstone samples are slightly different. Combined with Table 5 and Figure 15, it can be seen that the fracture inclination is 90°; the total damage value corresponding to the peak strength of single fracture sandstone sample is the largest. When the fracture inclination is 60°, the total damage value corresponding to the peak strength of the T-shaped fracture sandstone sample is the largest; when the fracture inclination is 0°, the total damage value corresponding to the peak strength of the Y-shaped fractured sandstone sample is the largest. Overall, the fracture inclinations do

not affect the total damage trend of the multi-shape fractured sandstone samples, but only affects their numerical value.

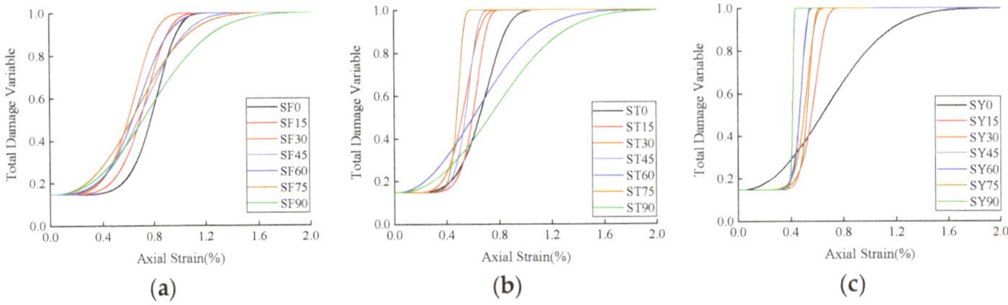

Figure 15. Loaded damage evolution curves of the multi-shape fractured samples: (**a**) single fracture samples, (**b**) T-shaped fracture samples, and (**c**) Y-shaped fracture samples.

5. Conclusions

The strength, deformation, and damage evolution characteristics of the multi-shape fractured sandstone under hydro-mechanical coupling were systematically studied by conducting hydro-mechanical coupling tests on fractured sandstone samples at different inclinations (α = 0°, 15°, 30°, 45°, 60°, 75°, and 90°) under a confining pressure of 10 MPa and a water pressure of 3 MPa. The main conclusions are as follows:

(1) Based on the stress–strain curve, the volumetric strain response method is proposed by integrating the advantages of the volumetric strain method, lateral strain method, crack volumetric strain method, and lateral strain response method, and the σ_{cc} and σ_{ci} values of the multi-shape fractured sandstone samples are determined. By introducing standard deviation (SD) and the coefficient of variation (CoV), the dispersion degree of σ_{cc} and σ_{ci} obtained by various methods was analyzed, and the rationality of the method is verified.

(2) The peak strength of the single fractured sandstone samples is the lowest at 75° and the highest at 0°, the peak strength of the T-shaped fracture sandstone samples is the lowest at 60° and the highest at 15°, and the peak strength of the Y-shaped fractured sandstone samples is the lowest at 60° and the highest at 75°. The crack closure stress and crack initiation stress of the multi-shape fractured sandstone samples are different, as well as the angles when the damage stress and peak strength (σ_c) reach their peak values.

(3) There are no significant differences in the stress ratios for intact sample without water pressure, the intact sample with water pressure, and the prefabricated fracture samples, indicating that the stress ratio is almost unaffected by the shape and inclination of the fractures in the rocks.

(4) The samples, in descending order of peak strength, are the intact sample without water pressure, intact sample with water pressure, and all the samples with prefabricated fractures, which fully shows that the weakening effect of water has less of an influence on the strength than the prefabricated fractures. The elastic modulus of the intact sample without water pressure is the largest, and the single fracture samples are most affected by water and fractures. In addition, the Poisson's ratio of the multi-shape sandstone samples fluctuates significantly with the increase in the fracture inclination.

(5) A superposition or coupling damage constitutive model, considering the water effect, micro-flaws and macrocracks, is proposed, which is in good agreement with the experimental stress–strain curves and numerical simulation curves, and is scientifically reasonable. When the fracture inclinations are 90°, 60°, and 0°, respectively, the total damage values corresponding to the peak strength of the single, T-shaped, and Y-shaped fracture sandstone samples are the largest.

Author Contributions: Conceptualization, Y.Z. and X.W.; methodology, Q.G. and Z.Z.; validation, Y.Z., X.W. and Q.G.; formal analysis, X.W. and L.T.; investigation, Z.Z.; resources, M.C.; writing original draft preparation, Y.Z., X.W. and Q.G.; writing—review and editing, Q.G. and Z.Z.; supervision, M.C.; project administration, M.C.; funding acquisition, Q.G. and M.C. All authors have read and agreed to the published version of the manuscript.

Funding: This research was funded by the National Key Research and Development Program of China: 2021YFC3001301, National Natural Science Foundation of China: 51974014 and 52074020, Fundamental Research Funds for the Central Universities: FRF-DRY-20-032.

Institutional Review Board Statement: Not applicable.

Informed Consent Statement: Not applicable.

Data Availability Statement: Not applicable.

Conflicts of Interest: The authors declare that they have no conflict of interest.

References

1. Gao, F.; Xiong, X.; Xu, C.; Zhou, K. Mechanical property deterioration characteristics and a new constitutive model for rocks subjected to freeze-thaw weathering process. *Int. J. Rock Mech. Min. Sci.* **2021**, *140*, 104642. [CrossRef]
2. Zhu, Z.N.; Tian, H.; Jiang, G.S.; Cheng, W. Effects of high temperature on the mechanical properties of Chinese marble. *Rock Mech. Rock Eng.* **2018**, *51*, 1937–1942. [CrossRef]
3. Liu, Y.; Dai, F. A damage constitutive model for intermittent jointed rocks under cyclic uniaxial compression. *Int. J. Rock Mech. Min. Sci.* **2018**, *103*, 289–301. [CrossRef]
4. Yilmaz, I. Influence of water content on the strength and deformability of gypsum. *Int. J. Rock Mech. Min. Sci.* **2010**, *47*, 342–347. [CrossRef]
5. Vasarhelyi, B.; Ván, P.J.E.G. Influence of water content on the strength of rock. *Eng. Geol.* **2006**, *84*, 70–74. [CrossRef]
6. Kou, M.; Liu, X.; Tang, S.; Wang, Y. 3-D X-ray computed tomography on failure characteristics of rock-like materials under coupled hydro-mechanical loading. *Theor. Appl. Fract. Mech.* **2019**, *104*, 102396. [CrossRef]
7. Wang, H.L.; Xu, W.Y.; Shao, J.F. Experimental researches on hydro-mechanical properties of altered rock under confining pressures. *Rock Mech. Rock Eng.* **2014**, *47*, 485–493. [CrossRef]
8. Wang, L.; Liu, J.F.; Pei, J.L.; Xu, H.N.; Bian, Y. Mechanical and permeability characteristics of rock under hydro-mechanical coupling conditions. *Environ. Earth Sci.* **2015**, *73*, 5987–5996. [CrossRef]
9. Song, Z.; Cheng, Y.; Tian, X.; Wang, J.; Yang, T. Mechanical properties of limestone from Maixi tunnel under hydro-mechanical coupling. *Arab. J. Geosci.* **2020**, *13*, 402. [CrossRef]
10. Wang, Q.; Hu, X.; Zheng, W.; Li, L.; Xu, C. Mechanical properties and permeability evolution of red sandstone subjected to hydro-mechanical coupling: Experiment and discrete element modelling. *Rock Mech. Rock Eng.* **2021**, *54*, 2405–2423. [CrossRef]
11. Kawamoto, T.; Ichikawa, Y.; Kyoya, T. Deformation and fracturing behaviour of discontinuous rock mass and damage mechanics theory. *Int. J. Numer. Anal. Methods Geomech.* **1988**, *12*, 1–30. [CrossRef]
12. Swoboda, G.; Shen, X.P.; Rosas, L. Damage model for jointed rock mass and its application to tunnelling. *Comput. Geotech.* **1998**, *22*, 183–203. [CrossRef]
13. Zhang, T.; Xu, W.; Huang, W.; Wu, G. Experimental study on mechanical properties of multi-layered rock mass and statistical damage constitutive model under hydraulic-mechanical coupling. *Eur. J. Environ. Civ. Eng.* **2020**. [CrossRef]
14. Chen, W.L.; Li, N. Damage model of the rock mass medium with intermittent cracks. *Chin. J. Geotech. Eng.* **2000**, *22*, 430–434.
15. Xu, P.; Yang, S.Q. A fracture damage constitutive model for fissured rock mass and its experimental verification. *Arab. J. Geosci.* **2017**, *10*, 164. [CrossRef]
16. Chen, S.; Qiao, C. Composite damage constitutive model of jointed rock mass considering crack propagation length and joint friction effect. *Arab. J. Geosci.* **2018**, *11*, 283. [CrossRef]
17. Martin, C.D.; Chandler, N.A. The progressive fracture of Lac du Bonnet granite. *Int. J. Rock Mech. Min. Sci. Geomech. Abstr.* **1994**, *31*, 643–659. [CrossRef]
18. Brace, W.F.; Paulding, B.W.; Scholz, C. Dilatancy in the fracture of crystalline rocks. *J. Geophys. Res.* **1966**, *71*, 3939–3953. [CrossRef]
19. Hoek, E.; Bieniawski, Z.T. Brittle fracture propagation in rock under compression. *Int. J. Fract.* **1965**, *1*, 137–155. [CrossRef]
20. Hallbauer, D.K.; Wagner, H.; Cook, N.G.W. Some observations concerning the microscopic and mechanical behaviour of quartzite samples in stiff, triaxial compression tests. *Int. J. Rock Mech. Min. Sci. Geomech. Abstr.* **1973**, *10*, 713–726. [CrossRef]
21. Zhao, X.G.; Cai, M.; Wang, J.; Ma, L.K. Damage stress and acoustic emission characteristics of the Beishan granite. *Int. J. Rock Mech. Min. Sci.* **2013**, *64*, 258–269. [CrossRef]
22. Lajtai, E.Z. Brittle fracture in compression. *Int. J. Fract. Mech.* **1974**, *10*, 525–536. [CrossRef]
23. Nicksiar, M.; Martin, C.D. Evaluation of methods for determining crack initiation in compression tests on low-porosity rocks. *Rock Mech. Rock Eng.* **2012**, *45*, 607–617. [CrossRef]

24. Zhao, X.G.; Wang, J.; Ma, L.K.; Su, R.; Cai, M.; Wang, G.B. Acoustic emission behaviors of the Beishan granite under uniaxial and triaxial compression conditions. In Proceedings of the 3rd ISRM SINOROCK Symposium, Shanghai, China, 17–21 June 2013; CRC Press/Balkema: London, UK, 2013; pp. 90–95.
25. Huang, D.; Li, X.Q. Numerical simulation research on characteristic strength of marble based on development of microcrack. *Rock Mech. Rock Eng.* **2017**, *38*, 253–262.
26. Ruth, D.; Ma, H. On the derivation of the Forchheimer equation by means of the averaging theorem. *Transp. Porous Med.* **1992**, *7*, 255–264. [CrossRef]
27. Zhao, C.; Xing, J.Q.; Niu, J.L.; Ma, C.C. Experimental study on crack propagation of precrack rock-like samples under hydro-mechanical coupling. *Chin. J. Rock Mech. Eng.* **2019**, *38*, 2823–2830.
28. Lemaitre, J. A continuous damage mechanics model for ductile fracture. *J. Eng. Mater. Technol.* **1985**, *107*, 83–89. [CrossRef]
29. Lemaitre, J. How to use damage mechanics. *Nucl. Eng. Des.* **1984**, *80*, 233–245. [CrossRef]
30. Lemaitre, J.; Chaboche, J.L. *Mechanics of Solid Materials*; Cambridge University Press: Cambridge, UK, 1990; pp. 159–168.
31. Yang, T.H.; Liu, J.; Zhu, W.C.; Elsworth, D.; Tham, L.G.; Tang, C.A. A coupled flow-stress-damage model for groundwater outbursts from an underlying aquifer into mining excavations. *Int. J. Rock Mech. Min. Sci.* **2007**, *44*, 87–97. [CrossRef]
32. Weibull, W. A statistical distribution function of wide applicability. *J. Appl. Mech.* **1951**, *18*, 293–297. [CrossRef]
33. Drucker, D.C.; Prager, W. Soil mechanics and plastic analysis or limit design. *Q. Appl. Mech.* **1952**, *10*, 157–165. [CrossRef]
34. Zou, Y.G.; Wang, X.J.; Li, S.C.; Zeng, Q.; Wang, X.M.; Liu, J. Mechanical Properties and Damage Evolution Constitutive Model of Copper Oxide Rocks under Chemical Corrosion. *Min. Res. Dev.* **2021**, *40*, 56–61.
35. Pan, J.L.; Cai, M.F.; Li, P.; Guo, Q.F. A damage constitutive model of rock-like materials containing a single crack under the action of chemical corrosion and uniaxial compression. *J. Cent. South Univ.* **2021**, *29*, 486–498. [CrossRef]

minerals

MDPI

Article

Strain Energy Release and Deep Rock Failure Due to Excavation in Pre-Stressed Rock

Peng Xiao, Diyuan Li * and Quanqi Zhu

School of Resources and Safety Engineering, Central South University, Changsha 410083, China; xiaopengaizhanghuimin@csu.edu.cn (P.X.); quanqi_zhu@csu.edu.cn (Q.Z.)
* Correspondence: diyuan.li@csu.edu.cn

Abstract: Deep rock engineering is in a high pre-stressed state before excavation. In this research, a method to calculate the release of strain energy caused by excavation in pre-stressed rock is proposed. The normal stress release after excavation leads to a reduction in strain energy in rock specimens. The influence of excavation height and width on strain energy release is inconsistent under vertical loading. When the height of the hole is 1 mm, the strain energy release is large, and the increase in height of hole leads to a slow increase in the strain energy release. When the width of the hole is 1 mm, the strain energy release is very small, and the increase in the width of the hole leads to an increasingly faster release of strain energy. This strain energy release exponentially increases with the increase in the lateral pressure coefficient, showing a trend in the second power of the lateral pressure coefficient. Moreover, the tunnel failure caused by excavation under high stress is obtained by a numerical calculation. The failure modes of the deep tunnel model are strain rockbursts caused by tangential stress concentrations and spalling caused by normal stress release, which is also observed in the failure mode of the actual tunnel.

Keywords: deep engineering; tunnel excavation; high ground stress; strain energy release; rockburst

Citation: Xiao, P.; Li, D.; Zhu, Q. Strain Energy Release and Deep Rock Failure Due to Excavation in Pre-Stressed Rock. *Minerals* **2022**, *12*, 488. https://doi.org/10.3390/min12040488

Academic Editor: Yosoon Choi

Received: 24 March 2022
Accepted: 14 April 2022
Published: 16 April 2022

1. Introduction

With the large-scale development of deep underground projects, such as deep mines [1,2], deep buried traffic tunnels [3,4], and hydropower station chambers [5,6], many ground pressure disasters have been reported in deep projects [7–9]. Therefore, the study of deep ground pressure disasters has become a popular topic in the field of rock engineering [10]. Many disasters often occur in deep engineering, such as rockbursts, slabbing, large-scale collapse and large deformation [11–13]. Many scholars explain the mechanism of these disasters from many aspects, such as in situ stress [14,15], strength theory [16–18], rock fault [19,20], the energy storage of rock [21,22], engineering structure [23,24], dynamic disturbance [25–28], stress monitoring [29], and so on [30–34]. Among them, the high ground stress is one of the most important causes of ground pressure disasters. Many studies show that most ground pressure disasters occur in high-stress areas [18,35,36].

In the study of the influence of ground stress on ground pressure disasters, Hu and Gong et al. found that the tangential stress concentration after excavation led to spalling and rockburst through the true triaxial testing of specimens with holes [37–39]. Tao et al. found that the release of normal stress on the free face after excavation caused damage to rock through numerical simulation [40–42]. Many engineering practices show that a faster excavation speed leads to more serious surrounding rock damage [43]. This is because the faster the excavation results, the faster the release of normal stress on the excavation surface. Therefore, the normal stress release on the excavation surface of surrounding rock by excavation needs to be considered.

Su et al. found that the excavation in high-stressed rock mass led to the release of radial stress and the concentration of tangential stress [44]. Considering the initial stress

state, there are some differences between the excavation after loading and the loading after excavation. For the loading after excavation, e.g., the laboratory test of a rock specimen with holes, there is no release of normal stress during the loading process, since there is no normal stress around the holes to begin with. For excavation after loading, it is more suitable to describe the complete process of deep engineering excavations. As shown in Figure 1, normal stress exists around the roadway before excavation, and excavation will lead to the release of normal stress around the roadway and the concentration of tangential stress.

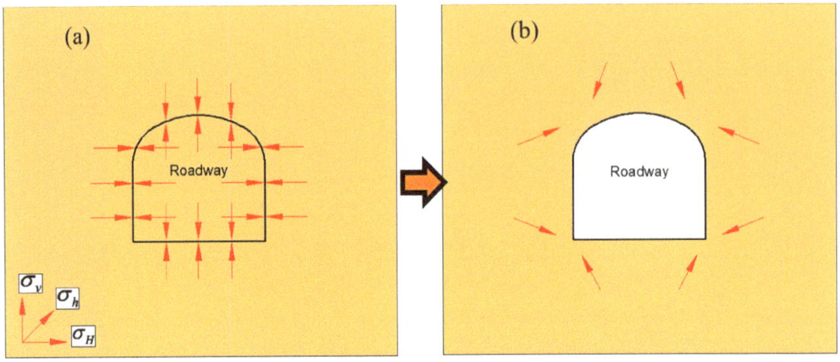

Figure 1. The stress field evolution caused by excavation in deep engineering. (**a**) Stress state before excavation. (**b**) Release of normal stress and concentration of tangential stress after excavation.

It is important to calculate the released strain energy after excavation, which can provide an important basis for energy absorption support and rockburst prediction. At present, there are relatively few studies on the normal stress release caused by excavation in deep pre-stressed rock. The tests of excavation in pre-stressed rock are difficult in the laboratory. This research proposes a method to calculate the strain energy release caused by excavation. The released strain energy caused by excavation in pre-stressed rock is calculated through laboratory tests and numerical calculations. Moreover, the influence of excavation height, width and lateral pressure on the strain energy release are analyzed. Finally, the deep tunnel failure due to excavation in pre-stressed rock is reproduced by means of a numerical simulation.

2. Calculation of Strain Energy Release by Excavation in Pre-Stressed Rock

2.1. Methodology

It is difficult to measure the released strain energy directly caused by excavation, but it can be indirectly obtained by another way. As shown in Figure 2, the intact specimen and the specimen with holes are compressed by a test machine and subjected to the same deformation l_0, except that the second specimen includes a hole of size $w \times h$, the size of the two specimens is the same: W is width of specimen, H is height, T is thickness, F is external load, l is displacement of load, w is the width of hole, and h is the height of hole.

During loading, the strain energy is stored inside the two specimens. In this paper, the U_{Int} represents the strain energy of the intact specimen, which is calculated by Equation (1); U_{Hole} represents the strain energy of the hole, which is calculated by Equation (2); $U_{Aft\text{-}ex}$ represents the strain energy of the specimen after excavation, which is calculated by Equation (3); and $U_{Spe\text{-}ho}$ represents the strain energy of the specimen with a hole, which is calculated by Equation (4). Due to the same deformation of the two specimens, the stress state of the intact specimen after excavating a hole will eventually be the same as that of the specimen with a hole. If the $U_{Aft\text{-}ex}$ and $U_{Spe\text{-}ho}$ are equal, there is no strain energy release

after excavation; otherwise, there is a strain energy release after excavation. The actual released strain energy U_{Rele} can be calculated by Equation (5):

$$U_{Int} = \int_0^{l_0} F_1 dl \quad (1)$$

$$U_{Hole} = \frac{hw}{HW} U_{Int} \quad (2)$$

$$U_{Aft-ex} = U_{Int} - U_{Hole} \quad (3)$$

$$U_{Spe-ho} = \int_0^{l_0} F_2 dl \quad (4)$$

$$U_{Rele} = U_{Int} - U_{Hole} - U_{Spe-ho} \quad (5)$$

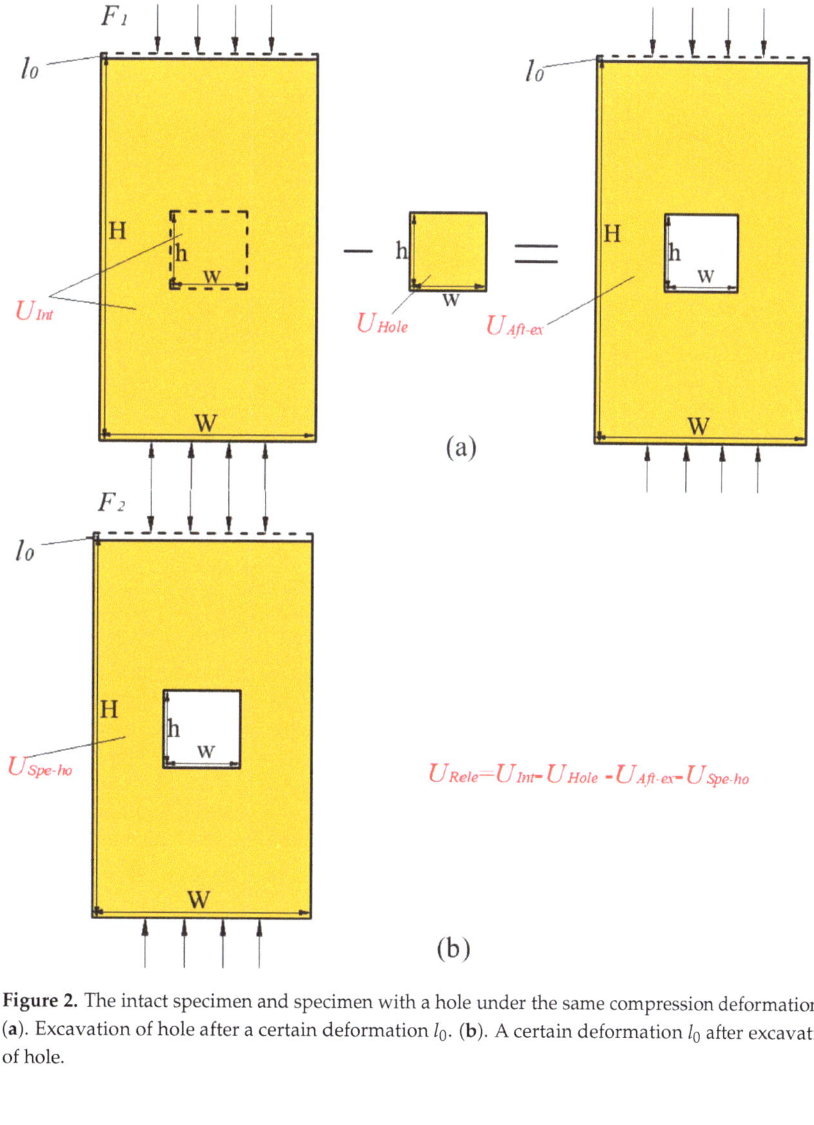

Figure 2. The intact specimen and specimen with a hole under the same compression deformation l_0. (**a**). Excavation of hole after a certain deformation l_0. (**b**). A certain deformation l_0 after excavation of hole.

2.2. Strain Energy Release by Excavation in Laboratory Test

As shown in Figure 3, the specimen is marble, its size is 150 mm × 120 mm × 20 mm ($H \times W \times T$), and a rectangular hole is prefabricated in the geometric center of specimen. In order to ensure the integrity and homogeneity of specimen, a whole marble plate was selected for processing, and the surface of specimen was smooth without obvious defects visible to the naked eye. The rectangular hole was made by professional water jet cutting mechanical equipment OMAX 2626. There are two types of specimens: intact specimen (S-Intact) and specimen with rectangle hole (S-Hole). The specimen sizes are shown in Table 1. Instron-1346 hydraulic servo control testing machine was used for uniaxial compression test. The loading strain rate was 2×10^{-5}, which belonged to quasi-static loading. The loading curve with deformation was selected as 0.453 mm, equivalent to 0.3% strain to ensure that all specimens were always in elastic stage. The load versus displacement curves of two specimens are shown in Figure 4. On the whole, the stress–strain curves of two specimens showed the characteristics of plastic elastic deformation.

Table 1. The size of specimen and hole.

Specimens	High/mm	Width/mm	Thick/mm	Hole Size/mm
S-Intact	149.7	119.1	21.0	0
S-Hole	149.3	119.2	21.6	22 × 22

Figure 3. Intact specimen and specimen with rectangle hole.

The strain energy of each part was calculated, as shown in Figure 4 and Equations (1)–(5); the strain energy of each part is shown in Table 2. If the stored energy in the hole is subtracted, is the residual strain energy of intact specimen equal to the strain energy of specimen with hole? It can be seen from Table 2 that the residual strain energy $U_{Aft\text{-}ex}$ of the intact specimen after excavating the hole is greater than that $U_{Spe\text{-}ho}$ of the specimen with the hole. Since the stress state of the intact specimen after excavating the hole will eventually be same as that of the specimen with hole, this indicates that there is strain energy release around the hole after excavation. The actual released strain energy U_{Rele} can be calculated by U_{Int}-U_{Hole}-$U_{Spe\text{-}ho}$.

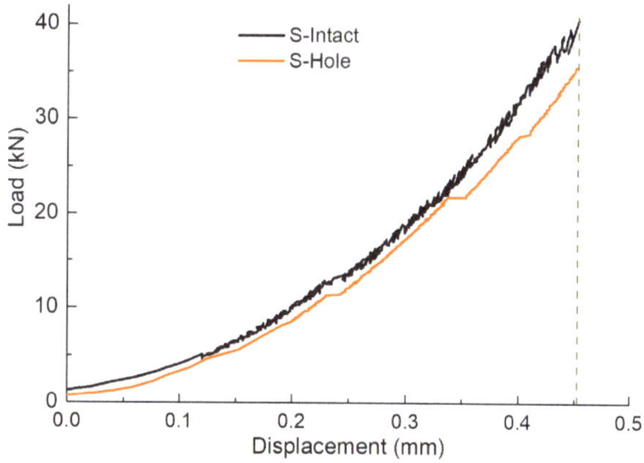

Figure 4. Load versus displacement curve of four specimens.

Table 2. The strain energy of each part.

Specimen	U_{Int}/J	U_{Hole}/J	$U_{Aft\text{-}ex}$/J	$U_{Spe\text{-}ho}$/J	U_{Rele}/J
S-Intact	6.848	0.184	6.664	---	0.703
S-Hole	---	---	---	5.961	

The strain energy change in the intact specimen from loading to excavation and then to stability (stress balance) is shown in Figure 5; points 1 to 2 represent the gradual increase in strain energy during loading of intact specimen. Point 2 represents stop loading. Point 3 represents the residual strain energy $U_{Aft\text{-}ex}$ of intact specimen after excavating hole, the stress around hole is not balanced. Point 4 represents the equilibrium state after the normal stress around the hole is released and the tangential stress is concentrated. It can be seen from Figure 5 that the strain energy of intact specimen gradually decreases from point 2 to point 3 and then to point 4. This also means that the deep pre-stressed rock changes from the stable state before excavation to the unstable state after excavation, and finally to the stable state.

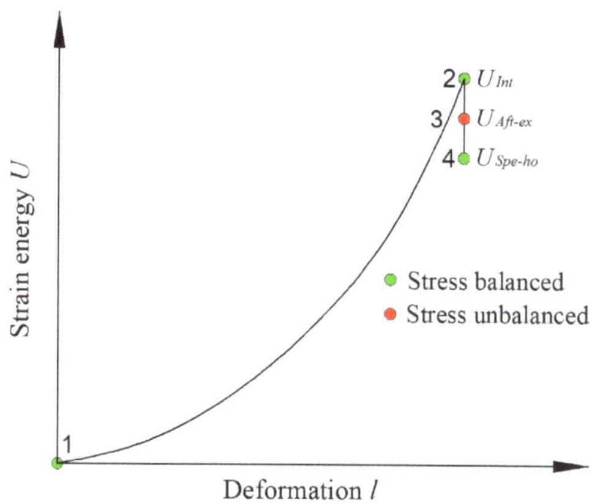

Figure 5. Strain energy change in specimen at different stages.

2.3. *Strain Energy Release by Excavation in Numerical Analysis*

As shown in Figure 6, two elastic material models were built in a code ELFEN. The height H of specimen is 150 mm, width W is 120 mm, thickness T is 20 mm, and the size of hole is the same as that in Section 2. In order to obtain a more obvious law of strain

energy release under different states, the model was set to elastic, and no failure criterion was adopted in the model, which means that plastic deformation and failure did not occur. The model properties are shown in Table 3. The elastic modulus of the specimen was obtained from the load displacement curve of the intact specimen in Figure 4, which was about 5.01 GPa. The specimen was placed between two pressure platens. The loading rate was about 6.7×10^{-3} s^{-1}, which could be considered as a quasi-static loading. When the displacement of the pressure platen reached 0.453 mm, the loading ended.

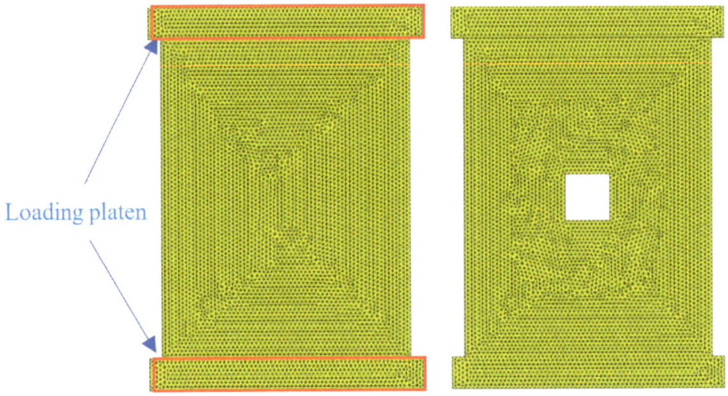

Figure 6. Two models of intact specimen and specimen with rectangle hole.

Table 3. Material properties adopted in model.

Mechanical Parameters	Specimen	Loading Platen
Young's modulus (E, GPa)	5.01	211
Poisson's ratio (ν)	0.33	0.29
Density (ρ, kg/m^3)	2700	7840
Normal penalty (P_n, N/mm^2)	5010	211,000
Tangential penalty (P_t, N/mm^2)	501	21,100
Friction (γ)	0	0
Mesh element size (mm)	2	2
Contact type	Node edge	Node edge

The load versus displacement curve of the two specimens is shown in Figure 7. On the whole, the stress–strain curves of two specimens show the characteristics of elastic deformation. The strain energy of each part was calculated, as shown in Figure 7 and Equations (1)–(5); the strain energy of each part is shown in Table 4. Because there is no initial compaction stage, a specimen of the elastic materials accumulates more strain energy than elastic–plastic materials under the same displacement. The residual strain energy of the intact specimen after excavating the hole was still greater than that of specimen with hole. This showed that the excavation in pre-stressed elastic materials also led to the release of strain energy.

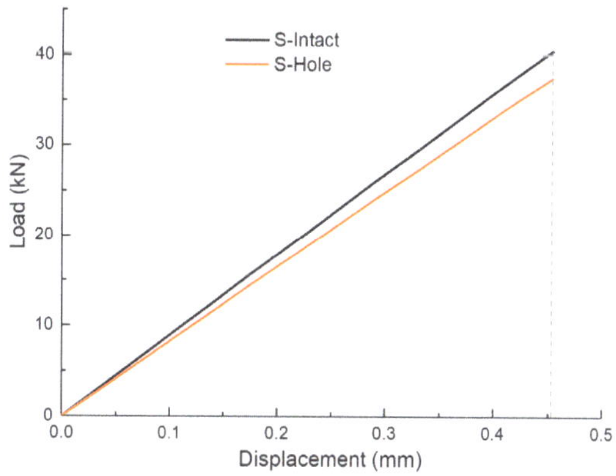

Figure 7. Load versus displacement curve of two models.

Table 4. The strain energy of each part.

Specimen	Displacement/mm	Load/kN	U_{Int}/J	U_{Hole}/J	$U_{Aft\text{-}ex}$/J	$U_{Spe\text{-}ho}$/J	U_{Rele}/J
S-Intact	0.453	40.56	9.186	0.247	8.939	---	0.434
S-Hole		37.55	---	---	---	8.505	

3. Effect of Excavation Height and Width on Strain Energy Release

3.1. Effect of Excavation Height

Through the previous analysis, it was found that excavating holes in pre-stressed rock led to the release of the strain energy of the specimen. In this section, the influence of height h of rectangle hole on energy release is analyzed. As shown in Figure 8, if the width of 22 mm is unchanged, the height of rectangle hole gradually changes from 1 mm to 22 mm. The specimen size, loading method and material parameters are consistent with Section 2.3. The common tunnel section size in some deep mines is about 4.5 m × 4.3 m, the maximum size of the hole in the test is 22 mm × 22 mm, and the scale of the test to the real condition is 1:200.

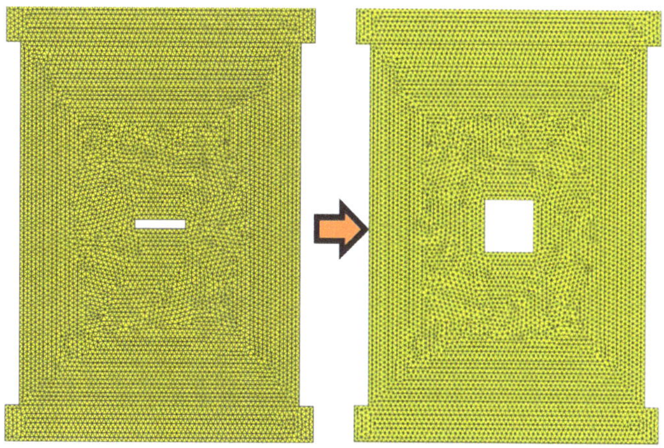

Figure 8. The height h of hole increases from 1 mm to 22 mm.

Table 5 shows the load of specimens with holes that have different height and strain energy in each part. Figure 9 is the strain energy release caused by different heights h of holes. With the increase in height h, the load of the specimen with holes under the same

displacement gradually decreases. When the height of hole is 1 mm, the strain energy release is large. In addition, the increase in height h will lead to the slow increase in strain energy release.

Table 5. The load of specimens with holes that have different height and strain energy in each part.

h/mm	Displacement/mm	Load/kN	U_{Int}/J	U_{Hole}/J	$U_{Aft\text{-}ex}$/J	$U_{Spe\text{-}ho}$/J	U_{Rele}/J
0		40.56		---	---	---	0
1		39.10		0.011	9.175	8.856	0.319
4		38.81		0.045	9.141	8.791	0.350
8	0.453	38.46	9.186	0.090	9.096	8.712	0.384
12		38.18		0.135	9.051	8.648	0.403
16		37.92		0.180	9.007	8.590	0.417
22		37.55		0.247	8.939	8.504	0.435

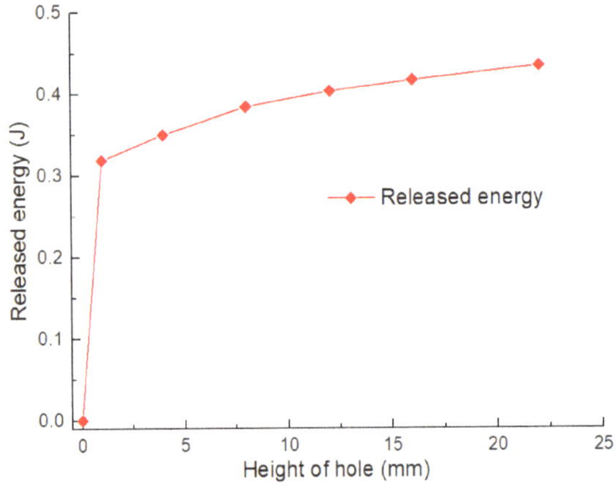

Figure 9. Strain energy release caused by different heights h of excavation holes.

3.2. Effect of Excavation Width

In addition to the excavation height, the influence of excavation width w on strain energy release is analyzed. As shown in Figure 10, if the height of 22 mm remains unchanged, then the width of hole gradually changes from 1 mm to 22 mm. The specimen size, loading method and material parameters are consistent with Section 2.3.

Table 6 is the load of specimens with holes that have different height and strain energy in each part. Figure 11 is the strain energy release caused by different heights h of hole. With the increase in the width of the hole, the load of the specimen with holes under the same displacement gradually decreases. When the width of the hole is 1 mm, the released strain energy is very small, and the increase in the width of the hole will lead to an increasing release of strain energy.

Table 6. The strain energy of each part.

w/mm	Displacement/mm	Load/kN	U_{Int}/J	U_{Hole}/J	$U_{Aft\text{-}ex}$/J	$U_{Spe\text{-}ho}$/J	U_{Rele}/J
0		40.56		---	---	---	0
1		40.49		0.011	9.175	9.172	0.003
4		40.31		0.045	9.141	9.129	0.012
8	0.453	39.90	9.186	0.090	9.096	9.038	0.058
12		39.37		0.135	9.051	8.918	0.133
16		38.73		0.180	9.007	8.772	0.234
22		37.55		0.247	8.939	8.504	0.435

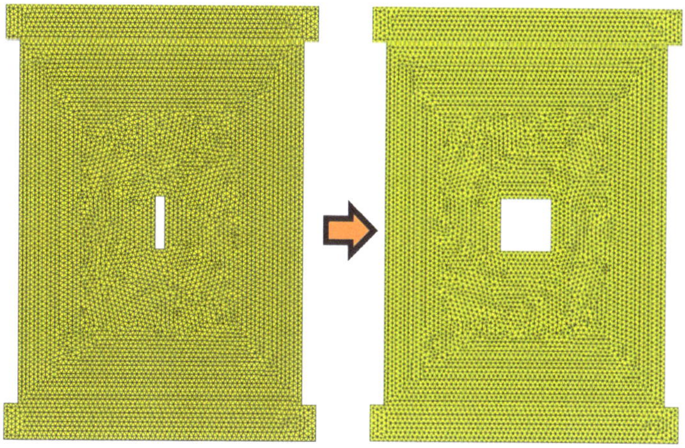

Figure 10. The width of hole w increases from 1 mm to 22 mm.

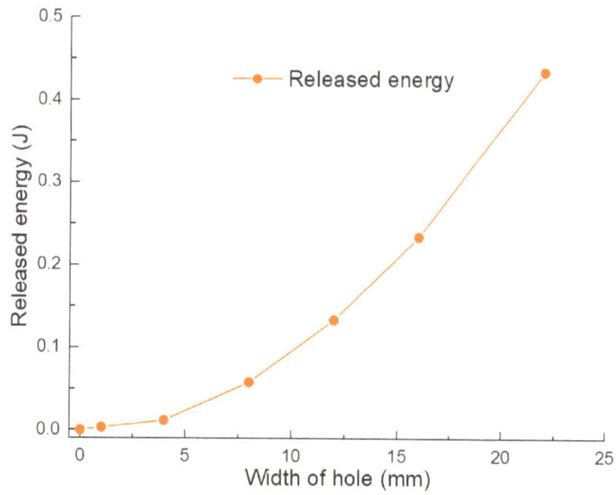

Figure 11. Strain energy release caused by different width w of excavation hole.

3.3. Equation for Released Strain Energy Caused by Excavation

In previous studies, it is mentioned that the excavation of a hole in pre-stressed rock will lead to strain energy release. The strain energy release per unit thickness is given by Equation (6) [18]:

$$U_e = \frac{\pi\sigma^2 A^2}{4E} \tag{6}$$

Considering the specimen thickness, the released strain energy caused by holes is calculated by Equation (7). The maximum principal stress of the intact model in Table 6 is 16.9 MPa, T is 20 mm, E is 5.01 GPa. When the width w of hole is 22 mm, the theoretical value of strain energy release can be calculated as 0.433 J according to Equation (8), which is very close to 0.435 J in Table 6, indicating that Equation (7) is reliable:

$$U_{Rele} = \frac{\pi\sigma^2 w^2 T}{4E} \tag{7}$$

$$U_{Rele} = \frac{\pi \times 16.9^2 \times 10^{12} \times 22^2 \times 20 \times 10^{-9}}{4 \times 5.01 \times 10^9} = 0.433\ \text{J} \tag{8}$$

Combined with Figures 9 and 11, the strain energy releases caused by different height and width of holes are shown in Figure 12. When the size of height or width is less than 16 mm, the strain energy release caused by height and width is very different. When the height and width reach 22 mm, the shapes of the two holes are square with the same side length, and the released strain energy is the same.

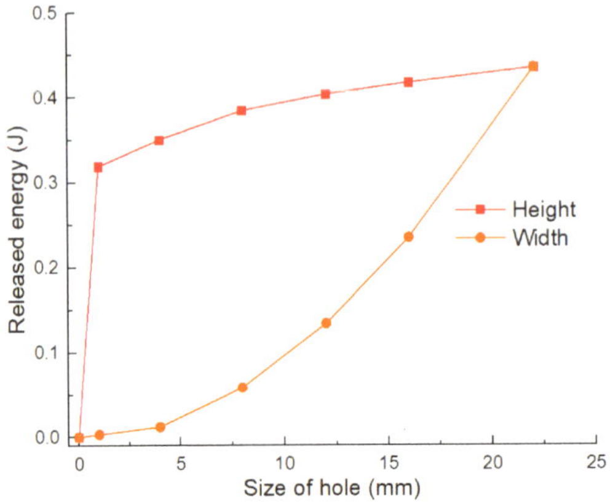

Figure 12. Strain energy release caused by different height and width of excavation holes.

Since the strain energy release is discontinuous when the height of the hole is from 0 to 1 mm, the part from 1 mm to 22 mm is considered, which can represent holes with various height width ratio. Considering the influence of hole height, Equation (9) is obtained by modifying Equation (7). As shown in Figure 13, the fitting accuracy is high. The theoretical value of strain energy release caused by height and width of holes can be calculated by Equation (9):

$$U_{Rele} = \frac{\pi\sigma^2 w^2 T}{4E}(0.78 + 0.22\frac{h}{w}) \tag{9}$$

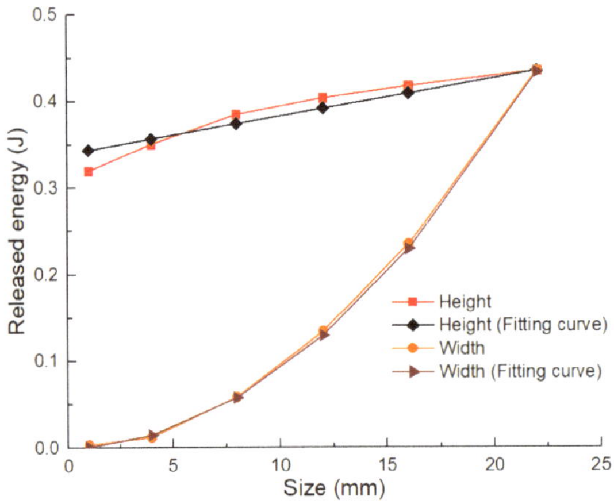

Figure 13. The fitting curve of Equation (9). If the width or height of 22 mm are maintained, then another size changes from 1 mm to 22 mm.

4. Effect of Lateral Pressure on Strain Energy Release

4.1. Model Construction and Results

Due to the influence of the Earth structure motion, the horizontal stress in most areas is higher than the vertical stress. The influence of lateral pressure on strain energy release should be analyzed, and the ratio of horizontal stress σ_x to vertical stress σ_y is defined as the lateral pressure coefficient λ. As shown in Figure 14, in addition to two pressure platens arranged at the upper and lower ends of specimen, another two pressure platens are arranged on the left and right sides of specimen. The load in the vertical direction is Fy, and the load in the horizontal direction is Fx. A positive displacement indicates that the specimen is compressed in this direction, and a negative displacement indicates that the specimen expands in this direction. The size and material parameters of the specimen are consistent with Section 2.3. The rectangle hole of specimen with holes is 22 mm × 22 mm.

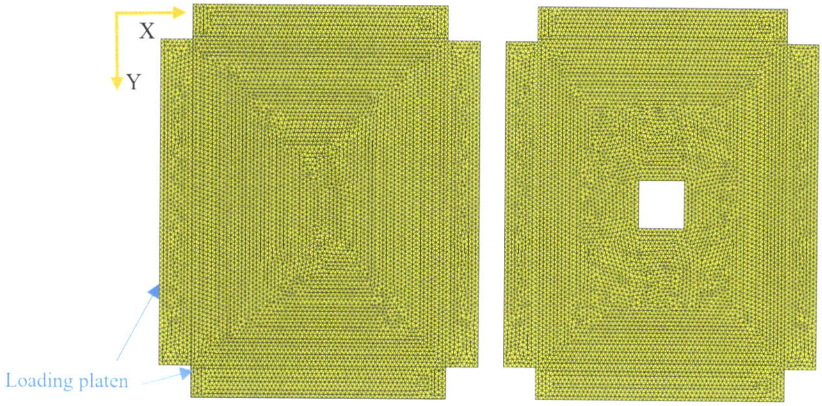

Figure 14. The lateral pressure coefficient λ increases from 0.1 to 1.35 under biaxial loading.

The lateral pressure coefficient is for the intact specimen. The vertical stress should be maintained at 16.9 MPa, then the corresponding horizontal stress is calculated according to the lateral pressure coefficient. Different lateral pressures of intact specimens are achieved by changing the displacement of the pressure platen in the horizontal and vertical directions. Firstly, the intact specimen is loaded, and then the specimen with holes is loaded with the same displacement. The displacement of the specimen with hole is the same as that of the intact specimen in the Y and X directions. The lateral pressure coefficient increases from 0.1 to 1.35.

The displacement and load corresponding to different lateral pressure coefficients are shown in Table 7. As can be seen from Table 7, when the lateral pressure coefficient is 0.1 to 0.4, the displacement Dx in the horizontal direction is negative, which is caused by the expansion in the X direction of the specimen under compression in Y direction. With the increase in lateral pressure coefficient, the displacement in the X direction gradually increases, and the displacement in Y direction gradually decreases. This is because the increase in horizontal stress will lead to an increase in vertical stress caused by the Poisson effect. Therefore, in order to reach the same vertical stress σ_y, it is necessary to reduce the displacement in Y direction. In addition, under the same displacement, the load in the Y direction of the specimen with holes decreases slowly with the increase in the lateral pressure.

Table 7. Displacement and load in vertical and horizontal directions under different lateral pressure coefficients.

λ	Displacment/mm		Intact Specimen				Specimen with Hole	
	Dy	Dx	Fy/kN	Fx/kN	σ_y/MPa	σ_x/MPa	Fy/kN	Fx/kN
0.10	0.431	−0.142	40.56	5.07	16.90	1.69	37.06	3.27
0.25	0.398	−0.088		12.67		4.22	36.93	10.26
0.40	0.365	−0.033		20.28		6.76	36.80	17.26
0.60	0.322	0.039		30.42		10.14	36.66	26.44
0.80	0.277	0.112		40.56		13.52	36.42	35.84
1.00	0.232	0.186		50.70		16.90	36.19	45.24
1.20	0.189	0.257		60.84		20.28	36.05	54.43
1.35	0.156	0.312		68.44		22.82	35.97	61.52

4.2. Effect of Lateral Pressure

The strain energy of each part is calculated by Table 7 and Equations (1)–(5), as shown in Table 8 and Figure 15. As shown in Table 8, the strain energy of intact specimens and specimens with holes first decreases and then increases with the increase in lateral pressure coefficient, and the minimum value appears when the lateral pressure coefficient is 0.4–0.6. This is because under the same vertical load, the increase in lateral displacement reduces the displacement in vertical direction, resulting in the decrease in vertical strain energy. When the lateral pressure coefficient is greater than 0.6, the strain energy in the horizontal direction increases exponentially.

Table 8. The strain energy of each part under different lateral pressure coefficients λ.

λ	U_{Int}/J	U_{Hole}/J	$U_{Aft\text{-}ex}$/J	$U_{Spe\text{-}ho}$/J	U_{Rele}/J
0.10	8.38	0.23	8.15	7.76	0.40
0.25	7.51	0.20	7.31	6.90	0.41
0.40	7.07	0.19	6.88	6.43	0.45
0.60	7.12	0.19	6.93	6.42	0.51
0.80	7.89	0.21	7.68	7.05	0.63
1.00	9.42	0.25	9.17	8.41	0.76
1.20	11.65	0.31	11.34	10.40	0.94
1.35	13.84	0.37	13.47	12.40	1.07

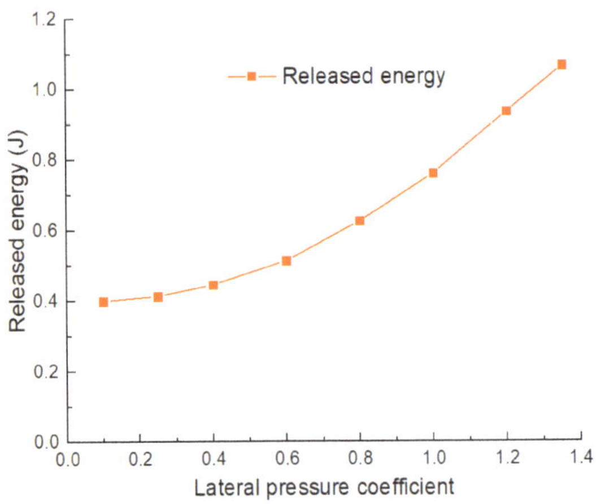

Figure 15. Strain energy release caused by different lateral pressure coefficient λ.

As shown in Figure 15, the strain energy release increases exponentially with the increase in lateral pressure, showing a trend of the second power of lateral pressure coefficient. This shows that the vertical stress remains unchanged, and the increase in horizontal stress will lead to the increase in strain energy release. Similar to the second power shown in Figure 12, this is because the release of strain energy in the horizontal direction is the second power of the horizontal stress, which can be calculated according to Equation (9).

5. Rock Failure Caused by Excavation

5.1. Numerical Method and Model

It is difficult to simulate the excavation in rock mass under high stress through indoor physical experiments, especially the failure process caused by excavation. Fortunately, the numerical analysis method can reproduce the excavation and failure process in rock mass under high stress. The coupled finite and discrete element method (FDEM) is used for the research, and the simulation is carried out by the code ELFEN. The method can realize the real simulation of the brittle failure process of materials by mixing finite and discrete elements and introducing the principle of fracture mechanics [19].

The Mohr–Coulomb criterion with a tension cut-off yield criterion was chosen as the failure criterion of the rock material in this study. Compared with the traditional Mohr-Coulomb criterion, the modified criterion can better describe both the shear and tensile failure of rock material, as shown in Figure 16.

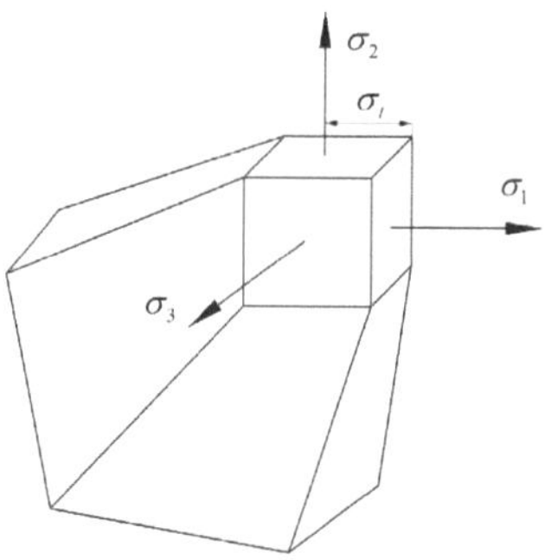

Figure 16. Mohr–Coulomb with tension cut-off yield surface.

The Mohr–Coulomb criterion with a tension cut-off yield criterion combines the Mohr-Coulomb yield criterion and the Rankine tensile yield criterion. The Mohr-Coulomb yield criterion is used to judge shear failure and is described by Equation (10):

$$\tau = c - \sigma_n \tan \varphi \quad (10)$$

where τ is the shear stress, c is the cohesion, σ_n is the normal pressure, and φ is the friction angle.

The Rankine tensile yield criterion is used to judge tensile failure and is described by Equation (11):

$$\sigma_i - \sigma_t = 0 \quad \mathrm{i} = 1,\ 2,\ 3 \quad (11)$$

where σ_i is each principal stress, σ_t is the tensile strength.

The cohesion of material decreases after plastic strain occurs, and the tensile strength is softened by the decrease in cohesion, as shown in Equation (12). This ensures that there is always normal stress on the failure shear surface:

$$\sigma_t \leq c(1 - \sin \varphi)/\cos \varphi \quad (12)$$

The fracture energy G_f is an important parameter for fracture development. For a material, the more energy that is released, the greater the generated crack area. This can be calculated by uniaxial tensile test, and refers to the energy required to generate continuous cracks per unit area. The unit is J/m^2, also written as N/m. It is defined as:

$$G_f = \int \sigma du = \int \sigma \varepsilon(s) ds \tag{13}$$

where σ is the tensile stress, and u is the tensile displacement.

As shown in Figure 17, a model with side length of 20 m is first established. Then, according to the principal stress value of the −720 m buried depth in a mine [14], the maximum horizontal principal stress 33 MPa and vertical principal stress 20 MPa are applied at the boundary of the model, as shown in Equation (14). Finally, a tunnel with a side length of 4 m is excavated in the model center. The material properties are shown in Table 9.

$$\begin{cases} \sigma_H = 0.043h + 1.433 \\ \sigma_v = 0.027h + 0.838 \end{cases} \tag{14}$$

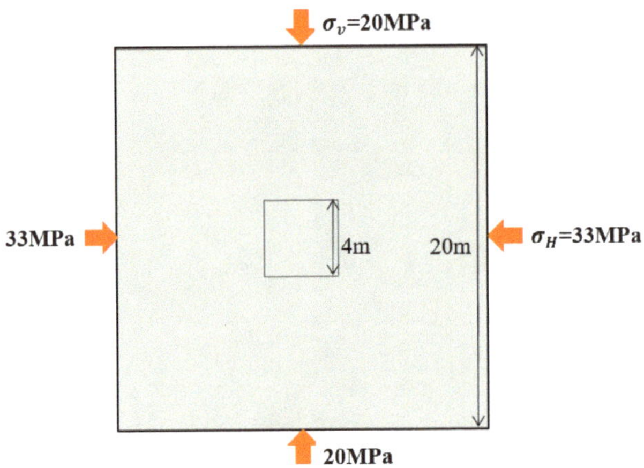

Figure 17. Loading schematic of model.

Table 9. Material properties adopted in model.

Mechanical Parameters	Model
Young's modulus (E, GPa)	20.7
Poisson's ratio (ν)	0.23
Density (ρ, kg/m^3)	2790
Friction angle (φ)	30°
Cohesion (c, MPa)	10
Tensile strength (σ_t, MPa)	2
Fracture energy (G_f, N/m)	4

5.2. Rock Failure Caused by Excavation

As shown in Figure 18, the surrounding rock of the tunnel suffered failure after excavation. The excavation caused tangential stress concentration and normal stress release. It is known that the maximum tangential stress appears at the roof and bottom, which is the reason for the V-shaped rockburst failure of the roof and bottom. In addition, normal stress release caused the tensile spalling failure of the sidewall. Similar damage was also observed in a deep tunnel of a gold mine, as shown in Figure 19.

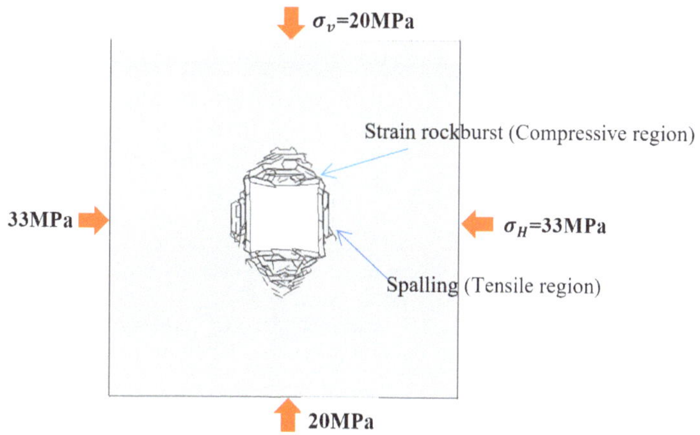

Figure 18. Tunnel failure caused by excavation under pre-stressed state.

Figure 19. Tunnel failure at buried depth −830 m of a gold mine. (**a**) rockburst and rock bolts falling in roof; (**b**) spalling in sidewall.

Figure 19 shows the tunnel failure with a buried depth −830 m in a gold mine. The principal stresses in three directions are marked in the figure, and the maximum horizontal principal stress is perpendicular to the sidewall. It can be seen that the roof and sidewall are damaged, similar to the failure mode in Figure 18. The rockburst failure in the roof is caused by the concentration of tangential stress, while the spalling failure in the sidewall is caused by the release of normal stress.

6. Conclusions

The main conclusions of this article are listed as follows:

(1) Excavation in pre-stressed rock leads to a change in the original stress field, which is manifested in the release of normal stress on the free surface and the concentration of tangential stress. In this research, a method is proposed to calculate the strain energy release caused by the excavation by loading an intact specimen and a specimen with holes. The energy change in the intact specimen from loading to excavation, and then to stability (stress balance) causes a gradual increase in strain energy during loading, a reduction in strain energy after excavation, and strain energy release after excavation.

Both the excavation and normal stress release after excavation lead to a reduction in strain energy in the rock specimen.

(2) The influence of excavation height and width on strain energy release is inconsistent under vertical loading. When the height of hole is 1 mm, the strain energy release is large, and the increase in the height of the hole will lead to a slow increase in strain energy release. When the width of hole is 1 mm, the strain energy release is very small, and the increase in width of hole will lead to an exponential release of strain energy. Through equation fitting, the theoretical Equation (9) of strain energy release caused by holes with different height and width is obtained, and the fitting accuracy is high.

(3) The strain energy release increases exponentially with the increase in the lateral pressure coefficient, showing a trend of the second power of lateral pressure coefficient. This shows that the vertical stress remains unchanged, and the increase in horizontal stress will lead to the increase in strain energy release.

(4) The tunnel failure mode caused by excavation under high stress is obtained by numerical calculation. The failure modes of the tunnel are strain rockbursts caused by tangential stress concentration and spalling caused by normal stress release, which are also observed in the failure mode of the actual tunnel. For an underground mine exposed to high horizontal stresses, the single excavation height should be reduced to reduce the rate of horizontal strain energy release. In addition, it is recommended that energy-absorbing materials are used to absorb the strain energy release caused by excavation. The support design can be based on the strain energy release of surrounding rock and the amount of energy absorption by the support structure.

Author Contributions: Conceptualization, P.X.; methodology, P.X.; software, P.X.; writing—review and validation, P.X., D.L. and Q.Z.; funding acquisition, D.L. All authors have read and agreed to the published version of the manuscript.

Funding: This research was funded by the National Natural Science Foundation of China (Grant No. 52074349).

Institutional Review Board Statement: Not applicable.

Informed Consent Statement: Not applicable.

Data Availability Statement: The data presented in this study are available on request from the corresponding author.

Conflicts of Interest: The authors declare no conflict of interest.

References

1. Li, X.; Zhou, J.; Wang, S.; Liu, B. Review and practice of deep mining for solid mineral resources. *Chin. J. Nonferrous Met.* **2017**, *27*, 1236–1262.
2. Shi, Q.; Pan, J.; Wang, S.; Liu, S.; Mishra, B.; Seitz, S. Field Monitoring of Delayed Coal Burst in an Advancing Entry of a Deep Coal Mine. *Min. Metall. Explor.* **2021**, *38*, 2417–2431. [CrossRef]
3. Feng, G.-L.; Feng, X.-T.; Xiao, Y.-X.; Yao, Z.-B.; Hu, L.; Niu, W.-J.; Li, T. Characteristic microseismicity during the development process of intermittent rockburst in a deep railway tunnel. *Int. J. Rock Mech. Min. Sci.* **2019**, *124*, 104135. [CrossRef]
4. Lu, M.; Grov, E.; Dahle, H.; Qiao, H.; Wen, B.; Chen, Z.; Zhao, Q. Rock Support Design For Special Lighting Caverns In High In-Situ Stress Rock Mass. *Chin. J. Rock Mech. Eng.* **2008**, *27*, 35–41.
5. Shan, Z.-G.; Yan, P. Management of rock bursts during excavation of the deep tunnels in Jinping II Hydropower Station. *Bull. Eng. Geol. Environ.* **2010**, *69*, 353–363. [CrossRef]
6. Liang, Z.; Xue, R.; Xu, N.; Li, W. Characterizing rockbursts and analysis on frequency-spectrum evolutionary law of rockburst precursor based on microseismic monitoring. *Tunn. Undergr. Space Technol.* **2020**, *105*, 103564. [CrossRef]
7. Li, X.; Gong, F.; Tao, M.; Dong, L.; Du, K.; Ma, C.; Zhou, Z.; Yin, T. Failure mechanism and coupled static-dynamic loading theory in deep hard rock mining: A review. *J. Rock Mech. Geotech. Eng.* **2017**, *9*, 767–782. [CrossRef]
8. Li, C.C.; Mikula, P.; Simser, B.; Hebblewhite, B.; Joughin, W.; Feng, X.; Xu, N. Discussions on rockburst and dynamic ground support in deep mines. *J. Rock Mech. Geotech. Eng.* **2019**, *11*, 1110–1118. [CrossRef]
9. Lu, C.-P.; Liu, Y.; Zhang, N.; Zhao, T.-B.; Wang, H.-Y. In-situ and experimental investigations of rockburst precursor and prevention induced by fault slip. *Int. J. Rock Mech. Min. Sci.* **2018**, *108*, 86–95. [CrossRef]

10. Zhou, J.; Li, X.; Mitri, H.S. Evaluation method of rockburst: State-of-the-art literature review. *Tunn. Undergr. Space Technol.* **2018**, *81*, 632–659. [CrossRef]
11. He, M.; Sousa, R.L.; Müller, A.L.; Vargas, E., Jr.; Sousa, L.R.; Chen, X. Analysis of excessive deformations in tunnels for safety evaluation. *Tunn. Undergr. Space Technol.* **2015**, *45*, 190–202. [CrossRef]
12. Wu, Z.; Wu, S.; Cheng, Z. Discussion and application of a risk assessment method for spalling damage in a deep hard-rock tunnel. *Comput. Geotech.* **2020**, *124*, 103632. [CrossRef]
13. Li, C.C. Principles and methods of rock support for rockburst control. *J. Rock Mech. Geotech. Eng.* **2021**, *13*, 46–59. [CrossRef]
14. Miao, S.; Cai, M.; Guo, Q.; Huang, Z. Rock burst prediction based on in-situ stress and energy accumulation theory. *Int. J. Rock Mech. Min. Sci.* **2016**, *83*, 86–94. [CrossRef]
15. Li, P.; Cai, M.; Guo, Q.; Miao, S. In Situ Stress State of the Northwest Region of the Jiaodong Peninsula, China from Overcoring Stress Measurements in Three Gold Mines. *Rock Mech. Rock Eng.* **2019**, *52*, 4497–4507. [CrossRef]
16. Feng, F.; Li, X.; Rostami, J.; Peng, D.; Li, D.; Du, K. Numerical Investigation of Hard Rock Strength and Fracturing under Polyaxial Compression Based on Mogi-Coulomb Failure Criterion. *Int. J. Geomech.* **2019**, *19*, 04019005. [CrossRef]
17. Zhou, Z.; Li, Z.; Gao, C.; Zhang, D.; Wang, M.; Wei, C.; Bai, S. Peridynamic micro-elastoplastic constitutive model and its application in the failure analysis of rock masses. *Comput. Geotech.* **2021**, *132*, 104037. [CrossRef]
18. Xiao, P.; Li, D.; Zhao, G.; Liu, H. New criterion for the spalling failure of deep rock engineering based on energy release. *Int. J. Rock Mech. Min. Sci.* **2021**, *148*, 104943. [CrossRef]
19. Feng, F.; Li, X.; Rostami, J.; Li, D. Modeling hard rock failure induced by structural planes around deep circular tunnels. *Eng. Fract. Mech.* **2019**, *205*, 152–174. [CrossRef]
20. Wei, C.; Zhang, C.; Canbulat, I.; Huang, W. Numerical investigation into impacts of major fault on coal burst in longwall mining—A case study. *Int. J. Rock Mech. Min. Sci.* **2021**, *147*, 104907. [CrossRef]
21. Gong, F.; Yan, J.; Li, X.; Luo, S. A peak-strength strain energy storage index for rock burst proneness of rock materials. *Int. J. Rock Mech. Min. Sci.* **2019**, *117*, 76–89. [CrossRef]
22. Jiang, Q.; Feng, X.-T.; Xiang, T.-B.; Su, G.-S. Rockburst characteristics and numerical simulation based on a new energy index: A case study of a tunnel at 2500 m depth. *Bull. Eng. Geol. Environ.* **2010**, *69*, 381–388. [CrossRef]
23. Qiu, S.; Feng, X.; Zhang, C.; Xiang, T. Estimation of rockburst wall-rock velocity invoked by slab flexure sources in deep tunnels. *Can. Geotech. J.* **2014**, *51*, 520–539. [CrossRef]
24. He, B.-G.; Zelig, R.; Hatzor, Y.H.; Feng, X.-T. Rockburst Generation in Discontinuous Rock Masses. *Rock Mech. Rock Eng.* **2016**, *49*, 4103–4124. [CrossRef]
25. Xiao, P.; Li, D.; Zhao, G.; Zhu, Q.; Liu, H.; Zhang, C. Mechanical properties and failure behavior of rock with different flaw inclinations under coupled static and dynamic loads. *J. Cent. South Univ.* **2020**, *27*, 2945–2958. [CrossRef]
26. Qiu, J.; Li, X.; Li, D.; Zhao, Y.; Hu, C.; Liang, L. Physical Model Test on the Deformation Behavior of an Underground Tunnel Under Blasting Disturbance. *Rock Mech. Rock Eng.* **2021**, *54*, 91–108. [CrossRef]
27. Zhou, Z.; Cai, X.; Li, X.; Cao, W.; Du, X. Dynamic Response and Energy Evolution of Sandstone Under Coupled Static-Dynamic Compression: Insights from Experimental Study into Deep Rock Engineering Applications. *Rock Mech. Rock Eng.* **2020**, *53*, 1305–1331. [CrossRef]
28. Han, Z.; Li, D.; Zhou, T.; Zhu, Q.; Ranjith, P.G. Experimental study of stress wave propagation and energy characteristics across rock specimens containing cemented mortar joint with various thicknesses. *Int. J. Rock Mech. Min. Sci.* **2020**, *131*, 104352. [CrossRef]
29. Skrzypkowski, K. Case Studies of Rock Bolt Support Loads and Rock Mass Monitoring for the Room and Pillar Method in the Legnica-Gogow Copper District in Poland. *Energies* **2020**, *13*, 2998. [CrossRef]
30. Jiang, J.; Su, G.; Zhang, X.; Feng, X.-T. Effect of initial damage on remotely triggered rockburst in granite: An experimental study. *Bull. Eng. Geol. Environ.* **2020**, *79*, 3175–3194. [CrossRef]
31. Chen, G.; He, M.; Fan, F. Rock Burst Analysis Using DDA Numerical Simulation. *Int. J. Geomech.* **2018**, *18*, 04018001. [CrossRef]
32. Lin, M.-Q.; Zhang, L.; Liu, X.-Q.; Xia, Y.-Y.; Zhang, D.-J.; Peng, Y.-L. Microscopic analysis of rockburst failure on specimens under gradient stress. *Rock Soil Mech.* **2020**, *41*, 2984–2992.
33. Rui, Y.; Zhou, Z.; Lu, J. A novel AE source localization method using clustering detection to eliminate abnormal arrivals. *Int. J. Min. Sci. Technol.* **2022**, *32*, 51–62. [CrossRef]
34. Xiao, P.; Li, D.; Zhao, G.; Liu, M. Experimental and Numerical Analysis of Mode I Fracture Process of Rock by Semi-Circular Bend Specimen. *Mathematics* **2021**, *9*, 1769. [CrossRef]
35. Read, R. 20 years of excavation response studies at AECL's Underground Research Laboratory. *Int. J. Rock Mech. Min. Sci.* **2004**, *41*, 1251–1275. [CrossRef]
36. Kang, S.S.; Ishiguro, Y.; Obara, Y. Evaluation of core disking rock stress and tensile strength via the compact conical-ended borehole overcoring technique. *Int. J. Rock Mech. Min. Sci.* **2006**, *43*, 1226–1240. [CrossRef]
37. Hu, X.; Su, G.; Chen, G.; Mei, S.; Feng, X.; Mei, G.; Huang, X. Experiment on Rockburst Process of Borehole and Its Acoustic Emission Characteristics. *Rock Mech. Rock Eng.* **2019**, *52*, 783–802. [CrossRef]
38. Gong, F.; Wu, W.; Li, T.; Si, X. Experimental simulation and investigation of spalling failure of rectangular tunnel under different three-dimensional stress states. *Int. J. Rock Mech. Min. Sci.* **2019**, *122*, 104081. [CrossRef]

39. Qiu, P.-Q.; Ning, J.-G.; Wang, J.; Hu, S.-C.; Li, Z. Mitigating rock burst hazard in deep coal mines insight from dredging concentrated stress: A case study. *Tunn. Undergr. Space Technol.* **2021**, *115*, 104060. [CrossRef]
40. Tao, M.; Li, X.; Li, D. Rock failure induced by dynamic unloading under 3D stress state. *Theor. Appl. Fract. Mech.* **2013**, *65*, 47–54. [CrossRef]
41. Feng, F.; Li, X.; Luo, L.; Zhao, X.; Chen, S.; Jiang, N.; Huang, W.; Wang, Y. Rockburst response in hard rock owing to excavation unloading of twin tunnels at great depth. *Bull. Eng. Geol. Environ.* **2021**, *80*, 7613–7631. [CrossRef]
42. Ren, F.; Chang, Y.; He, M. A systematic analysis method for rock failure mechanism under stress unloading conditions: A case of rock burst. *Environ. Earth Sci.* **2020**, *79*, 370. [CrossRef]
43. Huang, R.Q.; Wang, X.N.; Chan, L.S. Triaxial unloading test of rocks and its implication for rock burst. *Bull. Eng. Geol. Environ.* **2001**, *60*, 37–41. [CrossRef]
44. Su, G.; Hu, L.; Feng, X.; Wang, J.; Zhang, X. True triaxial experimental study of rockburst process under low frequency cyclic disturbance load combined with static load. *Chin. J. Rock Mech. Eng.* **2016**, *35*, 1309–1322.

minerals

Article

Study on the Stress Variation Law of Inclined Surrounding Rock Roadway under the Influence of Mining

Qian Jia [1], Hai Wu [1,2,*], Tao Ling [3], Kai Liu [3], Weiwei Peng [3], Xu Gao [4] and Yanlin Zhao [1,*]

1 School of Resources, Environment and Safety Engineering, Hunan University of Science and Technology, Xiangtan 411201, China; gj_wkjq@163.com
2 Work Safety Key Lab on Prevention and Control of Gas and Roof Disasters for Southern Goal Mines, Hunan University of Science and Technology, Xiangtan 411201, China
3 Railway No. 5 Bureau Group First Engineering Co., Ltd., Changsha 411104, China; lingtao2022@126.com (T.L.); liukai15@crecg.com (K.L.); pengweiwei@crecg.com (W.P.)
4 China Occupational Safety and Health Association, Beijing 100013, China; gaoxu987202@163.com
* Correspondence: wuhai@hnust.edu.cn (H.W.); yanlin_8@163.com (Y.Z.)

Abstract: This study takes three roadways with similar burial depths in different strata in Xieqiao Mine of Huainan as its research object. This study involves the observation and analysis of borehole stress gauge data under the influence of mining pressure. The observation data show that: (1) under the influence of mining, the high-wall vertical stress increases as the distance from the roadway surface increases, and the peak point is at 6 m. The increment value of vertical stress at the low side has a maximum value at 8 m and a peak value at 14 m. The increase value of horizontal stress of the high side has two peaks, which are 4 m and 6 m, respectively. The increment of horizontal stress in low walls is also about 8 m. (2) The mining influence range of working face mining is about 150 m. Mining influence distance can be divided into three stages: 0–25 m, 25–60 m, and beyond 60 m. The increase of vertical and horizontal stress caused by mining increases sharply within 25 m from the working face. (3) The buried depth of the roadway has an influence on the range of mining influence and the increase of mining stress caused by working face mining. The more the buried depth of the roadway increases, the greater the range of mining influence and the increased value of mining stress. (4) After roadway excavation, the surface deformation of roadway surrounding rock reduces the increase of mining stress near the roadway surface. The mutual verification between the analysis results and theoretical calculation results is helpful to roadway support design and advanced support design of the working face.

Keywords: roadway surrounding rock; mining influence; stress increases; distribution characteristics

Citation: Jia, Q.; Wu, H.; Ling, T.; Liu, K.; Peng, W.; Gao, X.; Zhao, Y. Study on the Stress Variation Law of Inclined Surrounding Rock Roadway under the Influence of Mining. *Minerals* **2022**, *12*, 499. https://doi.org/10.3390/min12050499

Academic Editor: Tuncel M. Yegulalp

Received: 7 March 2022
Accepted: 14 April 2022
Published: 19 April 2022

1. Introduction

In recent years, as the mining of mineral resources has continued to extend to the deep, geological conditions have become increasingly complex, mining has become more difficult, and rock strata where the roadway is located have become more stressed. The surrounding rock of the inclined roadway is prone to heterogeneous deformation. As the buried depth of the roadway increases, so too does the deformation of the surrounding rock of the inclined rock roadway. Therefore, it is more and more important to study the deformation law of the surrounding rock of deeply inclined rock roadways.

Scholars at home and abroad have conducted extensive research on the deformation law of inclined roadways. Zhao et al. [1] found that the stress distribution characteristics of roadway surrounding rock in inclined coal seam are complex and diverse. Chang et al. [2] revealed the stability control mechanism of surrounding rock in deep rock roadway by analyzing the stress evolution characteristics and deformation and failure laws of surrounding rock after deep rock roadway excavation. Chen et al. [3] analyzed the stress field and displacement field of surrounding rock in two stages of the gob-side excavation and

mining of deep inclined coal seam. Chen et al. revealed the deformation and failure characteristics of the surrounding rock of this kind of roadway, and proposed stability control technology. Wang et al. [4] established the mechanical model of the overlying structure in the gob-side excavation of deep inclined coal seam by a theoretical analysis method. In terms of numerical simulation, Wei et al. [5] established numerical calculation models of different rock strata dip angles, and calculated and analyzed the stress change process of surrounding rock of inclined coal seam under dynamic pressure. Wu et al. [6] used UDEC discrete finite element simulation software to establish the stress model of roadways in deep inclined strata. Gao et al. [7] adopted the FLAC3D numerical analysis method to obtain the asymmetric deformation law of the surrounding rock of soft rock roadways in the inclined coal seam. Regarding the combination of numerical simulation and field measurement, Sun et al. [8] concluded that the obtuse angle of deep inclined strata roadway sections and the inclined direction of strata are key components of asymmetric deformation and failure. Zhang et al. [9] analyzed the stress distribution law, displacement change, and deformation failure characteristics of surrounding rock in the process of extremely close coal seam mining with upper and lower mining goaf. Fu et al. [10] studied and concluded that, under the influence of the double mining stress of the adjacent working face and the own working face, the mine pressure appeared sharply and the roadway maintenance was difficult. Chen et al. [11] revealed the mechanism of asymmetric deformation and failure after roadway excavation in the steeply inclined coal seam and analyzed the distribution characteristics of the stress field and plastic zone after roadway excavation. Through the numerical simulation and monitoring of roadway surrounding rock deformation and surrounding rock stress during mining, Wang et al. [12] judged the deformation law of roadway surrounding rock under the influence of working face mining and the influence range of advanced dynamic pressure. Using a combination of numerical simulation, theoretical analysis, and field measurement, Li et al. [13] determined the loosening range and plastic zone range of a coal roadway, and comprehensively analyzed the asymmetric deformation characteristics of coal and rock under the environment of the broken surrounding rock. Yang et al. [14] studied the deformation and failure mechanism of surrounding rock in the steeply inclined roadway, and found that the deformation and failure of steeply inclined soft and hard interlayer roadways are asymmetrical, with obvious bias effect, characterized by the mutual staggered deformation characteristics of roof siding and floor heave, with serious floor heave and easy roof collapse. Luo et al. [15] studied the roof stress and deformation characteristics of steeply inclined thin coal seam working faces. Wang et al. [16,17] studied the mechanical characteristics of roadway surrounding rock and the influence of surrounding rock deformation and the plastic zone, revealing the intrinsic mechanical nature of the influence of maximum horizontal principal stress direction on the stability of roadway surrounding rock. Zhao et al. [18] analyzed the mechanical characteristics and size effect of narrow coal pillars in the gob-side excavation of inclined coal seams. As the pillar width increases, the peak values of vertical stress and horizontal stress gradually increase. Sun et al. [19] concluded that the stress concentration of a roadway through inclined coal seam is distributed along the layer. At the same time, the research on the stress distribution of the roadway surrounding rock of inclined coal seam is also focused on the combination of theoretical analysis, field measurement, numerical simulation, and field investigation. Wang et al. [20] integrated field investigation, theoretical analysis, numerical calculation, downhole test and measurement, etc. The distribution and propagation of floor stress under mining conditions and its relationship with surrounding rock stress, dynamic deformation and failure characteristics of floor dark inclined shaft disturbed by mining process, and the instability mechanism and control method of floor roadway are systematically studied.In this way, Wang et al. [21–24] systematically studied the deformation and stress evolution of the surrounding rock of roadways in inclined coal seam by combining multiple means. In terms of field measurement, Shi et al. [25] analyzed the law of mining pressure in the fully mechanized caving roadways along goaf through field measurement, providing an analytical basis for determining roadway support mode. Han et al. [26] used the field

observation method to analyze the change of internal stress of small coal pillars and the coal body at the working face, the influence range of leading abutment pressure, and the deformation and failure law of small coal pillar and coal roadway. Song et al. [27] analyzed the distribution rule of advance abutment pressure on the working face by using the field measurement method, and obtained the influence range of advance abutment pressure. According to the monitoring data, Zhang et al. [28] analyzed how the stress change of surrounding rock can be divided into four change modes: gradual change, sudden jump, oscillation, and compound change. However, these studies are not comprehensive enough in data analysis.

In this paper, the field measurement method is adopted to analyze the data of borehole stress gauge under the action of mining pressure for three tunnels with similar burial depth in different layers. This is undertaken in order to reveal the secondary distribution of roadway surrounding rock stress under the influence of mining pressure, and to provide a basis for roadway support design and advanced support at the end of working face.

2. Geological Survey of Roadway and Station Layout

Three roadways with similar burial depths in different strata of Xieqiao Mine of the Huainan Mining Group are taken as research objects. A lot of original data were obtained by one-month observation of mine pressure on the working face of 11318 (Section 2.1), 11613 (Section 2.2), and 12521 (Section 2.3) roadways in Xieqiao Mine.

2.1. Geological Survey and Station Layout of 11318 Roadway

The section of the 11318 roadway is a straight wall with a rectangular width of 4.3 m and a height of 2.8 m, with a dip angle of 10~15°. The coal thickness is 1.5~3.6 m and the average coal thickness is 2.75 m. The old roof of the coal seam is sandstone, partial sections have mudstone false roof (or phase turns into gangue), and the direct bottom is mudstone. Ground elevation is +20.8~+27.6 m and working face elevation is −511~−620 m. The measuring instrument is a vibrating string-type borehole stress gauge, the diameter of which is 62 mm. A total of 14 borehole stress gauges were installed to monitor the increase of vertical stress on both sides of the roadway under the influence of distance from the working face. Six small high side coal pillars are installed in the roadway, and the monitoring positions are 1 m, 2 m, 3 m, 4 m, 5 m, and 6 m, respectively. Eight are installed in the low side solid coal side of the roadway, and the monitoring positions are 2 m, 4 m, 6 m, 8 m, 10 m, 12 m, 14 m, and 15 m, respectively, as show in Figures 1 and 2.

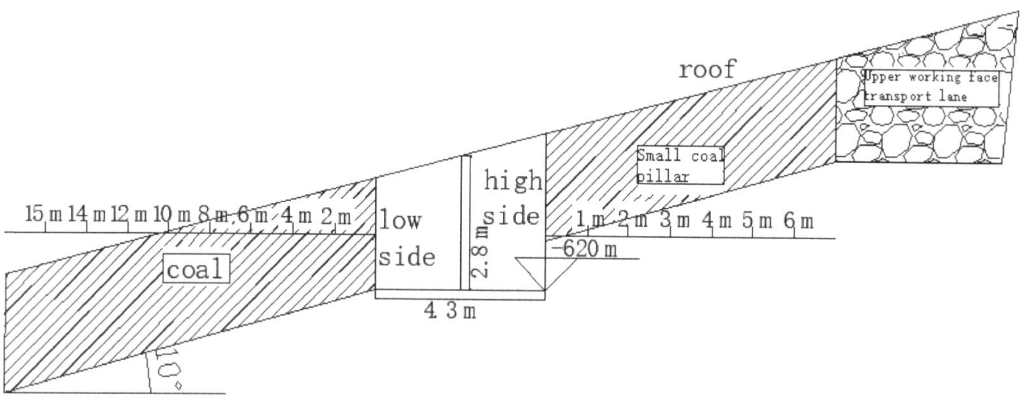

Figure 1. 11318 Schematic diagram of section layout of measuring station.

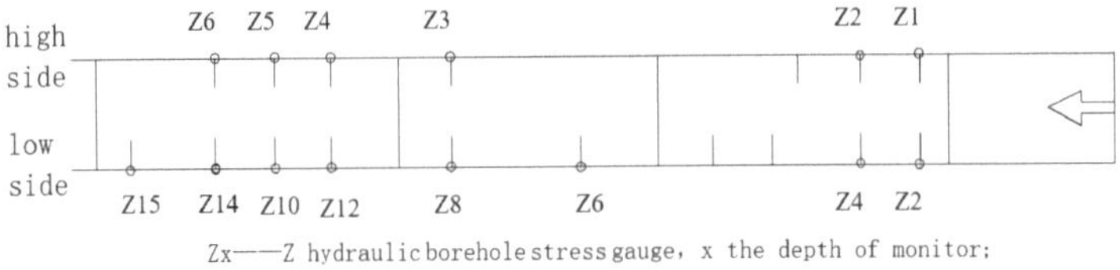

Figure 2. 11318 Graphic layout of measuring station.

2.2. Geological Survey and Station Layout of 11613 Roadway

The working face where the roadway is located has an elevation of −608 m~−723 m, a strike length of 1932 m, and an inclination length of 205 m. The roadway section is 5.2 m × 2.8 m, with an inclination angle of 10~15° and an average inclination angle of 12°. The old roof of the coal seam is medium sandstone, the local phase is sandy mudstone, and the direct roof is mudstone and coal. A total of 24 vibrating string-typeborehole stress gauges are installed within the observation range. Twelve high side small coal pillars are installed in the roadway to monitor the high side vertical and horizontal stress, respectively, and the monitoring depth is 1 m, 2 m, 3 m, 4 m, 5 m, and 6 m. Twelve of them are installed in the solid coal side of low side to monitor the vertical and horizontal stress of low side, respectively, and the monitoring depth is 2 m, 4 m, 6 m, 8 m, 10 m, and 12 m, as show in Figures 3 and 4.

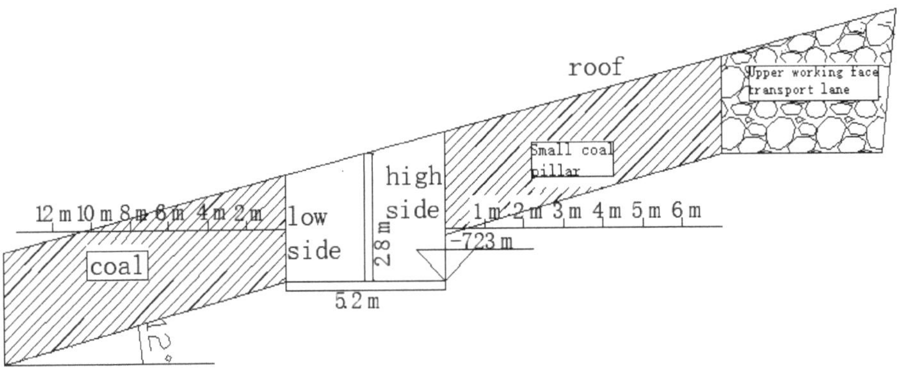

Figure 3. 11613 Schematic diagram of section layout of measuring station.

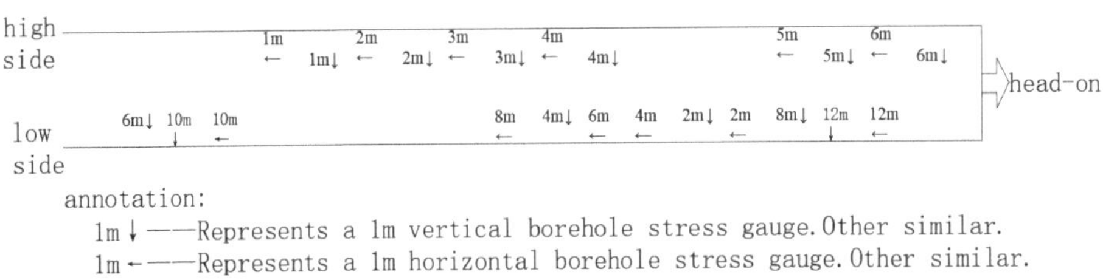

Figure 4. 11613 Graphic layout of measuring station.

2.3. The Basic Condition of 12521 Roadway and Layout of Measuring Station

The working face where the roadway is located has a large fluctuation from east to west from the roof and floor of the coal seam. Generally, it is high at both ends and low in the middle. The occurrence of coal strata is 180~200° < 10~18°, the roadway section is 5.0 m × 3.0 m, and the average dip angle is 14.5°. The coal is black, powdery, and massive. A layer of gangue is commonly developed in the coal seam, and the partial gangue phase turns into a false top. The thickness is 0~0.4 m, with an average of 0.2 m. The thickness

of the coal seam is 1.5~3.0 m, with an average of 2.4 m. Ground elevation +22.3~+24.9 m, working face elevation −642~−720 m. A total of 32 vibrating string drill stress gauges are installed within the observation range. Twelve high side small coal pillars are installed in the roadway to monitor the vertical and horizontal stress, respectively, and the monitoring depth is 1 m, 2 m, 3 m, 4 m, 5 m, and 6 m. Twenty of them are installed in the solid coal side of the low side to monitor the vertical and horizontal stress of the low side, respectively, and the monitoring depth is 2 m, 4 m, 6 m, 8 m, 10 m, 12 m, 14 m, 16 m, 18 m, and 20 m, as show in Figures 5 and 6.

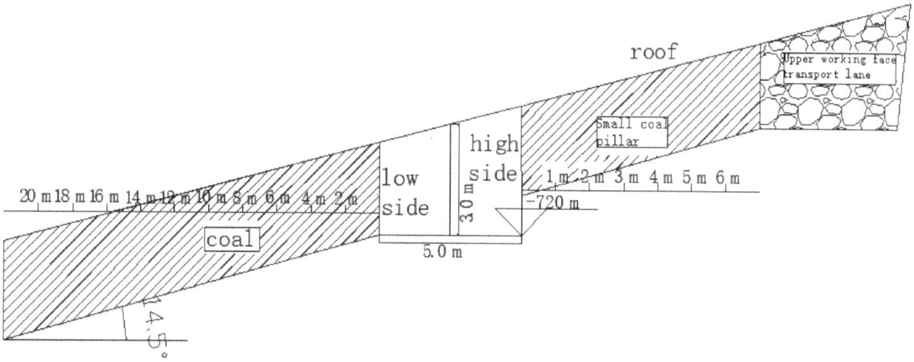

Figure 5. 12521 Schematic diagram of section layout of measuring station.

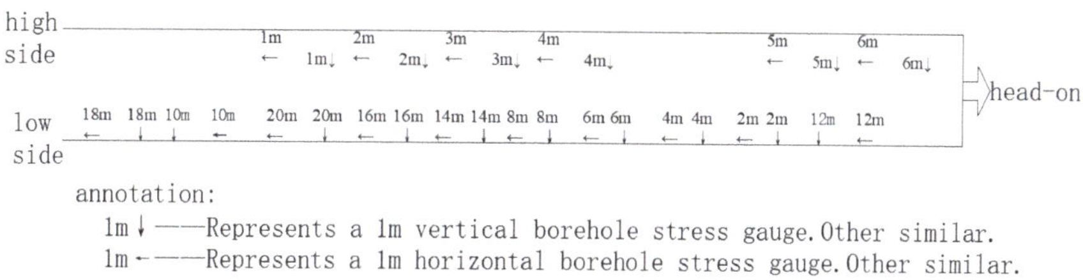

Figure 6. 12521 Graphic layout of measuring station.

All the borehole stress gauges mentioned above are zlGH-40 and produced by the Shandong Rosel Company, as shown in Figure 7. The instrument works through a vibrating string sensor inside the sensor. When measuring, the stress gauge is loaded into the drilling hole, and the pressure head is pressed close to the rock wall by twisting the rotating rod. When the borehole is under pressure, the pressure head will measure the frequency change of the vibration string sensor caused by the pressure. According to the calibration value in advance, the stress change value can be converted.

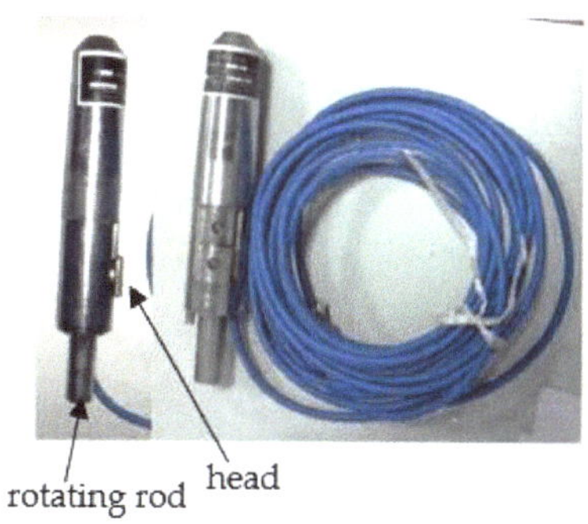

Figure 7. Drill hole stress gauge.

3. Stress Secondary Distribution Theory of Mining Roadway Surrounding Rock

After the roadway is excavated in the rock mass, the surrounding rock of the roadway will inevitably redistribute its stress. Generally, the increased tangential stress after the change of both sides of the roadway is called the support pressure. Support pressure is an important component of mine pressure. In the development from the outer side of the roadway to the deep, the bearing capacity of coal and rock mass increases obviously as it is away from the roadway surface. Within a certain width from the edge of the coal body, the bearing capacity of the coal body and support pressure is in the limit equilibrium state. The stress distribution diagram is shown in Figure 8.

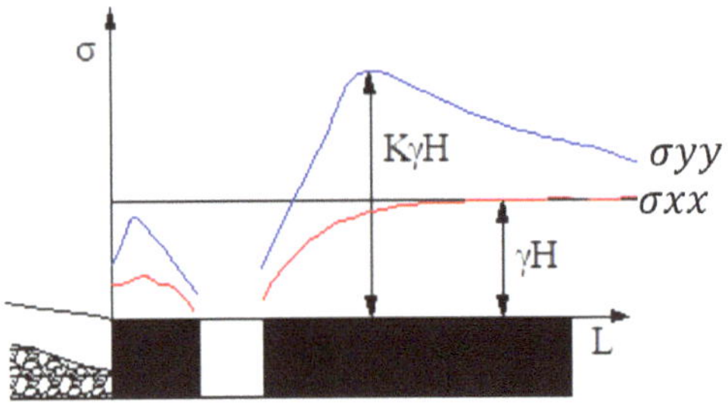

Figure 8. Stress distribution diagram of mining roadway surrounding rock.

4. Analysis of Distribution Law of Stress Increment of Roadway Surrounding Rock under the Influence of Mining

ORIGIN data processing software was used to analyze the stress increase measured value of the monitoring points on the two sides of the three roadways. The comparative analysis charts of horizontal and vertical stress increment of monitoring points in two sides of roadway under the influence of mining were obtained, as shown in Figures 9–12.

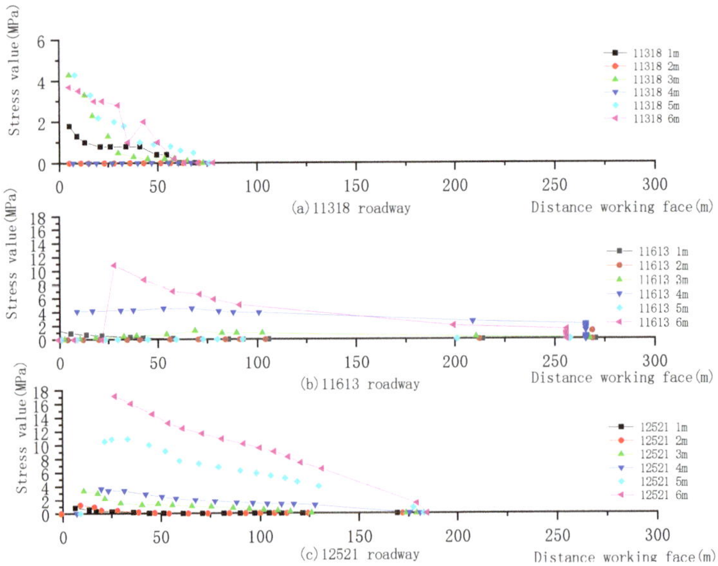

Figure 9. Comparison analysis diagram of vertical stress increment at different depth measuring points in high wall. (**a**) 11318 roadway; (**b**) 11613 roadway;(**c**) 12521 roadway.

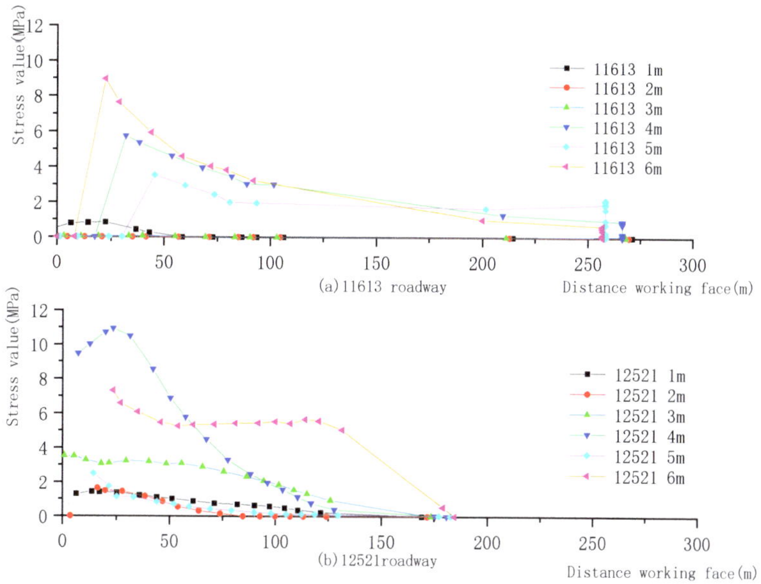

Figure 10. Comparison analysis diagram of horizontal force increment at different depth measuring points in high wall. (**a**) 11613 roadway; (**b**) 12521 roadway.

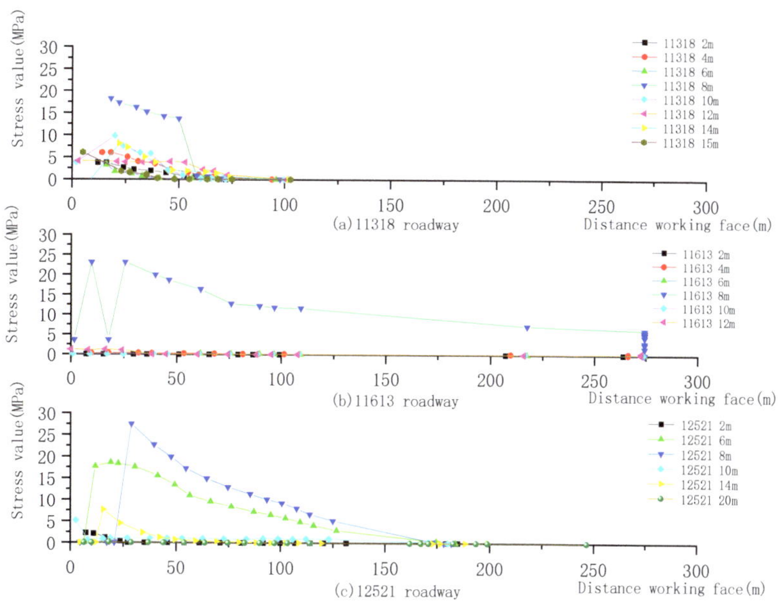

Figure 11. Comparison analysis diagram of vertical stress increment at different depth measuring points in low wall. (**a**) 11318 roadway; (**b**) 11613 roadway; (**c**) 12521 roadway.

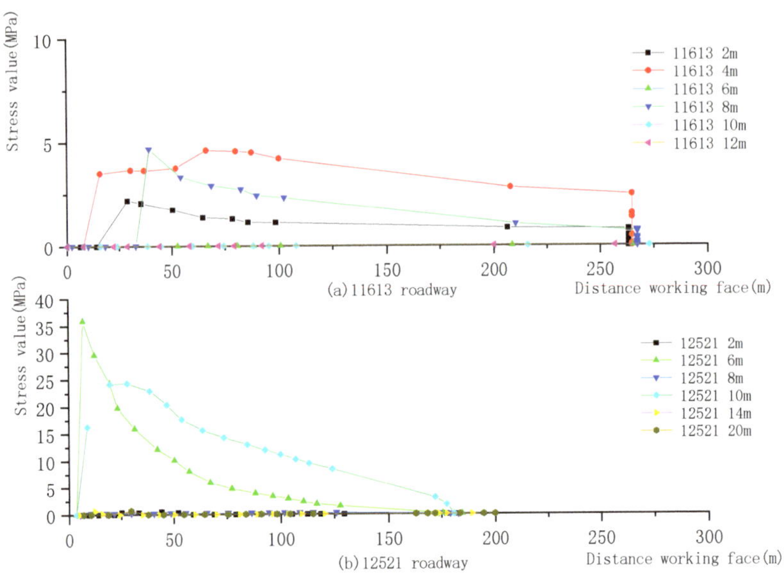

Figure 12. Comparison diagram of horizontal stress at different depth measuring points at low side. (**a**) 11613 roadway; (**b**) 12521 roadway.

4.1. Contrastive Analysis of Distribution Law of Increased Value of Roadway High Side Internal Stress during the Monitoring Period

Figure 9 shows that the increase of vertical stress at all measuring points in the high wall of the monitoring roadway shows a trend of increasing gradually with the decrease of mining distance. However, in Figure 9a, the stress at 6 m away from the roadway surface drops and rises again at 35 m away from the working face, but it still shows a trend of increasing gradually with the decrease of the mining distance. On the whole, the increase value of vertical stress at most monitoring points within 3 m from the roadway surface is small, and the maximum value is about 4 MPa. Figure 9b,c are consistent with such an overall trend. In Figure 9a, the increase value of vertical stress at 1 m away from the roadway surface is larger than that at 2 m, but it is still small compared with the whole. In Figure 9a, the increase of vertical stress at a distance of 3 m from the roadway surface is small when the distance is greater than 30 m from the working face, and increases rapidly when the distance is less than 30 m. Except for the point 3 m away from the roadway surface in Figure 9a, the other points all conform to the conclusion that the increment of vertical stress of most monitoring points within 3 m away from the roadway surface is small. However, the increase values of vertical stress at the monitoring point 4 m away from the roadway surface in Figure 9a and 5 m away from the roadway surface in Figure 9b are almost 0. Nevertheless, on the whole, the increase value of vertical stress at monitoring points 4 m, 5 m, and 6 m away from the roadway surface is larger than that within 3 m away from the roadway surface. Additionally, the added value at 6 m is almost greater than that at other parts. Among them, the added value of vertical stress at 6 m monitoring points in 12521 roadway reaches the maximum value of 17 MPa. In terms of the increased rate of monitoring data, mining influence distance can be divided into three stages: 0~25 m, 25~60 m, and beyond 60 m.

Although some curves in Figure 10 show a certain degree of decline or fluctuation when the mining distance is small, on the whole, the increased value of horizontal stress in the high wall of the roadway increases with the decrease of the mining distance. In the range of 1 m, 2 m, and 3 m away from the roadway surface, the increment of horizontal stress is small on the whole. Except for the small increment of horizontal stress at the measuring point at 5 m in Figure 10b, the increment of horizontal stress at 4 m, 5 m, and

6 m is large. In Figure 10a, the stress increase at 6 m is generally greater than that at other monitoring points. In Figure 10b, the stress increase at the 6 m measuring point is greater than that at other measuring points when the mining distance is greater than 70 m; when the mining distance is less than 70 m, it is only less than that at the 4 m measuring point, and the stress increase still remains at a large level. In general, the stress increase at the 6 m measuring point is at a large level throughout the whole process. In terms of the increased rate of monitoring data, mining influence distance can be divided into three stages: 0~25 m, 25~60 m, and beyond 60 m.

4.2. *Contrastive Analysis of Distribution Law of Increased Value of Roadway Low Side Internal Stress during the Monitoring Period*

Figure 11 shows that the stress added value of measuring points in the middle part of the low side of the three roadways does not change much (some of them fail). From the monitoring results, the stress added value of 8 m monitoring points in the low side of the three roadways is larger than that of other monitoring points, among which the stress added value of 12521 roadway is the largest (27.5 MPa), followed by 11613 (22.5 MPa), then 11318 (18 MPa). In terms of the increased rate of monitoring data, mining influence distance can be divided into three stages: 0~25 m, 25~60 m, and beyond 60 m.

Figure 12 shows that the increment of most of the horizontal stress monitoring points in the low side of the two roadways is relatively small (basic failure). The increase of horizontal stress at 2 m, 4 m, and 8 m monitoring points in 11613 roadway is small, while that at 6 m and 10 m monitoring points in 12521 roadway is large, reaching 25 MPa and 35 MPa, respectively.

4.3. *Change Law of Stress Increment of Two Sides of the Roadway under the Same Mining Distance*

The stress meter measures the actual value of stress, so it cannot directly measure the added value of stress. In order to obtain the change rule of the added value of stress in both sides of roadway under the influence of the same mining distance, the measured stress value should be modified. The correction method is as follows: select the minimum distance between the roadway monitoring station and the working face when the layout is completed as the reference distance, and the measured stress at the reference distance is the reference stress; subtract the reference stress from the measured stress at the other distances to obtain the stress increment, and the stress increment at the reference distance is 0. The minimum distance is selected as the distance (80 m) between 11318 roadway survey station layout and the working face. The three roadways were all taken as the reference distance of 80 m from the working face, the stress at their respective reference distance was measured, and then the stress at the remaining distance was modified accordingly. After modification, the variation of the added value of internal stress in both sides of the roadway under the influence of the same mining distance is obtained, as shown in Figures 13–16.

Figure 13 shows that under the influence of the same mining distance, the increase of vertical stress at the monitoring points in the high side of the three roadways shows the following characteristics: (1) the increasing value of vertical stress appears in the distance from the roadway surface, and shows a trend of increasing with the increase of the distance from the roadway surface; (2) the added value of 6 m monitoring points within the monitoring range of different roadways is the largest; (3) the maximum value of stress added value 12521 > 11613 > 11318; and (4) the increment of vertical stress in the inner shallow part of the high wall is relatively small. When the distance from the working face is less than 25 m, the increment of vertical stress begins to increase slightly.

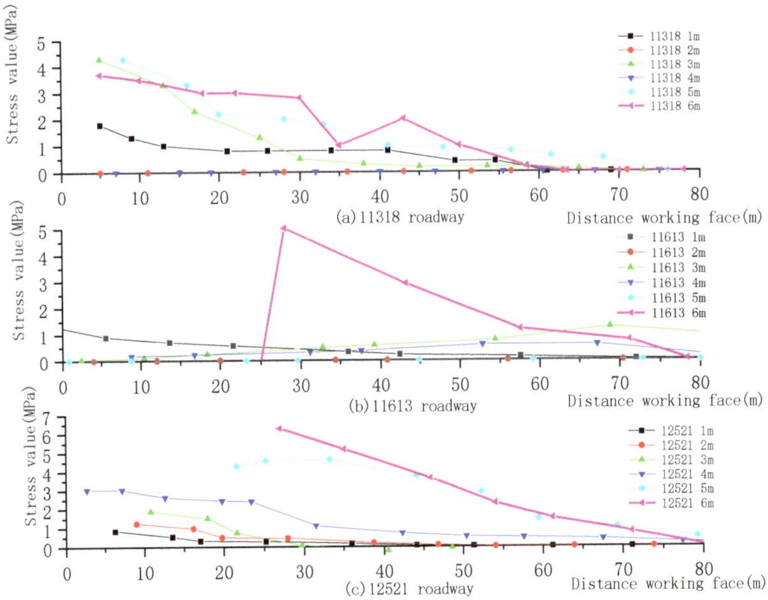

Figure 13. Contrast analysis diagram of high side vertical stress. (**a**) 11318 roadway; (**b**) 11613 roadway; (**c**) 12521 roadway.

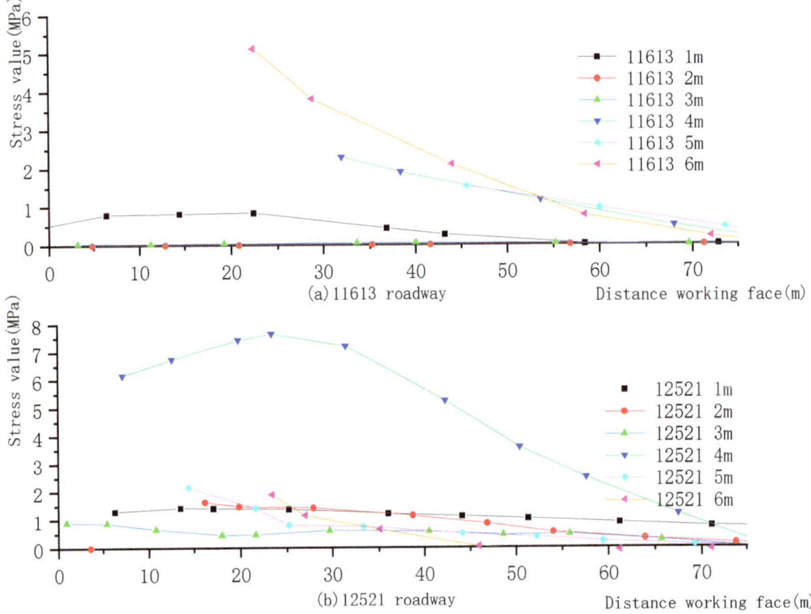

Figure 14. Comparison analysis diagram of high lateral stress. (**a**) 11613 roadway; (**b**) 12521 roadway.

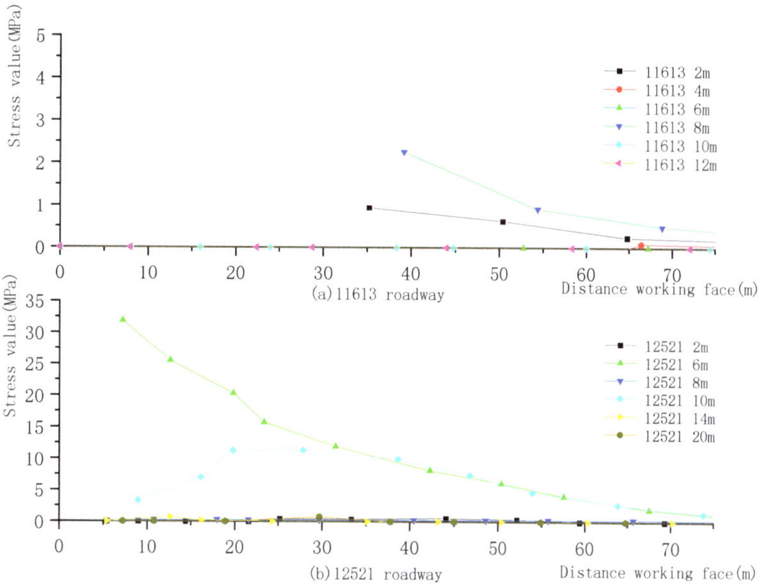

Figure 15. Comparison analysis diagram of horizontal stress increment of low side within 75 m. (**a**) 11613 roadway; (**b**) 12521 roadway.

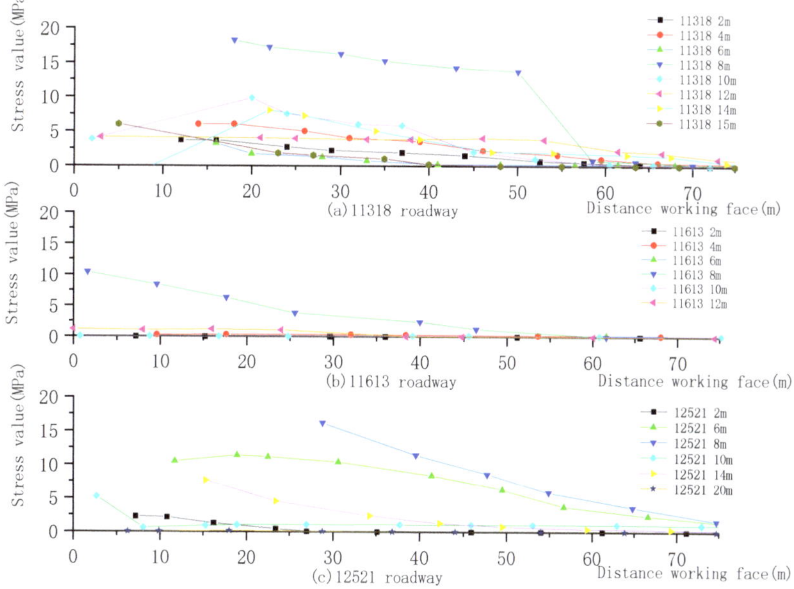

Figure 16. Comparison analysis diagram of vertical stress increment of low side within 75 m. (**a**) 11318 roadway; (**b**) 11613 roadway; (**c**) 12521 roadway.

Figure 14 shows that, under the influence of the same mining distance, the horizontal stress increment of monitoring points on the high side of the roadway shows the following characteristics: (1) the stress increment of 11613 roadway is mainly concentrated at 4 m and 6 m measuring points, and that of 12521 roadway is mainly concentrated at 4 m. On the whole, the stress increment is mainly concentrated at a large distance from the roadway

surface. (2) The stress increase of other horizontal stress increment monitoring points is relatively small.

Figure 15 shows that under the influence of the same mining distance, the increase of horizontal stress at measuring points of the lower wall of 11613 roadway is mainly concentrated at measuring points of 2 m and 8 m. The increase of horizontal stress of 12521 roadway's low side measuring point is mainly concentrated at 6 m and 10 m measuring point, and the increase value is greater than that of 11613 roadway. The whole is concentrated at measuring points 6–10 m away from the roadway surface. Figure 16 shows that, under the influence of the same mining distance, the added vertical stress value of the monitoring points in the low side of the roadway shows the following characteristics: (1) the increase value of vertical stress at 8 m monitoring points within the monitoring range is the largest. Combined with the increasing trend, the maximum value is 12521 > 11318 > 11613. (2) The increase value of vertical stress decreases from 8 m to the adjacent measurement point, but it is larger than other monitoring points.

4.4. Comparison of Mining Stress Increment of Two Sides of Different Roadways under 25 m Lead Distance

The increase of stress in the two sides when the monitoring points of the three roadways are 25 m away from the working face is calculated, and the distribution diagram of vertical stress increases in the two sides when the distance is 25 m away from the working face is obtained, as shown in Figure 17.

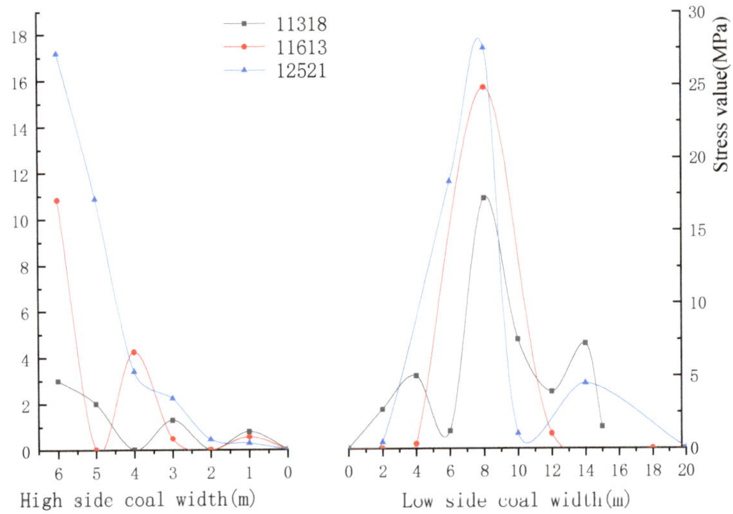

Figure 17. Comparison diagram of vertical stress increment of high and low sides at 25 m lead distance.

Figure 17 shows that, at a distance of 25 m from the working face, as the distance from the roadway surface increases—although the added value of vertical stress in the three roadways fluctuates—it shows an overall trend of increasing first and then decreasing. The peak value is reached at 8 m, and then the vertical stress increment begins to fluctuate and decline, and the second peak value appears at 14 m. Although the increase value of high side vertical stress fluctuates partly, on the whole it increases with the increase of the depth of the monitoring point, and reaches its maximum value at 6 m.

5. Discussion

(1) The influence of buried depth on the added value of mining stress. The increased ranges of the stress added value of the monitoring points in the two sides of the three

roadways are different under the influence of mining. It can be seen from Figure 15 that the increase order of stress added value is 12521 > 11613 > 11318. Additionally, the buried depth of the three roadways is 12521 approximately 660 m, 11613 approximately 630 m, and 11318 approximately 530 m. This shows that buried depth plays a decisive role in the increase of mining-induced stress: the greater the buried depth, the greater the increase of mining-induced stress.

(2) The influence of the setting time of monitoring points on the added value of mining stress. The setting and observation times of the monitoring points on the two sides of the three roadways are different. The observation time of 11613 is the longest, followed by 12521, and then 11318 is the shortest. The distance between observation point 11613 and the working face has not changed for a long time after the installation, but the stress increase in both sides has increased significantly after the installation. However, the final total stress increase is not greater than that of other tunnels. This shows that the installation time does not affect the later stress increment.

(3) Influence of working face distance on the increase of mining stress. According to the observation results, the influence range of mining caused by working face mining is about 150 m. When the distance affected by mining is less than 75 m, the increase rate of stress added in roadway is accelerated. When the mining distance is less than 25 m, the stress increase increases sharply. The stress increment of different roadways is basically the same, but the increment increment increases with the increase of buried depth.

(4) When the distance from the working face is less than 10 m, many sensors are damaged due to the influence of mining, which is an important reason for the disappearance of many sensor data. Some damage is due to the influence of mining stress, and the stress meter cannot measure the stress after the surrounding rock is broken.

(5) Influence of roadway deformation on the added value of mining stress. From the increase range of stress increment at monitoring points and distance of roadway surface. After high side deformation, the increase value of stress is smaller than that of coal pillar 3 m away, and the deformation of roadway surface releases the deformation energy of surrounding rock.The deformation of a roadway's low wall causes the peak value of stress increase to move to the depth of the surrounding rock, and the peak value of vertical stress increase is about 8 m away from the roadway surface.

6. Conclusions

Under the influence of working face mining, the stress increment in the two sides of the roadway shows the following characteristics.

(1) In terms of the increased range of stress added value, the high side vertical stress increases with the increase of distance from the roadway surface, and the peak point is at 6 m. Meanwhile, the low side vertical stress added value has a maximum value at 8 m and a peak value at 14 m. The increased value of horizontal stress of the high side has two peaks, which are 4 m and 6 m, respectively. The increment of horizontal stress in the low side is also around 8 m.

(2) The mining influence range of working face mining is about 150 m, and the mining influence distance can be divided into three stages: 0–25 m, 25–60 m, and 60 m. The increase of vertical and horizontal stress caused by mining influence increases sharply within 25 m away from the working face.

(3) The buried depth of the roadway influences the range of mining influence and the increase of mining stress caused by working face mining. The greater the extent to which the buried depth of the roadway increases, the greater the range of mining influence and the increase of mining stress.

(4) After roadway excavation, the surface deformation of roadway surrounding rock reduces the increase of mining stress near the roadway surface.

Author Contributions: Data curation, Q.J. and H.W.; Investigation, T.L. and K.L.; Project administration, W.P. and X.G.; Resources, Y.Z. All authors have read and agreed to the published version of the manuscript.

Funding: Funded by the National Natural Science Foundation of China (No. 52074117, No. 51774133).

Data Availability Statement: Data available on request due to privacy restrictions.

Conflicts of Interest: The authors declare no conflict of interest.

References

1. Zhao, K.; Li, L. Analysis and development of coal resource security in China. *Coal Eng.* **2018**, *50*, 185–189.
2. Chang, J.; Xie, G. Mechanical characteristics and stability control of deep roadway surrounding rock. *J. China Coal Soc.* **2009**, *34*, 881–886.
3. Chen, X.; Wang, M. Study on deformation characteristics and control technology of surrounding rock in deep inclined coal seam along with goaf excavation. *J. Min. Saf. Eng.* **2015**, *32*, 485–490. [CrossRef]
4. Wang, M.; Bai, J.; Wang, X.; Chen, B.; Han, Z. Study on stability and control of the overlying structure of deep inclined coal seam along goaf. *J. Min. Saf. Eng.* **2015**, *32*, 426–432. [CrossRef]
5. Wei, S.; Song, J.; Jing, H. Numerical analysis of surrounding rock stress distribution in dynamic pressure roadway of inclined coal seam. *Met. Mine* **2012**, 37–41.
6. Wu, H.; Zhang, N.; Wang, W.; Peng, G. Simulation and control technology of roadway deformation characteristics in deep inclined strata. *J. Hunan Univ. Sci. Technol. (Nat. Sci. Ed.)* **2013**, *28*, 6–12.
7. Gao, Y. Study on defamation law and surrounding rock control of soft rock roadway in gently inclined Coal seam. *Min. Saf. Environ. Prot.* **2017**, *44*, 53–56, 59.
8. Sun, X.; Zhang, G.; Cai, F.; Yu, S. Mechanism and control countermeasures of asymmetric deformation of roadway in deep inclined strata. *Chin. J. Rock Mech. Eng.* **2009**, *28*, 1137–1143.
9. Zhang, X.; Chang, J. Study on stress and failure characteristics of surrounding rock in extremely close seam mining with upper and lower goaf. *J. Min. Saf. Eng.* **2014**, *31*, 506–511.
10. Fu, Y. Deformation and failure mechanism of roadway surrounding rock undermining stress and grouting reinforcement technology. *Coal Min.* **2017**, *22*, 34–39. [CrossRef]
11. Chen, J.; Yan, R.; Liu, K. Asymmetric deformation mechanism of roadway in steeply inclined extra-thick coal seam. *J. Coal* **2018**, *43*, 3007–3015. [CrossRef]
12. Wang, H.; Yang, C.; Tian, X. Study on surrounding rock deformation law and support technology of mining-influenced roadway with small coal pillar in goaf. *Coal Eng.* **2020**, *52*, 53–57.
13. Li, L.; Lai, X.; Li, Y.; Lin, H. Monitoring of roadway stress and deformation under a broken surrounding rock. *J. Xi'an Univ. Sci. Technol.* **2010**, *30*, 24–28. [CrossRef]
14. Yang, F.; Chen, W.; Zheng, P.; Wu, G.; Yuan, J.; Yu, J. Study on deformation and failure mechanism and support technology of steeply inclined soft and hard interbedded roadway. *Rock Soil Mech.* **2014**, *35*, 2367–2374, 2425. [CrossRef]
15. Lu, D. Study on roof stress distribution and deformation characteristics of steeply inclined thin coal seam mining. *Coal Eng.* **2015**, *47*, 84–87.
16. Wang, W.; Yuan, C.; Yu, W.; Zhao, Y.; Peng, W.; Wu, H.; Peng, G.; Qu, Y. Control technology of surrounding rock reserved deformation in deep high-stress roadway. *J. China Coal Soc.* **2016**, *41*, 2156–2164.
17. Xie, G.; Li, C.; Wang, L. Mechanical characteristics and engineering practice of roadway surrounding rock stress shell. *J. China Coal Soc.* **2016**, *41*, 2986–2992.
18. Zhao, P.; Li, G.; Li, S.; Liu, L.; Wang, X.; Jia, Y.; Wang, H. Dimensional effect analysis of mechanical characteristics of coal pillar in inclined thick coal seam along goaf. *J. Min. Saf. Eng.* **2019**, *36*, 1120–1127. [CrossRef]
19. Sun, Z.; Gao, K.; Cui, G.; Sun, Y. Stress distribution and asymmetric deformation law of deep inclined coal seam crossing roadway. *Coal Mine Saf.* **2017**, *48*, 61–64. [CrossRef]
20. Wang, W.; Yuan, Y.; Yu, W.; Chen, X. Failure mechanism and control of floor dark deviated well-undermining influence. *J. China Coal Soc.* **2014**, *39*, 1463–1472.
21. Wang, P.; Zhou, G.; Zou, C.; Wang, H.; Wang, C. Asymmetrical support technology of gob-side roadway in inclined coal seam. *Met. Mine* **2014**, 70–73.
22. Wang, X.; Bai, J.; Wang, M. Uneven instability mechanism and control technology of deeply inclined rock roadway under the influence of weak surface. *J. Min. Saf. Eng.* **2015**, *32*, 544–551. [CrossRef]
23. Qi, F. *Deformation Mechanism and Control Technology of Surrounding Rock Along Gob Roadway in Fully Mechanized Caving*; China University of Mining and Technology: Xuzhou, China, 2016.
24. Chen, S.; Song, C.; Guo, Z.; Wang, J.; Wang, Y. Mechanical mechanism and control countermeasures of asymmetric deformation of deep dynamic pressure roadway. *Coal J.* **2016**, *41*, 246–254. [CrossRef]
25. Shi, P.; Xu, S.; Chen, Z. Study on the law of mining pressure in goaf excavation with fully mechanized caving. *Mine Press. Roof Manag.* **2004**, 32–33, 31–118.

26. Han, C.; Zhang, K.; Xu, X.; Li, D.; Xie, P. Study on the failure law and reasonable size of section small coal pillar. *J. Min. Saf. Eng.* **2007**, 370 373.
27. Song, G.; Zhang, K. Study on the law of mining pressure in mining roadway of inclined coal seam. *Min. Saf. Environ. Prot.* **2010**, *37*, 14–16, 19, 91.
28. Zhang, L.; Kou, J.; Dou, M.; Huang, Z. Monitoring and research of surrounding rock stress under the influence of goaf. *China Saf. Sci. Technol.* **2015**, *11*, 33–39.

minerals

MDPI

Article

Assessment of Machine Learning Models for the Prediction of Rate-Dependent Compressive Strength of Rocks

Ziquan Yang [1,2], Yanqi Wu [3,*], Yisong Zhou [2], Hui Tang [2] and Shanchun Fu [2]

1 School of Civil Engineering and Architecture, Wuhan University of Technology, Wuhan 430070, China; civillhy@163.com
2 School of Civil Engineering, Xinyang College, Xinyang 464000, China; yisongzhou2022@163.com (Y.Z.); myzj1990@163.com (H.T.); chenhg2022@163.com (S.F.)
3 School of Civil Engineering, Southeast University, Nanjing 211189, China
* Correspondence: yanqiwu@seu.edu.cn

Abstract: The prediction of rate-dependent compressive strength of rocks in dynamic compression experiments is still a notable challenge. Four machine learning models were introduced and employed on a dataset of 164 experiments to achieve an accurate prediction of the rate-dependent compressive strength of rocks. Then, the relative importance of the seven input features was analyzed. The results showed that compared with the extreme learning machine (ELM), random forest (RF), and the original support vector regression (SVR) models, the correlation coefficient R^2 of prediction results with the hybrid model that combines the particle swarm optimization (PSO) algorithm and SVR was highest in both the training set and the test set, both exceeding 0.98. The PSO-SVR model obtained a higher prediction accuracy and a smaller prediction error than the other three models in terms of evaluation metrics, which showed the possibility of the model as a rate-dependent compressive strength prediction tool. Additionally, besides the static compressive strength, the stress rate is the most important influence factor on the rate-dependent compressive strength of the rock among the listed input parameters. Moreover, the strain rate has a positive effect on the rock strength.

Keywords: machine learning; rock; rate-dependent compressive strength; SVR; random forest

Citation: Yang, Z.; Wu, Y.; Zhou, Y.; Tang, H.; Fu, S. Assessment of Machine Learning Models for the Prediction of Rate-Dependent Compressive Strength of Rocks. *Minerals* **2022**, *12*, 731. https://doi.org/10.3390/min12060731

Academic Editors: Diyuan Li, Zhenyu Han, Xin Cai, Shijie Xie and Paris Xypolias

Received: 17 April 2022
Accepted: 5 June 2022
Published: 8 June 2022

1. Introduction

Strain rate is one of the most important factors affecting the dynamic properties of rocks [1]. Especially for engineering projects involving blasting and excavation, the stress waves generated by blasting are different from the static loads acting on rocks and are complex dynamic processes. In the blasting analysis of rock tunnels, the dynamic properties of rocks show obvious strain rate dependence, and the effect of strain rate should be further considered [2–4]. Many studies have shown that the mechanical properties of rock materials change significantly as the strain rate increases [5,6]. Therefore, it is of great theoretical and practical importance to study the compressive strength of rocks under different strain rate conditions. Moreover, understanding the rate dependence of rock strength is of great importance to rock engineering design and construction [7,8].

As a key indicator of the mechanical properties of rocks, the current methods for determining the compressive strength of rocks rely on laboratory tests such as static compression tests and the Split Hopkinson Pressure Bar (SHPB) test [9,10]. These test methods investigate the rate dependence of the compressive strength of rocks at low and high strain rates, respectively. However, these methods are cumbersome and time-consuming to operate. Moreover, it is difficult to use these test methods to directly study the compressive strength of rocks at moderate strain rates. The study of the mechanical properties of rocks at moderate strain rates can be of great help in understanding the mechanism of excavation-induced geohazards (e.g., rock bursts) [11–13]. Some scholars have attempted to modify the test setup to control the intermediate strain rate [14], but it

remains a challenging task. This is because there are many influencing factors, and these factors are complexly coupled with each other. Although numerical simulations have been applied in this area [15–17], the accurate prediction of rock strength depends on a reasonable intrinsic model and reliable model calibration. Therefore, there is an urgent need for a simple, fast, and reliable intelligent method to predict the compressive strength of rocks at different strain rates and to facilitate understanding of the rate dependence of rock strength.

With the development of artificial intelligence, some intelligent methods such as machine learning have been used to solve some complex engineering problems. In recent years, machine learning techniques such as artificial neural network (ANN), support vector machine (SVM), random forest (RF), extreme learning machine (ELM), and other models have been employed to predict concrete compressive strength and have achieved good prediction behavior [18–25]. Similarly, some intelligent models have been introduced to predict the mechanical properties and stability of rocks, with significant progress being made by numerous scholars [26–34]. For example, Ebrahim, et al. [35] developed a model tree approach to predict the uniaxial compressive strength and elastic modulus of carbonate rocks, which provided better prediction results. Li, et al. [36] utilized the Least squares support vector machine model to predict the compressive strength and shear strength of rocks, and the correlation coefficient R^2 of the predicted results exceeded 0.99. Yılmaz, et al. [37] employed an ANN to predict the compressive strength and elastic modulus of rocks. Compared with the conventional statistical model, the ANN network obtained a higher prediction accuracy. Ehsan, et al. [38] proposed a particle swarm optimization (PSO) algorithm-based ANN model with four input parameters including point load index, Schmidt hammer rebound number, P-wave velocity, and dry density, to predict the uniaxial compressive strength of rocks. Compared with the conventional ANN, the hybrid optimized model reached higher prediction accuracy with $R^2 = 0.97$. Gupta et al. [39] employed five machine learning models on 170 samples to predict rock strength, and the proposed density-weighted least squares twin support vector regression (SVR) model showed the best predictive performance compared with the other four models in terms of evaluation indicators. Hany et al. [40] developed two models, random forest and functional network, to predict the unconstrained compressive strength of rocks using six parameters: drilling torque, weight on bit, mud pumping rate, stand-pipe pressure, drill string rotating speed, and the rate of penetration. The model results showed that the developed RF and functional network models can provide accurate uniaxial compression strength estimations from drilling data in real-time. The application of the model saves time and costs and provides data support and guidance to improve well stability.

Although many successful applications have been achieved using methods such as empirical formulas [41] and intelligent models to estimate the compressive strength of rocks [42,43], it is undeniable that the relevant research has mainly focused on static compressive strength, less research has been performed on the rate-independent compressive strength of rocks, and more research is needed to improve the understanding of this problem. In particular, further analysis is required to understand the compressive strength of rocks under rate independence to improve the ability to predict the rate dependence of rock strength. Fortunately, these successful applications provide a research base and important guidance for the further extension of machine learning model applications to rock dynamics. To this end, this paper attempts to propose a hybrid model which combines a PSO algorithm and SVR to predict the rate-independent compressive strength of rocks. In addition, a comparative analysis with three other models (ELM, random forest, and SVR) is performed, providing new insights into the rate-dependence of rock strength.

2. Method and Models

2.1. Extreme Learning Machine

ELM is an algorithm based on a single hidden layer feedforward neural network, in which the input weights and biases are randomly assigned, and the output weights are

calculated using the Moore–Penrose generalized inverse within the framework of the least-squares criterion [44–46]. Therefore, ELM has the advantage of fast convergence and is less likely to fall into local extremes than traditional neural networks based on gradient descent learning theory [47,48]. The network architecture of ELM is shown in Figure 1. Given a dataset containing N arbitrary samples (x_i, t_i), the number of nodes in the input layer is n and the number of nodes in the output layer is m, where $x_i = [x_{i1}, x_{i2}, \cdots, x_{in}] \in R^n$, $t_i = [t_{i1}, t_{i2}, \cdots, t_{im}] \in R^m$. For a neural network with an activation function $g(x)$ and a single hidden layer with K hidden nodes, the expression is shown below.

$$\sum_{i=1}^{K} \beta_i g(\langle w_i, x_i \rangle + b_i) = o_j \; j = 1, 2, \cdots, N \tag{1}$$

where $w_i = [w_{i1}, w_{i2}, \cdots, w_{in}]^T$ is the weight vector between the ith hidden node and the input node; $\beta_i = [\beta_{i1}, \beta_{i2}, \cdots, \beta_{in}]^T$ is the weight vector between the ith hidden node and the output node; b_i is the bias of the ith hidden node; $\langle w_i, x_i \rangle$ is the inner product of w_i and x; o_j is the output value.

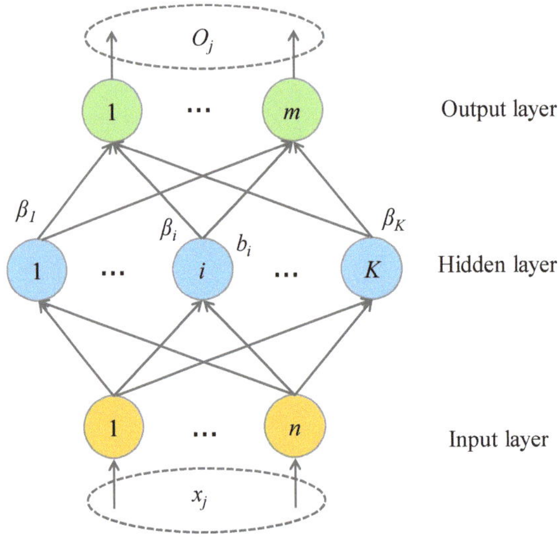

Figure 1. Extreme Learning Machine Network Structure.

It is known that the learning goal of a single hidden layer neural network is to minimize the error in the output, i.e., there exists β_i, w_i, and b_i satisfying the following conditions.

$$\sum_{i=1}^{K} \beta_i g(\langle w_i, x_i \rangle + b_i) = t_j \; j = 1, 2, \cdots, N \tag{2}$$

Equation (2) is expressed in matrix form as follows.

$$H\beta = T \tag{3}$$

2.2. Random Forest

Random forest is a variant of the bagging integration algorithm to improve the model. The random forest uses decision trees as the base learner and builds a random forest model by integrating several decision trees, while the random forest introduces a random selection of feature attributes in the training process of the decision trees [49,50]. Based on this mechanism, random forest inherits the advantages of sample perturbation from the bagging integration algorithm and improves on it by introducing the perturbation strategy

of a random selection of attributes. Therefore, for the same data set, the double randomness of the random forest provides a better generalization ability and overfitting resistance. As shown in Figure 2, the algorithm principle of random forest can be expressed as follows.

$$Y = \frac{1}{N}\sum_{i=1}^{N} F_i(X) \tag{4}$$

where X is the input feature vector, Y is the prediction result, and N is the number of regression tree models built.

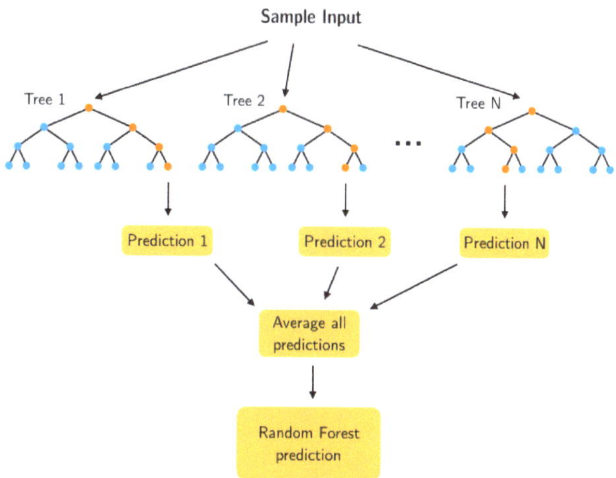

Figure 2. Schematic diagram of random forest model [51].

2.3. Support Vector Regression

Support vector machines can effectively solve classification and complex nonlinear regression problems [52]. When they are applied to regression problems, the basic idea is to find an optimal classification surface that minimizes the error of all training samples from that classification surface [53–55]. Assuming that there is a set of training samples $\{(x_i, y_i), i = 1, 2, \cdots, l\} \in (R^n \times R)$, a linear regression function is established in the high-dimensional feature space as follows.

$$f(x) = w\phi(x) + b \tag{5}$$

where $\phi(x)$ is a nonlinear mapping function. By finding w as well as b according to the structural risk minimization principle, the problem of solving the regression function is transformed into the optimization problem of the function [56,57].

$$\begin{cases} \min \frac{1}{2}\|w^2\| + C\sum_{i=1}^{l}(\xi_i + \xi_i^*) \\ s.t. \begin{cases} y_i - w\phi(x) - b \le \varepsilon + \xi_i \\ -y_i + w\phi(x) + b \le \varepsilon + \xi_i^* \\ \xi_i, \xi_i^* \ge 0 \end{cases} \end{cases} \tag{6}$$

where C is the penalty factor. A larger C indicates a larger penalty for samples with a training error greater than ε; ε specifies the error requirement of the regression function, and a smaller ε indicates a smaller error of the regression function. ξ_i and ξ_i^* are slack variables.

This optimization problem can be solved by introducing the Lagrange function and transforming it into a pairwise form. The regression function can eventually be expressed as:

$$f(x) = \sum_{i=1}^{l} (\alpha_i - \alpha_i^*) K(x_i, x) + b \tag{7}$$

where $K(x_i, x) = \phi(x_i)\phi(x)$ is the kernel function. In this paper, the radial basis function (RBF) with a wide convergence domain is selected as the kernel function.

2.4. Support Vector Regression with Particle Swarm Optimization

For the SVR model with RBF function, the combined values of C and g have a significant effect on the predictive ability of the model. Some intelligent algorithms such as grid search, genetic algorithm, and PSO algorithm, are used to optimize the model. Considering the advantages of the PSO algorithm with fewer parameters and higher efficiency, a PSO-based SVR model is proposed to improve the model performance of the original SVR. The PSO algorithm can be represented as follows. In addition, the training and testing process of the hybrid model is shown in Figure 3.

$$\begin{cases} v_i^{k+1} = \omega \cdot v_i^k + c_1 r_1 (pbest_i^k - x_i^k) + c_2 r_2 (gbest_i^k - x_i^k) \\ x_i^{k+1} = x_i^k + v_i^{k+1} \end{cases} \tag{8}$$

where ω is the initial weight, k is the number of iterations, v_i^k and x_i^k are the velocity and position vectors of the particle, respectively, c_1 and c_2 are the learning factors, r_1 and r_2 are arbitrary values between [0, 1], $pbest_i^k$ is the best position passed by the i-th particle, and $gbest_i^k$ is the global best position.

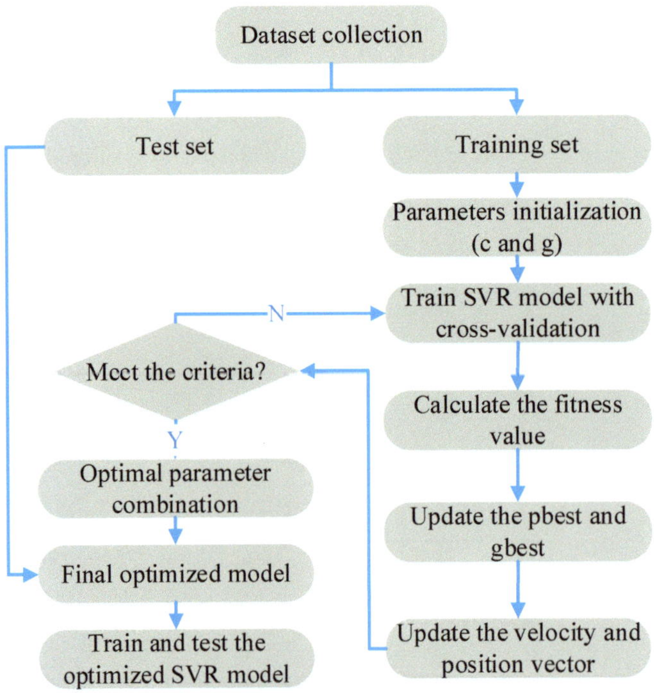

Figure 3. Flowchart of the proposed hybrid model.

3. Dataset Description

A total of 164 datasets were collected from the literature [5]. Each data set consists of seven input parameters which include length, diameter, grain size, bulk density, P-wave velocity, strain rate, and static compressive strength (SCS). The output is the rate-dependent compressive strength (CS) of the rocks. The distribution between each input parameter and output parameter is shown in Figure 4. To further understand the characteristics of the input parameters, the distribution characteristics of these parameters are listed in Table 1. For model training and testing, 131 sets of data were randomly selected for the training of the model, and the remaining data set was used as the test set. For comparison purposes, the training set and test set will be fixed after the training set has been randomly selected by the first model. In this way, the training and test sets of the four models are identical. The model training phase used 10-fold cross-validation.

Table 1. Distribution characteristics of the parameters.

Parameter	Length	Diameter	Grain Size	Bulk Density	P-Wave Velocity	Strain Rate	SCS	CS
Unit	mm	mm	mm	kg/m^3	m/s	s^{-1}	MPa	MPa
Max	70	70	3.5	2850	6651	223	212	352.71
Min	10	2.5	0.03	2278	2437	0.000005	28.6	30.03
Mean	40.43	41.95	0.75	2479.68	3542.99	59.48	87.55	128.14
Median	49.77	49.50	0.23	2384.00	3031.00	56.00	71.91	101.42
Standard deviation	15.81	18.15	1.07	186.21	1231.89	46.85	56.39	80.23
Coefficient of variation	0.39	0.43	1.43	0.08	0.35	0.79	0.64	0.63
Kurtosis	−0.74	−0.55	1.76	−1.04	0.12	1.16	−0.81	−0.32
Skewness	−0.76	−0.62	1.75	0.59	1.33	0.90	0.75	0.91
Pearson correlation coefficient	−0.22	0.13	0.54	0.71	0.78	0.16	0.89	1

In addition, three evaluation metrics are introduced for quantification when assessing and comparing the predictive performance of the models. The definitions of these evaluation metrics are listed in Table 2 [58–61].

Table 2. Definitions of evaluation indexes.

Evaluation Metrics	Definition
Correlation coefficient	$R^2 = \frac{(n\sum_{i=1}^{n}(O_eO_p) - \sum_{i=1}^{n} O_e \sum_{i=1}^{n} O_p)2}{[n(\sum_{i=1}^{n} O_e^2) - (\sum_{i=1}^{n} O_e)^2][n(\sum_{i=1}^{n} O_p^2) - (\sum_{i=1}^{n} O_p)^2]}$
Mean absolute error	$MAE = \frac{1}{n}\sum_{i=1}^{n}\lvert O_e - O_p\rvert$
Mean absolute percentage error	$MAPE = \frac{1}{n}\sum_{i=1}^{n}\left\lvert\frac{O_e - O_p}{O_e}\right\rvert \times 100\%$

Where O_e and O_p are the true and predicted result of the rate-dependent compressive strength, respectively.

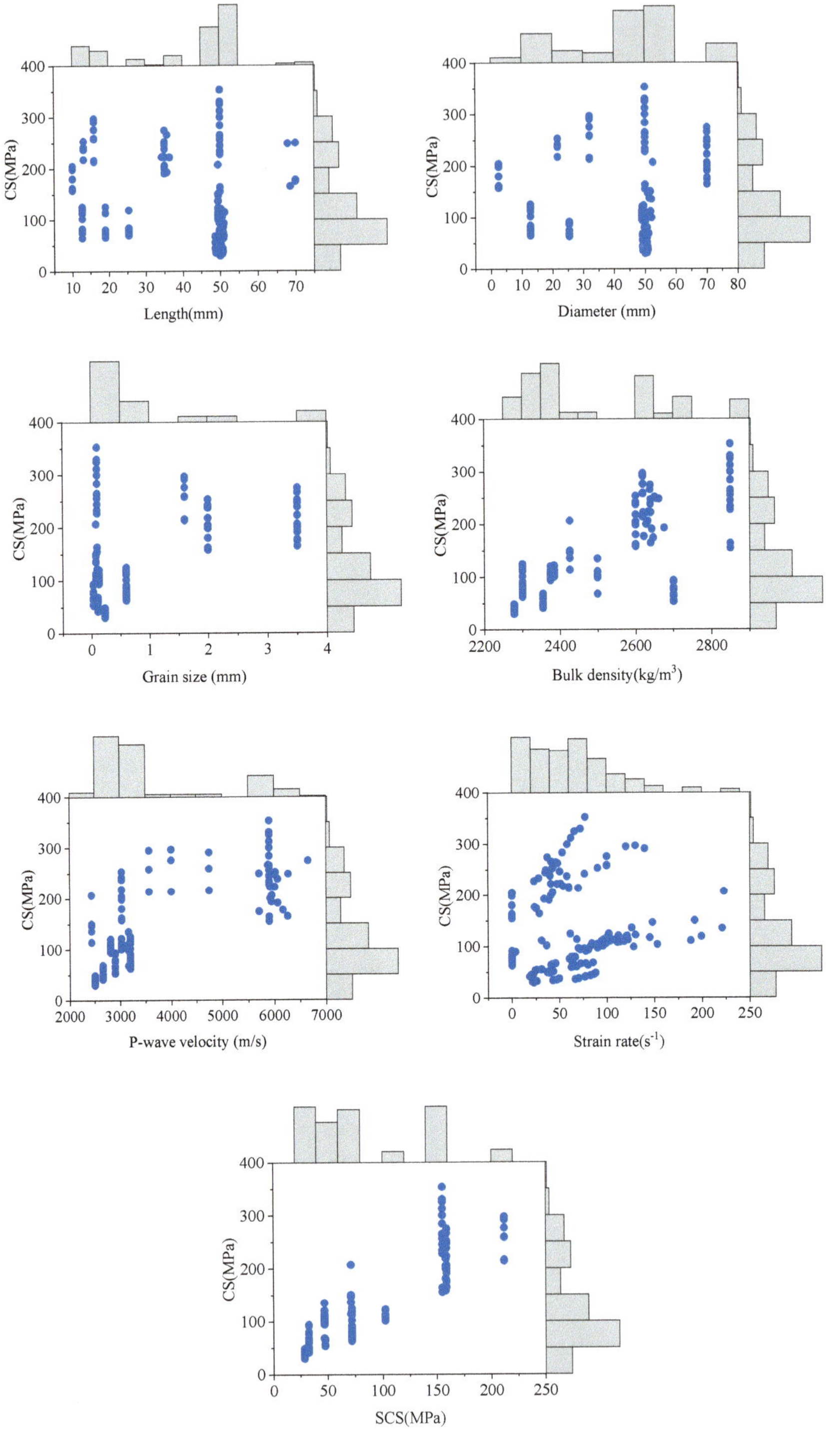

Figure 4. Scatterplot and histogram of the distribution of the input and output parameters.

4. Model Performance Comparison

After training and testing, the prediction results of the four models are shown in Figure 5. Moreover, a linear fitting was performed between the predicted results and the actual values. According to the correlation coefficient R^2 provided in Figure 5, it can be seen that among the four models, the PSO-SVR model achieves the best prediction performance, followed by the random forest, SVR, and ELM model. Specifically, the PSO-SVR model presents the best prediction results both in the training and testing phases. For both the training and test sets, the correlation coefficient R^2 of the models exceeded 0.98.

The box line plots of the relative prediction errors for the testing samples are shown in Figure 6. Compared with the ELM model, the mean prediction relative errors of the other three models were within 10%. For the PSO-SVR model, the mean value of the prediction error reached a minimum of 7.944%. To facilitate further comparison of the prediction performance of different models, Table 3 lists the evaluation metrics for the training and testing phases of each model. For the PSO-SVR model, the correlation coefficient R^2 is the highest and is close to 1. The other two error metrics are smaller than the other three models, which confirm the high accuracy, reliability, and generalization ability of the PSO-SVR model in predicting the rate-dependence compressive strength of rocks.

Table 3. Evaluation indicators of four models.

Model	MAE	MAPE/%	R^2
ELM training	12.531	10.620	0.947
ELM test	15.573	12.664	0.946
RF training	9.429	8.411	0.972
RF test	10.312	9.007	0.979
SVR training	10.218	9.871	0.969
SVR test	11.04145	9.278841	0.972
PSO-SVR training	4.9511	4.718	0.992
PSO-SVR test	10.052	7.944	0.980

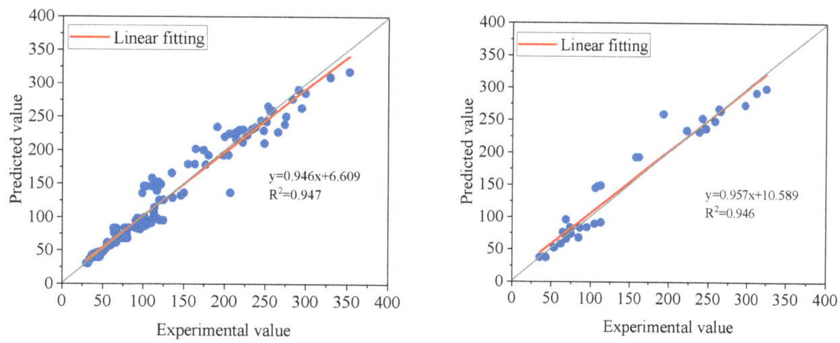

(**a**) ELM model prediction results: training set (left), test set (right)

Figure 5. *Cont.*

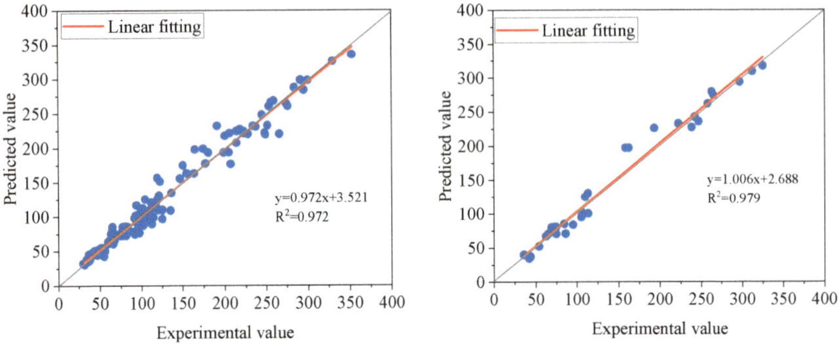

(**b**) RF model prediction results: training set (left), test set (right)

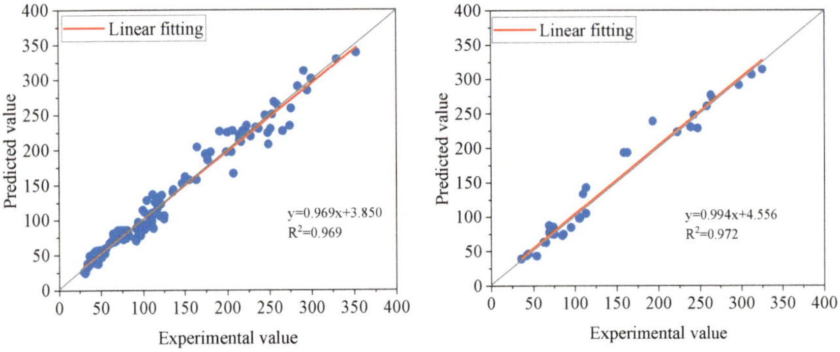

(**c**) SVR model prediction results: training set (left), test set (right)

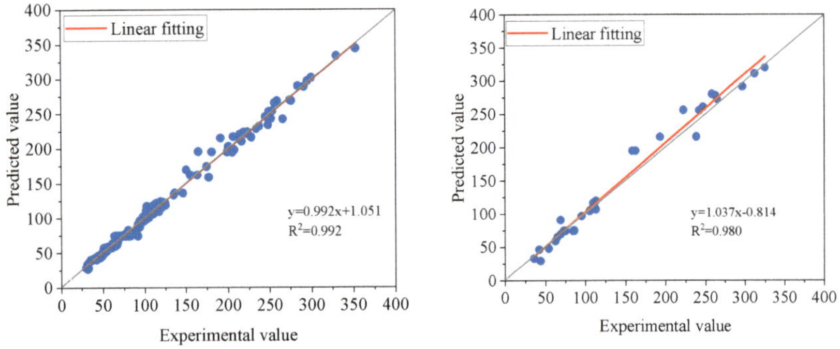

(**d**) PSO-SVR model prediction results: training set (left), test set (right)

Figure 5. Prediction results of three models. (**a**) ELM model prediction results; (**b**) RF model prediction results; (**c**) SVR model prediction results; (**d**) PSO-SVR model prediction results.

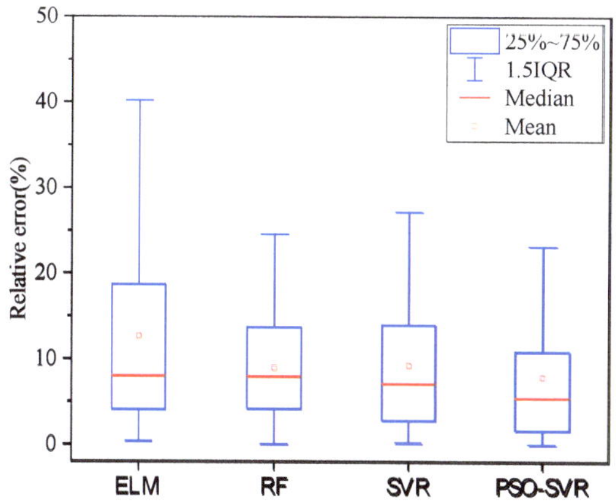

Figure 6. Relative error of model prediction results for the test set.

5. Relative Importance of Input Parameters

Analyzing the influence degree and relative importance of each influencing parameter on the output results is important for the prediction of the results. In particular, it is important to understand the positive or negative effect of each influencing parameter on the output. Figure 7 shows the relative importance of each influence parameter on the output results. It can be seen that among the seven influence parameters listed, the static compressive strength is the most important parameter, with a relative importance of more than 75%. It is followed by the strain rate, whose relative importance is about 15%. The remaining five parameters have a smaller degree of influence on the output results. Further, the effect of each influencing parameter on the output results is analyzed in detail and presented in Figure 8. Each scatter in Figure 8 represents a rock sample in the data set. The red color indicates that the parameter is positive for the output, while the blue color indicates that it is negative for the output. It can be observed that the three parameters, static compressive strength, strain rate, and bulk density, have a significant positive effect on the rate-independent compressive strength of rocks. Their increase leads to an increase in compressive strength.

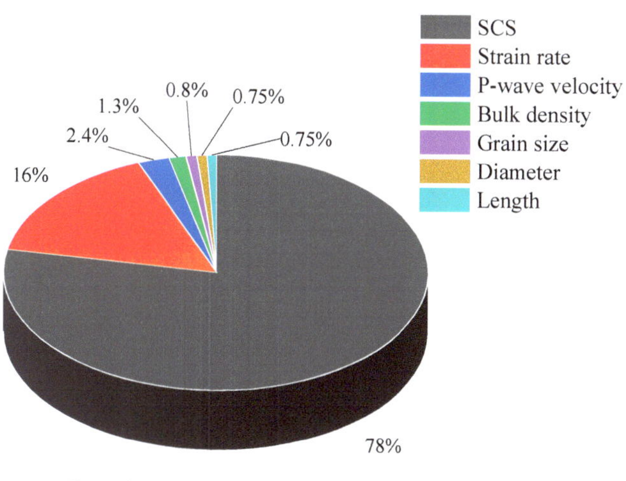

Figure 7. Three-dimensional pie chart of the relative importance of parameters to output.

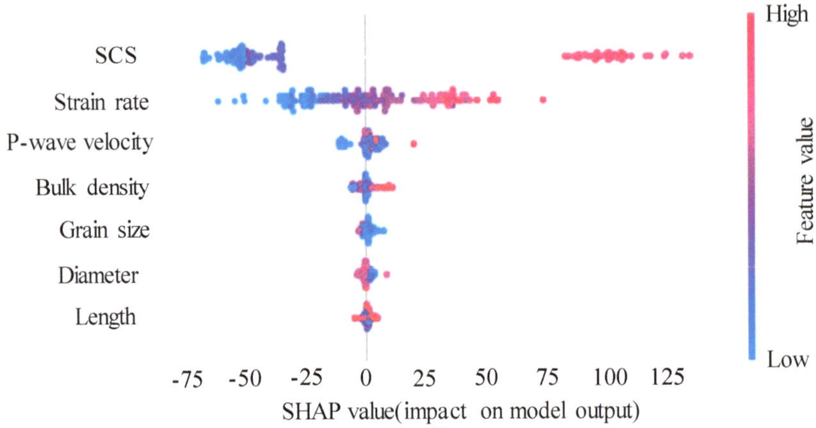

Figure 8. SHAP summary plot of compressive strength.

6. Conclusions

In this work, four intelligent machine learning models were introduced to predict the rate-independent compressive strength of rocks. The main findings are summarized below.

(1) All four machine learning models presented in this paper can effectively achieve a fast and rough estimation of the rate-independent compressive strength of rocks for a given combination of input parameters. Compared with ELM, the average relative prediction errors of the random forest, SVR, and PSO-SVR models were all within 10%, while the PSO-SVR model reached a minimum average relative error of 7.944% for the test set.

(2) The PSO-SVR model could capture the complex nonlinear mapping between multiple inputs and outputs more accurately than the other two models in terms of the evaluation metrics, and its prediction performance is superior to the other three methods.

(3) Among the seven input parameters mentioned, the static compressive strength and the strain rate are the two most important variables for the rate-independent compressive strength of rocks. The three parameters, static compressive strength, strain rate, and bulk density, have significant positive effects on the rate-independent compressive strength of rocks. Their increase leads to an increase in compressive strength.

Author Contributions: Z.Y.: Writing revised manuscript, writing response to comments; Y.W.: methodology, writing—original draft, conceptualization, formal analysis; Y.Z.: data curation, resources; H.T.: validation, data curation; S.F.: funding acquisition, visualization. All authors have read and agreed to the published version of the manuscript.

Funding: Key Scientific and Technological Research Projects of Henan Province (222102210306).

Data Availability Statement: Not applicable.

Conflicts of Interest: The authors declare no conflict of interest.

References

1. Zaid, M.; Sadique, M.R.; Samanta, M. Effect of unconfined compressive strength of rock on dynamic response of shallow unlined tunnel. *SN Appl. Sci.* **2020**, *2*, 2131. [CrossRef]
2. Zaid, M.; Sadique, M.R. A Simple Approximate Simulation Using Coupled Eulerian–Lagrangian (CEL) Simulation in Investigating Effects of Internal Blast in Rock Tunnel. *Indian Geotech. J.* **2021**, *51*, 1038–1055. [CrossRef]
3. Zaid, M.; Shah, I.A. Numerical Analysis of Himalayan Rock Tunnels under Static and Blast Loading. *Geotech. Geol. Eng.* **2021**, *39*, 5063–5083. [CrossRef]
4. Zaid, M.; Sadique, M.R.; Alam, M.M. Blast analysis of tunnels in Manhattan-Schist and Quartz-Schist using coupled-Eulerian–Lagrangian method. *Innov. Infrastruct. Solut.* **2021**, *6*, 69. [CrossRef]

5. Wei, M.; Meng, W.; Dai, F.; Wu, W. Application of machine learning in predicting the rate-dependent compressive strength of rocks. *J. Rock Mech. Geotech. Eng.* 2022, *in press*. [CrossRef]
6. Yang, L.; Wang, G.; Zhao, G.-F.; Shen, L. A rate- and pressure-dependent damage-plasticity constitutive model for rock. *Int. J. Rock Mech. Min. Sci.* **2020**, *133*, 104394. [CrossRef]
7. Xie, S.J.; Han, Z.Y.; Chen, Y.F.; Wang, Y.X.; Zhao, Y.L.; Lin, H. Constitutive modeling of rock materials considering the void compaction characteristics. *Arch. Civ. Mech. Eng.* **2022**, *22*, 60. [CrossRef]
8. Xie, S.J.; Lin, H.; Chen, Y.F.; Yong, R.; Xiong, W.; Du, S.G. A damage constitutive model for shear behavior of joints based on determination of the yield point. *Int. J. Rock Mech. Min. Sci.* **2020**, *128*, 104269. [CrossRef]
9. Bieniawski, Z.T.; Bernede, M.J. Suggested methods for determining the uniaxial compressive strength and deformability of rock materials: Part 1. Suggested method for determining deformability of rock materials in uniaxial compression. *Int. J. Rock Mech. Min. Sci. Geomech. Abstr.* **1979**, *16*, 138–140. [CrossRef]
10. Han, Z.; Li, D.; Li, X. Experimental study on the dynamic behavior of sandstone with coplanar elliptical flaws from macro, meso, and micro viewpoints. *Theor. Appl. Fract. Mec.* **2022**, *120*, 103400. [CrossRef]
11. Zhong, K.; Zhao, W.; Qin, C.; Gao, H.; Chen, W. Mechanical Properties of Roof Rocks under Superimposed Static and Dynamic Loads with Medium Strain Rates in Coal Mines. *Appl. Sci.* **2021**, *11*, 8973. [CrossRef]
12. Duan, K.; Ji, Y.; Wu, W.; Kwok, C.Y. Unloading-induced failure of brittle rock and implications for excavation-induced strain burst. *Tunn. Undergr. Space Technol.* **2019**, *84*, 495–506. [CrossRef]
13. Xie, S.; Han, Z.; Hu, H.; Lin, H. Application of a novel constitutive model to evaluate the shear deformation of discontinuity. *Eng. Geol.* **2022**, *304*, 106693. [CrossRef]
14. Whittington, W.R.; Oppedal, A.L.; Francis, D.K.; Horstemeyer, M.F. A novel intermediate strain rate testing device: The serpentine transmitted bar. *Int. J. Impact Eng.* **2015**, *81*, 1–7. [CrossRef]
15. Du, H.-b.; Dai, F.; Xu, Y.; Liu, Y.; Xu, H.-n. Numerical investigation on the dynamic strength and failure behavior of rocks under hydrostatic confinement in SHPB testing. *Int. J. Rock Mech. Min. Sci.* **2018**, *108*, 43–57. [CrossRef]
16. Huang, X.; Qi, S.; Xia, K.; Shi, X. Particle Crushing of a Filled Fracture during Compression and Its Effect on Stress Wave Propagation. *J. Geophys. Res. Solid Earth* **2018**, *123*, 5559–5587. [CrossRef]
17. Duan, K.; Li, Y.; Wang, L.; Zhao, G.; Wu, W. Dynamic responses and failure modes of stratified sedimentary rocks. *Int. J. Rock Mech. Min. Sci.* **2019**, *122*, 104060. [CrossRef]
18. Wu, Y.; Zhou, Y. Hybrid machine learning model and Shapley additive explanations for compressive strength of sustainable concrete. *Constr. Build. Mater.* **2022**, *330*, 127298. [CrossRef]
19. Ahmad, A.; Ahmad, W.; Aslam, F.; Joyklad, P. Compressive strength prediction of fly ash-based geopolymer concrete via advanced machine learning techniques. *Case Stud. Constr. Mater.* **2022**, *16*, e00840. [CrossRef]
20. Güçlüer, K.; Özbeyaz, A.; Göymen, S.; Günaydın, O. A comparative investigation using machine learning methods for concrete compressive strength estimation. *Mater. Today Commun.* **2021**, *27*, 102278. [CrossRef]
21. Tang, F.; Wu, Y.; Zhou, Y. Hybridizing Grid Search and Support Vector Regression to Predict the Compressive Strength of Fly Ash Concrete. *Adv. Civ. Eng.* **2022**, *2022*, 3601914. [CrossRef]
22. Ray, S.; Haque, M.; Rahman, M.M.; Sakib, M.N.; Al Rakib, K. Experimental investigation and SVM-based prediction of compressive and splitting tensile strength of ceramic waste aggregate concrete. *J. King Saud Univ.-Eng. Sci.* 2021, *in press*. [CrossRef]
23. Ahmad, W.; Ahmad, A.; Ostrowski, K.A.; Aslam, F.; Joyklad, P.; Zajdel, P. Application of Advanced Machine Learning Approaches to Predict the Compressive Strength of Concrete Containing Supplementary Cementitious Materials. *Materials* **2021**, *14*, 5762. [CrossRef] [PubMed]
24. Sonebi, M.; Cevik, A.; Grünewald, S.; Walraven, J. Modelling the fresh properties of self-compacting concrete using support vector machine approach. *Constr. Build. Mater.* **2016**, *106*, 55–64. [CrossRef]
25. Saha, P.; Debnath, P.; Thomas, P. Prediction of fresh and hardened properties of self-compacting concrete using support vector regression approach. *Neural Comput. Appl.* **2020**, *32*, 7995–8010. [CrossRef]
26. Meng, F.; Jing, S.; Sun, X.; Wang, C.; Liang, Y.; Pang, D. A New Approach of Disaster Forecasting Based on Least Square Optimized Neural Network. *Geofluids* **2020**, *2020*, 8882241. [CrossRef]
27. Acar, M.C.; Kaya, B. Models to estimate the elastic modulus of weak rocks based on least square support vector machine. *Arab. J. Geosci.* **2020**, *13*, 590. [CrossRef]
28. Ceryan, N.; Okkan, U.; Kesimal, A. Prediction of unconfined compressive strength of carbonate rocks using artificial neural networks. *Environ. Earth Sci.* **2013**, *68*, 807–819. [CrossRef]
29. Ceryan, N.; Okkan, U.; Samui, P.; Ceryan, S. Modeling of tensile strength of rocks materials based on support vector machines approaches. *Int. J. Numer. Anal. Methods Geomech.* **2013**, *37*, 2655–2670. [CrossRef]
30. Fattahi, H. Application of improved support vector regression model for prediction of deformation modulus of a rock mass. *Eng. Comput.* **2016**, *32*, 567–580. [CrossRef]
31. Mohamad, E.T.; Armaghani, D.J.; Momeni, E.; Abad, S.V.A.N.K. Prediction of the unconfined compressive strength of soft rocks: A PSO-based ANN approach. *Bull. Eng. Geol. Environ.* **2015**, *74*, 745–757. [CrossRef]
32. Rezaei, M.; Majdi, A.; Monjezi, M. An intelligent approach to predict unconfined compressive strength of rock surrounding access tunnels in longwall coal mining. *Neural Comput. Appl.* **2014**, *24*, 233–241. [CrossRef]

33. Yagiz, S. Predicting uniaxial compressive strength, modulus of elasticity and index properties of rocks using the Schmidt hammer. *Bull. Eng. Geol. Environ.* **2009**, *68*, 55–63. [CrossRef]
34. Armaghani, D.J.; Mohamad, E.T.; Momeni, E.; Monjezi, M.; Narayanasamy, M.S. Prediction of the strength and elasticity modulus of granite through an expert artificial neural network. *Arab. J. Geosci.* **2015**, *9*, 48. [CrossRef]
35. Ghasemi, E.; Kalhori, H.; Bagherpour, R.; Yagiz, S. Model tree approach for predicting uniaxial compressive strength and Young's modulus of carbonate rocks. *Bull. Eng. Geol. Environ.* **2018**, *77*, 331–343. [CrossRef]
36. Li, W.; Tan, Z. Research on Rock Strength Prediction Based on Least Squares Support Vector Machine. *Geotech. Geol. Eng.* **2017**, *35*, 385–393. [CrossRef]
37. Yılmaz, I.; Yuksek, A.G. An Example of Artificial Neural Network (ANN) Application for Indirect Estimation of Rock Parameters. *Rock Mech. Rock Eng.* **2008**, *41*, 781–795. [CrossRef]
38. Momeni, E.; Armaghani, D.J.; Hajihassani, M.; Amin, M.F.M. Prediction of uniaxial compressive strength of rock samples using hybrid particle swarm optimization-based artificial neural networks. *Measurement* **2015**, *60*, 50–63. [CrossRef]
39. Gupta, D.; Natarajan, N. Prediction of uniaxial compressive strength of rock samples using density weighted least squares twin support vector regression. *Neural Comput. Appl.* **2021**, *33*, 15843–15850. [CrossRef]
40. Gamal, H.; Alsaihati, A.; Elkatatny, S.; Haidary, S.; Abdulraheem, A. Rock Strength Prediction in Real-Time While Drilling Employing Random Forest and Functional Network Techniques. *J. Energy Resour. Technol.* **2021**, *143*, 093004. [CrossRef]
41. Xie, S.J.; Lin, H.; Chen, Y.F.; Wang, Y.X. A new nonlinear empirical strength criterion for rocks under conventional triaxial compression. *J. Cent. South Univ.* **2021**, *28*, 1448–1458. [CrossRef]
42. Farhadian, A.; Ghasemi, E.; Hoseinie, S.H.; Bagherpour, R. Prediction of Rock Abrasivity Index (RAI) and Uniaxial Compressive Strength (UCS) of Granite Building Stones Using Nondestructive Tests. *Geotech. Geol. Eng.* **2022**, *40*, 3343–3356. [CrossRef]
43. Amirkiyaei, V.; Ghasemi, E.; Faramarzi, L. Estimating uniaxial compressive strength of carbonate building stones based on some intact stone properties after deterioration by freeze–thaw. *Environ. Earth Sci.* **2021**, *80*, 352. [CrossRef]
44. Liang, N.; Huang, G.; Saratchandran, P.; Sundararajan, N. A Fast and Accurate Online Sequential Learning Algorithm for Feedforward Networks. *IEEE Trans. Neural Netw.* **2006**, *17*, 1411–1423. [CrossRef]
45. Guang-Bin, H. Learning capability and storage capacity of two-hidden-layer feedforward networks. *IEEE Trans. Neural Netw.* **2003**, *14*, 274–281. [CrossRef] [PubMed]
46. Ding, S.; Zhao, H.; Zhang, Y.; Xu, X.; Nie, R. Extreme learning machine: Algorithm, theory and applications. *Artif. Intell. Rev.* **2015**, *44*, 103–115. [CrossRef]
47. Tran, V.Q. Machine learning approach for investigating chloride diffusion coefficient of concrete containing supplementary cementitious materials. *Constr. Build. Mater.* **2022**, *328*, 127103. [CrossRef]
48. Wan, D. Research on Extreme Learning of Neural Networks. *Chin. J. Comput.* **2010**, *33*, 279–287.
49. Han, Q.; Gui, C.; Xu, J.; Lacidogna, G. A generalized method to predict the compressive strength of high-performance concrete by improved random forest algorithm. *Constr. Build. Mater.* **2019**, *226*, 734–742. [CrossRef]
50. Li, H.; Lin, J.; Lei, X.; Wei, T. Compressive strength prediction of basalt fiber reinforced concrete via random forest algorithm. *Mater. Today Commun.* **2022**, *30*, 103117. [CrossRef]
51. He, B.; Lai, S.H.; Mohammed, A.S.; Sabri, M.M.; Ulrikh, D.V. Estimation of Blast-Induced Peak Particle Velocity through the Improved Weighted Random Forest Technique. *Appl. Sci.* **2022**, *12*, 5019. [CrossRef]
52. Wu, Y.; Li, S. Damage degree evaluation of masonry using optimized SVM-based acoustic emission monitoring and rate process theory. *Measurement* **2022**, *190*, 110729. [CrossRef]
53. Li, N.; Nguyen, H.; Rostami, J.; Zhang, W.; Bui, X.-N.; Pradhan, B. Predicting rock displacement in underground mines using improved machine learning-based models. *Measurement* **2022**, *188*, 110552. [CrossRef]
54. Bardhan, A.; Kardani, N.; GuhaRay, A.; Burman, A.; Samui, P.; Zhang, Y. Hybrid ensemble soft computing approach for predicting penetration rate of tunnel boring machine in a rock environment. *J. Rock Mech. Geotech. Eng.* **2021**, *13*, 1398–1412. [CrossRef]
55. Liu, B.; Wang, R.; Guan, Z.; Li, J.; Xu, Z.; Guo, X.; Wang, Y. Improved support vector regression models for predicting rock mass parameters using tunnel boring machine driving data. *Tunn. Undergr. Space Technol.* **2019**, *91*, 102958. [CrossRef]
56. Jueyendah, S.; Lezgy-Nazargah, M.; Eskandari-Naddaf, H.; Emamian, S.A. Predicting the mechanical properties of cement mortar using the support vector machine approach. *Constr. Build. Mater.* **2021**, *291*, 123396. [CrossRef]
57. Wu, Y.; Zhou, Y. Prediction and feature analysis of punching shear strength of two-way reinforced concrete slabs using optimized machine learning algorithm and Shapley additive explanations. *Mech. Adv. Mater. Struct.* **2022**, 1–11. [CrossRef]
58. Song, H.; Ahmad, A.; Farooq, F.; Ostrowski, K.A.; Maślak, M.; Czarnecki, S.; Aslam, F. Predicting the compressive strength of concrete with fly ash admixture using machine learning algorithms. *Constr. Build. Mater.* **2021**, *308*, 125021. [CrossRef]
59. Chen, H.; Li, X.; Wu, Y.; Zuo, L.; Lu, M.; Zhou, Y. Compressive Strength Prediction of High-Strength Concrete Using Long Short-Term Memory and Machine Learning Algorithms. *Buildings* **2022**, *12*, 302. [CrossRef]
60. Farooq, F.; Amin, M.N.; Khan, K.; Sadiq, M.R.; Javed, M.F.; Aslam, F.; Alyousef, R. A Comparative Study of Random Forest and Genetic Engineering Programming for the Prediction of Compressive Strength of High Strength Concrete (HSC). *Appl. Sci.* **2020**, *10*, 7330. [CrossRef]
61. Nafees, A.; Javed, M.F.; Khan, S.; Nazir, K.; Farooq, F.; Aslam, F.; Musarat, M.A.; Vatin, N.I. Predictive Modeling of Mechanical Properties of Silica Fume-Based Green Concrete Using Artificial Intelligence Approaches: MLPNN, ANFIS, and GEP. *Materials* **2021**, *14*, 7531. [CrossRef] [PubMed]

minerals

Article

Effects of Accumulated Damage on the Dynamic Properties of Coal Measures Sandstone

Ziheng Sha [1], Hai Pu [1,2,*], Junce Xu [1], Hongyang Ni [1,3] and Shiru Guo [1]

1 State Key Laboratory for Geomechanics & Deep Underground Engineering, China University of Mining & Technology, Xuzhou 221116, China; zhsha@cumt.edu.cn (Z.S.); xujunce@cumt.edu.cn (J.X.); nhyang@cumt.edu.cn (H.N.); guoshiru@cumt.edu.cn (S.G.)
2 College of Mining Engineering and Geology, Xinjiang Institute of Engineering, Urumqi 830091, China
3 LaMcube, FRE2016, CNRS, University of Lille, 59000 Lille, France
* Correspondence: haipu@cumt.edu.cn

Abstract: The coupling effect of accumulated damage and impact load substantially affects the integrity of the surrounding rock structure in deep coal mining engineering, which has inhibited safe and effective coal mining. Therefore, dynamic compression tests were performed on coal measures sandstone specimens with accumulated damage using the SHPB device. The effects of a high strain rate and accumulated damage on the sandstone's mechanical behavior and damage evolution were investigated. The results reveal that accumulated damage has a considerable impact on specimen stress–strain curves and lowers dynamic compressive strength and deformation modulus substantially. The sandstone failure mode looks to be shear failure from a macroscopic perspective, while it appears to be intergranular fracture between the mineral particles from a microscopic perspective. The macroscopic and microscopic failure mechanisms of the sandstone specimens likewise conformed to the energy absorption law. The accumulated damage factor and the accumulated damage correction coefficient were presented in order to construct a statistical damage constitutive model of rocks based on the Weibull distribution. This model provides a decent description of the effects of accumulated damage and the strain rate on sandstone's mechanical behavior, with parameters that are all of evident physical significance.

Keywords: accumulated damage; SHPB; mechanical behavior; failure modes; constitutive model

Citation: Sha, Z.; Pu, H.; Xu, J.; Ni, H.; Guo, S. Effects of Accumulated Damage on the Dynamic Properties of Coal Measures Sandstone. *Minerals* **2022**, *12*, 810. https://doi.org/10.3390/min12070810

Academic Editors: Diyuan Li, Zhenyu Han, Xin Cai, Shijie Xie and Gianvito Scaringi

Received: 17 May 2022
Accepted: 22 June 2022
Published: 25 June 2022

Publisher's Note: MDPI stays neutral with regard to jurisdictional claims in published maps and institutional affiliations.

1. Introduction

Deep mining engineering has more engineering risks than shallow mining engineering, such as frequent and high-intensity rockburst, massive mine pressure bumps, and considerable deformation of the roadway surrounding rock [1–3]. It could be owing to the deep rock mass's complicated geological environment of "three highs and one disturbance" [4,5]. As a result, the deep roadway's surrounding rock mass is frequently subjected to significant impact loads [2,5]. At the same time, the deep surrounding rock mass will be damaged to varying degrees after excavation due to the more complex stress field in the deep mine roadway (Figure 1) [1,4,6]. Therefore, comprehending the rock mechanics under the coupling action of accumulated damage and impact load is critical for deep mining engineering.

Over the last few decades, rock dynamics research has advanced significantly. Researchers began by studying the uniaxial compression characteristics of rocks under impact load [7–9], then looked at the tensile performance of rocks under impact load [10,11], and finally came to a conclusion that, in most cases, the impact load dramatically increases the strength of rocks, which is manifested as the strain rate effect. The impact of high temperatures on the mechanical characteristics of rocks under impact loads has been thoroughly investigated using a background in deep mining engineering [12,13]. Similarly, the effects of coupled static and dynamic conditions have also been extensively studied [14,15]. The freeze–thaw cycle has a significant impact on the dynamic characteristics of rocks in

open-pit mining and geotechnical engineering in cold and wet environments [16,17]. Some researchers studied the evolution of rocks' mechanical behavior under cyclic impact loads in order to better understand the dynamic damage characteristics of rocks [18,19]. They discovered that as the cyclic impact number increases, the dynamic strength and deformation capacity of rocks weaken, reflecting the gradual accumulation of fatigue damage. In recent years, the dynamic fracture characteristics of rocks in different types of damage states have also been studied in depth by a number of scholars. Peng et al. [20] evaluated the static and dynamic mechanical behavior of granite at various burial depths and discovered that the initial damage variables of the rock specimens grew significantly at first and then gradually dropped as the burial depth increased. Wang et al. and Li et al. [21,22] inflicted initial damage on Fangshan granite with the aid of microwave radiation. Microwave damage was quantified using CT values to investigate further the effect of the initial damage on dynamic rock fragmentation. Yan et al. [23,24] examined the dynamic cracking behavior and energy evolution of flawed rocks, considering that rocks in subsurface rock engineering are initially subjected to static tectonic stress or gravity stress and then to dynamic loading. In truth, the investigation of accumulated damage is not restricted to rocks [25]. The nature of the accumulated damage is that it alters the microstructural parameters of rocks, such as porosity, pore size, and pore distribution, causing changes in the physical properties of the rocks, which strongly affect their mechanical characteristics [26–28]. Therefore, the concern regarding accumulated stress damage caused by deep mining engineering requires immediate attention.

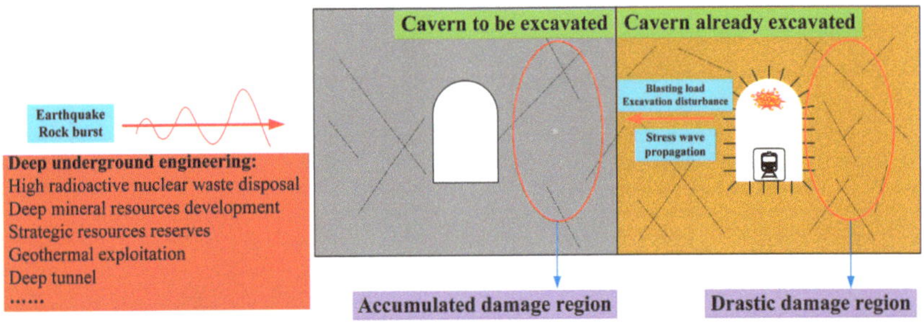

Figure 1. Overview of rock dynamics and accumulated damage problems in deep underground engineering.

The constitutive model of rocks has become a difficult research task due to the discontinuity and non-uniformity of rock materials [2,29]. Numerous academics have contributed to rock dynamics by doing extensive theoretical research. The Jh-2 model [30] is frequently used to describe the dynamic response of rocks. It is a phenomenological constitutive model that has no correlation with the microstructure of rocks and ignores their fundamental features. The classical TCK model was developed by Taylor et al. [31]. They assumed the distribution and development law of micro-cracks in rocks from the microscopic perspective and characterized the rocks' behavior. Liu et al. [32] developed a dynamic damage constitutive model for jointed rock masses based on the TCK theory. To characterize the damage induced by micro-cracks and joints, they proposed a mesoscopic damage variable and a macroscopic damage tensor. The G–K model [33] has also been widely accepted by scholars. It can effectively describe the process of rock properties evolving as a result of cracks in the rock. The rock dynamic constitutive relationship created using the 3D scatterplot method, proposed by Zhou et al. [34], takes into account the scenario of unstable strain rate loading and may provide more plentiful strain rate information. Scholars address more intricate situations while studying building rock material models. For example, Li and Shi [35] proposed a rock dynamic material model based on the D-P strength criterion and the Johns–Cook model for high confining pressures. Wu et al. [36] and Ke et al. [17] proposed a model for dynamic strength degradation of rocks that takes into account the

number of F–T cycles. The initial accumulated damage effect of rocks has not yet been included in theoretical rock dynamics studies.

In this work, the SHPB test system was used to evaluate the performance of coal measures sandstone after it was subjected to initial accumulated damage under the action of an impact load. Sandstone deformation and failure modes, as well as energy absorption patterns, were analyzed in detail. Finally, a statistical damage constitutive model that considers accumulated damage and the strain rate effect has been developed employing the accumulated damage factor and the accumulated damage correction coefficient. The sensitivities of the corresponding parameters were evaluated to measure the relative physical significance.

2. Material and Methods

2.1. Material Characterization

The rock material employed in this research is coal measures sandstone, sourced from the Sanhejian coal mine, Xuzhou, Jiangsu province, China, at a depth of 800 m, as indicated in Figure 2. In this location, thick Jurassic–Cretaceous strata overlay the coal measures strata. Disc-type rock specimens were treated for dynamic testing according to the test concept. Simultaneously, a portion of the sandstone was turned into standard cylinders with a diameter of 50 mm and a height of 100 mm for quasi-static tests. The specimens were machined to suit ISRM specifications [37] (Figure 2d). The sandstone specimens with apparent faults or cracks on the surfaces were removed after processing. The remaining specimens were then ultrasonically tested with an HC-U81 ultrasonic detector, and any sandstone specimens with anomalous acoustic values were discarded. Finally, the remaining specimens were employed to conduct quasi-static and dynamic tests, ensuring that sandstone specimen dispersion remained within the controlled range.

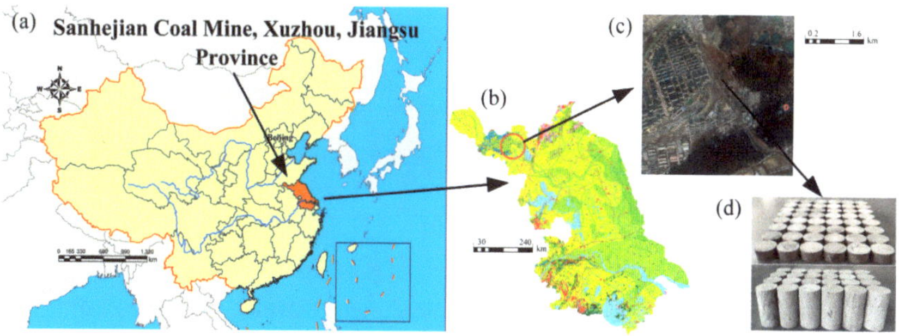

Figure 2. Coal measures sandstone sampling sites: (**a**) Location in China; (**b**) geological sketch of Jiangsu province (from the National Geological Archive of China); (**c**) satellite view of the Sanhejian coal mine; (**d**) part of sandstone specimens.

The SEM images and the XRD graph of coal measures sandstone are shown in Figures 3 and 4, respectively. The SEM images show some tiny internal porosities and mineral grains on the surface of the sandstone, indicating that the selected sandstone may be sensitive to an accumulated damage effect. The mineralogical composition of the sandstone is represented by an XRD graph. According to the XRD graph, the sandstone used in this study contains a large proportion of quartz, along with a slight presence of berlinite, clinochlore, kaolinite, and nacrite.

2.2. Physical and Quasi-Static Tests

The physical and quasi-static properties of the rock are critical for describing material qualities. The sandstone specimens were dried for 24 h in an electric drying oven at 85 °C, following which the moisture content of the sandstone in its natural condition could be

determined. According to several observations and estimates, sandstone has a density of 2589 kg/m^3, a P-wave velocity of 3682 m/s, and a natural moisture content of 1.13%.

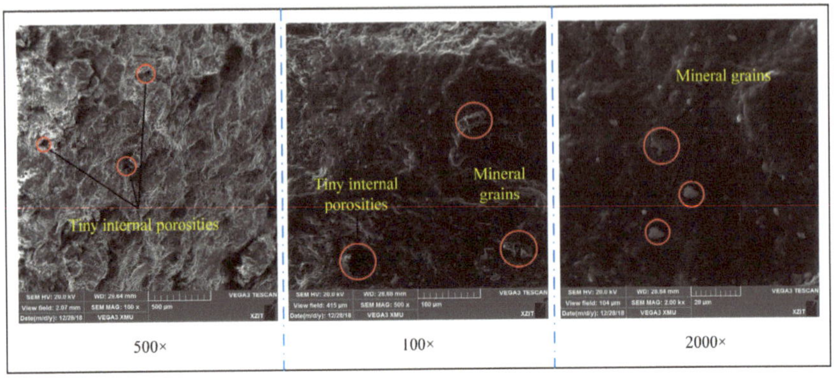

Figure 3. Scanning electron microscopy (SEM) analysis of sandstone.

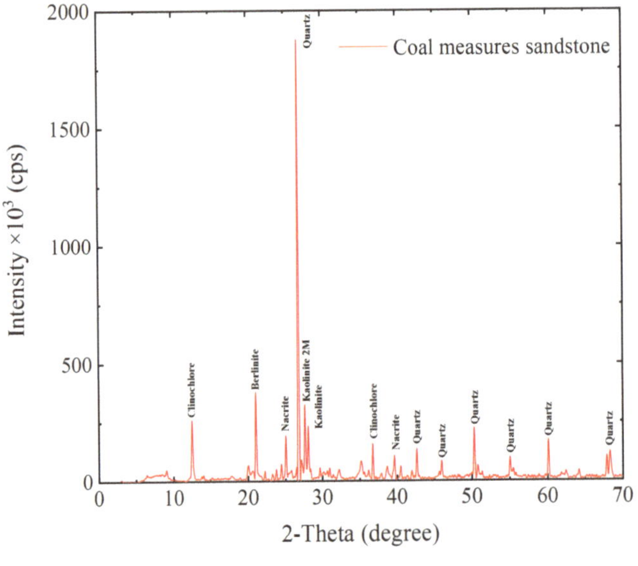

Figure 4. X-ray diffraction (XRD) study of sandstone.

MTS816 was used to conduct quasi-static tests on the sandstone, including uniaxial compression tests and Brazilian tests (Figure 5). Strain gauges were fitted to the center of the specimens before the tests to measure the transverse strain in a direction perpendicular to the loading force. The uniaxial compression stress–strain curves are shown in Figure 6. The stress–strain curves exhibit a tiny bending state when the stress level is low. The stress–strain curves become straighter as the stress increases to a level high enough to destroy the specimens. Shear and tensile failures are responsible for specimen failure. This type of rock is known as a plastic-elastomer [38]. The specimens had an average uniaxial compressive strength of 22.46 MPa, an average elastic modulus of 1.91 GPa, and an average Poisson's ratio of 0.14 after three sets of uniaxial compression testing. The lateral stress–lateral strain graphs in Figure 7 show the typical tensile failure characteristics. The sandstone's average indirect tensile strength is 2.38 MPa. The physical and mechanical properties of the sandstone are summarized in Table 1.

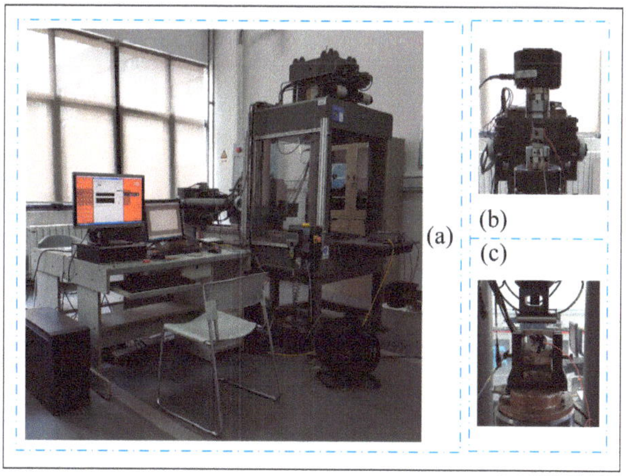

Figure 5. Quasi-static test methods: (**a**) MTS816 system; (**b**) uniaxial compression tests; (**c**) Brazilian tests.

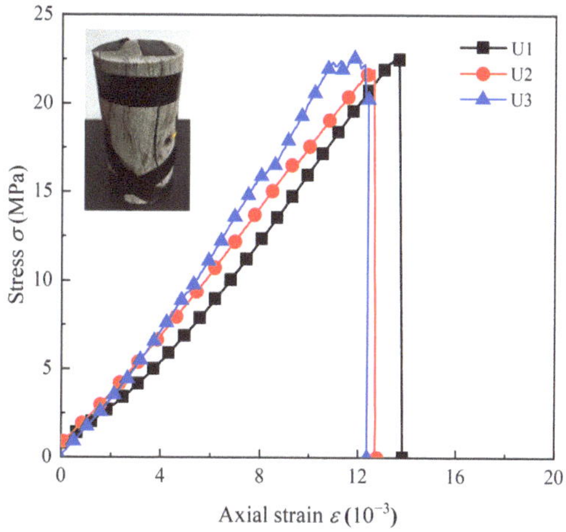

Figure 6. Uniaxial compression stress–strain curves.

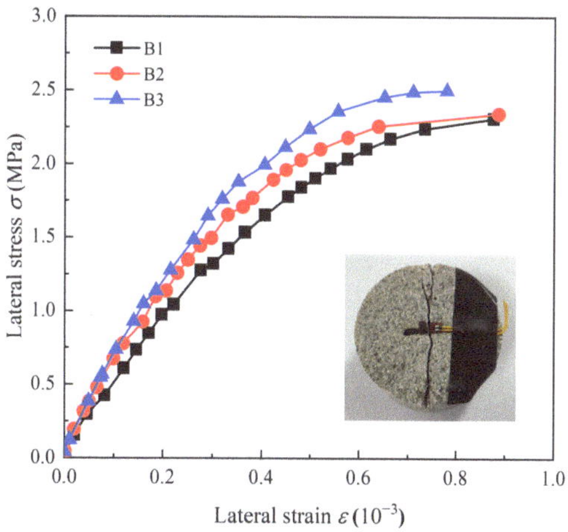

Figure 7. Indirect tensile stress–strain curves.

Table 1. Basic physical and mechanical properties of sandstone.

ρ/(kg·m^{-3})	V_P/(m·s^{-1})	ω/%	E/GPa	σ_u/MPa	σ_T/MPa	μ
2589	3682	1.13	1.91	22.46	2.38	0.14

Notes: ρ, density; V_P, P-wave velocity; ω, natural moisture content; E, elasticity modulus; σ_u, uniaxial compressive strength; σ_T, tensile strength; μ, Poisson's ratio.

2.3. Dynamic Testing Facilities

The experimental investigation on coal measures sandstone was carried out at Central South University using a 50 mm diameter split Hopkinson pressure bar (SHPB) testing system, as shown in Figure 8.

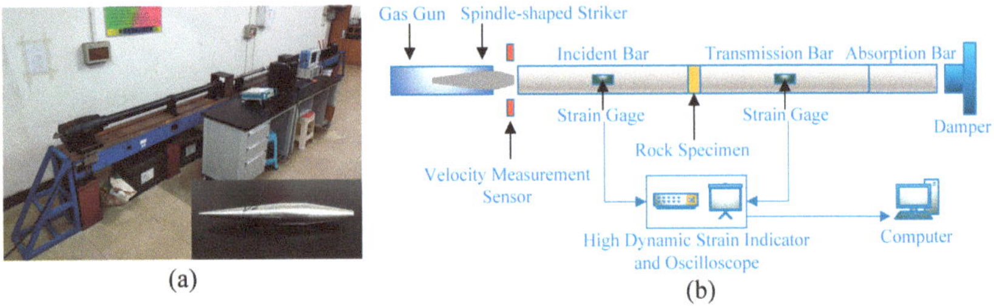

Figure 8. Split Hopkinson pressure bar test system: (**a**) SHPB device, Central South University; (**b**) schematic diagram.

The specific parameters of the SHPB device are detailed by Li et al. [39] and Qiu et al. [40]. The striker is shaped similar to a spindle to obtain the half-sine incident stress wave along the direction of the bars. It prevents fragile materials from being damaged prematurely during the tests. This particularly formed striker permits rock specimens to be loaded at a constant strain rate for as long as possible, which is critical in dynamic tests [41,42]. Two sets of strain gauges are attached to the middle of the bar system, and the wave data can be collected by the SDY2107A high dynamic strain indicator and the DL850E oscilloscope. The following part of the research is carried out using the first complete set of waveforms acquired during testing. Another key step before the tests is to lubricate the end faces of the specimens and the bar system with a thin layer of lubricant to reduce frictional effects. Many researchers have investigated the impact of various length-to-diameter ratios on test results, recommending a range of 0.5 to 1 as the best length-to-diameter ratio [43]. The size of the rock specimen should be more than 10 times the maximum diameter of the rock particles, according to the ISRM-suggested method for SHPB testing [37], and the inertia effect and dispersion effects should be avoided as much as feasible. For this study, disc-type sandstone specimens with a length-to-diameter ratio of 0.5 were used, i.e., 25 mm length and 50 mm diameter. The dynamic properties of rock with strain rates between 10^1 and 10^3 s^{-1} can be obtained using the SHPB test system, which covers the range of high strain rates seen in deep mining engineering.

Two essential assumptions underpin the SHPB experimental technique. The one-dimensional stress wave theory is the initial assumption [42]. The stress wave is thought to propagate in a one-dimensional direction in the bars. The stress wave is unaffected by the dispersion effect throughout the operation, and the cross-section of the bars is always flat. The specimen's stress uniformity assumption is the second hypothesis [44]. The stress and strain in the specimen are equally distributed during stress wave propagation, and the stress and strain of the two end faces of the specimen are equal. The stress equilibrium condition can be verified by the relationship between the incident wave, the reflected wave, and the transmitted wave. For a typical compression test in Figure 9, the sum of the incident wave and the reflected wave is in high agreement with the transmitted

wave, proving that the forces at both ends of the sandstone specimen are approximately equivalent. Equation (1) is naturally obtained using these two assumptions:

$$\begin{cases} \sigma(t) = \frac{AE}{2A_s}[\varepsilon_I(t) + \varepsilon_R(t) + \varepsilon_T(t)] \\ \varepsilon(t) = \frac{C}{l_s}\int_0^t[\varepsilon_I(t) - \varepsilon_R(t) - \varepsilon_T(t)]dt \\ \dot{\varepsilon}(t) = \frac{C}{l_s}[\varepsilon_I(t) - \varepsilon_R(t) - \varepsilon_T(t)] \end{cases} \quad (1)$$

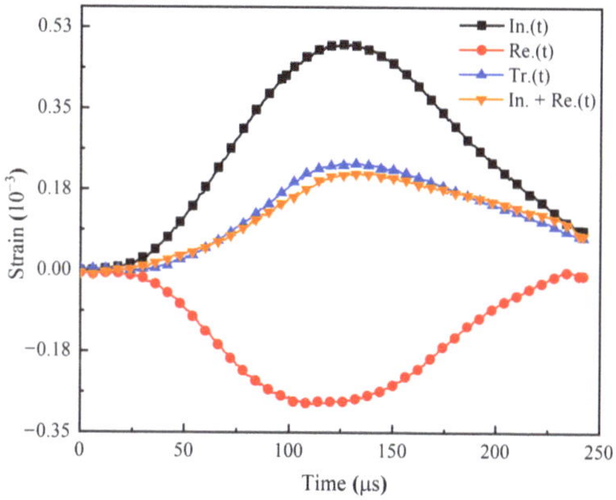

Figure 9. Dynamic stress equilibrium for a typical compression test.

2.4. Definition of Accumulated Damage

Micro-cracks begin to develop within the rock when damage is produced in the specimen. After that, as the damage within the rock progressively intensifies, the cracks gradually expand to penetrate the rock, eventually leading to rock destruction. The rock will deform when it is subjected to external loads. At the same time, energy, namely strain energy, will gradually accumulate inside the sample. The rock sample will be destroyed once the accumulated strain energy exceeds the limit value. So, we define the strain energy density as the amount of energy stored per unit volume of the rock. The strain energy density can reflect rock failure caused by the nature and stress state of the rock material itself [45–47]. The strain energy density can be calculated by Equation (2):

$$U = \frac{dW}{dV} = \int_0^{\varepsilon} \sigma d\varepsilon \quad (2)$$

where U denotes the strain energy density; W denotes the strain energy; V denotes the specimen volume; and σ and ε denote the stress and the corresponding strain, respectively.

Uniaxial compression tests apply different loads to rock specimens to form additional accumulated damage. The rock specimen will be broken when its strength reaches its maximum value, and the strain energy density of the specimen will be the highest. The ratio of the corresponding strain energy density under a specific load to the maximum strain energy density is called the accumulated damage factor (Figure 10), which is expressed by Equation (3):

$$D = \frac{U_{\varepsilon}}{U_{\text{MAX}}} \quad (3)$$

where D denotes the accumulated damage factor, U_{ε} denotes the strain energy density when strain is ε, and U_{MAX} denotes the maximum strain energy density. After calculation, $U_{MAX} = 140.7 \times 10^3 J \cdot m^{-3}$.

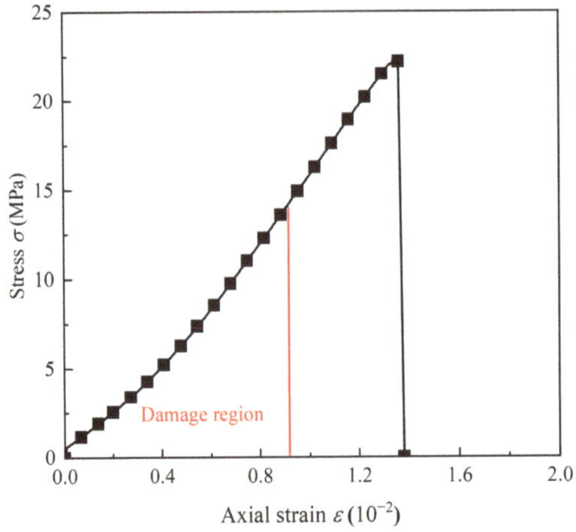

Figure 10. Characterization of the damage region.

Different sandstone specimens were subjected to axial pressures of 40%, 50%, 60%, 70%, 80%, and 90% of uniaxial compressive strength to control different accumulated damage degrees. The corresponding accumulated damage factors were calculated, as shown in Table 2. Figure 11 demonstrates that the damage factor increases linearly with the growth of the axial load.

Table 2. Axial stress and the corresponding accumulated damage factor.

σ/MPa	8.99	11.23	13.48	15.73	17.97	20.22
D	0.1893	0.2764	0.3789	0.4939	0.6283	0.7891

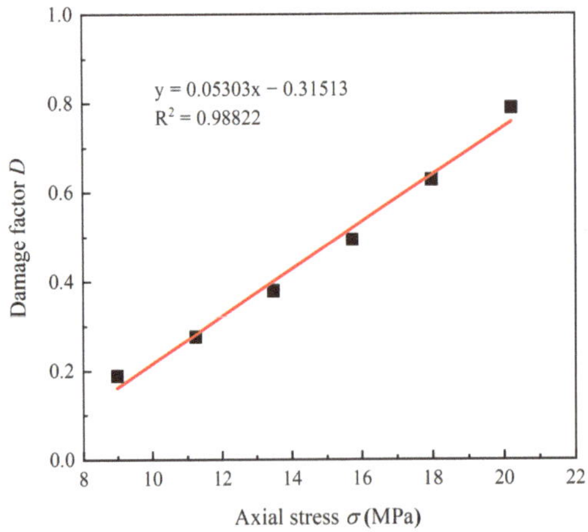

Figure 11. Relationship between the accumulated damage factor and axial stress.

2.5. Dynamic Testing Scheme

Dynamic tests were performed consecutively under seven different accumulated damage situations in this study. The specific test scheme is as follows:

1. Initial accumulated damage tests. The MTS816 is used to apply various levels of accumulated damage to selected coal measures sandstone specimens. The accumulated damage factor is calculated to measure the extent of the sandstone damage.

2. Dynamic tests of sandstone. The SHPB test equipment is used to perform impact tests on sandstone to investigate the coupling effect of a high strain rate and the accumulated damage. There are five different strain rates and seven different accumulated damage conditions in this study. In each condition, only one test is conducted. In the event of a significant irregularity in the test results, a second test under the identical settings is required to ensure reliability.
3. Observation of sandstone failure modes. An SEM device is used to observe sandstone's microstructure after failure and compare it with the macroscopic failure morphology. Finally, the effects of high strain rates and accumulated damages on the sandstone failure mode are explored.

2.6. Dynamic Testing Facilities

In order to explore the microscopic morphology modes of the fracture surface after sandstone failure, the TESCAN VEGA3 scanning electron microscope system (SEM) of the Xuzhou University of Engineering is employed (Figure 12). Several small fragments with a diameter of about 10 mm in the middle of the fracture surface are selected for observation. The samples are well prepared by spraying the surfaces of the observed fragments with gold using the SBC-12 small ion sputtering apparatus.

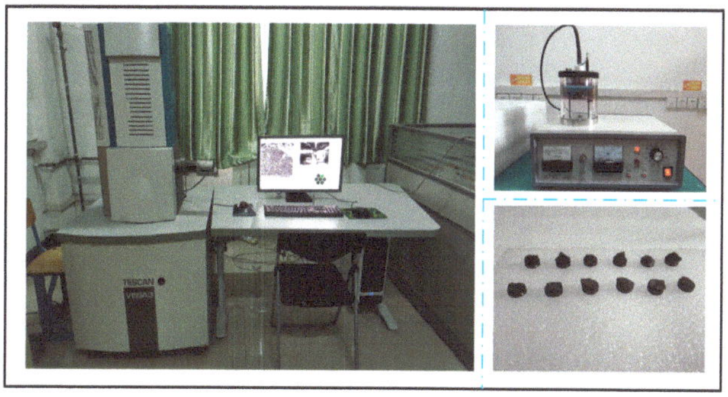

Figure 12. TESCAN VEGA3 scanning electron microscopy (SEM) test system.

3. Results and Discussion

3.1. Dynamic Properties of Coal Measures Sandstone

In this study, five similar high strain rates are obtained by setting five different impact pressures to investigate the dynamic mechanical properties of coal measures sandstone under impact loading conditions. Figure 13 illustrates the impacts of the strain rate on the stress–strain curves of the sandstone specimens without accumulated damage. The stress–strain curves with high strain rates show certain similarities, which means they can all be roughly classified into three phases. The first phase is the linear elastic phase, in which the dynamic stress–strain curves exhibit a straight upward tendency. In this phase, the stress wave propagated from the incident bar into the sandstone specimen will propagate many times inside the specimen and, finally, the strain and stress will reach a uniform state. The second phase is the plastic deformation phase, which begins at the end of the linear elastic phase and ends when the stress of the specimen reaches the peak value. In this phase, the sandstone specimen undergoes primarily plastic deformation. The third phase is called the strain-softening phase. In this phase, the sandstone specimen still has specific strength until it is completely destroyed. For some rocks, there is still a stage prior to the linear elastic phase, in which the rock specimen undergoes the compaction of the initial internal crack, as shown in the corresponding static tests [48]. However, the initial crack compaction phase is almost negligible under impact loading.

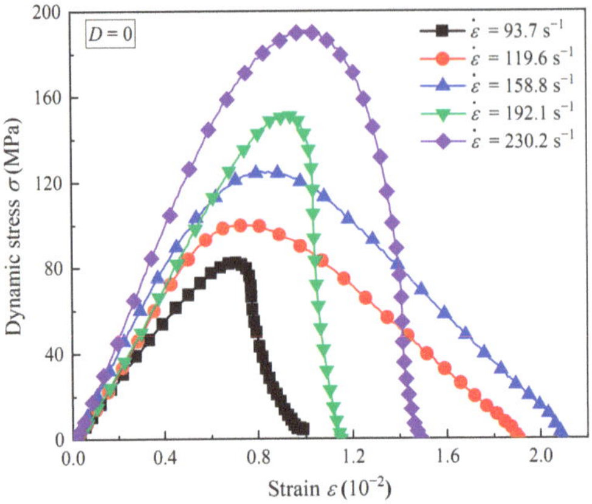

Figure 13. Dynamic stress–strain curves of sandstone at different strain rates.

The dynamic strength and the deformation modulus of sandstone are available using the dynamic stress–strain curves (Figure 14). As the strain rate rises from 93.7 s^{-1} to 230.2 s^{-1}, the peak stress progressively increases from 82.8 MPa to 190.9 MPa. From static tests, sandstone strength is 22.5 MPa, which indicates that the high strain rate significantly contributes to the sandstone strength. The deformation modulus is calculated by the ratio of the peak stress of the sandstone specimen to the strain at the peak stress. It is adopted to describe the sandstone deformation characteristics. The deformation modulus of sandstone increases from 11.8 GPa to 19.3 GPa as the strain rate rises from 93.7 s^{-1} to 230.2 s^{-1}. Under impact loading conditions, sandstone specimens' strength and deformation modulus have a similar variation trend, but the increased amplitude is different. Specifically, when the strain rate experiences the same variation range, the sandstone strength increases by 130.56%, while the deformation modulus of sandstone increases by 63.01%. Namely, the sandstone strength is more sensitive to strain rates than to deformation characteristics.

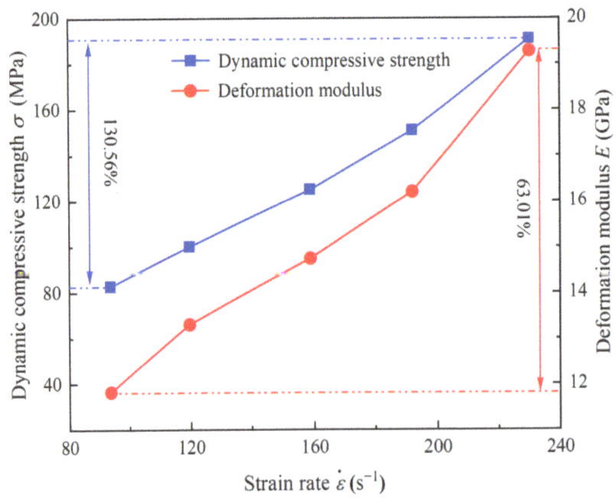

Figure 14. Variation in dynamic strength and the deformation modulus of sandstone with strain rates.

3.2. *Effects of Accumulated Damage on the Stress–Strain Curves*

For the purpose of investigating the effects of accumulated damage on the dynamic behavior of sandstone under five kinds of similar strain rates, dynamic stress–strain curves of sandstone specimens at different strain rates and accumulated damages are first obtained, as shown in Figure 15. We use accumulated damage factors to distinguish the extent of the accumulated damage. The accumulated damage factor D ranges from 0 to 1. The larger the

D, the higher the accumulated damage degree of the sandstone specimen, and the more the number of microscopic cracks. When *D* is closer to 1, these microscopic cracks are likely to show up in the form of larger macroscopic cracks.

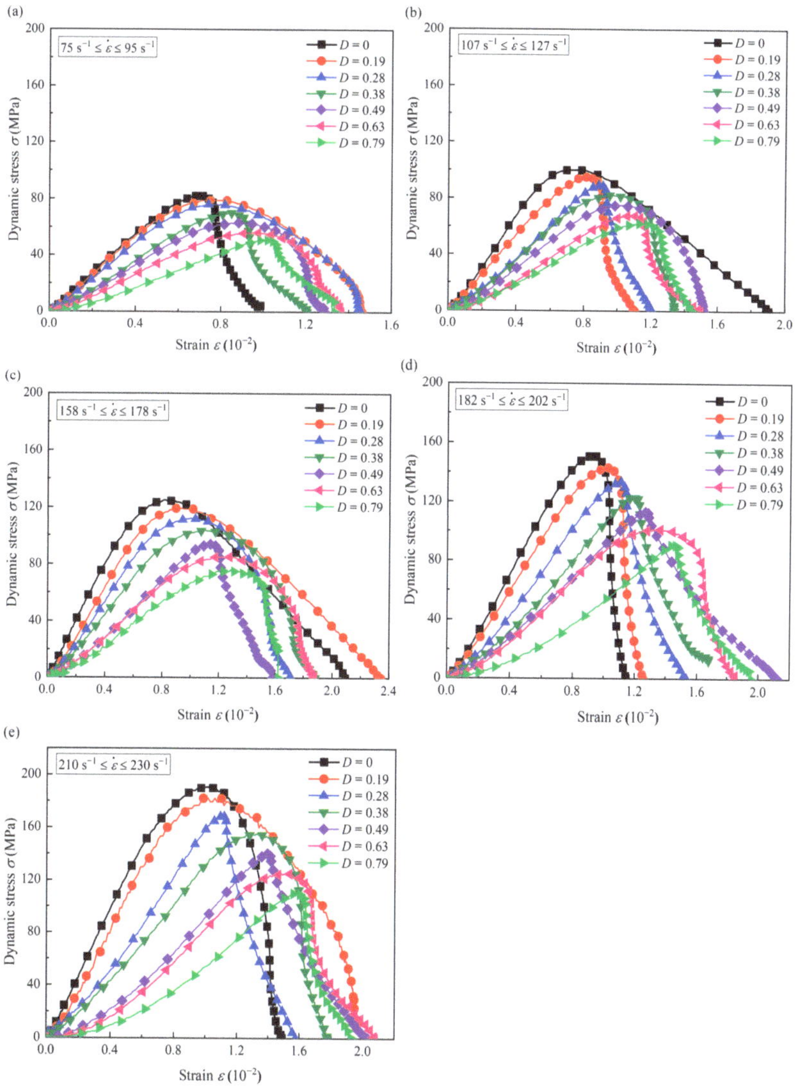

Figure 15. Dynamic stress–strain curves of sandstone with different accumulated damages when strain rates between (**a**) 75 s^{-1} and 95 s^{-1}, (**b**) 107 s^{-1} and 127 s^{-1}, (**c**) 158 s^{-1} and 178 s^{-1}, (**d**) 182 s^{-1} and 202 s^{-1}, and (**e**) 210 s^{-1} and 230 s^{-1}.

As described in the previous section, the stress–strain curves of sandstone specimens with accumulated damages at different strain rates reveal a certain similarity in the overall shape. The linear elastic phase, the plastic deformation phase, and the strain softening phase are well reflected in these stress–strain curves, indicating that their mechanical behaviors are similar. However, the initial phases show differences as the accumulated damage increases. Specifically, when the specimen has accumulated damage, the first phase

in the stress–strain curve is no longer a linear elastic phase but a crack closure phase. In this phase, the stress–strain curves are nonlinear and tend to be concave.

The stress–strain curves of the sandstone at different strain rates when the accumulated damage is at its maximum are given in Figure 16. It is noticed that these stress–strain curves all have more obvious crack closure phases. More cracks in the sandstone specimens cause the appearance of the crack closure phase. When the accumulated damage factor is more significant, the accumulated damage degree of the specimen is greater, and the micro-cracks in the sandstone specimens gradually increase. These micro-cracks may produce macroscopic cracks visible to the naked eye due to the penetration. The increase in internal cracks leads to the crack closure phase of the samples in dynamic tests, and this stage will gradually grow longer with the increase in accumulated damage. Simultaneously, the high strain rate will profoundly impact the process, as evidenced by the fact that the initial crack closure phase of the sandstone specimens becomes progressively longer as the strain rate increases, which can be observed in Figure 15.

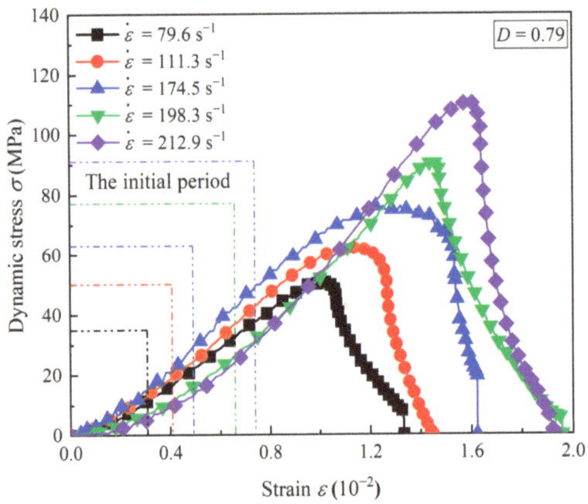

Figure 16. Dynamic stress–strain curves of damaged sandstone at different strain rates.

3.3. Effects of Accumulated Damage on the Compressive Strength

The effects of the strain rate and the accumulated damage on the dynamic compressive strength of sandstone are compared and investigated using the dynamic stress–strain curves of sandstone specimens in Figure 15. Figure 17 plots the variation in sandstone dynamic compressive strength at different strain rates and accumulated damage conditions and the increase factor of dynamic strength under the same accumulated damage factor. It can be seen that under each strain rate, the increase in the damage factor causes a gradual decrease in the dynamic compressive strength, suggesting that the accumulated damage inflicts excellent damage on the sandstone specimens. For instance, when the strain rate is the maximum, with the increase in the accumulated damage factor from 0 to 0.79, the dynamic compressive strength decreases from 190.9 MPa to 110.2 MPa, with a decrease of 42.3%. Another significant feature is that when D is less than 0.3, the dynamic compressive strength of sandstone with a similar strain rate decreases slowly. When D exceeds 0.3, the decreasing trend of the peak stress at a similar strain rate becomes more pronounced, indicating that the high strain rate will increase the change rate of sandstone strength. Furthermore, when the specimens experience the same accumulated damage, the dynamic compressive strength gradually rises with the increasing strain rate, suggesting that strain rate and the accumulated damage have an essential role in the sandstone behavior at the same time. The most likely explanation for the decrease in sandstone strength with the increasing damage factor is that the existence of accumulated damage leads to more micro-cracks. As loading proceeds, the transverse cracks in the sandstone specimens gradually close, while the vertical cracks continue to expand faster. These micro-cracks may also

interpenetrate and eventually form macroscopic cracks, thus leading to a sharp reduction in the bearing capacity of sandstone. Therefore, the damage pattern of rock specimens is more likely to consist of two modes: tensile fracture and shear fracture. This situation is more evident under a high strain rate than in the quasi-static tests. Figure 17b exhibits how the increased amplitude of sandstone strength varies with the accumulated damage factor within the same range of strain rates. When the accumulated damage factor is 0, the higher impact velocity increases the sandstone peak stress by 130%, while the strength increases by 115% at a damage factor of 0.79. In other words, the increase in accumulated damage leads to a gradual reduction in the increased amplitude of sandstone strength at different strain rates.

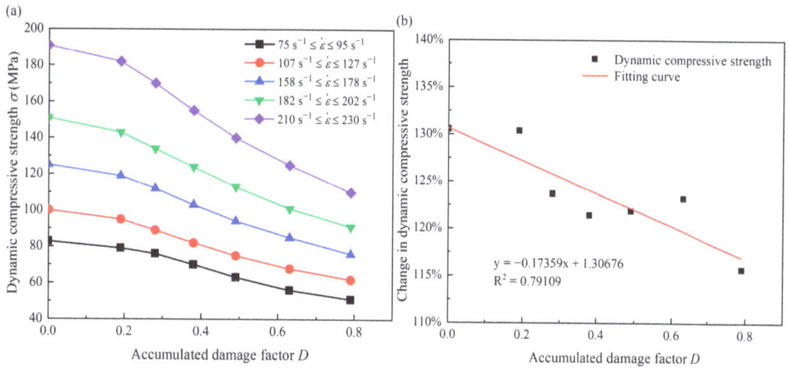

Figure 17. Variation in (**a**) the dynamic compressive strength of sandstone and (**b**) change in sandstone strength with accumulated damage.

The dynamic increase factor is evaluated to compare the variability in sandstone's dynamic strength and quasi-static strength. The DIF is defined as the ratio of the dynamic and quasi-static test values for sandstone, representing the increase in rock properties under impact load [9,44,49]. Figure 18 depicts the variation in the DIF with the strain rate. For each accumulated damage factor, the DIF is distinguished into the medium-strain-rate part and the high-strain-rate part. For the medium strain rate, the rate of increase of the DIF is significantly lower than that of the high strain rate. The accumulated damage reduces the influence of the strain rate on the DIF.

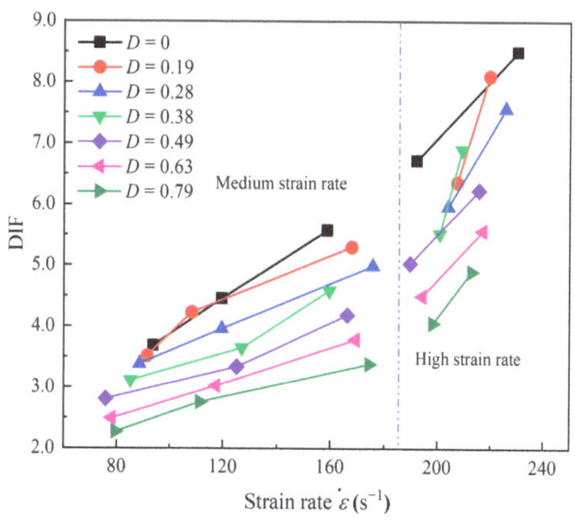

Figure 18. Dynamic increase factor (DIF) of dynamic compressive strength.

3.4. Effects of Accumulated Damage on the Deformation Modulus

In addition to affecting the extreme bearing capacity of the sandstone, the accumulated damage also influences the ability of the specimens to resist deformation under high strain rates. The deformation modulus is defined as the ratio of dynamic peak strength to the corresponding critical strain. Figure 19 shows the variation in sandstone deformation modulus with different strain rates and accumulated damage. Generally speaking, the dynamic deformation modulus and compressive strength have similar forms. As the strain rate is approximately 220 s^{-1}, with the accumulated damage factor increasing from 0 to 0.79, the deformation modulus of sandstone specimens decreases from 19.3 GPa to 7.1 GPa, with a decrease rate of 63.2%, which is more significant than the decrease rate of sandstone compressive strength under the same condition. However, when the strain rate is the lowest, the decrease in the deformation modulus becomes 56.8%, which implies that the strain rate causes a specific impact on the changing of the sandstone deformation modulus. Figure 19b shows how the increased amplitude of the sandstone deformation modulus varies with the accumulated damage factor within the same range of strain rates. When there is no accumulated damage to the sandstone specimen, the deformation modulus of sandstone increases by 63% as the strain rate turns from minimum to maximum. With the increase in the accumulated damage, the increase in the sandstone deformation modulus decreases. As the accumulated damage factor reaches 0.79, the growth in the sandstone deformation modulus decreases to 38%. The increased amplitude of the sandstone deformation modulus is reduced by almost 40% when the accumulated damage factor rises from 0 to 0.79, far more than that of compressive strength. It thus indicates that compared with the compressive strength, the deformation modulus of sandstone is more susceptible to accumulated damage.

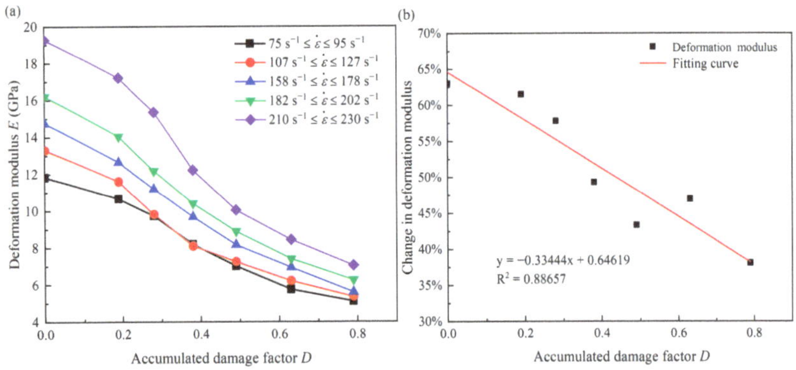

Figure 19. Variation in (**a**) the deformation modulus of sandstone and (**b**) change in deformation modulus with accumulated damage.

Meanwhile, the DIF of the deformation modulus has also been calculated [44]. The deformation modulus increases remarkably along with the strain rate compared with the results of quasi-static tests. The increasing trend of the deformation modulus is similar to compressive strength and can also be divided into medium-strain-rate and high-strain-rate parts. When the samples do not experience accumulated damage, the DIF varies from 4.9 to 8.1. The DIF varies between 2.1 and 3.0 when the accumulated damage to the sandstone specimens is the largest (Figure 20).

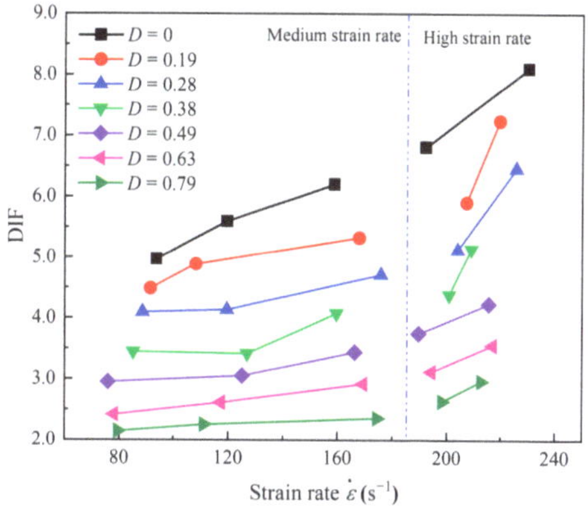

Figure 20. Dynamic increase factor (DIF) of the deformation modulus.

4. Deformation and Failure Modes of Coal Measures Sandstone

In addition to the quantitative properties of materials, rock failure behavior and fragment establishment are also essential to obtain the impacts of a high strain rate and accumulated damage on failure modes.

4.1. Macroscopic Failure Modes

A high-speed camera is used to capture the process of testing sandstone specimens using the SHPB device. Figure 21 displays the failure states of a sandstone specimen in an experiment. The axial crack first appears on the specimen during the compressive test and gradually expands (Figure 21a). After that, some secondary cracks occur near the axial cracks, and most of the secondary cracks are shear cracks (Figure 21b,c). It is due to the reason that the specimen has accumulated damage before being subjected to impact load. The accumulated damage reduces the internal friction angle of the specimen, leading to shear failure. Finally, these cracks together cause the complete destruction of the sandstone specimen (Figure 21d).

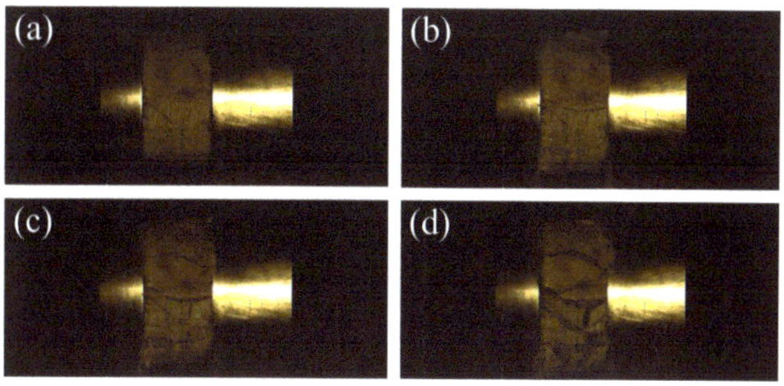

Figure 21. Failure process of a sandstone specimen at 209.3 s^{-1}: (**a**–**d**) represent four moments of failure in chronological order.

Figure 22 shows the fragments of sandstone specimens after dynamic tests at a similar strain rate. For compressive load, Figure 22 proves that the fragment sizes of specimens decrease significantly with increasing accumulated damage. Tensile fracture is dominant when the sandstone specimen has not experienced the initial accumulated damage. The failure of the specimen begins with the peeling of its peripheral radial part (Figure 22a). The fragments reveal multiple huge block shapes when the accumulated damage factor reaches 0.38, showing that the specimens are failing due to tensile and shear fracture, both of which contribute significantly to the eventual failure (Figure 22b). As the accumulated damage

increases, the volume of the wedge-shaped block decreases gradually, smaller pieces appear, and the degree of the sandstone failure increases further (Figure 22c). The appearance of more wedge-shaped blocks represents that the angle between the fracture surface and the principal stress direction of sandstone samples increases gradually, which implies that the internal friction angle decreases gradually with the increase in the accumulated damage degree suffered by sandstone. Another possible reason for this failure pattern is that the increase in accumulated damage leads to more pores or micro-cracks in the sandstone, thus reducing the cohesion of sandstone specimens. In other words, the failure mode of sandstone under impact load depends not only on the high strain rate but also on the cohesion and the internal friction angle. In general, compressive failure is characterized by the cracks expanding along the axial direction of the specimen until the specimen is destroyed. The raised strain rate causes overall particle fragmentation failure to occur in the specimens. Furthermore, because of the accumulated damage, the rock particles are destroyed many times, resulting in the formation of fine-grained powder.

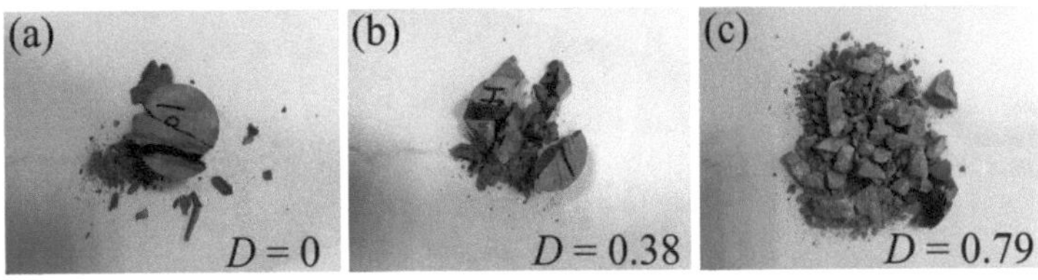

Figure 22. Fragment establishment of sandstone specimens with different accumulated damages: (**a**) $D = 0$, (**b**) $D = 0.38$, and (**c**) $D = 0.79$.

4.2. *Microscopic Failure Modes*

The microscopic morphology of sandstone after dynamic failure is observed by the SEM test system. Figure 23 shows the micromorphological characteristics of sandstone specimens with accumulated damage before and after dynamic tests.

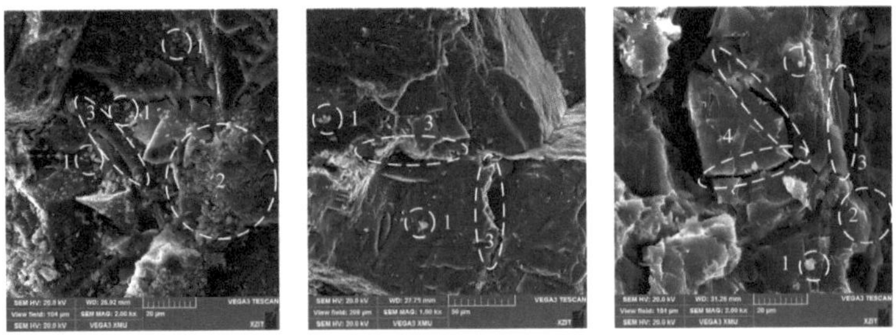

(a) (b) (c)

Figure 23. The microscopic failure characteristics of sandstone with accumulated damage: (**a**) sandstone specimen not subjected to impact load; (**b**) sandstone specimen at a low impact velocity; (**c**) sandstone specimen at a high impact velocity (1: tiny mineral particles; 2: rough surface; 3: intergranular fracture; 4: transgranular fracture).

It can be seen that there are many tiny mineral particles and some rough portions on the surface of the sandstone specimen with accumulated damage but not subjected to impact load (Figure 23a). The existence of accumulated damage also causes intergranular fracture in the sandstone specimen. When the sandstone specimen is coupled with a low strain rate and accumulated damage, the sandstone failure surface is relatively comparable to that of the specimen not subjected to impact load. However, more intergranular fractures appear (Figure 23b). When the sandstone specimen suffers a greater strain rate, a new failure mode, transgranular fracture, is exhibited on the failure surface of the sandstone

(Figure 23c). In other words, accumulated damage mainly causes intergranular fracture between the mineral particles, while a higher strain rate mainly causes transgranular fracture inside the mineral particles.

4.3. Energy Absorption Law

In order to understand the energy absorption law of sandstone samples, the specific energy absorption (SEA) factor is calculated [50].

$$\begin{aligned} W_I &= \frac{A_0 C_0}{E_0}\int \sigma_I{}^2 dt \\ W_R &= \frac{A_0 C_0}{E_0}\int \sigma_R{}^2 dt \\ W_T &= \frac{A_0 C_0}{E_0}\int \sigma_T{}^2 dt \\ W_L &= W_I - (W_R + W_T) \\ F_{SEA} &= \frac{W_L}{V_S} \end{aligned} \tag{4}$$

where W_I, W_R, and W_T are the energy of the relevant waves; A_0, C_0, and E_0 are the cross-sectional area, the longitudinal wave velocity, and the elastic modulus of the incident bar and the transmission bar, respectively; σ_I, σ_R, and σ_T are the relevant stress signals; W_L is the energy absorbed by the sandstone specimen; and V_S is the specimen volume. Finally, the SEA factor F_{SEA} can be figured out according to Equation (4).

We assume that the stress wave is completely transmitted during the experiment, indicating that there is no energy loss at the interface between the specimens and the bar system. Thus, energy absorption factors of sandstone under different accumulated damages and strain rates are calculated, as shown in Figure 24. The results demonstrate that an increase in the strain rate can significantly enhance the value of the energy absorption factor. Further, the accumulated damage also increases the value of the energy absorption factor. The results correspond to the macro- and micro-failure patterns. When the strain rate and the accumulated damage reach a higher level, the sandstone specimen absorbs more energy. As a result, the failure mode of sandstone is manifested in the appearance of more and smaller wedge-shaped blocks. Due to the repeated destruction of sandstone specimens, many powdery sandstone particles will appear. Sandstone specimens absorb more energy at a microscopic level, leading to a gradual shift in the sandstone failure mode from intergranular to transgranular fracture.

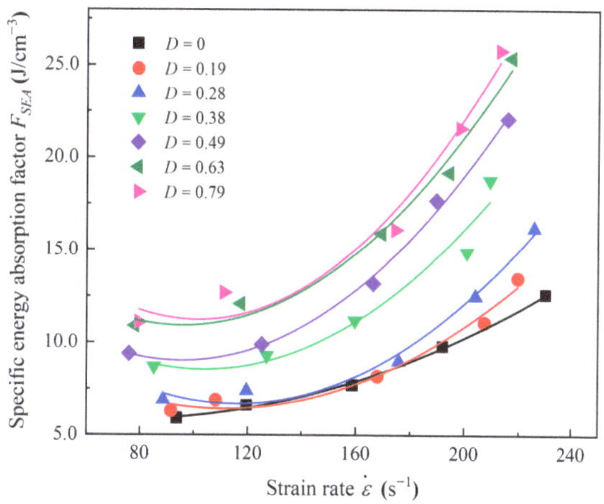

Figure 24. Specific energy absorption factors of sandstone.

5. Statistical Damage Constitutive Model Considering the Strain Rate Effect and the Accumulated Damage

Based on the Weibull distribution, a statistical damage constitutive model of coal measures sandstone is proposed by considering the accumulated damage and the strain rate effect.

5.1. Statistical Damage Constitutive Model

It is generally accepted that rock failure begins with internal micro-cracks, and with the increase in the external load, the micro-cracks gradually expand to macroscopic cracks, eventually leading to rock failure. Damage evolution of rock leads to the degradation of elasticity [51]. Assuming that [50,51] (1) the intact rock is a homogeneous and isotropic material; (2) each mesoscopic microelement of the rock is continuous and elastic before destruction; and (3) the rock intensity obeys the Weibull distribution, the probability distribution function is expressed by Equation (5):

$$P(F) = \frac{m}{F_0}\left(\frac{F}{F_0}\right)^{m-1} \exp\left[-\left(\frac{F}{F_0}\right)^{m}\right] \tag{5}$$

where F denotes the strength of the rock material microelement, F_0 denotes the mean value, and m denotes the shape parameter or uniformity index representing the microelement. Tang [52] points out that the larger the m, the higher will be the material uniformity.

From strain strength theory [53], F in the above formula can be replaced by ε, so Equation (5) changes to:

$$P(F) = \frac{m}{\varepsilon_0}\left(\frac{\varepsilon}{\varepsilon_0}\right)^{m-1} \exp\left[-\left(\frac{\varepsilon}{\varepsilon_0}\right)^{m}\right] \tag{6}$$

The rock damage variable D is specified as the ratio of the number of failure microelements and the number of all microelements N in the rock [51]. Then, the expression of the damage variable is:

$$D = \frac{N_D}{N} \tag{7}$$

When the load on the rock increases from σ to $\sigma + d\sigma$, the number of failure microelements increases to $NP(\sigma)d\sigma$. Therefore, when the external load increases from 0 to σ, the total number of failure microelements is:

$$N_D(\sigma) = \int_0^{\sigma} NP(x)dx = N\left\{1 - \exp\left[-\left(\frac{\varepsilon}{\varepsilon_0}\right)^{m}\right]\right\} \tag{8}$$

Substituting Equation (8) into Equation (7) yields:

$$D = 1 - \exp\left[-\left(\frac{\varepsilon}{\varepsilon_0}\right)^{m}\right] \tag{9}$$

At this point, the damage factor D is caused by the loads on the rock and its value varies from 0 to 1. ε_0 and m refer to the proportion parameter and the shape parameter of rock materials, respectively. According to Wang et al. [51], when m is the same, D increases as $\frac{\varepsilon}{\varepsilon_0}$ increases. When the stress level is the same, the greater the m, the earlier the appearance of the complete damage.

Based on the Lemaitre strain equivalence hypothesis [54,55], the deformation modulus of damaged materials is:

$$E_D = E_0(1 - D) \tag{10}$$

where E_D denotes the deformation modulus of the damaged rock and E_0 denotes the deformation modulus of the undamaged rock.

However, after the sandstone specimen has experienced accumulated damage, some micro-cracks have already appeared within the rock before the dynamic tests, which makes

the deformation modulus of the sandstone specimen with accumulated damage different from that of the non-damaged specimen. Assume that the accumulated damage factor is D_A. At the same time, the stress–strain curves of the sandstone specimens that have experienced accumulated damage show an obvious crack closure stage, which is significantly different from that of the complete sandstone specimens. From the analysis in the previous section, the existence of accumulated damage gradually reduces the internal friction angles and cohesion of the sandstone specimens. Therefore, $i = a\varepsilon^b$ is defined as the accumulated damage correction coefficient to represent the characteristics closely related to the reduction in cohesion and internal friction angles caused by accumulated damages. The parameters a and b represent the shape coefficients of the stress–strain curves. The actual damage deformation modulus of the specimen after experiencing accumulated damage is then expressed by the following equation:

$$E_D = iE_0(1 - D)(1 - D_A) = a\varepsilon^b E_0(1 - D)(1 - D_A) \tag{11}$$

Therefore, the stress–strain relationship of the specimen with accumulated damage is given by:

$$\sigma = E_D\varepsilon = aE_D(1 - D)(1 - D_A)\varepsilon^{b+1} \tag{12}$$

Substituting Equation (9) into Equation (12) yields the statistical damage constitutive model considering accumulated damage:

$$\sigma = E_D\varepsilon = aE_D\varepsilon^{b+1}(1 - D_A)\exp\left[-\left(\frac{\varepsilon}{\varepsilon_0}\right)^m\right] \tag{13}$$

Parameters a, b, ε_0, and m in Equation (13) are all functions of strain rate, which are given by:

$$\begin{aligned} a &= a(D_A, \dot{\varepsilon}) \\ b &= b(D_A, \dot{\varepsilon}) \\ \varepsilon_0 &= \varepsilon_0(D_A, \dot{\varepsilon}) \\ m &= m(D_A, \dot{\varepsilon}) \end{aligned} \tag{14}$$

Substituting Equation (14) into Equation (13) yields the statistical damage constitutive model considering the accumulated damage and the strain rate effect:

$$\sigma = E_D\varepsilon = a(D_A, \dot{\varepsilon})E_D\varepsilon^{b(D_A,\dot{\varepsilon})+1}(1 - D_A)\exp\left[-\left(\frac{\varepsilon(\dot{\varepsilon})}{\varepsilon_0(D_A, \dot{\varepsilon})}\right)^{m(D_A,\dot{\varepsilon})}\right] \tag{15}$$

5.2. Sensitivity Analysis of Parameters

The sandstone stress–strain curve with the accumulated damage of 0.79 and the strain rate of 111.3 s^{-1} is selected for the comparative study of parameter sensitivity. The experimental data are fitted using the constitutive model in the previous section, and the fitted curve is provided in Figure 25. The values of the four fitting parameters in the constitutive model can be obtained as follows:

$$\begin{aligned} a &= 2.16 \\ b &= 0.26 \\ \varepsilon_0 &= 0.013 \\ m &= 16 \end{aligned} \tag{16}$$

When parameter a is set to 1.50, 2.16, and 3.00, three different stress–strain curves are shown in Figure 26a. The graph reveals that parameter a mainly affects the peak stress of curves. The larger the value of parameter a, the greater the peak stress, thus increasing the deformation modulus. Figure 26b shows three kinds of stress–strain curves as parameter b is 0.10, 0.26, and 0.40, respectively. The value of parameter b mainly affects the pre-peak phases. The greater the value of parameter b, the more pronounced the crack closure

stage, which indicates that the value of parameter b is affected by accumulated damage. Figure 26c gives the stress–strain curves with different values of ε_0. It can be seen that parameter ε_0 has a significant effect on the peak stress and the corresponding peak strain in the stress–strain curve. The larger the ε_0, the greater the peak stress and the corresponding peak strain. From Figure 26d, parameters m and ε_0 have a similar effect. They all have a specific effect on the peak stress and the corresponding peak strain. However, the difference is that the existence of parameter m makes the post-peak phases change significantly, and it changes the shape of the curves. The larger the value of m, the steeper the falling phases.

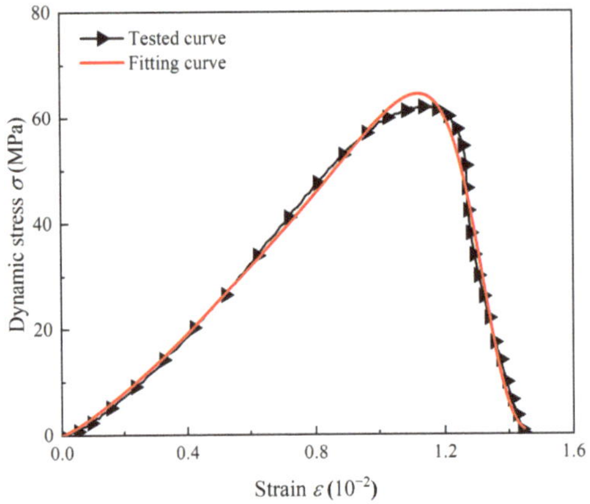

Figure 25. Comparison of tested and fitting stress–strain curves.

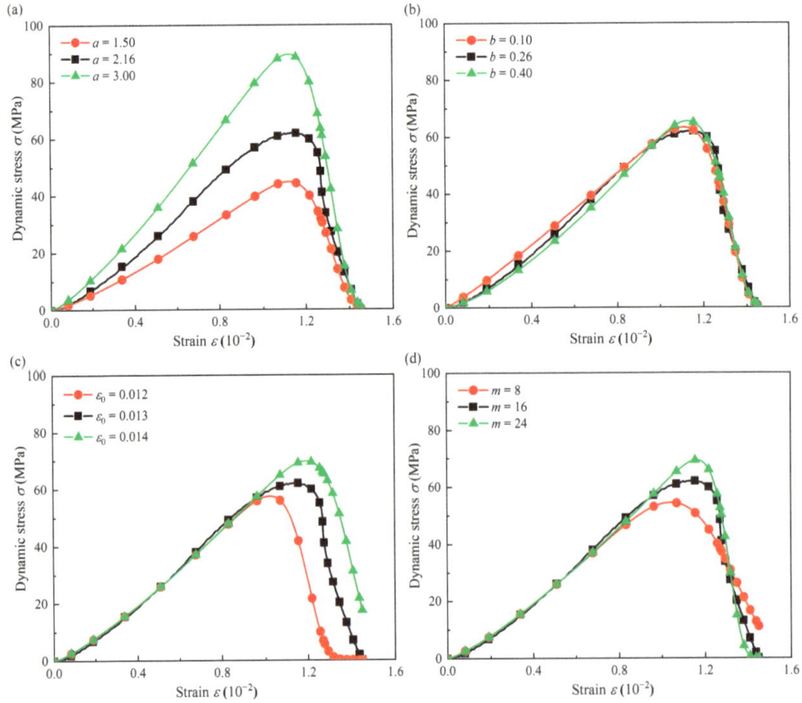

Figure 26. Effects of model parameters on the stress–strain curve: (**a**) parameter a; (**b**) parameter b; (**c**) parameter ε_0; (**d**) parameter m.

5.3. Validation and Discussion of the Model

According to the statistical damage constitutive model, the dynamic stress–strain curves of sandstone under different accumulated damages are fitted, and the tested and fitting curves are shown in Figure 27.

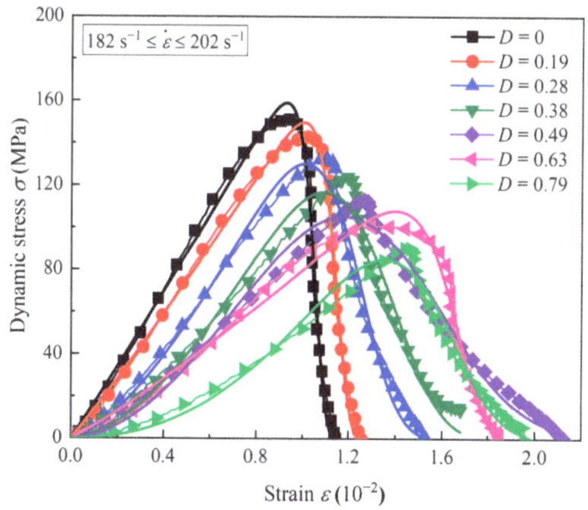

Figure 27. Comparison of tested and fitting stress–strain curves with different accumulated damages.

It can be seen that compared with the data obtained from the tests, the prediction results obtained from the constitutive model show a higher consistency. It is noteworthy that, as described in Section 3.2, the presence of accumulated damage gradually lengthened the crack closure phase of the stress–strain curve, which is well represented by this model. At the same time, the post-peak phase is also accurately predicted. In addition, it is worth emphasizing that there are four parameters in this model, all of which have their specific physical meanings. They all affect each phase in the stress–strain curves and are indispensable, which is also different from some previous studies [31,32,56,57]. In conclusion, the model presented in this paper can reasonably predict the influence of accumulated damage effects and strain rate effects, which indicates that the model has good applicability.

This research aims to highlight the effects of accumulated damage on deep rocks. In this study, coal measures sandstone is set as the main object of study for impact tests. It is well known that different types of rocks have distinct physical and mechanical properties, which makes it likely that the effects of accumulated damage on them will be highly varied. It is unknown whether the experimental results and the proposed statistical damage constitutive model in this study are applicable to other types of rocks. The next attempt is to apply accumulated damage to other types of rocks and alter the type of accumulated damage using different stress paths. Future research will offer fresh perspectives on estimating the accumulated damage impacts of deep in situ rock environments.

6. Conclusions

The coupling effect of accumulated damage and a high strain rate has a profound influence on deep mining engineering. In this study, six accumulated damage conditions were assigned to coal measures sandstone, and a statistical damage constitutive model for sandstone considering the strain rate effect and the accumulated damage was developed. The main conclusions are as set out below:

1. The stress–strain curves of the intact coal measures sandstone are classified into three phases: the linear elastic phase, the plastic deformation phase, and the strain-softening phase. The accumulated damages result in the crack closure phases in the dynamic curves of sandstone, and the greater the accumulated damage, the longer the crack closure phase.

2. The accumulated damage results in a significant reduction in the dynamic strength and the deformation modulus of the sandstone specimens. Compared with the compressive strength, the deformation modulus of sandstone is more susceptible to accumulated damage. The DIFs of strength and the deformation modulus are divided into two stages. The high-strain-rate phase has a faster change rate than the low- and medium-strain-rate phases.
3. Accumulated damage reduces the internal friction angle of sandstone, leading to tensile fracture and shear fracture in the macroscopic failure mode. From a microscopic perspective, accumulated damage leads to intergranular fracture between mineral particles, whereas a high strain rate induces transgranular fracture in mineral particles. The energy absorption law corresponds to the failure modes of sandstone specimens.
4. A statistical damage constitutive model of sandstone is established, which can well reflect the effect of the strain rate and accumulated damage on the dynamic response. Parameter a mainly affects the peak stress, while parameter b mainly affects the crack closure phase. The parameters ε_0 and m have certain effects on the peak stress and the deformation modulus, but parameter m also affects the strain-softening phase.

Author Contributions: Methodology, Z.S. and H.P.; investigation, J.X.; resources, H.N.; data curation, Z.S. and S.G.; writing—original draft preparation, Z.S., J.X. and H.N.; writing—review and editing, Z.S., S.G. and H.P.; visualization, Z.S. and J.X.; supervision, H.P. All authors have read and agreed to the published version of the manuscript.

Funding: This research was funded by the National Natural Science Foundation of China (No. 51974296 and No. 52061135111).

Data Availability Statement: Data available on request due to privacy restrictions.

Conflicts of Interest: The authors declare no conflict of interest.

References

1. He, M.C.; Xie, H.P.; Peng, S.P.; Jiang, Y.D. Study on Rock Mechanics in Deep Mining Engineering. *Chin. J. Rock Mech. Eng.* **2005**, *24*, 2803–2813.
2. Huang, L.X. Development and New Achievements of Rock Dynamics in China. *Rock Soil Mech.* **2011**, *32*, 2889–2900.
3. Qian, Q.H. Challenges Faced by Underground Projects Construction Safety and Countermeasures. *Chin. J. Rock Mech. Eng.* **2012**, *31*, 1945–1956.
4. Xie, H.P. Research Review of the State Key Research Development Program of China: Deep Rock Mechanics and Mining Theory. *J. China Coal Soc.* **2019**, *44*, 1283–1305.
5. Xia, K.W.; Wang, S.; Xu, Y.; Chen, R.; Wu, B.B. Advances in Experimental Studies for Deep Rock Dynamics. *Chin. J. Rock Mech. Eng.* **2021**, *40*, 448–475.
6. Rong, T.L.; Zhou, H.W.; Wang, L.J.; Ren, W.G.; Guo, Y.B. A Damage-Based Permeability Models of Deep Coal under Mining Disturbance. *Rock Soil Mech.* **2018**, *39*, 3983–3992.
7. Bauer, S.J.; Song, B.; Sanborn, B. Dynamic Compressive Strength of Rock Salts. *Int. J. Rock Mech. Min. Sci.* **2019**, *113*, 112–120. [CrossRef]
8. Duan, K.; Li, Y.C.; Wang, L.; Zhao, G.F.; Wu, W. Dynamic Responses and Failure Modes of Stratified Sedimentary Rocks. *Int. J. Rock Mech. Min. Sci.* **2019**, *122*, 104060. [CrossRef]
9. Mishra, S.; Khetwal, A.; Chakraborty, T. Dynamic Characterisation of Gneiss. *Rock Mech. Rock Eng.* **2019**, *52*, 61–81. [CrossRef]
10. Gong, F.Q.; Zhao, G.F. Dynamic Indirect Tensile Strength of Sandstone under Different Loading Rates. *Rock Mech. Rock Eng.* **2013**, *47*, 2271–2278. [CrossRef]
11. Jiang, W.; Lu, Y.B.; Jiang, X.Q.; Duan, J.Z. Experimental Study on Dynamic Direct Tensile Mechanical Properties of Mortar-Granite ITZ. *Chin. J. Rock Mech. Eng.* **2018**, *37*, 1905–1915.
12. Li, M.; Mao, X.B.; Cao, L.L.; Pu, H.; Mao, R.R.; Lu, A.H. Effects of Thermal Treatment on the Dynamic Mechanical Properties of Coal Measures Sandstone. *Rock Mech. Rock Eng.* **2016**, *49*, 3525–3539. [CrossRef]
13. Yin, T.B.; Li, X.B.; Yin, Z.Q.; Zhou, Z.L.; Liu, X.L. Study and Comparison of Mechanical Properties of Sandstone under Static and Dynamic Loadings after High Temperature. *Yanshilixue Yu Gongcheng Xuebao/Chin. J. Rock Mech. Eng.* **2012**, *31*, 273–279.
14. Gong, F.Q.; Li, X.B.; Liu, X.L.; Zhao, J. Experimental Study of Dynamic Characteristics of Sandstone under One-Dimensional Coupled Static and Dynamic Loads. *Chin. J. Rock Mech. Eng.* **2010**, *29*, 2076–2085.
15. Zhu, W.C.; Bai, Y.; Li, X.B.; Niu, L.L. Numerical Simulation on Rock Failure under Combined Static and Dynamic Loading during SHPB Tests. *Int. J. Impact Eng.* **2012**, *49*, 142–157. [CrossRef]

16. Han, T.L.; Shi, J.P.; Cao, X.S. Fracturing and Damage to Sandstone under Coupling Effects of Chemical Corrosion and Freeze–Thaw Cycles. *Rock Mech. Rock Eng.* **2016**, *49*, 4245–4255. [CrossRef]
17. Xu, J.C.; Pu, H.; Sha, Z.H. Mechanical Behavior and Decay Model of the Sandstone in Urumqi under Coupling of Freeze–Thaw and Dynamic Loading. *Bull. Eng. Geol. Environ.* **2021**, *80*, 2963–2978. [CrossRef]
18. Zhou, T.; Dong, S.L.; Zhao, G.F.; Zhang, R.; Wu, S.Y.; Zhu, J.B. An Experimental Study of Fatigue Behavior of Granite under Low-Cycle Repetitive Compressive Impacts. *Rock Mech. Rock Eng.* **2018**, *51*, 3157–3166. [CrossRef]
19. Tong, L.H.; Yu, Y.; Lai, S.K.; Lim, C.W. Dynamic Weakening of Sandstone Subjected to Repetitive Impact Loading. *Rock Mech. Rock Eng.* **2019**, *52*, 2197–2206. [CrossRef]
20. Peng, K.; Liu, Z.P.; Zou, Q.L.; Zhang, Z.Y.; Zhou, J.Q. Static and Dynamic Mechanical Properties of Granite from Various Burial Depths. *Rock Mech. Rock Eng.* **2019**, *52*, 3545–3566.
21. Li, X.; Wang, S.; Xu, Y.; Yao, W.; Xia, K.W.; Lu, G.M. Effect of Microwave Irradiation on Dynamic Mode-I Fracture Parameters of Barre Granite. *Eng. Fract. Mech.* **2020**, *224*, 106748. [CrossRef]
22. Wang, S.; Xu, Y.; Xia, K.W.; Tong, T.Y. Dynamic Fragmentation Of Microwave Irradiated Rock. *J. Rock Mech. Geotech. Eng.* **2021**, *13*, 300–310. [CrossRef]
23. Yan, Z.L.; Dai, F.; Liu, Y.; Du, H.B.; Luo, J. Dynamic Strength and Cracking Behaviors of Single Flawed Rock Subjected to Coupled Static-Dynamic Compression. *Rock Mech. Rock Eng.* **2020**, *53*, 4289–4298. [CrossRef]
24. Yan, Z.L.; Dai, F.; Zhu, J.B.; Xu, Y. Dynamic Cracking Behaviors and Energy Evolution of Multi-Flawed Rocks under Static Pre-Compression. *Rock Mech. Rock Eng.* **2021**, *54*, 5117–5139. [CrossRef]
25. Liu, K.; Xu, P.X.; Wang, F.; Jin, C.; Liu, Q.T.; Pang, H.P.; Xie, H.Z. The Accumulated Stress Damage and Residual Life Prediction of Unreinforced Concrete Pavement with Electric Heating Pipes. *Constr. Build. Mater.* **2021**, *278*, 122258. [CrossRef]
26. Ni, H.Y.; Liu, J.F.; Chen, T.; Chen, S.J.; Meng, Q. Bin Coal Permeability Prediction Method Based on the Microscopic Pore-Fracture Dual-Porosity Structure. *J. Pet. Sci. Eng.* **2022**, *211*, 110107. [CrossRef]
27. Jia, B.; Chen, Z.L.; Xian, C.G. Investigations of CO2 Storage Capacity and Flow Behavior in Shale Formation. *J. Pet. Sci. Eng.* **2022**, *208*, 109659. [CrossRef]
28. Jia, B.; Xian, C.G. Permeability Measurement of the Fracture-Matrix System with 3D Embedded Discrete Fracture Model. *Pet. Sci.* **2022**. [CrossRef]
29. Kim, T.; Jeon, S. Experimental Study on Shear Behavior of a Rock Discontinuity under Various Thermal, Hydraulic and Mechanical Conditions. *Rock Mech. Rock Eng.* **2019**, *52*, 2207–2226. [CrossRef]
30. Ai, H.A.; Ahrens, T.J. Simulation of Dynamic Response of Granite: A Numerical Approach of Shock-Induced Damage beneath Impact Craters. *Int. J. Impact Eng.* **2006**, *33*, 1–10. [CrossRef]
31. Taylor, L.M.; Chen, E.P.; Kuszmaul, J.S. Microcrack-Induced Damage Accumulation in Brittle Rock under Dynamic Loading. *Comput. Methods Appl. Mech. Eng.* **1986**, *55*, 301–320. [CrossRef]
32. Liu, H.Y.; Lv, S.R.; Zhang, L.M.; Yuan, X.P. A Dynamic Damage Constitutive Model for a Rock Mass with Persistent Joints. *Int. J. Rock Mech. Min. Sci.* **2015**, *75*, 132–139. [CrossRef]
33. Grady, D.E.; Kipp, M.E. The Micromechanics of Impact Fracture of Rock. *Int. J. Rock Mech. Min. Sci.* **1979**, *16*, 293–302. [CrossRef]
34. Zhou, Z.L.; Li, X.B.; Ye, Z.Y.; Liu, K.W. Obtaining Constitutive Relationship for Rate-Dependent Rock in SHPB Tests. *Rock Mech. Rock Eng.* **2010**, *43*, 697–706. [CrossRef]
35. Li, H.Y.; Shi, G.Y. A Dynamic Material Model for Rock Materials under Conditions of High Confining Pressures and High Strain Rates. *Int. J. Impact Eng.* **2016**, *89*, 38–48. [CrossRef]
36. Wu, A.J.; Deng, J.H.; Gu, X.; Hong, Y. Research on Mechanical Properties and Damage Evolution Law of Argillaceous Dolomite under Freeze-Thaw Cycles. *Rock Soil Mech.* **2014**, *35*, 3065–3072.
37. Dai, F.; Huang, S.; Xia, K.W.; Tan, Z.Y. Some Fundamental Issues in Dynamic Compression and Tension Tests of Rocks Using Split Hopkinson Pressure Bar. *Rock Mech. Rock Eng.* **2010**, *43*, 657–666. [CrossRef]
38. Cai, M.F.; He, M.C.; Liu, D.Y. *Rock Mechanics and Engineering*; Science Press: Beijing, China, 2013.
39. Li, X.B.; Zhou, T.; Li, D.Y.; Wang, Z.W. Experimental and Numerical Investigations on Feasibility and Validity of Prismatic Rock Specimen in SHPB. *Shock Vib.* **2016**, *2016*, 7198980. [CrossRef]
40. Qiu, J.D.; Li, D.Y.; Li, X.B.; Zhou, Z.L. Dynamic Fracturing Behavior of Layered Rock with Different Inclination Angles in SHPB Tests. *Shock Vib.* **2017**, *2017*, 7687802. [CrossRef]
41. Li, X.B.; Lok, T.S.; Zhao, J. Dynamic Characteristics of Granite Subjected to Intermediate Loading Rate. *Rock Mech. Rock Eng.* **2005**, *38*, 21–39. [CrossRef]
42. Gong, F.Q.; Si, X.F.; Li, X.B.; Wang, S.Y. Dynamic Triaxial Compression Tests on Sandstone at High Strain Rates and Low Confining Pressures with Split Hopkinson Pressure Bar. *Int. J. Rock Mech. Min. Sci.* **2019**, *113*, 211–219. [CrossRef]
43. Su, C.D. Testing Study on the Effect of Crystal Size and Specimen Length on the Rockburst Tendency. *Chin. J. Rock Mech. Eng.* **2004**, *23*, 3750–3753.
44. Zhang, Q.B.; Zhao, J. A Review of Dynamic Experimental Techniques and Mechanical Behaviour of Rock Materials. *Rock Mech. Rock Eng.* **2014**, *47*, 1411–1478. [CrossRef]
45. Sun, Q.; Li, S.C.; Feng, X.D.; Li, W.T.; Yuan, C. Study of Numerical Simulation Method of Rock Fracture Based on Strain Energy Density Theory. *Rock Soil Mech.* **2011**, *32*, 1575–1582.

46. Wen, Z.J.; Tian, L.; Jiang, Y.J.; Zuo, Y.J.; Meng, F.B.; Dong, Y.; Lin, G.; Yang, T.; Lv, D.W. Research on Damage Constitutive Model of Inhomogeneous Rocks Based on Strain Energy Density. *Chin. J. Rock Mech. Eng.* **2019**, *38*, 1332–1343.
47. Xie, H.P.; Ju, Y.; Li, L. Yun Criteria for Strength and Structural Failure of Rocks Based on Energy Dissipation and Energy Release Principles. *Chin. J. Rock Mech. Eng.* **2005**, *24*, 3003–3010.
48. Li, L.Y.; Xu, Z.Q.; Xie, H.P.; Ju, Y.; Ma, X.; Han, Z.C. Failure Experimental Study on Energy Laws of Rock under Differential Dynamic Impact Velocities. *J. China Coal Soc.* **2011**, *36*, 2007–2011.
49. Millon, O.; Ruiz-Ripoll, M.L.; Hoerth, T. Analysis of the Behavior of Sedimentary Rocks under Impact Loading. *Rock Mech. Rock Eng.* **2016**, *49*, 4257–4272. [CrossRef]
50. Wang, Z.L.; Shi, H.; Wang, J.G. Mechanical Behavior and Damage Constitutive Model of Granite under Coupling of Temperature and Dynamic Loading. *Rock Mech. Rock Eng.* **2018**, *51*, 3045–3059. [CrossRef]
51. Wang, Z.L.; Li, Y.C.; Wang, J.G. A Damage-Softening Statistical Constitutive Model Considering Rock Residual Strength. *Comput. Geosci.* **2007**, *33*, 1–9. [CrossRef]
52. Tang, C.A. Numerical Simulation of Progressive Rock Failure and Associated Seismicity. *Int. J. Rock Mech. Min. Sci. Geomech. Abstr.* **1997**, *34*, 249–261. [CrossRef]
53. Li, Y.W.; Jia, D.; Rui, Z.H.; Peng, J.Y.; Fu, C.K.; Zhang, J. Evaluation Method of Rock Brittleness Based on Statistical Constitutive Relations for Rock Damage. *J. Pet. Sci. Eng.* **2017**, *153*, 123–132. [CrossRef]
54. Lemaitre, J. Coupled Elasto-Plasticity and Damage Constitutive Equations. *Comput. Methods Appl. Mech. Eng.* **1985**, *51*, 31–49. [CrossRef]
55. Xie, H.P.; Ju, Y.; Dong, Y.L. Discussion on the "Elastic Modulus Method" in the Classical Damage Definition. *Mech. Eng.* **1997**, *19*, 1–5.
56. Xie, L.X.; Zhao, G.M.; Meng, X.R. Research on Damage Viscoelastic Dynamic Constitutive Model of Soft Rock and Concrete Materials. *Chin. J. Rock Mech. Eng.* **2013**, *32*, 857–864.
57. Zhu, J.J.; Li, X.B.; Gong, F.Q.; Wang, S.M. Dynamic Characteristics and Damage Model for Rock under Uniaxial Cyclic Impact Compressive Loads. *Chin. J. Geotech. Eng.* **2013**, *35*, 531–539.

minerals

Article

Experimental Test and Field Observations of an Electric Potential Monitoring Device for Dynamic Hazards during Mining Activities

Yue Niu [1,2,3,4,5], Enyuan Wang [2,4,5,*], Zhonghui Li [2,4,5,*], Tiancheng Shan [2,4,5], Mingfu Wang [6], Jiali Wang [6], Honghao Wang [7], Hongchang Liu [4], Junhai Ding [4], Jingkun Wang [4], Feiyang Luo [4] and Wei Shi [4]

1 State Key Laboratory for GeoMechanics and Deep Underground Engineering, China University of Mining and Technology, Xuzhou 221116, China; y.niu16@cumt.edu.cn
2 Key Laboratory of Coal Methane and Fire Control, Ministry of Education, China University of Mining and Technology, Xuzhou 221116, China; shantiancheng@cumt.edu.cn
3 School of Mechanics and Civil Engineering, China University of Mining and Technology, Xuzhou 221116, China
4 School of Safety Engineering, China University of Mining and Technology, Xuzhou 221116, China; 02201415@cumt.edu.cn (H.L.); 02203702@cumt.edu.cn (J.D.); 02201302@cumt.edu.cn (J.W.); 08204303@cumt.edu.cn (F.L.); 02203201@cumt.edu.cn (W.S.)
5 National Engineering Research Center for Coal Gas Control, China University of Mining and Technology, Xuzhou 221116, China
6 Xinzheng Coal Power Co., Ltd., Zhengzhou Coal Industry (Group) Co., Ltd., Xinzheng 451100, China; d.mingfu@163.com (M.W.); yuer_new@126.com (J.W.)
7 Yanghe Coal Industry Co., Ltd., Zhengzhou Coal Industry (Group) Co., Ltd., Xinmi 452370, China; wanghonghao617@163.com
* Correspondence: weytop@cumt.edu.cn (E.W.); leezhonghui@163.com (Z.L.)

Citation: Niu, Y.; Wang, E.; Li, Z.; Shan, T.; Wang, M.; Wang, J.; Wang, H.; Liu, H.; Ding, J.; Wang, J.; et al. Experimental Test and Field Observations of an Electric Potential Monitoring Device for Dynamic Hazards during Mining Activities. *Minerals* **2022**, *12*, 852. https://doi.org/10.3390/min12070852

Academic Editors: Diyuan Li, Zhenyu Han, Xin Cai, Shijie Xie and Abbas Taheri

Received: 20 May 2022
Accepted: 28 June 2022
Published: 3 July 2022

Abstract: The EP (electric potential) signals can be generated during the deformation and fracture process of coal and rock mass. Meanwhile, the EP response is closely related to its stress state and damage evolution, which is expected to be used in monitoring and coal and rock dynamic disaster hazards. Based on this, this paper developed an EP monitoring device for mining to continuously monitor the temporal response characteristics and spatial distribution of coal seam internal EP signals in real time. Further, the experimental tests were carried out, whose results showed that the device has high monitoring sensitivity and little error for the EP signals and can reveal the loading state and damage degree of the coal and rock specimens during the deformation and fracture process. Moreover, the tests and application of EP monitoring were carried out during mining activities in the field. The results showed that the EP signals fluctuate during the coal mining stage and remain relatively stable during the maintenance stage. When the abnormal mining stress or the coal cannon phenomenon occurs, the intensity of EP signals increases rapidly and fluctuates violently, which has precursory response information for the hazards of dynamic disasters. Considering the advantages of sensitive response and nearly non-destructive monitoring, the study results can provide key monitoring equipment and research basis for field testing the EP signals during the mining process, to monitor and forecast the hazards of coal and rock dynamic disasters.

Keywords: EP monitoring device; experimental test; field application; coal and rock dynamic disaster; monitoring and forecasting

1. Introduction

Coal is an important raw material for energy resources, which is widely utilized in many fields, including thermal power generation, metallurgical industry, and chemical production [1]. However, most of the coal resources are hidden in the deep strata underground, which need to be mined. In the process of coal mining activities, as the coal mass is extracted from the original coal stratum, the structure changes, breaking its original

stress balance state. It causes the deformation, damage, and instability in the coal and rock mass, leading to coal and rock dynamic disasters when reaching the critical conditions, such as rock burst, coal and gas outburst, roof collapse, etc. [2]. It is very important for real-time monitoring of the damage and failure process of coal and rock mass and accurate forecasting of its instability and failure [3]. This can reliably ensure the safety of coal mining activities, and contribute to avoiding miners' personal casualties, mechanical equipment damage, roadway deformation fracture, etc. [4].

The essence of coal and rock dynamic disasters is the damage and failure of coal and rock mass under the action of mining stress, which eventually leads to the loss of structural stability and dynamic failure of coal and rock [5]. The process belongs to the mechanical category [6]. However, coal mass is a material with significant inhomogeneity and anisotropy, which makes it difficult to accurately monitor the failure and its precursor information of the coal body using traditional mechanical methods [7,8]. Previous studies show that the deformation and rupture process of a coal body can generate significant electric potential (EP) signals, and the EP response is closely related to the damage evolution process and stress state of the coal body [9,10]. Studying the EP response characteristics of a coal body during damage and failure processes can indirectly monitor the damage evolution process of a coal body and identify the precursor information of instability and failure of coal and rock mass [11]. Eccles et al. observed abnormal signals of EP response before rock fracture under different experimental conditions [12]. Wang et al. found that the process of deformation and rupture of a coal body under loading results in charge separation through crack propagation, the triboelectric effect, the piezoelectric effect, and so on, and revealed the physical mechanism of the EP signals generated by the charge [13]. Niu et al. found that when the main fracture occurs inside gas-bearing coal, the EP intensity increases and mutates abruptly, verified by acoustic emission and crack propagation simultaneously [14].

Previous studies on the EP effect of coal and rock mass were mainly carried out using experimental study and theoretical analysis [9,15,16]. These studies have achieved fruitful results and have been widely recognized, and also clarified the potential of using the EP method to monitor and forecast coal and rock dynamic disasters [17]. However, the study on the EP effect of coal and rock lacks testing and application in the field during coal mining activities [18]. Compared with the laboratory study conditions, the interference of environmental signals is very complex in the process of field coal mining, and the change in mining stress and the evolution of coal damage are not controllable [19]. In order to guide the engineering practice, the experimental achievement must be tested and applied in the field, and it is the basis and premise for developing the monitoring equipment for EP signals of a coal body suitable for field coal mining activities [20]. The performance of the mine EP monitoring equipment must be guaranteed, and the validity of the test results should also be verified.

Therefore, in this paper, the mine EP monitoring device was developed independently, and the test effect of the device under laboratory experiment was tested, and then the application in the coal mining site was also carried out. The study results are expected to lay the equipment foundation and research basis for further monitoring and forecasting of coal and rock dynamic disasters with the EP monitoring method.

2. Development of the EP Monitoring Device for Mining

2.1. Monitoring Host

In this paper, the mine EP monitoring device was developed independently, and the host of this equipment is shown in Figure 1. The host surface includes a digital display screen, a control panel, multiple channel interfaces, etc. The host has a resolution of 0.01 mV for EP testing and a range of [−1500 mV,+1500 mV]. The sampling period is adjustable in the interval of 1~240 s. Each channel is connected to the electrode through a shielded cable and can measure EP signals independently. The host has a built-in large-capacity battery with a single power cycle of 30 days approximately, which can independently supply power

to the motherboard and the screen. The whole machine has an explosion-proof function and can be applied to downhole testing. The host can use multiple channels to monitor the EP signals of different measuring points synchronously, continuously, and in real time. The sheet-shaped copper electrode is utilized as a sensor to receive EP signals, with high conductivity and low impedance, to sensitively monitor the input EP signal.

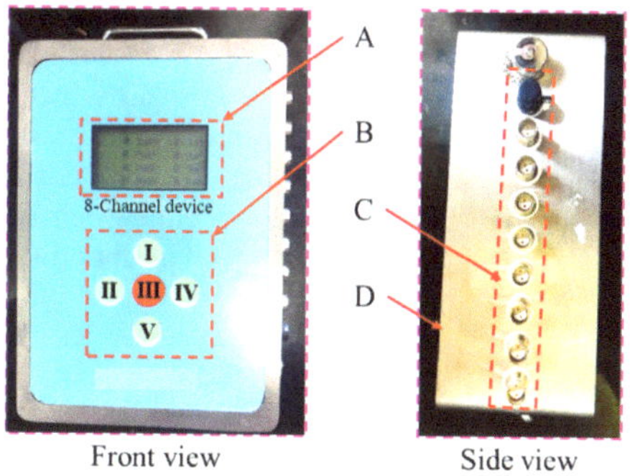

Figure 1. Physical photograph of the EP monitoring host for mining. I—project; II—subtraction; III—power Supply; IV-addition; V-choice; A—data display screen; B—operation panel; C—multichannel interface; D—explosion-proof enclosure.

This host can be regarded and equivalent to a special "voltmeter", and the process flow of measuring EP signal is shown in Figure 2. The electrode receives the EP signals and transmits them to the host through a shielded cable. After being amplified by the preamplifier in the host, the EP signals are converted from an analog/digital converter to digital signals, and then processed by the CPU and stored in memory. The stored data is transmitted to the data reading terminal by the wireless transmission module, and the display can show the measuring data in real time.

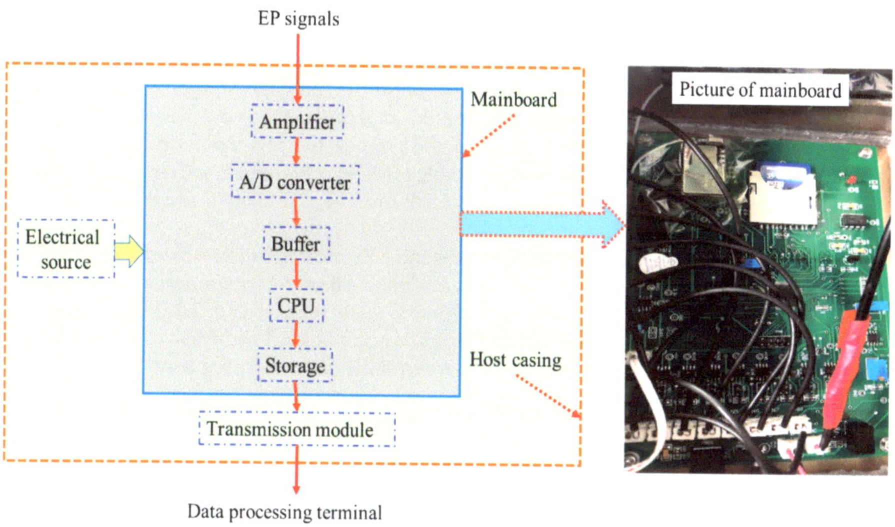

Figure 2. Schematic diagram of EP signal processing.

2.2. EP Data Analysis Software

As shown in Figure 3, the EP monitoring host is matched with self-developed EP data analysis software. The software imitates the Windows interface design with real-time

display, historical inquiry, data analysis, forecasting, and other functions. After acquiring the EP data recorded in the EP monitoring host, it can query the EP historical data of different ID hosts, different channels and different time periods, conduct statistical analysis, and issue early warning information in time.

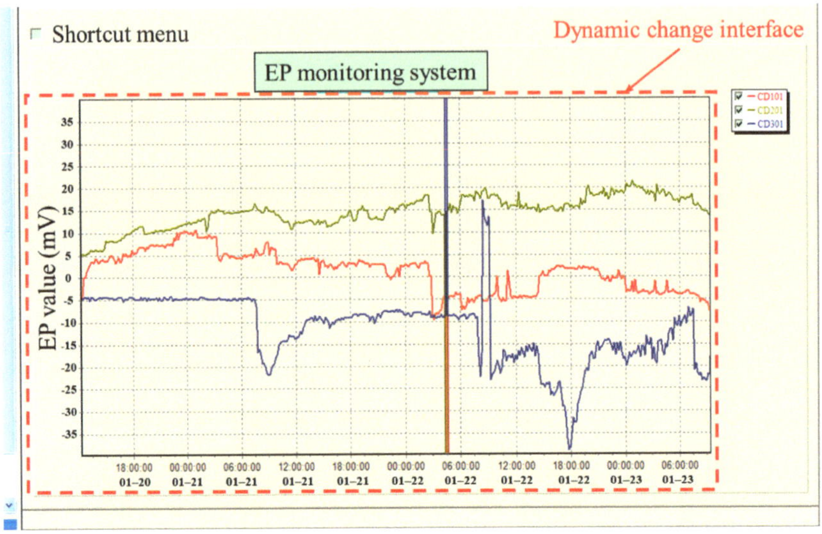

Figure 3. EP analysis software.

2.3. EP Monitoring System Inside Coal Seam

The purpose of developing the mine EP monitoring device is to test the EP signals in different zones inside the coal seam, analyze the correlation between the EP response and the coal and rock dynamic disasters, so as to locate and monitor the hazard zones where the disaster occurs. In order to achieve this goal, this paper designed and developed a multi-point test system for EP spatial distribution inside the coal seam in the field. It can realize the real-time monitoring of EP signals from measurement points at different locations and depths inside the coal seam and analyze the spatial distribution characteristics and dynamic evolution laws of the EP signals.

As shown in Figure 4, the test system consists of an EP monitoring host, push pipes, electrodes, shielding cables, inflatable capsules, flexible air conduits, etc. Among them, the pushing pipe is made of PVC, the length of the single unit is 1.5 m, and the different push pipe units can be connected together through the joint, which can realize pushing the push pipes to the depth of the coal seam. The inside of the push pipe unit is hollow, and a groove is opened in the middle of each unit, where capsules and electrodes can be placed. The electrode is selected as a thin sheet electrode made of copper, which has the advantages of sensitive response and corrosion resistance. The size of this electrode is 50mm × 30mm × 1.5 mm, which has a large coverage and moderate flexibility.

The test method of this system includes six steps. (1) An appropriate position in the coal seam underground is selected and the boreholes are drilled inside the coal mass. (2) The PVC pipes are used to transport electrodes to the designated positions in the boreholes, as well as cables, capsules, air conduits, etc. (3) The gas is filled into the capsule through the air conduits outside the borehole, which makes the capsule expand and deform. (4) The deformed capsule lifts the electrode fixed above the capsule, so that the electrode is fully contacted with the coal body on the borehole wall and the EP signals of the coal body are tested. (5) The electrodes are connected to the host outside borehole through shielded cables, and the host records the EP signals monitored in real time. (6) Multiple electrodes can be arranged in the same borehole to test the EP signals at different locations in the borehole.

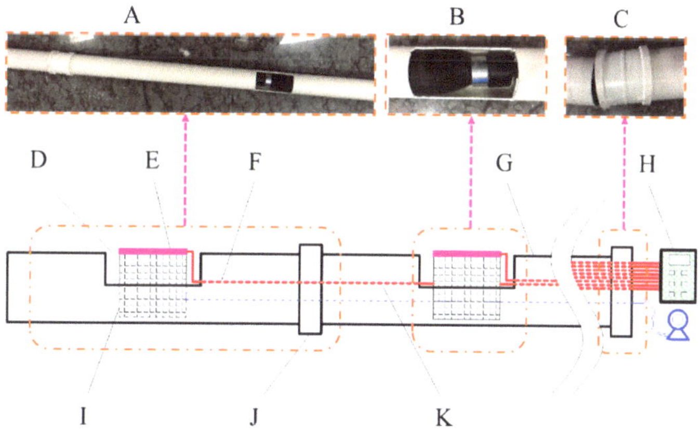

Figure 4. Structural diagram of the EP measuring system for mining. A—PVC push pipe; B—grooves; PVC; C—PVC push pipe joint; D—groove; E—electrode; F—shielding cable; G—PVC push pipe; H—EP host; I—inflatable capsule; J—PVC push pipe joint; K—flexible air conduits.

3. EP Monitoring Host Experimental Test Results

To verify the measuring effect of the EP monitoring host, the test work with two parts under laboratory conditions were carried out. On the one hand, the EP signals were emitted by the power supply to test the EP signal's response accuracy and its spatial distribution characteristics of the EP monitoring host [21]. On the other hand, the uniaxial compression experiments of coal and rock specimens were carried out to test the EP signal's response characteristics of the EP monitoring host during the deformation and fracture process of coal and rock mass [22]. Combined with the above experimental results, the test effect of the EP monitoring host was verified and analyzed.

3.1. Test Experiment of EP Signals Emitted by the Power Supply

3.1.1. Test Results of EP Response Accuracy

The EP signals were emitted by the power supply, which were recorded by the EP monitoring host, and the test results are shown in Figure 5. The EP value of the power supply input is defined as the input EP value, and the EP value measured by the EP monitoring host is defined as the receiving EP value. The results showed that the receiving EP value was highly consistent with the input EP value, and they were almost linearly increased, and the error rate was controlled within 1.5%. It indicates that the EP monitoring host has high sensitivity to the input EP signals, and the test results are reliable.

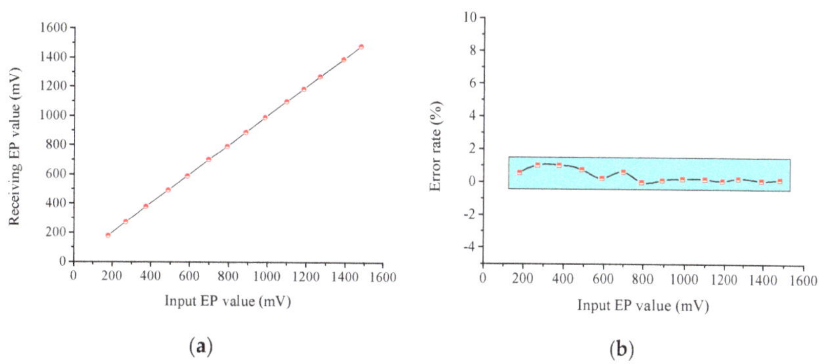

Figure 5. Experimental results of EP measuring emitted by power supply. (**a**) Value comparison of input EP and output EP; (**b**) error rate of comparison results.

3.1.2. Test Results of EP Spatial Distribution

As shown in Figure 6a, 1500 mL of NaCl solution with a concentration of 5% was injected into a bathtub with a diameter of 50 cm, and the response of the EP monitoring host to the EP spatial distribution in the solution was tested in the water bath experiments [23]. The power supply was connected to the input positive electrode and the input negative electrode, and an electrostatic field was excited by the power supply in the solution; thus, generating an EP difference between any two points. The EP monitoring host was connected to the receiving positive electrode and the receiving negative electrode, and the EP difference (i.e., the EP value) between the two electrodes was tested. At the same time, the receiving positive electrode and the receiving negative electrode were connected with the voltmeter to verify the test results of the EP monitoring host.

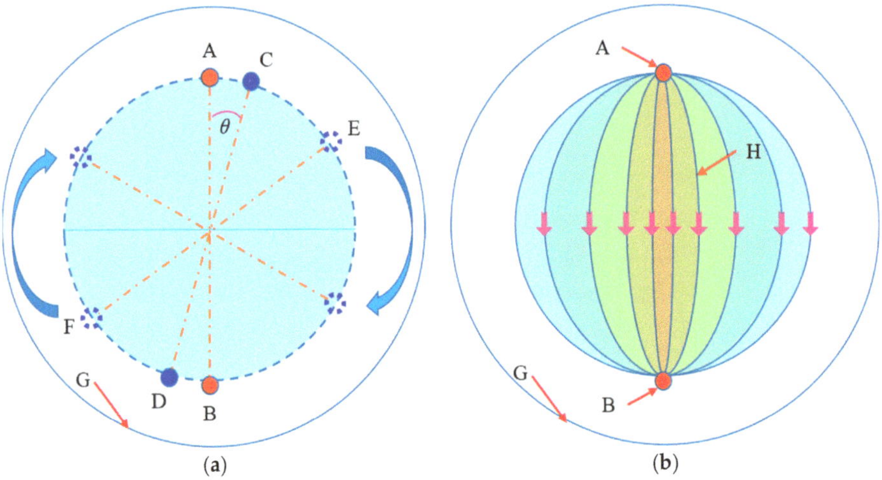

Figure 6. Schematic diagram of electrode arrangement for the water bathing experiment. A—input positive electrode; B—input negative electrode; C—receiving positive electrode; D—receiving negative electrode; E—potential position of receiving positive electrode after rotation; F—potential position of receiving negative electrode after rotation; G—solution boundary; H—electric field line diagram. (**a**) Schematic diagram of receiving electrode rotation relative to input electrodes. (**b**) Schematic diagram of electric field excited by input electrode.

As shown in Figure 6a, the positions of the input positive electrode and the input negative electrode were kept constant, and the positions of the receiving positive electrode and the receiving negative electrode were rotated synchronously along their midpoints to make a certain angle between the two sets of electrodes. As shown in Figure 6b, theoretically, as the angle increases gradually in the interval of [0°,90°], the distance between the receiving electrodes remains unchanged, but the EP difference between the electrodes will gradually decrease [24]. Similarly, the absolute value of the EP difference between the receiving electrodes should gradually increase as the angle gradually increases in the interval of [90°,180°].

The test results of the EP monitoring host and voltmeter under different electrode angles were recorded, and a Rose Diagram was drawn, shown in Figure 7. With the gradual increase in the angle, the EP signals tested by the EP monitoring host and the voltmeter tend to decrease and then increase, and reach the lowest value at an angle of 90° and the highest value at an angle of 180°. This is consistent with the theoretical analysis results, so the test results of the EP monitoring host can reflect the nonlinear distribution characteristics of the original EP signals in space.

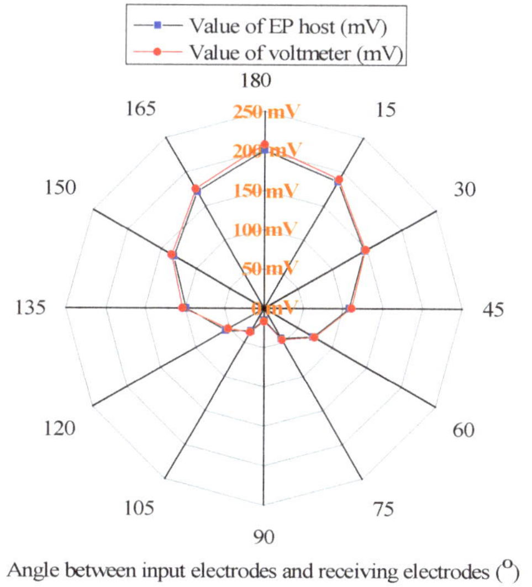

Figure 7. Experiment results of EP measuring in the water bathing environment.

3.2. *Experiment Results of the EP Signals Test under the Uniaxial Compression Process*

3.2.1. Test Results of Raw Coal Specimen

For mining activities, coal is regarded as a special rock, which contains more abundant pores and fracture structures. The coal's main component is organic matter, which can adsorb gas with stronger plasticity and weaker elastic modulus. The materials involved in coal rock dynamic disasters can be regarded as "coal", "coal rock", "coal-measure rock", or "coal and rock". In order to simulate the deformation and fracture process of a coal body in mining activities, the cylindrical raw coal specimens with the size of Φ50mm × 100 mm were made, and the uniaxial compression load was applied to the specimens until failure. The loading method is force control with rate of 100 N/s. The EP measuring points are arranged on the specimen surfaces, and the EP signals of the specimens during the loading process is tested using the EP monitoring host. The experimental results are shown in Figure 8.

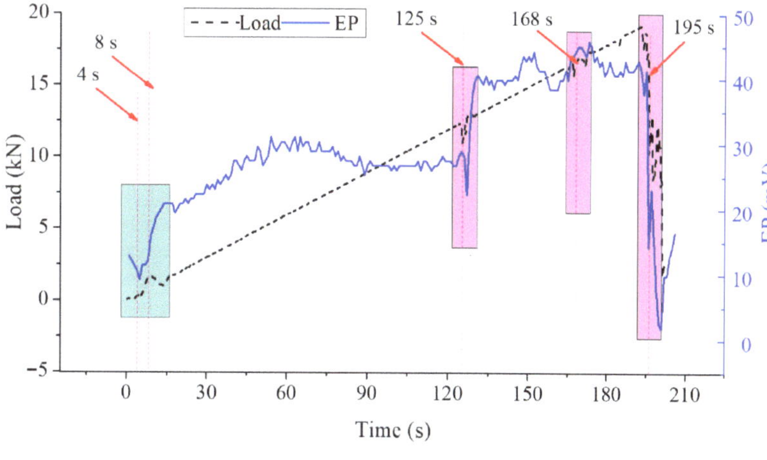

Figure 8. Experimental results of EP response on raw coal specimens under uniaxial compression.

It can be seen from Figure 8 that with the increase in load, the damage degree of the coal body was increasingly aggravated. When there was a small abrupt change in the

load, indicating that the specimen was severely damaged in the local position. When the load suddenly decreased to the residual value, it showed that the specimen structure lost stability and dynamic damage occurred. Meanwhile, with the increase in load, the EP signal intensity showed an increasing trend. When the specimen was severely damaged locally, the EP signals fluctuated several times.

In details, (1) in the early loading stage, there was a small abnormal fluctuation in the process of load increase, and the EP intensity also showed an abnormal response that fluctuated to a valley value and increased rapidly at the corresponding moments (4 and 8 s). This was because the raw coal specimen contained abundant pore and fissure structures, which would inevitably cause some damage to the internal structure of the specimen during the specimen preparation process. Even at a low load level, small damage and rupture might occur, resulting in fluctuations in EP signals [11]. (2) At 125 s, the load fluctuated abnormally again, and the EP intensity increased in a stepwise manner. At this time, under the action of higher load, the relatively serious deformation and rupture occurred inside the specimen, resulting in load fluctuation. At the corresponding time, the cracks inside the specimen expanded rapidly, resulting in a sudden increase in EP intensity. (3) When the loading process continued to 168 s, the load showed significant abnormal fluctuations again, resulting in violent fluctuations in the EP signals again. (4) At 195 s, the main fracture of the specimen occurred, and the load rapidly decreased to the residual level, where the EP signals showed a sudden drop mutation again. This was basically consistent with the previous laws of test results by using high-precision EP monitoring equipment suitable for laboratory conditions [14].

The experimental results indicate that the independently developed EP monitoring host can effectively measure the EP signals generated during the deformation and fracture process of the raw coal specimen. The response characteristics of the EP signals can reveal the stress state of the coal body and have cumulative enhanced response characteristics (i.e., "memory effect") for the damage and deformation of the coal body. When the specimen is severely damaged or even the main fracture occurs, the EP signals have abnormal response characteristics. It is manifested as "↑" and "↓" type mutations, which can monitor and forecast the instability and fracture of the coal body.

3.2.2. Test Results of the Concrete Specimen

The brittle fracture of raw coal occurs during the uniaxial compression. However, in most cases during mining activities, due to the limitation of the mining space and the effect of the lateral stress of the surrounding rock, the deformation and fracture of coal and rock often show strong plastic characteristics. Meanwhile, the concrete material has strong plasticity. Under the uniaxial compression test conditions, the concrete specimen can have obvious plastic deformation before reaching the failure strength without sudden fracture, which is more in line with the characteristics of actual deformation and fracture of coal and rock mass at the excavation and mining site. At the same time, the production process of concrete can better avoid the initial damage of the specimens. Previous studies have shown that concrete can also generate abundant EP signals during the deformation and fracture process [25,26]. In order to further simulate the deformation and fracture of coal and rock mass in mining space, the concrete specimens were prepared in this section, and the uniaxial compression test was carried out. Additionally, the EP signals generated at different positions during the deformation and fracture of the specimens were tested by the EP monitoring host. The concrete specimens are cubes with dimensions of 150 mm × 150 mm × 150 mm. The loading method is displacement control and the loading speed is 0.25 mm/min. The test results are shown in Figure 9.

Figure 9 shows that the EP signals are tested at different measuring points under the action of load, and the response characteristics of the EP signals are basically consistent, which show an increasing trend with the load increasing. Different from raw coal, concrete has strong plasticity and weak brittleness. The damage process of the specimen is relatively stable, and there is no significant load mutation characteristic, so the change in EP signals is relatively stable. When the load level is low, the specimen undergoes transverse expansion

deformation under the action of axial stress, and the damage degree is higher [27]. In this process, the abundant EP signals are generated due to crack propagation and friction between crack surfaces, and the EP intensity increases with the increase in fracture surfaces. At the later stage of loading, with the increase in the load level, the deformation and fracture of the specimen are intensified, and the cracks gradually converge and coalescence, forming the macroscopic fracture zones. The charges are constantly accumulating and neutralizing each other, which leads to a slow increase or even a decrease in the EP intensity too varying degrees, but the EP intensity is still at a high level on the whole.

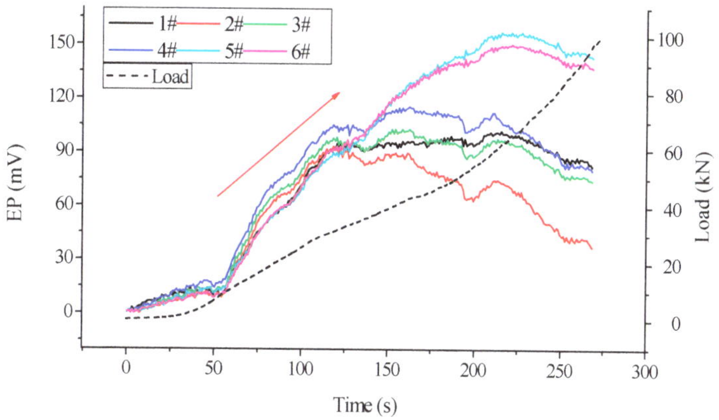

Figure 9. Experimental results of EP response on the concrete specimen under uniaxial compression.

There are also some differences in the EP response characteristics of different measuring points, which are closely related to the degree of deformation and rupture of the measuring points on the specimen surfaces. Among them, the 1~3# EP measuring points 0were arranged on the front surface of the specimen, and the 4~6# EP measuring points were arranged on the rear surface of the specimen. The average value and standard deviation of EP intensity at different measuring points are shown in Figure 10. The EP response of 1~3# was relatively weak, which showed that the average value and standard deviation of EP were lower, and the weakest was 2#. Relatively, the EP response of 4~6# was stronger, which showed that the average value and standard deviation of EP were higher. It meant that the EP response was more intense, and the strongest was 8#. It could be seen from the physical maps of the specimen surfaces with deformation and fracture (the yellow dotted line indicates the crack propagation trajectory) that the crack propagation was less near the 1~3# EP measuring points on the front surface, and the plane near the 2# EP measuring point was the most complete. Meanwhile, the crack propagation near the 6~8# EP measuring points on the rear surface was greater in number and larger in scale (the crack lengths were longer and the widths were wider), and the crack propagation was the most abundant near the 8# measuring point. The crack propagation was the macroscopic manifestation of the internal local severe damage. In the region with abundant crack propagation, the damage localization of the specimen developed more intensely, which led to a higher mean value for the EP signals of the measuring point at this position. Additionally, the larger the standard deviation was, the more intense the EP response was. Therefore, the EP monitoring host could effectively monitor the EP signals generated during the deformation and fracture of materials such as coal and rock and could identify the damage degree of different local positions of the specimen.

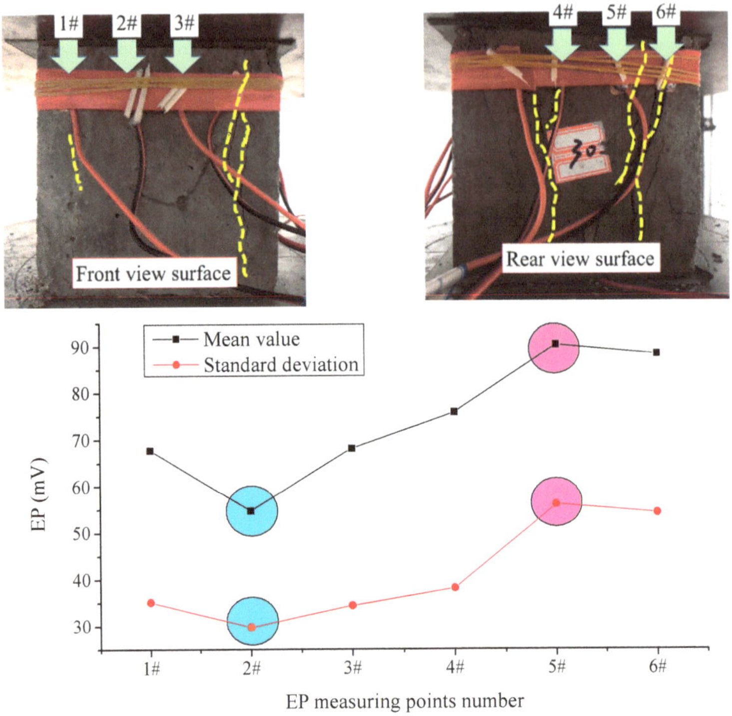

Figure 10. Statistics of mean and standard deviation of the EP signal at different measuring points of the concrete specimen.

4. Field Test and Application Results of the EP Monitoring Device during Mining Activities

4.1. Survey of the Mining Coal Seam in the Field

In coal mining activities, as the coal body is mined, a suspended area (i.e., "goaf") is formed in the original position, resulting in a change in the stress distribution of the coal seam, which imposes an "additional loading stress" on the coal body, similar to Section 3.2 [28]. At the same time, the coal mining process uses the machine to cut the coal body, so that the coal body is broken and gas is pushed out, which further disturbs the stability of the coal seam. Under the action of mining stress, the deformation and fracture of the coal body develop to a certain extent and even have the hazard of dynamic disasters such as coal and gas outburst and roof collapse [29].

In order to further verify the application effect of the EP monitoring device for mining in the field, this paper used the EP monitoring system in Section 2.3 to test the EP signals in the mining coal seam of Xuehu Coal Mine in Shangqiu City, Henan Province, China. The test address was the return air roadway (belonging to the coal roadway) of the 25050 fully mechanized mining face in the NO. 25 Mining Area, where the EP measuring points were arranged. The NO. 25050 Ming Area has a strike length of 1200 m, an inclination length of 160 m, an average inclination angle of 11°, an average coal thickness of 2.3 m, and a height of 720–750 m from the ground. It is located in the outburst threat area, with a large gas content, and the absolute gas emission volume of the excavation is 3.0–1.0 × 10^3 m^3/min. There are many small faults and magmatic rocks intrusion locally, which have an important impact on the stress distribution of the mining coal seam and the gas migration. Moreover, it has high pressure and serious slats, and it is easy to form stress anomalous areas locally, resulting in a broken roof, development of cracks, and gas enrichment, which has the danger of inducing coal and rock dynamic disasters. It is necessary to strengthen the monitoring

of coal and rock dynamic hazards in the mining face. The arrangement of EP measuring points in the field coal mining face is shown in Figure 11.

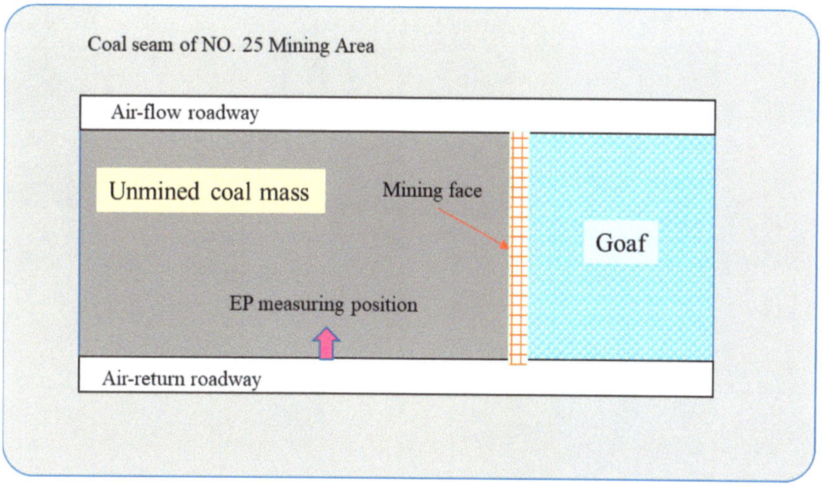

Figure 11. Schematic diagram of EP measuring layout in the coal seam of the NO. 25 Mining Area.

4.2. Response Characteristics of EP Signals in the Coal Mining Process

4.2.1. "Mining-Maintenance" Process and EP Response in the Coal Seam

Generally, coal mining activities are carried out intermittently. In Xuehu Coal Mine, the coal mining activities are carried out continuously for 20 h on the mining face each day, followed by a four-hour pause for equipment maintenance during the coal mining process. After the completion of maintenance, coal mining activities continue to carry out circularly.

Taking 27–30 August as an example, the response characteristics of the EP signals during maintenance and coal mining were analyzed, shown in Figure 12. (1) The maintenance stage was carried out from 9:00 to 14:00 on 28 August, during which the EP signals quickly entered a quiet period from a relatively fluctuating state and remained stable. Additionally, the EP signals quickly left the relatively quiet state after entering the next coal mining cycle and fluctuated significantly. (2) The mining process entered the maintenance stage again from 9:00 to 14:00 on 29 August, and the EP signals showed a trend of "fluctuation–calm–fluctuation" again. (3) From 9:00 to 14:00 on 30 August, the mining process entered the maintenance stage again, and the EP signals of the measuring point were not as calm as the previous maintenance stage, and the EP signals fluctuated greatly.

The EP signals of the coal body change with the coal mining activities, which is mainly caused by the change in the coal body stress distribution due to coal cutting by the coal mining machine [30]. Theoretically, the stress state and structure of the coal seam in front of the working face are relatively stable during the maintenance stage, so the EP signals should be in a relatively stable state, and the fluctuation of the EP signals during coal mining is more significant. During the coal mining process, the coal mining face is constantly moving forward, while the EP measuring points are fixed, so the distance between the EP measuring points and the coal mining face is constantly decreasing. With the continuous decrease in this distance, the coal body at the EP measuring points is increasingly affected by the coal mining activities, so the EP intensity shows an overall increasing trend. On 29 August, the EP measuring point was about 10 m away from the coal mining face. The coal near the measuring point was greatly affected by the pressure relief effect of the coal mining face, and the coal at the measuring point had a large deviatoric stress difference, resulting in severe damage and fracture. Therefore, even in the maintenance stage, the damage state of the coal body at the EP measuring points was still in a slow evolution and development process, and the EP signals had a relative fluctuating response.

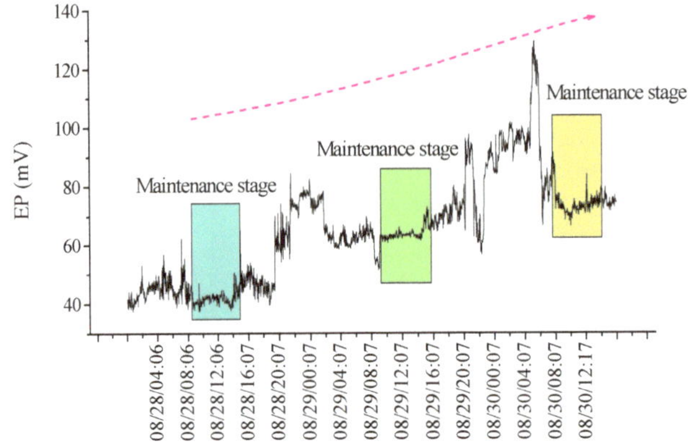

Figure 12. Influence on EP response during mining activities by maintenance.

4.2.2. Mining Stress Anomaly and EP Response

The geological structures of an underground coal seam are complex, which are often rich in small faults, collapse columns, water-bearing layers, gas abundance zones, and so on, resulting in the distribution of stress anomalies in coal and rock mass during mining [31]. When the mining stress is abnormal, the damage to the coal body is more severe under the same mining influence, and it is easy to induce dynamic damage phenomena of the coal body.

On 21 August, the mining stress of the coal body was abnormal. Correspondingly, at 18:30 on 21 August, the EP intensity suddenly increased, with an increase of 1.5–6.2 times for the original value. The mean and standard deviation for the daily EP signals of coal body were counted, as shown in Figure 13. Before 21 August, the EP intensity and standard deviation increased rapidly, and after the occurrence of stress anomaly, the EP intensity continued to increase and the standard deviation decreased after reaching the peak, which meant that the EP intensity and standard deviation showed the abnormal response characteristics with types of "↗" and "∧", respectively. The above test results showed that when the stress abnormality occurred, the stress distribution of the coal body near the measurement point changed from a relatively balanced state to an unstable state, and the damage was more severe, resulting in an increase in the EP strength and the increase in the dispersion of the EP signals, and the heterogeneity of the coal body damage evolution was enhanced [32]. As the stress distribution of the coal body gradually transitioned from "abnormal" to "new equilibrium" state, the loading state of the coal body was more stable, and even though the EP signals continued to increase, the dispersion of EP signals decreased, resulting in a decrease in the EP standard deviation.

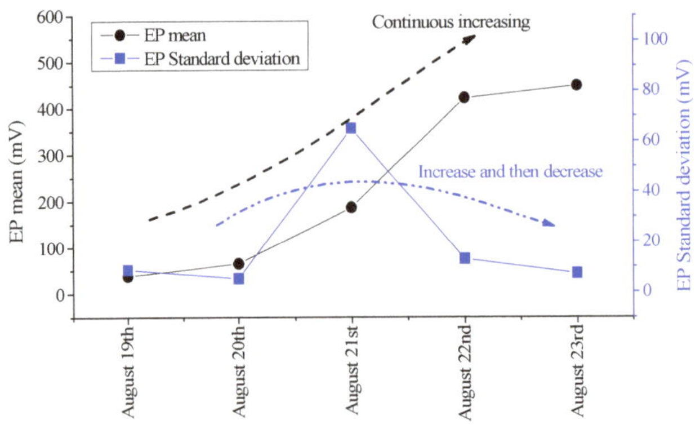

Figure 13. Response results of EP signals before and after stress abnormality occurrence.

Therefore, when the mining stress is abnormal, the EP signals have a rapid and greatly enhanced abnormal response characteristic. The increase in EP strength is the most intuitive reflection of the mining stress anomalies, while the EP standard deviation is more able to identify the continuous "anomalous" and "new equilibrium" state of the mining stress distribution. The EP monitoring results have significant anomalous response characteristics to the coal mining stress anomaly and can monitor the stress state and damage process inside the coal and rock body.

4.2.3. Coal Cannon Phenomenon and EP Response

Coal is damaged and deformed under the action of mining activities, and the elastic energy is continuously accumulated. When the accumulation of elastic energy reaches the critical condition, due to roof fracture, rock vibration, and so on, the coal body occurs in a small range of sudden fracture and emits a loud sound [32]. It is called the "coal cannon" phenomenon, belonging to a classical type of coal dynamic failure. When the energy released by the coal cannon is large, especially when the continuous high-energy coal cannon phenomenon occurs and cannot be effectively controlled, it will pose a threat to the safety of personnel and equipment in the mining space, which is one of the early warning signs of coal and rock dynamic disaster hazards.

As shown in Figure 14, a continuous high-energy coal cannon phenomenon occurred on 20 September, resulting in a significant crack with 1.2 m length and 0.5 m width on the floor of the roadway with a large destructive force. Before the occurrence of the coal cannon, the EP intensity was already at a high level up to 159.7 mV and showed a certain increasing trend. At 21:30 on 18 September, the EP response began to reveal precursory information, showing a significant abnormal response characteristic of "plunge-stabilization-swell-stabilization-decrease", and rapidly decreased after 9:00 on 19 September. Compared with the moment when the coal cannon occurred, the precursory information of the EP signals was about 2 days earlier and lasted about 1 day. When the coal cannon occurred, the EP intensity was close to the peak, and then the EP signals fluctuated violently during the continuous occurrence of the coal cannon. After the occurrence of the coal cannon, the EP intensity decreased rapidly again and remained relatively stable afterwards.

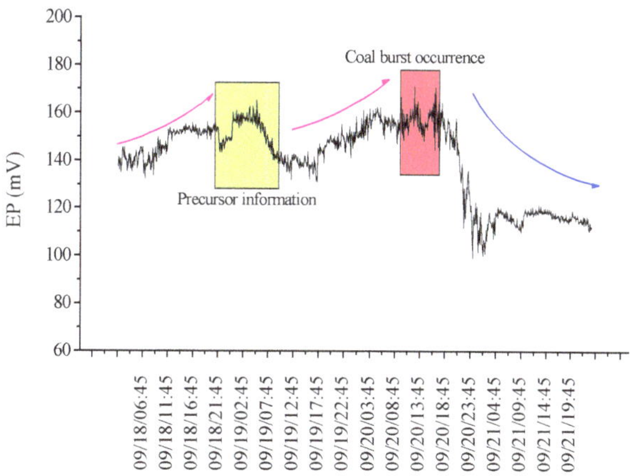

Figure 14. EP measuring results around the coal burst occurrence.

Before the occurrence of the coal cannon, the strain energy continued to accumulate by a large amount, leading to an increase in the elastic deformation and plastic damage of the coal body and a subsequent increase in EP strength [30]. This induced the abnormal response characteristics of the EP signals during the inoculation process of the coal cannon, which showed the precursory information of sustained increasing and violent fluctuation, and the EP intensity was close to the peak level when the coal cannon occurred. After the

occurrence of the coal cannon, the mining stress level decreased rapidly with the release of energy, which was similar to the residual stress state in the uniaxial compression experiment (shown in Figure 11). At this time, the hazard of the coal and rock dynamic disaster was significantly reduced, and the EP intensity decreased rapidly and remained at a relatively stable level.

4.3. Application Prospect of the EP Monitoring Method in Forecasting of Coal and Rock Dynamic Disaster Hazards

According to the above analysis, the EP signals can be generated in the mining process of coal seam, and the response characteristics of EP signals are closely related to the stress state and damage evolution process of coal and rock mass. The EP signals have the response characteristics of the strength rising to a high value for coal and rock dynamic disaster hazards (i.e., mining stress anomaly or coal cannon phenomenon), which can reflect the loading and damaging state of specimen, and reveal the hazard degree of the coal and rock dynamic disasters [9,10]. Therefore, through monitoring the EP signals inside the coal seam during mining activities, it is possible to identify the abnormal response characteristics, monitor and forecast the coal and rock dynamic hazards, which are prevented and controlled by taking targeted measures. This provides a new engineering application method for fine forecasting and precise control of coal and rock dynamic disasters at the coal mining sites.

Compared with the previous conventional monitoring methods, such as the stress monitoring method, electromagnetic radiation method, and microseismic method, the EP monitoring has its own unique advantages [33–35]. Taking electromagnetic radiation as an example, it is a non-contact monitoring method. It monitors the electromagnetic radiation signal generated during the process of coal deformation and fracture, but it is easily disturbed by the electromagnetic induction phenomenon caused by the start, stop, and operation of large-scale electromechanical equipment, such as a shearer and scraper. This is unavoidable during the daily mining work. As a comparison, the EP monitoring needs to contact the electrode with the coal, and the electrode should avoid direct contact with electromechanical equipment, the metal anchor net, and wet coal, which effectively avoids the interference of the electromagnetic induction phenomenon. Similarly, the acoustic emission signal is easily disturbed by the noise generated during the operation of large electromechanical equipment. Due to the wide frequency spectrum of the acoustic emission signal generated in the mining activities, there is no widely adaptive filtering method to eliminate it. Usually, the acoustic emission sensor is arranged on the bolt to receive the acoustic emission signal generated inside the coal body.

For the EP monitoring method, on the one hand, the EP monitoring has the advantages of accurate response and strong anti-interference. The EP signals are not only sensitive to the large-scale fracture event of the coal body, but also respond to the deformation of the coal body (i.e., the strain field), which can reveal the evolution process of the coal body damage and failure from a deep level [30]. The EP signals are obtained by direct contact tests between the electrode and the coal body, and coal is one of the typical poor conductors, which is less disturbed by the work of electromechanical equipment. On the other hand, the process of EP monitoring is simple, requiring only the construction of ordinary boreholes with a small workload and negligible impact on production, which is similar to "non-destructive monitoring". The EP monitoring system can push the electrodes to the specified borehole depth through the pushing device, which can test the EP signals of the same borehole with different borehole depths in real time and continuously at high frequency, and realize the location monitoring of the EP spatial distribution law [10].

In this paper, the mine EP monitoring device was independently developed, which was tested and applied during mining activities. Consequently, some preliminary achievements and progress were made. However, the study on the EP monitoring technology and system of coal and rock gas dynamic disasters still needs to be carried out in depth. The authors' research group has independently developed EP monitoring instruments,

matching facilities, and analysis software for coal seam mining on site. The next step is to monitor the coal seam EP signals online. Through artificial intelligence theory and big data analysis method, sensitive indicators are set up to carry out intelligent monitoring and forecasting of coal and rock dynamic failure and its induced disasters. The study results are not only applicable to the coal mining face, but also play a monitoring role in the tunneling of coal and rock roadway. Considering the technical advantages and low cost of the EP monitoring, this technical method has broad engineering study prospects.

5. Conclusions

(1) The host device, data analysis software, and monitoring system of the EP monitoring device for mining were independently developed. The EP monitoring host has the function of synchronous, continuous, and real-time monitoring for the EP signals at the different measuring points. Further, the data analysis software can record the EP data of different acquisition channels and different test periods. The coal seam EP monitoring system can monitor the EP signals of multiple measuring points in real time, analyzing the spatial distribution characteristics.

(2) Multiple sets of laboratory experiments were carried out with EP monitoring host. The EP monitoring host has high sensitivity and little error for the EP signals. It shows the reliable test results and can reflect the nonlinear distribution characteristics of the EP signals in space. The experimental results of coal and rock specimen loading failure show that the EP monitoring host can effectively measure the EP signals generated by the deformation and rupture process of the specimens. The response characteristics of EP signals can reveal the stress state and damage degree of a specimen, to forecast its instability and fracture.

(3) The field tests and applications were carried out in coal mining faces, and the relationship between EP response and the hazard of coal and rock dynamic disasters was analyzed. The results show that the EP signals fluctuate greatly during coal mining and remain relatively stable during the maintenance stage. When the mining stress is abnormal, the EP signal intensity and standard deviation show the abnormal response characteristics with types of "↗" and "∧", respectively. Before the occurrence of the coal cannon, the EP signals show precursory information of increased intensity and violent fluctuation and were close to the peak level when the coal cannon occurred. After that, the hazard of disaster decreases, and the EP intensity remains at a relatively stable level.

(4) The EP monitoring method has the technical advantages of sensitive response, strong anti-interference, and approximate non-destructiveness, which can reveal the loading and damaging state of coal mass and indicates the hazard degree of coal and rock dynamic disasters. By identifying the abnormal response characteristics of EP monitoring results, it is expected to provide a new engineering application method for fine forecasting and precise prevention of coal and rock dynamic disasters in coal mining sites.

Author Contributions: Conceptualization, E.W. and Z.L.; methodology, Y.N.; software, Y.N.; formal analysis, Y.N.; investigation, H.W.; resources, E.W. and Z.L.; writing—original draft preparation, Y.N.; writing—reviewing and editing, H.L., J.D., J.W. (Jingkun Wang), F.L. and W.S.; supervision, J.W. (Jiali Wang); project administration, M.W.; funding acquisition, E.W., Z.L. and T.S. All authors have read and agreed to the published version of the manuscript.

Funding: This research is supported by the National Science Foundation for Young Scientists of Jiangsu Province (BK20210504), National Natural Science Foundation of China, China (52104234, 51934007, 52074280), and the Fundamental Research Funds for the Central Universities (2021QN1104).

Conflicts of Interest: The authors declare that they have no known competing financial interests or personal relationships that could have appeared to influence the work reported in this paper.

References

1. Zhou, H.; Bhattarai, R.; Li, Y.; Si, B.; Dong, X.; Wang, T.; Yao, Z. Towards sustainable coal industry: Turning coal bottom ash into wealth. *Sci. Total Environ.* **2021**, *804*, 149985. [CrossRef] [PubMed]
2. Batugin, A.; Kolikov, K.; Ivannikov, A.; Ignatov, Y.; Krasnoshtanov, D. Transformation of the geodynamic hazard manifestation forms in mining areas. *Int. Multidiscip. Sci. GeoConf. SGEM* **2019**, *19*, 717–724.
3. Khan, M.; He, X.; Farid, A.; Song, D.; Li, Z.; Tian, X.; Ni, M. A novel geophysical method for fractures mapping and risk zones identification in a coalmine, Northeast, China. *Energy Rep.* **2021**, *7*, 3785–3804. [CrossRef]
4. Wolf, K.-H.A.; van Bergen, F.; Ephraim, R.; Pagnier, H. Determination of the cleat angle distribution of the RECOPOL coal seams, using CT-scans and image analysis on drilling cuttings and coal blocks. *Int. J. Coal Geol.* **2008**, *73*, 259–272. [CrossRef]
5. Lu, J.; Zhang, D.; Huang, G.; Li, X.; Gao, H.; Yin, G. Effects of loading rate on the compound dynamic disaster in deep underground coal mine under true triaxial stress. *Int. J. Rock Mech. Min. Sci.* **2020**, *134*, 104453. [CrossRef]
6. Wang, C.; He, B.; Hou, X.; Li, J.; Liu, L. Stress–Energy Mechanism for Rock Failure Evolution Based on Damage Mechanics in Hard Rock. *Rock Mech. Rock Eng.* **2019**, *53*, 1021–1037. [CrossRef]
7. Kuchibhatla, S.V.; Karakoti, A.; Bera, D.; Seal, S. One dimensional nanostructured materials. *Prog. Mater. Sci.* **2007**, *52*, 699–913. [CrossRef]
8. Bordas, S.; Rabczuk, T.; Zi, G. Three-dimensional crack initiation, propagation, branching and junction in non-linear materials by an extended meshfree method without asymptotic enrichment. *Eng. Fract. Mech.* **2008**, *75*, 943–960. [CrossRef]
9. Li, Z.; Zhang, X.; Wei, Y.; Ali, M. Experimental Study of Electric Potential Response Characteristics of Different Lithological Samples Subject to Uniaxial Loading. *Rock Mech. Rock Eng.* **2020**, *54*, 397–408. [CrossRef]
10. Niu, Y.; Wang, E.; Li, Z.; Gao, F.; Zhang, Z.; Li, B.; Zhang, X. Identification of Coal and Gas Outburst-Hazardous Zones by Electric Potential Inversion During Mining Process in Deep Coal Seam. *Rock Mech. Rock Eng.* **2022**, *55*, 3439–3450. [CrossRef]
11. Li, Z.; Niu, Y.; Wang, E.; Liu, L.; Wang, H.; Wang, M.; Ali, M. Experimental Study on Electric Potential Response Characteristics of Gas-Bearing Coal During Deformation and Fracturing Process. *Processes* **2019**, *7*, 72. [CrossRef]
12. Eccles, D.; Sammonds, P.; Clint, O. Laboratory studies of electrical potential during rock failure. *Int. J. Rock Mech. Min. Sci.* **2005**, *42*, 933–949. [CrossRef]
13. Wang, E.; Li, Z.; Liu, Z.; Li, Y.; Song, X. Experimental study on surface potential effect of coal under load. *Chin. J. Geophys.* **2009**, *52*, 1318–1325. [CrossRef]
14. Niu, Y.; Wang, C.; Wang, E.; Li, Z. Experimental Study on the Damage Evolution of Gas-Bearing Coal and Its Electric Potential Response. *Rock Mech. Rock Eng.* **2019**, *52*, 4589–4604. [CrossRef]
15. Takeuchi, A.; Lau, B.W.S.; Freund, F.T. Current and surface potential induced by stress-activated positive holes in igneous rocks. *Phys. Chem. Earth* **2006**, *31*, 240–247. [CrossRef]
16. Revil, A.; Ehouarne, L.; Thyreault, E. Tomography of self-potential anomalies of electrochemical nature. *Geophys. Res. Lett.* **2001**, *28*, 4363–4366. [CrossRef]
17. Niu, Y.; Song, X.; Li, Z.; Wang, E.; Zhang, X.; Wang, M.; Dong, M. Study on EP response to compression-induced damage of coal-measure rock and the mechanism behind such response. *J. Geophys. Eng.* **2019**, *16*, 242–252. [CrossRef]
18. Li, D.; Wang, E.; Ju, Y.; Wang, D. Laboratory Investigations of a New Method Using Pressure Stimulated Currents to Monitor Concentrated Stress Variations in Coal. *Nonrenew. Resour.* **2020**, *30*, 707–724. [CrossRef]
19. Revil, A. On charge accumulation in heterogeneous porous rocks under the influence of an external electric field. *GEOPHYSICS* **2013**, *78*, D271–D291. [CrossRef]
20. Roubinet, D.; Irving, J. Discrete-dual-porosity model for electric current flow in fractured rock. *J. Geophys. Res. Solid Earth* **2014**, *119*, 767–786. [CrossRef]
21. Czabanski, R.; Horoba, K.; Wrobel, J.; Matonia, A.; Martinek, R.; Kupka, T.; Jezewski, M.; Kahankova, R.; Jezewski, J.; Leski, J. Detection of Atrial Fibrillation Episodes in Long-Term Heart Rhythm Signals Using a Support Vector Machine. *Sensors* **2020**, *20*, 765. [CrossRef] [PubMed]
22. Zhang, X.; Li, Z.; Wang, E.; Li, B.; Song, J.; Niu, Y. Experimental investigation of pressure stimulated currents and acoustic emis-sions from sandstone and gabbro samples subjected to multi-stage uniaxial loading. *Bull. Eng. Geol. Environ.* **2021**, *80*, 7683–7700. [CrossRef]
23. Djozan, D.; Farajzadeh, M.A.; Sorouraddin, S.M.; Baheri, T. Molecularly imprinted-solid phase extraction combined with sim-ultaneous derivatization and dispersive liquid–liquid microextraction for selective extraction and preconcentration of meth-amphetamine and ecstasy from urine samples followed by gas chromatography. *J. Chromatogr. A* **2012**, *1248*, 24–31.
24. Lei, R.; Shi, Y.; Ding, Y.; Nie, J.; Li, S.; Wang, F.; Zhai, H.; Chen, X.; Wang, Z.L. Sustainable high-voltage source based on triboelectric nanogenerator with a charge accumulation strategy. *Energy Environ. Sci.* **2020**, *13*, 2178–2190. [CrossRef]
25. Li, D.; Wang, E.; Kong, X.; Zhao, S.; Kong, Y.; Wang, X.; Wang, D.; Liu, Q. Mechanical properties and electromagnetic radiation characteristics of concrete specimens after exposed to elevated temperatures. *Constr. Build. Mater.* **2018**, *188*, 381–390.
26. Wang, X.; Wang, E.; Liu, X. Damage characterization of concrete under multi-step loading by integrated ultrasonic and acoustic emission techniques. *Constr. Build. Mater.* **2019**, *221*, 678–690. [CrossRef]
27. Shen, R.; Chen, T.; Li, T.; Li, H.; Fan, W.; Hou, Z.; Zhang, X. Study on the effect of the lower limit of cyclic stress on the mechanical properties and acoustic emission of sandstone under cyclic loading and unloading. *Theor. Appl. Fract. Mech.* **2020**, *108*, 102661. [CrossRef]

28. Wold, M.; Connell, L.; Choi, S. The role of spatial variability in coal seam parameters on gas outburst behaviour during coal mining. *Int. J. Coal Geol.* **2008**, *75*, 1 14. [CrossRef]
29. Wang, K.; Du, F. Coal-gas compound dynamic disasters in China: A review. *Process Saf. Environ. Prot.* **2020**, *133*, 1–17. [CrossRef]
30. Niu, Y.; Li, Z.; Wang, E.; Shen, R.; Cheng, Z.; Gao, X.; Zhang, X.; Wang, H.; Ali, M. Study on characteristics of EP responsing to coal mining. *Eng. Fract. Mech.* **2020**, *224*, 106780. [CrossRef]
31. Waclawik, P.; Ptacek, J.; Grygar, R. Structural and stress analysis of mining practice in the Upper Silesian Coal Basin. *Acta Geodyn. Geomater.* **2013**, *10*, 170. [CrossRef]
32. Xue, Y.; Liu, J.; Ranjith, P.; Liang, X.; Wang, S. Investigation of the influence of gas fracturing on fracturing characteristics of coal mass and gas extraction efficiency based on a multi-physical field model. *J. Pet. Sci. Eng.* **2021**, *206*, 109018. [CrossRef]
33. Yu, L.; Ignatov, Y.; Ivannikov, A.; Khotchenkov, E.; Krasnoshtanov, D. *Common Features in the Manifestation of Natural and Induced Geodynamic Events in the Eastern Regions of Russia and China*; IOP Publishing: Bristol, UK, 2019; p. 012004.
34. Hu, S.; Wang, E.; Liu, X. Spatiotemporal multifractal characteristics of electromagnetic radiation in response to deep coal rock bursts. *Nat. Hazards Earth Syst. Sci.* **2014**, *14*, 2089–2103. [CrossRef]
35. Batugin, A.; Wang, Z.; Su, Z.; Sidikovna, S.S. Combined support mechanism of rock bolts and anchor cables for adjacent roadways in the external staggered split-level panel layout. *Int. J. Coal Sci. Technol.* **2021**, *8*, 659–673. [CrossRef]

minerals

MDPI

Article

Integrated Collaborative Control of "Shielding-Filling-Grouting" of 1 km Deep Large-Section Roadways: A Case Study

Wei Zhang [1,2], Yandong Zhang [2,*], Yanchao Zhu [2], Jiajia Tang [3], Longtao Cheng [4] and Zhiliang Suo [2]

1 State Key Laboratory of Coal Resources and Safe Mining, China University of Mining and Technology, Xuzhou 221116, China; zhangwei@cumt.edu.cn
2 School of Mines, China University of Mining and Technology, Xuzhou 221116, China; yanchaozhu@cumt.edu.cn (Y.Z.); suozhiliang@cumt.edu.cn (Z.S.)
3 Guqiao Coal Mine, Huainan Mining Industry (Group) Co., Ltd., Huainan 232150, China; jiajiatang@cumt.edu.cn
4 Kouzidong Coal Mine, China Coal Xinji Energy Co., Ltd., Fuyang 236008, China; chenglongtao1983@163.com
* Correspondence: zhangyandong@cumt.edu.cn

Abstract: Effective control of deformation failure of surrounding rock in deep roadway has become an important prerequisite for the safe and efficient development of deep coal resources. In this study, the field measurement of the study area's in-situ stress was carried out for the specific engineering geological conditions of the KCM −967 m level west-wing main track roadway. The west-wing main track roadway's full-section deformation failure features were summarized and analyzed, and the main roadway's surrounding rock nonlinear deformation failure mechanism was revealed from the perspective of elastoplastic mechanics. Based on that, a set of highly targeted integrated collaborative control technology of "shielding-filling-grouting" system was proposed. The industrial field test revealed that, after the above integrated collaborative control scheme was adopted, there was no strong deformation failure on the surface of the main roadway surrounding rock and deep rock mass, which played the role of active and passive support collaborative control, reduced the subsequent repeated repair and maintenance workload of the roadway, and satisfied the needs of long-term safe and efficient production of the mine. The results obtained provide a reference for the control of surrounding rock of deep and large-section roadways in other mining areas.

Keywords: 1 km-deep mine; large-section roadway; nonlinear deformation failure; in-situ stress measurement; integrated collaborative control

Citation: Zhang, W.; Zhang, Y.; Zhu, Y.; Tang, J.; Cheng, L.; Suo, Z. Integrated Collaborative Control of "Shielding-Filling-Grouting" of 1 km Deep Large-Section Roadways: A Case Study. *Minerals* **2022**, *12*, 854. https://doi.org/10.3390/min12070854

Academic Editor: Abbas Taheri

Received: 8 June 2022
Accepted: 30 June 2022
Published: 4 July 2022

1. Introduction

A gradual exhaustion of shallow coal resources worldwide resulted in the growing number of kilometer-deep coal mines [1], including over sixty such mines in China [2]. Noteworthy is that the depth of newly excavated roadways increases at a rate of 8~12 m per year [3]. The typical "high in-situ stress, high geothermal, high karst water pressure, and strong mining disturbance" occurrence environment of deep coal rock leads to a series of problems, such as frequent roof fall accidents of deep roadway, significant deformation failure of roadway walls, common roadway floor heave, and severe damage of roadway support components [4–6]. One of the conditions for ensuring the safety and continuity of production is to maintain the cross-section of the excavation, which is closely related to the deformation conditions of surrounding rocks [7]. Therefore, effective control of deformation failure of surrounding rocks in deep roadways has become an important prerequisite for safe and efficient development of deep coal resources.

Compared to medium- and shallow-depth roadways, in-situ stresses of deep-depth ones continuously grow, and the soft rock characteristics of coal rock become increasingly

significant under the action of high stresses [8,9]. For instance, the maximum principal stress of some kilometer-deep mines can reach more than 40 MPa, significantly exceeding most of the coal rock uniaxial compressive strength in coal measure strata, resulting in serious deformation failure of roadway surrounding rock [10–12]. In addition to the high stress effect, the deep roadway surrounding rock stress field will be further redistributed under the action of geological structure and mining-induced stress, resulting in discrete stress degradation and stress concentration. Currently, scholars have studied the ultra-high strength bolt, the grouting modified material, and its mechanical performance applied to the control of surrounding rocks in deep roadways [13–17]. The support-modification-pressure relief collaborative control method of deep roadways and the modification technology of high-pressure splitting grouting of deep roadways were put forward from the technical point of view [18,19]. The deformation failure mechanism of roadways with high stresses and large mining depth was analyzed [20]. The above studies enriched the research results of deformation control of surrounding rock in deep roadways. However, due to the complex geological conditions of deep roadways, the technology and method of deformation control of surrounding rocks in deep roadways still need to be explored.

The Kouzidong Coal Mine (KCM) in the city of Fuyang, China, is a typical 1 km deep mine, which is located in the west of the Huainan Coalfield, with the mine's first production level elevation of −967 m. As a key point, the roadway in the KCM, namely, the west-wing main track roadway, is responsible for important production auxiliary tasks of the whole mine. Nevertheless, the main roadway of the west-wing track has a large depth, the stress environment of surrounding rock in the roadway is complicated, surrounding rock exhibits poor physical and mechanical properties, and the roadway shows typical features of large deformation, strong rheology, and sudden change after excavation. As a result, some areas were exposed to severe deformation failures, thus affecting operation of the roadway and mining safety. In this study, the KCM west-wing main track roadway was taken as the engineering background. Field measurements of the in-situ stress in the study area were carried out, and the difficulties in control of the surrounding rock in the 1 km deep roadway were analyzed. Based on the original support scheme of the west-wing main track roadway, the main roadway's surrounding rock full-section deformation failure features and mechanical mechanism were revealed. Given this, a 1 km deep, large-section roadway "shielding-filling-grouting" integrated collaborative control technology system was put forward. The industrial field tests were carried out. The proposed technology system enhanced the effect of control of the surrounding rock in the roadway, and satisfied the practical needs of safe and efficient mining production. The relevant research results can provide a reference for other similar deformation control of the surrounding rock in deep roadways.

2. In-Situ Stress Measurement of West-Wing Main Track Roadway and Difficulty in the Surrounding Rock Control in Roadways

2.1. Engineering Background of the Main Roadway in the West-Wing Track

2.1.1. Stratigraphic Characteristics of the Study Area

The main roadway of the west-wing track of the KCM had a depth of −950 m, being adjacent to the main roadway of the west-side belt conveyor haulage in the north, the tail joint roadway of the west-wing main track roadway in the west, and the Fufeng fault in the south. Fine sandstone, quartz sandstone, siltstone, mottled mudstone, mudstone, coal, carbonaceous mudstone, and other coal rock strata were exposed during roadway excavation. The roadway of the study area was excavated in the roof of the #8 coal seam and the roadway surrounding is mainly composed of mudstone and fine sandstone. The comprehensive lithology column of the study area stratum is shown in Figure 1.

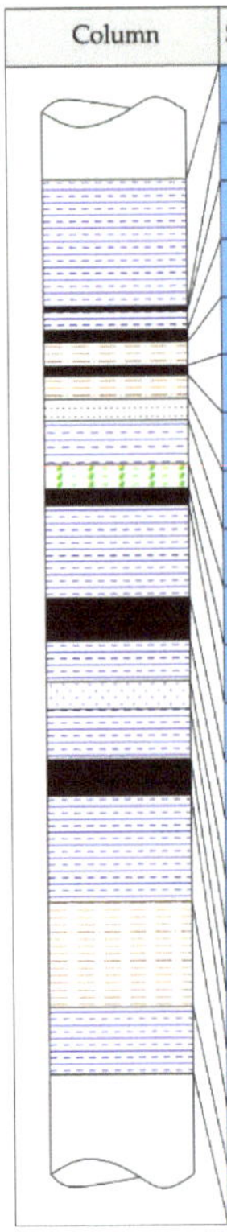

Succession	Thickness /m	Gross thickness/m	Lithology
1	6.39	6.39	Mudstone
2	0.54	6.93	Coal seam
3	0.94	7.87	Mudstone
4	0.64	8.51	Coal seam
5	1.19	9.70	Sandy mudstone
6	0.50	10.20	Coal seam
7	1.19	11.39	Sandy mudstone
8	1.09	12.48	Siltstone
9	2.18	14.66	Mudstone
10	1.24	15.90	Carbonaceous mudstone
11	0.74	16.64	Coal seam
12	4.70	21.34	Mudstone
13	2.13	23.47	#9 Coal seam
14	2.08	25.55	Mudstone
15	1.39	26.94	Fine sandstone
16	2.08	29.02	Mudstone
17	1.88	30.90	#8 Coal seam
18	5.15	36.05	Mudstone
19	5.27	41.32	Sandy mudstone
20	3.38	44.70	Mudstone

Figure 1. Strata comprehensive column of the study area.

2.1.2. Original Support Scheme of the Main Roadway

The section of the main roadway of the west-wing track is a semi-circular arch with an inclined wall, and the combined support of "bolt mesh cable shotcrete" (see Figure 2) is adopted. The main technical parameters are as follows:

(1) Roof support: Eleven bolts and seven anchor cables are symmetrically arranged in each row of the main roadway roof; the row spacing between bolts is 800 × 800 mm, the size is Φ22 × L2500 mm, the material is high-strength rebar; the row spacing between anchor cables is 1200 × 1600 mm, the size is Φ21.8 × L9200 mm, the material is steel strand; the steel strip is H-shaped steel strip welded with Φ12 mm round steel; the anchor mesh is made by reinforcement mesh, with specification Φ6 × 1000 × 2000 mm; the strength grade of shotcrete is C20, and the initial shotcrete thickness is 30~50 mm.

(2) Wall support: The two walls of the main roadway are symmetrically arranged with six bolts in each row, while other support parameters are the same as the roof support.

(3) Floor support: No reinforcement measures are taken for the floor of the main roadway.

2.2. In-Situ Stress Field Measurement and Difficulties in the Roadway Surrounding Rock Control

In-situ stress is one of the root causes of rock mass deformation failure in underground engineering. In-situ stress measurement is an effective method to determine and evaluate the state of underground coal rock stress, providing basic parameters and theoretical basis for scientific design and decision-making of underground engineering [21–24]. Therefore, to handle the stress field distribution of the KCM −967m level, the field measurements of in-situ stresses in the area near the −967 m level west-wing main track roadway (G1, G2, and G3 stations, respectively) [25] were conducted considering the specific construction condi-

tions of the underground site comprehensively. The specific measurement method was the hollow inclusion stress relief method, and the measuring instrument was a KX-81 hollow inclusion triaxial in-situ stress meter developed by the Institute of Geomechanics, Chinese Academy of Geological Sciences. Table 1 lists in-situ stress specific measurement results, while Figure 3 shows azimuths of horizontal principal stress at different stations. Herein, σ_1, σ_2, and σ_3 are the maximal, medium, and minimum principal stresses, respectively; α is the dip angle; σ_{Hmax} and σ_{Hmin} are the maximum and minimum horizontal principal stresses, respectively; σ_V is the vertical principal stress; β and γ are the maximum and minimum horizontal principal stress azimuths, respectively; k_1 is the ratio of the maximum horizontal principal stress to the vertical principal stress; k_2 is the maximum-to-minimum horizontal principal stress ratio.

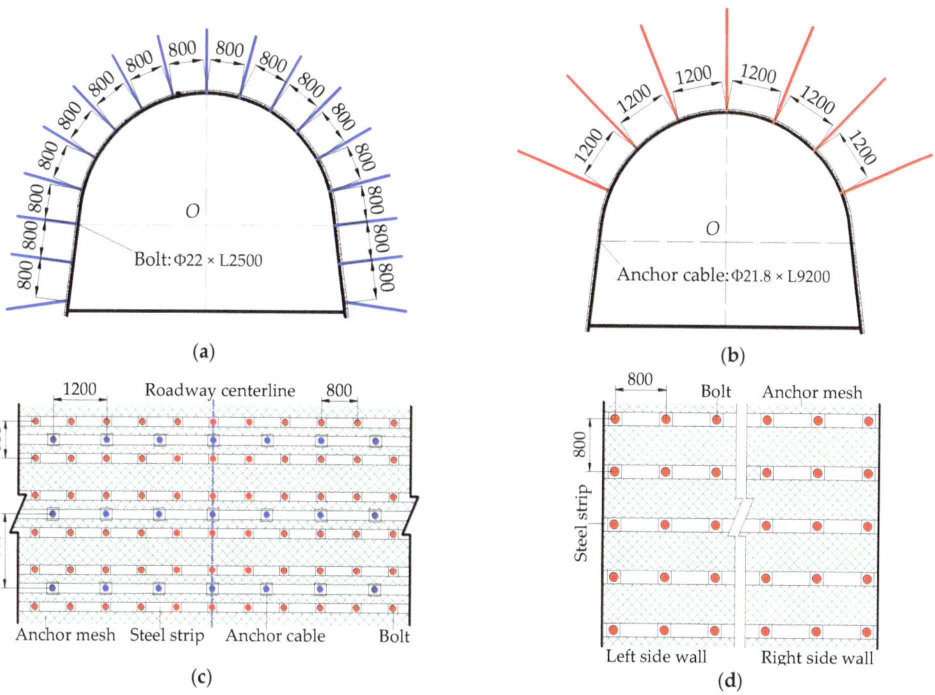

Figure 2. West-wing main track roadway original support scheme (Unit: mm): (**a**) bolt support section; (**b**) Anchor cable support section; (**c**) Roadway roof support; (**d**) Roadway wall support.

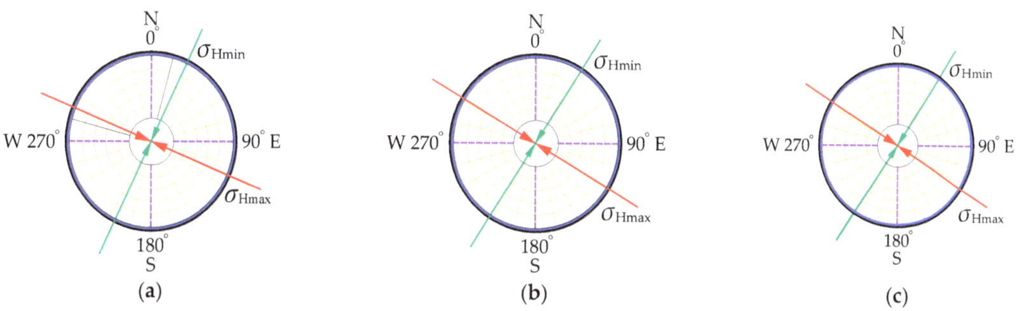

Figure 3. Azimuths of horizontal principal stresses at different stations: (**a**) G1 station; (**b**) G2 station; (**c**) G3 station.

Table 1. In-situ stress measurement results obtained at different stations.

Station	Depth (m)	σ/MPa	α	σ_{Hmax}/MPa	σ_{Hmin}/MPa	σ_V/MPa	β	γ	k_1	k_2
G1	960.50	σ_1 = 30.15 σ_2 = 27.64 σ_3 = 21.01	9.01° −69.54° −22.32°	30.15	21.05	27.60	N 112.54° E	N 25.43° E	1.09	1.43
G2	958.50	σ_1 = 32.81 σ_2 = 20.58 σ_3 = 14.01	3.08° −57.31° 39.01°	31.97	14.04	21.40	N 120.71° E	N 32.57° E	1.49	2.28
G3	958.30	σ_1 = 36.10 σ_2 = 24.32 σ_3 = 16.78	1.48° −69.47° 18.54°	37.27	20.63	24.30	N 122.29° E	N 33.28° E	1.53	1.81

According to the results of on-site in-situ stress measurement, the KCM −967 m level stress field distribution had the following features:

(1) Maximum and minimum depths of the three stations for monitoring of in-situ stress were 960.50 and 958.30 m, respectively, belonging to a typical 1 km deep environment. This implied that the coal rock was in high stress state. The maximum principal stresses of all three stations exceeded 30 MPa (up to 36.10 MPa) and their dip angles of maximum principal stress were below 10°. Additionally, dip angles of medium and minimum principal stresses varied widely, suggesting that the primary rock stress field of the study area was mainly structural stress in the approximately horizontal direction.

(2) After the coordinate system transformation, the maximum horizontal principal stresses in all three stations exceeded 30 MPa (30.15~37.27 MPa), with the azimuth angles ranging between N 112.54° E~N 122.29° E; the minimum horizontal principal stresses ranged from 14.04 to 21.05 MPa, with the azimuth angles between N 25.43° E~N 33.28° E; the vertical principal stresses ranged from 21.40 to 27.60 MPa.

(3) The maximum horizontal principal stresses of all three stations were larger than their corresponding vertical principal stresses. The respective ratio k_1 ranged from 1.09 to 1.53, demonstrating that the paleotectonic interaction in the study area was relatively strong; the measured maximum-to-minimum horizontal principal stress ratio k_2 ranged from 1.43 to 2.28, demonstrating that the in-situ stress in this area had obvious orientation.

In summary, the in-situ stress born by the KCM −967 m level roadway surrounding rock showed typical horizontal structural stress features. In particular, the maximum horizontal principal stress was as high as 37.27 MPa, which significantly exceeded the uniaxial compressive strength of the coal rock in this area, greatly reducing the self-bearing capacity of roadway surrounding rock and seriously affecting the original support effect during the roadway excavation. Besides the dynamic pressure effect, the high-stress roadway surrounding rock nonlinear deformation failure would be more severe, promoting the original support failure. Therefore, it was expedient to carry out the effective integrated reinforcement control of roadway surrounding rock on the basis of original support, in order to improve the guaranteed capacity of safe and efficient production of the mine.

3. Analysis of the West-Wing Main Track Roadway Surrounding Rock Deformation Failure Pattern and Mechanism

3.1. Main Roadway Surrounding Rock Full-Section Deformation Failure Pattern

When the original support scheme was adopted for the −967 m level west-wing main track roadway, the roadway surrounding rock presented the following full-section nonlinear deformation failure features, as shown in Figure 4:

(1) The roadway roof surrounding rock was cracked and separated, and the damage zone extended continuously along the central axis of the main roadway, accompanied by signs of bending and sinking. The shallow bolt (anchor cable) of roof had obvious bending dislocation. Therefore, deformation of shallow surrounding rock was greater than that of deep surrounding rock.

(2) The two walls of the roadway were damaged, and the surrounding rock was obviously broken unevenly and bulged into the roadway, especially in the corner. Deformation with dominating shear expansion failures also occurred in the side profile; the shotcrete layer of the inclined wall section in some areas peeled off obviously. The metal mesh was exposed, with bending and breaking tendency, while the bolt (anchor cable) at the side profile was also deformed.

(3) The roadway floor deformation lasted for a long time and had poor structural integrity. The breakthrough gap for strain energy could be easily filled, increasing the loose range of the floor-surrounding rock, with a serious floor heave in some sections.

(a) (b) (c)

Figure 4. Deformation failure of the surrounding rock in the main roadway. (**a**) the roof; (**b**) the wall; (**c**) the floor.

3.2. Mechanism of Nonlinear Deformation Failure of Surrounding Rock in Main Roadway

Excavation of the main roadway would inevitably violate the primary stress equilibrium and stable state of the rock mass. Therefore, the internal stress state of the rock surrounding the roof and the two walls would be redistributed. The gravity-induced stresses would transfer to the deep rock strata of the two walls, and the horizontal stresses would transfer to the deep rock strata of the roof. After exceeding the stress limits of the rocks surrounding the roof and the two walls, the integrity of the main roadway surrounding rock would be violated, and irregular cracks would appear. Meanwhile, the main roadway surrounding rock's mechanical performance gradually transformed from shallow brittleness to deep plasticity, with strong non-linear, non-smooth plastic deformation and rapid convergence of deformation, and the bearing capacity of surrounding rock is greatly reduced [26,27]. The −967 m level west-wing main track roadway had large depth, high in-situ stress, and poor physical and mechanical properties. The structural distribution in such roadway surrounding rock comprised elastic, plastic, and damage zones [28,29], as shown in Figure 5.

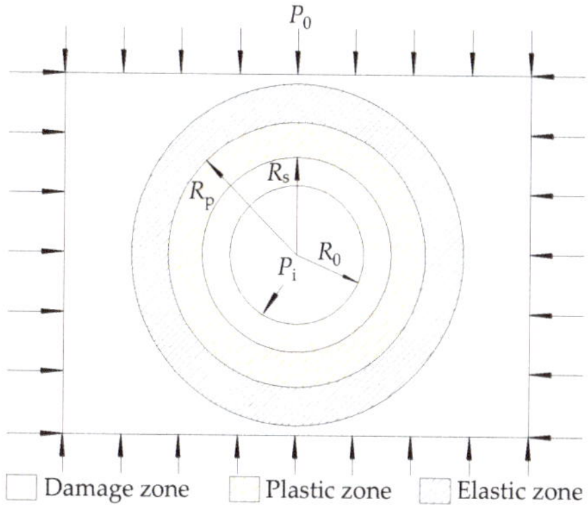

Figure 5. Zones in the main roadway surrounding rock.

The inelastic (damage and plastic) zones can be described as in [30,31]:

$$R_p = \frac{R_0}{t^*}\left(\frac{\frac{2}{\varepsilon^*+1}\left(P_0 + \frac{\sigma_c+\beta B_0}{\varepsilon^*-1}\right)t^{*\varepsilon^*-1} - \frac{2(\sigma_c-\sigma_c{}^*+\beta B_0)}{\varepsilon^{*2}-1}}{P_i + \frac{\sigma_c{}^*}{\varepsilon^*-1}}\right)^{\frac{1}{\varepsilon^*-1}} \quad (1)$$

$$R_s = t^* R_p \quad (2)$$

where $t^* = [\beta B_0/(\sigma_c - \sigma_c{}^* + \beta B_0)]^{1/2}$; $B_0 = (1+\mu)[(\varepsilon^*-1)P_0 + \sigma_c]/(\varepsilon^*+1)$; $\varepsilon^* = (1+\sin\varphi)/(1-\sin\varphi)$, R_p is the plastic zone radius, m; R_s is the damage zone radius, m; R_0 is the roadway's theoretical radius, m; P_0 is the primary rock stress, MPa; σ_c is the rock uniaxial compressive strength, MPa; β is brittleness coefficient; $\sigma_c{}^*$ is the uniaxial compression residual strength of the rock, MPa; φ is the internal friction angle, °; P_i is the support stress, MPa; μ is Poisson's ratio.

According to Equations (1) and (2), the main roadway surrounding rock plastic zone radius R_p and the damage zone radius R_s are positively correlated with the theoretical radius of the roadway R_0 and the primary rock stress P_0, being negatively correlated with the support stress P_i. Thus, the structural distribution of the main roadway surrounding rock can be related to the original physical and mechanical properties of the rock mass, stress environment, and support mode.

4. The "Shielding-Filling-Grouting" Integrated Collaborative Control Technology System for the West-Wing Main Track Roadway

According to the elastoplastic analysis results of Section 3.2, the main causes of the full-section nonlinear deformation failure of the west-wing main track roadway surrounding rock were as follows: (i) low strength of the main roadway surrounding rock, (ii) the complex horizontal structural stress environment, (iii) the insufficient original "bolt mesh cable shotcrete" support scheme strength, and (iv) the lack of effective control measures of the main roadway floor. Given this, an integrated collaborative control technology system of "shielding-filling-grouting" was proposed, including the U-shaped steel secondary shed passive support technology, the backwall rapid-setting material filling and control technology, and grouting reinforcement with combined deep and shallow holes (GR-CDSH) support technology.

4.1. U-Shaped Steel Secondary Shed Passive Support Technology

The U-shaped steel shed has the advantages of strong bearing capacity, good stability, and high compressibility, which can improve the shallow stress environment of the roadway surrounding rock and improve the crushing surrounding rock's residual strength by passively restraining the rock mass deformation [32,33]. The combination of passive and active supports can give full play to the bearing performance of the two types of support and greatly improve the overall supporting strength of the surrounding rock structure. Based on the combined support of "bolt mesh cable shotcrete", the U-shaped steel secondary shed passive support (see Figure 6) was implemented for the west-wing main track roadway. The basic mechanical performance of the U-shaped steel beam is summarized in Table 2.

Table 2. Basic mechanical performance parameters of the U-shaped steel shed.

Steel Grade	Specification	Yield Stress (MPa)	Tensile Trength (MPa)	Elongation (%)
20 MnK	36 U	400	550	18

The parameters of passive support technology for U-shaped steel secondary shed were as following:

A three-section shed manufactured from 36 U-shaped steel beams was adopted for the secondary shed passive support, the clear width of arch foundation was 5500 mm, the pure

height of arch foundation was 4530 mm, the pure width of the bottom corner was 5940 mm, and the shed middle-to-middle spacing was 650 mm; the overlapping length of shed beam and shed leg was 700 mm; the clamp was made of the 36 U-shaped steel beams, the clamp bolt specification was Φ30 × L160 mm, with the torque of no less than 300 N·m. The filling thickness reserved in the main roadway section was 250~350 mm.

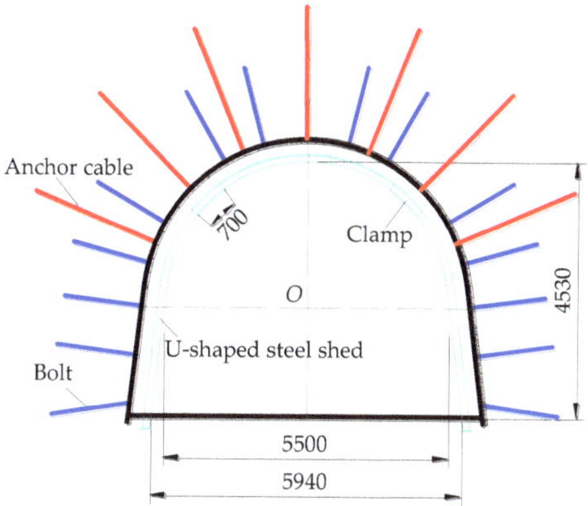

Figure 6. Secondary shed passive support scheme (Unit: mm).

4.2. Backwall Rapid-Setting Material Filling and Control Technology

Backwall filling can compact the gap between the main roadway surrounding rock and the U-shaped steel shed, generating a common mechanical bearing system between the surrounding rock, filling body, and support structure, in order to improve the overall coupling effect [34]. Based on the above combined support of "bolt mesh cable shotcrete" and secondary shed passive support, the automatic filling system produced by the Ruhr Group, Germany, was used to fill the setting material behind the wall of the west-wing main track roadway (see Figure 7), so as to further control the main roadway surrounding rock deformation. The filling material was prepared from cement, slag, and yellow sand. Table 3 shows the relationship between the curing time and strength.

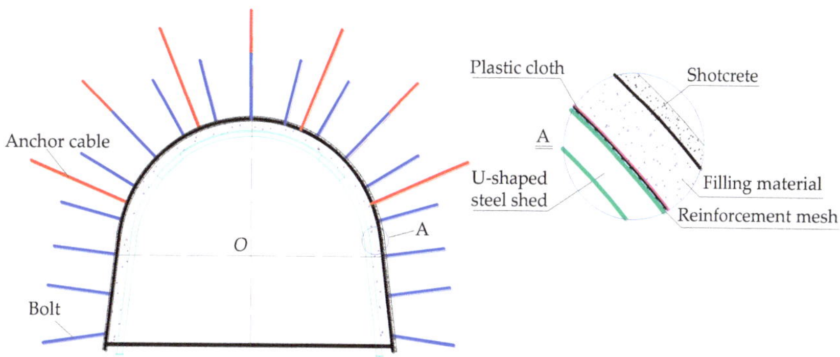

Figure 7. Backwall filling compaction control scheme.

Table 3. Correlation of filling material's curing time with the sample strength.

Time (h)	**5**	**12**	**24**	**48**
Compressive strength (MPa)	5	10	15	20

The construction process of filling and compacting quick-setting materials behind the wall was as follows:

(1) First, lay the reinforcement mesh behind the U-shaped steel shed, and then lay the plastic cloth from the shed top to both sides;
(2) At the starting and end positions of filling, stack woven bags filled with sand to form isolation belts, and arrange a filling window on the top of the U-shaped steel shed and two shoulder sockets, respectively;
(3) The equipment is used to mix the mixture with water to form concrete; then, it is transferred to the filling place for backwall filling by concrete pump;
(4) The head of the filling pipe is fixed by ladder. First, fill the side profile of the main roadway (the two walls are filled alternately), and then fill the top of the main roadway;
(5) When the material reaches the filling window position, the filling is completed and stopped. Use woven bags or plastic cloth to block the shoulder socket window tightly, and then use 10# iron wire to fix it, and then fill the top window;
(6) The hose is adopted for concrete delivery pump with sufficient compressive capacity (the pressure resistance shall be no less than 10 MPa), and its length can be subdivided into 10, 15, and 20 m long sections. The maximum delivery distance shall not exceed 150 m.

4.3. Grouting Reinforcing Technology with Combined Deep and Shallow Holes (GR-CDSH)

Grouting reinforcement is a strengthening method of injecting slurry into the crushing surrounding rock to induce physical and chemical interaction, which can effectively improve the bearing capacity of the surrounding rock and control the plastic deformation. According to the specific geological conditions of the west-wing main track roadway, based on the above combined support of "bolt mesh cable shotcrete", secondary shed passive support, and the backwall rapid-setting material filling and control technology, it was proposed to use the grouting pipe to implement the grouting reinforcement with combined deep and shallow holes (GR-CDSH) for the main roadway surrounding rock, as shown in Figure 8. It was envisaged to further improve the overall stability of the main roadway surrounding rock. The technical parameters of GR-CDSH were as follows:

(1) GR parameters of roof: Each row of main roadway roof was symmetrically arranged with three deep grouting holes with a depth of 8000 mm, and the row spacing was 2600 × 9600 mm, grouting pressure was 4.5 MPa; three shallow grouting holes with a depth of 3000 mm were symmetrically arranged with a row spacing of 1600 × 4800 mm, and the grouting pressure was 2.0 MPa.
(2) GR parameters of the two walls: Each row of the main roadway's two walls was symmetrically arranged with two deep grouting holes with a depth of 8000 mm, and the grouting pressure was 3.5 MPa; four shallow grouting holes with a depth of 3000 mm were arranged symmetrically. Other reinforcement parameters were the same as roof support.
(3) GR parameters of floor: As an important transportation, ventilation, and pedestrian passage of the KCM, the main roadway floor of the west-wing main track roadway was required to lay tracks to ensure the safety of material transportation. Therefore, the stability of the main roadway floor should be strictly controlled to avoid large floor heave deformation. Three shallow grouting holes with a depth of 3000 mm were symmetrically arranged in the main roadway floor, and the spacing and grouting pressure of these were the same as that of the GR parameters of roof.

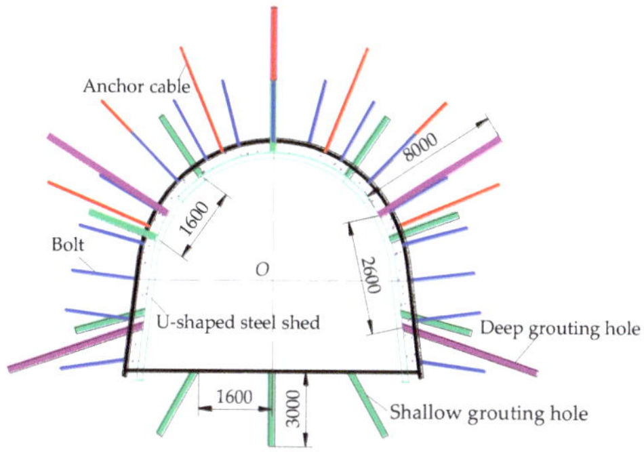

Figure 8. The GR-CDSH scheme (Unit: mm).

5. Effect Analysis of the Industrial Field Test

5.1. Field Observation Scheme Design

5.1.1. Borehole Peeping of Deep Failure Characteristic of the Surrounding Rock in the Main Roadway

Three stations (see Figure 9) were arranged at an equal distance of 20 m in the main roadway test section, and a YTJ20 borehole peep instrument was used to examine the failure of deep surrounding rock in the main roadway. There were seven Φ32 mm peep holes in each station section, including three holes in the roof, two in the two walls, and two in the floor. The roof hole depth was 8000 mm, while depths of the side profile and bottom holes were 5000 mm.

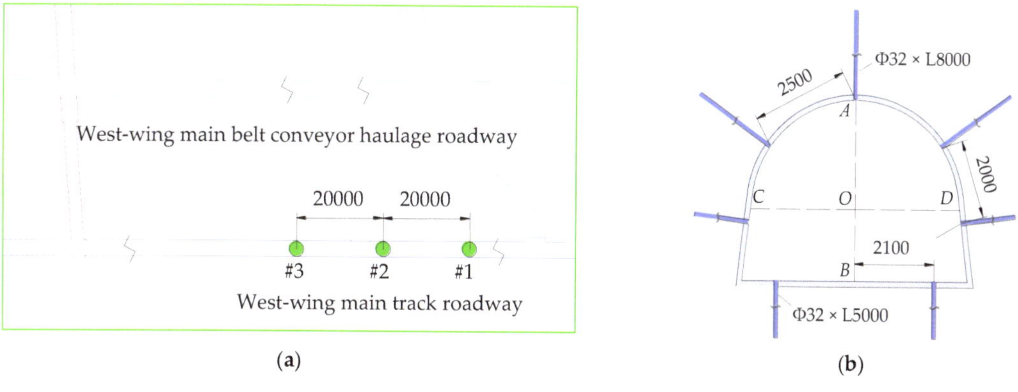

Figure 9. Observation scheme of deep failure of the surrounding rock (Unit: mm): (**a**) station arrangement positions; (**b**) peep borehole parameters.

5.1.2. Field Observation of Surface Displacement of the Surrounding Rock in the Main Roadway

The number and location of stations for monitoring of surface displacement of surrounding rock in the main roadway were the same as those of borehole peeping stations. The measuring base points were arranged in the center of the main roadway's roof, floor, and the two walls. The "cross-measurement method" was used for continuous monitoring for three months, and the surrounding rock deformation of the main roadway was recorded, as shown in Figure 9b. On the basis of the displacements of the excavation contour in the surrounding rock, it is possible to estimate the deformation of the future excavation performed in similar geological and mining conditions [35].

5.2. Field Test

Figure 10 shows the west-wing main track roadway surrounding rock deep's failure characteristic borehole peeping results.

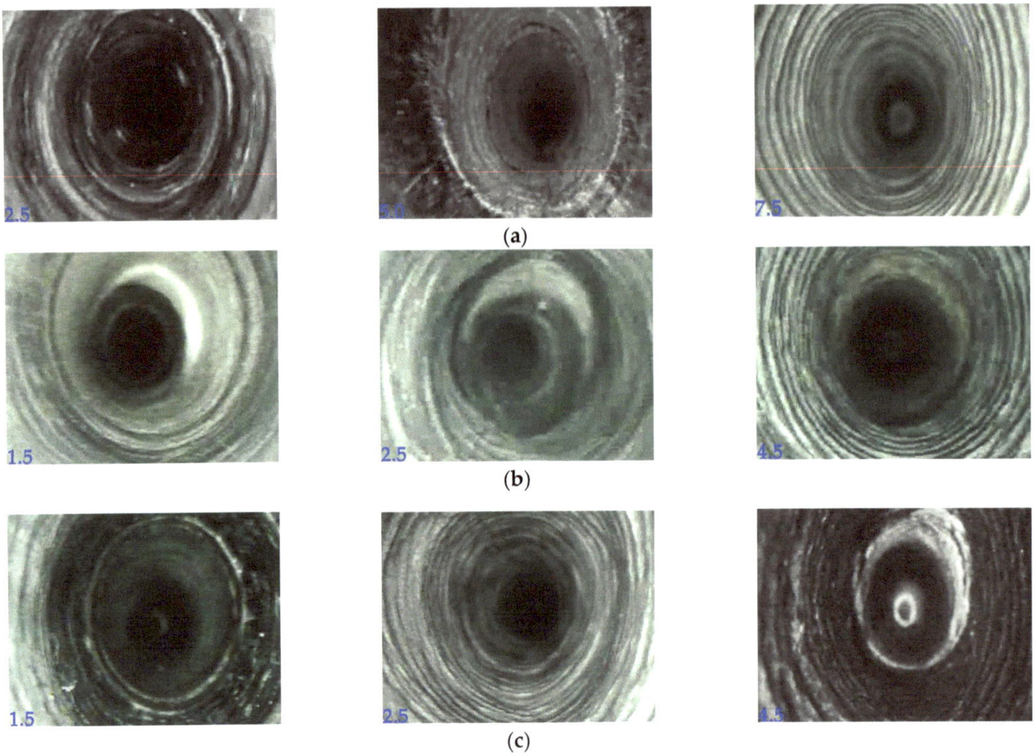

Figure 10. The main roadway-surrounding rock observations via peep boreholes (Unit: m): (**a**) roof; (**b**) two walls; (**c**) floor.

It can be seen from the peep borehole observations that the grouting slurry could effectively fill the surrounding rock cracks and bond them into a relatively complete structure. The cracking and fracturing of the main roadway surrounding rock were inhibited significantly. The overall stability and integrity of the surrounding rock were good. Although a small amount of microcracks occurred occasionally, there were no obvious traces of separation, dislocation, fracture, or other phenomena. It can be seen that the "shielding-filling-grouting" integrated collaborative control technology system achieved good application results in the west-wing main track roadway.

Figure 11 exhibits trends of surface displacement of the surrounding rock in the west-wing main track roadway obtained by field observation. As observed, the west-wing main track roadway #1 station roof and floor, as well as left- and right-wall displacements, finally reached 58, 46, 33, and 32 mm to reach stability. The #2 station's roof and floor, as well as left- and right-wall displacements finally reached 54, 44, 30, and 33 mm to reach stability; #3 station's roof and floor, as well as left- and right- wall displacements reached 51, 43, 32, and 36 mm to reach stability. The average changes in the displacements of rock surrounding the roof, floor, and two walls were only 54, 44, 32, and 34 mm, respectively. The overall deformation of the surrounding rock was small, suggesting that the integrated collaborative control scheme of "shielding-filling-grouting" could significantly reduce the deformation failure of the surrounding rock in the main roadway and effectively maintain the overall

stable state of the main roadway's surrounding rock, satisfying the requirements of the mine's long-term exploitation.

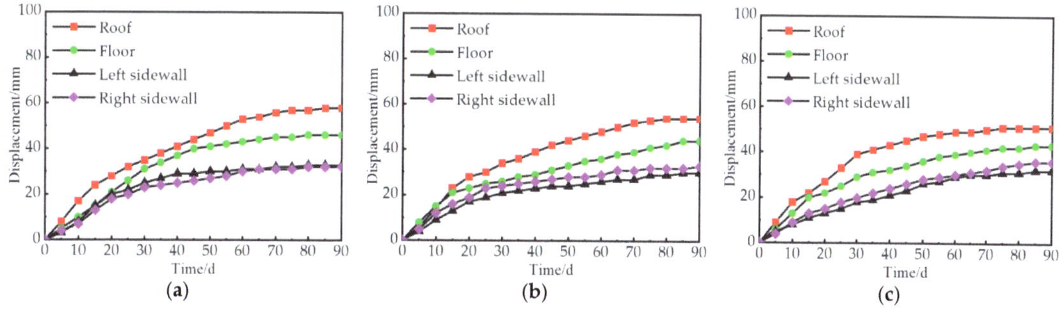

Figure 11. Surface displacement of the main roadway obtained by field observation: (**a**) #1 station curve; (**b**) #2 station curve; (**c**) #3 station curve.

6. Conclusions

Using a case study of the KCM (Fuyang, China), we analyzed its −967 m level west-wing main track roadway, as a typical 1 km deep large-section roadway. The results obtained made it possible to draw the following conclusions:

(1) Field measurement of the in-situ stress in the study area revealed that the in-situ stress state was dominated by the horizontal in-situ stress, showing typical structural stress field features. When the original "bolt mesh cable shotcrete" combined support scheme was adopted for the main roadway, the roadway surrounding rock presented full-section nonlinear deformation failure features, breaching the requirements of mine safety production.

(2) The analysis of elastoplastic mechanism of deformation failure of the surrounding rock in the main roadway revealed that the main roadway surrounding rock structure distribution was related to the original physical and mechanical properties of the rock mass, stress environment, and support mode. The main reasons for the above failure mode were the low strength of the main roadway surrounding rock, the complex stress environment of the horizontal structure, the insufficient support scheme strength of the original "bolt mesh cable shotcrete", and the lack of effective control measures for the main roadway floor.

(3) Aiming at the specific engineering geological conditions of the west-wing main track roadway and considering the advantages and disadvantages of the main roadway's original support scheme, a set of targeted "shielding-filling-grouting" integrated collaborative control technology system was proposed. It mainly included the U-shaped steel secondary shed passive support technology, the quick-setting material filling and compaction control technology behind the wall, and GR-CDSH support technology.

(4) The industrial field test revealed that when the integrated collaborative control scheme of "shielding-filling-grouting" was adopted in the main roadway, the grouting slurry could effectively fill the surrounding rock's crack and bond with it to form a relatively complete structure, achieving a sound grouting effect. The average changes in the displacements of rock surfaces surrounding the main roadway roof, floor, and two walls were 54, 44, 32, and 34 mm, respectively. There were no severe deformation failures in the surrounding rock. The proposed scheme achieved strength improvement and structural compensation of the weak position of the main surrounding rock, provided the active and passive support collaborative control, guaranteed the long-term stability of the roadway, and ensured the safety of staff and equipment, satisfying the requirements of mine safety and efficient production.

Author Contributions: Conceptualization, W.Z., L.C. and Y.Z. (Yandong Zhang); Data curation, J.T. and L.C.; Funding acquisition, W.Z.; formal analysis, Y.Z. (Yanchao Zhu) and Z.S.; Investigation, J.T. and L.C.; Methodology, Y.Z. (Yandong Zhang); Resources, W.Z.; Supervision, W.Z.; Writing—original draft, J.T.; Writing—review and editing W.Z., Y.Z. (Yandong Zhang), J.T. and Y.Z. (Yanchao Zhu). All authors have read and agreed to the published version of the manuscript.

Funding: This work was financially supported by the National Natural Science Foundation of China, grant number 51874278, the "Six Talent Peaks" Project of Jiangsu Province, grant number JNHB-087, and the Independent Research Project of State Key Laboratory of Coal Resources and Safe Mining, grant number SKLCRSM2020X04.

Conflicts of Interest: The authors declare no conflict of interest.

References

1. Xie, H.; Wang, J.; Wang, G.; Ren, H.; Liu, J.; Ge, S.; Zhou, H.; Wu, G.; Ren, S. New ideas of coal revolution and layout of coal science and technology development. *J. China Coal Soc.* **2018**, *43*, 1187–1197.
2. Zhang, N.; Li, X.; Zheng, X.; Xue, F. Current status and challenges of deep coal mining. In Proceedings of the National Symposium on Mining Technology of Kilometer Deep Coal Mine, Tai'an, China, 25 July 2013.
3. He, M.; Xie, H.; Peng, S.; Jiang, Y. Study on rock mechanics in deep mining engineering. *Chin. J. Rock Mech. Eng.* **2005**, *24*, 2803–2813.
4. He, M.; Wang, X.; Liu, W.; Yang, S. Numerical simulation on asymmetric deformation of deep soft rock roadway in Kongzhuang Coal Mine. *Chin. J. Rock Mech. Eng.* **2008**, *27*, 673–678.
5. Shen, B. Coal mine roadway stability in soft rock: A case study. *Rock Mech. Rock Eng.* **2014**, *47*, 2225–2238. [CrossRef]
6. Liu, W.; Guo, Z.; Wang, C.; Niu, S. Physico-mechanical and microstructure properties of cemented coal Gangue-Fly ash backfill: Effects of curing temperature. *Constr. Build. Mater.* **2021**, *299*, 124011. [CrossRef]
7. Skrzypkowski, K.; Zagórski, K.; Zagórska, A.; Apel, D.B.; Wang, J.; Xu, H.; Guo, L. Choice of the arch yielding support for the preparatory roadway located near the fault. *Energies* **2022**, *15*, 3774. [CrossRef]
8. Hou, C.; Wang, X.; Bai, J.; Meng, N.; Wu, W. Basic theory and technology study of stability control for surrounding rock in deep roadway. *J. China Univ. Min. Technol.* **2021**, *50*, 1–12.
9. Xie, H.; Gao, M.; Zhang, R.; Peng, G.; Wang, W.; Li, A. Study on the mechanical properties and mechanical response of coal mining at 1000 m or deeper. *Rock Mech. Rock Eng.* **2019**, *52*, 1475–1490. [CrossRef]
10. Kang, H. "Strata control and intelligent mining technology in deep coal mines with depth more than 1000 m" Guest editor of the album to readers. *J. China Coal Soc.* **2020**, *45*, 1211–1212.
11. Ma, Q.; Tan, Y.; Liu, X.; Zhao, Z.; Fan, D. Mechanical and energy characteristics of coal-rock composite sample with different height ratios: A numerical study based on particle flow code. *Environ. Earth Sci.* **2021**, *80*, 1–14. [CrossRef]
12. Jing, H.; Wu, J.; Yin, Q.; Wang, K. Deformation and failure characteristics of anchorage structure of surrounding rock in deep roadway. *Int. J. Min. Sci. Technol.* **2020**, *30*, 593–604. [CrossRef]
13. Fu, Y.; Ju, W.; Wu, Y.; He, J.; Jiao, J. Study on principle application of energy absorption and bump reduction of high impact toughness rock bolt. *Coal Sci. Technol.* **2019**, *47*, 68–75.
14. Yang, J.; Yang, S. Experimental study on reinforcement of deep soft roadway by using ACZ-1 modified grouting material. *Saf. Coal Mines* **2017**, *48*, 60–64.
15. Yu, Y.; Qin, Z.; Wang, X.; Zhang, L.; Chen, D.; Zhu, S. Development of modified grouting material and its application in roadway repair engineering. *Geofluids* **2021**, *2021*, 1–15. [CrossRef]
16. Zhang, H.; Zhou, R.; Liu, S.; Zhu, Y.; Wang, S.; Wang, J.; Guan, X. Enhanced toughness of ultra-fine sulphoaluminate cement-based hybrid grouting materials by incorporating in-situ polymerization of acrylamide. *Constr. Build. Mater.* **2021**, *292*, 123421. [CrossRef]
17. Oggeri, C.; Oreste, P.; Spagnoli, G. Creep behaviour of two-component grout and interaction with segmental lining in tunnelling. *Tunn. Undergr. Space Technol.* **2022**, *119*, 104216. [CrossRef]
18. Kang, H.; Jiang, P.; Huang, B.; Guan, X.; Wang, Z.; Wu, Y.; Gao, F.; Yang, J.; Cheng, L.; Zheng, Y.; et al. Roadway strata control technology by means of bolting-modification-destressing in synergy in 1000 m deep coal mines. *J. China Coal Soc.* **2020**, *45*, 845–864.
19. Zhang, Z.; Kang, H.; Jiang, Z.; Li, W.; Jiang, P.; Cai, R.; Zhu, Y.; Wang, J. Study and application of high-pressure splitting grouting modification technology in coalmine with depth more than 1000 m. *J. China Coal Soc.* **2020**, *45*, 972–981.
20. Huang, B.; Zhang, N.; Jing, H.; Kan, J.; Meng, B.; Li, N.; Xie, W.; Jiao, J. Large deformation theory of rheology and structural instability of the surrounding rock in deep mining roadway. *J. China Coal Soc.* **2020**, *45*, 911–926.
21. Zhang, G.; Zhu, W.; Zhao, P. In-situ stress measurements and analysis of action of geological structures of deep coal mines in Xuzhou. *Chin. J. Geotech. Eng.* **2012**, *34*, 2318–2324.
22. Qin, W.; Gui, S. Field study on ground stress distribution laws in Huainan Mining Area. *Saf. Coal Mines* **2013**, *44*, 185–188.
23. Yang, R.; Chen, J.; Xue, H.; Cheng, C.; Wang, W. Research of influencing factors of coalmine ground stress test precision by stress relieving method. *China Min. Mag.* **2014**, *23*, 136–139.

24. Wu, H.; Zhao, G.; Ma, S. Failure behavior of horseshoe-shaped tunnel in hard rock under high stress: Phenomenon and mechanisms. *Trans. Nonferr. Met. Soc. China* **2022**, *32*, 639–656. [CrossRef]
25. Wang, Q. Study on the Relationship between Kou Zidong Mine Deep Crustal Stress Distribution and Roadway Layout. Master's Thesis, Anhui University of Science and Technology, Huainan, China, 2014.
26. Shen, M.; Chen, J. *Rockmass Mechanics*, 1st ed.; Tongji University Press: Shanghai, China, 2006; pp. 142–184.
27. Podkopaiev, S.; Gogo, V.; Yefremov, I.; Kipko, O.; Iordanov, I.; Simonova, Y. Phenomena of stability of the coal seam roof with a yielding support. *Min. Miner. Depos.* **2019**, *13*, 28–41. [CrossRef]
28. Zhao, G.; Zhang, X.; Wang, C.; Meng, X. Mechanical analysis and numerical simulation for deep and shallow bearing structures of soft and broken roadway surrounding rock. *J. China Coal Soc.* **2016**, *41*, 1632–1642.
29. Wang, H.; Zhang, D.; Deng, D.; Jiang, Y.; Liu, Y. Stress distribution characteristics of roadway surrounding rock damaged zone under non-hydrostatic pressure. *J. China Coal Soc.* **2020**, *45*, 3717–3725.
30. Yuan, W.; Chen, J. Analysis of plastic zone and damaged zone of roadway in softened rock stratum. *J. China Coal Soc.* **1986**, *11*, 77–86.
31. Zuo, J.; Hong, Z.; Yu, M.; Liu, H.; Wang, Z. Study on gradient support model and classification control of borken surrounding rock. *J. China Univ. Min. Technol.* **2022**, *51*, 1–11.
32. Wang, Q.; Xie, W.; Jing, S.; Zang, L.; Zhang, N. Research on U-shape steel frame and anchor cable collaborative support mechanism and loading law of roadway under dynamical pressure impact. *J. China Coal Soc.* **2015**, *40*, 301–307.
33. Rotkegel, M.; Szot, Ł.; Fabich, S. The analysis of selected methods of the yielding of a circular arch support made of V profiles. *Arch. Min. Sci.* **2020**, *65*, 521–550.
34. Xie, W.; Jing, S.; Wang, T.; Ren, Y.; Zhang, N. Structural stability of U-steel support and it's control technology. *Chin. J. Rock Mech. Eng.* **2010**, *29*, 3743–3748.
35. Skrzypkowski, K. Case studies of rock bolt support loads and rock mass monitoring for the room and pillar method in the legnica-głogów copper district in poland. *Energies* **2020**, *13*, 2998. [CrossRef]

minerals
MDPI

Article

Investigation of the Deformation Failure Occurring When Extracting Minerals via Underground Mining: A Case Study

Xuanting Liu [1,2,*], Congxin Chen [1,2], Xiumin Liu [1,2], Kaizong Xia [1,2] and Tianlong Wang [1,2]

1 State Key Laboratory of Geomechanics and Geotechnical Engineering, Institute of Rock and Soil Mechanics, Chinese Academy of Sciences, Wuhan 430071, China
2 University of Chinese Academy of Sciences, Beijing 100049, China
* Correspondence: liuxuanting18@mails.ucas.ac.cn

Abstract: Metal mines mined using the sublevel caving method often exhibit various environmental problems on the ground surface. This can affect the safety of the production process in the mining area. A numerical model using Universal Distinct Element Code has been established to investigate deformation failure giving rise to this underground mining. The calculations are combined with in-situ monitoring data collected over a period of 10 years. The results indicate that the strata movement in the footwall can be divided into two stages: an arch caving development stage and post arch caving development stage. Mining disturbance is the main cause of caving in the arch caving development stage. The overlying strata experience 'caving–stability–caving', leading to the formation of an 'arch-shaped' caving pattern. In the post arch caving development stage, flexural-toppling deformation occurs in the strata in the direction of the collapse pit due to the tectonic stress present and high dip angles of the discontinuities. Through-going failure surface has been analyzed by studying the plastic state and displacement of elements. The formation of through-going failure surface is related to the flexural-toppling deformation and stress concentration caused by mining activities. Based on the different failure mechanisms, an efficient partition has been proposed such that the footwall can be divided into a stable zone, flexural-toppling failure zone, compression and slipping-toppling failure zone, and shear-slipping failure zone. The results are a useful reference when applied to the Chengchao Iron Mine and other similar metal mines.

Keywords: metal mine; underground orebody excavation; sublevel caving method; strata movement; mining disturbance; deformation failure; numerical simulation

Citation: Liu, X.; Chen, C.; Liu, X.; Xia, K.; Wang, T. Investigation of the Deformation Failure Occurring When Extracting Minerals via Underground Mining: A Case Study. *Minerals* **2022**, *12*, 1025. https://doi.org/10.3390/min12081025

Academic Editor: Abbas Taheri

Received: 29 June 2022
Accepted: 9 August 2022
Published: 15 August 2022

1. Introduction

Mineral resources are important for the continued development of society. Naturally, therefore, as development becomes more rapid, demand for mineral resources also increases rapidly. As a result, mining activity is gradually moving to deeper depths below ground. The caving method is widely adopted in many metal mines due to its high efficiency and relatively low operating costs [1–3]. Notable examples include the Kiirunavaara Iron Mine in Kiruna City, Norrbotten Province, Sweden, Chengchao Iron Mine in Ezhou City, Hubei Province, China, and Jinshandian Iron Mine in Huangshi City, Hubei Province, China. Compared to metal mines mined using the filling method, sublevel caving leads to the large-scale collapse of the strata overlying the mine. This can lead to unexpected disasters resulting from the formation of collapse pits and cracks and damage to buildings and facilities on the ground surface [4,5]. Due to the lack of research on stress–strain behavior when extracting minerals by underground mining, a large amount of surface subsidence was observed at the Kiirunavaara mine in Sweden. The surface disturbance was so significant that it affected parts of Kiruna city [6,7]. Another representative example is the Verkhnekamsk mine in Perm, Russia. Due to errors at the design stage and in the production process, two mines were destroyed and flooded, and the destruction affected the

city limits [8]. In other words, the sublevel caving method can lead to serious environmental and safety problems.

Numerous scholars have conducted research on the movement of strata due to coal mining activity, and some important results have been accomplished [9–15]. However, metal ores tend to be located in areas with complex geological conditions. This makes the problem of strata movement around metal mines more complicated; the research is still in the exploratory stage. Scholars have tried to address the problem using a variety of research methods, e.g., theoretical analysis, in situ monitoring, and numerical simulation. In terms of theoretical analysis, Hoek [16] used a theoretical approach to model the progressive failure (break angle) of the hanging wall in the Grangesberg Iron Mine in Bergslagen, Sweden. Based on this work, Brown et al. [17] and Lupo [18,19] made further attempts to provide an improved theoretical model. Baryakh et al. proposed a series of methods to evaluate subsidence and fracture due to underground mining [20–22]. Although such theoretical models are based on clear physical concepts (and are easy to use), a large number of idealized assumptions are made during their derivation. Unfortunately, some of these may be inconsistent with the complex geological conditions found in the field. In terms of in situ monitoring, photographic methods (using, for example, unmanned aerial vehicles), interferometric synthetic aperture radar, and global positioning systems (GPSs) have been used by many scholars to measure the deformation of the ground surface [23–25]. Although in situ monitoring is often limited by external conditions, it does reflect the most realistic assessment of the ground surface displacement occurring. In terms of numerical simulations, Sitharam et al. [26] used the concept of representative elementary volume in fast lagrangian analysis of continua (FLAC) to study the deformation and stress characteristics in the Kiirunavaara Iron Mine in Kiruna, Sweden. Rybak et al. explored the change in the stress–strain behavior and the possibility of development of deformation changes with numerical simulation using FLAC3D software (5.01 version, ITASCA Consulting Group, Minneapolis, MN, USA) [27]. Villegas et al. [28] used particle flow code (PFC) to study the effect of rock caving on the stability of the hanging wall. Hamdi et al. [29] used the finite–discrete element method (FDEM) to study the influences of the persistence and spacing of the discontinuities on the deformation of the ground surface of the hanging wall. Yu et al. show that natural and human-made geodynamic events associated with mining operations reflect the global geodynamic processes [30]. Senz, E. and Menéndez-Pidal, I. et al. also studied topics related to that described herein [31,32].

Most scholars focus on the hanging wall because the strata movement there is more extensive than that in the footwall. The Chengchao Iron Mine is a metal mine which is mined using the sublevel caving method. Moreover, we have been monitoring this iron mine for over 10 years. Our results show that, in this mine, the extent of the failure in the footwall is more serious than that in the hanging wall. Various scholars have utilized qualitative and limit equilibrium methods to analyze this phenomenon, but the mechanism responsible for the extensive damage in the footwall is still not fully understood [33–35]. Considering that many significant buildings and facilities are located in the footwall of the mining area, it is imperative that the deformation and failure mechanism should be studied further.

According to the aforementioned analysis, it can be noted that the deformation failure occurring when extracting minerals via underground mining is a very topical issue. Therefore, the purpose of this study was to investigate the strata movement and failure mechanism in the footwall of the Chengchao Iron Mine, and to achieve this, it was necessary to address the following tasks: the geological conditions in the Chengchao Iron Mine were analyzed and a reasonable two-dimensional (2D) numerical model was established using Universal Distinct Element Code (UDEC) and the model was then used to simulate the complete orebody-mining procedure. The results of the numerical simulations were then combined with the in situ monitoring data to investigate stratal movement in the footwall from multiple perspectives. As a result, the stratal movement was divided into two different stages. The formation of through-going failure surface was related to

the flexural-toppling deformation and stress concentration caused by mining activities. The footwall was subsequently divided into different zones according to the different failure mechanisms occurring. The results obtained should serve as a useful guide to the deformation occurring in the Chengchao Iron Mine and other similar metal mines.

2. Engineering Situation

2.1. Engineering Geology

The Chengchao Iron Mine is located in Ezhou in Hubei Province, China. The mining area is about 2300 m long in the east–west direction and 800 m wide in the north–south direction. The mine produces about 2.8 million tons of ore per year, making it the third largest underground metal mine in China (after the Jingtieshan Iron Mine in Jiayuguan City, Gansu Province, China and Meishan Iron Mine in Nanjing City, Jiangsu Province, China).

Many scholars have discussed the engineering geological conditions in the Chengchao Iron Mine region in detail [36,37]. Therefore, in this section, only some of the most important issues of relevance to the numerical simulations are briefly discussed. The mining area can be divided into eastern and western areas by exploration line no. 15 (Figure 1). In the eastern area, line no. 4 was chosen for further study. The geological profile along this line is shown in Figure 2. The rock mass in the hanging wall is mainly comprised of hornstone (the metamorphic product of mudstone or sandy shale). The strata overlying the orebody are mainly composed of marble and diorite. In fact, as shown in Figure 2, the marble is sandwiched between diorite and granite and has a 'funnel' shape in the vertical profile. The width of the marble varies from 60 to 80 m. In Figure 1, it appears as a narrow stripe aligned in the ESE–WNW direction. According to on-site drilling data, the marble contains fissures that are highly developed.

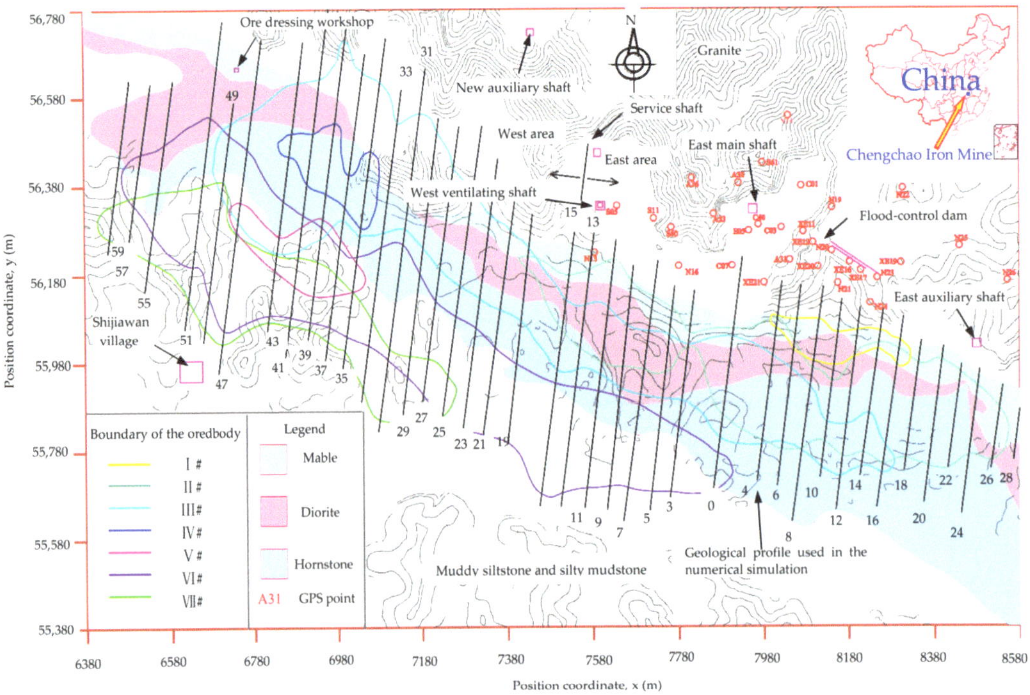

Figure 1. Distribution of the rock masses and orebodies in the Chengchao Iron Mine region [5] (numbers near the exploratory lines are the names of the geological profiles).

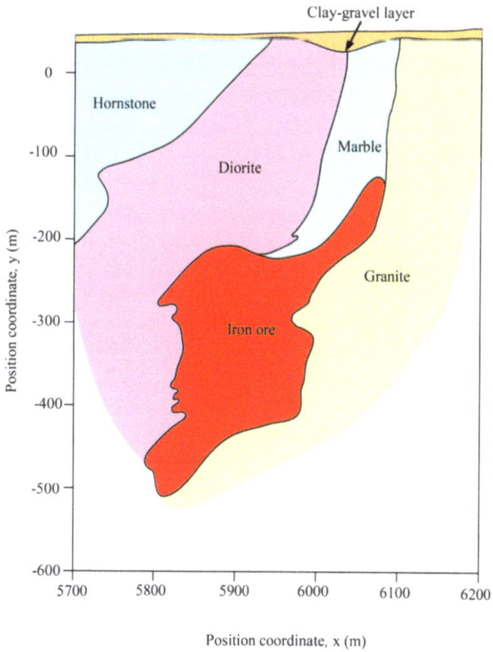

Figure 2. Geological profile along line no. 4 in the study area [5].

The footwall in the mining area mainly consists of granite, and discontinuities are widely developed in the surrounding rock mass. Two dominant discontinuity sets (J1–J2) were found in the footwall through investigation of the geological outcrops and mining roadways, as illustrated in Figure 3. The first discontinuity set, J1, has an orientation of 74°∠82°. It is the most developed and shows high persistence, so it is simply referred to here as 'one fully persistent discontinuity'. The discontinuity surfaces are rough, planar, and brown. The second discontinuity set, J2, has an orientation of 209°∠82°. It is developed and has medium persistence (the persistence of discontinuities is about 0.5). The discontinuity surfaces are rough and iron-stained. Some discontinuities in other orientations were also observed. However, compared to J1 and J2, these are not well-developed.

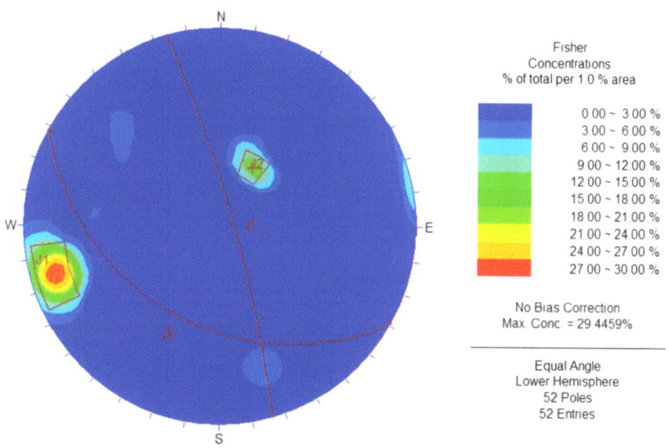

Figure 3. Pole plot of the discontinuities in the footwall (red rectangles denote the most developed region of the dominant discontinuity sets).

2.2. Mining Situation

Several orebodies can be found in the region (see Figure 1). The orebodies in the eastern part of the Chengchao Iron Mine (i.e., nos. I, II, III, and IV) have strikes in the NWW direction and dips in the SSW direction. The plans drawn up to mine the area specified that sublevel caving should be adopted to mine the orebody. This mining method utilizes a true caving technique, in that the aim of mining activity is to induce free displacement of the country rock overlying an orebody. The main conceptual components of sublevel caving are demonstrated in Figure 4. Mining progresses downwards in an orebody, with each old production horizon level being progressively eliminated as mining proceeds. Generally, the production horizon level consists of a series of sub-levels, and the phase height is between 60 to 80 m. The phase height is chosen to be 70 m and the sublevel interval varies from 10 to 12 m in the Chengchao Iron Mine. To improve the production capacity of the mine, the sublevel interval is increased to 14 m when mining the −218 to −358 m levels. Moreover, below the −358 m level, the sublevel interval is further increased to 17.5 m. The orebody in the eastern area was first mined in 1969. After more than 50 years of mining, the mined-out areas shown in Figure 5 have been formed. Further details of the mining schedule are presented in Table 1. As can be seen, the −80, −150, −220, −290, −360, and −430 m levels have been mined out. At present (up to June 2021), the −463 and −480.5 m levels are still being excavated.

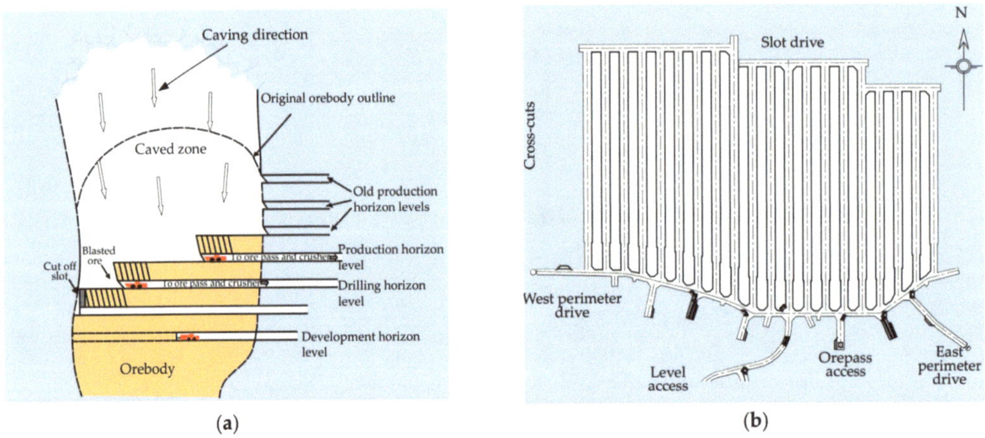

Figure 4. (**a**) Diagrammatic long section of sublevel caving; (**b**) diagrammatic plan of sublevel caving [38].

Table 1. Extent of the mining activity at different times.

Mining Level (m)		Mining Date		Orebody Number	Mining Scope (Exploration Line Numbers)
Start	**End**	**Beginning**	**Ending**		
−16.0	−96.0	1969	1978	I,II	4 to 16; 20 to 26
−106.0	−150.0	1979	1984	II	10 to 24
−160.0	−206.0	1984	1990	II	9 to 22
−218.0	−260.0	1990	1997	II	11 to 22
−260.0	−302.0	1998	2002	II	13 to 22
−302.0	−358.0	2002	2008/07	III	15 to 16
−358.0	−375.0	2007/05	2012/02	III	15 to 16
−375.0	−393.0	2009/05	2013/12	III	15 to 16
−393.0	−410.5	2012/03	2017/03	III	15 to 14
−410.5	−430.0	2014/07	2018/07	III	15 to 12
−430.0	−447.5	2016/06	2018/08	III	15 to 10
−447.5	−463.0	2018/11	-	-	-
−463.0	−480.0	2019/10	-	-	-

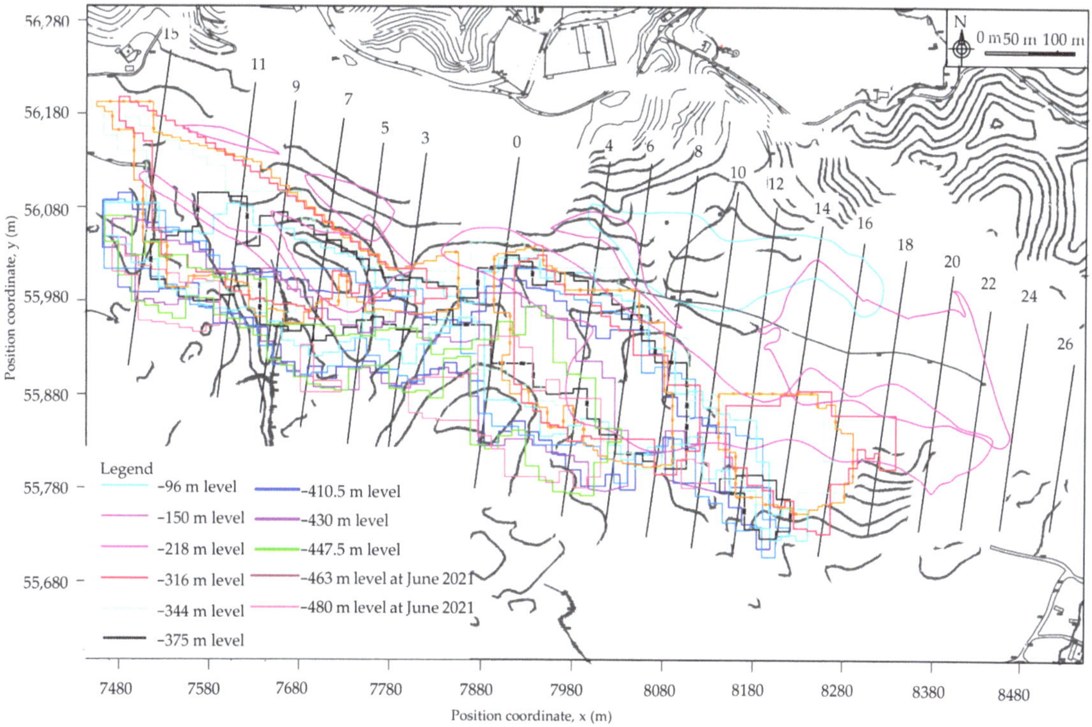

Figure 5. Distribution of the mined-out areas (numbers near the exploratory lines are the names of the geological profiles).

3. Numerical Modeling

3.1. Numerical Method

The application of continuum theory in rock mechanics is limited due to the existence of discontinuities [39]. To solve this problem, discrete element methods were employed in media containing discontinuities. Based on the monitoring data, the area affected by strata movement in the Chengchao Iron Mine is about 2500 m in length and 1500 m in width. It is most reasonable to use three-dimensional numerical simulation software to explore the strata movement in the eastern area of the Chengchao Iron Mine. However, it is difficult for 3D numerical simulation in the case of a large geometrical domain requiring fine mesh discretization. The orebody in the study area is essentially strip-shaped, with a length along the strike direction of about 2000 m, and width along the normal direction of about 300 m. The lithology changes little along the strike direction of the ore body. Because of this geometry, the strata movement can be simplified to a 2D plane strain problem along the strike direction of the orebody. In particular, the discrete element software package UDEC (4.0 version, ITASCA Consulting Group, Minneapolis, MN, USA) has achieved good results in geotechnical engineering applications in recent years [40,41]. The orebody in the study area is essentially 'strip-shaped', with little change in lithology. This means that the rock masses can be treated as a simple 2D plane strain problem. Hence, a 2D numerical UDEC method is used in this study.

3.2. Modeling and In Situ Stress Field

Figure 6 shows the numerical model established using UDEC (which is based on the geological profile shown in Figure 2, i.e., that appropriate to exploration line no. 4). As can be seen, the model is 1700 m long and 650 m wide. The model aims to accurately reproduce

the behavior of the hornstone, diorite, marble, granite, and orebody in the region. The boundary conditions of the model are as follows: the top surface is free, and the velocity of the bottom face is confined in the x- and z-directions. The velocity on two sides of the numerical model is confined in the x-direction.

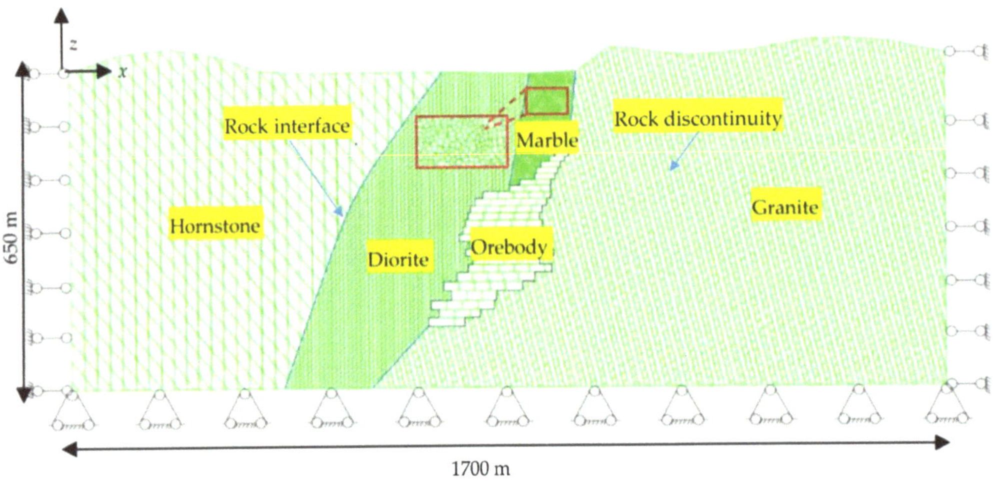

Figure 6. UDEC numerical model corresponding to the geological profile along line no. 4 (green lines denote discontinuities; blue lines denote the interfaces between different rock masses; red rectangles denote the partial enlarged detail of Voronoi blocks).

To ensure a stable response from the computational model, a modified FISH function "ZONK.fis" is utilized. More specifically, "ZONK.fis" can detect the interior boundaries of the excavation and gradually release the forces thereon. The release factor R is the key parameter that controls this process and ranges from 0 to 1. An R value of 0.2 is chosen herein, indicating that there are five relaxation stages and 20% of the force applied to an excavation boundary is released in each one. This function in UDEC is mainly applied to the excavation of underground spaces. Considering the complexity of the mined-out area, an advance of the working face of 20 m is regarded as a mining step. In this paper, the mining of the orebody in the −150 to −480 m levels is simplified into 86 steps. Table 2 illustrates a relationship between excavation steps and mining levels in the numerical model. As shown in Table 2, mining the −150 m level orebody corresponds to the first step in the numerical simulation; mining the −232 m level orebody corresponds to the eleventh to fifteenth steps in the numerical simulation, etc.

The granite in the numerical model features two main sets of discontinuities: one that is of high persistence and steeply dipping and one that is of medium persistence and gently dipping. The fissures in the marble are highly developed and it is difficult to describe the distribution characteristics of the discontinuities. Therefore, Voronoi blocks are utilized in the UDEC model to account for the irregular nature of the discontinuities. The specific geometric parameters and distributions of the discontinuities in the model are listed in Table 3.

The Chengchao Iron Mine has suffered from intensive tectonic movement, and the in situ stress test results can be summarized as follows:

1. The direction of the maximum principal stress σ_1 in the mining area is consistent with the strike of the orebody (N 85° to 75° W). The measured value of σ_1 is 1.1704 to 1.6296 γz, with a mean average value of 1.274 γz (where γ and z are the unit weight of rock mass and the depth below ground surface).

2. The direction of the intermediate principal stress σ_2 is found to be perpendicular to the strike of the orebody. In the deep parts, the intermediate principal stress is close to the vertical stress caused by the effect of gravity on the rock masses.
3. The direction of the minimum principal stress σ_3 is found to be perpendicular to the strike of the orebody. The measured value of σ_3 is 0.59 to 0.926 γz, with a mean average value of 0.758 γz.

Table 2. Relationship between excavation steps and mining levels in the numerical model.

Levels (m)	Corresponding Steps in the Numerical Model																		
	1	2	3	4	5	6	7	8	9	10	11	12	13	14	15	16	17	18	19
150	*																		
160		*																	
171.5			*																
183				*															
194.5					*														
206						*	*												
218								*	*	*									
232											*	*	*	*	*				
246																*	*	*	*
—	20	21	22	23	24	25	26	27	28	29	30	31	32	33	34	35	36	37	38
246	*																		
260		*	*	*	*	*													
274							*	*	*	*	*								
288												*	*	*	*	*			
302																	*	*	*
—	39	40	41	42	43	44	45	46	47	48	49	50	51	52	53	54	55	56	57
302	*	*																	
316			*	*	*	*	*												
330								*	*	*	*	*							
344													*	*	*	*	*	*	
358																			*
—	58	59	60	61	62	63	64	65	66	67	68	69	70	71	72	73	74	75	76
358	*	*	*																
375.5				*	*	*	*	*	*										
393										*	*	*	*	*					
410.5															*	*	*	*	*
—	77	78	79	80	81	82	83	84	85	86									
428	*	*	*																
445.5				*	*	*	*												
463								*	*										
480										*									

Note: * denotes the mining step in the numerical simulation.

Table 3. Geometric parameters and distributions of the discontinuities.

Lithology	Density (kg/m^3)	Deformation Modulus (GPa)	Poisson's Ratio (GPa)	Cohesion (MPa)	Tensile Strength (MPa)	Friction Angle (°)	Dilation Angle (°)
Granite	2600.00	16.09	0.22	0.80	0.16	30.20	0
Marble	2700.00	12.05	0.28	0.30	0.10	27.00	0
Hornstone	2600.00	22.50	0.25	1.84	0.18	31.80	0
Diorite	2630.00	25.31	0.26	0.80	0.16	31.00	0
Ore body	3923.00	22.41	0.21	2.48	0.25	38.10	0

Note: MV represents mean value; MRD the maximum random deviation from the mean.

In the numerical model (Figure 6), the x-axis corresponds to the strike direction of the No. 4 exploration line, the y-axis is perpendicular to exploration line No. 4, and the z-axis points vertically upwards. A modified FISH function 'Stress.field' is used in this paper. This function in UDEC is mainly applied to apply stress to each zone. The in situ stress field in the numerical model therefore takes the following form:

$$\sigma_x = k_x \gamma z \tag{1}$$

$$\sigma_y = k_y \gamma z \tag{2}$$

$$\sigma_z = \gamma z \tag{3}$$

where k_x and k_y are the horizontal pressure coefficients along the x- and y-directions, respectively. The actual values of k_x and k_y are 1.0 and 1.4, respectively.

3.3. Calibration of the Rock Mass Parameters

The Mohr–Coulomb constitutive model was adopted to describe the mechanical behavior of the rock masses. The presence of joints changes the mechanical parameters of rock massed and intact rock. Based on many laboratory tests on small specimens (50 mm × 100 mm), the mechanical parameters of the intact rock were determined. The mechanical parameters of the rock mass could be calculated using the Hoek–Brown failure criterion [42,43]. More specifically, the value of *GSI*, the Hoek–Brown constant (m_i), and the disturbance factor of the rock mass (*D*) were determined based on geological investigations and excavation characteristics. Then, the cohesion and friction angle of the rock mass were obtained by fitting the non-linear Hoek–Brown strength parameters using RocLab software (1.010 version, Rocscience, Toronto, ON, Canada). Finally, the mechanical parameters of rock mass were determined in Table 4.

Table 4. Mechanical parameters of the rock masses.

Lithology	Density (kg/m^3)	Bulk Modulus (GPa)	Shear Modulus (GPa)	Cohesion (MPa)	Tensile Strength (MPa)	Friction Angle (°)	Dilation Angle (°)
Granite	2600.00	8.38	6.82	0.80	0.16	30.20	0
Marble	2700.00	8.45	4.36	0.30	0.10	27.00	0
Hornstone	2600.00	15.00	9.00	1.84	0.18	31.80	0
Diorite	2630.00	18.00	10.00	0.80	0.16	31.00	0
Ore body	3923.00	12.90	9.26	2.48	0.25	38.10	0

A Coulomb slip model incorporating residual strength was used for the discontinuities. The strength parameters of the discontinuities were determined by laboratory testing. Compared to the strength parameters, it is difficult to determine the normal stiffness (k_n) and shear stiffness (k_s) of the discontinuities. Although UDEC does provide expressions to calculate the values of k_n and k_s,

$$k_n = m\left[\frac{K+\left(\frac{4}{3}\right)\cdot G}{\Delta z_{\min}}\right],\ 1 \le m \le 10 \tag{4}$$

$$k_s = k_n\left(\frac{G}{E}\right) \tag{5}$$

these parameters have been found to adopt a large range of values.

The monitoring data pertaining to the ground–surface displacement are used to conduct back-analysis to derive more reliable values for the stiffness parameters in this paper. More specifically, monitoring points A31 and C03 are located near the east main shaft (Figure 1), and their locations could be found in the UDEC model. The cumulative horizontal deformation of monitoring points A31 and C03 from March 2012 to March 2017 was 397.77 cm and 119.23 cm, and it corresponds to the excavation stage from −393.0 m to −410.0 m levels in the numerical model. By referring to Equations (4) and (5), many back-

analysis calculations were performed to simulate the mining of −393.0 m to −410.0 m levels in the numerical model. When the stiffness parameters of discontinuities listed in Table 5 were specified, the error rates of measuring points A31 and C03 between the calculated horizontal displacement and the measured horizontal displacement from March 2012 to March 2017 were minimal (Table 6); therefore, the stiffness parameters of discontinuities listed in Table 5 were used herein.

Table 5. Mechanical parameters of the discontinuities.

No.	Discontinuity Type	Normal Stiffness (GPa/m)	Shear Stiffness (GPa/m)	Friction Angle (°)	Cohesion (MPa)	Tensile Strength (MPa)	Residual Cohesion (MPa)
I	Rock interface	10.0	5.1	30	0.2	0.1	0.1
II	Rock discontinuity	24.0	12.4	30	0.4	0.1	0.2

Note: Rock interface denotes the interface between different rock masses; rock discontinuity represents a discontinuity in the rock mass.

Table 6. Back analysis of the calculated and in situ monitored horizontal deformations.

Monitoring Point	Horizontal Deformation (March 2012 to March 2017)		Error (cm)	Percentage Error (%)
	Calculated Value (cm)	Monitored Value (cm)		
A31	360.78	397.77	36.99	9.30
C03	102.34	119.23	16.89	14.2

The Voronoi approach was used to calibrate marble, and the mechanical properties of marble were determined by many Voronoi blocks. In the Voronoi approach, an elastic model is used to describe the mechanical behavior of the Voronoi blocks, and a Coulomb slip model is adopted to describe the contact between them. Thus, the microparameters of Voronoi blocks are composed of block properties and contact properties (Table 7). More specifically, the deformation behavior of each Voronoi block is controlled by the bulk moduli of the block, its shear moduli, and the normal and shear stiffnesses of the contact. Rock failure is controlled by the occurrence of contact failure. Shear or tensile failure can occur if the shear or tensile stress exceeds the contact strength. Thus, the strength of rock mass is controlled by cohesion, friction angle, and tensile strength of the contacts.

Table 7. Microparameters used in the UDEC Voronoi model used to describe the marble.

Voronoi Block Size (m)	Block Properties			Contact Properties				
	Unit Weight (kN/m^3)	Bulk Moduli (GPa)	Shear Moduli (GPa)	Normal Stiffness (GPa/m)	Shear Stiffness (GPa/m)	Friction Angle (°)	Cohesion (MPa)	Tensile Strength (MPa)
5	27.0	8.45	4.36	14.30	5.58	27.20	0.38	0.08

The process of calibrating microparameters is illustrated in Figure 7. As shown in Figure 7, the first step is to establish calibration model of direct tensile tests and compression tests. Figure 8a,c illustrates models of the compression test and tensile test. To reduce boundary effects, the dimensions of the test samples used should be at least 10 times the size of the Voronoi block. Thus, rectangular test samples measuring 50 m × 100 m were adopted in this work (as the average size of Voronoi block is 5 m). Secondly, a typical value of each microparameter was assigned to the test samples. Then, a series of numerical tests in compression and direct tension were conducted iteratively until the simulated test results were consistent with the macroparameters of the marble, as listed in Table 4 [44].

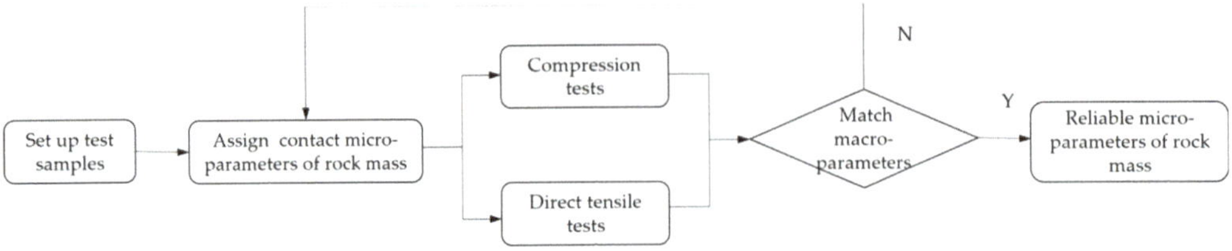

Figure 7. Flowchart summarizing the method of calibrating microparameters.

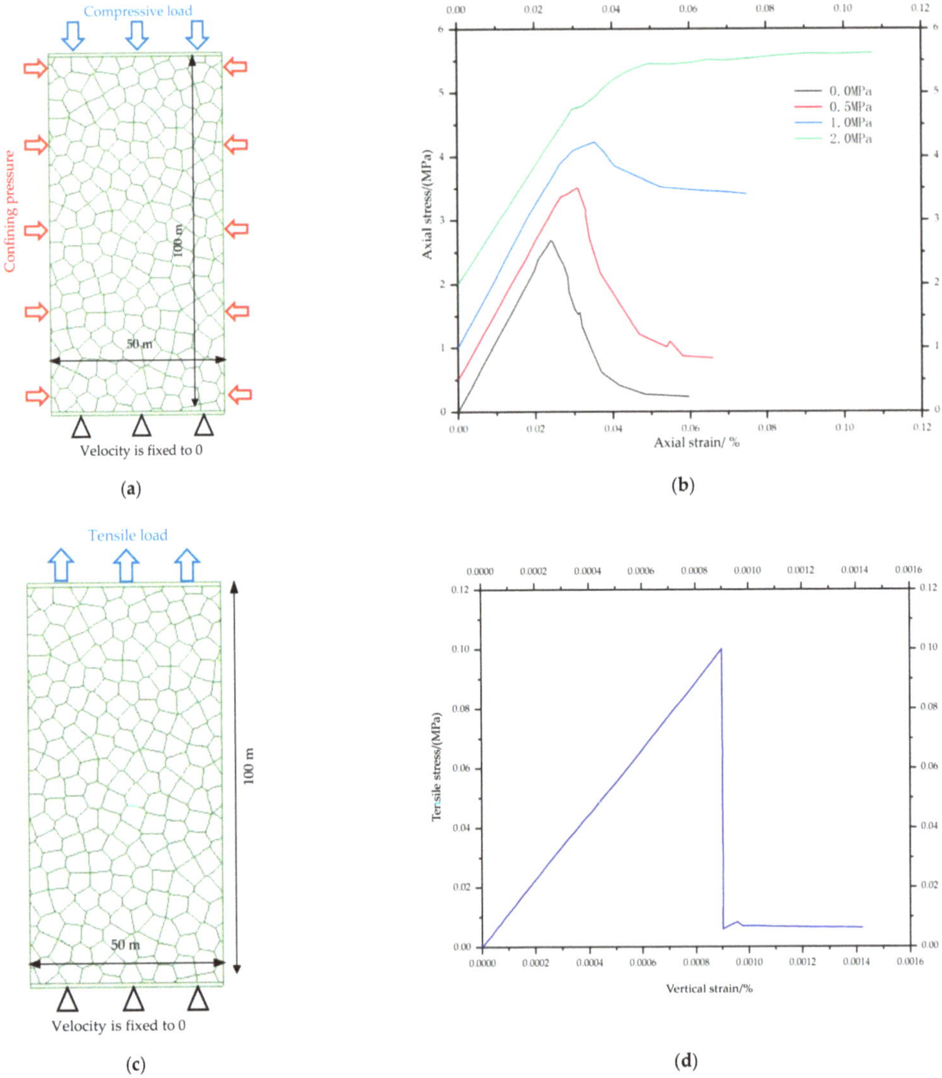

Figure 8. Calibration of the UDEC Voronoi model. (**a**) Compression test model; (**b**) axial stress–strain curves obtained from unconfined and confined compression tests (i.e., using confining pressures of 0, 0.5, 1.0, and 2.0 MPa); (**c**) tensile test model; and (**d**) the results of a direct tensile test showing the stress–strain curve obtained.

More specifically, the deformation moduli and shear stiffness of the block contacts were obtained by conducting a series of unconfined compression tests. As for the friction angle and cohesion of the contacts, they can be determined by conducting a series of confined compression tests. Based on a series of confined compression tests in Figure 8a, the strength envelope can be obtained from a regression analysis of the compressive strength for the various confining pressures. On this basis, the friction angle (φ) and cohesion of the contacts can be obtained from the gradient (m) and vertical intercept (b) of the strength envelope curve with reference to Equations (6) and (7) [45]:

$$\varphi = \arcsin\left(\frac{m-1}{m+1}\right) \tag{6}$$

$$c = b\left(\frac{1-\sin\varphi}{2\cos\varphi}\right) \tag{7}$$

The tensile strength of the contacts could also be acquired by a series of tensile tests in Figure 8c. The final deformation and strength characteristics obtained are shown in Figure 8b,d. Finally, the most appropriate microparameters for the blocks and contacts in the UDEC Voronoi model were calibrated (Table 7).

4. Results

When metal mines are mined using sublevel caving, the strata do not move very much in the early mining stages. As the orebody is continuously mined, however, the overlying strata continuously fall, and collapse pits are formed on the ground surface. It was also found that there is an obvious difference in the movement of the strata before and after the formation of the collapse pits. The formation of a collapse pit leads to stress release, and this accelerates the rate at which the strata move. The calculation results suggest that the movement of the strata in the footwall in the eastern part of the Chengchao Iron Mine can be divided into two stages. These are referred to as the arch caving development stage (ACDS) and post arch caving development stage (PACDS). The latter stage, PACDS, can be further subdivided into a flexural-toppling stage and deep rock mass failure stage.

4.1. Arch Caving Development Stage

The orebody in the eastern area of the Chengchao Iron Mine is 'olive-shaped' (i.e., narrow at both ends and wide in the middle). This means that the mined-out area formed in the initial mining stage is mainly located below the marble (see Figure 2). The disturbance caused by mining leads to the redistribution of the stress in the surrounding rock, and hence displacement. This induces the arch of the overlying strata to cave towards the mined-out area.

Figure 9 illustrates the caving process that occurs in the marble. In these Figures, the blue lines denote shear failure, and the red lines denote tensile failure. After the -150 m level is mined, the overlying strata in the mined-out area suffer 'arch-shaped' damage. More specifically, shear failure mainly occurs at the feet of the arch and tensile failure mainly occurs at its vault (Figure 9a). Although these areas are damaged (to different degrees), the overlying stratum remains stable, as a complete failure surface is not formed. The damage in the overlying stratum continues to expand further as the mined-out area continues to increase. When the -160 m level is completely mined, the marble starts to cave as a complete failure surface is formed between the vault and arch feet (Figure 9b). As the mining depth deepens, the overlying stratum caves several times. Caving occurs as the -171.5, -194.5, and -206 m levels are mined (Figure 9c,e,f), and the failure process follows the same pattern that occurred previously. The arch caving process gradually migrates to the ground surface so that a collapse pit is formed after the -218 m level is mined (Figure 9g).

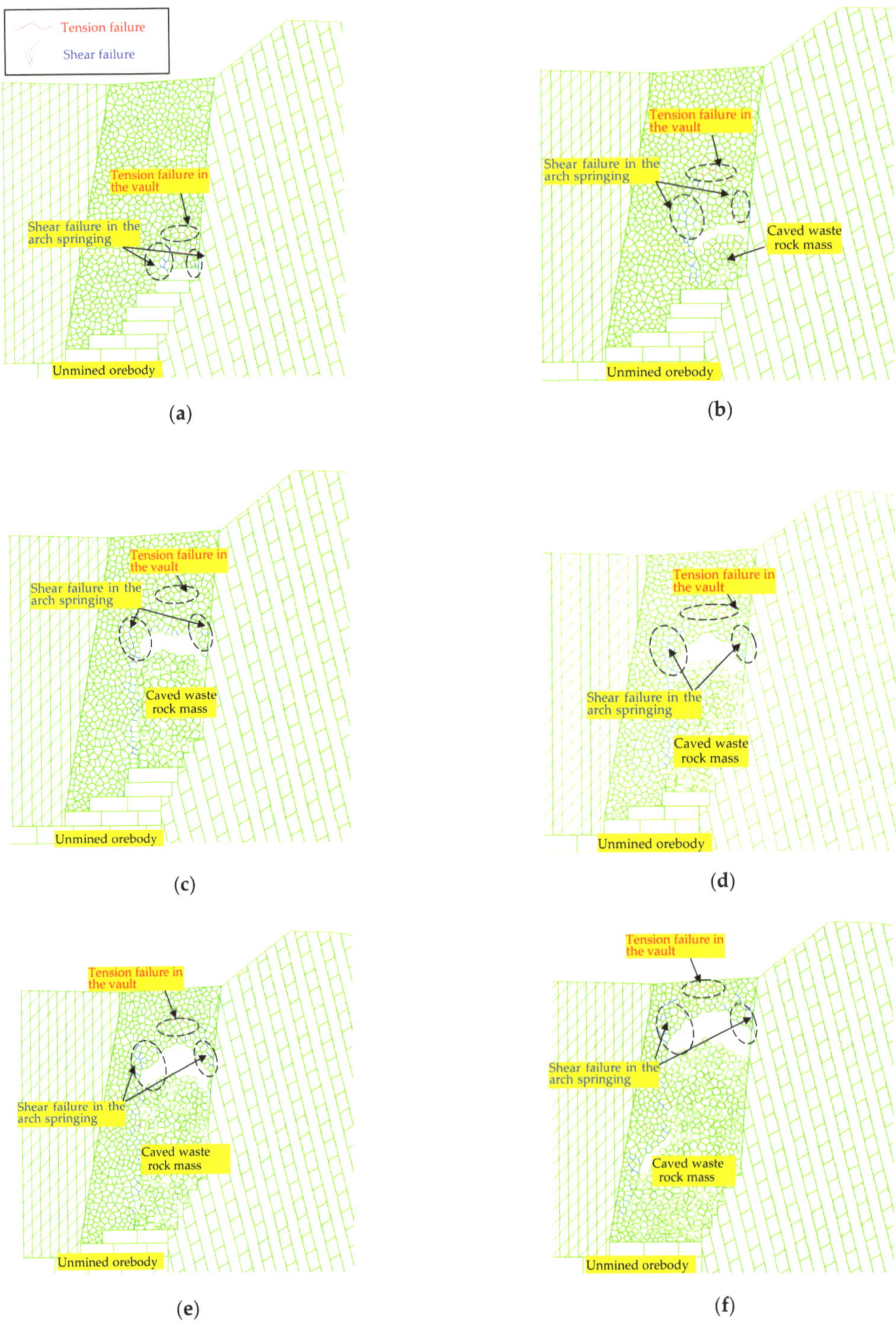

Figure 9. *Cont.*

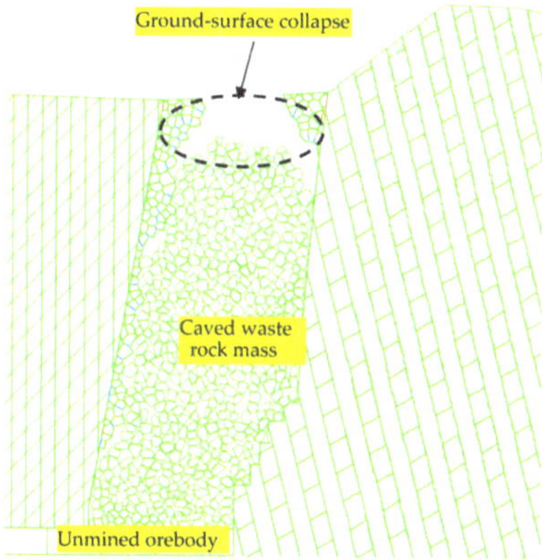

(g)

Figure 9. The seven stages shown correspond to depths of: (**a**) −150 m, (**b**) −160 m, (**c**) −171.5 m, (**d**) −183 m, (**e**) −194.5 m, (**f**) −206 m, and (**g**) −218 m. Red lines denote tensile failure; blue lines denote shear failure.

Figure 10 can be obtained using FISH function for the calculation results. Figure 10 presents a graph of the volume of the caved rock generated as the mining activity moves to progressively deeper levels (the thickness is taken as 1). As can be seen, the caving volume changes dynamically as the mining progresses. In fact, the caving process can be observed following three stages: slow caving, stable caving, and rapid caving. In other words, the caving process has a pattern that can be described as 'caving–stability–caving'.

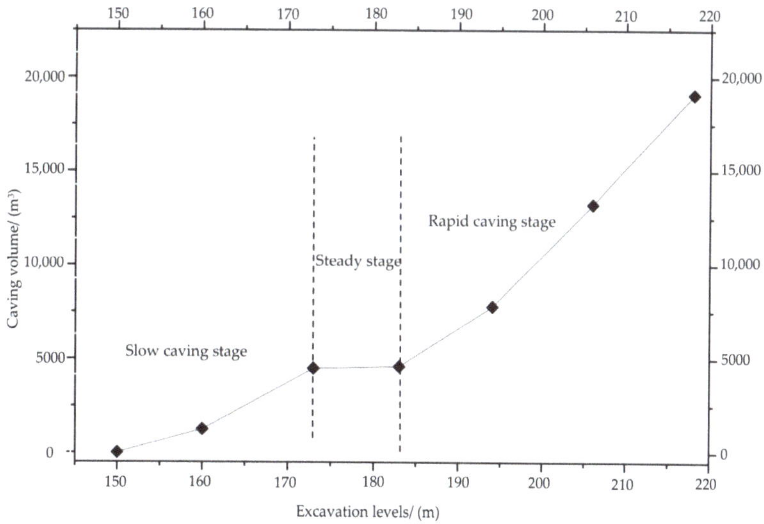

Figure 10. Plot of the caving volume of the overlying marble as a function of excavation level.

In the slow caving stage (−150 to −171.5 m levels), the overlying strata experience caving on a small scale, and so the caving volume is small. In the steady stage (−171.5 to −183 m levels), although there are a few individual unstable rock masses experiencing caving towards the mined-out area, the overlying rock mass does not exhibit instability overall. The caving volume is close to constant at this stage. As the mining depth increases, the disturbance caused by the mining activity continues to lead to instability

in the overlying strata, and then caving occurs again. In the rapid caving stage (−194.5 to −206 m levels), the overlying stratum experiences large-scale caving. The caving arch expands rapidly towards the ground surface and eventually forms a collapse pit.

The caving process outlined above is supported by the in situ monitoring data. Figure 11a illustrates the distributions of the ground surface collapse pits in the eastern area of the Chengchao Iron Mine at different times. In the 1990s (when the −218 m level was being mined), the ground surface collapse pits marked with blue lines in Figure 11a were discovered near exploration line no. 4. The range of the collapse pits subsequently continued to increase, leading to the present situation shown in Figure 11b.

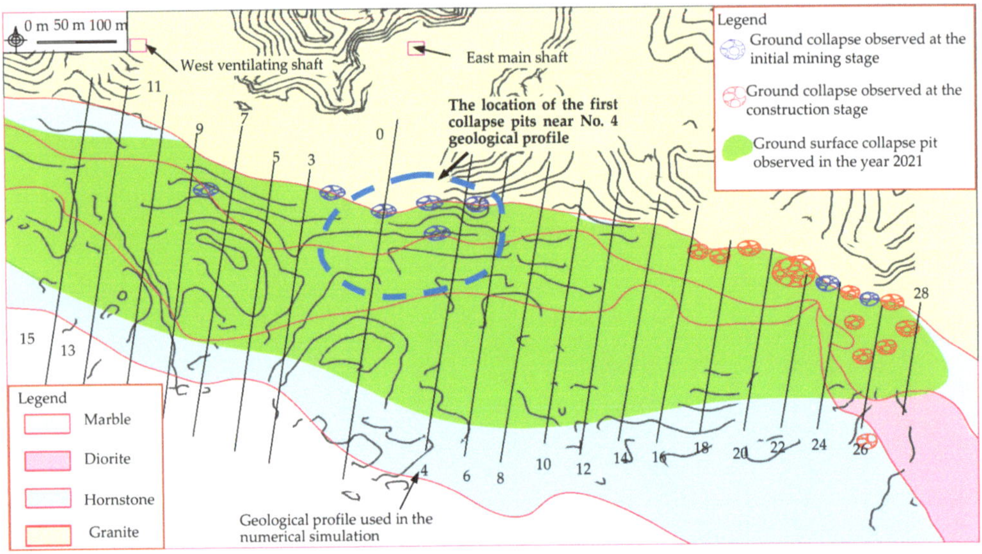

(**a**)

(**b**)

Figure 11. (**a**) Distribution of the ground surface collapse pits in the eastern area of the Chengchao Iron Mine at different mining stages (numbers near the exploratory lines are the names of the geological profiles); (**b**) photograph of the collapsed ground (taken on 26 August 2021).

4.2. Post Arch Caving Development Stage

4.2.1. Flexural-Toppling Stage

There are two groups of developed discontinuities in the footwall. Among these, those that are steeply dipping and have an orientation of 74°∠82° are the most developed. When a collapse pit is formed, a free face is generated in the footwall, at which deformation and

flexural toppling can occur. Figure 12 illustrates the plastic failure zones and displacement vectors in the footwall when the −232, −288, and −358 m levels are mined, respectively.

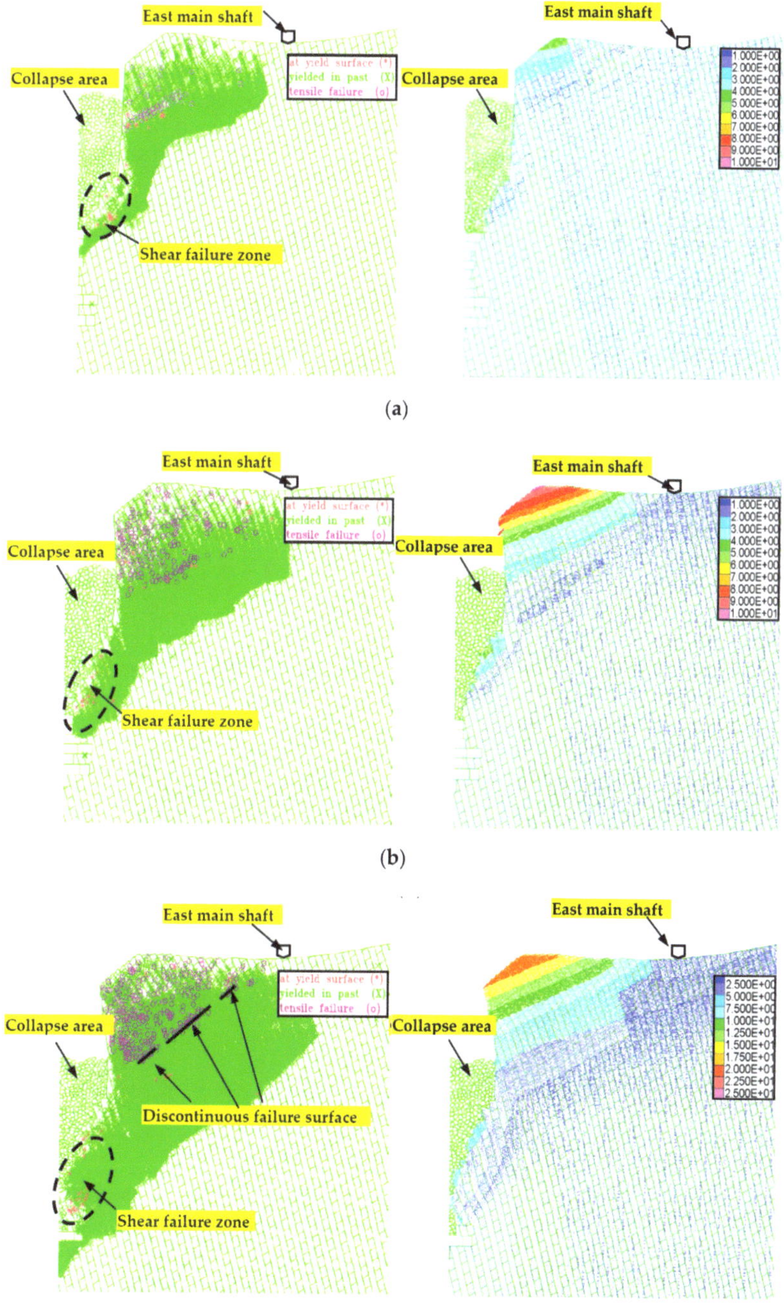

Figure 12. Plastic failure zones and displacement vector diagrams for the footwall at different excavation levels. The three stages shown correspond to depths of: (**a**) −232 m, (**b**) −288 m, and (**c**) −358 m.

After the −232 m level is mined, the displacement of the ground surface is mainly in the horizontal direction and points towards the collapse area (Figure 12a). The stratal displacement exhibits obvious layering; more specifically, the displacement of the shallow strata is greater than that of the deep strata. Flexural-toppling deformation is observed in the footwall near the collapse area. Continued mining activity leads to further stress release in the footwall. This extends the flexural-toppling deformation further and to deeper regions (Figure 12b). After the −358 m level is mined, the area of plastic failure is not obvious in the deep rock masses, though the flexural-toppling deformation has extended to the east main shaft. More specifically, the failure surface is discontinuous. In other words, a through-going failure surface is not produced in the footwall (Figure 12c). This may be because the caved rock mass (or mining waste) provides a certain amount of support to the footwall rock mass, thus limiting the occurrence of flexural-toppling failure [46]. Although a shear failure zone is observed around the mined-out area, it is not very well-developed.

The above results can be confirmed by carrying out on-site investigations of the surface cracks in the area and examining the GPS monitoring data. Figure 13c illustrates the distribution of major cracks in the area near the east main shaft. Four groups of major tension cracks were visible in the area in 2012. The crack shown in Figure 13a, which is 140 m long and 2 m wide, is the most developed. Unsurprisingly, these tension cracks constitute a serious threat to the safe production of ore from the mine. The GPS data shown in Figure 14 make it clear that the horizontal displacement is much larger than the vertical displacement near the east main shaft. This can, to a certain extent, be seen as a sign that flexural-toppling deformation is taking place.

(a)

(b)

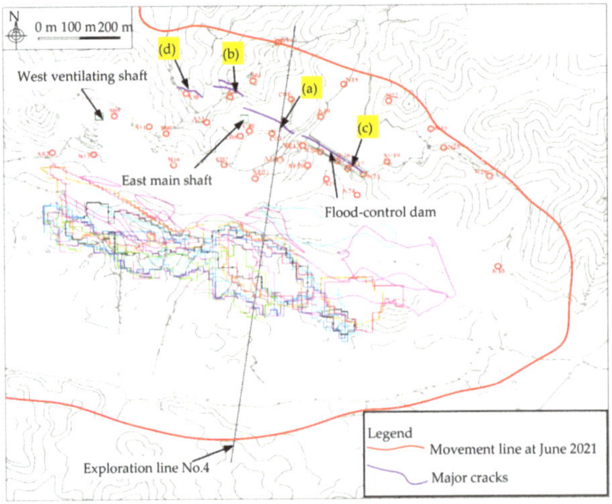

(c)

Figure 13. *Cont.*

(d)

(e)

Figure 13. (**a**) Major crack at the east main shaft, (**b**) major crack near the east main shaft, (**c**) distribution of cracks in the area near the east main shaft, (**d**) major crack near the flood-control dam, and (**e**) major crack near the west ventilating shaft.

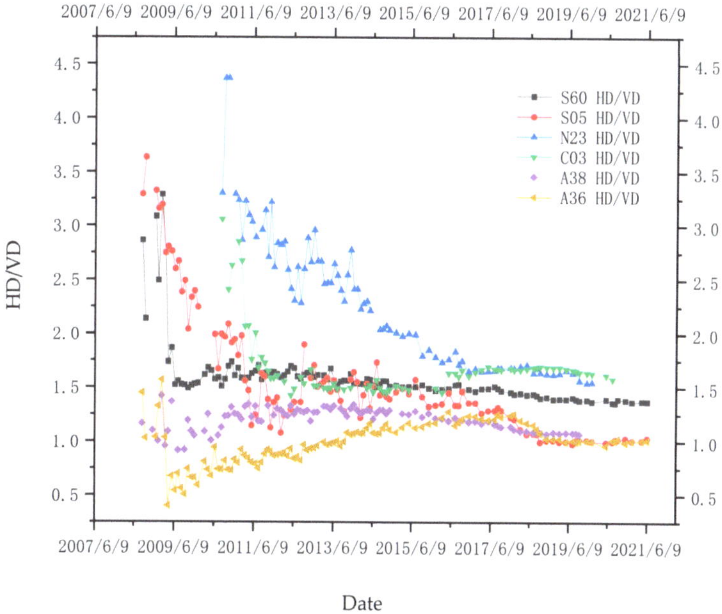

Figure 14. Plots of the HD/VD ratio as a function of time at various GPS monitoring points (HD stands for horizontal displacement; VD stands for vertical displacement).

4.2.2. Deep Rock Mass Failure Stage

Based on the above analysis, it is obvious that flexural-toppling deformation took place during the mining of the −218 to −358 m levels. Figure 15 illustrates the daily change in the displacement of certain representative monitoring points plotted as a function of time (HDR refers to the horizontal displacement rate and VDR refers to the vertical displacement rate).

The displacements of monitoring points E05, S60, and A53 first started to accelerate in 2009, which can be regarded as marking the beginning of the flexural-toppling deformation process. It is worth noting that the above monitoring points also underwent a second period of acceleration in 2017, when the ground surface deformation was in a rapid stage of growth. The occurrence of this second acceleration stage may imply that there is some special feature of the failure mechanism acting in the footwall at this point. If so, the calculation results may be able to clarify the reason for this phenomenon.

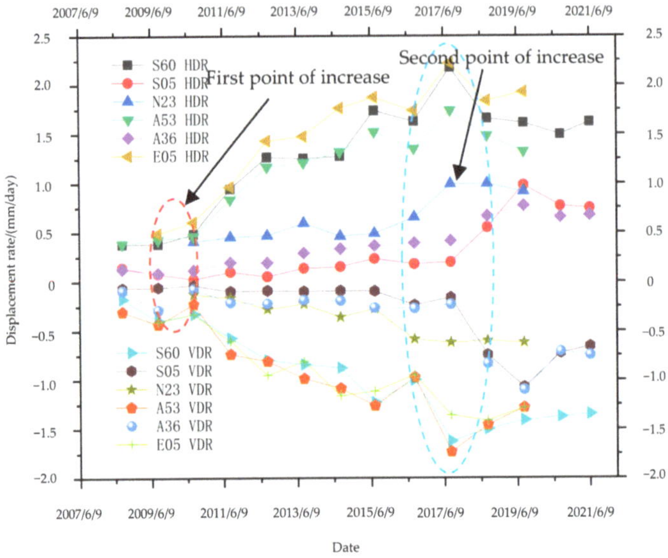

Figure 15. Graphs showing the rate of change of displacement as a function of time at representative monitoring points (HDR denotes the horizontal displacement rate; VDR the vertical displacement rate).

Figure 16 illustrates the plastic failure zones in the footwall when the −375.5, −393, and −410.5 m levels were mined. The caved waste rock exerts a supporting effect on the footwall rock mass, but this effect is weakened as the caved rock flows into the newly mined-out levels. After the −375.5 m level is mined, the flexural-toppling failure of the footwall rock mass is further intensified, and a flexural-toppling failure surface is formed in the deep rock mass (red line in Figure 16a). At the same time, a shear-slip failure surface is formed adjacent to the mined-out area due to the stress concentration formed by the mining disturbance (blue line in Figure 16a). By the time the orebody in the −393 m level has been excavated, the flexural-toppling and shear-slip failure surfaces can be seen to have expanded further, as shown in Figure 16b. One interesting observation is that a compression and slipping-toppling failure surface is formed between the flexural-toppling failure surface and shear-slip failure surface (cyan line in Figure 16; further details of this surface are given later). Under the continued disturbance caused by mining, the compression and slipping-toppling failure surface develops further and eventually forms a through-going failure surface (Figure 16c).

Two monitoring points have been established on each side of the through-going failure surface to monitor the velocity thereof. As shown in Figure 16c, monitoring point A is located between the collapse pit and the through-going failure surface, and monitoring point B is located outside the through-going failure surface. Figure 17 illustrates the velocity of monitoring points A and B as a function of the increasing number of simulation steps. As shown in Figure 17, a slow increase in velocity can be observed at monitoring point B. This means that the area outside the through-going failure surface is less affected by mining disturbance. Compared to monitoring point B, there are three important inflection points, Av_1, Av_2, and Av_3, on the velocity curve of monitoring point A. These are observed after 500,000, 1,750,000, and 3,700,000 steps of the numerical calculation, which correspond to mining of the −218 m, −288 m, and −410.5 m levels, respectively.

More specifically, stage I corresponds to the arch caving development stage. The values of velocity are very small at this stage. The inflection point Av_1 is caused by stress release due to overlying strata caving to the surface. Stage II corresponds to the flexural-toppling stage. The velocity increases slowly and approximately linearly. Continuous deformation

mainly occurs at this stage, and the rock mass is stable. Stage III corresponds to the failure of the deep rock mass. The velocity increases faster, which means that the rock mass is in its transition from the stable to unstable stages. Under the continued disturbance caused by mining, the velocity suddenly increases after Av_3. This means the through-going failure surface is eventually formed, and the rock mass is unstable after mining the −410.5 m level.

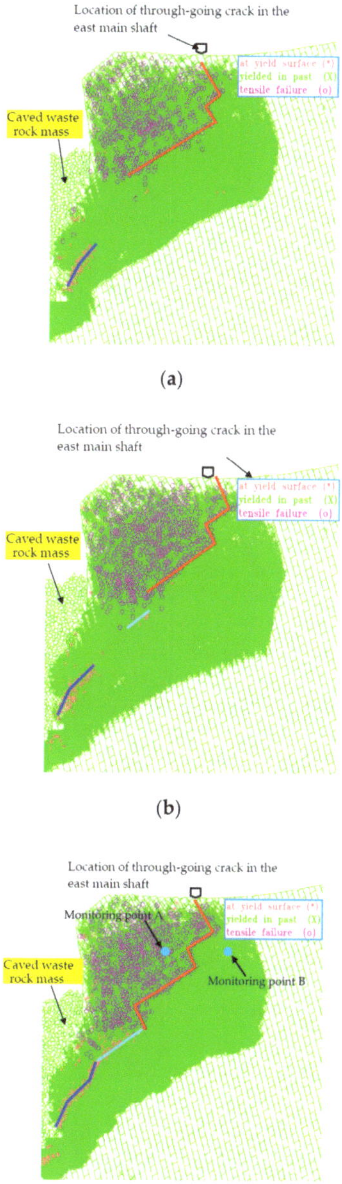

Figure 16. Plastic failure zones in the footwall corresponding to different excavation levels. The three stages shown correspond to depths of: (**a**) −375.5 m, (**b**) −393 m, and (**c**) −410.5 m. Red lines denote the flexural-toppling failure surface; cyan lines denote the compression and slipping-toppling failure surface; and blue lines denote the shear-slipping failure surface.

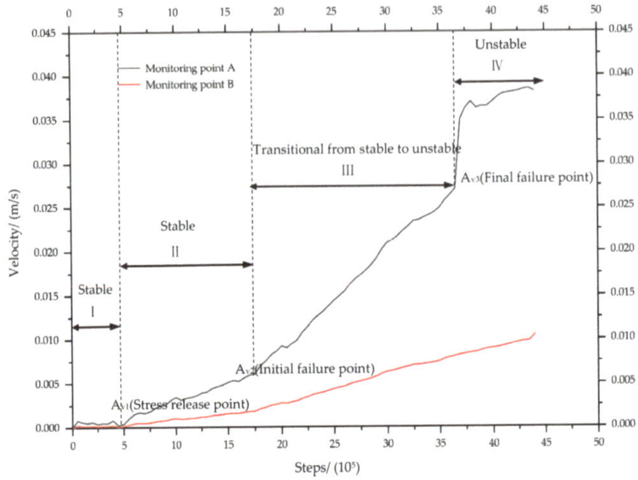

Figure 17. Graphs showing the velocity of monitoring points A and B as a function of simulation steps.

The industrial use of the east main shaft was eventually abandoned because of the through-going crack in the ground surface (Figure 13a). Due to a shortage of land resources in the mining area, the through-going crack was backfilled, and the east main shaft was again used in 2017. The through-going crack again became exposed to the surface at some point in 2020. Damage was then caused to buildings and facilities on the ground surface. The zone between the collapse pit and through-going failure surface is called the 'failure zone', and the parts of the footwall that remain undisturbed form the 'stable zone'. Monitoring points were set up on both sides of the through-going crack in May 2021 in order to monitor the deformation of the crack. The results show that the through-going crack is currently expanding at a rate of about 2 cm per month (Figure 18). Thus, the existence of the through-going failure surface allows the rock mass in the failure zone to move towards the mined-out area at a much faster rate than the rock mass in the stable zone.

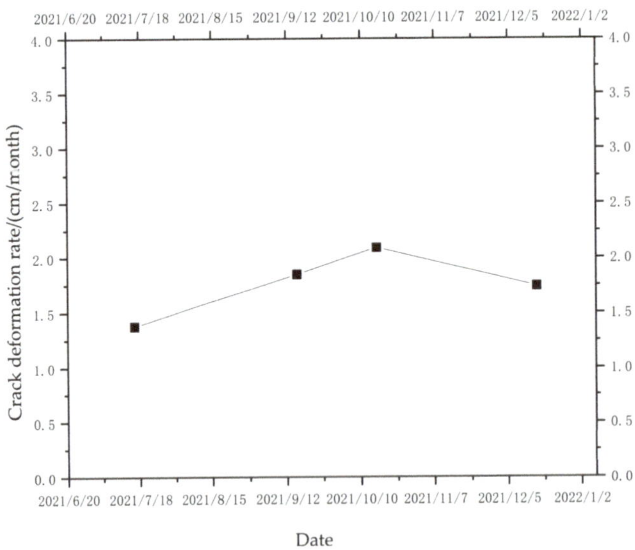

Figure 18. Graphs showing the crack deformation rate as a function of time.

5. Discussion

The footwall can be divided into a stable zone, flexural-toppling failure zone (FTFZ), shear-slipping failure zone (SSFZ), and compression and slipping-toppling failure zone (CSTFZ) according to the failure mechanism operating (Figure 19a). The details are as follows:

1. The FTFZ is located above the flexural-toppling failure surface, which corresponds to the red area in Figure 19a. The formation of the collapse pit causes stress to be released and provides a free surface for the strata to move in the footwall [47]. The existence of highly dipping discontinuities and tectonic stress further induces the occurrence of flexural-toppling deformation. Initially, the flexural-toppling failure surface is not obvious, due to the support provided to the footwall by the caved waste rock. The enlargement of the mined-out area allows the caved waste rock to continuously flow into these areas, which reduces the support it provides to the footwall. Finally, the flexural-toppling failure surface is formed in the deep rock mass after the −373 m level is mined out.
2. The SSFZ is located above the shear-slip failure surface (blue area in Figure 19a). The SSFZ is formed by the concentration of stress due to mining, its existence inducing the rock mass to slide towards the mined-out area. Svartsjaern et al. [48] observed a similar phenomenon in the Kiirunavaara Iron Mine in Sweden. In the early mining stages, the stress concentration has a less obvious effect because the mining depth is shallow and the mined-out area is relatively small. As the mining depth increases, an obvious shear-slip failure surface is formed near the mined-out area in the footwall (by the time the −375.5 m level is mined).
3. The CSTFZ is located above the compression and slipping-toppling failure surface (the cyan area in Figure 19a). The formation of this zone is controlled by the FTFZ and SSFZ. The rock mass in this zone is subjected to compressive forces from the rock mass in the FTFZ. The SSFZ induces it to slip towards the mined-out area. Mining activity further induces the formation of a compression and slipping-toppling failure surface. After mining the −410 m level, the CSTFZ develops further to form a through-going failure surface.
4. The stable zone is outside the through-going failure surface (black area in Figure 19a). As the rock mass in the stable zone is relatively far away from the mined-out area, it is less affected by the mining activity. The through-going failure surface marks the interface between the failure and stable zones. Once the through-going failure surface is formed, the ground surface in the failure zone undergoes a rapid increase in its rate of deformation, and serious damage is caused to the buildings and facilities on the surface (Figure 19b,c). Although microcracks may be observed in the buildings and facilities on the ground surface, the movement of the strata is generally not obvious [49,50]. Mining is a dynamic process, and the rock masses near the through-going failure surface that are currently in the stable zone may be absorbed into the failure zone in the future.

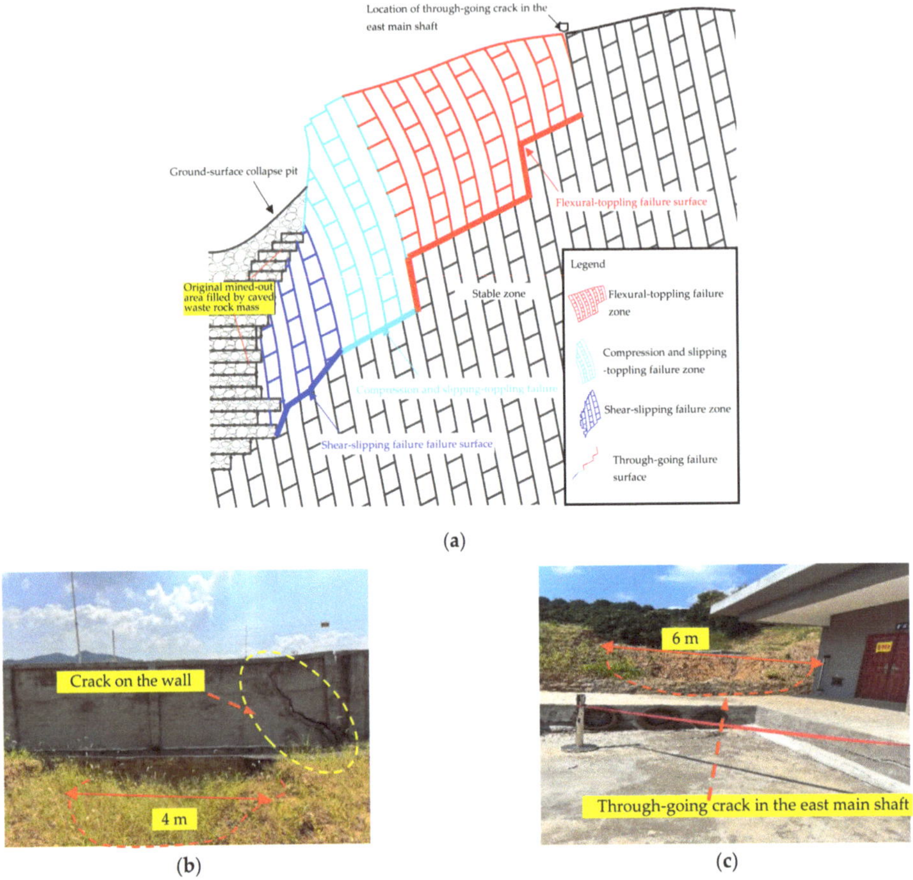

Figure 19. (**a**) Schematic diagram showing the different zones in the rock mass of the footwall. (**b**) Damaged buildings and facilities near the east main shaft. (**c**) The through-going crack in the east main shaft.

6. Conclusions

By analyzing the deformation failure occurring when extracting minerals by underground mining, the following conclusions could be drawn:

1. The movement of the strata in the footwall in the eastern area of the Chengchao Iron Mine seems to occur in different stages. Depending on whether the overlying strata have caved to the ground surface or not, the movement of the footwall strata can be divided into two stages called ACDS and PACDS. The PACDS can be further subdivided into a flexural-toppling stage and deep rock mass failure stage according to the failure mechanism operating.
2. In the ACDS, 'arch-shaped' caving occurs due to the nearby mining activity which leads to a redistribution of the stress in the overburden. Shear failure occurs in the feet of the arch, and tension failure occurs in the vault thereof. The caving process can be divided into three consecutive stages involving slow caving, stable caving, and then rapid caving.
3. In the PACDS, the strata undergo flexural-toppling deformation in the direction of the collapse pit driven by the tectonic stress and presence of steeply dipping discontinuities. The flexural-toppling failure of the rock mass intensifies as the mining activity continues. A flexural-toppling failure surface is formed after the −375.5 m

level is mined out. As the mining activity moves to greater depths, the compression and slipping-toppling failure surface develops further, eventually forming a through-going failure surface. Based on the different failure mechanisms, the footwall can be divided into four zones: a stable zone, flexural-toppling failure zone, compression and slipping-toppling failure zone, and shear-slipping failure zone.

4. The through-going failure surface forms the interface between the failure zone and stable zone. The failure surface is exposed on the ground surface in the form of a well-developed macroscopic crack. After the through-going failure surface is formed, the ground surface in the failure zone undergoes a rapid increase in its rate of deformation. When this occurs, serious damage may be suffered by the buildings and facilities on the ground surface.

Author Contributions: Conceptualization, X.L. (Xuanting Liu) and C.C.; data curation, K.X.; funding acquisition, C.C. and K.X.; Investigation, X.L. (Xiumin Liu); methodology, X.L. (Xuanting Liu) and K.X.; writing—original draft, X.L. (Xuanting Liu) and K.X.; writing—review & editing, X.L. (Xiumin Liu) and T.W. All authors have read and agreed to the published version of the manuscript.

Funding: This work was funded by the Young Scholar Fund of National Natural Science Foundation of China, grant number 42002292.

Data Availability Statement: Data are contained within the article.

Acknowledgments: The authors thank the Chengchao Iron Mine of the Wuhan Iron and Steel (Group) Corporation for providing the information. The authors also thank Zaiqiang Ren, Zaobing Wang, Jiabing Wang, and Zhian He for their assistance in collecting the in situ data from the Chengchao Iron Mine.

Conflicts of Interest: The authors declare no conflict of interest.

References

1. Brady, B.H.G.; Brown, E.T. *Rock Mechanics for Underground Mining*, 3rd ed.; Springer: Berlin, Germany, 2006; pp. 347–368.
2. Khodayari, F.; Pourrahimian, Y.; Liu, W.V. Production Scheduling with Horizontal Mixing Simulation in Block Cave Mining. *J. Min. Sci.* **2019**, *55*, 789–803. [CrossRef]
3. Zarate, E.U.; Pourrahimian, Y.; Boisvert, J. Optimizing block caving drawpoints over multiple geostatistical models. *Int. J. Min. Min. Eng.* **2018**, *34*, 55–74. [CrossRef]
4. Deng, Y.Y.; Chen, C.X.; Xia, K.Z.; Pang, H.S.; Sun, C.Y.; Yang, K.Y.; Zheng, X.W. Investigation on the distribution characteristics of ground cracks in the Chengchao Iron Mine. *Environ. Earth Sci.* **2019**, *78*, 280. [CrossRef]
5. Xia, K.Z.; Chen, C.X.; Zheng, Y.; Zhang, H.N.; Liu, X.M.; Deng, Y.Y.; Yang, K.Y. Engineering geology and ground collapse mechanism in the Chengchao Iron-ore Mine in China. *Eng. Geol.* **2019**, *249*, 129–147. [CrossRef]
6. Edelbro, C.; Sjöberg, J.; Malmgren, L.; Dahnér-Lindqvist, C. Prediction and follow-up of failure and fallouts in footwall drifts in the Kiirunavaara mine. *Can. Geotech. J.* **2012**, *49*, 546–559. [CrossRef]
7. Henry, E.; Dahnér-Lindqvist, C. Footwall stability at the LKAB's Kiruna sublevel caving operation, Sweden. In Proceedings of the 3rd International Conference and Exhibition on Mass Mining, Brisbane, Australia, 29 October–2 November 2000.
8. Kashnikov, Y.A.; Ermashov, A.O.; Efimov, A.A. Geological and Geomechanical Model of the Verkhnekamsk Potash Deposit Site. *J. Min. Inst.* **2019**, *237*, 259–267. [CrossRef]
9. Simmons, B.S.; Wempen, J.M. Quantifying relationships between subsidence and longwall face advance using DInSAR. *Int. J. Min. Sci. Technol.* **2021**, *31*, 91–94. [CrossRef]
10. Huang, P.; Zhang, J.X.; Yan, X.J.; Spearing, A.; Li, M.; Liu, S.W. Deformation response of roof in solid backfilling coal mining based on viscoelastic properties of waste gangue. *Int. J. Min. Sci. Technol.* **2021**, *31*, 279–289. [CrossRef]
11. Brücker, C.; Preuße, A. The future of underground spatial planning and the resulting potential risks from the point of view of mining subsidence engineering. *Int. J. Min. Sci. Technol.* **2020**, *30*, 93–98. [CrossRef]
12. Wang, H.Z.; Zhang, D.S.; Wang, X.F.; Zhang, W. Visual Exploration of the Spatiotemporal Evolution Law of Overburden Failure and Mining-Induced Fractures: A Case Study of the Wangjialing Coal Mine in China. *Minerals* **2017**, *7*, 35. [CrossRef]
13. Marschalko, M.; Yilmaz, I.; Křístková, V.; Fuka, M.; Bednarik, M.; Kubečka, K. Determination of actual limit angles to the surface and their comparison with the empirical values in the Upper Silesian Basin (CzechRepublic). *Eng. Geol.* **2012**, *124*, 130–138. [CrossRef]
14. Marshalko, M.; Yilmaz, I.; Kubechka, K.; Bushal, T.; Bednarik, M.; Druza, M.; Bendova, M. Utilization of ground subsidence caused by underground mining to produce a map of possible land-use areas for urban planning purposes. *Arab. J. Geosci.* **2015**, *8*, 579–588. [CrossRef]

15. Marschalko, M.; Bednárik, M.; Yilmaz, I.; Bouchal, T.; Kubečka, K. Evaluation of subsidence due to underground coal mining: An example from the Czech Republic. *Bull. Eng. Geol. Environ.* **2012**, *71*, 105–111. [CrossRef]
16. Hoek, E. Progressive caving induced by mining an inclined orebody. *Trans. Inst. Min. Metall. Sect. A Min. Technol.* **1974**, *83*, A133–A139.
17. Brown, E.T.; Ferguson, G.A. Prediction of progressive hanging-wall caving, Gaths mine, Rhodesia. *Trans. Inst. Min. Metall. Sect. A Min. Technol.* **1979**, *88*, A92–A105.
18. Lupo, J.F. Evaluation of Deformations Resulting from Mass Mining of an Inclined Orebody. Ph.D. Thesis, Colorado School of Mines, Denver, CO, USA, 1996.
19. Lupo, J.F. Progressive failure of hanging wall and footwall Kiirunavaara Mine, Sweden. *Int. J. Rock Mech. Min. Sci.* **1997**, *34*, 184.e1–186.e11. [CrossRef]
20. Baryakh, A.A.; Tennyson, L.O.; Samodelkina, N.A. Assessment of horizontal deformations in undermined areas. *Min. Inf. Anal. Bull.* **2021**, *11*, 5–18. [CrossRef]
21. Baryakh, A.A.; Fedoseev, A.K.; Lobanov, S.Y.; Matveenko, V.P.; Plekhov, O.A. Deformations and fracture of rock strata during deep level potash mining. In Proceedings of the 22nd Winter School on Continuous Media Mechanics, Moscow, Russia, 22–26 March 2021.
22. Baryakh, A.A.; Samoldekina, N.A. Geomechanical estimation of deformation intensity above the flooded potash mine. *J. Min. Sci.* **2018**, *53*, 630–642. [CrossRef]
23. Hastaoğlu, K.Ö.; Gül, Y.; Poyraz, F.; Kara, B.K. Monitoring 3D areal displacements by a new methodology and software using UAV photogrammetry. *Int. J. Appl. Earth Obs. Geoinf.* **2019**, *83*, 101916. [CrossRef]
24. Modeste, G.; Doubre, C.; Masson, F. Time evolution of mining-related residual subsidence monitored over a 24-year period using InSAR in southern Alsace, France. *Int. J. Appl. Earth Obs. Geoinf.* **2021**, *102*, 102392. [CrossRef]
25. Villegas, T.; Nordlund, E.; Dahnér-Lindqvist, C. Hangingwall surface subsidence at the Kiirunavaara mine, Sweden. *Eng. Geol.* **2011**, *121*, 18–27. [CrossRef]
26. Sitharam, T.G.; Sridevi, J.; Shimizu, N. Practical equivalent continuum characterization of jointed rock masses. *Int. J. Rock Mech. Min. Sci.* **2001**, *38*, 437–448. [CrossRef]
27. Rybak, J.; Khayrutdinov, M.M.; Kuziev, D.A.; Kongar-Syuryun, C.B.; Babyr, N.V. Prediction of the geomechanical state of the rock mass when mining salt deposits with stowing. *J. Min. Inst.* **2022**, *253*, 61–70. [CrossRef]
28. Villegas, T.; Nordlund, E. Numerical analyses of the hangingwall failure due to sublevel caving: Study case. *Int. J. Min. Min. Eng.* **2013**, *4*, 201–223.
29. Hamdi, P.; Stead, D.; Elmo, D.; Toyra, D. Use of an integrated finite/discrete element method-discrete fracture network approach to characterize surface subsidence associated with sub-level caving. *Int. J. Rock Mech. Min. Sci.* **2018**, *103*, 55–67. [CrossRef]
30. Yu, L.; Ignatov, Y.; Ivannikov, A.; Khotchenkov, E.; Krasnoshtanov, D. Common features in the manifestation of natural and induced geodynamic events in the eastern regions of Russia and China. In *IOP Conference Series: Earth and Environmental Science*; IOP Publishing: Bristol, UK, 2019; Volume 324, p. 012004.
31. Sanz Pérez, E.; Sanz Riaguas, C. Identification and Investigation of Subsidence Areas to Mitigate Karstic Risks in Urbanized Areas of Madrid, Spain: A Case Study. *Sustainability* **2021**, *13*, 7716. [CrossRef]
32. Sanz Pérez, E.; Menéndez Pidal, I.; Lomoschitz, A.; Galindo-Aires, R. The Pico de Navas slump (Burgos, Spain): A large rocky landslide caused by underlying clayey sand. *J. Iber. Geol.* **2016**, *42*, 55–68. [CrossRef]
33. Song, X.G.; Chen, C.X.; Xia, K.Z.; Yang, K.Y.; Chen, S.; Liu, X.M. Analysis of the surface deformation characteristics and strata movement mechanism in the main shaft area of Chengchao Iron Mine. *Environ. Earth Sci.* **2018**, *77*, 335. [CrossRef]
34. Xia, K.Z.; Chen, C.X.; Yang, K.Y.; Zhang, H.N.; Pang, H.S. A case study on the characteristics of footwall ground deformation and movement and their mechanisms. *Nat. Hazards* **2020**, *104*, 1039–1077. [CrossRef]
35. Xia, K.Z.; Chen, C.X.; Liu, X.M.; Zheng, Y.; Fu, H. Ground movement mechanism in tectonic stress metal mines with steep structure planes. *J. Cent. South Univ.* **2017**, *24*, 2092–2104. [CrossRef]
36. Cheng, G.W.; Chen, C.X.; Ma, T.H.; Liu, H.Y.; Tang, C.A. A Case Study on the Strata Movement Mechanism and Surface Deformation Regulation in Chengchao Underground Iron Mine. *Rock Mech. Rock Eng.* **2016**, *50*, 1011–1032. [CrossRef]
37. Deng, Y.Y.; Chen, C.X.; Xia, K.Z.; Yang, K.Y.; Sun, C.Y.; Zheng, X.W. Investigation on the characteristics of overlying strata caving in the Chengchao Iron Mine, China. *Environ. Earth Sci.* **2018**, *77*, 362. [CrossRef]
38. Brunton, I.D.; Fraser, S.J.; Hodgkinson, J.H.; Stewart, P.C. Parameters influencing full scale sublevel caving material recovery at the Ridgeway gold mine. *Int. J. Rock. Mech. Min Sci.* **2010**, *47*, 647–656. [CrossRef]
39. Han, Z.Y.; Li, D.Y.; Zhou, T.; Zhu, Q.Q.; Ranjith, P.G. Experimental study of stress wave propagation and energy characteristics across rock specimens containing cemented mortar joint with various thicknesses. *Int. J. Rock Mech. Min. Sci.* **2020**, *131*, 104352. [CrossRef]
40. Cao, S.; Song, W.D.; Deng, D.; Lei, Y.K.; Lan, J.Q. Numerical simulation of land subsidence and verification of its character for an iron mine using sublevel caving. *Int. J. Min. Sci. Technol.* **2016**, *26*, 327–332. [CrossRef]
41. Zhang, D.; Bai, J.B.; Yan, S.; Wang, R.; Meng, N.K.; Wang, G.Y. Investigation on the Failure Mechanism of Weak Floors in Deep and High-Stress Roadway and the Corresponding Control Technology. *Minerals* **2021**, *11*, 1408. [CrossRef]
42. Hoek, E.; Carranza-Torres, C.T.; Corkum, B. Hoek–Brown failure Criterion–2002 Edition. In Proceedings of the Fifth north American Rock Mechanics Symposium, Toronto, ON, Canada, 7–10 July 2002.

43. Xia, K.Z.; Chen, C.X.; Wang, T.L.; Zheng, Y.; Wang, Y. Estimating the geological strength index and disturbance factor in the Hoek–Brown criterion using the acoustic wave velocity in the rock mass. *Eng. Geol.* **2022**, *306*, 106745. [CrossRef]
44. Gao, F.Q.; Stead, D.; Coggan, J. Evaluation of coal long-wall caving characteristics using an innovative UDEC Trigon approach. *Comput. Geotech.* **2014**, *55*, 448–460. [CrossRef]
45. Sun, C.Y.; Chen, C.X.; Zheng, Y.; Zhang, W.; Liu, F. Numerical and theoretical study of bi-planar failure in footwall slopes. *Eng. Geol.* **2019**, *260*, 105234. [CrossRef]
46. Svartsjaern, M.; Saiang, D. Discrete element modelling of footwall rock mass damage induced by sub-level caving at the Kiirunavaara Mine. *Minerals* **2017**, *7*, 109. [CrossRef]
47. Cheng, G.W.; Chen, C.X.; Li, L.C.; Zhu, W.C.; Yang, T.H.; Dai, F.; Ren, B. Numerical modelling of strata movement at footwall induced by underground mining. *Int. J. Rock Mech. Min. Sci.* **2018**, *108*, 142–156. [CrossRef]
48. Svartsjaern, M.; Saiang, D.; Nordlund, E.; Eitzenberger, A. Conceptual numerical modeling of large-scale footwall behavior at the Kiirunavaara Mine, and implications for deformation monitoring. *Rock Mech. Rock Eng.* **2016**, *49*, 943–960. [CrossRef]
49. Pang, H.S.; Chen, C.X.; Xia, K.Z.; Deng, Y.Y.; Zhang, C.Q.; Sun, C.Y. A methodology based on strain analysis for identifying potential discontinuous deformation zones in sublevel caving mines. *Eng. Geol.* **2020**, *279*, 105872. [CrossRef]
50. Deng, Y.Y.; Chen, C.X.; Xia, K.Z.; Sun, C.Y.; Chen, S.; Zhang, Y.P.; Zheng, X.W. A field investigation on the distribution characteristics of structure cracks in the Chengchao Iron Mine, China. *Arab. J. Geosci.* **2020**, *13*, 231. [CrossRef]

minerals

Article

Full-Field Deformation and Crack Development Evolution of Red Sandstone under Impact and Chemical Erosion

Junce Xu [1], Hai Pu [1,2,*] and Ziheng Sha [1]

1 State Key Laboratory for Geomechanics & Deep Underground Engineering, China University of Mining & Technology, Xuzhou 221116, China
2 College of Mining Engineering and Geology, Xinjiang Institute of Engineering, Urumqi 830091, China
* Correspondence: haipu@cumt.edu.cn

Abstract: Coal mine reuse involves complex environments such as chemical erosion and dynamic perturbation. Therefore, the effect of chemical erosion on the dynamic behavior of the red sandstone was studied by split Hopkinson pressure bar (SHPB) tests under the strain rates of 70~125 s^{-1}. The full-field deformation of the sample was then recorded through high-speed 3D digital image correlation (3D-DIC) technique. The dynamic deformation characteristics, especially the lateral strain, were extracted by averaging the lateral strain field by pixels. Also, the fracture behavior was investigated based on the evolution of strain localization in the strain field. The results indicated that the deformation field evolution of the sample is controlled by the chemical erosion effect and the loading strain rate. The chemical erosion lowers the stress threshold for strain localization and accelerates its expansion rate, which is closely related to the dynamic strength degradation of the sample. In contrast, the loading strain rate increases the dynamic strength but advances the occurrence of strain localization and shortens the time to the peak stress. The normalized stress thresholds for the initiation and development of cracks inside the sample under dynamic loading are reduced by chemical erosion, with the two thresholds dropping to 10%~30% and 20%~70% of the peak stress, respectively. The minimum thresholds for the initiation and development of cracks inside the red sandstone under dynamic loading are 11% and 24% of the peak stress, respectively.

Keywords: red sandstone; chemical erosion; strain rate; SHPB; 3D-DIC

Citation: Xu, J.; Pu, H.; Sha, Z. Full-Field Deformation and Crack Development Evolution of Red Sandstone under Impact and Chemical Erosion. *Minerals* **2022**, *12*, 1038. https://doi.org/10.3390/min12081038

Academic Editor: Giang Nguyen

Received: 28 July 2022
Accepted: 16 August 2022
Published: 18 August 2022

1. Introduction

As part of the global effort to reduce carbon dioxide emissions, out-of-life coal mines, especially deep ones, are valuable resources for potential clean energy storage and carbon sequestration [1,2]. However, the utilization of these mines inevitably involves consideration of complex geological environments as well as various loading effects. For example, an abandoned mine was reconstructed as an underground pumped storage hydropower plant [3]. In this case, the impact loading effects on the dynamic behavior of the surrounding rocks should be evaluated [4]. Meanwhile, the presence of low valence iron or manganese ions significantly lowers the pH of the mine water as a result of aeration [5]. Acid mine water not only erodes the equipment, but also affects the stability of the surrounding rock, which affects the operating efficiency and safety of the whole system. Therefore, research on the dynamic mechanical behavior of rock in complicated environments plays an important role in repurposing out-of-life mines.

Numerous studies have been carried out to reveal the deformation and failure characteristics of rocks in chemical environments, and have yielded many helpful results [6–10]. The most important factors affecting the mechanical properties of rocks are the pH and the duration of immersion in solutions [8]. The chemical composition, ion concentration, and acidity of the solution also have a significant effect on the rock failure [6]. Acidic solutions, in comparison to alkaline solutions, aggravate the degradation of rock microstructure by

dissolving minerals [9,11]. Consequently, the dissolution kinetics of rock minerals is also a key factor affecting their damaged properties [12]. However, dynamic loading is more sensitive to microstructural changes in rocks than quasi-static loading [10,13]. For example, the strength loss of rocks subjected to chemical erosion is greater under dynamic loading than under static loading [14]. In addition, chemical erosion reduces the dynamic elastic modulus and fracture toughness of rocks [15]. The differences in dynamic properties are mainly due to the changes in microstructure and mineral elements caused by chemical erosion [16]. These studies are essential for understanding the mechanical properties of rocks under chemical erosion. However, unlike static loading, it is not possible to directly evaluate sample deformation and fracturing evolution under impact loading using the strain gauges, acoustic emissions (AE) or CT methods [17].

In order to understand the dynamic behavior and fracture process of rocks, it is necessary to closely monitor the deformation and evolution of the strain field [18,19]. Meanwhile, the strain localization evolution is crucial for predicting the structural decay of the rock [20]. Recently, with the progress of computer technology, the digital image correlation (DIC) technique has been developed and successfully applied to rock mechanics [21–27]. The DIC method is done by tracing the regional changes in gray-scale features between two successive images before and after deformation [28]. Some studies have found that the fracture behavior of rocks under static loading is closely related to the strain field development by the DIC method [29]. For example, Li et al. [21] analyzed the deformation and fracture process of marble using the 2D-DIC method, and found that strain localization affects the development of cracks; Tang et al. [30] investigated the effect of confining pressure on the progressive fracture behavior of the rock using the 3D-DIC system, and concluded that strain localization develops rapidly in the post-peak due to the crack propagation. In addition, several studies have used the DIC method to investigate the behavior of rock materials under impact loading. Sharafisafa et al. [31] explored the dynamic fracture patterns of rock-like materials under impact loading using the 3D-DIC method; Xing et al. [20] employed the 3D-DIC technique in impact tests and discovered strain recovery in the post-peak phase. These efforts demonstrated that the DIC method is suitable for studying the deformation and failure of rocks under impact loading. Although the DIC method has been used to analyze rock deformation and fracture processes, little research has focused on impact loading conditions, particularly chemical erosion.

Therefore, in this study, a series of dynamic tests were performed on sandstone samples with chemical erosion. The effects of chemical erosion on deformation characteristics and fracture behavior were also investigated using the high-speed 3D-DIC technique. Finally, the stress thresholds for crack development in the sample under the influence of chemical decay and loading strain rate were studied. This study can be a reference for the use of abandoned mines under complex geological conditions.

2. Material and Methods

2.1. Materials and Sample Preparation

The red sandstone used in this study was obtained from a coal mine in Shandong, China. According to the ISRM recommendations, the rock blocks were processed into standard samples with a diameter of 50 mm and an aspect ratio of 1. All samples were taken from the same rock block and sampled in the same direction. Then the P-wave velocity of the samples was measured, and those with similar velocities were selected for further testing. Finally, a total of 60 samples were selected and divided into four subgroups, numbered D1 to D4, for impact testing. Five samples with a length and diameter of 50 × 100 mm were selected for physical and mechanical tests, with the results listed in Table 1. In addition, X-ray diffraction (XRD) tests were performed to determine the mineral composition of the sample. The results revealed that the sandstone is mainly composed of quartz, orthoclase, chlorite, and muscovite, with minor amounts of calcite and albite (Figure 1).

Table 1. Static physical and mechanical parameters of the red sandstone.

V_s/(m/s)	V_p (m/s)	ρ_d/(g·cm^{-3})	ρ_{sat}/(g·cm^{-3})	E_s/(GPa)	σ_s/(MPa)
1247.2	2218.6	2.42	2.49	9.35	63.5

Notes: V_P, P-wave velocity; V_S, S-wave velocity; ρ_d, dry density; ρ_{sat}, saturation density; E_s, elastic modulus; σ_s, static peak strength.

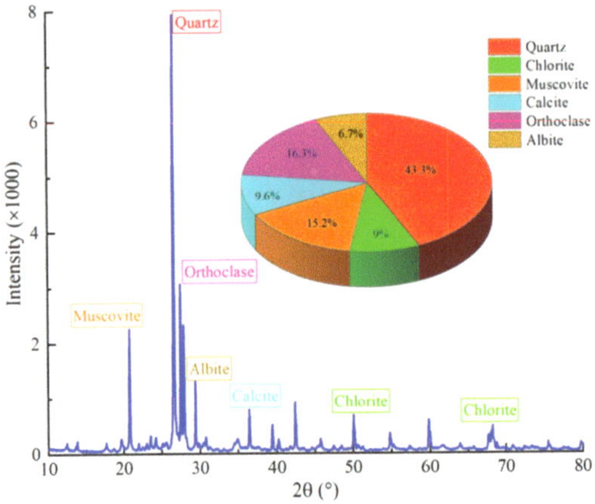

Figure 1. XRD result for the red sandstone.

In order to simulate the erosion process induced by underground mine water, an acidic solution with a pH of 3 was configured by means of hydrochloric acid, sodium chloride, potassium bisulfate, and pure water [4]. Four groups of the sample were immersed in the solution for 0 (D1), 7 (D2), 14 (D3), and 21 (D4) days, respectively. The chemical solution was changed every 7 days to accelerate the erosion rate. After all samples were immersed, they were dried and then subjected to impact loading. In this test, there were three samples for each loading strain rate and erosion condition, so a total of 60 samples were prepared for the five loading strain rates. As shown in Figure 2, XRD tests were also performed on the soaked samples.

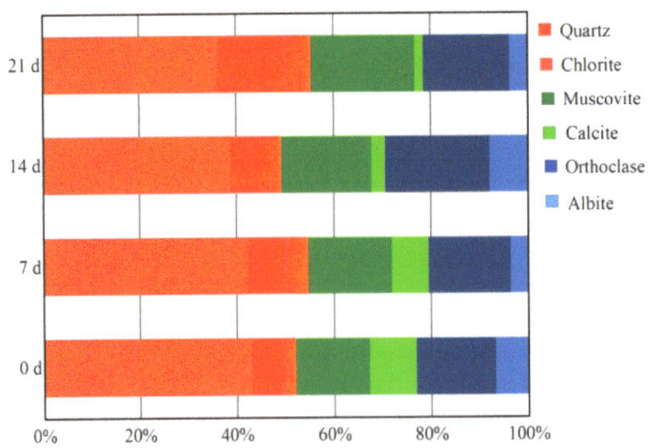

Figure 2. Variation of sample components under the chemical erosion effect.

2.2. Experimental System

As shown in Figure 3, the dynamic impact tests were performed using a SHPB system, which consists of a gas gun, a cylindrical striker, an incident bar, a transmitted bar, and an absorption bar. In the experimental system, the bars and the striker are made of high-strength Cr40 alloy steel with diameter, P-wave velocity, yield strength and elastic modulus

of 50 mm, 5400 m/s, 800 MPa and 208 GPa, respectively. To reduce the effect of friction between the bar and the sample on the experimental results, molybdenum disulfide was adopted as a lubricant [13]. Also, a round rubber pad with a diameter of 15 mm and a thickness of 2 mm was chosen as the pulse shaper [32]. When the high-pressure gas drives the cylindrical striker against the incident bar, a compressive stress wave is generated. A portion of the compressive wave can propagate through the sample to the transmitted bar. At this time, strain signals are generated in the bar, and can be measured by strain gauges on the incident and transmitted bars. Based on the one-dimensional stress wave propagation theory and the stress equilibrium assumption [33], the three-wave analysis method was used in this work to calculate the load and displacement of the samples. Therefore, the dynamic loads $P(t)$, strain rate $\dot{\varepsilon}(t)$ and strain $\varepsilon(t)$ were calculated by Equations (1)–(3) [32].

$$P_1(t) = A_r E_0[\varepsilon_I(t) + \varepsilon_R(t)],\ P_2(t) = A_r E_0 \varepsilon_T(t) \tag{1}$$

$$\dot{\varepsilon}(t) = \frac{C_0}{L}[\varepsilon_I(t) - \varepsilon_R(t) - \varepsilon_T(t)] \tag{2}$$

$$\varepsilon(t) = \int_0^t \dot{\varepsilon}(t)dt = -2\frac{C_s}{L}\int_0^t \varepsilon_R(t)dt \tag{3}$$

where A_r is the ratio of the cross-sectional area of the bar to the sample; E_0, C_s, and L are the elastic modulus of the bar, the P-wave velocity, and the length of the sample, respectively; $\varepsilon_I(t)$, $\varepsilon_R(t)$, and $\varepsilon_T(t)$ refer to the incident, reflected, and transmitted strain, respectively. In addition, the striker impact velocity was tested by a laser measurement system.

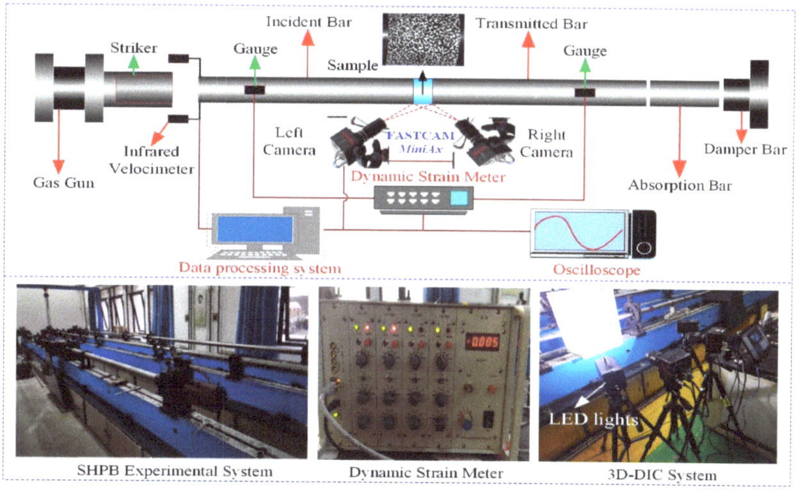

Figure 3. SHPB and 3D-DIC systems.

In addition, a 3D-DIC system was employed to capture the dynamic fracture process, full-field displacement, and strain evolution of the sample. The 3D-DIC system mainly consists of two high-speed complementary metal-oxide semiconductors (CMOS) cameras, two light-emitting diodes (LED), a set of camera tripods, and a computer for images acquisition. As presented in Figure 3, two 75 W LEDs provided about 15,000 lumens of illumination to improve image quality. Two high-speed CMOS cameras (FastCam AX200 developed by Photron Corporation, Tokyo, Japan) were selected in this work. The resolution was set to 256 × 256 pixels with a frame rate of 100,000 fps.

Before testing, the surface of the sample was first sprayed with matte white paint as the base color. Then, random spots were created with matte black paint, which finally formed a typical scatter pattern, as shown in Figure 3. The quality of the scatter pattern is an important carrier of deformation information on the sample surface and has a significant

effect on the accuracy and reliability of the measurement results [28]. In general, a good scatter pattern should have the characteristics of randomness, isotropy, and high contrast, i.e., the pattern should have no directional deviation, and show only black and white. During the test, the camera window covers 60 × 60 mm^2, and the region of interest in the target view is about 48 × 34 mm^2. So, the pixel size is 0.234 mm corresponding to the physical size of the sample. The accuracy of 3D-DIC is controlled to 5% of the pixel [20], which is about 11.7 μm, corresponding to 0.0234% of the engineering strain, thus meeting the accuracy requirements. In addition, the influence of rock fragments and human factors on the camera position during the experiment should be considered.

2.3. 3D-DIC Basic Principles

The DIC method can be divided into 2D-DIC and 3D-DIC. 2D-DIC measurement has some errors due to the influence of out-of-plane deformations [28]. At the same time, 2D-DIC is not suitable for curved surfaces. For instance, the measurement error for the deformation of a cylindrical sample is about 21% [20]. Therefore, the 3D-DIC method was developed to overcome the limitations of the 2D-DIC method. The 3D-DIC method uses binocular stereo vision technology to determine the actual 3D position of each point on the object by capturing the images from different angles with two cameras simultaneously [34]. As illustrated in Figure 4, two 3D spatial points K_1 and K_2 are projected as a single image point k on the plane I in the same projection ray (single camera), indicating an infinite number of 3D points corresponding to point k on the camera. Conversely, both planes can determine the unique point K_1 in space from the projection points (k, k_1). Similarly, the projection point (k, k_2) corresponds to the unique point K_2 in space. If the camera is calibrated, the exact spatial coordinates of the target points K_1 and K_2 can be obtained by triangulation.

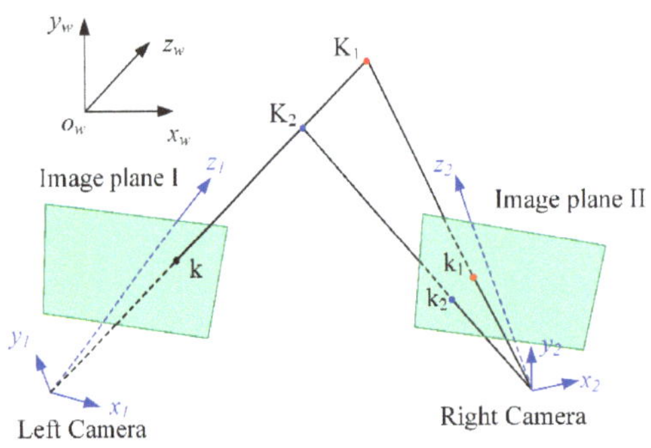

Figure 4. Principle of the binocular stereo vision in 3D-DIC method.

Another important component of the 3D-DIC method is digital image correlation [23]. High-speed cameras take a series of photographs during the testing, with the undeformed image as the reference image and the distorted image as the target image. As shown in Figure 5, the images are divided into small subsets of windows, also called regions of interest (ROI). In the reference image, a subset of (2m + 1) × (2m + 1) pixels with center C was selected. Then a subset with the maximum similarity to this reference subset was found in the target image. To evaluate the intensity similarity between the reference subset and the target subset, the zero-mean normalized sum-of-squares difference (ZNSD) criterion was usually used to search and determine the position of the corresponding target subset [28]. As a result, the centers C and C_0 can be considered the corresponding points. Once the displacement vector of the centers (white arrow) was determined, the pixel points in the deformation subset can be obtained from the reference subset of pixel points by first- or second-order shape functions.

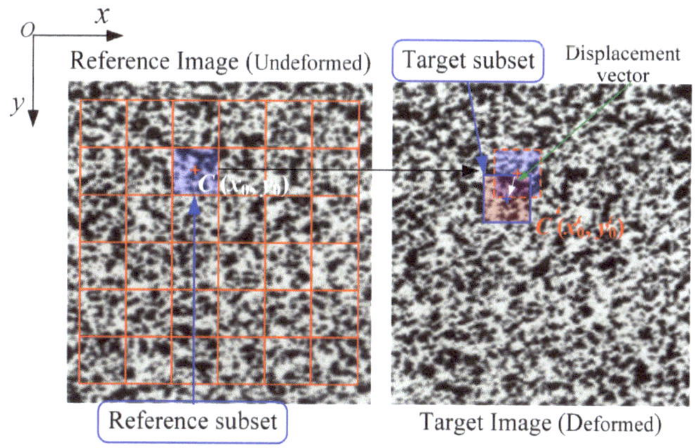

Figure 5. Schematic diagram of the original subset and the corresponding deformation subset.

In addition to the temporal alignment described above, the stereo alignment of the images, i.e., the calibration of the coordinates and transformation matrices of the two cameras in space, is an important part of the 3D-DIC alignment process. For this purpose, the stereo alignment of the images can be performed using either the polar linear constraint method or the iterative least squares method [34], which provides a 3D marker of the object surface at each stage.

3. Experimental Results and Analysis

In this work, the dynamic changes in the deformation and strain fields of the sample were evaluated using two high-speed CMOS cameras and 3D-DIC software. In addition, the dynamic fracture process of the sandstone samples was investigated.

3.1. Dynamic Stress-Strain Behavior

The stress-strain curves calculated using Equations (1)–(3) are presented in Figure 6. It can be seen that the dynamic strength decreases with increasing erosion time, while gradually increases with the strain rate. At a loading strain rate of about 120 s^{-1}, the dynamic strength of the sample decreased by 21.2% from 119.1 to 93.9 MPa with increasing immersion time from 0 to 21 days. However, when the loading strain rate was increased from 75 to 125 s^{-1}, the dynamic strength of the sample (14 d) increased from 75.9 to 102.5 MPa, which is about 35%. The reaction of the H ions in the solution with the minerals of the sample leads to structural degeneration, resulting in a loss of dynamic strength. As erosion increases, the stress-strain curves also exhibited a typical early compression phase. This effect was caused by dissolved minerals such as calcite in acidic liquids (Figure 2). It is worth noting that the stress-strain curves were calculated from the strain gauge signals and do not reveal the dynamic damage process of the samples. Therefore, to study the dynamic damage process of the sample, it is necessary to obtain the evolution characteristics of the strain and displacement fields during the dynamic loading.

3.2. Displacement and Strain Field Characteristics

The variations of the stress-strain curves are the external reflection of the erosion damage. The stress-strain curves obtained from Equations (1)–(3), however, cannot reflect the actual deformation process of the sample. Therefore, the deformation process of the sample was recorded by the high-speed 3D-DIC method, where ROI corresponds to a resolution of 145 × 209 pixels. Figure 7 shows the strain field of the 3D-DIC compared with the images from the CMOS cameras. It can be seen that the strain characteristics during loading are closely related to the crack development on the surface of the sample. Consequently, the

3D-DIC approach is able to collect credible data on the surface deformation of the sample during dynamic loading.

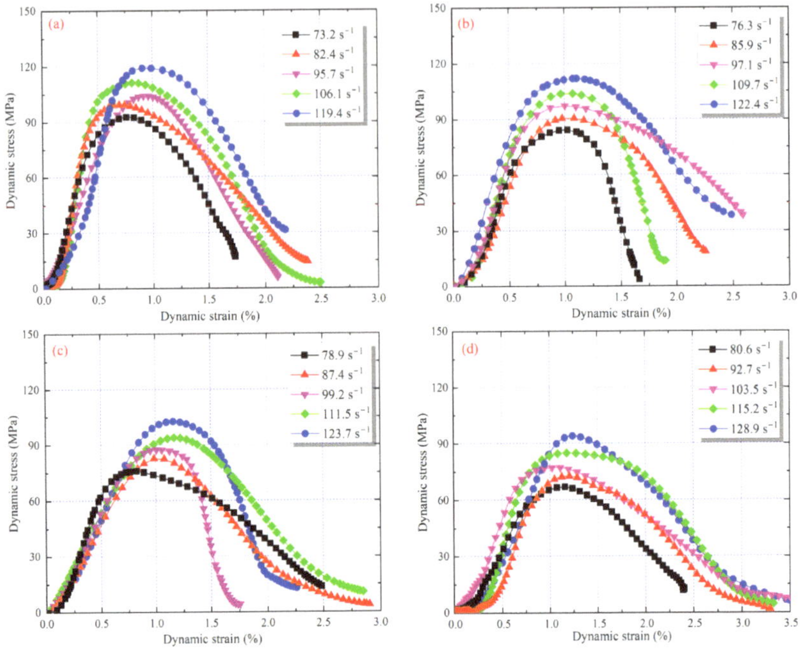

Figure 6. Dynamic stress-strain curves with various immersion time as measured by strain gauges: (**a**) 0 d; (**b**) 7 d; (**c**) 14 d; (**d**) 21 d.

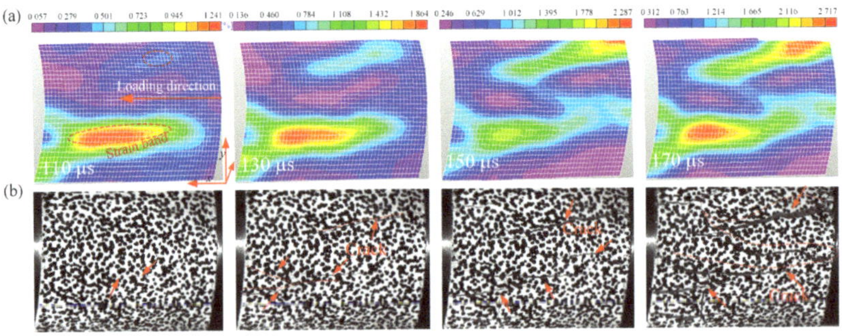

Figure 7. Strain filed by DIC (**a**) and the corresponding CMOS cameras images (**b**).

3.2.1. Evolution of the Displacement Field

To obtain the variation relationship of the displacement field with loading, eight representative stages were selected for the sample. By identifying the initial strain field on the surface, the strain measured by 3D-DIC, and the stress recorded by the strain gauges were synchronized to match the stress and displacement images. The evolution of the displacement field with the loading (120 s^{-1}) under erosion for 21 d was shown in Figure 8. Due to the limited frame rate of the high-speed cameras, only one image was captured at 10 μs intervals. Figure 8 shows an obvious compressive deformation field (compression in DIC is positive) on the right side of the sample upon contact with the striker. As the loading increases, the deformation field gradually shifts to the left side, indicating that the compressive stress wave propagates in the x-direction. At t = 50 μs, the displacement concentration phenomenon appears in the upper right corner of the sample. This is due to

the horizontal expansion deformation originating from both edges. Then, the horizontal deformation on both sides of the sample gradually expanded toward the center. Therefore, when the stress reached its peak, a pronounced shear displacement zone was formed in the center region of the sample. When the strain rate decreased to 93 s^{-1} (Figure 9), the time to peak stress increased to 110 μs, indicating that the time to peak stress is negatively correlated with the strain rate. When the stress increased to 35% of the peak stress, the displacement concentration appeared at the position of the lower right corner of the sample. As the loading time increased, the displacement concentration gradually developed toward the center. After the peak stress, the displacement concentration gradually expanded, and visible shear cracks appeared on the surface. In general, the higher the strain rate, the lower the stress threshold of displacement concentration and the shorter the time required to reach the peak stress.

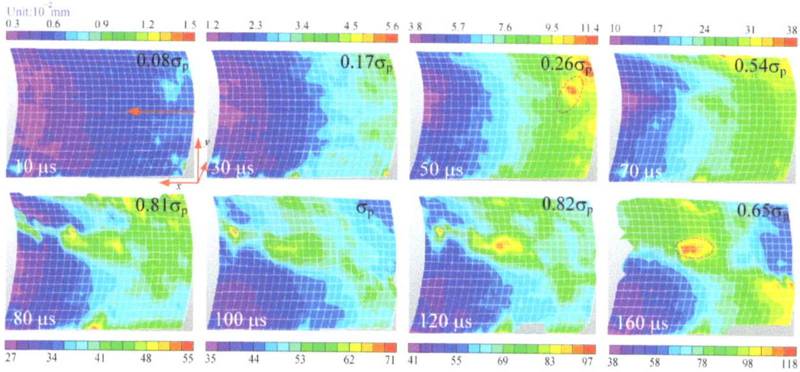

Figure 8. Displacement field evolution of the sample (21 d) at a strain rate of 128 s^{-1}.

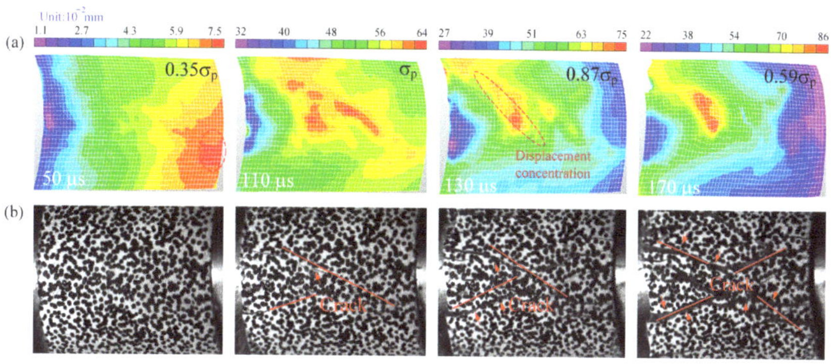

Figure 9. Displacement field evolution of the sample (21 d) at a strain rate of 92 s^{-1}. (**a**) DIC images; (**b**) CMOS cameras images.

In addition, Figure 10 shows the evolution of the displacement field for the eroded 7 d sample with a strain rate of 122 s^{-1}. Compared to the 21 d eroded samples, the occurrence of the edge displacement concentration is 52% of the peak stress, indicating that chemical erosion lowered the stress threshold for the occurrence of the displacement concentration in the sample. The reaction of H ions with calcite in the sandstone sample is the main reason for the phenomenon described above. After the peak stress, visible cracks appeared on the sample surface, similar to other erosion samples. In other words, visible microcracks hardly appear on the surface of the sample before it reaches the peak. In addition, the time required for the sample to reach the peak stress depends on the erosion effect and loading conditions, and therefore does not exhibit a clear pattern. When a large deformation occurs

on the surface of the sample, it leads to a spalling of the scatter points and thus to the appearance of defects in the displacement field.

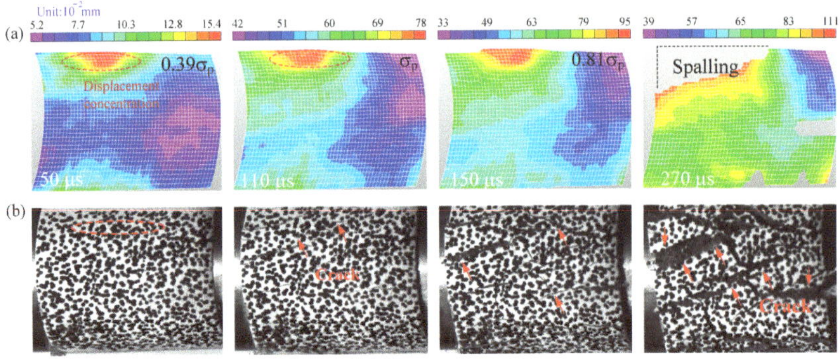

Figure 10. Displacement field evolution of the sample (7 d) at a strain rate of 119 s^{-1}. (**a**) DIC images; (**b**) CMOS cameras images.

3.2.2. Evolution of the Strain Field

Chemical erosion caused by acid mine water significantly affects the surface displacement field (Figures 8–10). Unlike the displacement field, the strain field can provide the process of strain localization distribution developing with the loading time [18], and thus reflect the time-domain characteristics of cracking from the side. Therefore, the strain field of the sample was determined by the 3D-DIC method. Figure 11 shows the evolution of the strain field of the sample (7 d) with a strain rate of 120 s^{-1}. It is obvious that chemical erosion has a significant effect on the strain localization process. The strain localization first appeared at 50 μs in the horizontal strain field of the sample when the loading stress was 39% of the peak stress. As the stress increased, more horizontal strain localization generated from the top and bottom edges and then expanded toward the center of the sample. When the peak stress was reached, shear strain localization formed at the surface due to the different stiffness inside the sample. The strain localization gradually penetrates into the surface and forms a larger shear strain band as the loading increases. After reaching the peak stress, the surface strain band continues to develop and coalesce. This is attributed to the continued release of elastic energy stored during compression. In addition, the development of the shear strain band is also an external manifestation of the increased rate of strength degradation. Moreover, the local strain decreased after cracks appeared on the surface of the sample in Figure 11, a phenomenon that is difficult to observe in dynamic stress-strain curves. Meanwhile, the decrease in local strain during the unloading stage indicated that the red sandstone belongs to the class II rock under dynamic impact conditions, i.e., the post-peak strain decreases, and the fracture extension is not stable. Hence, the strains determined by the DIC method, and the strain gauges differ from each other. The main reason for this discrepancy is that the strains measured by DIC and strain gauges come from two different data sources [20].

In addition, Figure 12 presents the strain fields at 120 s^{-1} in the y-direction (Syy), x-direction (Sxx), and xy-direction (Sxy) on the surface of the sample eroded for 21 days. The stress threshold for strain localization decreased from 39% to 28% of the peak stress compared to the non-eroded samples, while strain localization developed earlier at the horizontal edges due to radial expansion. After that, the horizontal strain localization occurs from both sides of the ROI and gradually moves to the center during loading. Then, a shear strain band appeared on the surface at 120 μs and connected the previous horizontal strain bands. Not until 130 μs did horizontal cracks appear in the horizontal strain field; in other words, the surface was cracked at the post-stress peak stage. However, as can be seen in Figure 6, there is a phase of crack development before the peak characterized by a slow decrease in elastic modulus, suggesting that the cracks initially develop inside the

rock. When the stress drops to 48% of the peak stress, visible shear cracks appear at 160 μs in the ROI. In contrast, at t = 170 μs, a vertical strain band appears at ROI and transitions into a tensile crack at 230 us. In fact, the tensile wave generated by the reflection at the interface is the cause of this spalling crack [17]. Thus, with increasing chemical corrosion, the crack development on the sample surface was more intense, which can be attributed to the deterioration of the internal structural strength.

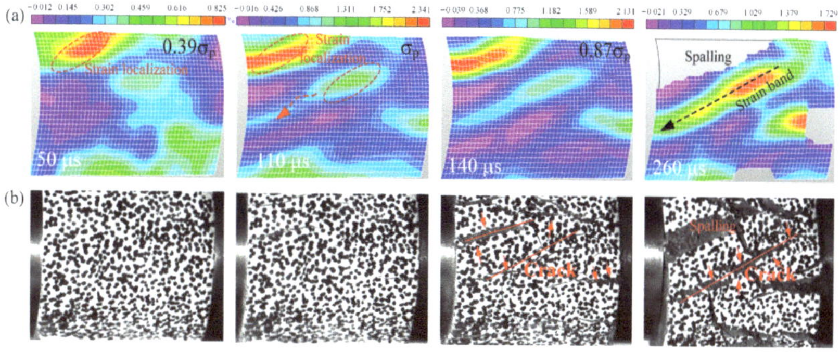

Figure 11. Strain field evolution of the sample (7 d) at a strain rate of 120 s^{-1}. (**a**) DIC images; (**b**) CMOS cameras images.

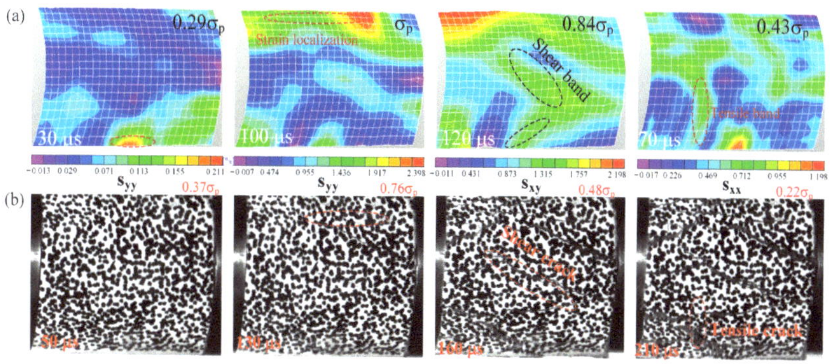

Figure 12. Strain field evolution (21 d) at a strain rate of 120 s^{-1}. (**a**) DIC image; (**b**) CMOS cameras images; Sxx, Syy and Sxy refer to the strain fields in x, y and xy directions, respectively.

In summary, the 3D-DIC method is able to present well the displacement and strain information on the surface of the sample during impact. Under all corrosion conditions, the horizontal strain localization first appears at the edge of the sample in the y-direction of the strain field before the peak stress, and then the strain band gradually expands toward the center. After the peak stress, shear strain bands appear on the surface of the sample and gradually lead to shear cracks on the surface. The weaker the chemical erosion, the higher the stress threshold required for strain localization; the higher the strain rate, the lower the stress threshold required for strain localization and the shorter the time to peak stress.

3.3. *Difference between DIC and Strain Gauges*

Figure 13 shows the stress and strain results obtained with the 3D-DIC and strain gauges. Compared to the results from the strain gauges on the bar, the strain in the DIC initially increases faster and more, but once the peak stress is reached, the strain from the DIC gradually begins to decrease (Figure 13a). The data sources between the DIC and the strain gauges are responsible for the fundamental difference. The stress measured by strain gauges is based on strain rates to calculate the time integral, which

makes two assumptions, namely, the one-dimensional propagation of stress waves and the homogeneity assumption [17]. These assumptions ignore the inertial and interfacial friction effects of the bar, resulting in low theoretically calculated axial strains. Therefore, the dynamic modulus measured with strain gauges is larger than the actual one. In addition, it cannot be neglected that the DIC measurement is the surface strain. Figure 13b shows the stress-strain curves derived from DIC and strain gauges, where the stresses are all derived from the strain gauge results. Based on the brittle damage characteristics of the rock, the DIC results produce a type II strain, while the strain gauge results are type I strain. Moreover, the results measured with strain gauges show that both elastic modulus and peak strain increase with increasing strain rate. However, the stress-strain curves from the DIC method show that the elastic modulus of the rock is almost independent of the strain rate, while the peak strain shows an increasing trend with the strain rate. The elastic modulus determined by the 3D-DIC method is approximately 12.49 GPa in Figure 13b. Clearly, proper strain measurement is important to evaluate the deformation behavior and quantify the change in stress threshold during fracture development of the rock.

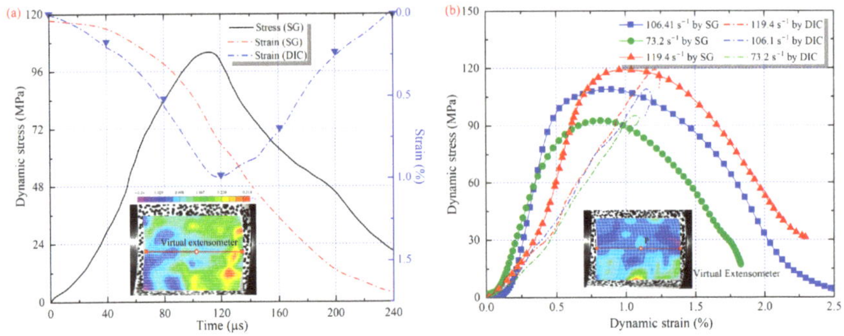

Figure 13. Results from strain gauges and 3D-DIC. (**a**) Stress and strain (**b**) Stress-strain curve.

3.4. Stress Thresholds for Cracks Development

The initiation and progression of cracks induced by stress are the fundamental causes of rock failure, so identifying the stages of crack development is critical to preventing rock engineering disasters. However, measuring or determining the stress threshold for crack development is difficult because the transverse strain cannot be measured with strain gauges during impact tests. Therefore, a high-speed 3D-DIC method was used in this experiment to characterize the stress thresholds of cracks development in sandstone under different strain rates and erosion conditions. As shown in Equation (4), the volumetric strain of the sample under uniaxial compression is the sum of the axial, lateral, and radial strains, with the lateral strain equal to the radial strain. Alternatively, the volumetric strain may consist of two components, the elastic volumetric strain and the volumetric strain due to crack propagation (crack strain) [35], mathematically described by Equation (5). The elastic volumetric strain can be determined from the elastic modulus, Poisson's ratio, and dynamic axial stress using Equation (6). The crack strain during dynamic compression is then given by Equation (7).

$$\varepsilon_v = \varepsilon_1 + \varepsilon_2 + \varepsilon_3 = \varepsilon_1 + 2\varepsilon_3 \quad (4)$$

$$\varepsilon_v = \varepsilon_{ve} + \varepsilon_{vc} \quad (5)$$

$$\varepsilon_{ve} = (1 - 2v)\sigma_{\mathrm{dy}}/E \quad (6)$$

$$\varepsilon_{vc} = \varepsilon_v - \varepsilon_{ve} \quad (7)$$

where ε_1, ε_2 and ε_3 are axial, lateral and radial strains, respectively; ε_v, ε_{ve} and ε_{vc} refer to volumetric strain, elastic volumetric strain and crack strain, respectively; v, E and σ_{dy} are

Poisson's ratio, dynamic elastic modulus and dynamic stress, respectively. As can be seen in Equation (6), to determine the elastic volumetric strain, the dynamic elastic modulus and Poisson's ratio must first be determined. Xing et al. [20] defined the dynamic modulus by the ratio of stress and strain in the elastic stage using the 3D-DIC method. The 3D-DIC results in this study showed that the dynamic elastic modulus did not vary with increasing loading strain rate. The dynamic elastic modulus of the sample with different erosion is shown in Table 2. In addition, the dynamic Poisson's ratio in the elastic phase can be determined from Equation (8) [17]:

$$\nu = \frac{1}{2}\frac{r_v^2 - 2}{r_v^2 - 1} \tag{8}$$

with r_v the ratio of P-wave and S-wave velocities. The variation of Poisson's ratio with erosion was calculated as displayed in Figure 14, and it was found that v does not vary significantly with erosion. This result is similar to the results from Liu et al. [10]. In this test, the dynamic Poisson's ratio was taken to be about 0.21.

Table 2. Dynamic elastic modulus E_d by DIC and static elastic modulus E_s of the sample.

Method	0 d	7 d	14 d	21 d
E_d (3D-DIC)	12.49 GPa	11.52 GPa	10.58 GPa	9.32 GPa
E_s (Static)	9.35 GPa	9.03 GPa	8.63 GPa	8.04 GPa

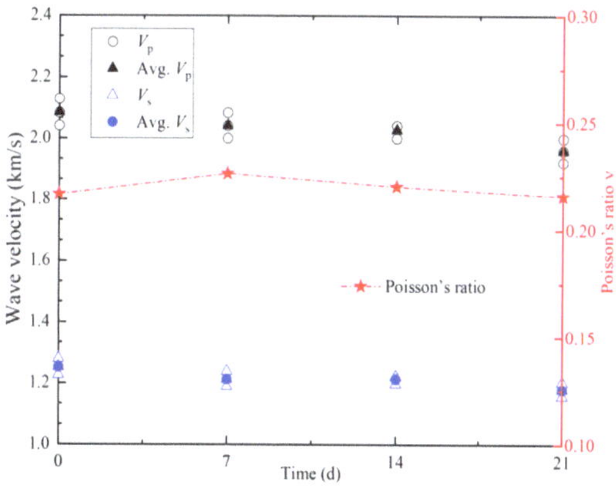

Figure 14. Variation of the wave velocity and the Poisson's ratio of the sample with corrosion.

Furthermore, the axial strain (ε_1) and lateral strain (ε_3) of the sample were extracted in the axial (S_{xx}) and lateral (S_{yy}) strain fields. As illustrated in Figure 15, the strain of each pixel in the axial and radial strain fields was extracted in the ROI. As a result, the axial and lateral strains during impact can be obtained according to Equation (9) [35].

$$\begin{cases} \varepsilon_1 = \left(\sum_{r=1}^{145}\sum_{c=1}^{209} \varepsilon_1(r,c)\right)/(145 \times 209)\ (S_{xx}) \\ \varepsilon_3 = \left(\sum_{r=1}^{145}\sum_{c=1}^{209} \varepsilon_3(r,c)\right)/(145 \times 209)\ (S_{yy}) \end{cases} \tag{9}$$

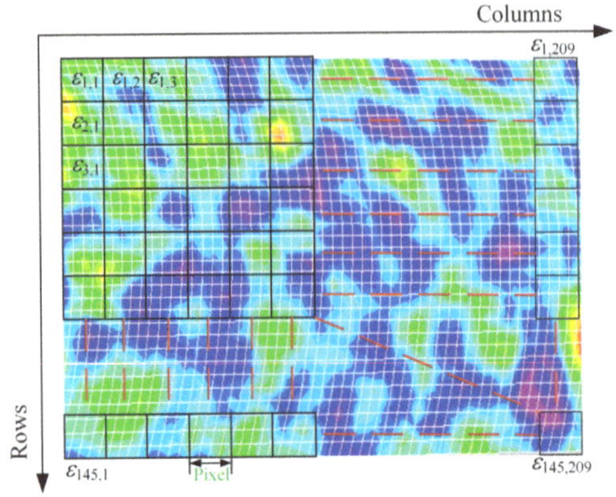

Figure 15. Schematic diagram of axial and radial strain based on pixel points.

The stress threshold diagrams for crack development in samples under different conditions are shown in Figure 16. Similar to the static test results, there are four main stages of crack development in the sample under dynamic loading, namely the crack closure (I), stable crack development (II), unstable crack development (III), and peak stress (IV). The crack closure stage corresponds to the gradual reduction of the crack strain to zero, during which the stress wave undergoes its first propagation. The corresponding point on the stress-strain curve is also the stress threshold for stable crack development (σ_{ci}). However, as soon as the crack strain increases to zero, it immediately starts to fall into the stable crack development stage, which means that there is no stable stage of elastic deformation in the dynamic impact process. Liang et al. [36] predicted this phenomenon by increasing the strain rate in the static tests; Xing et al. [17] observed and proved to this phenomenon. Nevertheless, it is undeniable that this phase still exhibits a strong linear variation of elastic modulus and a relatively slow crack development rate. Moreover, in the stage (II), there was no strain localization on the surface of the sample, let alone a crack. When strain localization occurs on the surface, it corresponds to the unstable crack development stage (III). At this stage, the tensile strain localization is initially generated from the edge in the S_{XX} strain field. Therefore, the occurrence of strain localization is an indication that stable crack development has accumulated to a high degree. When the stress reaches its peak, a shear strain band appears on the surface of the sample; moreover, obvious shear cracks appear on the sample in the failure phase after the peak. In addition, the stress threshold for stable/unstable crack development varied with the increase of erosion. Hence, the normalized stress thresholds for different conditions were summarized, as shown in Figure 17.

As can be seen in Figure 17, the normalized stress thresholds for the development of stable/unstable cracks in sandstone decrease with increasing strain rate. Previous studies have shown that the stress thresholds for the development of stable cracks under static loading are generally in the range of 30% to 50% of the peak stress, while the stress threshold for unstable cracks is in the range of 70% to 80%. However, in this study, these two thresholds drop to 10%–30% and 20%–70% of the peak stress under dynamic loading, respectively. In other words, the dynamic loading increases the strength of the sample but decreases the thresholds for the development of stable/unstable cracks, which is a significant difference from the static loading. Moreover, the normalized stress threshold for stable crack development decreased with increasing erosion at the approximate strain rate. At an approximate strain rate of 75 s^{-1}, the stress threshold for stable cracks development was 31.5 MPa and the normalized stress threshold was 0.34 for the uneroded sample. In comparison, the stress threshold for stable cracks development decreased by 33.1% to 22.1 MPa and the normalized stress threshold decreased by 14% to 0.29 for the sample eroded for 14 d. Similarly, the stress threshold for the development of unstable cracks also exhibited the same trend. The chemical reaction of some mineral components with the acidic

solution (e.g., Equation (10)) is the main responsible for the above phenomena. In addition, the difference between the stress thresholds for the development of stable/unstable cracks reflects the capacity of the quasi-elastic phase, obviously, the quasi-elastic phase continues to decrease with increasing erosion. The quasi-elastic phase also decreases with increasing strain rate. Thus, the sensitivity of the dynamic behavior of the sandstone to strain rate increases with chemical erosion. The microstructural changes in sandstone caused by chemical erosion are the essential reasons for variations in the loading response.

$$\begin{cases} SiO_2 + 2H_2O \rightarrow H_4SiO_4 \\ CaCO_3 + 2H^+ \rightarrow Ca^{2+} + H_2O + CO_2 \\ Fe_2O_3 + 6H^+ \rightarrow 2Fe^{3+} + 3H_2O \\ KAlSi_3O_8 + 4H^+ + 4H_2O \rightarrow K^+ + Al^{3+} + 3H_4SiO_4 \\ NaAlSi_3O_8 + 4H^+ + 4H_2O \rightarrow Na^+ + Al^{3+} + 3H_4SiO_4 \end{cases} \quad (10)$$

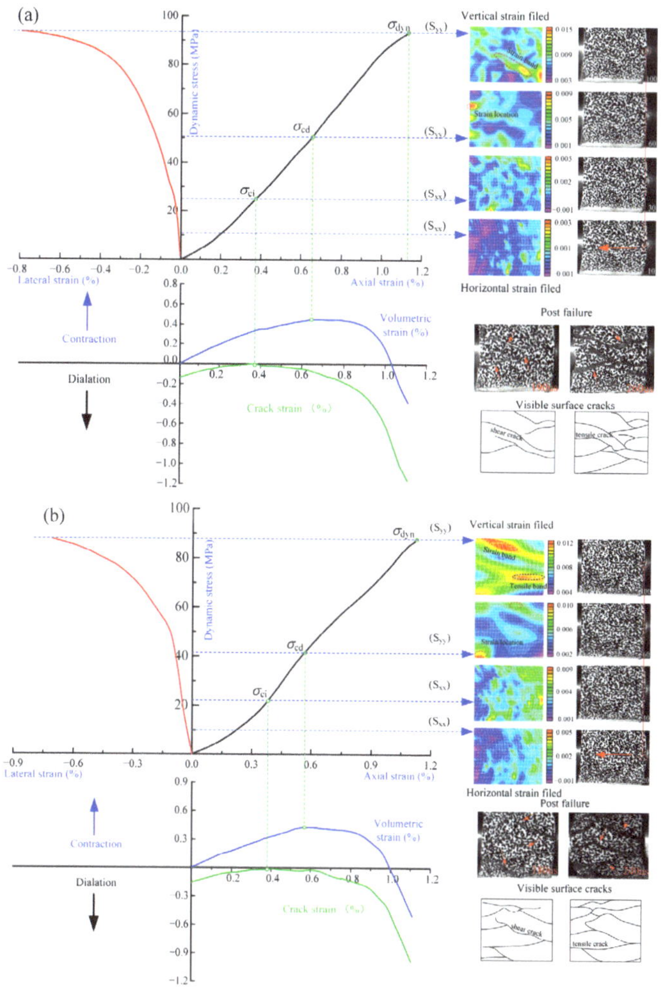

Figure 16. Pre-peak stress strain curves and stress thresholds for crack development with corresponding strain field. (**a**) 7d at 97.1 s^{-1}; (**b**) 14d at 99.2 s^{-1}.

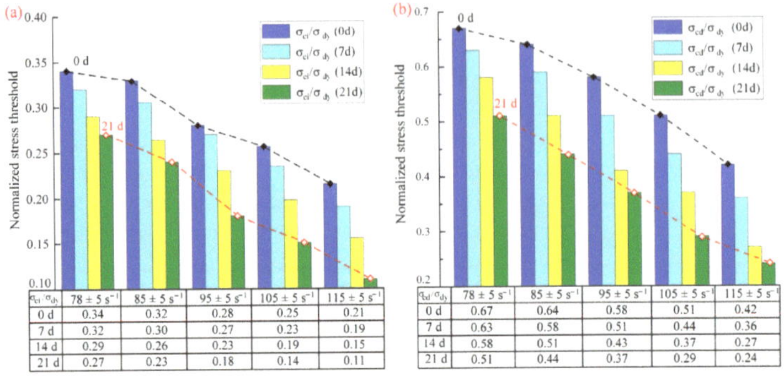

Figure 17. The stress thresholds for the development of stable (**a**) and unstable (**b**) cracks in samples under impact loading.

4. Discussions

Mine water at depth is often acidic due to the coal-associated minerals, such as reduced iron or manganese, come into contact with water and oxygen [37]. When coal mines are converted for other purposes, such as the construction of pumped storage plants [3], underground reservoirs [38], etc., they are inevitably affected by impact loading. The effects of impact loading on the mechanical response of rock have already been studied in detail, focusing on dynamic strength [32], modulus [10] and deformation [13]. However, little has been reported on the evolution of the overall deformation process during impact testing. In particular, for rocks subjected to chemical erosion, the effect of impact on the full-field evolution of rock fracture has not been well studied.

In this study, the evolution of the deformation and strain fields of the red sandstone under impact loading and chemical erosion conditions was revealed using the high-speed 3D DIC technique, and the variation of stress thresholds for the development of cracks was analyzed. At approximate loading strain rates, the degree of structural deterioration of the sandstone determines the development of strain localization at the surface, which in turn affects the fracture process of the rock. The longer the chemical erosion continues, the earlier the strain is localized at the surface of the sandstone and the lower the stress threshold for crack development. At the same time, the increase in strain rate leads to an increase in the strength of the sample, but the concentration of strain bands occurs earlier on the surface. It is worth mentioning that it is difficult to reveal the volumetric strain around the whole rock with two high-speed CMOS cameras [18]. In order to fully identify the dynamic fracture process of the rock, it is necessary to understand the evolution and distribution mechanism of crack development inside the rock. Currently, some techniques such as CT, AE, and digital volume correlation (DVC) are able to measure the internal deformation field [20,34]. However, these techniques are not yet mature enough for use in SHPB testing.

5. Conclusions

In this study, SHPB tests were performed on the red sandstone with different erosion effects, and full-field deformation was measured using the high-speed 3D-DIC method. The stress thresholds for the development of cracks in the sample under dynamic loading were investigated. The main conclusions are as follows:

1. The evolution of the displacement and strain fields of the red sandstone was affected by chemical erosion and impact loading. The higher the loading rate, the greater the dynamic strength, while the stress threshold for strain localization was lower and the time to reach the stress peak was shorter. Moreover, chemical erosion accelerated the strain localization process and lowered the stress threshold for the occurrence of strain localization. Similarly, the stress threshold for displacement localization on the sample surface decreased with increasing erosion.

2. Under the influence of the erosion effect, the horizontal strain localization first occurred at the edge of the sample in the y-direction of the strain field, and then the strain localization gradually extended toward the center of the sample before reaching the peak stress. In the unloading phase, shear strain bands began to appear on the sample, and shear cracks gradually formed on the surface. No visible cracks appeared on the surface of the sample until the peak stress was reached.
3. Normalized stress thresholds for stable/unstable crack development were both lowered by chemical erosion. After an erosion period of 21 d, the stress threshold for the development of stable crack in the sample decreased from 0.34 to 0.27. In addition, the strain rate accelerated the gestation and development of cracks in the rock. Compared to the static test, the two thresholds for crack development decreased to 10–30% and 20–70% of the peak stress, respectively.

Author Contributions: Methodology, J.X. and H.P.; investigation, Z.S.; resources, J.X.; data curation, J.X. and Z.S.; writing—original draft preparation, J.X. and Z.S.; writing—review and editing, J.X. and H.P.; visualization, J.X. and Z.S.; supervision, H.P. All authors have read and agreed to the published version of the manuscript.

Funding: This research was funded by the National Natural Science Foundation of China (No. 51974296, No. 52061135111).

Data Availability Statement: Data available on request due to privacy restrictions.

Conflicts of Interest: The authors declare no conflict of interest.

References

1. Pu, H.; Xu, J. Research Status and Progress of Geothermal Energy Development and Utilization from Closed/Abandoned Coal Mines. *J. China Coal Soc.* **2022**, *47*, 2243–2269.
2. Menéndez, J.; Fernández-Oro, J.M.; Galdo, M.; Loredo, J. Transient simulation of underground pumped storage hydropower plants operating in pumping mode. *Energies* **2020**, *13*, 1781. [CrossRef]
3. Menéndez, J.; Fernández-Oro, J.M.; Loredo, J. Economic feasibility of underground pumped storage hydropower plants providing ancillary services. *Appl. Sci.* **2020**, *10*, 3947. [CrossRef]
4. Zhang, C.; Wang, F.; Bai, Q. Underground space utilization of coalmines in China: A review of underground water reservoir construction. *Tunn. Undergr. Space Technol.* **2021**, *107*, 103657. [CrossRef]
5. Farr, G.; Sadasivam, S.; Watson, I.A.; Thomas, H.R.; Tucker, D. Low enthalpy heat recovery potential from coal mine discharges in the South Wales Coalfield. *Int. J. Coal Geol.* **2016**, *164*, 92–103. [CrossRef]
6. Ma, T.; Ding, W.X.; Wang, H.Y.; Chen, G.X.; Chen, H.J.; Yan, Y.Y. Dissolution characteristics and mechanical properties of limestone with different mineral composition contents eroded by acid chemical solution. *Chin. J. Geotech. Eng.* **2021**, *43*, 1550–1557.
7. Lin, Y.; Zhou, K.; Gao, R.; Li, J.; Zhang, J. Influence of chemical corrosion on pore structure and mechanical properties of sandstone. *Geofluids* **2019**, *2019*, 24–30. [CrossRef]
8. Li, H.; Zhong, Z.; Liu, X.; Sheng, Y.; Yang, D. Micro-damage evolution and macro-mechanical property degradation of limestone due to chemical effects. *Int. J. Rock Mech. Min. Sci.* **2018**, *110*, 257–265. [CrossRef]
9. Feng, X.; Chen, S.; Li, S. Effects of water chemistry on microcracking and compressive strength of granite. *Int. J. Rock Mech. Min. Sci.* **2001**, *38*, 557–568. [CrossRef]
10. Yongsheng, L.; Liu, W.; Dong, X. Dynamic mechanical properties and constitutive model of rock under chemical corrosion. *J. Yangtze River Sci. Res. Inst.* **2015**, *32*, 72–75.
11. Han, T.; Shi, J.; Chen, Y.; Li, Z.; Li, C. Laboratory Investigation on the Mechanical Properties of Sandstone Immersed in Different Chemical Corrosion under Freeze-thaw Cycles. *Acta Mech. Solida Sin.* **2017**, *38*, 503–520.
12. Ding, W.X.; Chen, J.P.; Xu, T.; Chen, H.J.; Wang, H.Y. Mechanical and chemical characteristics of limestone during chemical erosion. *Rock Soil Mech.* **2015**, *36*, 1825–1830.
13. Xu, J.; Pu, H.; Sha, Z. Mechanical behavior and decay model of the sandstone in Urumqi under coupling of freeze–thaw and dynamic loading. *Bull. Eng. Geol. Environ.* **2021**, *80*, 2963–2978. [CrossRef]
14. Li, G.; Yu, L.; Su, H.; Jing, H.; Zhang, T. Dynamic properties of corroded limestone based on SHPB. *Chin. J. Rock Mech. Eng.* **2018**, *37*, 2075–2085.
15. Yu, L.; Zhang, Z.; Wu, J.; Liu, R.; Qin, H.; Fan, P. Experimental study on the dynamic fracture mechanical properties of limestone after chemical corrosion. *Theor. Appl. Fract. Mech.* **2020**, *108*, 102620. [CrossRef]
16. Miao, S.; Cai, M.; Guo, Q.; Wang, P.; Liang, M. Damage effects and mechanisms in granite treated with acidic chemical solutions. *Int. J. Rock Mech. Min. Sci.* **2016**, *88*, 77–86. [CrossRef]

17. Xing, H.Z.; Zhang, Q.B.; Zhao, J. Stress Thresholds of Crack Development and Poisson's Ratio of Rock Material at High Strain Rate. *Rock Mech. Rock Eng.* **2018**, *51*, 945–951. [CrossRef]
18. Wang, Y.; Zhu, C.; Zhang, B.; Hou, Z. Full-Field Deformation Characteristics of Anisotropic Marble under Compression Revealed by 3D Digital Image Correlation. *Lithosphere* **2021**, *2021*, 1098235. [CrossRef]
19. Wang, Y.; Yang, H.; Zhu, C.; Gao, S. On the Fracture Evolution and Instability of Pyrite-Filled Marble Exposed to Freeze-Thaw-Compression Loads. *Lithosphere* **2021**, *2021*, 7599916. [CrossRef]
20. Xing, H.Z.; Zhang, Q.B.; Ruan, D.; Dehkhoda, S.; Lu, G.X.; Zhao, J. Full-field measurement and fracture characterisations of rocks under dynamic loads using high-speed three-dimensional digital image correlation. *Int. J. Impact Eng.* **2018**, *113*, 61–72. [CrossRef]
21. Li, D.; Zhu, Q.; Zhou, Z.; Li, X.; Ranjith, P.G. Fracture analysis of marble specimens with a hole under uniaxial compression by digital image correlation. *Eng. Fract. Mech.* **2017**, *183*, 109–124. [CrossRef]
22. Zhou, X.P.; Wang, Y.T.; Zhang, J.Z.; Liu, F.N. Fracturing Behavior Study of Three-Flawed Specimens by Uniaxial Compression and 3D Digital Image Correlation: Sensitivity to Brittleness. *Rock Mech. Rock Eng.* **2019**, *52*, 691–718. [CrossRef]
23. Solav, D.; Moerman, K.M.; Jaeger, A.M.; Genovese, K.; Herr, H.M. MultiDIC: An open-source toolbox for multi-view 3D digital image correlation. *IEEE Access* **2018**, *6*, 30520–30535. [CrossRef]
24. Golewski, G.L. Measurement of fracture mechanics parameters of concrete containing fly ash thanks to use of Digital Image Correlation (DIC) method. *Meas. J. Int. Meas. Confed.* **2019**, *135*, 96–105. [CrossRef]
25. Munoz, H.; Taheri, A. Specimen aspect ratio and progressive field strain development of sandstone under uniaxial compression by three-dimensional digital image correlation. *J. Rock Mech. Geotech. Eng.* **2017**, *9*, 599–610. [CrossRef]
26. Chen, C.; Xu, J.; Okubo, S.; Peng, S. Damage evolution of tuff under cyclic tension–compression loading based on 3D digital image correlation. *Eng. Geol.* **2020**, *275*, 105736. [CrossRef]
27. Munoz, H.; Taheri, A.; Chanda, E.K. Pre-Peak and Post-Peak Rock Strain Characteristics During Uniaxial Compression by 3D Digital Image Correlation. *Rock Mech. Rock Eng.* **2016**, *49*, 2541–2554. [CrossRef]
28. Zhao, J.; Sang, Y.; Duan, F. The state of the art of two-dimensional digital image correlation computational method. *Eng. Rep.* **2019**, *1*, e12038. [CrossRef]
29. Zhu, Q.; Li, D.; Han, Z.; Li, X.; Zhou, Z. Mechanical properties and fracture evolution of sandstone specimens containing different inclusions under uniaxial compression. *Int. J. Rock Mech. Min. Sci.* **2019**, *115*, 33–47. [CrossRef]
30. Tang, Y.; Okubo, S.; Xu, J.; Peng, S. Experimental Study on Damage Behavior of Rock in Compression–Tension Cycle Test Using 3D Digital Image Correlation. *Rock Mech. Rock Eng.* **2019**, *52*, 1387–1394. [CrossRef]
31. Sharafisafa, M.; Shen, L. Experimental Investigation of Dynamic Fracture Patterns of 3D Printed Rock-like Material Under Impact with Digital Image Correlation. *Rock Mech. Rock Eng.* **2020**, *53*, 3589–3607. [CrossRef]
32. Xu, J.; Pu, H.; Sha, Z. Experimental Study on the Effect of Brittleness on the Dynamic Mechanical Behaviors of the Coal Measures Sandstone. *Adv. Civ. Eng.* **2021**, *2021*, 6679333. [CrossRef]
33. Li, X. *Rock Dynamic Fundamentals and Applications*; Li, X., Ed.; Science Press: Beijing, China, 2014; ISBN 987-7-03-040425-1.
34. Xing, H.Z.; Zhang, Q.B.; Braithwaite, C.H.; Pan, B.; Zhao, J. High-Speed Photography and Digital Optical Measurement Techniques for Geomaterials: Fundamentals and Applications. *Rock Mech. Rock Eng.* **2017**, *50*, 1611–1659. [CrossRef]
35. Xing, H.; Wang, M.; Fan, P.; Wang, D. Grain-size effect on dynamic behavior of sandstone based on high-speed 3D-DIC technique. *Explos. Shock Waves* **2021**, *41*, 43–54.
36. Liang, C.; Li, X.; Wang, S.; Li, S.; He, J.; Ma, C. Experimental investigations on rate-dependent stress-strain characteristics and energy mechanism of rock under uniaixal compression. *Chin. J. Rock Mech. Eng.* **2012**, *31*, 1830–1838.
37. Pu, H.; Bian, Z.; Zhang, J.; Xu, J. Research on a reuse mode of geothermal resources in abandoned coal mines. *J. China Coal Soc.* **2021**, *46*, 677–687.
38. Zhang, K.; Gao, J.; Jiang, B.; Han, J.; Chen, M. Experimental study on the mechanism of water-rock interaction in the coal mine underground reservoir. *J. China Coal Soc.* **2019**, *44*, 3760–3772.

minerals

Article

Experimental Study on the Mechanical and Acoustic Emission Characteristics of Tuff with Different Moisture Contents

Wenlong Dong [1], Lijun Han [1,*], Lingdong Meng [1], Hexuan Zhu [1], Shuai Yan [1], Changyu Xu [1] and Yaning Dong [2]

1 State Key Laboratory for Geomechanics and Deep Underground Engineering, China University of Mining and Technology, Xuzhou 221116, China
2 Anhui Masteel Luohe Mining Co., Ltd., Hefei 231500, China
* Correspondence: hanlj666@163.com

Abstract: Rock materials are often affected by water in underground engineering. In this study, the mechanical and failure characteristics of tuff under different moisture content were studied using laboratory tests. The moisture content variation of tuff was studied in water absorption tests, and the mechanical and failure characteristics of tuff under different moisture contents were studied through uniaxial compression tests with a Micro-II acoustic emission (AE) control acquisition system. The results showed that the moisture content of tuff increases rapidly at the initial stage of water absorption tests and stabilizes after 180 h of immersion. According to the results of uniaxial compression tests, both uniaxial compressive strength and elasticity modulus decreased with the increase in moisture content. The AE parameters analyses showed that, when the moisture content increased, the accumulated AE counts and energy gradually decreased, and the "quiet period" at the initial stage of uniaxial compression tests lasted longer, and the RA (rise time⁄amplitude)–AF (AE counts⁄duration) distribution and the failure characteristics verified that the failure patterns evolved from shear failure to tensile failure. Scanning electron microscopy was used to observe the morphology of the fracture surface and analyze the influence of moisture content on the fracture characteristics of the tuff at the mesoscopic level. The results of this research can be used as a basis for studying the influence of water on tuff.

Keywords: moisture content; uniaxial compression test; rock acoustic emission; SEM; rock failure mode

Citation: Dong, W.; Han, L.; Meng, L.; Zhu, H.; Yan, S.; Xu, C.; Dong, Y. Experimental Study on the Mechanical and Acoustic Emission Characteristics of Tuff with Different Moisture Contents. *Minerals* **2022**, *12*, 1050. https://doi.org/10.3390/min12081050

Academic Editors: Abbas Taheri and Alison Ord

Received: 22 June 2022
Accepted: 18 August 2022
Published: 20 August 2022

1. Introduction

As underground projects such as tunnels, mining, and underground chambers are built in deeper and more complex geological environments, surrounding rocks will be more significantly affected by groundwater. The presence of groundwater increases the moisture content of rocks, and a large number of experiments and studies have shown that the increase in moisture content leads to a decrease in the strength and deformation properties of rocks, which will cause many rock engineering disasters [1–3]. Under the influence of groundwater, the engineering properties of the surrounding rock will be degraded to a certain extent, which will greatly affect the stability and safety of the projects [4]. Therefore, it is quite essential for the safety evaluation of underground engineering to carry out in-depth research on the deterioration of rock properties and failure characteristics under different moisture content.

Tuff is a common volcanic clastic rock, formed by the transport, scattering, cementation, and deposition of volcanic ash from volcanic eruptions, and it is widely distributed in southeastern China. Tuff is a widely existing rock in underground engineering, which often endangers the stability and safety of engineering under the influence of water. Therefore, it is of great meaning to study the influence of moisture content on the properties of tuff.

In recent years, acoustic emission (AE) technology has been widely used in geotechnical engineering monitoring as a nondestructive testing technology [5]. The AE phenomenon

refers to the elastic wave phenomenon generated by the rapid release of the local stress energy that propagates through the material. The AE signals collected by the AE collection device contain a large amount of information about the material cracks, and the amplitude, frequency, count, and energy obtained are widely used in the material failure process and failure characteristic analysis [6–9]. As a kind of brittle material, a large number of microcracks and fracture extensions will be generated inside the specimens during the process from initial loading to damage, and several AE events will be generated along with this process. By analyzing the AE characteristics parameters of rocks during the damage process, the failure characteristics of rocks under different moisture content can be effectively analyzed.

Previously, many scholars conducted numerous studies on the failure characteristics of rocks under different moisture content. Yao et al. [10] conducted uniaxial compression tests on coal, mudstone, and siltstone under different water absorption times, and they found that the uniaxial compressive strength, elastic modulus, and peak strain of coal and rock specimens all decreased to a certain extent as the moisture content increasing. Zhou et al. [11] analyzed the water influence on the mechanical behavior of rock in saturation and drying processes under static and dynamic tests. Vasarhelyi [12] studied the variations of porosity, uniaxial compressive strength, elastic modulus, and tensile strength of limestone at different water content. Yilmaz [13] studied the degradation of gypsum rock mechanical property parameters by uniaxial compression test under moisture content and obtained the law that the uniaxial compressive strength and elastic modulus of gypsum rock decreased with the growth of water content. Vishal et al. [14] studied the strength characteristics of saturated coal and rock mass and the initiation of internal micro-cracks under high-stress conditions. Yu et al. [15] studied the influence of moisture content on the mechanical properties and crack propagation of siltstone through uniaxial compression tests, and they proved that the moisture content was negatively linearly correlated with peak stress and elastic modulus. Roy et al. [16] conducted experimental research on the fracture and mechanical properties of saturated sedimentary rocks, and the results showed that saturability had a significant effect on the peak stress and fracture properties of sedimentary rocks. In addition, the Brazilian split strength and fracture toughness decreased with the increase in saturation.

AE technology can monitor the initiation and development of microcracks in the failure process of rock specimens, thereby revealing the evolution of internal damage and failure of the rock. In view of the advantage, AE technology has been used to conduct several studies on rock failure mechanism [17–23]. Liu et al. [8] studied the influence of strain rate on AE characteristic parameters and rock failure characteristics through experiments, and they obtained the law that the cumulative AE counts decreased as a power function with increasing strain rate. Chen et al. [24] used AE technology to study the law of fracture development during the hydraulic fracturing tests process, thereby revealing the fracture mechanism of fluid–solid interaction during hydraulic fracturing. Chen et al. [25] conducted uniaxial compression tests and AE monitoring on siltstones under different water saturation levels, and they studied the relationship between AE characteristic parameters and siltstone crack propagation with different moisture content. The results of the tests showed that the AE count rate decreased gradually during the compaction process as the moisture content of siltstone increased. Carpinteri et al. [26] conducted uniaxial compression tests with AE monitoring on concrete and rock specimens, revealing the law of energy absorption and release during the compression process of brittle materials. According to the study of AE parameters, it was proven that the heterogeneity of various properties of partly soaked coal specimens was more distinct than that of completely soaked and non-soaked coal specimens. Amann et al. [27] studied the brittle failure characteristics of clay using uniaxial compression tests and AE technology. The test results showed that the crack initiation threshold was about 30% of the peak stress, and the crack damage threshold was about 70% of the peak stress. Xiao et al. [28] used AE and infrared characterization methods to monitor coal samples with different moisture content

under uniaxial compression tests, revealing the overall and surface damage variables and their evolution. The test results showed that the internal damage of the coal specimens preceded the external damage, and the appearance of the AE signal was earlier than the abnormal infrared point. Rodriguez and Celestino [29] conducted splitting tests and AE monitoring on marble and granite, and the results showed that the internal cracks development processes and cracking types of the rock were mainly determined by the rock mineral composition.

In this article, tuff specimens with different moisture content were prepared and used to study the variation of mechanical and acoustic emission characteristics of tuff with moisture content through a uniaxial compression test and acoustic emission monitoring system. Then, scanning electron microscopy (SEM) was used to study the mesoscopic characteristics of tuff under different moisture content and verified the macroscopic failure characteristics at the micro level.

2. Materials and Methods

2.1. Specimen Preparation

The tuff specimens used in this study were taken from the Luohe Iron Mine located in Hefei, Anhui Province, China. In order to satisfy the requirements of uniaxial compression tests, the rock specimens were made into a cylinder of φ50 mm × H100 mm in accordance with the recommended standards of the International Society for Rock Mechanics (ISRM). Both ends of all specimens were ground to obtain a flat surface. There were 27 specimens in total, which required drying before testing. All specimens were placed in the oven and dried at 110 °C for 24 h, before being placed into a desiccator and weighed. The specimens were divided into nine groups of three specimens each and subjected to water absorption tests through free absorption. In order to obtain specimens with different moisture contents, the immersion time was set to 0 h, 6 h, 12 h, 24 h, 60 h, 180 h, 450 h, and 960 h, and the last group of specimens was forcibly saturated.

2.2. Experimental Equipment and Procedure

An MTS-816 electrohydraulic servomechanical test system was employed to carry out the uniaxial compression tests, with the loading rate set to 0.1 mm/min. Vaseline was applied to both ends of the specimens to eliminate the influence of end face constraints. A Micro-II AE acquisition system was used to monitor AE signals (seen in Figure 1), and the AE parameters were acquired by the AE sensor affixed on the surface of the specimens. A DH3816N static strain collection system was used to acquire strain data, as shown in Figure 2.

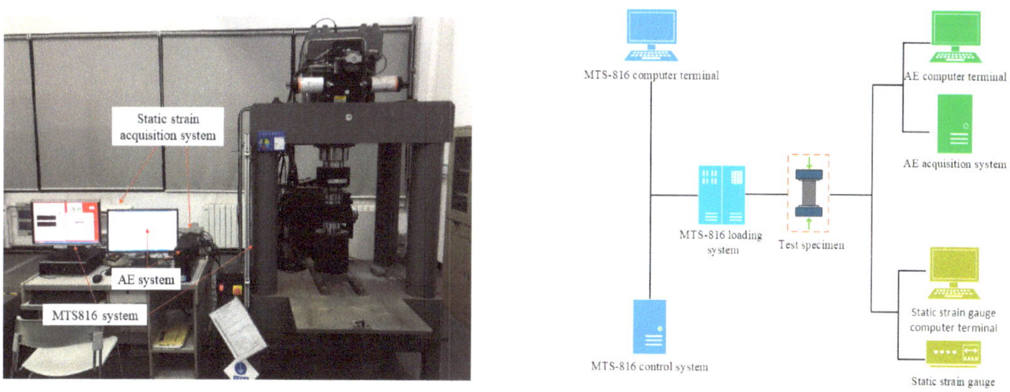

(a) (b)

Figure 1. Testing system: (a) diagram of testing system; (b) layout of testing system.

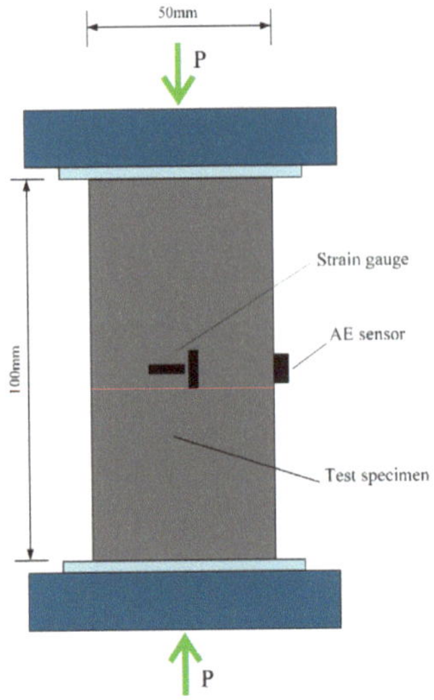

Figure 2. Sensor layout on the surface of specimen.

The acquisition parameters of the AE system for this study were set as shown in Table 1.

Table 1. Acquisition parameters of the AE system.

Parameter	MSPS	Threshold (dB)	HLT (μs)	HDT (μs)	PDT (μs)	Resonant Frequency (kHz)	Pre-Amplified (dB)
Value	1	40	1000	800	200	300	40

MSPS is the sampling rate; HLT, HDT, and PDT denote the hit lockout time, hit definition time, and peak definition time, respectively.

3. Experimental Results and Discussion

3.1. Relationship between Moisture Content and Immersion Time

The specimens were subjected to water absorption tests, and the moisture contents varied with the time of immersion. The saturated specimens were prepared by boiling the specimens for 6 h after immersion. The moisture content and average moisture content data of the specimens are shown in Table 2.

The relative relationship between average moisture content and immersion time is shown in Figure 3. With the increase in immersion time, the average moisture content showed an upward trend. During the immersion time from 0 h to 24 h, the moisture content increased rapidly, which was mainly due to the rapid water absorption on the surface of the specimens at the initial stage of immersion. From 24 h to 180 h, the moisture content still showed an upward trend with the accumulation of immersion time, but the rising rate dropped rapidly. This was mainly due to the slower rate of water infiltration into the interior of rock after the surfaces of the specimens were soaked by water. From 180 h to 960 h, the moisture content of the tuff specimens continued to increase with immersion time, but the rising rate was very small, and the moisture content was close to the saturation state in value. This was mainly because the pores that allow water to penetrate freely were basically occupied by water; thus, the water absorption rate of the rock became very slow. Accordingly, it was difficult for water to penetrate into the rock specimens during this stage. After being immersed in water for 960 h, the average moisture content of the specimens was 1.02%, and the average saturated moisture content was 1.05%, which were very close to each other.

Table 2. Results of water absorption test.

No.	Immersion Time	Dry Weight (g)	Wet Weight (g)	Moisture Content (%)	Average Moisture Content (%)
A-1		667.28		0	
A-2	0 h	658.17		0	0
A-3		649.84		0	
B-1		669.93	671.87	0.30	
B-2	6 h	662.31	664.10	0.27	0.28
B-3		652.79	654.55	0.26	
C-1		663.80	666.39	0.39	
C-2	12 h	671.15	673.90	0.41	0.40
C-3		661.26	663.91	0.4	
D-1		646.68	650.69	0.62	
D-2	24 h	647.25	651.07	0.59	0.60
D-3		658.74	662.69	0.6	
E-1		660.41	665.63	0.79	
E-2	60 h	656.28	661.60	0.81	0.81
E-3		644.91	650.20	0.82	
F-1		655.21	661.04	0.88	
F-2	180 h	647.26	653.02	0.90	0.90
F-3		655.95	661.92	0.91	
G-1		649.58	655.82	0.96	
G-2	450 h	644.32	650.63	0.99	0.97
G-3		656.28	662.71	0.97	
H-1		651.79	658.37	1.01	
H-2	960 h	647.66	654.27	1.02	1.02
H-3		649.57	656.33	1.04	
I-1		649.39	656.21	1.05	
I-2	Saturated	654.92	661.73	1.04	1.05
I-3		651.74	658.71	1.07	

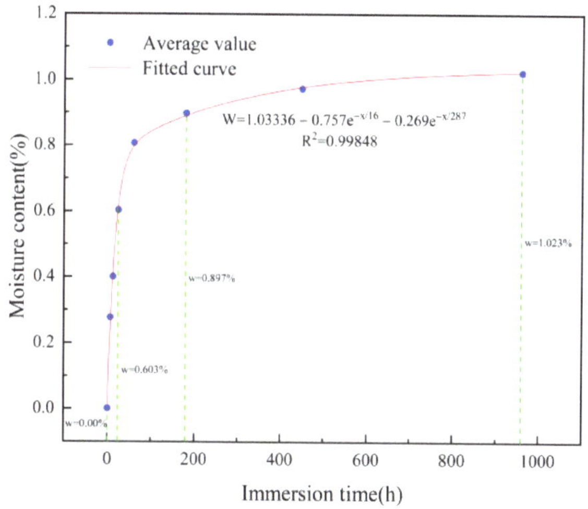

Figure 3. Relationship between moisture content and immersion time.

The relationship between the average moisture content and the immersion time obtained in the tests revealed the relationship of the sample moisture content with the immersion time, which could be fitted using Equation (1).

$$\omega = 1.03336 - 0.757e^{-x/16} - 0.269e^{-x/287}, \tag{1}$$

where ω denotes the average moisture content, and x represents the specimen immersion time. The correlation coefficient of the fitted function was R^2 = 0.99848, proving that the formula could accurately characterized the relationship.

3.2. Mechanical Behavior of Tuff under Different Moisture Content

The results of the uniaxial compression tests are listed in Table 3. As shown in Figure 4, the peak stress gradually declined with the increase in moisture content. The stress–strain curves of all specimens showed a concave upward trend in the initial loading stage, which was mainly due to the nonlinear deformation caused by the crack closure in the rock specimens. With the increase in load, the stress–strain curves showed an approximately linear growth, with a lower moisture content resulting in a higher growth rate. With continued loading, the stress–strain curve reached the peak stress after going through a downward concave section. In addition, both the peak stress and the corresponding strain gradually decreased as the moisture content increased.

Table 3. Results of uniaxial compression tests.

No.	Immersion Time	UCS (MPa)	Average UCS (MPa)	Elasticity Modulus (GPa)	Average Elasticity Modulus (GPa)	Poisson's Ratio	Average Poisson's Ratio
A-1		170.01		18.06		0.20	
A-2	0 h	170.53	170.05	17.96	17.95	0.18	0.19
A-3		169.61		17.82		0.19	
B-1		158.09		17.07		0.21	
B-2	6 h	157.57	157.59	16.93	16.97	0.24	0.23
B-3		157.11		16.90		0.23	
C-1		144.85		16.26		0.23	
C-2	12 h	148.10	146.39	16.56	16.41	0.26	0.25
C-3		146.22		16.41		0.27	
D-1		128.96		16.30		0.27	
D-2	24 h	127.36	127.84	16.11	16.05	0.26	0.27
D-3		127.20		15.73		0.27	
E-1		120.78		15.81		0.27	
E-2	60 h	119.89	119.83	15.66	15.69	0.29	0.27
E-3		118.82		15.60		0.26	
F-1		112.26		15.42		0.28	
F-2	180 h	111.35	112.16	15.48	15.39	0.30	0.28
F-3		112.88		15.26		0.26	
G-1		108.22		15.00		0.27	
G-2	450 h	107.73	107.27	15.22	15.19	0.30	0.29
G-3		105.87		15.34		0.30	
H-1		105.41		15.12		0.30	
H-2	960 h	105.16	104.88	14.94	15.09	0.29	0.30
H-3		104.07		15.20		0.31	
I-1		103.14		15.01		0.29	
I-2	Saturated	103.02	103.03	14.99	15.01	0.31	0.30
I-3		102.93		15.02		0.31	

The UCS refers to the uniaxial compressive strength.

In addition, the stress–strain curve presented different development trends under different moisture contents after peak stress. For the dry specimens, the peak stress obtained by the tests reached the maximum value, and the stress–strain curve had no obvious post-peak segment, indicating a brittle failure mode. The peak stress gradually decreased with the increase in moisture content, and the post-peak segments gradually emerged. Furthermore, a greater moisture content led to a more obvious post-peak segment. It can be seen from the results that, when the moisture content increased, the tuff specimen showed a certain bearing capacity after failure, which presented an obvious enhancement of plasticity and toughness.

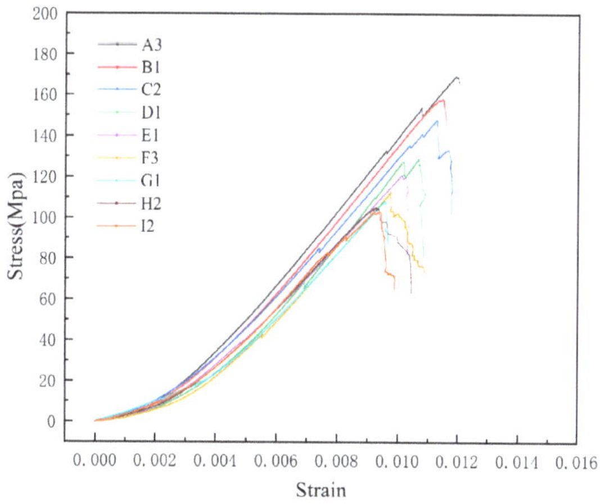

Figure 4. Stress–strain curve of specimens with different moisture content.

The variation law of peak stress and average UCS with moisture content is shown in Figure 5. The peak stress of the specimens showed a downward trend as the moisture content increased, but the decline rate was not maintained at a constant value. The peak stress decline rate was maintained at a relatively high level during the initial stage of increasing moisture content. However, as the moisture content continued to increase, the decline rate of the peak stress gradually dropped. The results indicate that the weakening effect of water on the strength of the tuff gradually decreased after the moisture content reached a certain level. The relative relationship between peak stress and moisture content could be fitted using Equation (2).

$$y = 17.128 + 158.365e^{-0.564x}, \tag{2}$$

where y is the uniaxial compressive strength, and x denotes the moisture content. The correlation coefficient of the fitted function was $R^2 = 0.98242$.

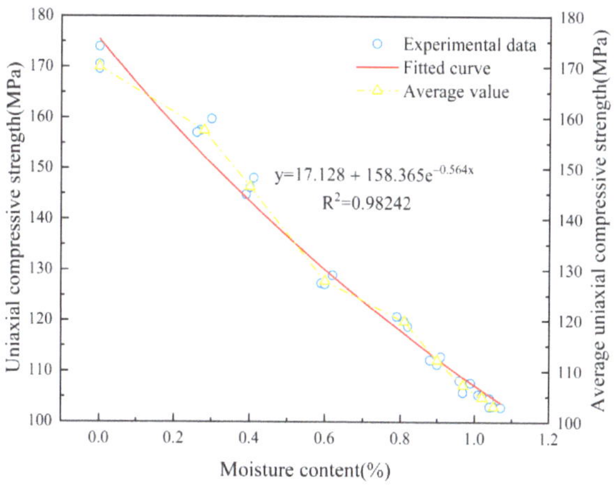

Figure 5. Relationship between uniaxial compressive strength and moisture content.

As shown in Figure 6, the average elastic modulus showed a downward trend as the moisture content increased, but the decline rate of the elastic modulus gradually dropped. When the specimens were dry, the average value of elastic modulus was 17.95 GPa, whereas, when the specimen was saturated, the average value of elastic modulus decreased to

15.01 GPa, with a decrease of 16.38%. This demonstrated that water produced an obvious decrease in the elastic modulus of the tuff. The relationship between the average value of the elastic modulus and the moisture content could be fitted using Equation (3).

$$y = 12.766 + 5.161e^{-0.767x}, \tag{3}$$

where y is the elastic modulus, and x denotes the moisture content. The correlation coefficient of the fitted function was $R^2 = 0.99331$.

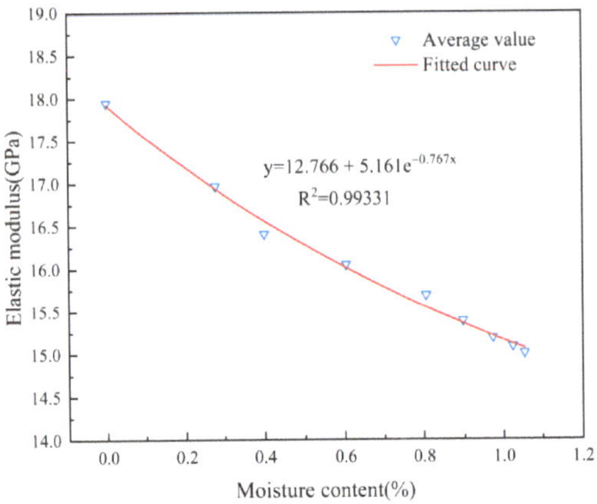

Figure 6. Relationship between elastic modulus and moisture content.

The Poisson's ratio of the tuff specimens gradually increased with the increase in moisture content (Figure 7). The average Poisson's ratio of the dry specimens was 0.19, whereas the average Poisson's ratio of the saturated specimens was 0.30, showing an increase of 57.89%. The results proved that the increase in moisture content caused a large increase in the Poisson's ratio of the tuff specimens. In addition, the Poisson's ratio was more affected by moisture content than the elastic modulus. The relationship between the average Poisson's ratio and the moisture content could be fitted using Equation (4).

$$y = 0.342 - 0.152e^{-1.126x}, \tag{4}$$

where y represents Poisson's ratio, and x denotes the moisture content. The correlation coefficient of the fitted function is $R^2 = 0.97451$.

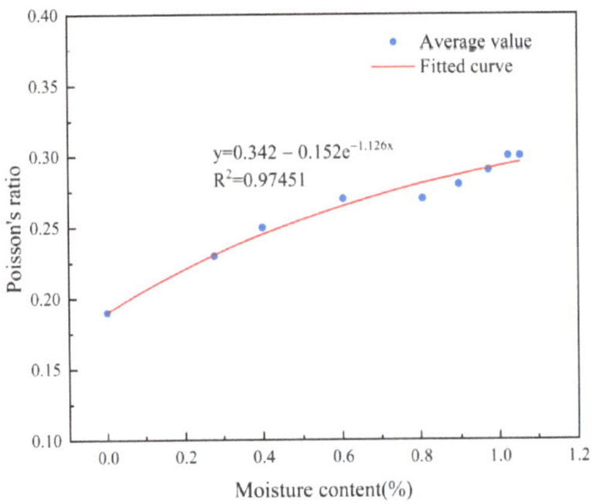

Figure 7. Relationship between Poisson's ratio and moisture content.

Many scholars have studied the influence of moisture content on the mechanical characteristics of other types of rocks, such as sandstone, red sandstone, marble, and siltstone, and found that the weakening effect of water will make the mechanical parameters such as uniaxial compressive strength and elastic modulus show similar change trends [25,30–32].

3.3. AE Counts and Energy Characteristics of Tuff with Different Moisture Content

The stress and AE parameter results with time under different moisture content in uniaxial compression tests are demonstrated in Figure 8, representing the nine different moisture states. The AE parameters collected during the tests included AE counts, amplitude, and energy, which showed a good correlation with stress evolution as a function of loading. During the uniaxial compression tests, the stress of the specimens was low at the initial loading stage, and the AE signals were rarely collected. With the increase in stress, the AE energy and accumulated counts increased gradually and showed a significant growth when the specimens were about to fail. This proved that the AE energy and accumulated counts had a good correlation with the crack propagation and failure of the specimens during loading.

During the initial loading period of uniaxial compression tests, there was a "quiet period" with few cracks being generated and expanding within the rock. AE events hardly occurred at this stage. When the moisture content of the specimens was 0, as shown in Figure 8a, the AE counts started to accumulate at about 151 s, and the AE events occurred earlier compared with the other specimens. As the moisture content increased, the AE events occurred later. When the specimens were saturated, as shown in Figure 8i, the AE counts started to accumulate at 246 s. The test results showed that the duration of the "quiet period" increased significantly with the moisture content during the uniaxial compression tests, which further confirmed that the initiation time of cracks within the tuff became later and later with the increase in moisture content.

The AE signals began to accumulate after the "quiet period". The AE counts basically rose steadily according to a certain slope before the peak stress, and the AE energy was also collected with the generation of AE events. However, both the growth rate of accumulated AE counts and the AE energy value were kept at a low level. When the stress reached the peak value, the accumulated AE counts and energy showed a very significant increase, which indicated that a large number of fracture events with high energy release occurred in the specimens when destroyed.

The information of AE parameters during the tests of the specimens with different moisture content showed different characteristics. With the increase in moisture content, the accumulated AE counts and energy value both presented a gradually decreasing trend. When the moisture content was low, the AE energy appeared more frequently with a higher value. However, as the moisture content increased, the occurrence of AE events became sparse, and the value of AE energy was relatively reduced, as shown in Figure 8. The analysis of AE parameters also revealed that a higher moisture content resulted in a lower AE energy and fewer accumulated AE counts at the peak stress. Compared with dry specimens, the accumulated AE counts of saturated specimens decreased by an order of magnitude, and the maximum AE energy also decreased by 32.86%. This was mainly due to the softening effect on the tuff specimens caused by water. Higher moisture content of the specimens led to less internal fracture generation, expansion, and penetration during the loading process, and less energy was released at the same time. Although the overall energy value showed a downward trend when the moisture content increased, this did not affect the phenomenon of sudden energy bursts when the specimens were destroyed.

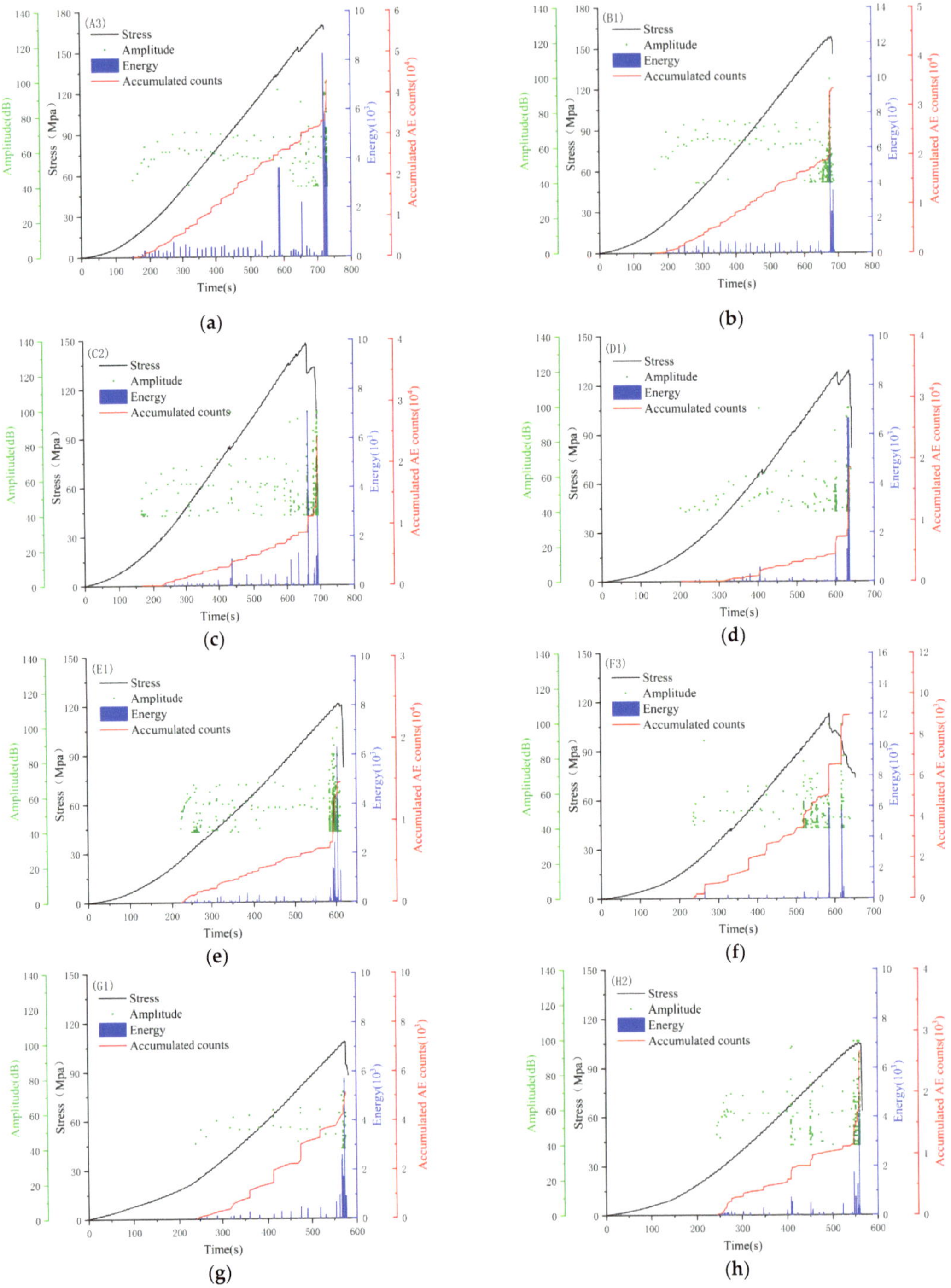

Figure 8. *Cont.*

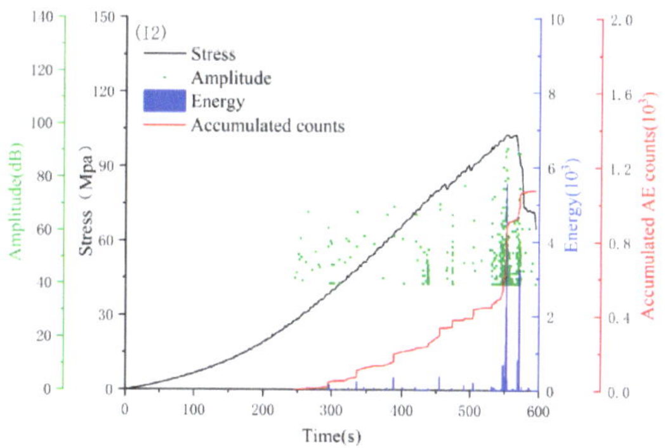

(i)

Figure 8. The evolution of stress, AE energy, accumulated AE counts, and amplitude with time under different moisture content. (**a–i**) Specimens under different moisture content.

3.4. RA–AF Distribution and Tuff Failure Characteristics under Different Moisture Content

The ratio of the rise time to the amplitude of an AE event was recorded as the RA value, and the ratio of the AE counts to the duration time was recorded as the AF value. The RA value and AF value are very important AE parameters to analyze the rock failure process. The combined analysis of RA and AF is often used to determine the fracture type in the analysis of rock fracture generation mechanisms [33–35]. Their definitions are shown in Equations (5) and (6).

$$\text{RA} = \text{Rise time}/\text{Amplitude}. \tag{5}$$

$$\text{AF} = \text{AE counts}/\text{Duration}. \tag{6}$$

The RA–AF distribution analysis was implemented in a rectangular coordinate system with RA value as the horizontal axis and AF value as the vertical axis. During the uniaxial compression tests, a lower RA value and a higher AF value are generated when tensile cracks occurred. In contrast, the generation of shear cracks results in a higher RA value and a lower AF value. A diagonal line was used to divide the rectangular coordinate system into a tensile crack region and a shear crack region, as shown in Figure 9.

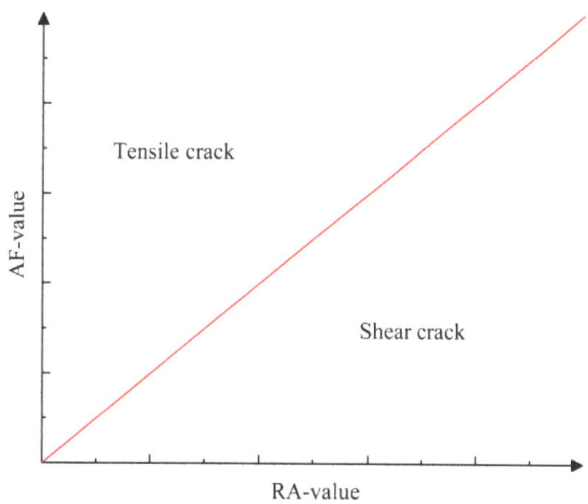

Figure 9. Relationship between crack types and RA–AF distribution.

The RA value and AF value were normalized and then drawn in the coordinate system, as shown in Figure 10. It can be seen from the RA–AF distribution that the RA–AF values

were mainly distributed along the horizontal axis when the moisture content was 0. The RA values were higher, and the AF values were lower, being mainly distributed in the shear crack region, as shown in Figure 10a. This phenomenon revealed that the cracks generated within the rock were mainly shear cracks when the rock specimens were dry. When the moisture content increased, the distribution of RA–AF values gradually transited from the shear crack region to the tension crack region. As shown in Figure 10c, when the moisture content reached 0.41%, RA–AF values were evenly distributed in the tensile and shear crack region. Therefore, both shear cracks and tension cracks were generated under this moisture content, and these two types of cracks were relatively close in number. For saturated specimens, the RA–AF values were mainly distributed along the longitudinal axis. The RA values were lower, and the AF values were higher, being mainly distributed in the tensile crack region, as shown in Figure 10i. This means that the majority of cracks generated inside the specimens were tension cracks when saturated.

The variation of RA–AF distribution was very significant at the initial stage of the rising process of moisture content. When the moisture content was higher than 0.6%, as shown in Figure 10e–i, the RA–AF distribution tended to be stable and mainly concentrated in the tensile crack region. The results showed that the cracks generated during the uniaxial compression tests were mainly tensile cracks when the moisture content of the specimens reached a certain level, and the crack type no longer changed with the continued increase in moisture content.

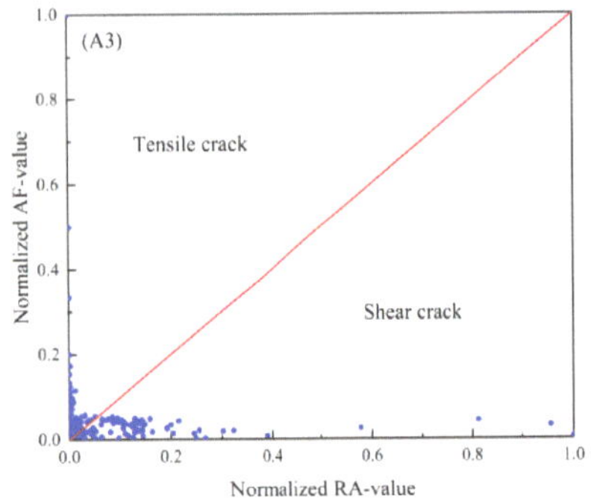

(a)

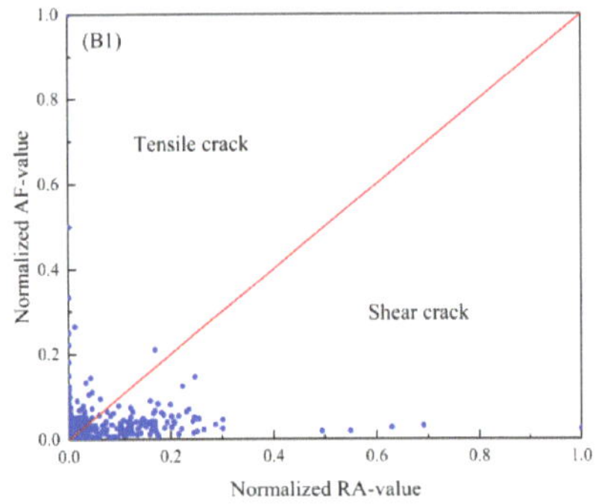

(b)

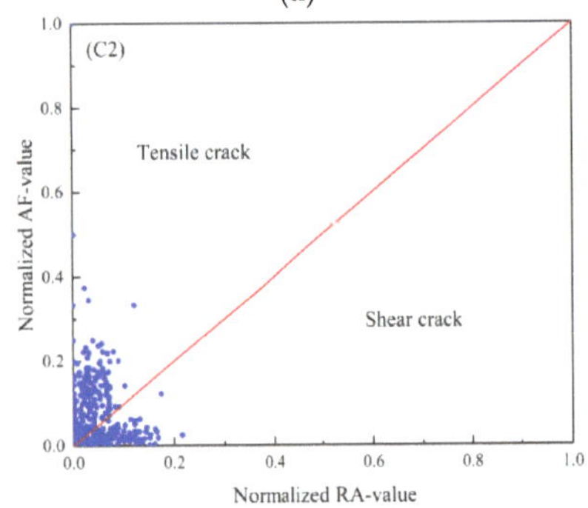

(c)

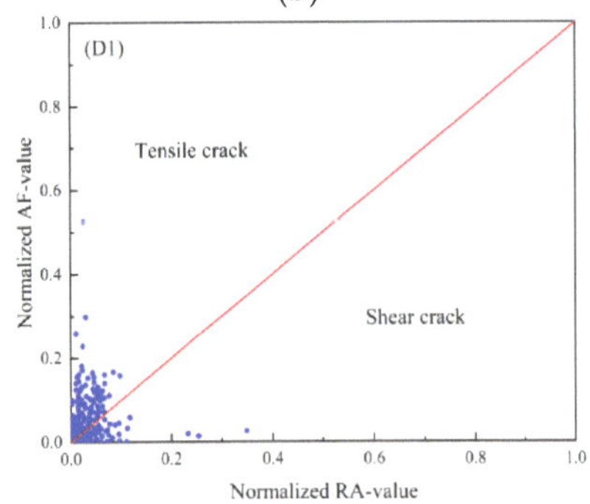

(d)

Figure 10. *Cont.*

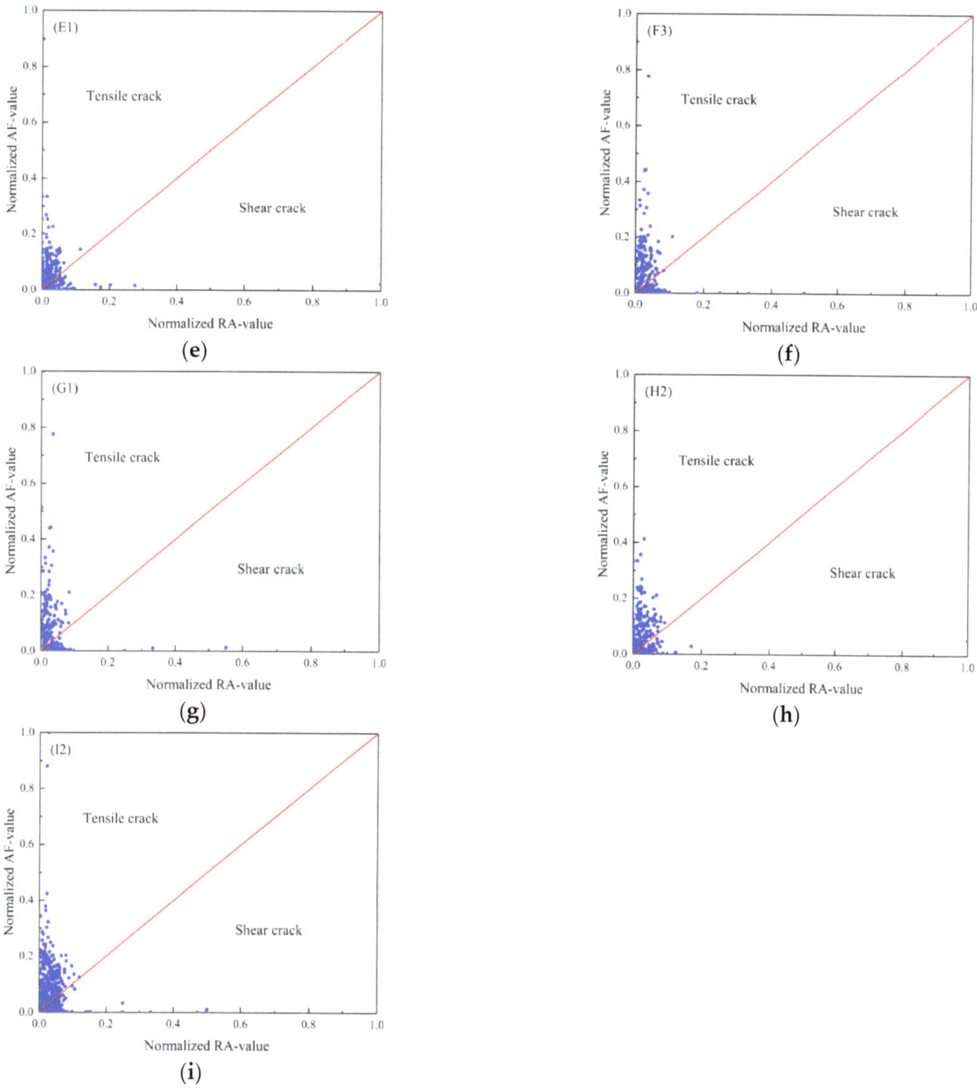

Figure 10. RA–AF distribution of tuff specimens with different moisture content. (**a**–**i**) Specimens under different moisture content.

The failure patterns of the tuff specimens after the tests are shown in Figure 11. The failure characteristics demonstrated that the crack types were mainly diagonal shear cracks when the moisture content of the specimens was low. The failure pattern was mainly manifested as shear slip along the oblique cracks, as shown in Figure 11a–c. During the process of increasing moisture content, the type of cracks gradually changed from shear cracks along the diagonal to tensile cracks along the longitudinal. As shown in Figure 11h,i, the cracks produced were mainly tensile cracks when the specimens failed under high moisture content, and the failure pattern was mainly manifested as splitting tensile failure. In addition, through the comparison of the test results, it was found that the tuff specimens had a higher level of fragmentation under high moisture content, and more fragments were produced when they were broken.

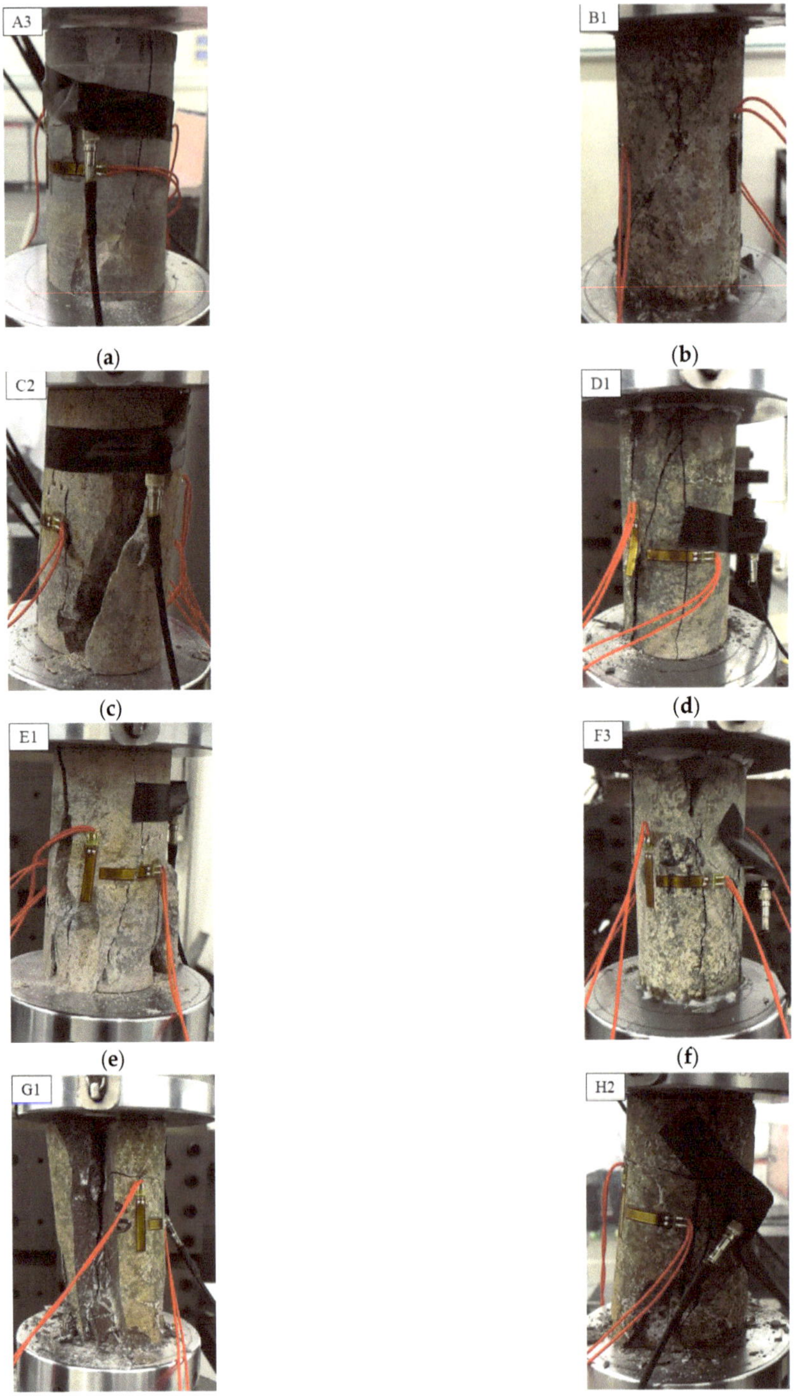

Figure 11. *Cont.*

(i)

Figure 11. Failure patterns of tuff specimens under different moisture content. (**a–i**) Specimens under different moisture content.

These experimental phenomena showed that the level of moisture content affected the failure patterns of the tuff specimens. As the moisture content increased, the tuff specimens gradually transitioned from shear failure to tensile failure, and the level of fragmentation gradually increased. This was mainly because of the increased infiltration of water into the rock with the increase in moisture content. Furthermore, the cementation between the mineral particles inside the specimens was weakened, which in turn caused a change in failure pattern and more severe damage when the rock specimens failed.

The fracture characteristics of the tuff specimens in the uniaxial compression tests under different moisture content were basically consistent with the fracture types obtained via the analysis of RA–AF distribution. This further proved that the RA–AF distribution could successfully determine the failure types of tuff specimens under different moisture content in the uniaxial compression test.

3.5. Mesoscopic Analysis of Failure Fracture under Different Moisture Content

In order to study the mesoscopic characteristics of tuff under different moisture content, scanning electron microscopy (SEM) was used to analyze the fracture mesoscopic morphology of the samples.

The damaged samples with 0% (A3), 0.62% (D1), and 1.04% (I2) moisture content were selected for SEM detection of the fracture surface, and the magnification was 2000×. As shown in Figure 12a–c, when the sample was dry, the failure section was relatively neat, and there were fewer fractures of holes and step zones. With the increase in moisture content, the failure mode tended to be complex, and the failure degree was higher, with more holes and cracks appearing. In addition, a higher moisture content led to more step zone cracks.

This is mainly because, with the increase in moisture content, water weakens the cementation of rock, resulting in a higher failure degree of tuff, and more holes and crack zones appear at the fracture surface at the mesoscopic level. This also explains the more serious macroscopic damage of tuff in the case of high moisture content at the microscopic level.

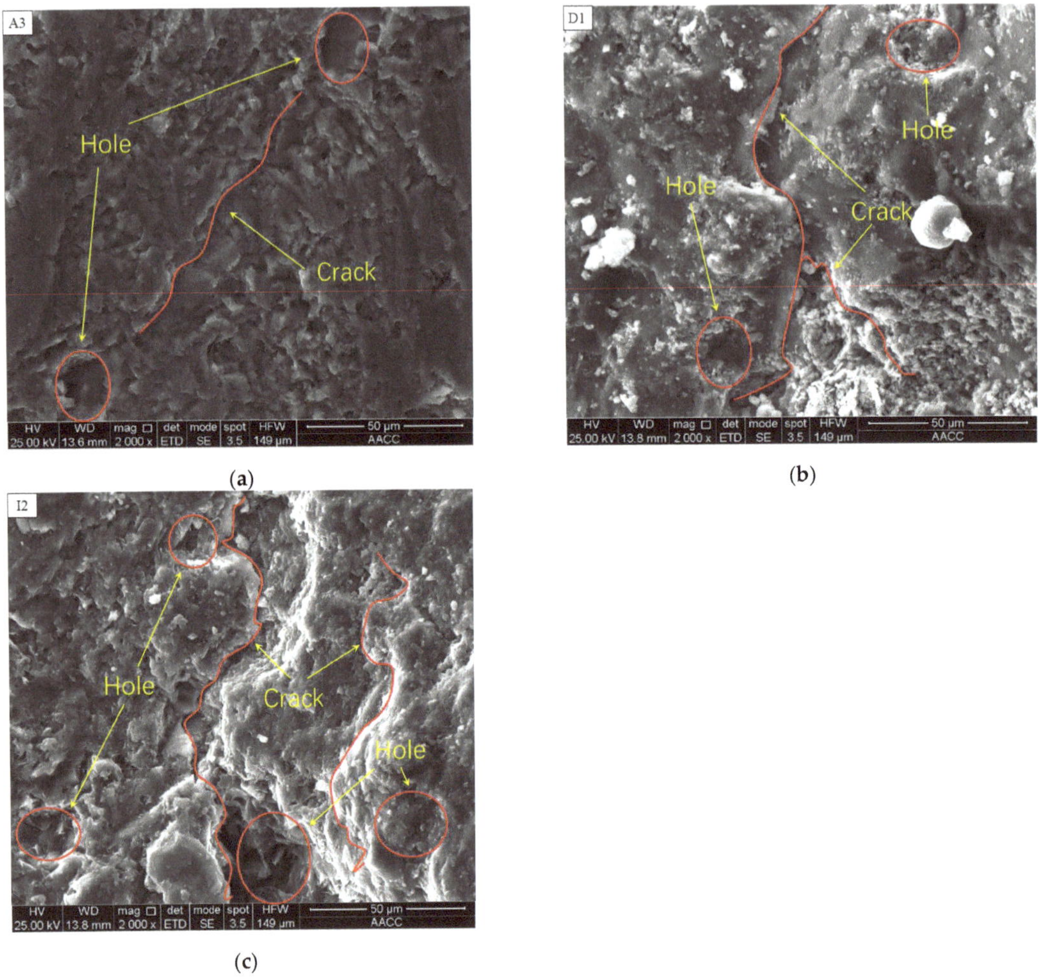

Figure 12. Mesoscopic fracture morphology of the samples under different moisture content according to SEM (2000×): (**a**) dry specimen; (**b**) specimen with 0.62% moisture content; (**c**) saturated specimen.

4. Conclusions

In this article, water absorption tests and uniaxial compression tests were carried out on tuff specimens to study the influence of moisture content on their mechanical and failure characteristics. The AE parameters obtained by the AE monitoring system were applied to analyze the characteristics of crack propagation and failure types in the rock under different moisture contents. The key findings obtained from this study are as follows:

(1) The moisture content increased with immersion time. At the initial stage of immersion, the moisture content rose rapidly (0–24 h). After 24 h of immersion, the increasing rate of moisture content dropped significantly. After 180 h of immersion, the moisture content gradually stabilized and approached saturation in numerical terms.

(2) The uniaxial compressive strength and elastic modulus of the tuff specimens both decreased with the increase in moisture content, reaching the lowest values when the specimens were saturated. The uniaxial compressive strength decreased rapidly at the initial stage of immersion, and the decrease rate slowed down as the moisture content increased. The variation of elastic modulus also showed a similar trend. When the moisture content increased, the stress–strain curves showed obvious post-peak

segments, which proved that the increase in moisture content led to an improvement in plasticity and toughness of the tuff. In addition, the Poisson's ratio increased with moisture content.

(3) Higher moisture content of the tuff specimens led to a longer "quiet period" experienced during the initial loading stage. In addition, the increase in moisture content also led to a decrease in the accumulated AE counts and energy. This proved that the generation and expansion of tuff rock cracks were delayed by the softening effect of water, and the number of internal crack events in the rock was relatively decreased. Furthermore, the energy release generated by internal crack events of the tuff was also reduced.

(4) When the moisture content of the rock was low (lower than 0.41%), the RA–AF values were mainly distributed along the horizontal axis, indicating that the crack type was dominated by shear crack, whereas, when the rock moisture content was high (higher than 0.41%), the RA–AF values became distributed along the vertical axis, indicating that the crack type was mainly tension crack. The failure morphology of the tuff specimens also further verified that the dominant crack type changed from shear crack to tensile crack with the increase in moisture content, while the disruption at the microscopic level was also more intense.

Author Contributions: Writing—original draft preparation, W.D.; writing—review and editing, L.H. and W.D.; methodology, L.M.; visualization, H.Z. and S.Y.; project administration, C.X.; funding acquisition, L.H. and Y.D.; supervision, L.H. All authors read and agreed to the published version of the manuscript.

Funding: This research was funded by National Natural Science Foundation of China (No. 51574223).

Data Availability Statement: Not applicable.

Acknowledgments: The authors thank Anhui Masteel Luohe Mining Co., Ltd. for the test samples. The authors also sincerely thank the State Key Laboratory for Geomechanics and Deep Underground Engineering, China University of Mining and Technology for providing the test equipment.

Conflicts of Interest: The authors declare no conflict of interest.

References

1. Kim, E.; Changani, H. Effect of water saturation and loading rate on the mechanical properties of Red and Buff Sandstones. *Int. J. Rock Mech. Min. Sci.* **2016**, *88*, 23–28. [CrossRef]
2. Vasarhelyi, B.; Van, P. Influence of water content on the strength of rock. *Eng. Geol.* **2006**, *84*, 70–74. [CrossRef]
3. Li, L.P.; Sun, S.Q.; Wang, J.; Yang, W.M.; Song, S.G.; Fang, Z.D. Experimental study of the precursor information of the water inrush in shield tunnels due to the proximity of a water-filled cave. *Int. J. Rock Mech. Min. Sci.* **2020**, *130*, 104320. [CrossRef]
4. Zhu, J.; Deng, J.H.; Chen, F.; Huang, Y.M.; Yu, Z.Q. Water Saturation Effects on Mechanical and Fracture Behavior of Marble. *Int. J. Geomech.* **2020**, *20*, 04020191. [CrossRef]
5. Ranjith, P.G.; Jasinge, D.; Song, J.Y.; Choi, S.K. A study of the effect of displacement rate and moisture content on the mechanical properties of concrete: Use of acoustic emission. *Mech. Mater.* **2008**, *40*, 453–469. [CrossRef]
6. Liu, X.X.; Wu, L.X.; Zhang, Y.B.; Liang, Z.Z.; Yao, X.L.; Liang, P. Frequency properties of acoustic emissions from the dry and saturated rock. *Environ. Earth Sci.* **2019**, *78*, 67. [CrossRef]
7. Shiotani, T.; Ohtsu, M.; Ikeda, K. Detection and evaluation of AE waves due to rock deformation. *Constr. Build. Mater.* **2001**, *15*, 235–246. [CrossRef]
8. Liu, X.; Liu, Z.; Li, X.; Gong, F.; Du, K. Experimental study on the effect of strain rate on rock acoustic emission characteristics. *Int. J. Rock Mech. Min. Sci.* **2020**, *133*, 104420. [CrossRef]
9. Lockner, D. The role of acoustic emission in the study of rock fracture. *Int. J. Rock Mech. Min. Sci. Geomech. Abstr.* **1993**, *30*, 883–899. [CrossRef]
10. Yao, Q.; Wang, W.; Zhu, L.; Xia, Z.; Tang, C.; Wang, X. Effects of moisture conditions on mechanical properties and AE and IR characteristics in coal–rock combinations. *Arab. J. Geosci.* **2020**, *13*, 615. [CrossRef]
11. Zhou, Z.L.; Cai, X.; Cao, W.Z.; Li, X.B.; Xiong, C. Influence of Water Content on Mechanical Properties of Rock in Both Saturation and Drying Processes. *Rock Mech. Rock Eng.* **2016**, *49*, 3009–3025. [CrossRef]
12. Vasarhelyi, B. Statistical analysis of the influence of water content on the strength of the miocene limestone. *Rock Mech. Rock Eng.* **2005**, *38*, 69–76. [CrossRef]

13. Yilmaz, I. Influence of water content on the strength and deformability of gypsum. *Int. J. Rock Mech. Min. Sci.* **2010**, *47*, 342–347. [CrossRef]
14. Vishal, V.; Ranjith, P.G.; Singh, T.N. An experimental investigation on behaviour of coal under fluid saturation, using acoustic emission. *J. Nat. Gas Sci. Eng.* **2015**, *22*, 428–436. [CrossRef]
15. Yu, L.; Yao, Q.; Li, X.; Wang, W.; Han, H.; Zhang, M. Experimental study of failure characteristics and fissure propagation in hydrous siltstone. *Arab. J. Geosci.* **2020**, *13*, 527. [CrossRef]
16. Roy, D.G.; Singh, T.N.; Kodikara, J.; Das, R. Effect of Water Saturation on the Fracture and Mechanical Properties of Sedimentary Rocks. *Rock Mech. Rock Eng.* **2017**, *50*, 2585–2600. [CrossRef]
17. Rudajev, V.; Vilhelm, J.; Lokajicek, T. Laboratory studies of acoustic emission prior to uniaxial compressive rock failure. *Int. J. Rock Mech. Min. Sci.* **2000**, *37*, 699–704. [CrossRef]
18. Backers, T.; Stanchits, S.; Dresen, G. Tensile fracture propagation and acoustic emission activity in sandstone: The effect of loading rate. *Int. J. Rock Mech. Min. Sci.* **2005**, *42*, 1094–1101. [CrossRef]
19. Cai, M.; Kaiser, P.K.; Morioka, H.; Minami, M.; Maejima, T.; Tasaka, Y.; Kurose, H. FLAC/PFC coupled numerical simulation of AE in large-scale underground excavations. *Int. J. Rock Mech. Min. Sci.* **2007**, *44*, 550–564. [CrossRef]
20. Ganne, P.; Vervoort, A.; Wevers, M. Quantification of pre-peak brittle damage: Correlation between acoustic emission and observed micro-fracturing. *Int. J. Rock Mech. Min. Sci.* **2007**, *44*, 720–729. [CrossRef]
21. Ohno, K.; Ohtsu, M. Crack classification in concrete based on acoustic emission. *Constr. Build. Mater.* **2010**, *24*, 2339–2346. [CrossRef]
22. Sun, X.; Xu, H.; Zheng, L.; He, M.; Gong, W. An experimental investigation on acoustic emission characteristics of sandstone rockburst with different moisture contents. *Sci. China Technol. Sci.* **2016**, *59*, 1549–1558. [CrossRef]
23. Yang, K.; Yan, Q.X.; Zhang, C.A.; Wu, W.; Wan, F. Study on mechanical properties and damage evolution of carbonaceous shale under triaxial compression with acoustic emission. *Int. J. Damage Mech.* **2021**, *30*, 899–922. [CrossRef]
24. Chen, L.H.; Chen, W.C.; Chen, Y.C.; Benyamin, L.; Li, A.J. Investigation of Hydraulic Fracture Propagation Using a Post-Peak Control System Coupled with Acoustic Emission. *Rock Mech. Rock Eng.* **2015**, *48*, 1233–1248. [CrossRef]
25. Chen, L.; Zhao, J.; Zheng, Z. Acoustic Emission Characteristics of Compressive Deformation and Failure of Siltstone under Different Water Contents. *Adv. Mater. Sci. Eng.* **2017**, *2017*, 4035487. [CrossRef]
26. Carpinteri, A.; Corrado, M.; Lacidogna, G. Heterogeneous materials in compression: Correlations between absorbed, released and acoustic emission energies. *Eng. Fail. Anal.* **2013**, *33*, 236–250. [CrossRef]
27. Amann, F.; Button, E.A.; Evans, K.F.; Gischig, V.S.; Blumel, M. Experimental Study of the Brittle Behavior of Clay shale in Rapid Unconfined Compression. *Rock Mech. Rock Eng.* **2011**, *44*, 415–430. [CrossRef]
28. Xiao, F.; He, J.; Liu, Z.; Shen, Z.; Liu, G. Analysis on warning signs of damage of coal samples with different water contents and relevant damage evolution based on acoustic emission and infrared characterization. *Infrared Phys. Technol.* **2019**, *97*, 287–299. [CrossRef]
29. Rodriguez, P.; Celestino, T.B. Application of acoustic emission monitoring and signal analysis to the qualitative and quantitative characterization of the fracturing process in rocks. *Eng. Fract. Mech.* **2019**, *210*, 54–69. [CrossRef]
30. Liu, H.L.; Zhu, W.C.; Yu, Y.J.; Xu, T.; Li, R.F.; Liu, X.G. Effect of water imbibition on uniaxial compression strength of sandstone. *Int. J. Rock Mech. Min. Sci.* **2020**, *127*, 104200. [CrossRef]
31. Huang, S.B.; He, Y.B.; Liu, G.F.; Lu, Z.X.; Xin, Z.K. Effect of water content on the mechanical properties and deformation characteristics of the clay-bearing red sandstone. *Bull. Eng. Geol. Environ.* **2021**, *80*, 1767–1790. [CrossRef]
32. Huang, Y.M.; Deng, J.H.; Zhu, J. An Experimental Investigation of Moisture-Induced Softening Mechanism of Marble Based on Quantitative Analysis of Acoustic Emission Waveforms. *Appl. Sci.* **2019**, *9*, 446. [CrossRef]
33. Sagar, R.V.; Prasad, B.K.R. A review of recent developments in parametric based acoustic emission techniques applied to concrete structures. *Nondestruct. Test Eval.* **2012**, *27*, 47–68. [CrossRef]
34. Meng, Y.Y.; Jing, H.W.; Liu, X.W.; Yin, Q.; Wei, X.C. Experimental and numerical investigation on the effects of bedding plane properties on the mechanical and acoustic emission characteristics of sandy mudstone. *Eng. Fract. Mech.* **2021**, *245*, 107582. [CrossRef]
35. Wang, H.J.; Liu, D.; Cui, Z.D.; Cheng, C.; Jian, Z. Investigation of the fracture modes of red sandstone using XFEM and acoustic emissions. *Theor. Appl. Fract. Mech.* **2016**, *85*, 283–293. [CrossRef]

minerals

MDPI

Article

Safe and Efficient Recovery Technique of Horizontal Isolated Pillar under Loose Tailings Backfill; A Case Study in a Zinc-Lead Mine

Fan Feng [1,2,*], Jiqing Zhang [2,4], Zhikai Yan [2], Yedan Wu [2,4], Yamin Song [3,*], Zhiwei Xie [2] and Xiang Cheng [1]

1 State Key Laboratory of Mining Response and Disaster Prevention and Control in Deep Coal Mines, Anhui University of Science and Technology, Huainan 232001, China
2 College of Energy and Mining Engineering, Shandong University of Science and Technology, Qingdao 266590, China
3 Don College of Shandong Jiaotong University, Jinan 250357, China
4 Qingdao Ruiyuan Engineering Group Co., Ltd., Qingdao 266555, China
* Correspondence: fengfan0213@sdust.edu.cn (F.F.); sym511@126.com (Y.S.)

Abstract: The presence of loose medium backfill above the horizontal pillar will technically hinder the efficient recovery of the pillar since the improper design and preserved roof protection layer height will potentially lead to casualties and equipment damage caused by large area collapse of filled tailings as well as roof fall accidents. In this study, a safe and efficient technique for the recovery of isolated pillars under loose tailings backfill was carried out via field investigation, theoretical analysis, numerical simulation, and analytic hierarchy process using the isolated pillars in the 855 middle sublevel of Hongling Zinc-Lead Mine, Chifeng, Inner Mongolia, as a practical engineering background. Current studies have revealed that the optimal scheme for an isolated horizontal pillar recovered via the cut-and-fill stoping of a drift vertical to ore body strike involves preserving a 1.0-m roof protection layer above the crown pillar combined with a spaced mining extraction sequence. This design minimizes ore dilution and losses during the pillar extraction process under safe operation. Our research results provide theoretical and technical support for the safe and efficient recovery of isolated pillars under loose tailings backfill in similar mines.

Keywords: isolated pillar; overlying backfill; roof protection layer height; recovery scheme; failure behavior; numerical simulation; analytic hierarchy process

Citation: Feng, F.; Zhang, J.; Yan, Z.; Wu, Y.; Song, Y.; Xie, Z.; Cheng, X. Safe and Efficient Recovery Technique of Horizontal Isolated Pillar under Loose Tailings Backfill; A Case Study in a Zinc-Lead Mine. *Minerals* **2022**, *12*, 1066. https://doi.org/10.3390/min12091066

Academic Editor: Yosoon Choi

Received: 4 August 2022
Accepted: 23 August 2022
Published: 24 August 2022

1. Introduction

The vast majority of metal mine workings feature preserved pillars, the ore volumes of which account for approximately 20 to 40% of their workable reserves [1–3]. Recovering remnant pillars can improve the utilization rate of mineral resources, extend mine life while shortening the underground operation line, aid in coordinating the sequence of operations, and improve mine management capacity. For example, Rybak [4] discusses the possibility of reducing the loss of minerals left in the pillars due to the use of a system with backfill; simulates in sufficient detail the impact of mining operations on the undermined rock mass, the possibility of developing subvertical disturbance, violation of the water-protective strata and the formation of subsidence on the Earth's surface. The technical conditions for mining remnant crowns and sill pillars are particularly complex, leading to more stringent safety and cost requirements [5,6].

There are two main upper load forms for horizontal isolated pillars: (1) in cases where the upper part of the pillar has been subjected to cut-and-fill stoping, the overlying uniform load is jointly produced by the gravitational loading of the overlying backfill and the ground pressure of the pit [7]; (2) when the ore body above the isolated horizontal pillar is extracted using either open stoping or caving, the overlying uniform load is

primarily induced by the weight of the collapsed surrounding rock or residual ore body in the goaf [8]. Under the first condition, the presence of loose medium backfill above the horizontal pillar will technically hinder the efficient recovery of the pillar. The high permeability of filled tailings results in the deposition of significant quantities of water at the bottom of the pit. A resulting column collapse will result in tailing leakage and loss, potentially leading to blockage of the operational space and roadway and, in serious cases, surface subsidence. Thus, a sufficiently thick roof protection layer must be preserved in advance to prevent casualties and equipment damage caused by large area collapse and roof fall accidents involving loose tailings backfill. It is worth noting that preserving a roof protection layer that is too thick will not ensure the safety of the roof and will further waste mineral resources, increasing ore damage and dilution. It follows that determining a reasonable roof protection layer thickness and stability in the recovery of pit crown pillars is an important and complex scientific research task [8–10]. The analysis and study of the mining stability of crown pillars under backfill in metal mines have significant theoretical and practical significance in improving the recovery rate of resources while achieving safe asnd efficient extraction.

It is clear from the relevant national and international literature that the size of the preserved roof protection layer of the isolated pillar is related to the technical conditions of mining (the reserve conditions; the size, occurrence, and stability of the crown and sill pillars; the grade and value of the ore; the dip of the ore body and the stability of the surrounding rock; the surface conditions; and the characteristics of the overlying uniform load distribution) and is inseparable from the main technical factors of crown pillar mining (the selected mining method and process parameters) [11,12]. For example, Liu [11] analyzed the stope span on the roof stability in the Sanshandao gold mine and proposed the rational stope structure parameters when recovering the horizontal pillar under backfill. The technical conditions of mining vary significantly by mine and, therefore, so do the recovery schemes. This variety has led to gaps in the literature; for example, there is not much research on the appropriate recovery technology for cases in which both the upper and lower parts of an isolated pillar comprise loose tailings backfill. In addition, the empirical formulas used in past studies to calculate the thickness of the roof protection layer of isolated pillars are potentially inaccurate and of limited general applicability [13]. Zhao et al. [13] calculated the size of the insulating pillar by using methods of limit span mechanism, the empirical equation method, and the limit equilibrium method. The results indicate the discreteness of the calculated pillar size is very large. With the rapid development of rock mechanics and computer science, numerical simulation methods have been increasingly used to analyze geotechnical engineering problems [14–16]. The authors conducted several studies to investigate the failure behavior of surrounding rocks in the underground mine by using the FLAC^{3D} (Finite Difference Method) and ELFEN (Finite/Discrete Element Method) software [17,18]. In addition, optimized fuzzy mathematics can be applied to transform various fuzzy or subjective factors involved in the preservation of the roof protection layer of an isolated pillar into a mathematical form to enable the quantification via evaluation indexing of individual schemes, allowing for more scientific, reasonable, and reliable scheme selection [19–21]. Thus, using classical theory, numerical simulation and optimized fuzzy mathematics to study the safe and efficient recovery technology of isolated pillars under loose tailings filling body will greatly improve the rationality and accuracy of the calculation results. To date, however, only a few relevant studies have applied such a combined approach.

Analyzing the above, it can be noted that the safe and efficient recovery technique of isolated horizontal pillars under loose tailings backfill is a topical issue. To achieve this, it is necessary to solve the following tasks: (1) the proposal of a compatible middle sublevel crown pillar recovery scheme; (2) the determination of rational mining sequence when recovering a crown pillar; and (3) the calculation of preserved roof protection layer height. The present study demonstrates that the optimal design for an isolated horizontal pillar recovered via the cut-and-fill stoping of a drift vertical to the ore body strike involves

preserving a 1.0-m roof protection layer above the crown pillar combined with a spaced mining extraction sequence. The research results are intended to provide theoretical support and scientific decision-making for the safe and efficient recovery of isolated pillars under loose tailings backfill in similar mines.

2. Engineering Background and Existing Problems

2.1. Basic Overview of the Mining Area

The Hongling Zinc-Lead Mine, Chifeng, is located at Haobugao Village, Wulandaba Township, Lindong Town, Bairin Left Banner, Chifeng City, Inner Mongolia Autonomous Region. The mining area has a total length of 5700 m. The ore body is reserved at an elevation range of 630–1110 m and has a maximum deepening of 480 m and an overall orientation of 59° N.E., with a tendency toward the northwest. The ore body has a dip of 60–80°, with an average dip of 75°, and a strike length of 1350 m. The ore zone is divided into five sublevels from east to west, with the main industrial ore body in the mining area located in Sublevel III. Sublevel III has a total length of 1350 m and a width of 100 m, tends toward the northwest, and has a dip of 80°. It circles three zinc ore bodies, namely 1, 1-1, and 2 and one iron ore body.

2.2. Existing Problems

In the mine, zinc and iron ores are recovered in the middle sublevel at 905 and 855 m from the south and west flanks using the shallow-hole shrinkage and sublevel open-stope methods. The surrounding rock of the hanging and footwall is marble and slate, respectively. The average thickness and dip of the ore body are 15 m and 75°, respectively. The pit has the following structural parameters: rib pillar of 18 m, stope room of 32 m, crown pillar of 10 m, and the width of the ore body between 10 and 40 m. The schematic of the mining method used in the 855 middle sublevel of the pit is shown in Figure 1.

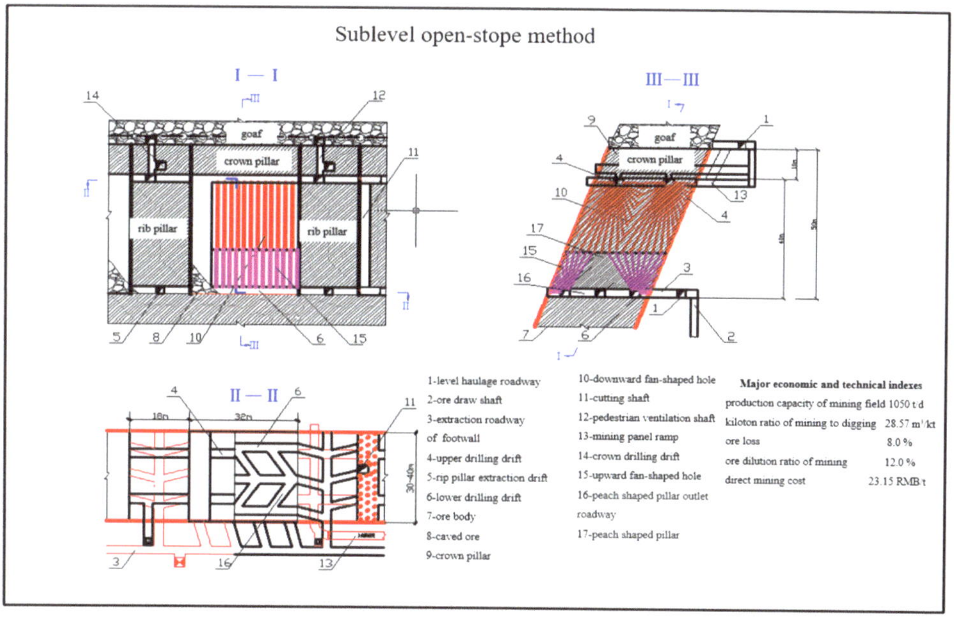

Figure 1. Schematic of mining method used in the 855 middle sublevel of the pit. (Done by the authors).

At present, the 855 m middle sublevel 1# ore body goaf has been governed, and part of the goaf in the 905 m middle sublevel is backfilled with 1:10 graded tailings (height of approximately 20 m), as shown in Figure 2. To reduce ore loss and extract additional ore

resources, the crown pillars of the 855 m middle sublevel (with a height of approximately 10 m, length along the strike direction of approximately 100 m, and containing approximately 30,000 tons of ore) are scheduled to be recovered; however, the rib pillars and the crown pillars corresponding to them will not be extracted for the time being, and the necessary support pillars will be preserved to maintain the stability of the pit in conjunction with the strength of the ore rock body itself. Full recovery of the crown pillars of the middle sublevel would cause a significant collapse of overlying loose tailings backfill and ore mixture, block the ore removal channel, and threaten the safety of downhole operators. To preserve the top protection ore, a solution to the technical problem of executing an economic and reasonable recovery scheme based on selecting a suitable roof protection layer thickness is urgently needed.

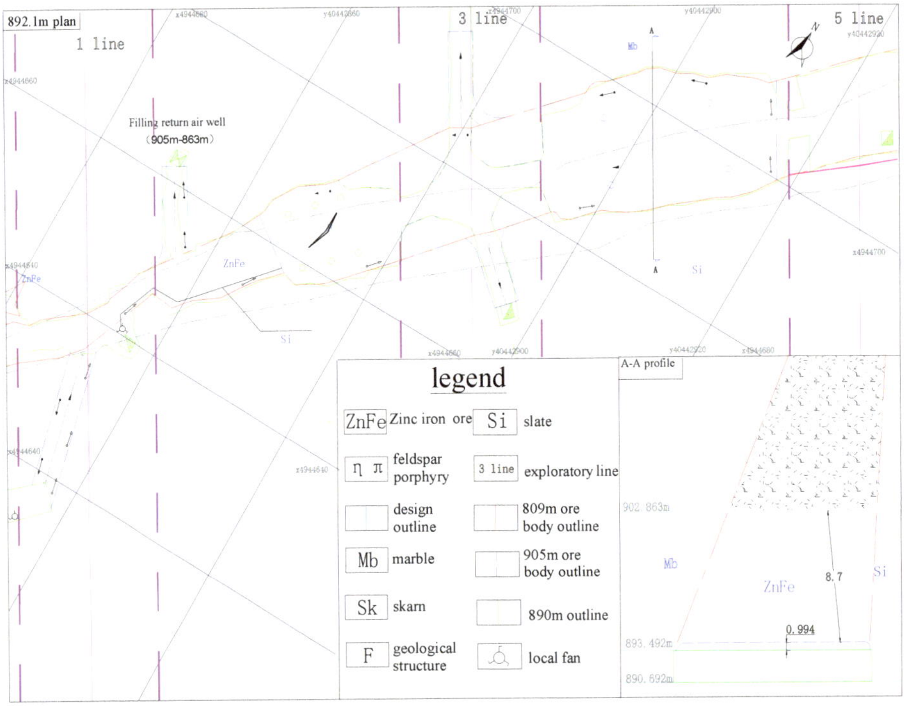

Figure 2. Plan of Lines 1–3 of 890 horizontal in 855 m middle sublevel.

2.3. Methods

Based on the engineering background and existing problems, this study presents a crown pillar recovery scheme applicable to the actual technical conditions in the Hongling Zinc-Lead Mine, Chifeng, Inner Mongolia, using the isolated horizontal pillar of the mine's 855 middle sublevel as the engineering background. Using the structure and size of the pit obtained through on-site research and the theoretical analysis, the uniform load value of the overlying backfill and the reasonable thickness range of the preserved roof protection layer of the crown pillar in the 855 middle sublevel were obtained. Numerical simulation and inversion analysis of different roof protection layer preservation and recovery schemes were then carried out using Fast Lagrangian Analysis of Continua (FLAC^{3D}) numerical simulation technology to obtain the displacement of the roof protection layer and the distribution pattern of the plastic zone within the test pit. Finally, the evaluation indices of the respective extraction scheme were quantified and analyzed using an analytic hierarchy process.

3. Proposed Middle Sublevel Crown Pillar Recovery Schemes

Based on the existing mining technology conditions and the stress environment around the crown pillars of the 855 middle sublevel of Hongling Zinc-Lead Mine, Chifeng, three crown pillar recovery schemes were proposed. The advantages, disadvantages, and applicable conditions for each scheme are listed in Table 1. It is worth noting that the isolated pillar of the 855 middle sublevel has a total thickness of 10 m and that the 5-m-thick ore body under it remains in a relatively stable state during extraction; accordingly, we limited our recovery scheme comparison to the isolated pillars of the upper (close to the loose tailings backfill) 5-m-thick ore body. In all three schemes, the cut and fill stoping are used to recover the pillars, but the pit structure, stope preparation, cut, and extraction operation differ significantly by proposal.

Table 1. Basic Characteristics and Advantages and Disadvantages of Three Crown Pillar Recovery Schemes.

Schemes	Scheme I	Scheme II	Scheme III
Name	Cut and fill stoping of drift along the ore body strike	Cut and fill stoping of drift vertical to the ore body strike	Cut and fill stoping of small drift along the ore body strike
Basic characteristics	Crown pillars are divided into two ore blocks along the ore body strike direction, and each block is further divided into stope rooms and pillars. The stope room has a length of 40 m, and the pillar has a width of 10 m. In the stope room, the drifts are arranged along the same direction of the ore body, and 3–5 drifts can be arranged at the same time.	Drifts are arranged for the crown pillars along the ore body tendency body without division of stope rooms and pillars. The width of the drifts can be 3–5 m, the length of the drift is the thickness of the crown pillar. If the length of the crown pillar is 100 m, more than 20 drifts need to be arranged.	The small blocks are arranged along the crown pillars, with a width of 20 m and a length of the thickness of the ore body, and 5 small blocks need to be arranged. Each small block is subdivided into stope rooms with a width of 15 m and point pillars with a width of 5 m. In the stope room, the drifts are arranged along the same direction of the ore body, and 3 drifts can be arranged at the same time.
Advantages	The pit structure is relatively simple, while the number of drifts is small, and the extraction and backfill operation does not need to be changed frequently, so there is less mutual interference; the stope preparation and cutting work are relatively simple.	The pit features a simpler structure, smaller exposed area of drifts, shorter maintenance time of drifts and higher ore recovery rate, in which the thickness of the roof protection layer required to be preserved can be reduced appropriately; the spaced extraction has a low impact on the ore recovery progress; multiple drifts can be extracted simultaneously, which improves mining efficiency.	Compared to both Schemes I and II, the pit drift features a smaller exposed area, better stability, shorter maintenance time, less thickness of roof protection layer required to be preserved, and a higher ore recovery rate.
Disadvantages	The length of each drift is larger, the exposed area of the roof is larger, and the thicker roof protection layer needs to be preserved, resulting in a higher ore loss rate; the non-spaced extraction method was adopted while the number of drifts extracted is only one, affecting the extraction progress inside the pit, thus leading to the extraction period delay; blind heading causes poor ventilation effect in the pit.	The length of the drift is the thickness of the ore body, more drifts need to be arranged in the pit, and the extraction work and the filling work need to be changed frequently, affecting the extraction progress to a certain extent; blind heading causes poor ventilation effect in the pit.	The pit structure is more complex, and the quantities of stope preparation and cutting works are larger; the non-spaced extraction method is adopted, while the number of drifts simultaneously extracted is only one, affecting the extraction progress and work efficiency in the pit, leading to the extraction period delay; blind heading causes poor ventilation effect in the pit.

In Scheme I, the pit features a simpler structure but longer drifts, resulting in a large, exposed area of extraction drifts; as a result, a thicker roof protection layer is required, and the ore loss rate is higher. Thus, Scheme I was excluded because of its poor performance in terms of residual deposit resource recovery. Under Scheme III, the pit has the smallest exposed area of drifts; however, the difference in this regard relative to Scheme II is minimal. In contrast, the pit under Scheme III has the disadvantages of a complex structure, a large workload for stope preparation and cutting, and a high strength required for transport roadway support, etc. Moreover, this scheme adopts non-spaced extraction, reducing extraction progress and work efficiency within the mining site. Scheme II features some of the advantages of Schemes I and III with a simple pit structure, small exposed area of drifts, and enhanced extraction efficiency. Therefore, following comprehensive comparison and analysis Scheme II was selected as the recovery scheme for the crown pillars of the 855 middle sublevel.

4. Theoretical Calculation of a Reasonable Size for the Preserved Roof Protection Layer of the Crown Pillars of the 855 Middle Sublevel

4.1. Calculation of the Overlying Uniform Load on the Roof Protection Layer

A site survey revealed that the surrounding rocks in the upper goaf would undergo failure and collapse under the double effects of mining-induced stress and excavation unloading, resulting in the deposition of waste rock at the bottom. Accordingly, the load on the upper part of the crown pillars in the 855 middle sublevel was determined to comprise two components: the loads from the loose tailings backfill and the waste rock falling from the surrounding rock caving from a height of 2–3 m, respectively. The load of the latter on the horizontal pillars is equivalent to the self-weight of the waste rock, that is, approximately 0.059 MPa. In calculating the overlying uniform load, the upper part of the roof protection layer was considered as the loose backfill, and according to Janssen's loose mass pressure theory [22,23], the uniform load of the loose tailings acting on the roof protection layer was determined to be 0.221 MPa. Thus, the total overlying uniform load on the roof protection layer was approximately 0.28 MPa.

4.2. Theoretical Analysis of the Reasonable Size of Preserved Roof Protection Layer

We calculated and analyzed the thickness of the roof protection layer using limit span theory, K.B. Rubeneita's method, B.H. Bogorlyubov's method, the length-to-width ratio plate-girder method, the load intersection line method, Protodyakonov's arch method, and limit equilibrium theory [13,24,25]. Given the limited space in this paper, the parameter definitions and the calculated top protection layer results obtained under each theory are summarized in Table 2.

Table 2. Theoretical calculation methods and results.

Theoretical Method	Calculation Formula	Definition of Parameters	Thickness of Roof Protection Layer (m)
Limit span theory	$h_p = W(\frac{\gamma_s W n}{2\sigma_t} + 3.3\sqrt{\frac{\gamma_s h_s n}{\sigma_c}})$	h_p—thickness of the roof protection layer, m; W—drift width, m; σ_t, σ_c—the tensile strength and compressive strength of the ore in the roof protection layer, respectively, MPa; γ_s—waste rock unit weight, kN/m^3; hs—the height of caved rock or backfill waste pile, m; n—the safety factor, by using the open stoping, taken as 1.2.	1.6

Table 2. *Cont.*

Theoretical Method	Calculation Formula	Definition of Parameters	Thickness of Roof Protection Layer (m)
K.B. Rubeneita method	$H = K\frac{0.25RB^2 + \sqrt{R^2B^2 + 800\sigma_s g}}{98\sigma_s}$	H-thickness of the roof protection layer, m; K—safety factor, taken as 1.2; γ_2—the rock unit weight of the roof protection layer, t/m^3; B—drift width, m; $\sigma_S = \sigma_{n5}/K_0K_3$ is the rock strength limit under bending conditions, K_0 = 2–3, K_3 = 7–10, $\sigma_{n5} = 0.1\sigma_c$, σ_c—the uniaxial compressive strength of rock, MPa; g—the uniform load above the roof protection layer, MPa.	0.6
B.H. Bogorlyubov method	$H = K\frac{\gamma_p W_0^2 + (\gamma_p W_0^2 + 16\sigma_{n5}P)^{\frac{1}{2}}}{q\sigma_{n5}}$	H—thickness of the roof protection layer, m; P—dynamic load formed due to blasting rock, $P = rH_Y K_n(K_c + K_{nep})/K_p$; H_Y—the gradient height, taken as 5 m; r—crater index, taken as 1.5; K_c, K_{nep}, K_n and K_p—the bench height decreasing coefficient, over-drilling coefficient, dynamic load factor and ore rock loose coefficient, respectively, when blasting is carried out, with K_c as 0.1, K_{nep} as 1.2, K_n as 2 and K_p as 1.3 according to previous studies;	1.5
Length-to-width ratio plate-girder method	$H_n = \frac{L_n}{8} \times \frac{\gamma L_n + \sqrt{\gamma^2 L_n^2 + 16\sigma(P + P_1)}}{\sigma}$	H_n—thickness of the roof protection layer, m; L_n—pit span (thickness of the ore body), m; γ—unit weight of the isolated pillar rock, kN/m^3; σ—permitted tensile stress of the isolated pillar rock, kN/m2; P_1—unit pressure of the waste rock and other additional loads on the isolated pillar, kN/m^2; P—dynamic load generated by blasting, $P = rH_Y K_n(K_c + K_{nep})/K_p$. Where: H_Y—the gradient height, taken as 5 m; r—crater index, taken as 1.5; K_c, K_{nep}, K_n and K_p—the bench height decreasing coefficient, over-drilling coefficient, dynamic load factor and ore rock loose coefficient, respectively, when blasting is carried out, with K_c as 0.1, K_{nep} as 1.2, K_n as 2 and K_p as 1.3 according to previous studies.	0.5
Load intersection line method	$h = \frac{KL}{2\tan\beta}$	h—thickness of roof protection layer, m; L—drift width, m; β—included angle between the load transfer line and the vertical center line of the goaf roof, taken as 30°, 32° and 35°; K—safety factor, taken as 1.2.	4.32–5.23
Protodyakonov arch method	$H_y = \frac{b + h \cdot \tan(45 - \varphi/2)}{f}$	H_y—thickness of the roof protection layer, m; b—half the width of the drift, m; h—maximum height of the pit, m; φ—angle of internal friction of the rock, °; f—Protodyakonov coefficient of the rock;	1.0
Limit equilibrium theory	$h \geq \sqrt{\frac{12q_0}{\pi^2[\sigma_t]} \cdot \frac{1}{\frac{3}{a^4} + \frac{2}{a^2b^2} + \frac{3}{b^4}} \cdot (\frac{1}{a^2} + \frac{\mu}{b^2})}$	h—thickness of the roof protection layer, m; σ_t—tensile strength of the ore body, MPa; a—length of the drift, m; b—width of the drift, m.	0.8

Note: The physical and mechanical parameters of the ore rock are listed in Table 3. The calculations assume a drift width of 5 m, a drift length (thickness of the ore body) of 15 m, and a waste rock unit weight of γ_1 = 26 KN/m^3.

Table 3. List of material parameters selected for the numerical simulation.

Lithology	Elastic Modulus (GPa)	Bulk Modulus (GPa)	Poisson's Ratio	Internal Friction Angle (°)	Density (g/cm^3)	Compressive Strength (MPa)	Tensile Strength (MPa)	Cohesive Force (MPa)
Marble (Surrounding rock of hanging wall)	11.64	8.43	0.27	47	2.663	7.86	0.7	1.2
Slate (Surrounding rock of foot wall)	18.81	10.81	0.21	49	2.77	17.02	0.9	1.5
Ore body	21.26	11.43	0.19	50	3.565	25.33	2	1.8

It was clear from the analysis results that the purely theoretical calculations produced highly discrete horizontal pillar thicknesses and, therefore, that subsequent numerical simulation was required to obtain a clearer roof protection layer thickness value. To this end, the theoretical results were categorized and summarized. For the convenience of grid division in the numerical simulation and to ensure operable field construction, the theoretical values in Table 1 were revised and summarized to obtain three schemes with different roof thicknesses for the preservation of the roof protection layer: Scheme I (0.5 m), Scheme II (1 m), and Scheme III (1.5 m). (It is noteworthy that the theoretical value derived from the load intersection line method had a large discrepancy from that derived from other theoretical calculations; therefore, its results were not considered for the time being).

5. Numerical Simulation and Inversion of Crown Pillar Extraction and Roof Protection Layer Schemes of the 855 Middle Sublevel

5.1. Numerical Modelling

The FLAC^{3D}-based numerical simulation involved the static analysis of a model with a total length of 100 m, a width of 75 m, a height of 75 m, and bottom and top elevations of +880 and +955 m, respectively, in the Z- direction. To simplify the model, we set the dip of the ore body at 90°. We fixed the X- and Y-axis displacements on the planes at both ends of the X- and Y-axis directions, respectively, of the model and the X-, Y-, and Z-direction displacements at the bottom boundary of the model. Our preliminary measurement of in-situ stress revealed that the maximum horizontal principal stress increased with the buried depth and that the orientation of the maximum horizontal principal stress remained consistent at different buried depths, and the fitting equations of the three principal stresses are presented, respectively, as follows:

$$\left.\begin{array}{l} \sigma_v = 0.0279h, \\ \sigma_{h\max} = 0.0256h + 10.202, \\ \sigma_{h\min} = 0.0233h + 1.4743. \end{array}\right\} \quad (1)$$

(1) According to the calculations, the vertical stress at +955 elevation was 4.19 MPa, the maximum horizontal principal stress was 14 MPa, and the minimum horizontal principal stress was 4.97 MPa. The models were divided into 25 groups, each featuring isolated pillars, roof protection layers, upper 20 m backfill of roof protection layer and lower 20 m backfill of isolated pillars, and surrounding rocks at the hanging and foot walls according to material characteristics, as shown in Figure 3.

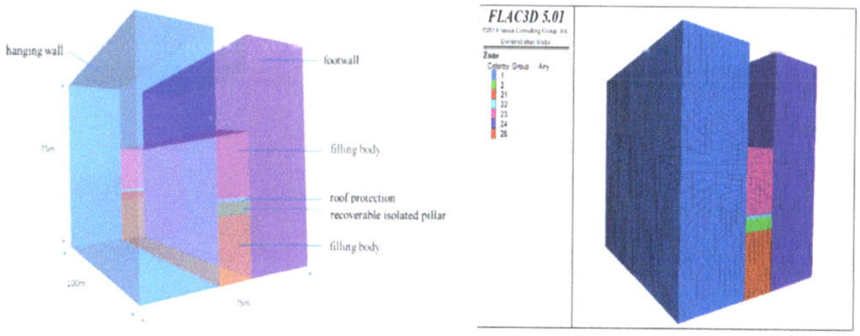

Figure 3. 3D numerical model of mine based on FLAC^{3D}.

5.2. Selection of Model Material Parameters

Based on the basic mechanical parameters of the rock masses obtained from a large number of indoor rock mechanics tests in the early stage, the rock solidity can be graded in combination with the structural surface reserve in the field, and the mechanical parameters of the rocks can be discounted using the Hoek-Brown strength criterion [26,27]. The discounted material parameters more accurately reflect the mechanical properties of the rock mass in the field. The material parameters of the ore body and surrounding rock obtained from the relevant literature [8,28] are listed in Table 3. A backfill scheme with an ash-sand ratio of 1:10 was used in backfilling the isolated ore based on the backfill material parameters listed in Table 4.

Table 4. Material parameters of backfills.

Proportioning	Bulk Modulus /MPa	Shear Modulus/MPa	Poisson's Ratio	Internal Friction Angle /°	Density/g/cm^3	Compressive Strength /MPa	Tensile Strength/MPa	Cohesive Force/KPa
1:10	80.9	58.2	0.21	36.9	1.82	1.07	0.1	143

5.3. Numerical Simulation Schemes and Steps

A total of six extraction schemes involving different combinations of roof protection layer preservation and extraction sequence were considered (during extraction, the crown pillar had a total height of 10 m; considering that the lower 5 m has been mined, numerical simulation was confined to extraction schemes involving crown pillars within the upper 5 m), as listed in Table 5. To reflect an actual mining situation, two extraction sequences were considered: mining at an interval of one strip and strip-by-strip mining. In both cases, three drifts were simultaneously extracted following an overall extraction sequence proceeding from the west to the east wing.

Table 5. Selection of numerical simulation schemes.

Scheme A	The thickness of roof protection layer is 0.5 m and strip-by-strip extraction is adopted	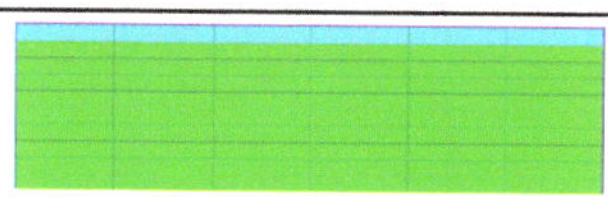
Scheme B	The thickness of roof protection layer is 0.5 m and spaced extraction is adopted	
Scheme C	The thickness of roof protection layer is 1.0 m and strip-by-strip extraction is adopted	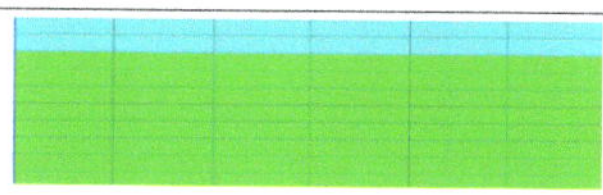
Scheme D	The thickness of roof protection layer is 1.0 m and spaced extraction is adopted	
Scheme E	The thickness of roof protection layer is 1.5 m and strip-by-strip extraction is adopted	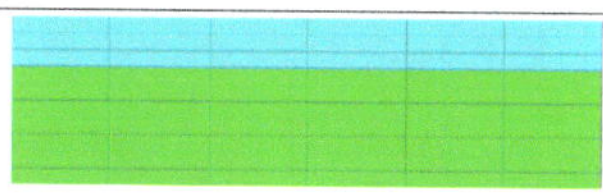
Scheme F	The thickness of roof protection layer is 1.5 m and spaced extraction is adopted	

Note: Blue indicates the preserved roof protection layer; green indicates the crown pillars to be extracted. Each grid has a width of 0.5 m.

5.4. Analysis and Discussions of Numerical Calculation Results

5.4.1. Analysis of Displacement Field

Figure 4 shows Z-direction displacement cloud pictures (plan views) of the roof protection layer for different periods under two extraction sequences (Schemes A and B) when the thickness of the roof protection layer is 0.5 m. Given the space limitations of this paper, only some of the extraction and backfilling stages are shown. The analysis found that, under both Schemes A and B, the displacement value at the hanging wall near the ore body was generally larger than that at the footwall during the overall extraction process (the displacement value of the roof protection layer near the hanging wall was approximately 1.14 cm; that near the footwall was approximately 0.76 cm). Because the strength of the surrounding rock at the hanging wall was lower, it was less able to support the roof protection layer, resulting in a larger sedimentation value. Analysis of the evolution rule of the displacement field of the roof protection layer at different stages revealed that the extraction sequence has a significant influence on the sedimentation development trend in the roof protection layer. When Step 2 was completed (for Drifts 1, 8, and 15 of Scheme A and Drifts 1, 3, and 5 of Scheme B), only a small increase in the displacement of the roof protection layer occurred at the corresponding drifts; when Step 11 was completed (for Drifts 6, 13, and 20 of Scheme A and Drifts 8, 10, and 12 of Scheme B), more obvious sedimentation occurred on the left side, middle, and right side of the roof protection layer in Scheme A. In contrast, a larger range of sedimentation occurred on the right side of Scheme B. When the last step was completed (for Drifts 7 and 14 of Scheme A and Drift 20 of Scheme B), the basic area of the displacement field of the roof protection layer was the same in both schemes: the maximum displacements under Schemes A and B were approximately 1.62 and 1.6 cm, respectively, and the sedimentation value under Scheme B was slightly smaller than that under Scheme A.

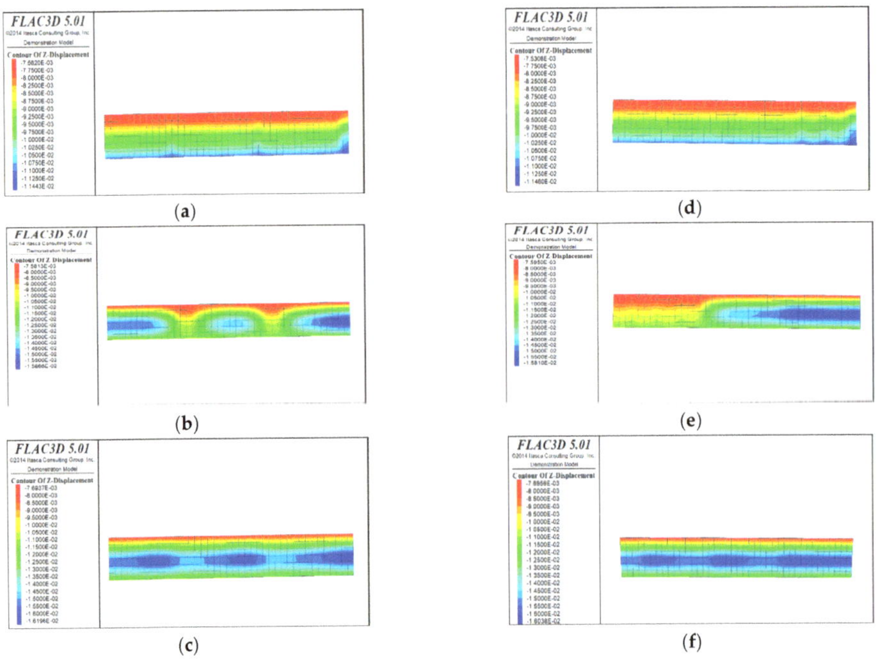

Figure 4. Z-direction displacement cloud pictures (plan views) for different periods under Schemes A and B with a roof protection layer thickness of 0.5 m. (**a**) After extraction and backfill of Drifts 1, 8, and 15 (Step 2). (**b**) After extraction of Drifts 6, 13, and 20 (Step 11). (**c**) After extraction and backfilling of Drifts 7 and 14 (last step). (**d**) After extraction and backfill of Drifts 1, 3, and 5 (Step 2). (**e**) After extraction of Drifts 8, 10, and 12 (Step 11). (**f**) After extraction and backfilling of Drift 20 (last step).

Figure 5 shows displacement clouds after all extraction and backfill for models with spaced extraction under different roof protection layer thicknesses (as Figure 4 includes Scheme B, only Schemes D and F are shown). An analysis of Figures 4f and 5a,b reveals that the maximum displacement sedimentation value generally occurs at the side of the roof protection layer near the footwall, regardless of the thickness of the roof protection layer. The maximum displacement sedimentation values under Schemes B, D, and F are 1.6, 1.38, and approximately 1.26 cm, respectively. It is seen that the maximum displacement sedimentation value monotonically decreases as the thickness of the roof protection layer increases, which indicates that increasing the thickness of this layer improves its stability. In addition, the displacement sedimentation value decreases by 0.22 cm when the thickness of the roof protection layer increases from 0.5 to 1 m, whereas increasing the thickness from 1 to 1.5 m decreases the displacement sedimentation value by only 0.12 cm, indicating that the decrease in amplitude of the displacement sedimentation value begins to decline when the thickness of the roof protection layer is increased to 1 m.

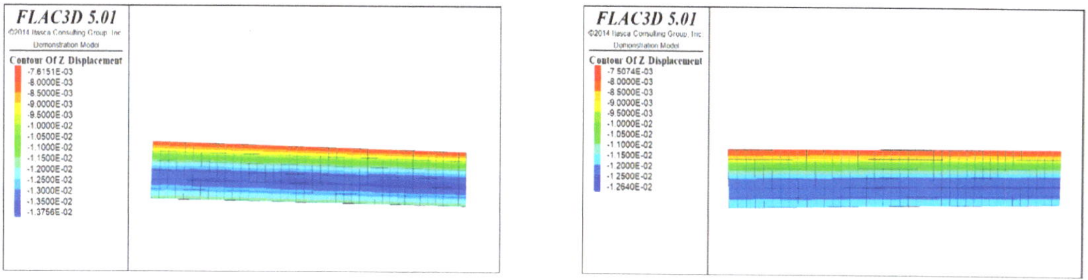

(**a**) (**b**)

Figure 5. Z-direction displacement cloud pictures (plan views) of the roof protection layer after the completion of extraction and backfilling under Schemes D and F. (**a**) Scheme D. (**b**) Scheme F.

5.4.2. Analysis of Plastic Zone

(1) Distribution characteristics of the plastic zones in the roof protection layer under different extraction sequences

Tables 6–9 list the evolution rules of the plastic zones in the roof protection layer during different periods under Schemes C and D with a roof protection layer thickness of 1 m. Note that the green, red and light blue colors indicate the plastic zones in each figure.

Table 6. Distributions of roof protection layer plastic zones under Schemes C and D (excavation stages 1 and 2).

		Excavation Stage 1 (Drifts 1, 8 and 15 of Scheme C; and Drifts 1, 3 and 5 of Scheme D)	**Backfill Stage 1 (Drifts 1, 8 and 15 of Scheme C; and Drifts 1, 3 and 5 of Scheme D)**	**Excavation Stage 2 (Drifts 2, 9 and 16 of Scheme C; and Drifts 7, 9 and 11 of Scheme D)**	**Backfill Stage 2 (Drifts 2, 9 and 16 of Scheme C; and Drifts 7, 9 and 11 of Scheme D)**
Scheme C	Distribution diagram of plastic zones in the roof protection layer observed from top to bottom	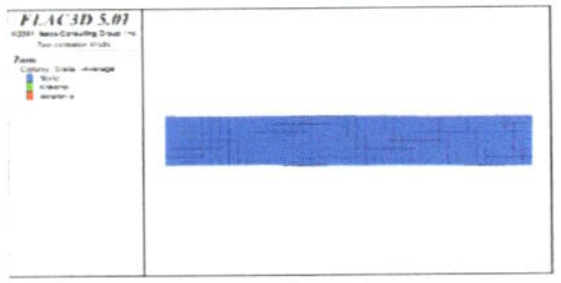	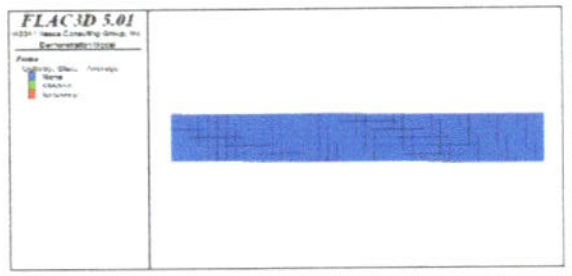	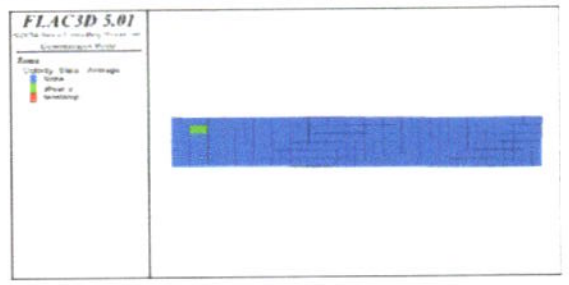	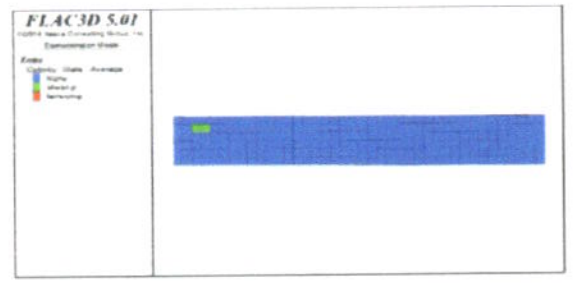
	Distribution diagram of plastic zones in the roof protection layer observed from bottom to top	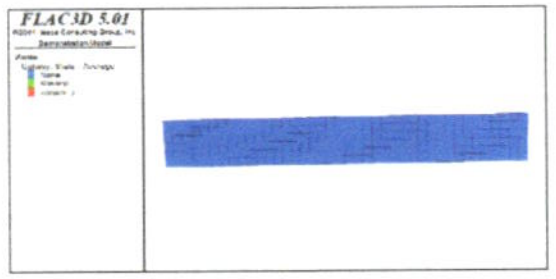	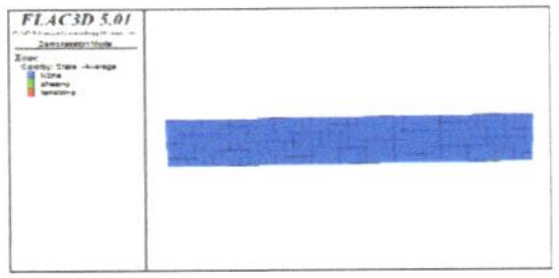	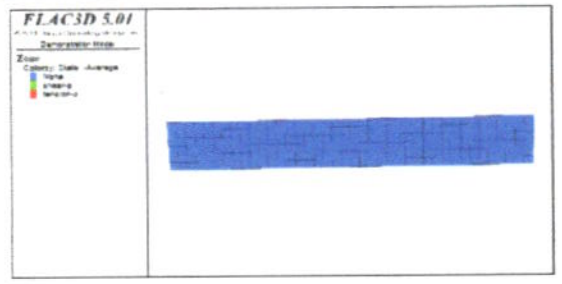	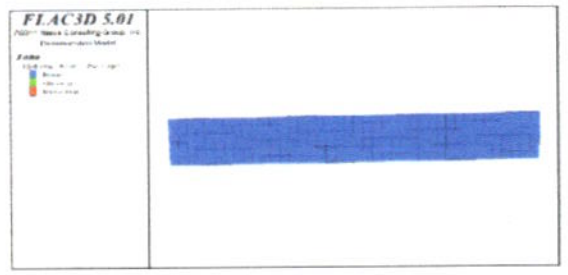
Scheme D	Distribution diagram of plastic zones in the roof protection layer observed from top to bottom	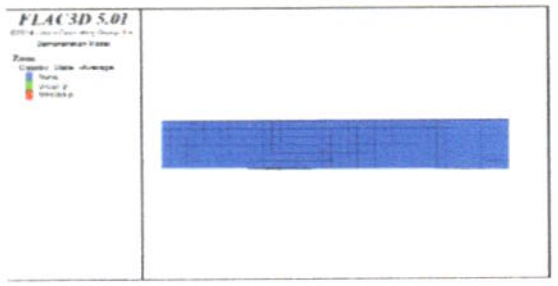	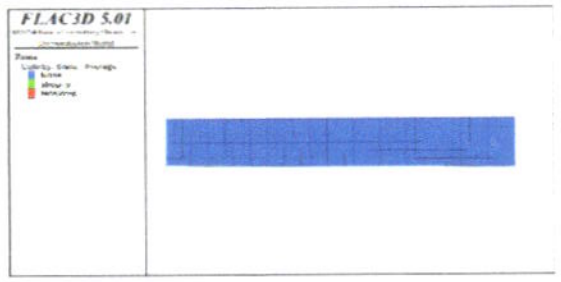	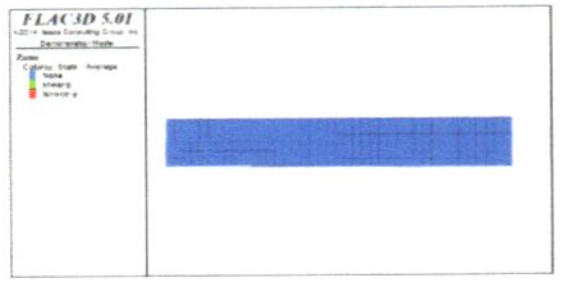	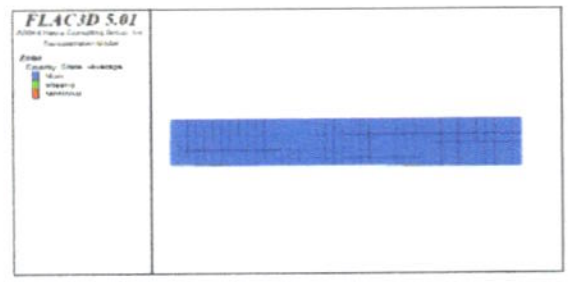
	Distribution diagram of plastic zones in the roof protection layer observed from bottom to top	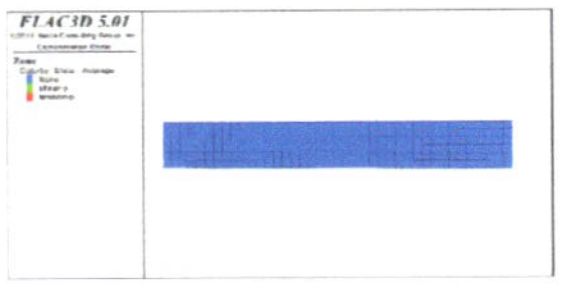	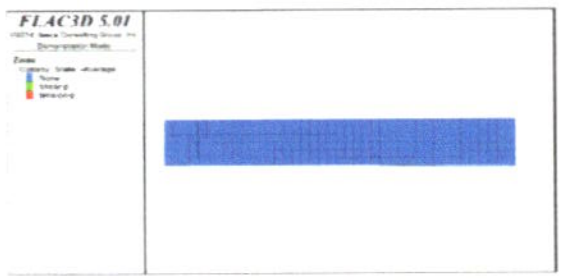	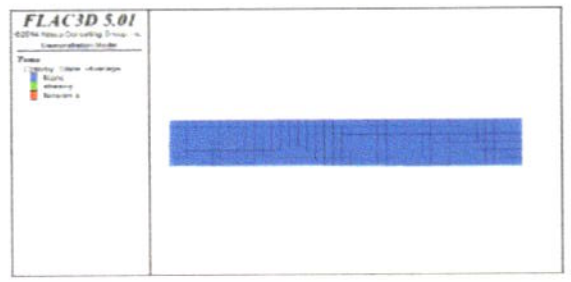	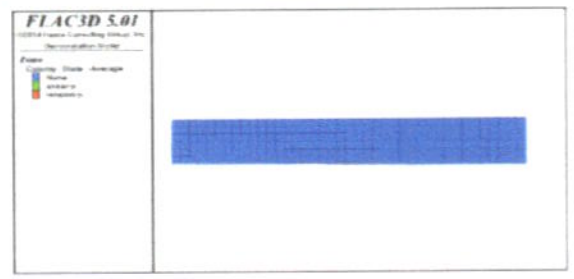

Table 7. Distributions of roof protection layer plastic zones under Schemes C and D (excavation stages 3 and 4).

		Excavation Stage 3 (Drifts 3, 10 and 17 of Scheme C; and Drifts 13, 15 and 17 of Scheme D)	**Backfill Stage 3 (Drifts 3, 10 and 17 of Scheme C; and Drifts 13, 15 and 17 of Scheme D)**	**Excavation Stage 4 (Drifts 4, 11 and 18 of Scheme C; and Drift 19 of Scheme D)**	**Backfill Stage 4 (Drifts 4, 11 and 18 of Scheme C;and Drift 19 of Scheme D)**
Scheme C	Distribution diagram of plastic zones in the roof protection layer observed from top to bottom	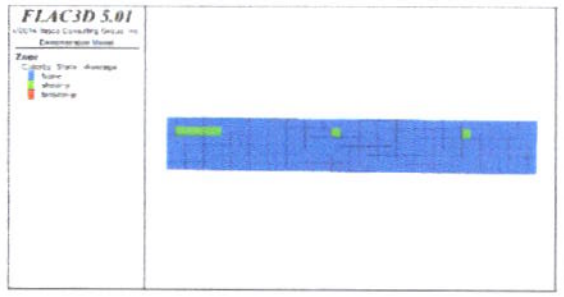	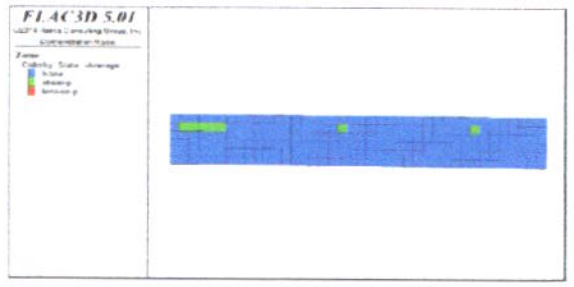	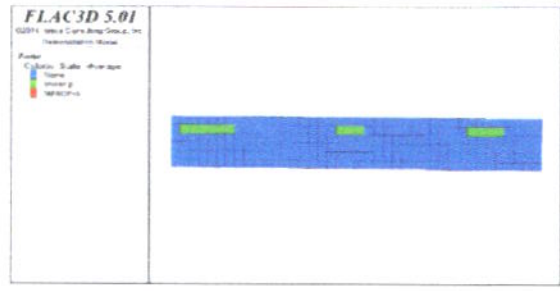	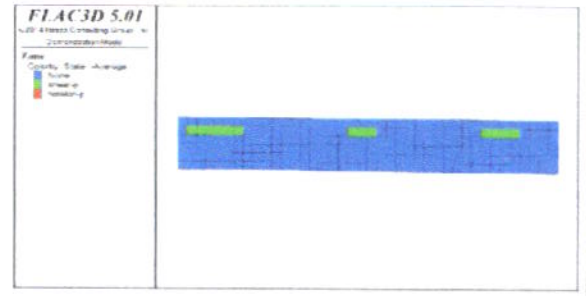
	Distribution diagram of plastic zones in the roof protection layer observed from bottom to top	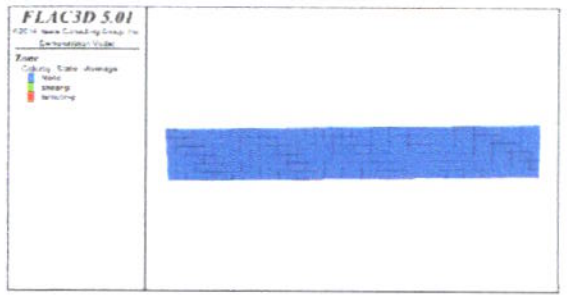	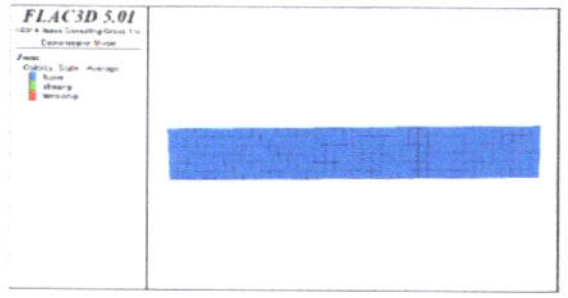	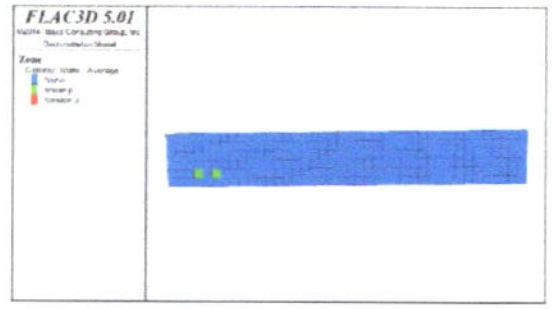	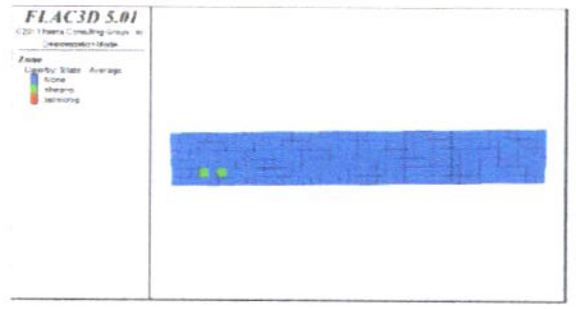
Scheme D	Distribution diagram of plastic zones in the roof protection layer observed from top to bottom	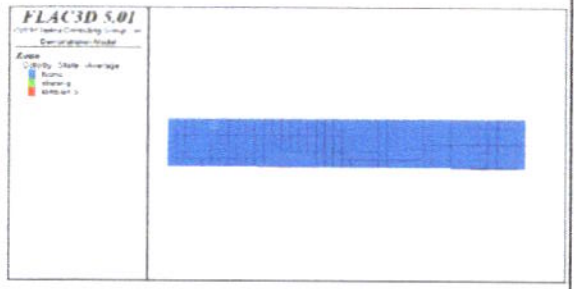	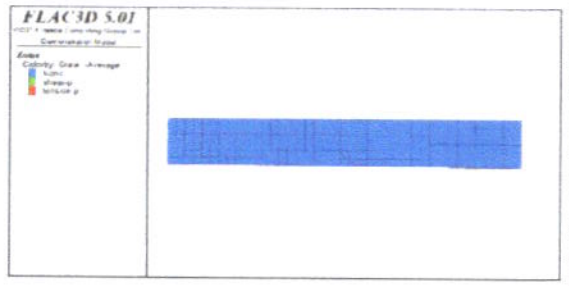	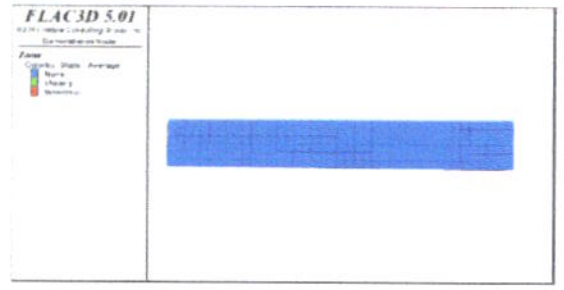	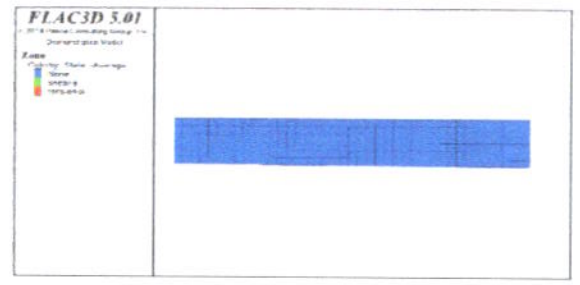
	Distribution diagram of plastic zones in the roof protection layer observed from bottom to top	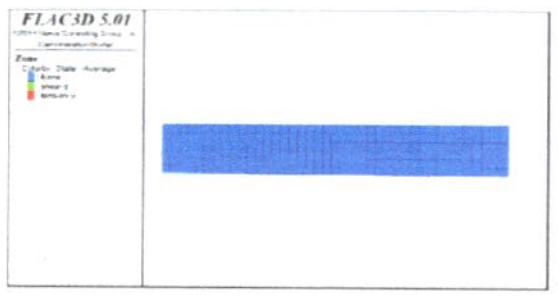	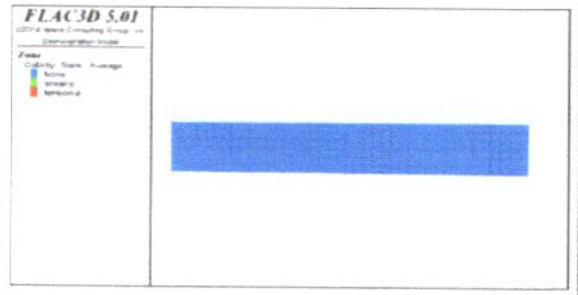	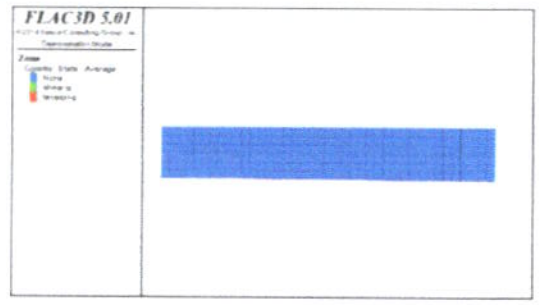	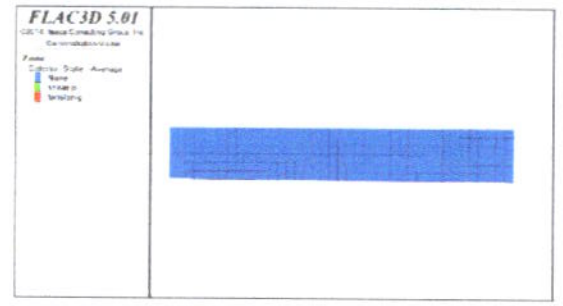

Table 8. Distributions of roof protection layer plastic zones under Schemes C and D (excavation stages 5 and 6).

		Excavation Stage 5 (Drifts 5, 12 and 19 of Scheme C; and Drifts 2, 4 and 6 of Scheme D)	Backfill Stage 5 (Drifts 5, 12 and 19 of Scheme C; and Drifts 2, 4 and 6 of Scheme D)	Excavation Stage 6 (Drifts 6, 13 and 20 of Scheme C; and Drifts 8, 10 and 12 of Scheme D)	Backfill Stage 6 (Drifts 6, 13 and 20 of Scheme C; and Drifts 8, 10 and 12 of Scheme D)
Scheme C	Distribution diagram of plastic zones in the roof protection layer observed from top to bottom	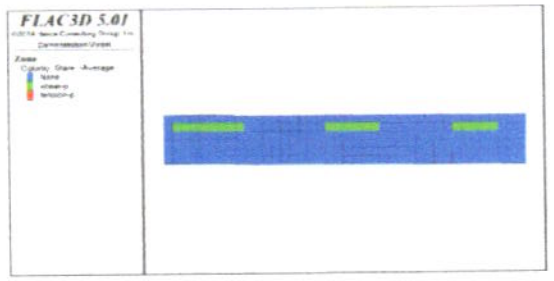	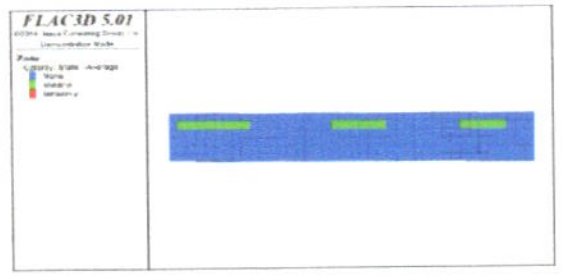	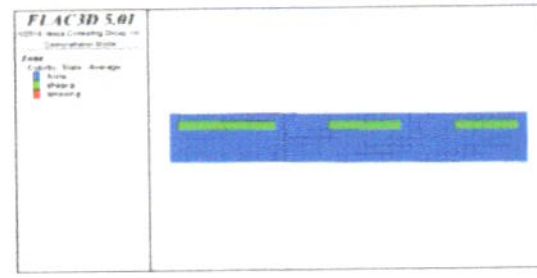	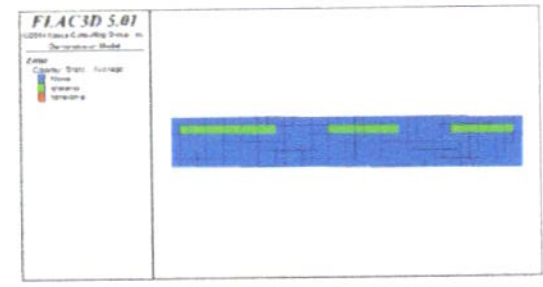
	Distribution diagram of plastic zones in the roof protection layer observed from bottom to top	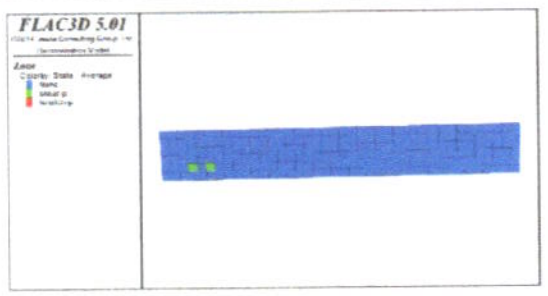	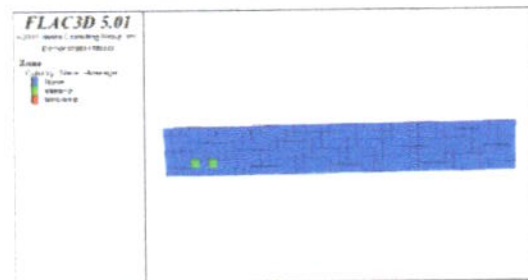	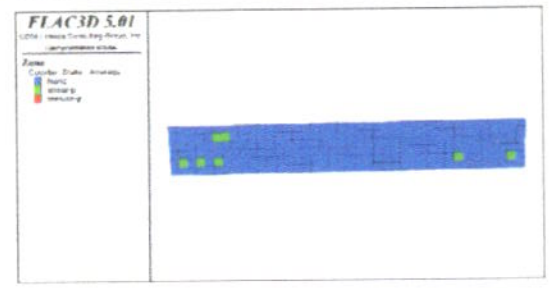	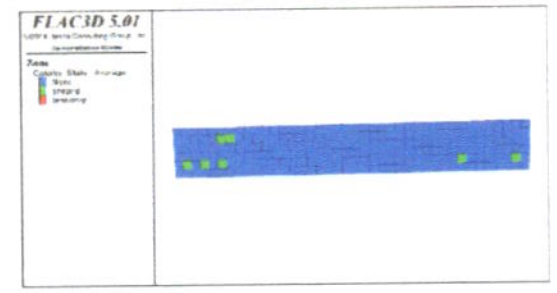
Scheme D	Distribution diagram of plastic zones in the roof protection layer observed from top to bottom	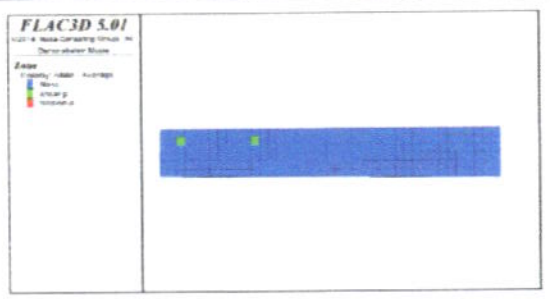	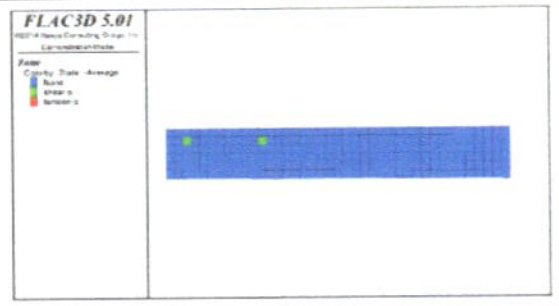	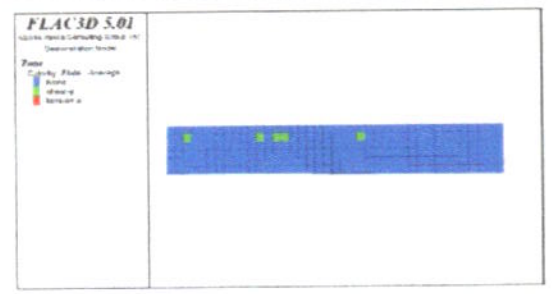	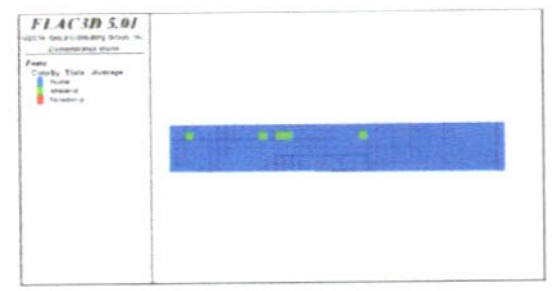
	Distribution diagram of plastic zones in the roof protection layer observed from bottom to top	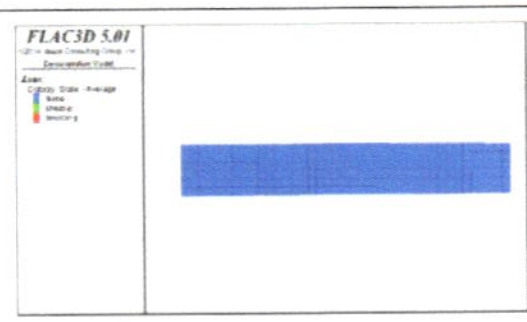	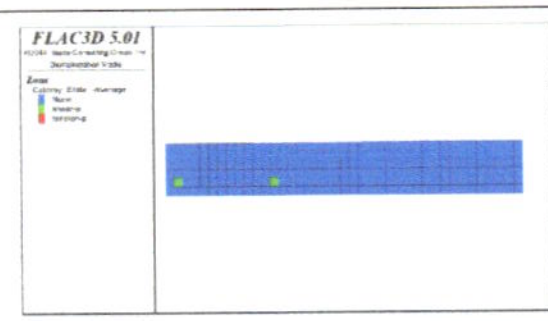	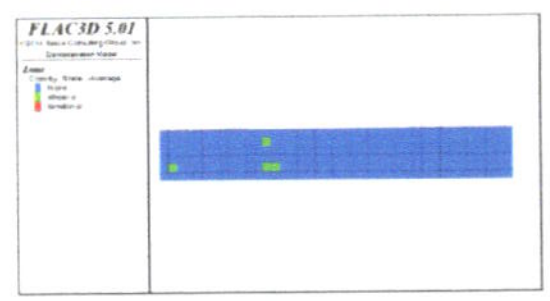	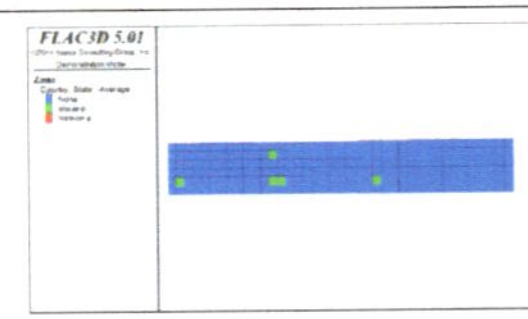

Table 9. Distributions of roof protection layer plastic zones under Schemes C and D (excavation stages 7 and 8).

		Excavation Stage 7 (Drifts 7 and 14 of Scheme C; and Drifts 14, 16 and 18 of Scheme D)	Backfill Stage 7 (Drifts 7 and 14 of Scheme C; and Drifts 14, 16 and 18 of Scheme D)	Excavation Stage 8 (mining completion in Scheme C; and Drift 20 of Scheme D)	Backfill Stage 8 (mining completion in Scheme C; and Drift 20 of Scheme D)
Scheme C	Distribution diagram of plastic zones in the roof protection layer observed from top to bottom	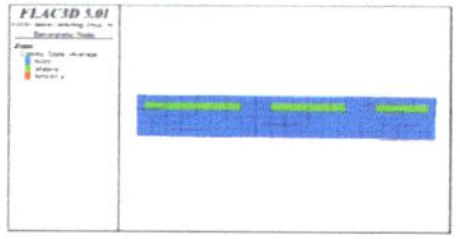	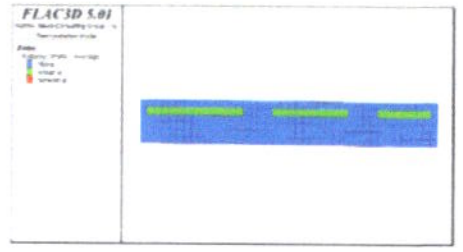		
	Distribution diagram of plastic zones in the roof protection layer observed from bottom to top	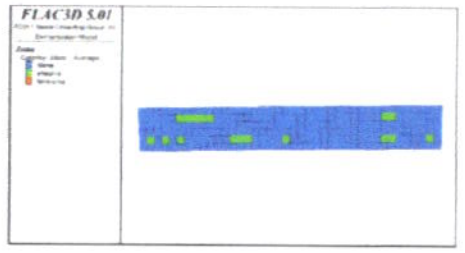	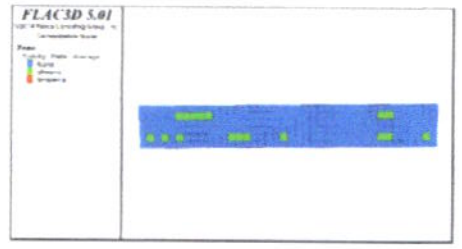		
Scheme D	Distribution diagram of plastic zones in the roof protection layer observed from top to bottom	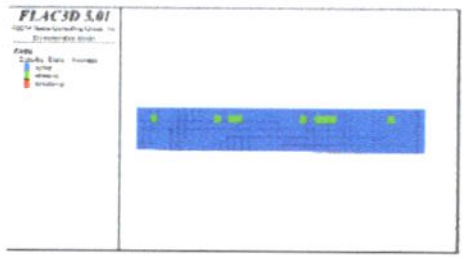	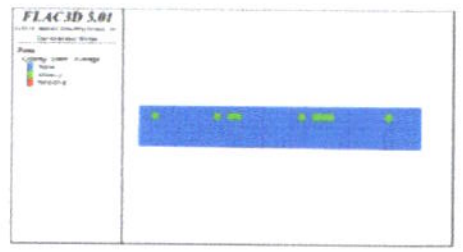	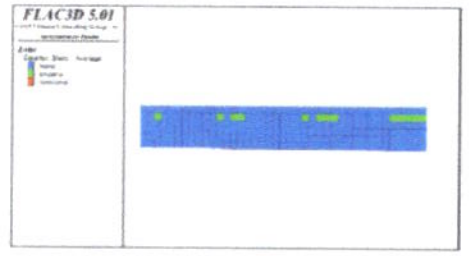	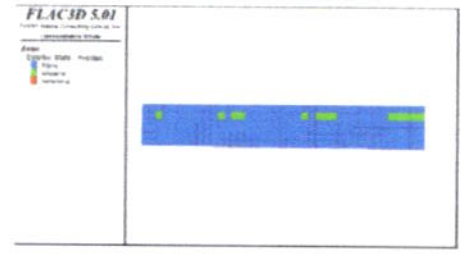
	Distribution diagram of plastic zones in the roof protection layer observed from bottom to top	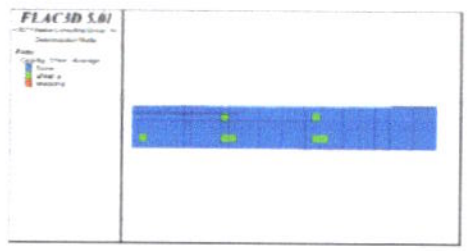	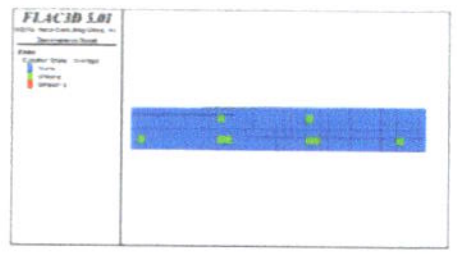	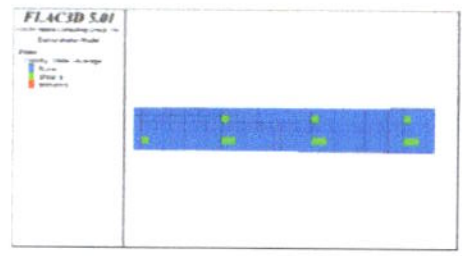	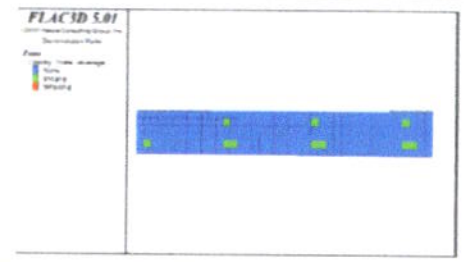

① Analysis of the plastic zones in the roof protection layer at each stage of Scheme C revealed the following: because one side of the drift was an isolated pillar and the other side was loose tailings backfill during mining, there was a plastic zone at the beginning stage; as the excavation and backfill continued, the plastic zones accumulated, with the plastic zones of the roof protection layer following an overall trend of steady growth as the excavation and backfill proceeded. Eventually, the failure range of the plastic zones of the roof protection layer accounted for 9.8% of the overall roof protection layer.

② The changes in the plastic zones of the roof protection layer during each stage under Scheme D revealed that the roof protection layer did not undergo failure during the mining of the odd-numbered drifts; at the same time, the plastic zones of the roof protection layer began to increase during the extraction of the even-numbered drifts. Ultimately, the plastic zones of the roof protection layer under Scheme D accounted for 4.7% of the overall roof protection layer. These results can be explained as follows. The odd-numbered drifts were bordered on each side by harder isolated pillars featuring ore bodies that could effectively support the roof protection layer during mining through a reasonable choice of drift size. The even-numbered drifts, by contrast, were bordered by loose tailings backfill, leading to the growth of roof protection layer plastic zones during mining.

③ Comprehensive analysis of Schemes C and D revealed significant differences between the two in terms of the percentage of plastic zones, with the former producing a higher final percentage than the latter. Although both sides of the pillar comprised relatively loose tailings backfills during drifts extraction in the later period under Scheme D, the plastic zones in the roof protection layer could be effectively controlled in the early stage of extraction, enhancing the overall stability of the layer in the later stage. By contrast, even when the strip-by-strip extraction process was applied under Scheme A, some plastic zones still appeared in the early stage of mining, resulting in the accumulation of plastic zones in the later extraction process and reducing the overall protection of the roof protection layer.

(2) Distribution characteristics of roof protection layer plastic zones with different roof protection layer thicknesses

Figures 6–8 show the distribution characteristics of the plastic zones produced at different roof protection layer thicknesses (Schemes B, D, and F) using the spaced extraction sequence following full completion of pit backfill and extraction. It is seen from the figures that plastic zones account for 60, 4.7, and 0.6% of the final overall layers under Schemes B, D, and F, respectively. The percentage of plastic zones showed a monotonically decreasing trend with increasing roof protection layer thickness. Under Scheme B, plastic zones appeared in most of the areas in the roof protection layer. They penetrated each other, indicating that complete failure had occurred in the roof protection layer. Under Schemes D and F, plastic zones appeared only within small ranges of the roof protection layer and failed to penetrate the upper and bottom parts of the layer (this was particularly true under Scheme F), indicating that the roof protection layers were relatively stable. These results indicate that thickening the roof protection layer reduces the likeliness of plastic zone appearance and is more conducive to the stability of the pit. Furthermore, increasing the thickness of the roof protection zone from 0.5 to 1 m decreased the plastic zone coverage by 55.2%, whereas an increase from 1 to 1.5 m resulted in a plastic zone decrease of only 4.3%, indicating that increasing the protection layer thickness beyond 1 m results in diminishing returns in terms of plastic zone coverage reduction.

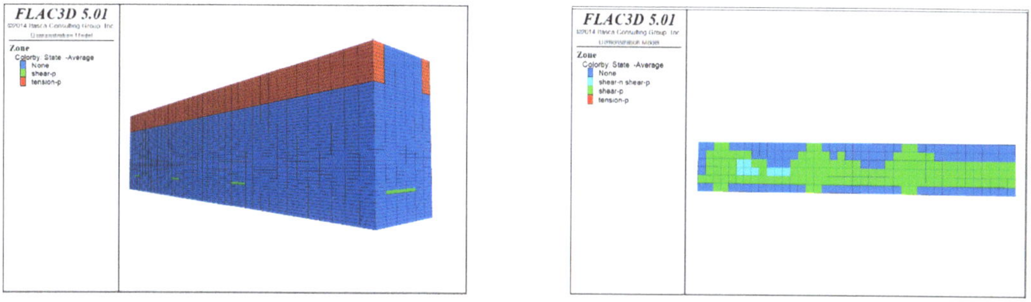

(**a**) (**b**)

Figure 6. Distribution of plastic zones in the roof protection layer under Scheme B. (**a**) 3D drawing. (**b**) plan of roof protection layer.

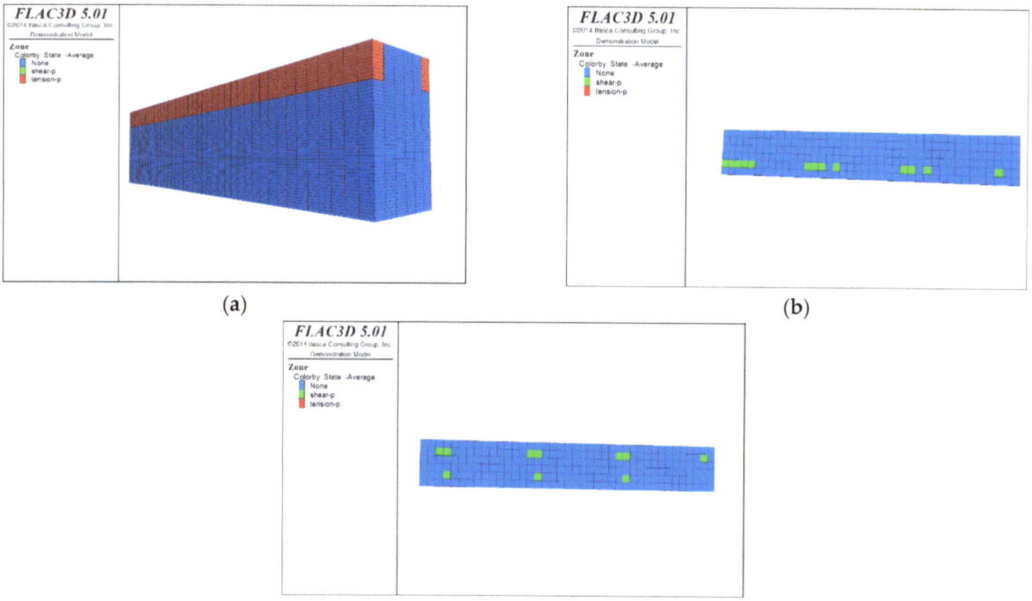

(**a**) (**b**) (**c**)

Figure 7. Distribution of plastic zones in the roof protection layer under Scheme D. Note that the heigt of each slice is 0.5m. (**a**) 3D drawing. (**b**) plan of roof protection layer (the first slice). (**c**) plan of roof protection layer (the second slice).

The roof protection layer plastic zone volumes and percentages obtained under different extraction schemes are listed in Table 10. As the numerically simulated plastic zones under Schemes A and B reach 82 and 60%, respectively, indicating that complete failure occurred in the roof protection layers, these two schemes can be excluded first. Although the differences among Schemes C, D, E, and F in terms of the percentage of plastic zones are not significant, Schemes C and D and E and F differ by 750 m^3 in terms of ore production. To fully evaluate the safety and economics (ore recovery rate) of pillar extraction under the four remaining schemes, we further optimized them using the analytic hierarchy process.

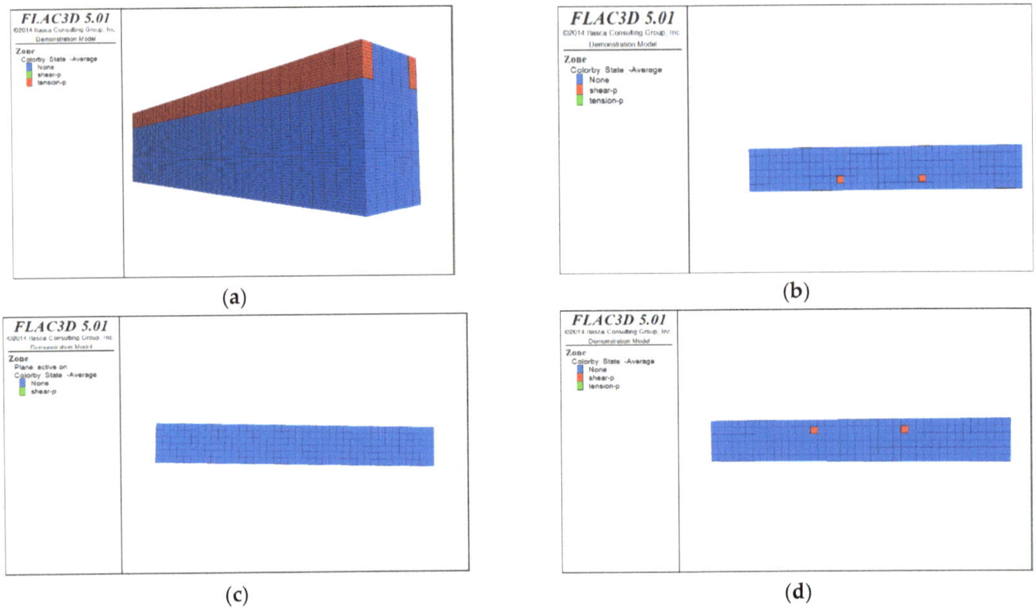

Figure 8. Distribution of plastic zones in the roof protection layer under Scheme F. Note that the heigt of each slice is 0.5m. (**a**) 3D drawing. (**b**) plan of roof protection layer (the first slice). (**c**) plan of roof protection layer (the second slice). (**d**) plan of roof protection layer (the thirdslice).

Table 10. Volumes and percentages of plastic zones under different extraction schemes.

	Scheme A	Scheme B	Scheme C	Scheme D	Scheme E	Scheme F
Volume of the plastic zones of the roof protection layer (m^3)	615.63	456.25	146.88	71.875	18.75	12.5
Total volume of pillars (m^3)	750	750	1500	1500	2250	2250
Percentage of plastic zones (%)	82	60	9.8	4.8	0.8	0.5

6. Optimization of the Schemes for the Preservation of Roof Protection Layer and Extraction Sequence Based on Analytic Hierarchy Process

The optimized fuzzy mathematics method was used to transform various fuzzy or subjective factors in the mining process into mathematical form as evaluation indices, which were then quantified and analyzed for each scheme, to produce a scheme selection approach that was scientific, reasonable, and reliable. The analytic hierarchy process was used to quantify and analyze the evaluation index achieved by each scheme [29,30], synthesize the safety and economy of the scheme, and use the results to determine an optimal preserved roof protection layer thickness and extraction sequence.

6.1. Establishment of Optimized Fuzzy Mathematics Model

In general, modeling using the application of the analytic hierarchy process to solve practical problems can be implemented through the following four steps:

Step 1: Hierarchical modelling

In applying the analytical hierarchy process to analyzing a decision-making problem, the problem should be organized so a hierarchical structure model can be built. Such

hierarchies can be divided into three categories: top hierarchies (target hierarchies), middle hierarchies (criterion hierarchies), and bottom hierarchies (scheme hierarchies). The number of hierarchies used in a hierarchical structure will reflect the complexity of the problem and the level of detail required for analysis. In general, an unlimited number of hierarchies can be used, but the number of elements dominated by each element in a hierarchy should not exceed nine.

Step 2: Construction of judgment matrices for each hierarchy

The individual criteria within a criterion hierarchy will not necessarily have the same proportions of importance for target measurement; however, each will have a proportion of importance in the mind of the decision maker. Here, the integers from 1–9 and their reciprocals are used as scales to define judgment matrices (see Table 11).

Table 11. Definition of judgment matrix scales.

Scale	Meaning
1	Indicates that two factors are of equal importance
3	Indicates that the former factor is slightly more important than the latter
5	Indicates that the former is significantly more important than the latter
7	Indicates that the former is strongly more important than the latter
9	Indicates that the former is more extremely important than the latter
2,4,6 and 8	Indicate the middle value of the above adjacent judgments
Reciprocals	The importance ratio of the latter factor to the former factor is the reciprocal of the importance ratio of the former factor to the latter factor

Step 3: Single ranking of hierarchies and consistency checking

(1) A consistency index (CI) is calculated for matrices of dimension n.

(2) A consistency index RI corresponding to n is then obtained from Table 12.

Table 12. Average random consistency index.

n	1	2	3	4	5	6	7	8	9	10	11	12	13	14
RI	0	0	0.52	0.89	1.12	1.24	1.36	1.41	1.46	1.49	1.52	1.54	1.56	1.58

(3) Calculation of consistency ratio (CR)

The consistency of a judgment matrix is considered acceptable when CR < 0.1; otherwise, the judgment matrix should be corrected as appropriate.

Step 4: After a reasonable judgment matrix is derived, its weight vector ω is obtained using the geometric mean method (square root method). All weights in each row of the judgment matrix are multiplied to obtain the row product of all elements; this extraction is repeated four times for each row product to obtain ω for each judgment matrix.

Step 5: Equations 5 are applied to normalize the judgment matrices, and their weight vectors are multiplied to obtain the scheme membership of each scheme.

Step 6: To obtain a total ranking of hierarchies, the scheme membership of each scheme is ranked to obtain the ranking weight of each element—particularly for the elements of the schemes within a bottom hierarchy—to produce a target for the selection of schemes.

6.2. Determination of the Optimal Scheme for the Preservation of Roof Protection Layer and the Extraction Sequence

A membership matrix of roof protection layer preservation methods was developed based on the preservation method selection principle. Four indices were used to select the optimal scheme from Schemes C, D, E, and F. The selected indices were the maximum mean stress value along the Z direction in the roof protection layer, the maximum mean sedimentation value along the Z direction in the roof protection layer, the degree of plastic failure in the roof protection layer, and the lost ore tonnage following extraction, as shown in Table 13.

Table 13. Quantitative indicators of each top protection layer retention scheme.

Index Factor	Scheme C	Scheme D	Scheme E	Scheme F
Maximum stress mean value in the Z direction of the roof protection layer (MPa)	4.27	4.28	4.39	4.40
Maximum sedimentation mean value in the Z direction of the roof protection layer (cm)	1.31	1.26	1.22	1.21
Percentage of plastic failure in the roof protection layer (%)	9.8	4.8	0.8	0.5
Lost ore tonnage after extraction (%)	20	20	30	30

The weight matrix R of the four indices was obtained using the analytic hierarchy process:

$$R = \begin{pmatrix} 1 & 1/2 & 1/4 & 1/8 \\ 2 & 1 & 1/3 & 1/8 \\ 4 & 3 & 1 & 1/5 \\ 8 & 8 & 5 & 1 \end{pmatrix} \tag{2}$$

Using the obtained maximum eigenvalue, λ_{max} = 4.1189, the consistency index was derived as

$$CI = \frac{\lambda_{\max} - n}{n-1} = 0.0396, \tag{3}$$

where n = 4, i.e., $CR = CI/RI = 0 < 0.1$ (RI = 0.89 was selected from Table 11), satisfying the consistency test. The weight vector affecting the selection of the roof protection layer preservation scheme, ω = [0.06, 0.08, 0.20, 0.66], was then obtained using the root method.

Using Table 12, the eigenvector matrix was obtained as

$$y_{1-4} = \begin{bmatrix} 4.27 & 4.28 & 4.39 & 4.40 \\ 1.31 & 1.26 & 1.22 & 1.21 \\ 9.8 & 4.8 & 0.8 & 0.5 \\ 20 & 20 & 30 & 30 \end{bmatrix}. \tag{4}$$

This matrix for smaller and better indices (all four of the indices used here are smaller and better indices) was normalized using the following equation:

$$r_{ij} = \frac{\min y_{ij}}{y_{ij}}, \tag{5}$$

where r_{ij} is the index normalized, $\min y_{ij}$ is the minimum-valued index in each row, and the y_{ij} correspond to the indices in each row. Applying this normalization produced the following matrix:

$$R_{1-4} = \begin{bmatrix} 1 & 0.998 & 0.973 & 0.970 \\ 0.924 & 0.960 & 0.992 & 1 \\ 0.051 & 0.104 & 0.625 & 1 \\ 1 & 1 & 0.667 & 0.667 \end{bmatrix} \tag{6}$$

From this matrix, the scheme membership $S = \omega \times R_{1-4}$ = (0.8041, 0.8175, 0.7030, 0.7784) was obtained, corresponding to the following priority ranking: Scheme D > Scheme C > Scheme F > Scheme E.

Based on this ranking, Scheme D—preservation of a 1.0-m-thick roof protection layer combined with mining at an interval of one strip—was selected as the optimal roof protection layer preservation scheme for the 855 middle sublevel.

7. Conclusions

(1) Based on the existing technical conditions of mining and ore body reserve characteristics of the crown pillar in the 855 middle sublevel, the cut-and-fill stoping of drift vertical to ore body strike was proposed to be the optimal recovery scheme for the crown pillar in the 855 middle sublevel.

(2) Based on the results of a site survey and information query, a preliminary reasonable isolated layer preserved thickness was obtained using seven classical theories. In conformance with the existing technical conditions of mining, three potential roof protection layer thicknesses were proposed: 0.5 m, 1 m, and 1.5 m.

(3) Based on the numerical simulation results, it was found that for each scheme, the displacement field of the roof protection layer and the change rule of the plastic zones were analyzed, the results revealing that the extraction sequence and the thickness of the roof protection layer both have a significant influence on the stability of the pit. By mining at intervals of one strip, the range of the plastic zones in the roof protection layer can be effectively controlled in the early stage of mining, which enhances the overall stability of the roof protection layer in the later stages. It was further shown that the proportion of plastic zones in the roof protection layer monotonically decreases as the thickness of the protection layer increases. However, there are significant reductions in the changes in sedimentation value and proportion of plastic zones once the roof protection layer thickness exceeds 1 m.

(4) Based on the scheme-ranking membership vector produced in the analytic hierarchy process, the optimal extraction scheme for the preservation of the roof protection layer in the 855 middle sublevel was determined to be cut-and-fill stoping of the drift vertical to ore body strike combined with a preserved roof protection layer thickness of 1 m and a spaced mining sequence.

Author Contributions: Conceptualization, F.F. and J.Z.; methodology, F.F.; software, Z.Y.; investigation, F.F. and Y.S.; resources, Y.W.; data curation, Z.Y.; writing—original draft preparation, F.F. and X.C.; writing—review and editing, Z.X.; All authors have read and agreed to the published version of the manuscript.

Funding: Funded by the National Natural Science Foundation of China (No. 52004143), the open fund for Laboratory of Mining Response and Disaster Prevention and Control in Deep Coal Mines (No. SKLMRDPC21KF06).

Data Availability Statement: Data available on request due to privacy restrictions.

Conflicts of Interest: The authors declare no conflict of interest.

References

1. Li, X.B.; Qiu, J.D.; Zhao, Y.Z.; Chen, Z.H.; Li, D.Y. Instantaneous and long-term deformation characteristics of deep room-pillar system induced by pillar recovery. *Trans. Nonferrous Met. Soc. China* **2020**, *30*, 2775–2791. [CrossRef]
2. Kaiser, P.K.; Kim, B.H.; Bewick, R.P.; Valley, B. Rock mass strength at depth and implications for pillar design. *Min. Technol.* **2014**, *120*, 170–179. [CrossRef]
3. Hu, G.J.; Yang, T.H.; Zhou, J.R.; Yu, Q.L.; Xie, L.K.; Liu, H.L.; Zhao, Y. Mechanism of surrounding rock failure and crack evolution rules in branched pillar recovery. *Minerals* **2017**, *7*, 96. [CrossRef]
4. Rybak, J.; Khayrutdinov, M.M.; Kuziev, D.A.; Kongar-Syuryun, C.B.; Babyr, N.V. Prediction of the geomechanical state of the rock mass when mining salt deposits with stowing. *J. Min. Inst.* **2022**, *253*, 61–70. [CrossRef]
5. Christopher, M.; Michael, G. Preventing roof fall fatalities during pillar recovery: A ground control success story. *Int. J. Min. Sci. Technol.* **2017**, *27*, 107–113.
6. Wang, F.F.; Ren, Q.Y.; Chen, B.; Zou, P.; Peng, Z.J.; Hu, W.J.; Ma, Z. Numerical investigation on safe mining of residual pillar in goaf: A case study of Panlong lead–zinc mine. *Geotech. Geol. Eng.* **2020**, *38*, 4269–4287. [CrossRef]
7. Ley, G.M.M.; Steed, C.M. Recovery of pillars under backfill. *World Min. Express* **1999**, *15*, 40–45.

8. Xia, K.Z.; Chen, C.X.; Liu, X.M.; Song, X.G.; Jiang, X.W.; Zhou, Y.C. Rheological mechanical model of pillar-protective roof supporting system in gypsum mines. *Rock Soil Mech.* **2017**, *38*, 2923–2930. (In Chinese)
9. Jiang, L.C.; Yang, C. Safety thickness of bauxite goaf protective roof in hard clay roof. Journal of Central South University. *Sci. Technol.* **2020**, *51*, 445–452.
10. Auvary, C.; Homand, F.; Sorgi, C. The aging of gypsum in underground mines. *Eng. Geol.* **2004**, *74*, 183–196. [CrossRef]
11. Liu, Z.Y. Stability Analysis of Mining of Roof and Bottom Pillars under the Backfill in Xinli Area of Sanshandao Gold Mine. Master's Thesis, Central South University, Changsha, China, 2014. (In Chinese).
12. Zhu, H.X.; Han, L.J.; Meng, Q.B.; Tian, M.L.; Meng, L.D.; Zhao, Z.L. Determining the minimum thickness of a crown pillar and optimizing the cross-sectional size of point pillars for the safe exploitation of a subsea gold mine. *Arab. J. Geosci.* **2021**, *14*, 1275. [CrossRef]
13. Zhao, X.D. Stability analysis of insulating pillar of excavation of Chambishi copper mine in depth. *Chin. J. Rock Mech. Eng.* **2010**, *29*, 2616–2622.
14. Ajayi, S.A.; Ma, L.Q.; Spearing, A.J.S. Ground Stress Analysis and Automation of Workface in Continuous Mining Continuous Backfill Operation. *Minerals* **2022**, *12*, 754. [CrossRef]
15. Huang, Y.G.; Yang, W.X.; Li, Y.Y.; Guo, W.B. Spatial distribution characteristics of plastic failure and grouting diffusion within deep roadway surrounding rock under three-dimensional unequal ground stress and its application. *Minerals* **2022**, *12*, 296. [CrossRef]
16. Safonov, A.; Filippi, M.; Mašín, D.; Bruthans, J. Numerical modeling of the evolution of arcades and rock pillars. *Geomorphol.* **2020**, *365*, 107260. [CrossRef]
17. Chen, S.J.; Feng, F.; Wang, Y.J.; Li, D.Y.; Huang, W.P.; Zhao, X.D.; Jiang, N. Tunnel failure in hard rock with multiple weak planes due to excavation unloading of in-situ stress. *J. Cent. South Univ.* **2020**, *27*, 2864–2882. [CrossRef]
18. Feng, F.; Li, X.B.; Rostami, J.; Peng, D.X.; Li, D.Y.; Du, K. Numerical investigation of hard rock strength and fracturing under polyaxial compression based on Mogi-Coulomb failure criterion. *Int. J. Geomech.* **2019**, *19*, 04019005. [CrossRef]
19. Ebrahim, G.; Mohammad, A.; Kourosh, S. An intelligent approach to predict pillar sizing in designing room and pillar coal mines. *Int. J. Rock Mech. Min. Sci.* **2014**, *65*, 86–95.
20. Saffet, Y.; Candan, G. Application of fuzzy inference system and nonlinear regression models for predicting rock brittleness. *Expert Syst. Appl.* **2020**, *37*, 2265–2272.
21. Hyunjun, I.; Huongdoo, J.; Erkan, T.; Micah, N. Long- and short-term strategies for estimation of hydraulic fracturing cost using fuzzy logic. *Minerals* **2022**, *12*, 715.
22. Sławomir, B.; Stanisław, P. Numerical study of pressure on dams in a backfilled mining shaft based on PFC3D code. *Comput. Geotech.* **2015**, *66*, 230–244.
23. Janssen, H.A. Versuche über Getreidedruck in Silozellen. *Z. Des. Ver. Dtsch. Ing.* **1895**, *39*, 1045–1049.
24. Feng, F.; Li, X.B.; Chen, S.J.; Peng, D.X.; Bian, Z. Rational size and stability analysis of horizontal isolated pillars in deep mining from caving to filling method. *E3S Web Conf.* **2021**, *303*, 01040. [CrossRef]
25. Swift, G.M.; Reddish, D.J. Stability problems associated with an abandoned ironstone mine. *Bull. Eng. Geol. Environ.* **2002**, *61*, 227–239.
26. Hoek, E.; Brown, E.T. The Hoek–Brown failure criterion and GSI—2018 edition. *J. Rock Mech. Geotech. Eng.* **2019**, *11*, 445–463. [CrossRef]
27. Yang, J.H.; Dai, J.H.; Yao, C.; Jiang, S.H.; Zhou, C.B.; Jiang, Q.H. Estimation of rock mass properties in excavation damage zones of rock slopes based on the Hoek-Brown criterion and acoustic testing. *Int. J. Rock Mech. Min. Sci.* **2020**, *126*, 104192. [CrossRef]
28. Li, X.B.; Peng, D.X.; Feng, F.; Li, X.S. Stability analysis of horizontal insulating pillar in deep mining from caving to filling method on the basis of refined plate theory. *J. China Univ. Min. Technol.* **2019**, *48*, 484–494.
29. Philip-Mark, S.; Christos, R.; Francis, P. A Methodology combining IDEF0 and weighted risk factor analysis for the strategic planning of mine reclamation. *Minerals* **2022**, *12*, 713.
30. Fatemeh, K.; Ataallah, B.; Jafar, A.S. Mineral processing plant site selection using integrated fuzzy cognitive map and fuzzy analytical hierarchy process approach: A case study of gilsonite mines in Iran. *Miner. Eng.* **2020**, *147*, 106143.

minerals

MDPI

Article

Failure Behavior of Cuboid Granite Sample with a Circular Hole beneath a Bonding Fracture under Biaxial Compression

Diyuan Li, Jingtai Jiang and Feihong Gao *

School of Resources and Safety Engineering, Central South University, Changsha 410083, China
* Correspondence: feihong.gaoo@gmail.com

Abstract: Ore bodies incubating within fault zones are a common phenomenon in geological strata and pose a huge challenge for underground mining. To effectively exploit mineral resources, the layout of the mining roadway and the interaction between the roadway and geological fault must be considered. In this paper, a bonding fracture was formed on granite samples to simulate a closed fault, under which a circular hole was fabricated to simulate the roadway of the gold mine. We performed a biaxial compression test at a true-triaxial electrohydraulic servo testing system for granite samples with a combined fracture-hole structure. It is worth noting that the fracture inclination β and relative distance between fracture and hole L were taken into account. The digital image correlation (DIC) technique was used to observe the displacement and strain field evolution around the fracture-hole structure. Our results demonstrate that (1) the strength of the granite sample decreases with increasing bonding fracture dip angle β, and the displacement drops between the hanging wall and foot wall raised in both the horizontal and vertical displacement directions. Macroscopic cracks become dense, and the failure degree becomes severe around simulated fault areas. (2) With the increase in the distance L, the strength of the granite sample increases, the influence of the hole on the slip of the fracture plane is weakened, and the discontinuity of displacement becomes less obvious. (3) The maximum principal strain field quantitatively reveals the details of the crack initiation, propagation, and coalescence around the fracture-hole structure, and displacement nucleation is observed in the vertical displacement field.

Keywords: biaxial compression test; fracture-hole structure; crack propagation; strain evolution; digital image correlation

Citation: Li, D.; Jiang, J.; Gao, F. Failure Behavior of Cuboid Granite Sample with a Circular Hole beneath a Bonding Fracture under Biaxial Compression. *Minerals* **2022**, *12*, 1328. https://doi.org/10.3390/min12101328

Academic Editor: Giang Nguyen

Received: 23 August 2022
Accepted: 18 October 2022
Published: 20 October 2022

1. Introduction

During underground engineering, many collapse accidents occur in the process of roadway or tunnel excavation, which is generally caused by unreasonable construction and excavation schemes, or imperfect support measures around the fault zone [1–3]. Usually, accidents can be minimized by avoiding fault zones. However, when the ore body is to be mined in the fault area, it is inevitable to arrange roadways near the fault [4], as shown in Figure 1, Jiaojia Gold Mine is located in Laizhou City, Shandong Province, China. The actual geological characteristics are extremely complex, as abundant gold resources are detected underneath the fault zone, which poses a huge challenge for mineral resource extraction. To ensure the safe and efficient mining practice of Jiaojia gold mine resources and reduce roadway maintenance costs, it is necessary to pay attention to the interaction between geological faults and roadways [5]. Meanwhile, the intersection angle and relative distance [6,7] between roadways and faults are significant factors affecting roadway performance [8,9]. Hence, it is necessary to explore the influence of fault occurrence on roadway layout.

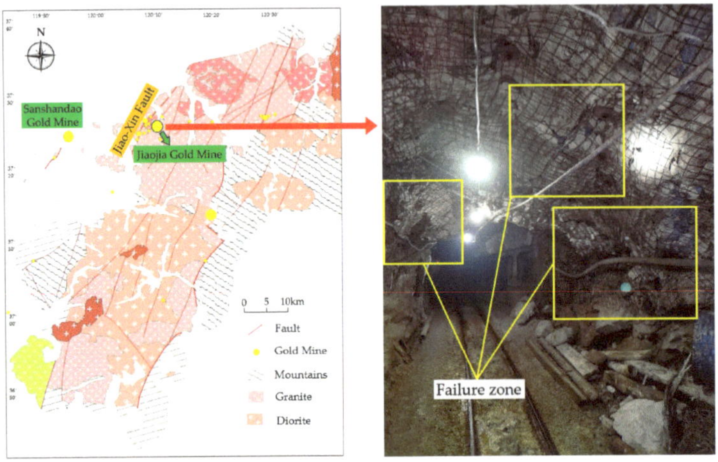

(a) (b)

Figure 1. Jiaojia Gold Mine in Jiaodong Peninsula. (**a**) Geological conditions of the Jiaojia Gold Mine; (**b**) failure zone of the roadway under the influence of the Jiao-Xin fault.

The surrounding rock stability and mechanical behavior of the roadway or tunnel construction have been widely studied. Some researchers carried out model tests to study the mechanical behavior of a circular tunnel with homogeneous strata [10,11] and analyzed the failure characteristics and strain field of the tunnel surrounding rock [12–16]. Precast cracks or faults have also been explored by a large volume of existing research [17–19]. However, due to the complex geological defects such as cracks and faults in the actual stratum, it is necessary to take these natural rock defects with roadways or tunnels into consideration to gain insight into their mutual effects. Among these existing studies, some studied the deformation characteristics and fracture evolution laws of fractured sandstone with prefabricated cavities [20,21]. Others used laboratory experiments and numerical analysis to study the failure characteristics of samples with combined defects of a single hole and oblique crack [8,21–23]. Many pieces of literature also studied the inclination of the fault and the distance between the fault and tunnel [24–28]. The existing research on the interaction between faults and roadways is limited to small-scale and numerical simulations. For field engineering practice, it is necessary to extend to carry out experimental research on large-scale structural planes.

Many researchers have conducted experiments on the impact of large-scale faults on tunnels or roadways based on engineering examples [2,12,29]. Some researchers have studied the influence of large faults on shallow tunnels and preventive measures [2,29,30], and carried out large-scale simulation tests to study the impact of faults on tunnels [12,31]. Some have used engineering examples and numerical simulations to study the impact of faults on mine shaft supports. Other studies have used numerical simulations to study the impact of faults on the stability of tunnels [9,32]. Although the above studies have considered the impact of large-scale structures on tunnels or roadways, which has certain practical engineering significance, experimental variables such as fault inclination and relative distance between fault and tunnel have not been considered, and have not been well explored in large-scale laboratory tests. To study the interaction between geological faults and roadways with varying fault occurrence, it is necessary to research the failure behavior of roadways beneath the simulated fault.

Therefore, we took the Jiaojia Gold Mine in Jiaodong Peninsula as the research background. As the lithology of the surrounding rock was mainly granite, granite was selected as the experimental material. The prefabricated hole and bonding fracture were used to simulate the fault and roadway. To better observe the deformation around the hole, a biaxial compression test was carried out, and the surface of the sample was observed

by the DIC equipment. The failure behavior of roadways under a simulated fault was explored. The fracture behavior and evolution were studied accordingly, which guided engineering practice.

2. Methodology

2.1. Sample Preparation

The engineering background of this article is the Jiaojia Gold Mine in the Jiaodong Peninsula, which is used to determine the basic mechanical parameters of the granite. The uniaxial compression test was performed on cylindrical samples with dimensions of Φ 50 mm × 100 mm [33]. The basic mechanical parameters are listed in Table 1. Granite samples for formal tests were made with a height of 150 mm, a length of 150 mm, and a width of 40 mm. Before the test, the rock plate was cut in half with a high-pressure water jet method and then glued together with high-strength epoxy resin to simulate a closed fault. The strength of the epoxy resin herein was approximately 12 MPa. In addition, a circular hole was fabricated beneath the oblique fracture to simulate the roadway, as shown in Figure 2.

Table 1. Basic physical and mechanical parameters of the granite sample.

Properties	Value
Young modulus E_s (GPa)	47.01
Poisson ratio ν	0.28
Uniaxial compressive strength σ_c (MPa)	125.87
Uniaxial tensile strength (MPa)	6.7
Longitudinal wave velocity V_p (m/s)	3692.9
Density ρ (kg/m^3)	2666.7

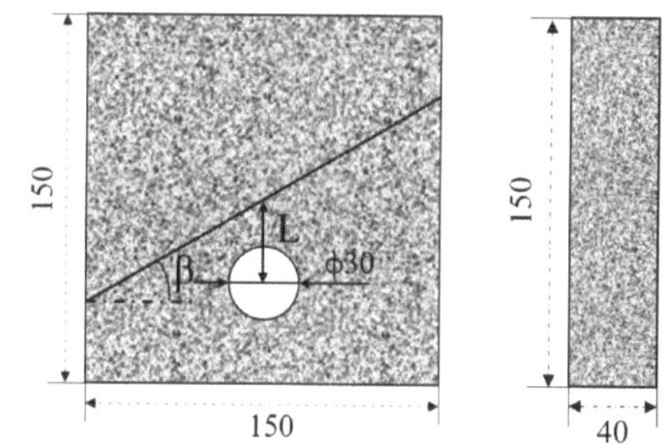

Figure 2. Schematic diagram of the granite sample in the biaxial compression test.

The purpose of this test is to study the influence of the fault dip angle and the distance between the roadway and the fault on the induced instability and failure of the surrounding rock by changing the variables β and L (as shown in Figure 1). Therefore, the samples can be divided into the following groups (as shown in Table 2):

(1) A-0, A-15, A-30, and A-45 are granite samples with fault dip angles of 0°, 15°, 30°, and 45°, and the distance L between the hole center and the fault is 40 mm.
(2) L-1, L-2, and L-3 are granite samples with distances L between the hole center and the fault of 30 mm, 35 mm, and 40 mm, respectively, and the fault dip angle is 30°. Considering the effect of the size effect, the hole bottom should not be too close to the sample boundary.
(3) C is the control group. The sample only contained a central hole without a prefabricated fault.
(4) O is the control group, the fault was not bonded with glue, and the fault dip angle was 30°. Considering that the stress propagation of the unbonded fault might be

blocked, the minimum distance L between the hole and the fault was 30 mm. The details regarding different types of samples are listed in Table 2.

Table 2. Morphology and mechanical parameters of granite samples.

Sample No.	Length/mm	Width/mm	Thickness/mm	Hole	Bonding Fracture	Angle	Distance between Fracture and Hole/mm
A-0	151.95	149.96	41.03	√	√	0°	40
A-15	149.45	149.66	40.56	√	√	15°	40
A-30	150.36	151.30	40.35	√	√	30°	40
A-45	153.12	152.22	41.02	√	√	45°	40
L-1	152.23	151.28	42.34	√	√	30°	30
L-2	149.36	150.34	39.86	√	√	30°	35
L-3	151.42	152.03	40.16	√	√	30°	40
C	151.42	151.30	41.12	√	-	-	-
O	150.35	152.21	42.01	√	×	30°	40

Moreover, the DIC technique was used to record the evolution process of the strain field and displacement field on the surface of the granite sample. Before the test begins, artificial speckle patterns should be printed on the surface of the sample to facilitate analysis and processing. Meanwhile, the low-speed camera of the CMOS sensor was placed 60 cm in front of the samples. The lamp was placed obliquely above the sample to supplement the illumination (as shown in Figure 3).

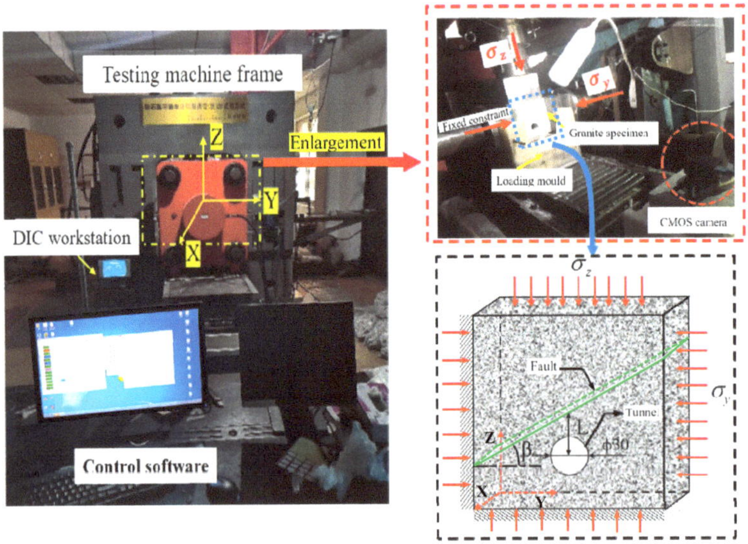

Figure 3. True-triaxial testing equipment and schematic diagram of the loading method for the sample.

The resolution of the camera is 4096 × 3000 pixels, with 8-bit gray digitalization. The camera automatically captures images at a frame rate of 10 frames per second. After filming, speckle images captured were analyzed using VIC-2D software. More details about the DIC technique and artificial speckle pattern can be found in the references [17,34,35]. The instrument used in this experiment is true triaxial equipment, which can realize loading stresses in one direction, two directions, and true-triaxial conditions. It can control the load, deformation, and displacement automatically. The experimental system is a digital closed-loop control system composed of a computer, control software, testing machine frames, and sensing element, as shown in Figure 3. The test device can apply a maximum force of 2000 kN in the X-axis and Y-axis directions and 3000 kN in the Z-axis direction [36–39].

2.2. *Experimental Procedure*

2.2.1. Determination of In Situ Stress

The three principal stresses at the depth of the Jiaojia Gold Mine are as follows: the maximum horizontal principal stress was up to 51.1 MPa (306.6 kN) and the vertical in-situ stress reached 27 MPa (162 kN). The details of the initial stresses are listed in Table 3. Loading to the initial set value at a loading rate of 1 kN/s was in the directions σ_y and σ_z [40].

Table 3. Initial in situ stress at the depth of the Jiaojia Gold Mine.

Depth (m)	σ_z (MPa)	σ_{ymax} (MPa)	σ_{ymin} (MPa)
1000	27	51.1	33.7

2.2.2. Loading Paths

To reduce the friction effect, petroleum jelly was smeared on four contact surfaces of sample boundaries. Four clamps were placed between the sample and loading end regions of the equipment to fix the granite sample, and one end region was fixed; the other end region was the loading face. After placing the sample and adjusting the test equipment, a preload of 2 kN was applied. When the preload reached 2 kN, the loading rate was fixed at 1 kN/s to a setting value [41].

The loading path is shown in Figure 4, and the specific loading procedure is described as follows:

(1) The σ_y and σ_z directions were loaded to 162 kN at a loading rate of 1 kN/s. The σ_z direction remained constant, and σ_y continued to be loaded to 306.6 kN.
(2) After the preset in situ stress was reached, σ_z maintained the stress level, and σ_y carried out graded loading at a rate of 1 kN/s. When the stress reached 306.6 kN, the stress was maintained for 2 min for every 60 kN increase. The purpose was to prevent the sample from suddenly failing.
(3) There was obvious damage and severe damage to the hole wall, the load was removed, the load of σ_y was first unloaded to σ_z, and then σ_y and σ_z were unloaded to 0.

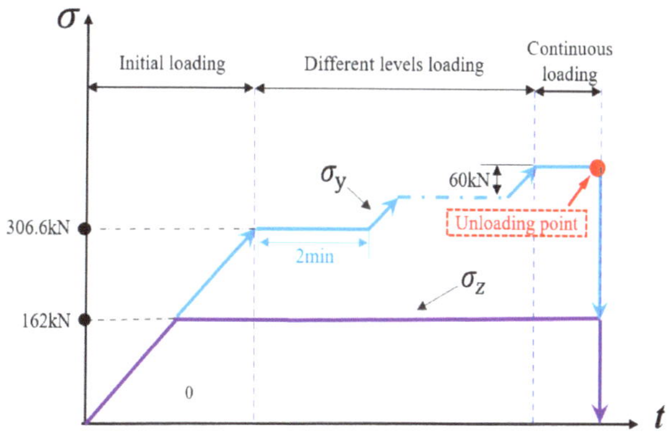

Figure 4. Schematic diagram of the loading path.

3. Results

3.1. *Mechanical Properties*

Biaxial compressive strength reflects the resistance of samples with bonding fracture, and it was represented by the peak strength of σ_y in this study. Take sample *L*-2 as an example. σ_y and σ_z reached the set vertical in situ stress (162.6 kN) with the same loading rate first. σ_z maintained stability, while σ_y maintained the original loading rate. Once the sample was damaged, σ_{ymax} (421.9 kN) was specified as the biaxial compressive

strength. The biaxial strength of granite samples with different geometric fracture patterns is summarized in Table 4.

Table 4. Mechanical parameters of different types of rock samples.

Initial Stress	A-0	A-15	A-30	A-45	L-1	L-2	L-3	C	O
σ_z/MPa	27.03	28.07	27.08	28.27	28.32	28.46	27.70	27.92	28.60
σ_y/MPa	91.24	85.17	69.76	53.08	62.72	70.32	77.69	84.09	92.90

Furthermore, Figure 5 shows the relationship between the biaxial compressive strength versus fracture dip angle (Figure 5a) and spacing distance L (Figure 5b). When the dip angles of the fracture were 0°, 15°, 30°, and 45°, the biaxial compressive strength of the rock sample was 91.24 MPa, 85.17 MPa, 69.76 MPa, and 53.08 MPa, respectively, which demonstrated a decreasing trend. In addition, when the relative distance between the hole and the fracture was 30 mm, 35 mm, and 40 mm, the biaxial compressive strength was 62.72 MPa, 70.32 MPa, and 77.69 MPa, respectively. The results demonstrated that with increasing distance L, the interaction between the fracture and the hole was weakened, and the strength of the rock sample increased.

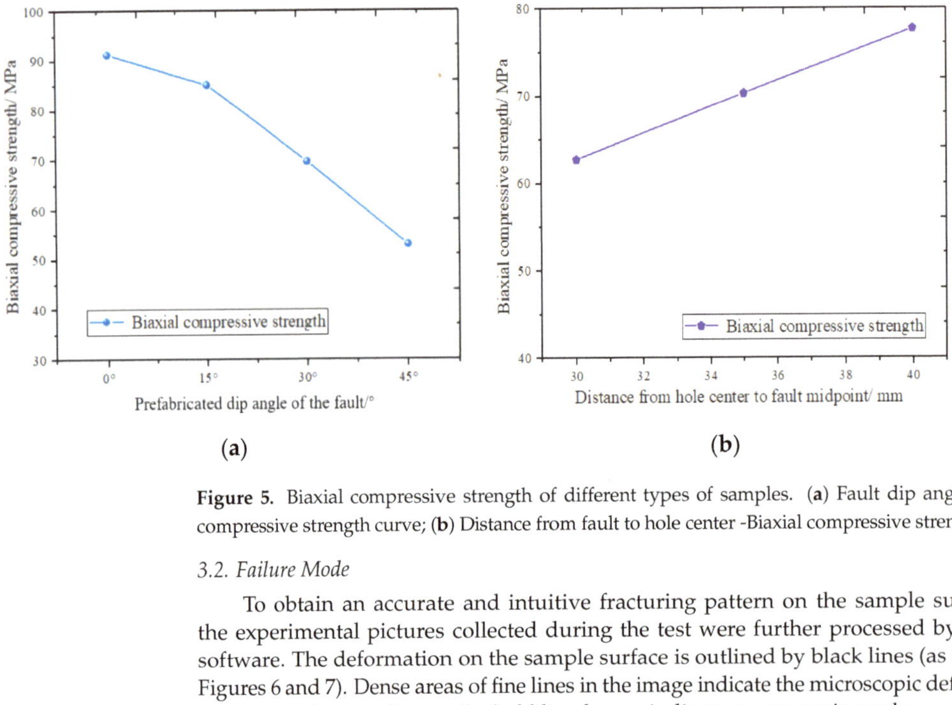

Figure 5. Biaxial compressive strength of different types of samples. (**a**) Fault dip angles-Biaxial compressive strength curve; (**b**) Distance from fault to hole center -Biaxial compressive strength curve.

3.2. Failure Mode

To obtain an accurate and intuitive fracturing pattern on the sample surface, all the experimental pictures collected during the test were further processed by the VIC software. The deformation on the sample surface is outlined by black lines (as shown in Figures 6 and 7). Dense areas of fine lines in the image indicate the microscopic deformation concentrated area, whereas the bold lined areas indicate macroscopic cracks.

Figure 6 shows the images of the final crack transfixion mode of different types of samples. Figure 6a,b show sample C with a hole and sample O with an unbonded fracture, respectively. The generating crack amount in these two types was small and mainly distributed at the footwall, indicating that the stress concentration generally occurred below the hole. In addition, there were no visible cracks around the unbonded fracture plane, indicating that the influence of the hole on the unbonded fracture plane was negligible.

The failure modes of granite samples with bonded fracture dip angles of 0°, 15°, 30°, and 45° are shown in Figure 6c–f. When the fracture dip angle was 0°, newborn cracks formed around the fracture plane end, and there were no obvious macrocracks generated around the fracture-hole structure, indicating that the fracture plane of 0° had a relatively

weak effect on the fracture-hole zone, which also helped verify the conclusion that the A-0 type had the largest strength. When the fracture dip angle was 15°, a macroscopic crack initiated from the middle of the fracture plane and extended to the hanging wall. At the same time, the roof and floor of the hole fractured and formed a V-shaped groove. When the fracture dip angle was 30°, the dominant crack was formed above the fracture plane and spread to the upper left direction. Several macroscopic cracks generated at the top of the hole extended continuously and tended to connect with the main crack on the other side of the fracture plane. When the crack was 45°, numerous macroscopic cracks were derived from the fracture plane toward the hanging wall. Among them, the dominant crack expanded from the middle of the fracture plane, and many secondary cracks were generated beside the main crack. A small number of cracks can be observed in the footwall and extend horizontally. Cracks were formed at the top of the hole and extended horizontally to the right.

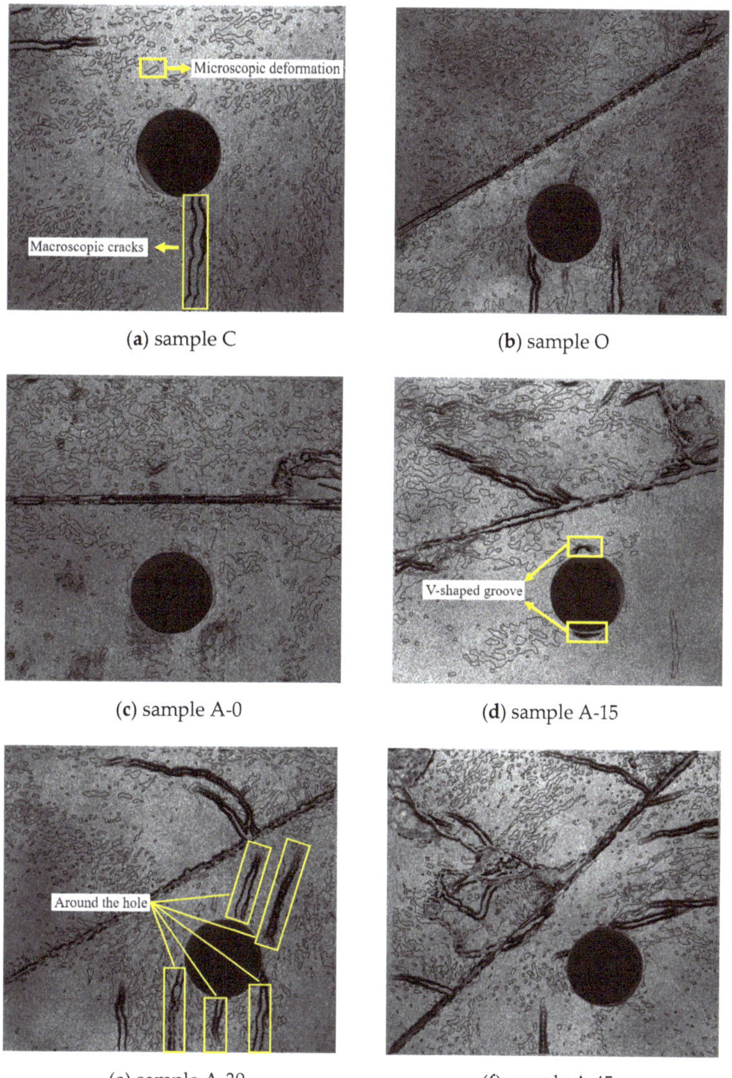

Figure 6. Fracture modes of granite samples with different bonding fracture dip angles.

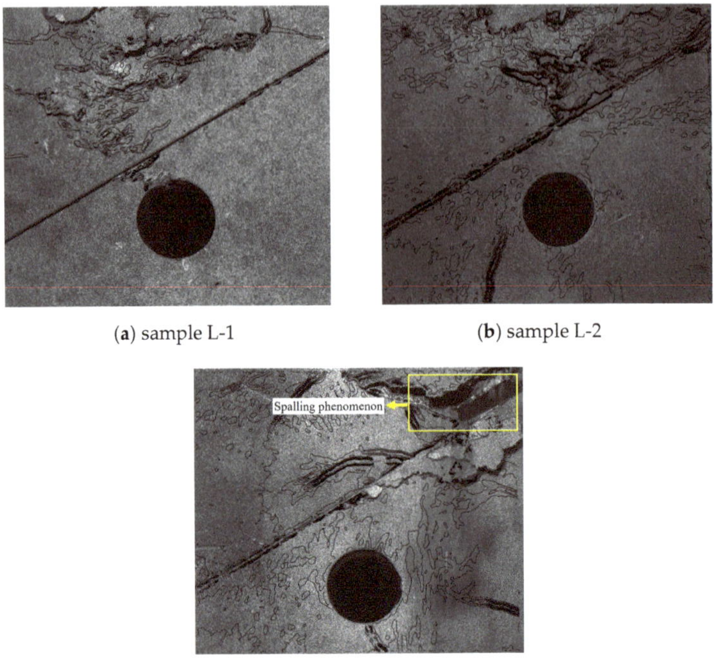

(a) sample L-1 (b) sample L-2

(c) sample L-3

Figure 7. Fracture modes of granite samples with bonding fracture at different spacing distances *L*.

The failure modes of rock samples with a spacing distance *L* ranging from 30 mm, 35 mm to 40 mm are shown in Figure 7a–c. It can be observed from Figure 7 that when the spacing distance was 30 mm, there were small-scale cracks between the hole and fracture plane, and they crossed the fracture plane. Finally, a divergent failure zone appeared on the hanging wall. When the spacing distance *L* was 35 mm, some cracks initiated from the fracture plane, and these cracks mainly gathered on only the hanging wall and extended to the boundary. Finally, the expansion and connection of macroscopic cracks led to the overall failure. When the distance between the hole and the fracture plane was 40 mm, cracks formed at the right end regions of the fracture zone while fracturing below the fracture plane. A crack was also formed under the hole and extended down to the bottom boundary of the sample.

In conclusion, when the spacing distance *L* is closed, the hole has a great influence on the fracture plane, and newborn cracks accumulate mainly in the middle region between the fracture and hole and further diffuse upward across the fracture plane. When the spacing distance *L* is far (*L* = 35 mm, 40 mm), the interaction between the fracture and hole is limited, and dominant macrocracks form in the hanging wall rather than dense microcracks.

4. Discussion

4.1. Discontinuity Analysis of Displacement

4.1.1. Fracture Slip Behavior

Static loads were applied in the Y and Z directions, and an LS camera was placed along the X direction to monitor the fracturing process. As shown in Figure 8 (A-0 and A-30 for examples), one platen was fixed, and the other end was subjected to axial static pressure. Figure 8a shows that when the fracture dip angle was 0°, the loading platen transmitted force to the granite sample from the upper and lower contact areas. Under such conditions, the contact area remained the same for the upper and lower parts, but the hole that existed

in the lower part resulted in different stress distributions. Two mass points along the fracture plane were analyzed. The displacement of two mass points could be decomposed into the slip distance along the fracture plane and vertical displacement resulting from the decomposed normal force. For the cases of fracture dip angles not equal to 0° (Figure 8b), the contact area of the upper part could be larger than that of the lower part, and the stress in the footwall, therefore, could be greater than that of the hanging wall. Furthermore, the footwall contained a hole, which further intensified the stress concentration around the hole. The asymmetry of the stress distribution led to the displacement difference along the fracture plane.

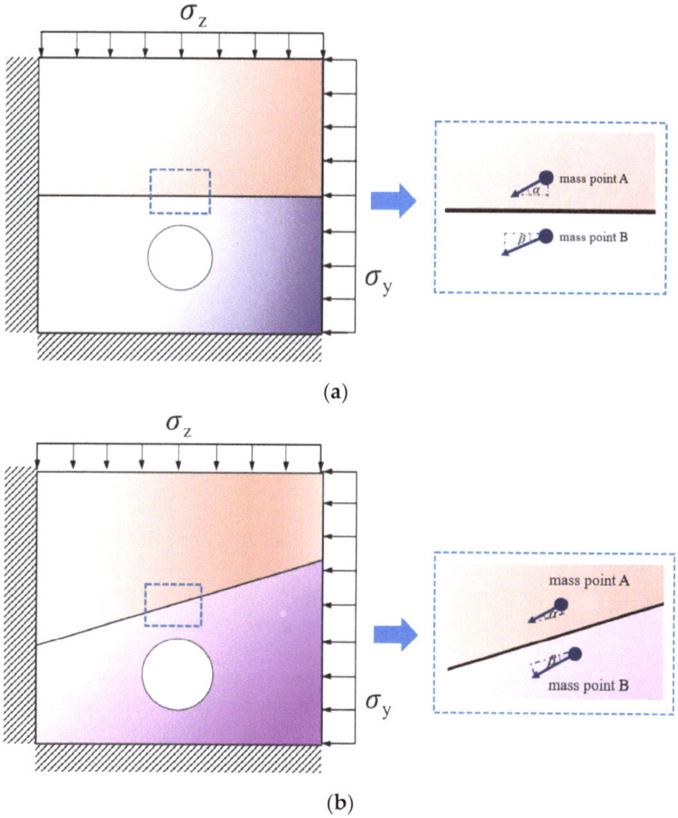

Figure 8. Loading mode diagram and sliding along the fracture plane for granite samples: (**a**) A-0; (**b**) A-30.

Furthermore, the displacement field on the sample surface can explain the failure mechanism. Taking the A-0 sample as an example, the displacement vector diagram was drawn using speckle data from VIC-2D software, as shown in Figure 9. This result indicated that the displacement vectors of the sample all inclined downward, which was related to the loading mode specified in Figure 8. This section analyzes the fracture slip value for different types of samples. The method to measure fracture slip is elaborated as follows:

(1) Determining the position of measurement points: points A and C were 2 cm away from the left and right boundaries, respectively, and point B was located directly above the hole.

(2) To analyze the fracture slip value of measurement points A, B, and C, reference points for each measurement point were taken above and below the fracture plane, as shown in Figure 9 (A_1, A_2, B_1, B_2, C_1, C_2). According to the displacement decomposition

method shown in Figure 8, the slip value along the fracture plane was derived for points A, B, and C.

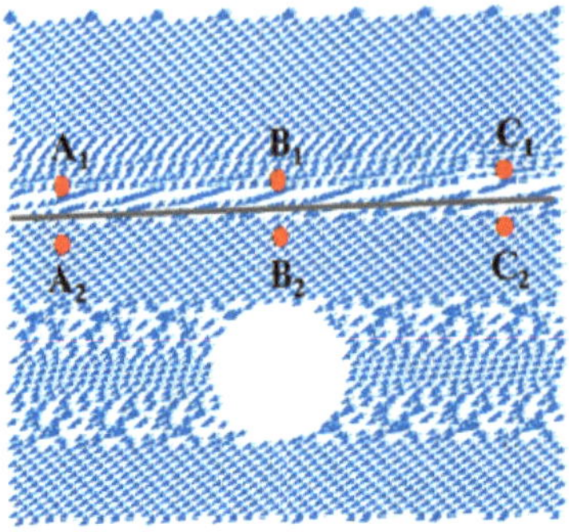

Figure 9. Displacement vector diagram of granite sample A-0 (σ_y = 90% σ_{ymax}).

After specifying the calculation method, the curves of the slip values for positions A, B and C are shown in Figures 10 and 11. Negative values on the plots indicate a downward slip along the fracture plane.

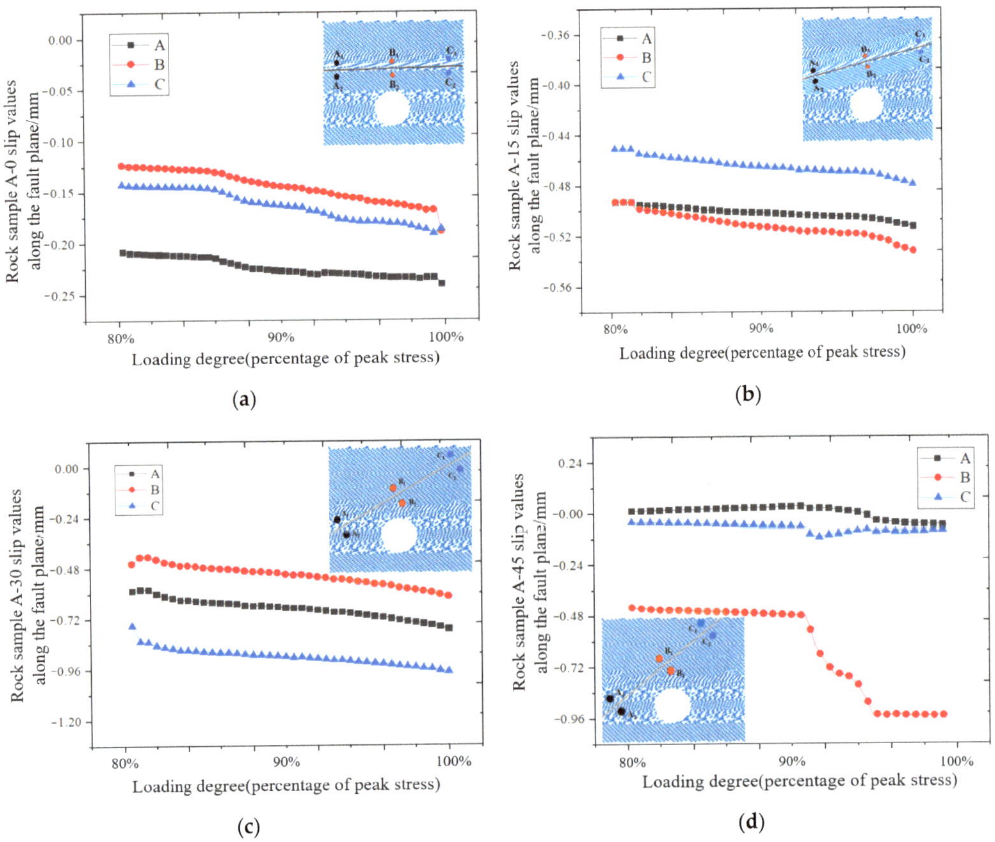

Figure 10. Schematic diagram of the relationship between fracture slip at measurement points and loading degree in group A. (**a**) sample A-0 slip values; (**b**) sample A-15 slip values; (**c**) sample A-30 slip values; (**d**) sample A-45 slip values.

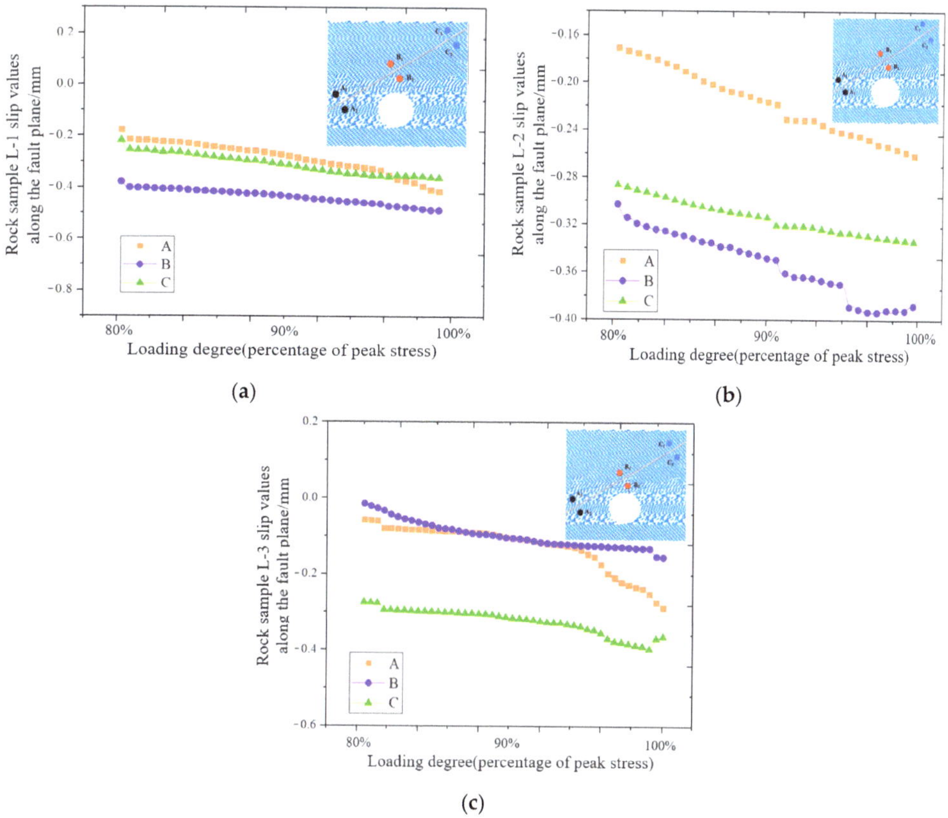

Figure 11. Schematic diagram of the relationship between fracture slip at points A, B, C, and loading degree in group *L*. (**a**) sample L-1 slip values; (**b**) sample L-2 slip values; (**c**) sample L-3 slip values; (**a**) sample A-45 slip values.

From Figures 10 and 11, it was concluded that for the A-0, A-15, and A-30 types, the slip values of the same locations (A, B, and C) generally increased with the increasing fracture dip angle (except for the measurement point B in Figure 10b-c, which were quite close). The behavior of sample A-45 was inconsistent with that of the other types of samples. The slip value of point B above the hole was significantly higher than that of points A and C, which were quite small and close to 0.

Figure 11 shows the relationship between slip values along the fracture plane and loading degree. The main characteristics were summarized as follows:

(1) When the fracture plane was close to the hole, the slip value at point B was the largest, followed by point C, and the slip value at point A away from the loading platen was the smallest. The slip values at points A and C of sample *L*-1 were similar, and the slip value at point B above the hole was larger, reaching 0.48 mm at peak stress. The slip value of point A at the left boundary of sample *L*-2 was the smallest, followed by point C at the right end, and the slip value for point B was the largest (less than *L*-1).

(2) When the fracture was far from the hole, the influence of the hole on fracture slip was reduced, the slip value of point B above the hole and point A away from the loading platen were small and similar, and point A increased obviously when the peak strength was reached. The slip value of position C near the loading platen was the largest.

4.1.2. Displacement Discontinuity

In the test process, fracture-dislocation occurred, and discontinuous displacement occurred on both sides of the fracture. To study the discontinuity of displacement on both sides of the fracture plane, a monitoring belt was arranged in the direction of the vertical fracture plane. The vertical fracture plane monitoring zone was arranged in the center of the opposite hole and perpendicular to both sides of the fracture plane, with a length of 10 mm.

Taking the sample A-0 as an example, the calculation method of the discontinuity of fracture displacement was elaborated. To analyze the displacement drops along the fracture at peak strength, the load-time curve of the granite sample A-0 was drawn in Figure 12 first, and then the displacement in the middle period of 950 s~1088 s was taken for analysis.

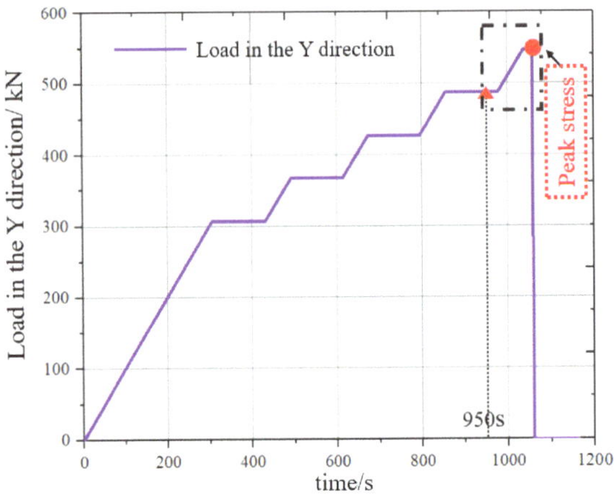

Figure 12. Load-time curve of sample A-0.

First, the horizontal and vertical displacement drops of the monitoring belt at different times are shown in Figures 13 and 14, respectively. In Figure 13a, the horizontal displacement line from M to N is depicted. The line breaking in the middle indicated the discontinuity of displacement for a certain time. Then, a series of horizontal displacement lines at a time range of 950 s~1088 s was deduced and drawn in Figures 13 and 14.

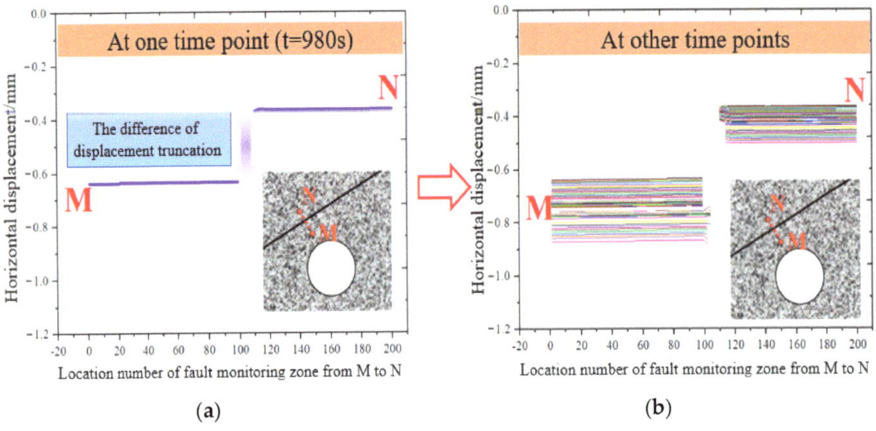

Figure 13. The horizontal displacement drops of granite sample A-0 during 950 s~1088 s. (**a**) The horizontal displacement drops when t=980 s; (**b**) The horizontal displacement drops when other time points.

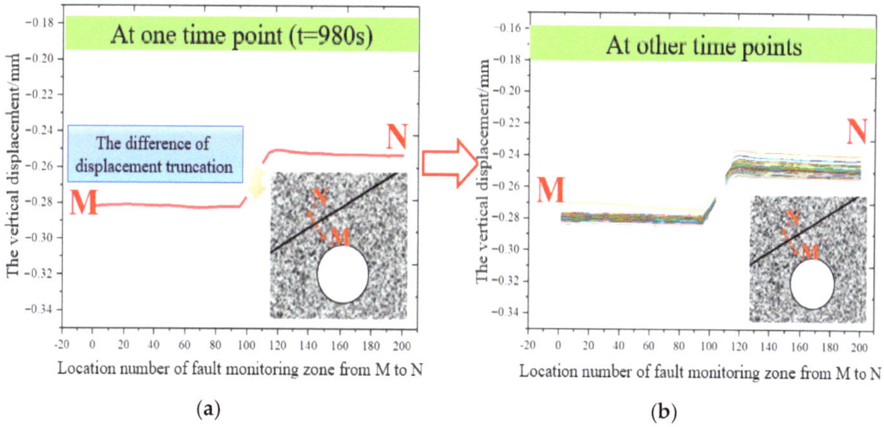

Figure 14. The vertical displacement drops of granite sample A-0 during 950 s~1088 s. (**a**) The vertical displacement drops when t=980 s; (**b**) The vertical displacement drops when other time points.

Second, calculate the displacement drops for every time point. Each displacement line is shown in Figures 13b and 14b. The scatter diagram of the drop values versus time range in 950 s~1088 s was plotted (as shown in Figure 15a,b). Figure 15a shows that the discontinuity of horizontal displacement slowly rises with time. In Figure 15b, an obvious linear relationship was not shown, but the displacement drops were maintained at 0.022–0.025 mm before the sample reached failure. To ensure accuracy, the mean value of the scattered points within this range was taken as the vertical displacement discontinuity value for sample A-0, as shown in Figure 15b.

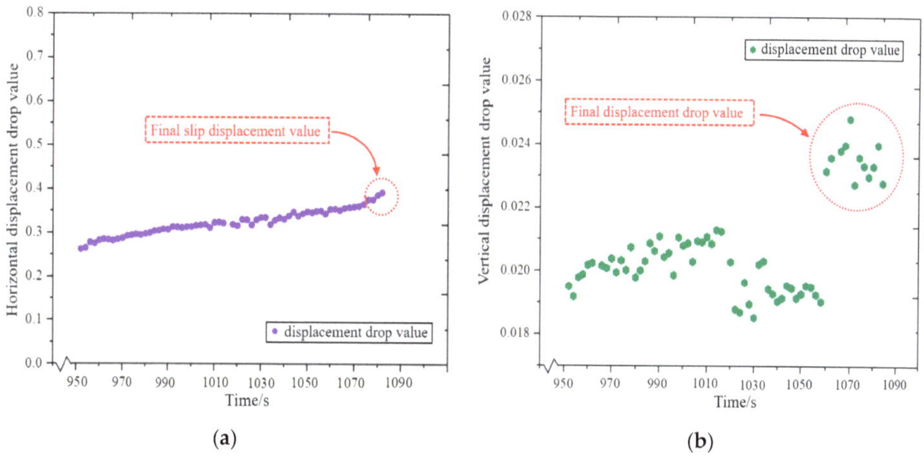

Figure 15. Final displacement drops value. (**a**) Final vertical drops in the horizontal direction in 950 s~1088 s (**b**). Final vertical drops in the horizontal direction in 950 s~1088 s.

Finally, the displacement drop value of the granite sample was calculated and compared. The relationship between the horizontal and vertical displacements dropped and different angles were obtained and are shown in Figure 16. As shown in Figure 16a, with increasing fracture dip angle, the displacement drops increased as well. The horizontal displacement drop value of the unbonded fracture (sample O) was 0.221 mm, while the horizontal displacement drop values of the 0°, 15°, 30°, and 45° bonded fractures were 0.393 mm, 0.554 mm, 1.282 mm, and 1.344 mm, respectively. According to Figure 16b, the vertical displacement drop value of the unbonded fracture was 0.051 mm, while the

displacement drop values of the 0°, 15°, 30°, and 45° bonded fractures were 0.024 mm, 0.248 mm, 0.594 mm, and 0.611 mm, respectively. Except for the unbonded fracture type O, the vertical displacement drop value generally demonstrated a rising trend with increasing angle. However, compared with the horizontal displacement, the drop value of the vertical displacement was reduced by half. This may be caused by the vertical direction Z being loaded to 162 kN and remaining at this level, but the Y-axis loading continues until the sample reaches failure, so the displacement drop in the horizontal direction is significantly greater than that in the vertical direction. Furthermore, Figure 16b shows that the vertical displacement drop of sample O is between sample 0° and sample 15°, mainly because the fracture dip angle of control group O was 30°, and the influence of the vertical load on the fracture displacement drops was lower than that of the fracture dip angle (30°).

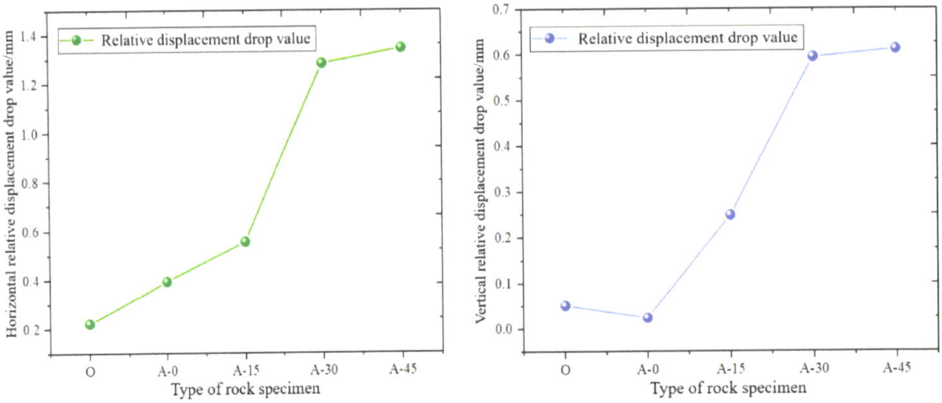

(**a**) (**b**)

Figure 16. The displacement truncation difference of different types of rock samples of group A; (**a**) Horizontal displacement; (**b**)Vertical displacement.

Figure 17 shows the displacement drop of different types in group *L*. The closer the distance between the fracture and the hole was, the more obvious the discontinuity of the displacement was, except for the unbonded type O (displacement drops value was 0.221 mm). The horizontal displacement drop values of the samples with spacing distances of 30 mm, 35 mm, and 40 mm were 0.555 mm, 0.265 mm, and 0.234 mm, respectively. The drop value characteristics for vertical displacement were similar to those of group A, with nearly half reduction for horizontal displacement. The drop values of the vertical displacement were 0.238 mm, 0.167 mm, and 0.150 mm, respectively. The distance between the fracture and hole was close, and the discontinuity of horizontal and vertical displacement was significantly greater than that of other granite types.

4.2. Evolution of the Deformation Field

4.2.1. Evolution of the Major Principal Strain Field

To analyze the major principal strain field on the sample surface under biaxial compression, several typical samples in groups A, *L*, and O were selected to discuss the evolution of fracturing for different sample types.

The evolution process of the major principal strain field of sample A-15 is shown in Figure 18. According to Figure 18, when the load reached 84.7% of the peak strength, strain concentration did not occur around the hole, only a slender crack appeared at the bottom boundary of the sample, and stress concentration occurred on the fracture plane. When the load reached 92.32% of the peak strength, there was no high strain around the hole, but strain began to concentrate at the left fracture end. While the load reached 98.39% of the peak strength, a high-strain concentration formed at both the roof and floor of the

hole, and a large-scale high-strain concentration also appeared in the hanging wall. When the strength peak was reached, regional failure occurred on the hanging wall, and the surface flaked.

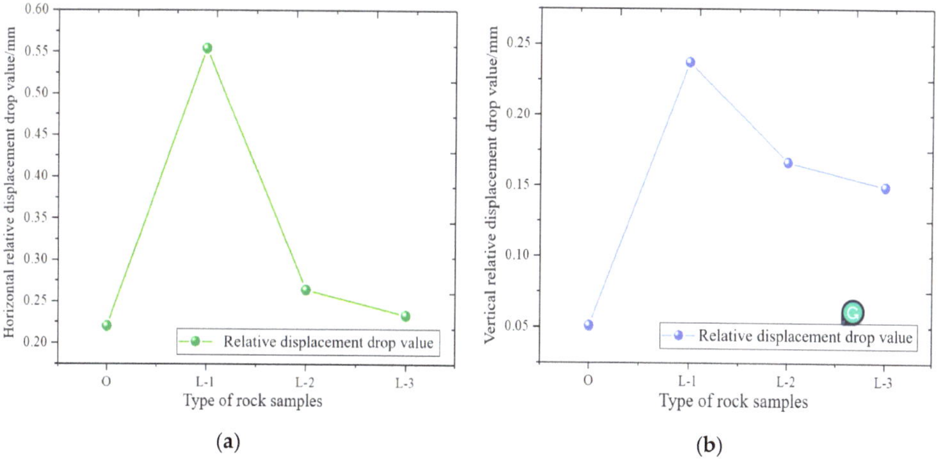

Figure 17. The displacement drops of different types of granite samples in group *L*. (**a**) The Horizontal displacement of samples; (**b**) The vertical displacement of samples.

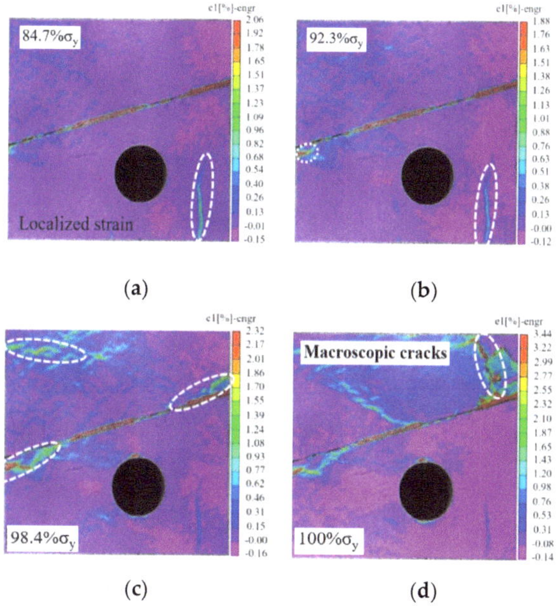

Figure 18. Evolution process of the major principal strain field of sample A-15. (**a**) 84.7%σ_y; (**b**) 92.3%σ_y; (**c**) 98.4%σ_y; (**d**) 100%σ_y.

The evolution process of the major principal strain field of sample A-30 is shown in Figure 19. When the load reached 84.2% of the peak strength, a high-strain concentration appeared in the lower left area of the footwall. A high-strain band initiated from the hole floor, which expanded along the loading direction. When the load reached 85.2% of the peak strength, the original high-strain zone expanded further. When the load reached 90.4% of the peak strength, the high-strain zone at the hole floor extended to the left boundary, and the strain zone appeared at the hole crown and tended to connect with the fracture

plane. When the sample reached its peak strength, the strain concentrated at the top and bottom of the hole, forming a V-shaped groove. The crack initiated from the hole crown and connected with the cracks formed at the left end of the fracture plane.

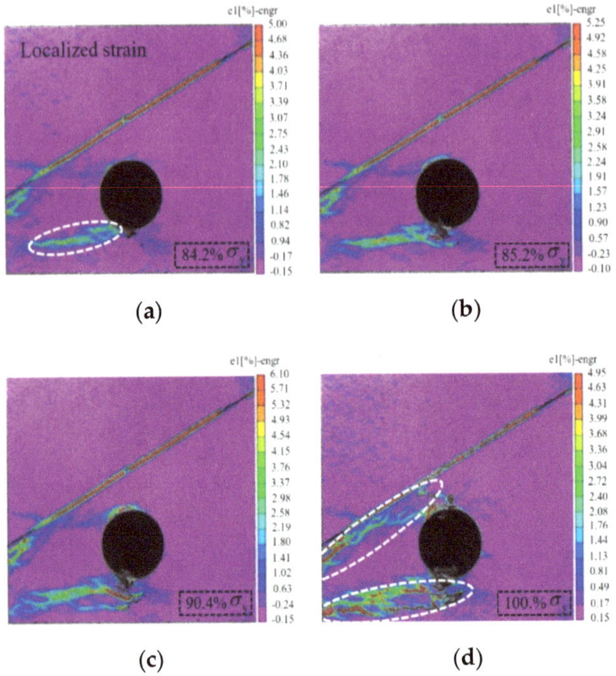

Figure 19. Evolution of the major principal strain field of sample A-30. (**a**) 84.2%σ_y; (**b**) 85.2%σ_y; (**c**) 90.4%σ_y; (**d**) 100%σ_y.

The evolution process of the major principal strain field of sample *L*-2 is shown in Figure 20. When the load reached 87.4% of the peak strength, little high strain appeared on both the top and bottom of the hole, and a crack initiated from the fracture plane and extended to the hanging wall. When the load reached 92.2% of the peak strength, the original high-strain zone around the hole expanded further, and a new crack was formed at the sample boundary and extended to the hole floor. Scattered high-strain zones are concentrated at the lower right end of the fracture plane. When the load reached 96.5% of the peak strength, the original crack extended further, and a new crack was generated at the bottom boundary of the sample. Meanwhile, a stress concentration area appeared at the upper right of the fracture plane. When the sample reached the peak strength, regional failure appeared on the upper right of the fracture plane, and a crack directly penetrated the fracture.

Figure 21 shows the evolution process of the major principal strain field of sample *L*-3. Figure 21 shows that when the load reached 87.7% of the peak strength, an obvious high-stress concentration occurred in the fracture plane, and small-scale high strain appeared above the hole. When the load reached 91.0% of the peak strength, ejection occurred at the hole floor, and a weak surface tangent to the hole crown appeared, which almost penetrated through the whole sample. When the load reached 95.7% of the strength, new cracks were generated from the hole floor and extended to the left boundary of the sample. When the load reached its peak strength, regional failure and cracks appeared in the right part of the hanging wall. The spalling phenomenon could be observed on the hole floor, accompanied by tensile cracks. The hole top and fracture surface formed a wedge block.

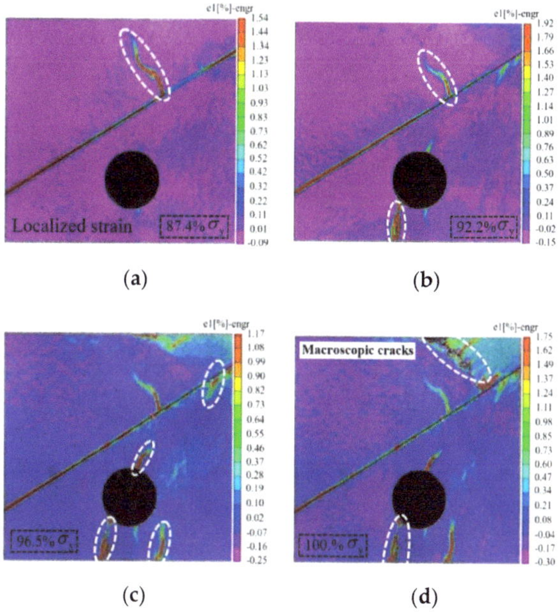

Figure 20. Evolution process of the major principal strain field of sample *L*-2. (**a**) 87.4%σ_y; (**b**) 92.2%σ_y; (**c**) 96.5%σ_y; (**d**) 100%σ_y.

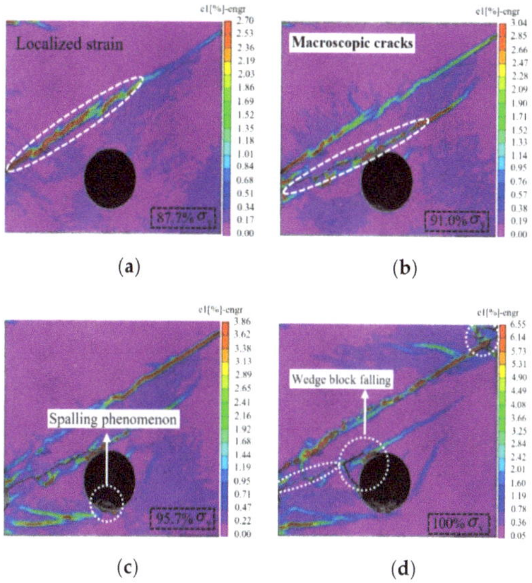

Figure 21. Evolution process of the major principal strain field of sample L-3. (**a**) 87.7%σ_y; (**b**) 91.0%σ_y; (**c**) 95.7%σ_y; (**d**) 100%σ_y.

The evolution process of the major principal strain field of control group O is shown in Figure 22. Figure 22 shows that when the load reached 87.30% of the peak strength, slender cracks along the vertical direction appeared at both the roof and floor of the hole, and an obvious high-stress concentration was observed at the fracture plane. When the load reached 89.36% of the peak strength, the granite sample was in a state of compression, and a floccule-like high-strain zone appeared on the right side of the footwall. When the load reached 93.28% of the strength, the cracks generated from the roof and floor of the

hole began to vanish gradually, and small-scale high strain began to appear at the right corner of the hanging wall. When the load reached 100% of the peak strength, the cracks initiated from the roof and floor of the hole closed completely, leaving only a belt crack derived from the oblique bottom of the hole. The high-strain concentration zone in the upper part of the fracture plane was enlarged and accompanied by the formation of several horizontal macrocracks, and the high-stress concentration also occurred in the left end of the fracture plane.

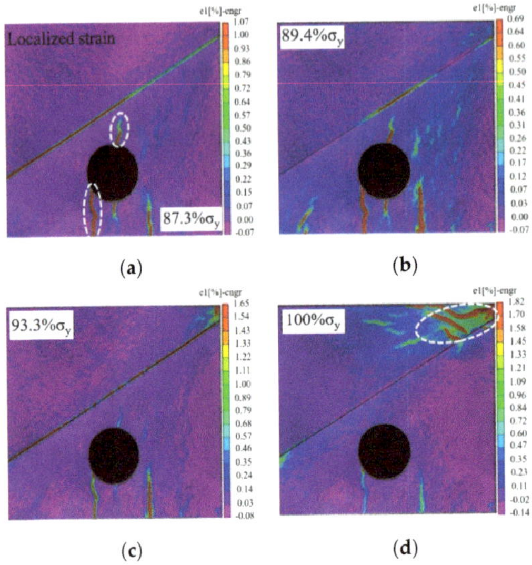

Figure 22. Evolution process of the major principal strain field of sample O. (**a**) 87.3%σ_y; (**b**) 89.4%σ_y; (**c**) 93.3%σ_y; (**d**) 100%σ_y.

For the failure process of the unbonded fracture O, the phenomenon that was different from other types of samples was that cracks were not generated near the fracture plane, and cracks still occurred around the hole. The failure of the sample was caused by the high-strain concentration at both ends of the fracture plane. The appearance of macroscopic cracks led to the fracturing and final failure of the granite material.

4.2.2. Displacement Field Characteristics

It was found that displacement nucleation existed in the vertical displacement field. More specifically, the vertical displacement between the fracture plane and the hole was nucleated, and the displacement value gradually decreased from the inside to the outside, as shown in Figure 23.

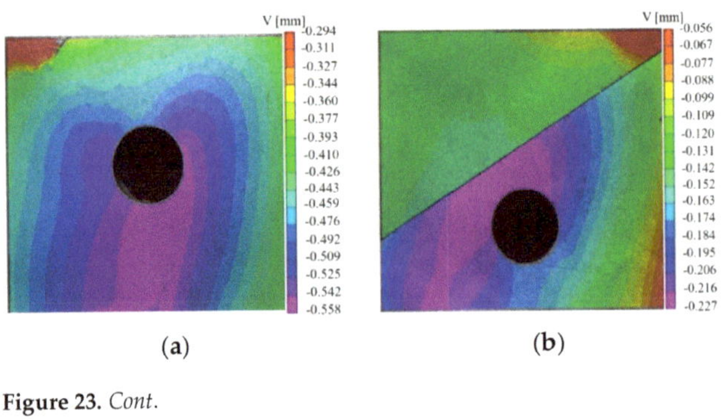

Figure 23. *Cont.*

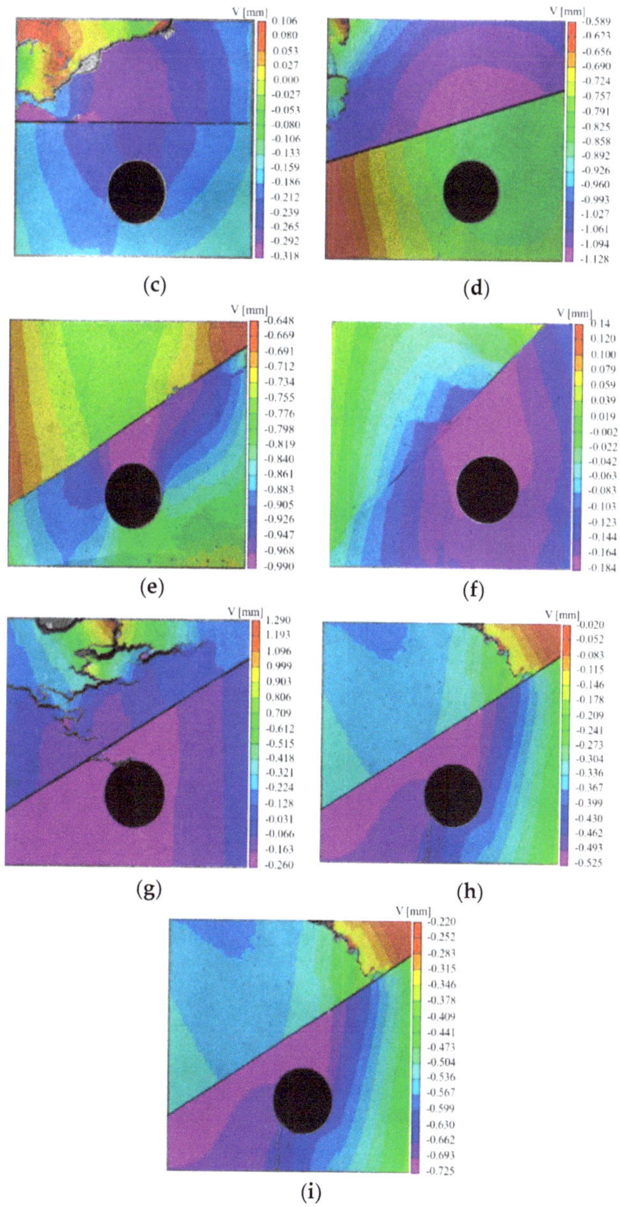

Figure 23. Nucleation phenomena of different types of granite samples; (**a**) Sample C (99.2%); (**b**) Sample O (97.5%); (**c**) Sample A-0 (96.2%); (**d**) Sample A-15 (99.3%); (**e**) Sample A-30 (94.6%); (**f**) Sample A-45 (96.6%); (**g**) Sample L-1 (99.8%); (**h**) Sample L-2 (98.9%); (**i**) Sample L-3 (98.3%).

The vertical displacement image of the granite sample with a hole is shown in Figure 23a. The phenomenon of displacement nucleation appeared below the hole, indicating that the microunits below the hole all moved downward. The vertical displacement images of granite samples with bonded fractures of 0°, 15° and 30°, 45°, and unbonded fractures are shown in Figure 23b–d. The red to purple in the legend indicates that the displacement value increased downward. Figure 23 shows that there was a purple nuclear displacement zone between the fracture and the hole, which meant that the vertical dis-

placement of the micro units within the fracture-hole structure demonstrated a downward movement under loading. Nucleation occurred when the vertical displacements reached 0.318 mm, 1.128 mm, 0.990 mm, and 0.184 mm. In group A, nucleation was more difficult to occur when the angle of the fracture was 15° and was most likely to occur when the angle of the fracture was 45°.

Figure 23g–i show the vertical displacement images of granite samples *L*-1, *L*-2, and *L*-3, respectively. The nucleation types of rock samples were centered on the region between the fracture and hole. Extending toward the left region of the footwall rather than spreading out similar to the types in group A, the nucleation types of the granite samples occurred when the vertical displacement reached 0.260 mm, 0.525 mm, and 0.726 mm. This indicated that with the increase in the spacing distance *L*, the micro units needed to move downward further in the vertical direction to form the nucleation.

5. Conclusions

The main conclusions are summarized as follows:

(1) Failure mode: With the increase in fracture dip angle, the number of macroscopic cracks also increases, and the failure degree is more severe. When the spacing distance *L* decreases, the circular hole has a great influence on the fracture plane, and the newly formed cracks accumulate in the middle region and further spread upward across the fracture plane. Therefore, to avoid roadway engineering accidents, it is necessary to avoid faults with large dip angles and keep a long distance from the fault as far as possible during roadway construction

(2) Fracture slip: when the fracture dip angle changes from 0° to 30°, the slip value at the measurement points (point A is far from the loading platen, point B is located above the hole, and point C is near the loading platen) along the fracture demonstrates less correlation. When the fracture dip angle is 45°, the slip value at the ends of the fracture plane (points A, B) is quite small, and the slip value of point B above the hole is significantly higher than that of the fracture ends. When the fracture is close to the hole, the slip value above the hole (point B) is the largest, and the slip value at point A far from the loading position is the smallest. When the fracture is far from the hole, the influence of the hole on fracture slip is reduced.

(3) Discontinuity of displacement: the displacement drops between the hanging wall and footwall increase with increasing fracture dip angle for both horizontal and vertical displacement. Furthermore, the displacement drop in the vertical direction is relatively small and is almost 50% lower than that in the horizontal direction. The closer the spacing distance between the fracture and the hole is, the more obvious the discontinuity of displacement. The closer the hole is to the fracture, the more likely the sample will be destroyed and the more serious the damage degree will be.

(4) The major principal strain field reveals the initiation, propagation, and coalescence of crack processes for different types of granite samples, and the displacement nucleation phenomenon is observed in the vertical displacement field.

Author Contributions: Conceptualization, D.L.; methodology, D.L.; F.G., J.J.; software, F.G.; validation, J.J., F.G.; formal analysis, J.J.; funding acquisition, D.L. All authors have read and agreed to the published version of the manuscript.

Funding: National Natural Science Foundation of China (NSFC) 52074349.

Data Availability Statement: The data presented in this study are available on request from the corresponding author.

Conflicts of Interest: The authors declare no conflict of interest.

References

1. Bruneau, G.; Tyler, D.B.; Hadjigeorgiou, J.; Potvin, Y. Influence of faulting on a mine shaft—A case study: Part I—Background and instrumentation. *Interna-Tional J. Rock Mech. Min. Sci.* **2003**, *40*, 95–111. [CrossRef]
2. Kun, M.; Onargan, T. Influence of the fault zone in shallow tunneling: A case study of Izmir Metro Tunnel. *Tunn.-Ling Undergr. Space Technol. Inc. Trenchless Technol. Res.* **2013**, *33*, 34–45. [CrossRef]
3. Zhao, K.; Janutolo, M.; Barla, G.; Chen, G. 3D simulation of TBM excavation in brittle rock associated with fault zones: The Brenner Exploratory Tunnel case. *Eng. Geol.* **2014**, *181*, 93–111. [CrossRef]
4. Panthee, S.; Singh, P.K.; Kainthola, A.; Singh, T.N. Control of rock joint parameters on deformation of tunnel opening. *J. Rock Mech. Geotech. Eng.* **2016**, *8*, 489–498. [CrossRef]
5. O'Rourke, M.J.; Liu, X. *Seismic Design of Buried and Offshore Pipelines, Monograph Series*; Multidisciplinary Center for Earthquake Engineering Research, (MCEER): New York, NY, USA, 2012.
6. Liu, Z.X.; Liu, J.Q.; Pei, Q.; Yu, H.T.; Li, C.C.; Wu, C.Q. Seismic response of tunnel near fault fracture zone under incident SV waves. *Undergr. Space* **2021**, *6*, 695–708. [CrossRef]
7. Zhang, Z.Q.; Chen, F.F.; Li, N.; He, M.M. Influence of Fault on the Surrounding Rock Stability for a Mining Tunnel: Distance and Tectonic Stress. *Adv. Civ. Eng.* **2019**, *2019*, 2054938. [CrossRef]
8. Sabagh, M.; Ghalandarzadeh, A. Numerical modelings of continuous shallow tunnels subject to reverse faulting and its verification through a centrifuge. *Comput. Geotech.* **2020**, *128*, 103813. [CrossRef]
9. Zhong, Z.L.; Wang, Z.; Zhao, M.; Du, X.L. Structural damage assessment of mountain tunnels in fault fracture zone subjected to multiple strike-slip fault movement. *Tunn. Undergr. Space Technol.* **2020**, *104*, 103527. [CrossRef]
10. Chu, B.L.; Hsu, S.C.; Chang, Y.L.; Lin, Y.S. Mechanical behavior of a twin-tunnel in multi-layered formations. *Tunn. Undergr. Space Technol.* **2007**, *22*, 351–362. [CrossRef]
11. Li, D.; Gao, F.; Han, Z.; Zhu, Q. Experimental evaluation on rock failure mechanism with combined flaws in a connected geometry under coupled static-dynamic loads. *Soil Dyn. Earthq. Eng.* **2020**, *132*, 106088. [CrossRef]
12. Huang, F.; Wu, C.; Ni, P.; Wan, G.; Karekal, S. Experimental analysis of progressive failure behavior of rock tunnel with a fault zone using non-contact DIC technique. *Int. J. Rock Mech. Min. Sci.* **2020**, *132*, 104355. [CrossRef]
13. Li, D.; Meng, L.; Han, Z.; Zhou, Z. Dynamic compressive mechanical properties of bedding sandstone with pre-existing hole. *Meitan Xuebao/J. China Coal Soc.* **2019**, *44*, 1349–1358.
14. Wang, S.; Ni, P.; Guo, M. Spatial characterization of joint planes and stability analysis of tunnel blocks. *Tunn. Undergr. Space Technol. Inc. Trenchless Technol. Res.* **2013**, *38*, 357–367. [CrossRef]
15. Yang, S.; Liu, X.; Li, Y. Experimental analysis of mechanical behavior of sandstone containing hole and fissure under uniaxial compression. *Chin. J. Rock Mech. Eng.* **2012**, *31*, 3539–3546.
16. Zhou, Y.; Zhu, Q.; Diyuan, L.I.; Chunde, M.A. Experimental study on mechanical and failure characteristics of sandstone with a filled elliptical hole. *J. Railw. Sci. Eng.* **2019**, *142*, 106780.
17. Li, D.Y.; Zhu, Q.Q.; Zhou, Z.L.; Li, X.B.; Ranjith, P.G. Fracture analysis of marble specimens with a hole under uniaxial compression by digital image correlation. *Eng. Fract. Mech.* **2017**, *183*, 109–124. [CrossRef]
18. Liu, J.P.; Li, Y.H.; Xu, S.D.; Xu, S.; Jin, C.Y.; Liu, Z.S. Moment tensor analysis of acoustic emission for cracking mechanisms in rock with a pre-cut circular hole under uniaxial compression. *Eng. Fract. Mech.* **2015**, *135*, 206–218. [CrossRef]
19. Pan, P.; Ding, W.; Feng, X.; Yao, H.; Zhou, H. Research on influence of pre-existing crack geometrical and material properties on crack propagation in rocks. *Chin. J. Rock Mech. Eng.* **2008**, *27*, 1882–1889.
20. Zhu, T.T.; Jing, H.W.; Su, H.J.; Yin, Q. Experimental investigation on mechanical behavior of sandstone with coupling effects under uniaxial compression. *Meitan Xuebao/J. China Coal Soc.* **2015**, *40*, 1518–1525.
21. Li, Y.; Liu, J.; Tang, X.; Yang, S.; Liu, D. Simulation analysis of mechanical properties of the soft rock mass with holes and fractures. *J. Min. Saf. Eng.* **2020**, *37*, 594–603.
22. Yin, Q.; Jing, H.W.; Su, H.J. Investigation on mechanical behavior and crack coalescence of sandstone specimens containing fissure-hole combined flaws under uniaxial compression. *Geosci. J.* **2018**, *22*, 825–842. [CrossRef]
23. Feng, F.; Li, X.; Li, D.; Chong, C. Numerical analysis of failure characteristics of circular hole specimen containing pre-existing flaw based on finite/discrete element method. *Yantu Lixue/Rock Soil Mech.* **2017**, *38*, 337–348.
24. Geng, P.; Yue, H.E.; Chuan, H.E.; Quan, Q.; Yan, Q. Research on reasonable aseismic fortified length for tunnel through fault fracture zone. *Chin. J. Rock Mech. Eng.* **2014**, *33*, 358–365.
25. Gu, Y.; Li, W.; Li, H. Experimental study on the relative distance between a tunnel and a fault and subsequent effects on the stability of the surrounding rock. *Mod. Tunn. Technol.* **2013**, *50*, 59–65.
26. Geng, P.; Chuan, W.U.; Tang, J.; Lin, L.I. Analysis of dynamic response properties for tunnel through fault fracture zone. *Chin. J. Rock Mech. Eng.* **2012**, *31*, 1406–1413.
27. Jiang, J.P.; Zhang, Y.S. Study on the influence of dip angles of normal fault on tunnel surrounding rocks in the fault footwall. *Caikuang Yu Anquan Gongcheng Xuebao/J. Min. Saf. Eng.* **2011**, *28*, 596–601.
28. Jiao, P.; Lai, H. Theoretical analysis on the influence of different dip angle reverse faults' dislocation on tunnel structure. *China Civ. Eng. J.* **2019**, *52*, 106–117.
29. Dalgic, S. Tunneling in fault zones, Tuzla tunnel, Turkey. *Tunn. Undergr. Space Technol.* **2003**, *18*, 453–465. [CrossRef]

30. Mohammadi, H.; Azad, A. Prediction of ground settlement and the corresponding risk induced by tunneling: An application of rock engineering system paradigm. *Tunn. Undergr. Space Technol.* **2021**, *110*, 103828. [CrossRef]
31. Han, Z.Y.; Li, D.Y.; Zhou, T.; Zhu, Q.Q.; Ranjith, P.G. Experimental study of stress wave propagation and energy characteristics across rock specimens containing cemented mortar joint with various thicknesses. *Int. J. Rock Mech. Min. Sci.* **2020**, *131*, 104352. [CrossRef]
32. Xiao, P.; Li, D.Y.; Zhao, G.Y.; Liu, M. Experimental and Numerical Analysis of Mode I Fracture Process of Rock by Semi-Circular Bend Specimen. *Mathematics* **2021**, *9*, 1769. [CrossRef]
33. Xiao, P.; Li, D.Y.; Zhu, Q.Q. Strain Energy Release and Deep Rock Failure Due to Excavation in Pre-Stressed Rock. *Minerals* **2022**, *12*, 488. [CrossRef]
34. Li, D.Y.; Zhang, C.X.; Zhu, Q.Q.; Ma, J.Y.; Gao, F.H. Deformation and fracture behavior of granite by the short core in compression method with 3D digital image correlation. *Fatigue Fract. Eng. Mater. Struct.* **2022**, *45*, 425–440. [CrossRef]
35. Li, D.Y.; Ma, J.Y.; Wan, Q.R.; Zhu, Q.Q.; Han, Z.Y. Effect of thermal treatment on the fracture toughness and subcritical crack growth of granite in double-torsion test. *Eng. Fract. Mech.* **2021**, *253*, 107903. [CrossRef]
36. Kun, D.U.; Li, X.B.; Li, D.Y.; Weng, L. Failure properties of rocks in true triaxial unloading compressive test. *Trans. Nonferrous Met. Soc. China* **2015**, *25*, 571–581.
37. Li, X.B.; Du, K.; Li, D.Y. True Triaxial Strength and Failure Modes of Cubic Rock Specimens with Unloading the Minor Principal Stress. *Rock Mech. Rock Eng.* **2015**, *48*, 2185–2196. [CrossRef]
38. Chen, J.; Feng, X. True triaxial experimental study on rock with high geostress. *Yanshilixue Yu Gongcheng Xuebao/Chin. J. Rock Mech. Eng.* **2006**, *25*, 1537–1543.
39. Han, Z.; Li, D.; Li, X. Dynamic mechanical properties and wave propagation of composite rock-mortar specimens based on SHPB tests. *Int. J. Min. Sci. Technol.* **2022**, *32*, 793–806. [CrossRef]
40. Gong, F.; Luo, Y.; Si, X.; Li, X. Experimental modelling on rockburst in deep hard rock circular tunnels. *Chin. J. Rock Mech. Eng.* **2017**, *36*, 1634–1648.
41. Han, Z.; Li, D.; Li, X. Effects of axial pre-force and loading rate on Mode I fracture behavior of granite. *Int. J. Rock Mech. Min. Sci.* **2022**, *157*, 105172. [CrossRef]

minerals

MDPI

Article

Effects of Freeze–Thaw Cycling on Dynamic Compressive Strength and Energy Dissipation of Sandstone

Bangbiao Wu [1], Ling Yang [1], Yong Mei [2,*], Yunhou Sun [2], Jinming Liu [2] and Jun Shen [2]

1 State Key Laboratory of Hydraulic Engineering Simulation and Safety, School of Civil Engineering, Tianjin University, Tianjin 300072, China
2 Institute of Defense Engineering, AMS, Beijing 100036, China
* Correspondence: meiyong1990@foxmail.com

Abstract: In this study, the dynamic compressive strength and dissipated energy of rocks were investigated under hydrostatic pressure after different freeze–thaw cycles (FTCs). A split Hopkinson pressure bar (SHPB) with a lateral confining pressure chamber was used for the dynamic testing of sandstone specimens. The results indicated that under a certain loading rate and hydrostatic pressure, both the dissipative energy and dynamic compression strength of rocks decreased with the increase in the number of FTCs. The hydrostatic pressure significantly increased the dissipated energy and dynamic compression strength of rocks, and the enhancement became more pronounced as the number of FTCs increased. By analyzing the energy and damage characteristics of the specimens, after 40 FTCs, the internal damage of the specimens became critical and desensitized the dissipated energy to the loading rate. Based on the testing results, an empirical function was proposed to describe how the dissipated energy was related to the number of FTCs, hydrostatic pressure, and loading rate. It was demonstrated that the dissipated energy had a negative linear relation with FTC, which was enhanced by confinement. The dissipated energy of the specimen corresponded to the dynamic compression strength according to a quadratic function.

Keywords: freeze–thaw cycling; hydrostatic pressure; loading rate; dissipated energy; dynamic compressive strength

Citation: Wu, B.; Yang, L.; Mei, Y.; Sun, Y.; Liu, J.; Shen, J. Effects of Freeze–Thaw Cycling on Dynamic Compressive Strength and Energy Dissipation of Sandstone. *Minerals* **2022**, *12*, 1331. https://doi.org/10.3390/min12101331

Academic Editor: Gianvito Scaringi

Received: 22 September 2022
Accepted: 18 October 2022
Published: 20 October 2022

1. Introduction

Freeze–thaw cycling (FTC) refers to the alternating phenomenon of the freezing and thawing of water contained in the surface and interior of a structural member and is the result of the phase transition between water and ice in the material caused by temperature change [1]. Studies indicated that freeze–thaw weathering has a significant influence on rock degradation and rock engineering stability in cold regions [2–6]. FTC weathering leads to the initiation of internal cracks in the rock, causing the reduction in rock strength, and eventually leads to the damage of the rock mass [2,3,7].

The effect of FTC on the petrophysical properties of rocks has been extensively studied. It was indicated that the number of FTCs had a certain influence on the change in rock porosity and that the P-wave velocity also showed a decreasing trend with the increase in the FTC number [8–10]. Park et al. [11] studied the petrophysical and mechanical properties of granite and sandstone under the effect of FTC and indicated that the specific heat capacity and Poisson's ratio of rock tended to decrease with the decrease in the freezing temperature. Thomachot and Matsuoka [12] found that the water content and the number of FTCs were the most important in the dilation of the materials. Tan et al. [13] obtained the effect of FTC on the mechanical properties of granite, including strength, deformation characteristics, elastic modulus, cohesive strength, and internal friction angle. Some studies indicated that the rock wave velocity decreased with the increase in the FTC number and that the decreasing trend was related to lithology [14,15].

Under the effect of FTC, the physical properties of rocks are weakened, so the mechanical properties of rocks also change greatly. Numerous studies showed that water content, loading rate, temperature, and the number of FTCs had important effects on the uniaxial compressive strength, P-wave velocity, and fracture of rocks [10,16–21]. Chen et al. [22] indicated that different initial water contents had an effect on the uniaxial compressive strength of rock after FTC. When the initial water content was above 70%, the compressive strength of the rock decreased sharply with the increase in the water content. Bayram [23] developed a statistical model to predict the degradation of uniaxial compression strength using intact tests of impact strength, modulus of elasticity, and water absorption. Amin Jamshidi [24] indicated a new petro-mechanical predictor parameter to predict the strength of granitic stones after FTC. Ren et al. [25] indicated that the damage caused by FTC was mainly concentrated on the side and edge area of rock specimens.

The above-mentioned studies mainly focused on the effect of FTC on the mechanical properties of rocks under static loading conditions. In practical engineering, rock masses are subjected to dynamic loads such as explosions and earthquakes. In recent years, the split Hopkinson pressure bar (SHPB) was widely used in the study of rock dynamic mechanics. Xia and Yao [26] gave a detailed introduction to the history of SHPB, working principle, wave shaping technology, and its application in rock dynamic tests. Some scholars used the SHPB system to systematically study various mechanical properties of rock materials, such as dynamic tensile properties, compressive properties, and fracture properties [26,27]. Numerous studies showed that the dynamic compression strength, dynamic tensile strength, and elastic modulus of rock decreased with the increase in the FTC number [28–32].

Material damage is a state instability phenomenon driven by energy, and energy dissipation is the fundamental cause of rock deformation and damage. Studying the energy dissipation laws of rocks in cold regions under different working conditions can effectively reveal the dynamic failure mechanism of rocks. Xie et al. [33] indicated that rock damage was associated with energy dissipation and release; therefore, the dissipated energy is often used to evaluate damage accumulation in rock materials [34]. Wang et al. [35] analyzed the stress–strain curve and energy evolution curve of the specimen, as well as the dynamic mechanical properties and energy index. A precise FTC rock damage model was established by Feng et al. [36]. Lundberg [37] performed dynamic compression tests on granite and limestone to evaluate the energy absorption of the specimens during crushing and indicated that the absorbed energy increased significantly when the load exceeded a certain level. Zhou et al. [32] established an increasing model of the relationship among porosity, absorbed energy, and the number of FTCs. Li et al. [38] applied dynamic loads to granite and measured the energy absorption; the results indicated that the size distribution of the fragments was determined by the energy absorption of the specimens. Deng et al. [39] further analyzed the effect of FTC on sandstone damage from the perspective of energy and found that the energy required to damage the specimen under dynamic impact load increased with the increase in the FTC number.

Recently, numerous studies were carried out on the effect of dynamic rock properties under FTC, but few researchers focused on the effect of hydrostatic confining pressure. However, in cold regions, deeply buried rock structures are also affected by FTC. Some studies showed that the physical and mechanical properties of rocks under the effect of hydrostatic pressure were quite different from those without hydrostatic pressure [40,41]. It is of great significance to study the dynamic mechanical properties and the energy dissipation of rock subjected to hydrostatic pressure.

In order to fill the gap, in this study, the dynamic compressive properties and energy dissipation of green sandstone specimens under the combined effects of freeze–thaw cycling, hydrostatic pressure, and loading rate were studied.

2. Specimen Preparation

2.1. Specimen Configuration and Basic Properties

Sedimentary rock is widely distributed on the surface of the Earth and is an important area for human activities such as resources and energy development, and underground engineering construction. In the current study, the rock specimens were made of green sandstone from Dazhou, Sichuan of China. The mineral composition of this rock was measured [42]; as shown in Figure 1, quartz and plagioclase were the main components of green sandstone. With reference to the dynamic test requirements of International Society for Rock Mechanics (ISRM), the specimens were made into discs with a nominal diameter of 50 mm and a thickness of 25 mm. The specimen end surfaces were polished to straightness of ±0.02 mm and a roughness of 0.5% over the full thickness [43], as shown in Figure 2. The mechanical and physical properties of green sandstone were measured in the previous study [44]. The density of green sandstone was 2.44 g/cm^3; the P-wave velocity and the static compressive strength were 3674 m/s and 62.97 MPa, respectively. The elastic modulus and Poisson's ratio were 7.83 GPa and 0.22, respectively.

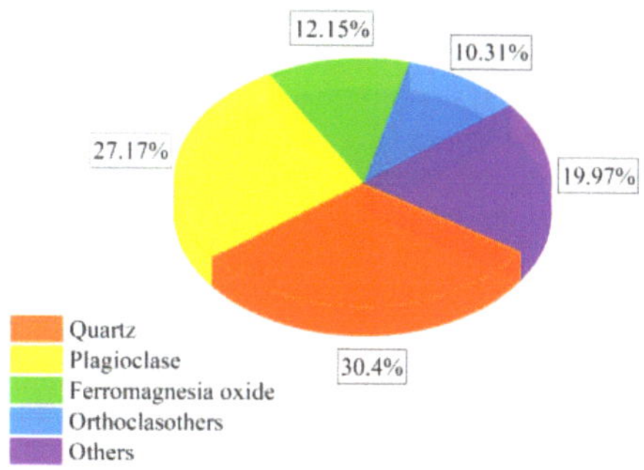

Figure 1. Schematic diagram of the mineral composition of green sandstone.

Figure 2. Photograph of disc specimens.

2.2. Freeze–Thaw Cycling Weathering

The green sandstone specimens were divided into six groups and prepared with 0, 10, 20, 30, 40, and 50 FTCs, respectively. At first, all specimens were dried for 48 h until the mass no longer changed and then cooled to room temperature naturally. All specimens were immersed in distilled water in a vacuum container for 48 h, and the surface of the water was at least 20 mm higher than the specimens. According to Chinese Standard for test methods of engineering rock mass (GB/T 50266-2013), the temperature range of the FTC process was set at −20–20 °C. The temperature control curve of the FTC process is shown in Figure 3. The fully saturated specimens were placed in the freeze–thaw box for 4 h under the condition of −20 °C; then, the frozen specimens were immersed in warm distilled water at 20 °C for 4 h. Altogether, this was one freeze–thaw cycle.

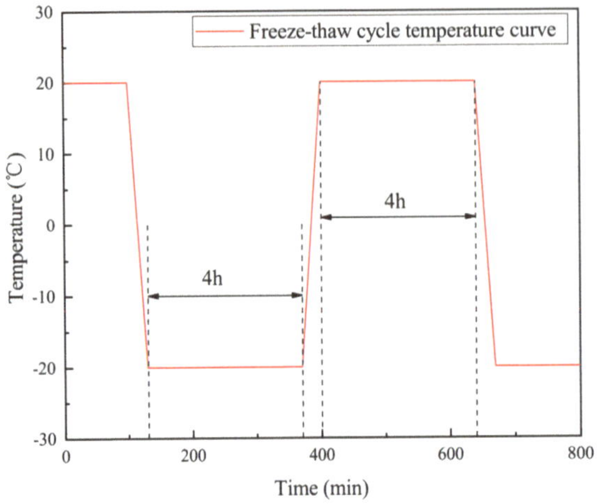

Figure 3. Temperature control curve of the freeze–thaw cycling process.

3. Experimental Techniques

3.1. Loading System

A split Hopkinson pressure bar (SHPB) with a lateral confining pressure chamber and an axial pressure cylinder was used for the dynamic testing of green sandstone specimens. The material of the bar was martensitic steel with Young's modulus of 211 GPa and a *P*-wave velocity of 5270 m/s. As the schematic diagram shows in Figure 4, the SHPB system could be divided into four parts: the bullet (impact bar), the incident bar, the transmission bar, and the absorption bar. The lengths of the bullet, incident bar, transmitted bar, and absorption bar were 0.35 m, 3.0 m, 1.8 m, and 0.5 m, respectively [44].

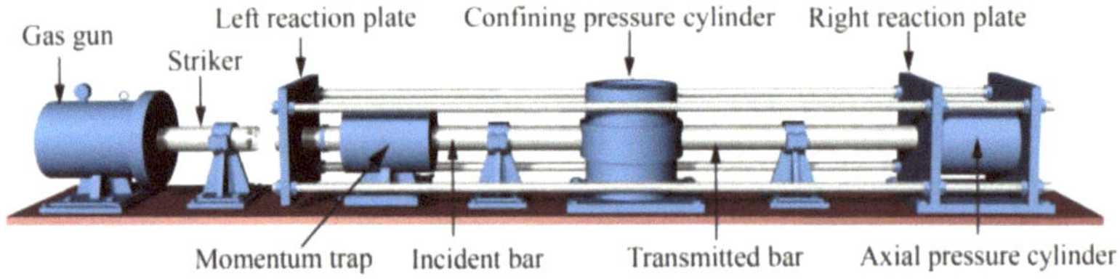

Figure 4. Schematic diagram of SHPB.

The confining pressure device could provide axial pressure and hydrostatic confining pressure, and the maximum confining pressure that the device could provide was 60 MPa. The working principle of the system was detailed in earlier studies. The strain gauges were attached on the surfaces of the incident bar and transmitted bar to measure the incident strain (ε_I), the reflected strain (ε_R), and the transmitted strain (ε_T).

According to the dynamic mechanic test method for rock proposed by ISRM, an effective test should satisfy the dynamic stress balance conditions:

$$\sigma_1 = \sigma_2 \tag{1}$$

where σ_1 and σ_2 are the loading stresses on the two ends of the rock specimen. In other words, all parts inside the specimen should satisfy the following formula:

$$\varepsilon_I(t) + \varepsilon_R(t) = \varepsilon_T(t) \tag{2}$$

where $\varepsilon_I(t)$ is the incidence strain, $\varepsilon_R(t)$ is the reflected strain, and $\varepsilon_T(t)$ is the transmitted strain. Moreover, the typical dynamic stress balance curve is shown in Figure 5.

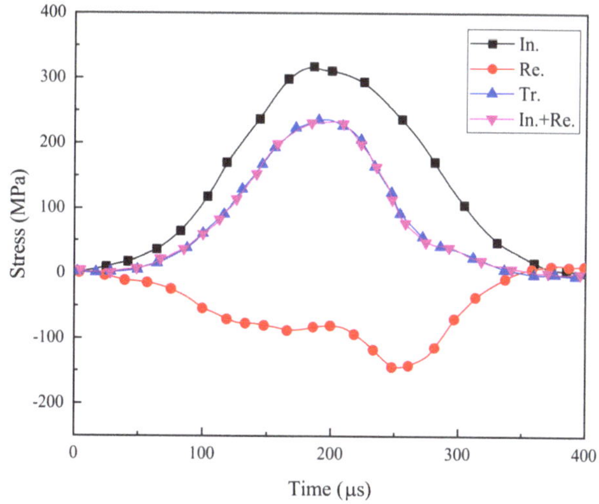

Figure 5. Stress balance verification.

On this basis, the dynamic compressive strength and strain of the specimens could be calculated using the three-wave method, and the calculation formula is as follows:

$$\sigma_s = \frac{P_1 + P_2}{2A_s} = \frac{AE}{2A_s}(\varepsilon_I + \varepsilon_R + \varepsilon_T) = \frac{AE}{A_s}\varepsilon_T \tag{3}$$

$$\varepsilon = \frac{C_0}{L_0}\int_0^t (\varepsilon_I - \varepsilon_R - \varepsilon_T)dt = -\frac{2C_0}{L_0}\int_0^t \varepsilon_R dt \tag{4}$$

$$\dot{\varepsilon} = \frac{C_0}{L_0}(\varepsilon_I - \varepsilon_R - \varepsilon_T) = -\frac{2C_0}{L_0}\varepsilon_R \tag{5}$$

where σ_s is the stress of the material, A_s is the cross-sectional area of the specimen, and L_0 is the length of the specimen. Along with the above equations, the stress–strain curve of the specimen material at the strain rate of $\dot{\varepsilon}$ could be obtained.

Figure 6 shows the relationship between the dynamic compressive stress and loading time. The linear slope of the fitting in the rising phase was defined as the loading rate. Then, the variation law of the compressive strength of the specimens with the loading rate under the effects of different freeze–thaw cycles and confining pressure could be obtained.

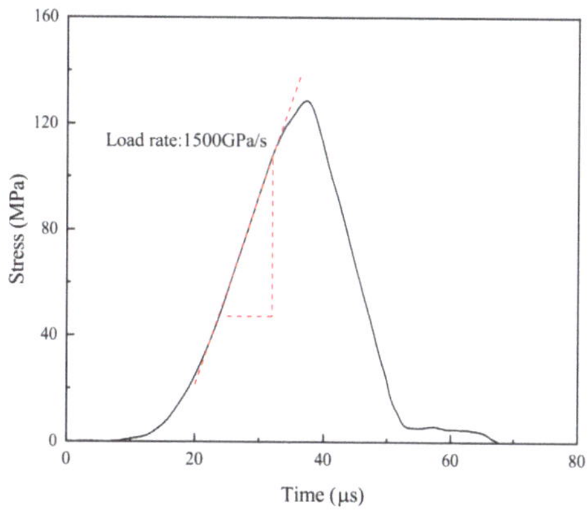

Figure 6. Calculation principle of loading rate.

In this study, three groups of different hydrostatic pressures (0 MPa, 2 MPa, and 4 MPa) were set. Among them, 0 MPa was the control test. Under each hydrostatic pressure, six groups of freeze–thaw cycle conditions (0 FTC, 10 FTC, 20 FTC, 30 FTC, 40 FTC, and 50 FTC) were set. Under each working condition, at least five tests with different loading rates were performed.

3.2. Single-Loading Technology

For a traditional split Hopkinson pressure bar, the reflected tensile wave can be reflected again at the impinging end of the incident bar to form a compression wave, causing the secondary loading or even multiple loading of the specimen, which is not conducive to the energy analysis of a specimen. To solve this problem, a single-loading technique was used in the current study, as shown in Figure 7. By setting the length of the prepared slit between the flange and the mass block, the prepared slit was closed when the reflected tension wave reached the impacting end of the incident bar. Due to the limitation of the mass block, the wave impedance at this end was much greater than the wave impedance of the incident bar, and the reflected tension wave remained a tension wave after reflection at this end [45]. The wave did not load the specimen in secondary compression or even multiple times, so the single loading of the specimen was achieved. The calculation of the prepared slit is shown below [46]:

$$S = \int_0^{T_0} \varepsilon_i c_0 dt \tag{6}$$

where ε_i is the strain in the incident bar and the integration region, T_0, is the loading pulse width.

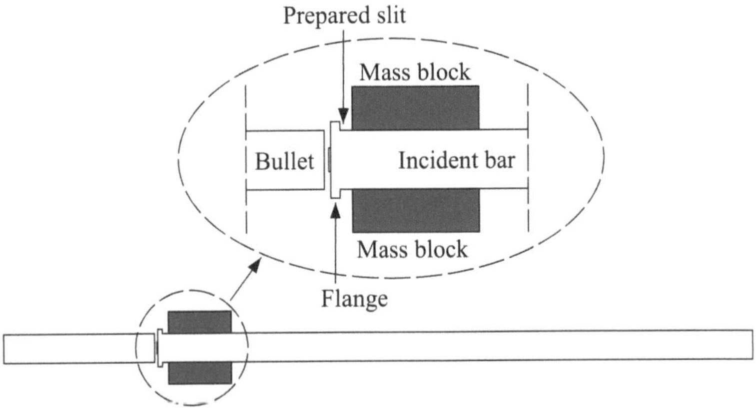

Figure 7. Schematic diagram of single-loading technology.

4. Results and Discussions

4.1. Stress–Strain Curve

Based on the information of the incident wave, reflected wave, and transmitted wave collected during the test, the stress–strain curve of each green sandstone specimen subjected to dynamic shock compression failure was obtained with the three-wave method. Figure 8 shows the dynamic compressive stress–strain curves of the specimens under the hydrostatic pressures of 0 MPa and 2 MPa, respectively, before FTC treatment. Under the same working conditions, the peak point of the stress–strain curve increased with the increase in the loading rate.

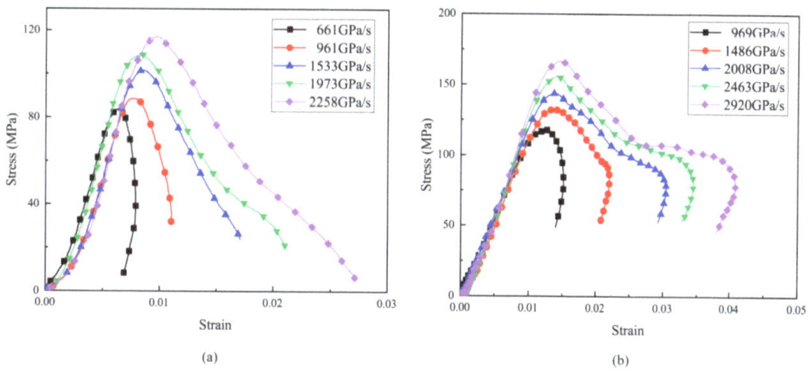

Figure 8. Dynamic compressive stress–strain curves of the specimen under different working conditions: (a) 0 FTC and 0 MPa; (b) 0 FTC and 2 MPa.

The stress–strain curves could be roughly divided into four stages: (I) compression of fissures, (II) linear deformation of rock, (III) yield of rock, and (IV) failure of rock. Typical schematic diagrams of different macroscopic failure stages of the specimens are also shown in Figure 9. The compression of fissures was the initial stage of the specimen loading process. Judging from the microstructural characteristics of the material, the green sandstone specimen had certain primary disconnected micro-cracks, and these micro-cracks were closed under pressure in this stage. In the stage of the linear deformation of rock, the stress–strain curve rose approximately linearly. The slope of the curve was defined as the elastic modulus of the specimen under this working condition. In the stage of the the yield of rock, with the increase in the stress, the slope of the curve decreased, and the local failure inside the specimen gradually developed; then, the crack ruptured through, which led to the generation of new cracks and the expansion of the original cracks.

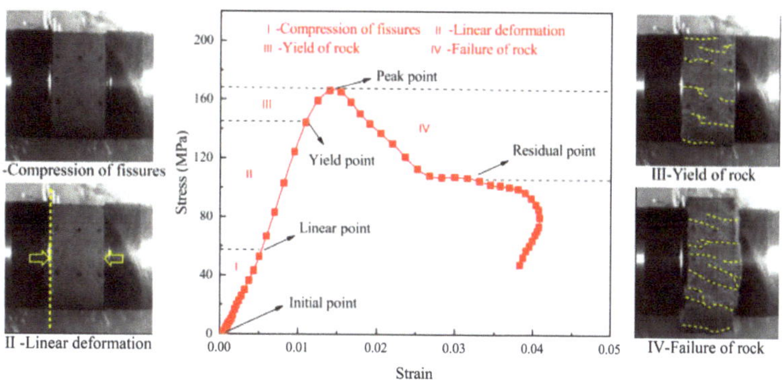

Figure 9. Schematic diagram of stress–strain curve and schematic diagrams of different macroscopic failure stages of the specimens.

After the specimens reached the peak point, the stress decreased rapidly. At this time, the stress–strain curve was closely related to the loading rate. As is shown in Figure 8a,b, the green sandstone specimens under confining pressure had an obvious strain-softening trend until reaching the residual strength point after the failure stage; then, they showed a significant rebound.

Under the condition of no confining pressure, the strain continued to increase after the specimens entered the failure stage, and there was no rebound stage, which indicated that the specimen was completely broken. When the loading rate was low (661 GPa/s in Figure 8a, 969 GPa/s in Figure 8b), the specimens remained intact after being impacted, and

the stress–strain curve showed a rebound trend immediately after the peak point. When the loading rate was high, the stress–strain curve continued to decrease with the increase in the strain after the peak point. At this time, the specimens still had a certain residual strength. Under the action of a large loading rate, the reduction rate of residual strength after the peaks decreased with the increase in the loading rate and appeared as a relatively gentle stage; then, the curve rebounded. Under the effect of hydrostatic pressure, the higher the loading rate was, the more obvious this stage was (2463 GPa/s and 2920 GPa/s in Figure 8b).

4.2. Dynamic Compressive Strength

Dynamic compressive strength is an important parameter to measure the mechanical properties of rock. Figure 10 shows the variation in dynamic compressive strength with the loading rate in green sandstone specimens under different hydrostatic pressure and FTCs. It could be seen that under the condition of a certain hydrostatic pressure, the dynamic compressive strength of the specimens increased approximately linearly with the loading rate. Under the condition of a constant loading rate, the dynamic compressive strength decreased with the increase in FTCs. The reason was that under the influence of FTC, the defects of the specimens continued to accumulate, and the impact resistance decreased.

Figure 11 shows the variation in dynamic compressive strength with the number of FTCs for two loading rates (1000 GPa/s and 2000 GPa/s) and different hydrostatic pressures. Under the conditions of the same loading rate and FTCs, the dynamic compressive strength of the specimen under the effect of hydrostatic pressure was significantly higher than that of the specimen without hydrostatic pressure, and the greater the hydrostatic pressure was, the greater the strength was. Under the conditions of 0 FTC and 1000 GPa/s, the strength of the specimen under 2 MPa hydrostatic pressure increased by 39.81% compared with that under 0 MP hydrostatic pressure, and the growth rate increased with the increase in FTCs. After 50 FTCs, the growth rate reached 70.47%. Under the conditions of 0 FTC and 2000 GPa/s, the strength of the specimen under 2 MPa hydrostatic pressure increased by 27.66% compared with that under 0 MPa hydrostatic pressure; after 50 FTCs, the growth rate reached 62.89%. The same trend could be found from the 2 MPa curve to the 4 MPa curve. It showed that the effect of hydrostatic pressure could significantly improve the dynamic compressive strength of the specimens, and the improvement effect was more obvious with more FTCs.

A linear fit analysis was performed on the data points, and the fitting parameters and coefficient of determination are shown in Table 1:

$$\sigma = kN + b \tag{7}$$

where N is the number of FTCs, σ is the dynamic compressive strength, and k and b are the fitting parameters. The slope, k, represents the rate of change in dynamic compression strength with N, and b represents the dynamic compression strength when the number of FTCs is 0.

The dynamic compression strength decreased with the increase in FTCs, and the rate of reduction decreased with the increase in hydrostatic pressure, which indicated that the hydrostatic pressure constraint could reduce the weathering damage to the specimen caused by FTC.

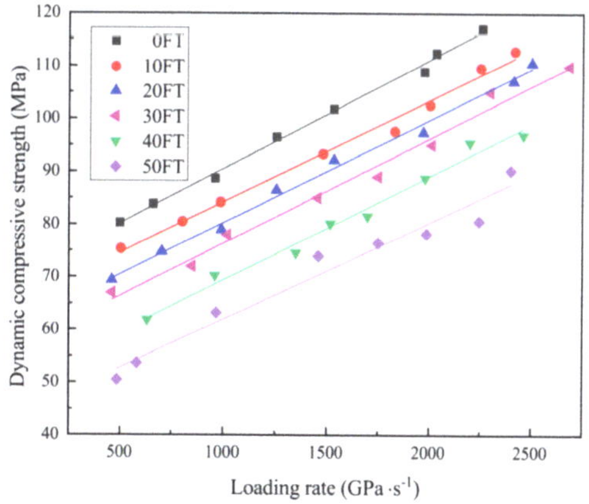

(a) 0MPa

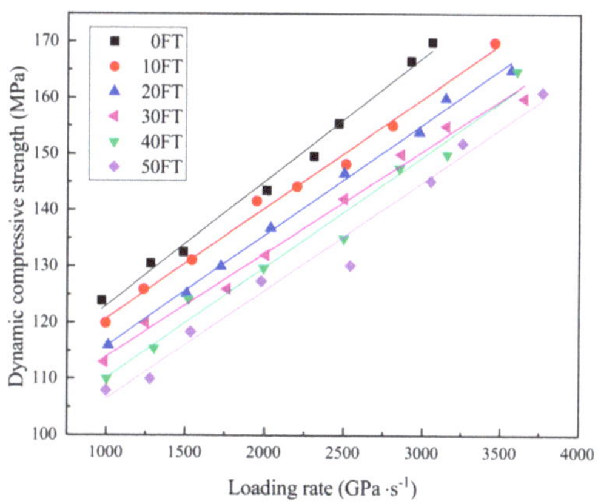

(b) 2MPa

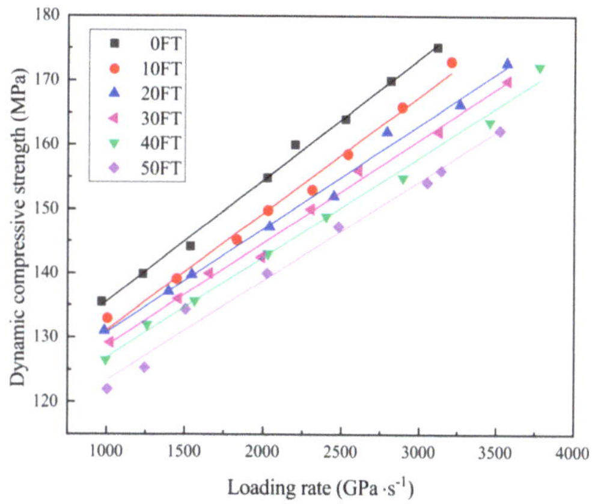

(c) 4MPa

Figure 10. Dynamic compressive strength of FT cycling-treated specimens subjected to different hydrostatic pressures: (**a**) 0 MPa; (**b**) 2 MPa; and (**c**) 4 MPa.

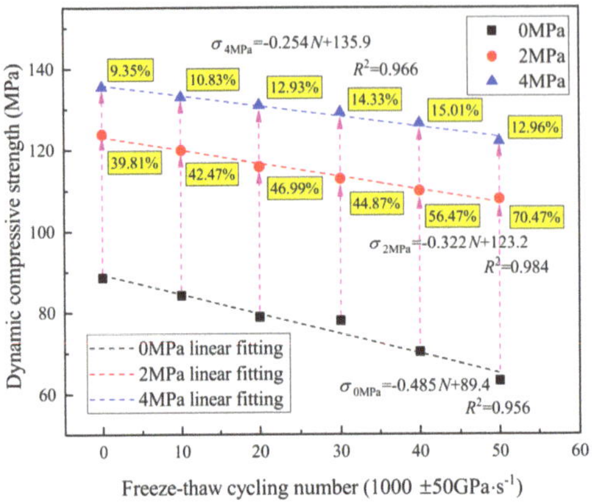

(a)

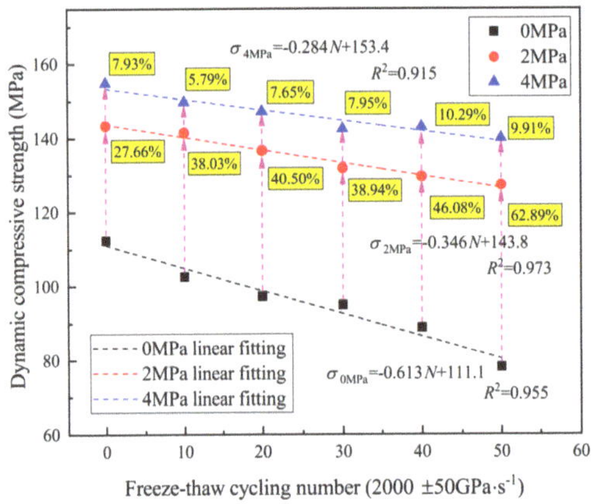

(b)

Figure 11. Dynamic compressive strength varied with freeze–thaw cycling under a certain loading rate (1000 GPa/s and 2000 GPa/s) and different hydrostatic pressures.

Table 1. Fitting parameters (k, b) and coefficient of determination (R^2).

Loading Rate (GP/s)	Hydrostatic Pressure (MPa)	k	b	R^2
1000 ± 50	0	−0.485	89.4	0.956
	2	−0.322	123.2	0.984
	4	−0.254	135.9	0.966
2000 ± 50	0	−0.613	111.1	0.955
	2	−0.346	143.8	0.973
	4	−0.284	153.4	0.915

4.3. Dissipated Energy

In dynamic experiments, the entire experimental system satisfies the law of conservation of energy. The energy in the whole SHPB system is divided into four parts: incident energy (U_I), transmission energy (U_T), reflection energy (U_R), and energy input on the specimen, which is also called absorption energy of specimen (U_S):

$$U_I(t) = \frac{A_0C_0}{E}\int \sigma_I^2(t)dt = A_0C_0E\int \varepsilon_I^2(t)dt \tag{8}$$

$$U_R(t) = \frac{A_0C_0}{E}\int \sigma_R^2(t)dt = A_0C_0E\int \varepsilon_R^2(t)dt \quad (9)$$

$$U_T(t) = \frac{A_0C_0}{E}\int \sigma_T^2(t)dt = A_0C_0E\int \varepsilon_T^2(t)dt \quad (10)$$

where E, C_0, and A_0 represent the elastic modulus, longitudinal wave velocity, and cross-sectional area of the incident bar (transmission bar), respectively. According to the law of conservation of energy, the energy (U_S) absorbed by the specimen is:

$$U_S = U_I - (U_R + U_T) \quad (11)$$

Previous studies indicated that the dissipated energy of the specimen accounts for about 95% of the energy input to the entire specimen [47]. Therefore, the energy absorbed by the specimen was used to approximate the dissipated energy (U_d):

$$U_d = U_S \quad (12)$$

Figure 12 shows the variation in the dissipated energy of the specimen with the loading rate under different hydrostatic pressures and FTCs. It could be seen that under the same working conditions, the dissipated energy of the specimen showed an approximately linear increasing trend with the loading rate. When the number of FTCs was certain, the internal damage of the specimen could be considered to be approximately the same, and the increase in the loading rate made the specimen more severely damaged, so the energy required was greater.

An empirical formula was proposed to fit the dissipated energy of rock related to the loading rate, the number of freeze–thaw cycles, and the confining pressure:

$$U_c^d = U_0^d\left[1-\alpha\frac{N}{N_0}\right]\left[1+\beta\ln\left(1+\frac{P}{P_0}\right)\right]\left[\left(\frac{\dot{\sigma}}{\dot{\sigma}_0}\right)^{\mu}\right] \quad (13)$$

where $\dot{\sigma}$ is the loading rate; N is the number of freeze–thaw cycles; P is the hydrostatic confining pressure; U_c^d is the dissipated energy of the specimen when the strain rate is $\dot{\sigma}$; P_0 and N_0 are the reference hydrostatic pressure and reference freeze–thaw cycles, where the values are both 1; and U_0^d is the dissipated energy of the specimen when the strain rate is $\dot{\sigma}_0$. In this study, the value of the dissipated energy of the specimen under the loading rate of 500 GPa/s without hydrostatic pressure and freeze–thaw cycles was taken as U_0^d. α, β, and μ are the fitting parameters related to the number of freeze–thaw cycles, hydrostatic pressure, and loading rate, respectively. Table 2 shows the fitting parameters (α, β, μ) and coefficient of determination (R^2). It could be seen that the coefficients of determination were all greater than 90%, so the fitting was reasonable, and the fitting curve is shown in Figure 12.

It should be noted here that when the hydrostatic pressure was 0 MPa, the change rate of dissipated energy under the conditions of 40 FTCs and 50 FTCs was obviously lower than that for other FTC numbers. It indicated that the internal damage of the specimens reached a large value after 40 FTCs, which desensitized the energy required for specimens to break. It could be analyzed from Table 2 that when the hydrostatic pressure was 0 MPa, parameter μ of the loading rate corresponding to 40 FTCs and 50 FTCs (0.835 and 0.922) was smaller than the correlation coefficient for 0 FTC (1.006). However, when the hydrostatic pressure was 2 MPa and 4 MPa, this pattern did not manifest. The reason was that the effect of hydrostatic pressure could offset a part of the damage to the specimens, so that the effect of loading rate on the dissipation energy of the specimen did not decrease. Under the effect of hydrostatic pressure, as the number of FTCs increased, the fitting parameter (β) of hydrostatic pressure decreased, indicating that FT weathering reduced the effect of hydrostatic pressure on dissipated energy.

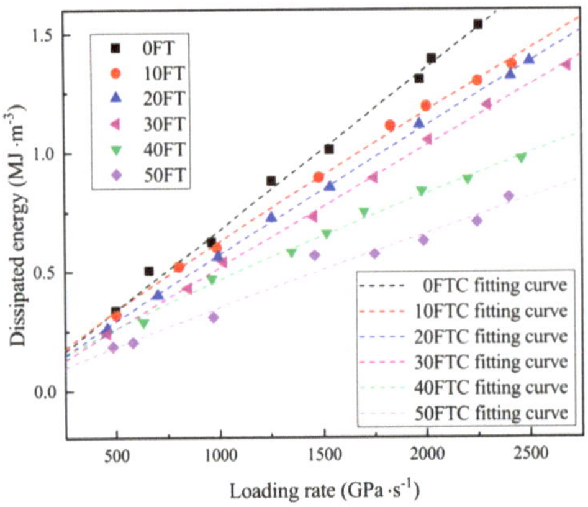

(a) 0MPa

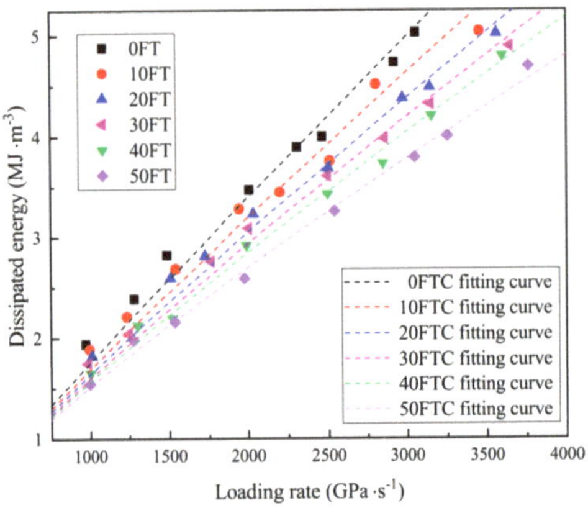

(b) 2MPa

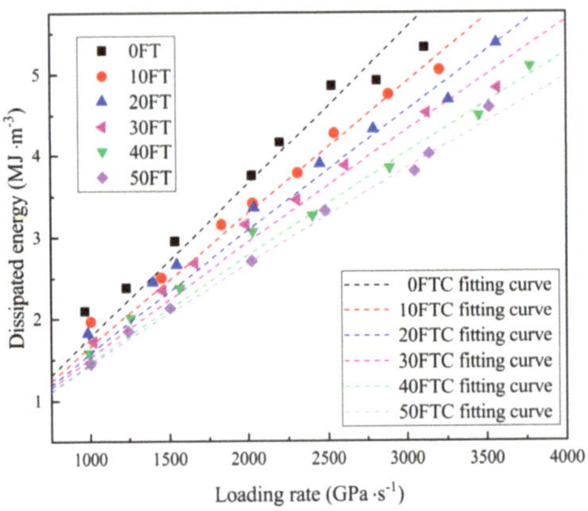

(c) 4MPa

Figure 12. Dissipated energy of specimen and fitting curves of FT cycling-treated specimen subjected to different hydrostatic pressures: (**a**) 0 MPa; (**b**) 2 MPa; and (**c**) 4 MPa.

Table 2. Fitting parameters (α, β, and μ) and coefficient of determination (R^2).

Confining Pressure (MPa)	FT Cycles	α	β	μ	R^2
0	0	0.00050	2.0	1.006	0.994
	10	0.00061	2.0	0.937	0.997
	20	0.00654	2.0	0.966	0.999
	30	0.00765	2.0	0.995	0.998
	40	0.00561	2.0	0.835	0.991
	50	0.00903	2.0	0.922	0.974
2	0	0.00170	1.561	0.950	0.949
	10	0.00169	1.553	0.920	0.956
	20	0.00162	1.560	0.896	0.983
	30	0.00153	1.552	0.881	0.989
	40	0.00141	1.561	0.865	0.992
	50	0.00124	1.556	0.832	0.994
4	0	0.00250	1.029	1.025	0.911
	10	0.00243	1.012	0.975	0.962
	20	0.00231	0.992	0.954	0.971
	30	0.00234	0.995	0.937	0.977
	40	0.00212	0.982	0.914	0.986
	50	0.00176	0.997	0.883	0.995

Figure 13 shows the variation law of the dissipated energy with the number of FTCs when the loading rate was about 1000 GPa/s and 2000 GPa/s. At a certain loading rate, the dissipated energy decreased with the increase in FTCs. As FTC progressed, the specimen was subjected to the frost heave effect of pore water, and the internal damage gradually increased, which was reflected in the increase in pores and the development of cracks. It could be seen from the experimental results that this damage was close to linear. Under the same conditions, the specimen needed to absorb less energy to break. A linear fit was performed on the data, and the fitting results are displayed in Table 3:

$$U^d = \eta N + \gamma \tag{14}$$

where N is the number of FTCs, U^d is the dissipated energy, and η and γ are the fitting parameters. The slope, η, represents the rate of change of dissipated energy with N, and γ represents the dissipated energy when the number of FTC is 0.

Under the two loading rates, with the increase in hydrostatic pressure, the absolute value of the correlation coefficient (η) of FTC increased, which indicated that the hydrostatic pressure enhanced the effect of FTC on the dissipated energy of the specimen. The dissipated energy of the specimens subjected to hydrostatic pressure was higher than that of the specimen not subjected to hydrostatic pressure under the same conditions. Hydrostatic pressure can enhance the dynamic mechanical properties of rock, limit the deformation of rock, and reduce the damage of rock caused by the load, so that more energy can be stored in the rock. Therefore, the rock specimens needed to absorb more energy to break. It is worth noting that under the conditions of 0 FTCs and 1000 GPa/s, the hydrostatic pressure was increased from 0 MPa to 2 MPa, and the dissipated energy increment was 210.7%. Under the condition of 50 FTCs, the dissipated energy increment was 400.1%. The same trend could also be seen under the hydrostatic pressure of 2000 GPa/s. The dissipated energy increment showed an increasing trend with the increase in FTCs. It showed that the effect of hydrostatic pressure could effectively improve the dissipation energy of the specimen, and the improvement effect was more obvious for the specimen subjected to more freeze–thaw cycles.

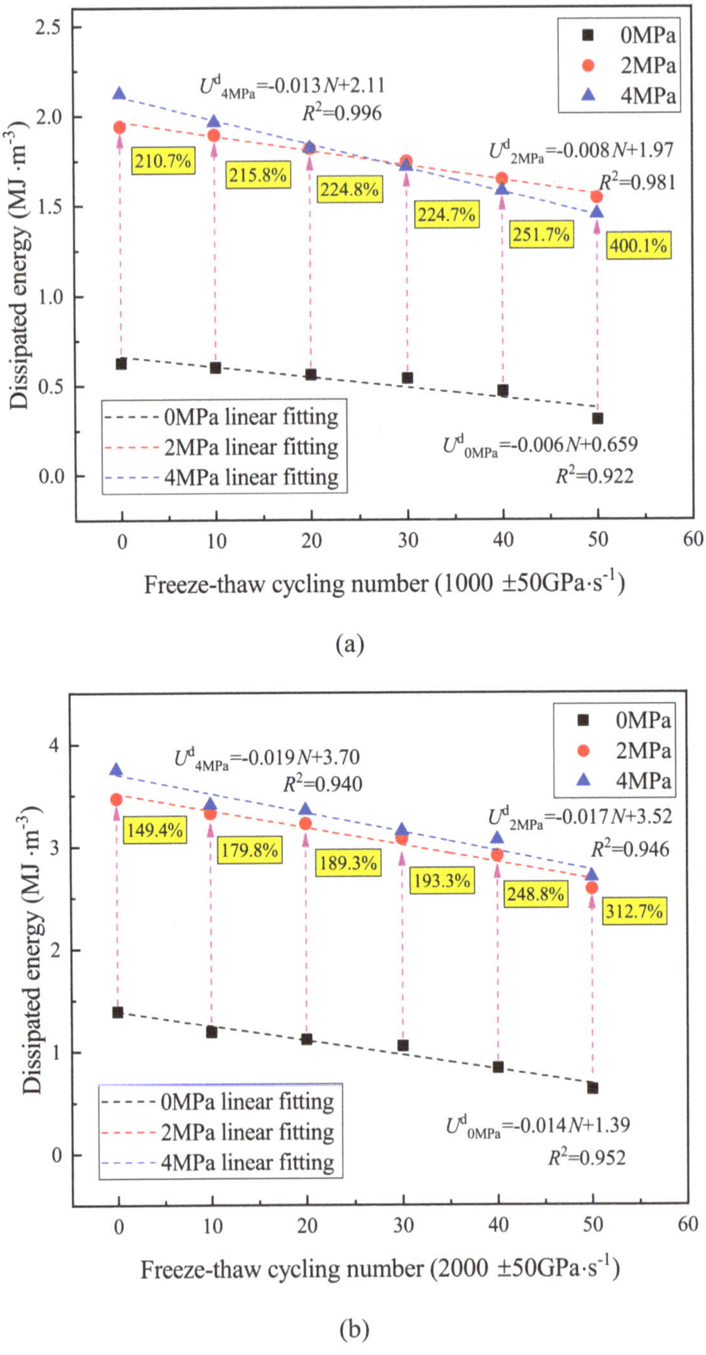

Figure 13. Dissipated energy varied with freeze–thaw cycling under a certain loading rate (1000 GPa/s and 2000 GPa/s) and different hydrostatic pressures.

Table 3. Fitting parameters (η, γ) and coefficient of determination (R^2).

Loading Rate (GP/s)	Hydrostatic Pressure (MPa)	η	γ	R^2
	0	−0.006	0.65	0.922
1000 ± 50	2	−0.008	1.97	0.981
	4	−0.013	2.11	0.996
	0	−0.014	1.39	0.955
2000 ± 50	2	−0.017	3.52	0.973
	4	−0.019	3.70	0.915

4.4. Dissipated Energy vs. Dynamic Compression Strength

Dynamic compression strength is an important parameter to describe the rock properties and is a key indicator of rock damage; rock fragmentation also directly effects the dissipation of energy. Therefore, the dynamic compression strength can be used to describe the energy dissipation of rock materials. As can be seen from Figure 14, the dissipated energy increased with the increase in dynamic compression strength and showed a certain trend. The following equation was obtained by fitting the data:

$$U^d = 3.192 \times 10^{-4}\sigma^2 - 0.0292\sigma + 0.852 \tag{15}$$

where U^d is the dissipated energy and σ is dynamic compression strength.

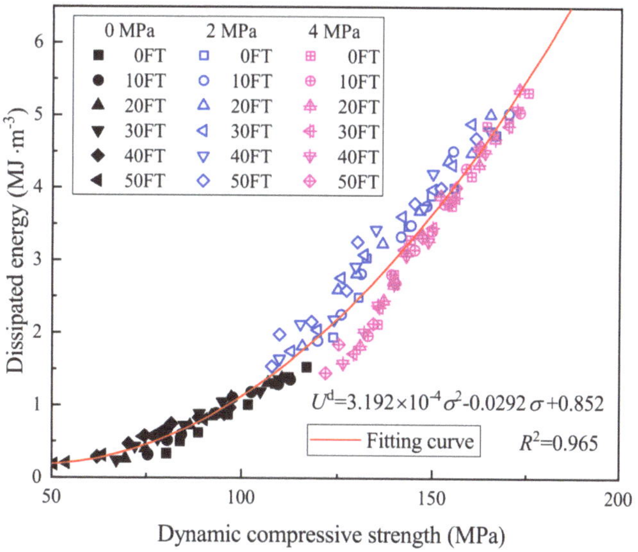

Figure 14. Dissipated energy vs. dynamic compression strength.

The dissipated energy varied as a quadratic function of the dynamic compression strength, and the trend was almost independent of the number of FTCs, hydrostatic pressure, and loading rate. During the destruction process, the rock needs to continuously absorb energy for the formation and development of internal cracks. When the absorbed energy reaches a certain value, the fracture throughs, and the rock is damaged, the energy dissipated; at this time, the rock reaches dynamic compression strength. For rock with certain internal damage and external force, the energy required to achieve damage was determined, so there was a correspondence between the energy and dynamic compression strength of rock.

4.5. Damage Characteristics Analysis

Studying the characteristics of a specimen after damage is important for the analysis of specimens subjected to FTC and external loading. For the specimens that were not subjected to hydrostatic pressure, the degree of damage was severe after loading, and the block sizes of the damaged specimens were well defined, so the average fragment size could be used as a relevant parameter to quantify the damage characteristics of the specimens. In the case of specimens subjected to hydrostatic pressure, the appearance of the damaged specimen remained intact due to the constraints of hydrostatic pressure, and the CT value could be used to quantify the internal damage of the specimens.

4.5.1. Average Fragment Size

A fragmentation analysis was carried out by sieving the fragmented rocks. The sieving device included nine sieves of 40 mm, 20 mm, 10 mm, 5 mm, 2 mm, 1 mm, 0.5 mm, 0.25 mm, and 0.075 mm. The average fragment size of the specimens after damage was indicated as δ, and the equation used is as follows:

$$\delta = \sum_{1}^{i} \delta_i = \sum_{1}^{i} \eta_i d_i \tag{16}$$

where i is the number of the sieve (i = 1, 2, 3, . . . ,10), η_i is the ratio of the mass of fragments to the initial mass of rock for each sieve, and d_i is the average value of the maximum and minimum particle sizes for each sieve.

The average fragment size reflected the size level of the specimen blocks after damage. A larger average fragment size indicated a larger proportion of large fragments of the specimen and a more complete specimen after damage. A smaller average fragment size indicated a larger proportion of small fragments of the specimen and a greater degree of damage to the specimen.

The calculated statistics for the average fragment size of specimens subjected to different FTCs and certain loading rates (1000 GPa/s and 2000 GPa/s) are shown in Figure 15. It could be seen that the average fragment size of specimens decreased with the increase in the number of FTCs under the same approximate loading rate conditions. After 40 FTCs, the average fragment size decreased rapidly, indicating that after 40 FTCs, the internal damage of the specimen increased dramatically, leading to a greater degree of fragmentation. As the loading rate increased, the average fragment size tended to decrease, which indicated that the sensitivity of the fragment size of sandstone specimens to the number of FTCs after damage was stronger at lower loading rates and relatively weaker at higher loading rates.

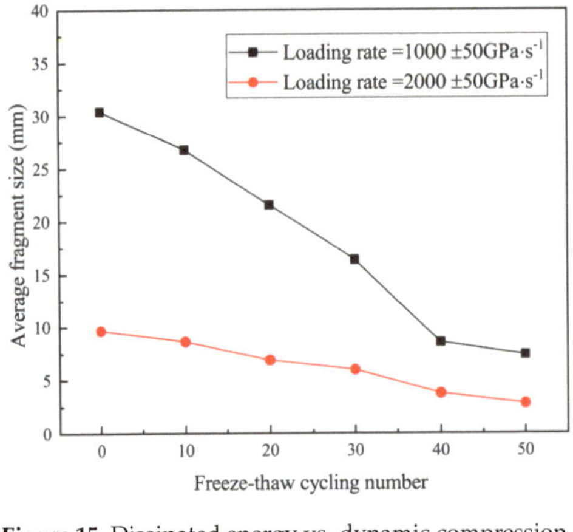

Figure 15. Dissipated energy vs. dynamic compression strength.

4.5.2. CT Value

CT (Computed Tomography) scanning is a technique that can realize the analysis of the internal construction of a specimen without damaging it. The principle is to reconstruct the internal construction according to the different transmission rates of different densities of matter to X-rays. In this paper, the CT value was used to characterize the relative value of the attenuation coefficient of the specimen to X-rays and was calculated as follows:

$$CT = \frac{\mu_c - \mu_w}{\mu_w} \times 1000 \quad (17)$$

where CT is the CT value of the specimen; μ_c is the attenuation coefficient of the specimen to X-rays, which can be measured using CT scanning; and μ_w is the attenuation coefficient of distilled water to X-rays.

The specimens that were not loaded were subjected to FTC and the CT values were measured as a control group. Figure 16 shows the CT value of the specimen subjected to different FTCs and the decay rate of the CT value vs. FTCs for the same approximate loading rate (1000 GPa/s). The CT value of the specimen decreased as the number of FTCs increased, indicating that the ability of the specimen to transmit X-rays was diminished. This is because the attenuation capacity of air for X-rays is lower than that of rock materials, and the more fractures are in the rock, the smaller the CT value is. With the increase in the number of FTCs under the effect of a similar loading rate, the volume of pores and fractures inside the rock specimen increased, and the degree of damage became more intense. It could be seen that the decay rate of the CT value was almost constant after 40 FTCs, indicating that the internal damage of the specimen reached a large value after 40 FTCs, which was consistent with the dissipated energy analysis reported above. The decay rate of the CT value for the specimens at 4 MPa was lower than that at 2 MPa, indicating that the effect of hydrostatic pressure could eliminate a part of the internal damage of the specimen and helped to improve the mechanical properties of the specimen.

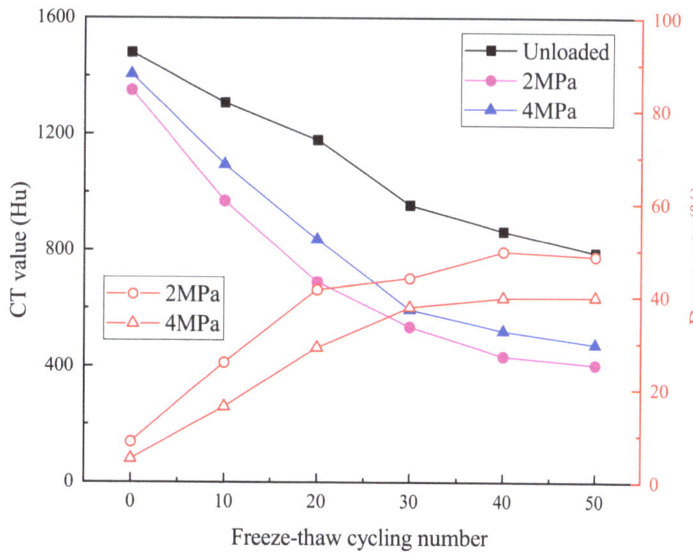

Figure 16. The CT value of specimen and decay rate of the CT value vs. different FTCs.

5. Conclusions

In this study, the dynamic compression strength and energy changes of saturated sandstone after FTC were systematically studied under different hydrostatic confining stresses. First, 0, 10, 20, 30, 40, and 50 FTC processing was performed on green sandstone

specimens. Then, dynamic compression tests on the specimens were performed with different loading rates under the hydrostatic pressures of 0 MPa, 2 MPa, and 4 MPa. By analyzing the stress–strain curve, dynamic compression strength, dispersion energy, and damage characteristics, the main conclusions were as follows:

(1) The dynamic failure process after FTC had obvious stages, and its stress–strain curve could be roughly divided into four stages: compression of fissures, linear deformation of rock, yield of rock, and failure of rock. Under different working conditions, the rock destruction phases were different. Under the effect of high loading rate and hydrostatic pressure, the stress–strain curve rebounded after a relatively gentle stage. This phenomenon became more obvious as the loading rate increased;

(2) The dynamic compression strength of rock increased linearly with the increase in the loading rate and decreased with the increase in FTCs. Under the same conditions of loading rate and FTC, the hydrostatic pressure could enhance dynamic compression strength, and this enhancement effect increased with the increase in the number of FTCs;

(3) The dissipated energy of sandstone specimens was affected significantly by FTC, hydrostatic pressure, and loading rate. The dissipated energy of sandstone specimens was negatively related to the number of FTCs, positively related to hydrostatic pressure, and positively related to the loading rate. An empirical equation of dissipated energy with freeze–thaw cycling, hydrostatic pressure, and loading rate was proposed, and the fitting parameters were provided. After 40 FT cycles, the internal damage of the specimens reached a larger value, which desensitized the energy required for specimens to reach breakage to the loading rate, but the sensitivity did not decrease under the effect of hydrostatic pressure. Freeze–thaw weathering reduced the effect of hydrostatic pressure on dissipated energy, and hydrostatic pressure enhanced the effect of FTC on the dissipated energy of the specimens;

(4) The dissipated energy of the specimen corresponded to the dynamic compression strength according to a quadratic function, and the correspondence was almost independent of the FTCs, loading rate, and hydrostatic pressure;

(5) The average fragment size after damage decreased with the increase in FTCs for specimens that were not subjected to hydrostatic pressure. For the specimens subjected to hydrostatic pressure, the appearance remained intact after loading, and the CT value decreased with the increase in the number of FTCs. Both phenomena indicated that FTC exacerbated the internal damage of rock. It was also found that 40 FTCs was an important number of FTCs for green sandstone whereby the internal damage reached a large value, which was consistent with the result given by the energy analysis.

Author Contributions: Methodology, B.W.; writing and experiments, L.Y.; resources, Y.M. and Y.S.; data curation, J.L. and J.S. All authors have read and agreed to the published version of the manuscript.

Funding: Natural Science Foundation of Tianjin City (No. 21JCZXJC00040).

Data Availability Statement: Data are available upon request due to privacy restrictions.

Conflicts of Interest: The authors declare no conflict of interest.

References

1. Park, J.; Hyun, C.U.; Park, H.D. Changes in microstructure and physical properties of rocks caused by artificial freeze-thaw action. *Bull. Eng. Geol. Environ.* **2015**, *74*, 555–565. [CrossRef]
2. Deprez, M.; De Kock, T.; De Schutter, G.; Cnudde, V. A review on freeze-thaw action and weathering of rocks. *Earth-Sci. Rev.* **2020**, *203*, 103143. [CrossRef]
3. Eppes, M.C.; Keanini, R. Mechanical weathering and rock erosion by climate-dependent subcritical cracking. *Rev. Geophys.* **2017**, *55*, 470–508. [CrossRef]
4. Liu, W.W.; Feng, Q.; Wang, C.X.; Lu, C.K.; Xu, Z.Z.; Li, W.T. Analytical solution for three-dimensional radial heat transfer in a cold-region tunnel. *Cold Reg. Sci. Technol.* **2019**, *164*, 102787. [CrossRef]

5. Luo, X.D.; Jiang, N.; Fan, X.Y.; Mei, N.F.; Luo, H. Effects of freeze-thaw on the determination and application of parameters of slope rock mass in cold regions. *Cold Reg. Sci. Technol.* **2015**, *110*, 32–37. [CrossRef]
6. Pudasaini, S.P.; Krautblatter, M. A two-phase mechanical model for rock-ice avalanches. *J. Geophys. Res.-Earth Surf.* **2014**, *119*, 2272–2290. [CrossRef]
7. Huang, S.B.; Liu, Y.Z.; Guo, Y.L.; Zhang, Z.L.; Cai, Y.T. Strength and failure characteristics of rock-like material containing single crack under freeze-thaw and uniaxial compression. *Cold Reg. Sci. Technol.* **2019**, *162*, 1–10. [CrossRef]
8. Khanlari, G.; Sahamieh, R.Z.; Abdilor, Y. The effect of freeze-thaw cycles on physical and mechanical properties of Upper Red Formation sandstones, central part of Iran. *Arab. J. Geosci.* **2015**, *8*, 5991–6001. [CrossRef]
9. Walbert, C.; Eslami, J.; Beaucour, A.L.; Bourges, A.; Noumowe, A. Evolution of the mechanical behaviour of limestone subjected to freeze-thaw cycles. *Environ. Earth Sci.* **2015**, *74*, 6339–6351. [CrossRef]
10. Yavuz, H. Effect of freeze-thaw and thermal shock weathering on the physical and mechanical properties of an andesite stone. *Bull. Eng. Geol. Environ.* **2011**, *70*, 187–192. [CrossRef]
11. Park, C.; Synn, J.H.; Shin, H.S.; Cheon, D.S.; Lim, H.D.; Jeon, S.W. An experimental study on the thermal characteristics of rock at low temperatures. *Int. J. Rock Mech. Min. Sci.* **2004**, *41*, 367–368. [CrossRef]
12. Thomachot, C.; Matsuoka, N. Dilation of building materials submitted to frost action. In Proceedings of the General Assembly of the European-Union-of-Geosciences, Vienna, Austria, 24–29 April 2005; p. 167.
13. Tan, X.J.; Chen, W.Z.; Yang, J.P.; Cao, J.J. Laboratory investigations on the mechanical properties degradation of granite under freeze-thaw cycles. *Cold Reg. Sci. Technol.* **2011**, *68*, 130–138. [CrossRef]
14. Del Rio, L.M.; Lopez, F.; Esteban, F.J.; Tejado, J.J.; Mota, M.I.; Gonzalez, I.; Ramos, A.; San Emeterio, J.L. Ultrasonic study of alteration processes in granites caused by freezing and thawing. In Proceedings of the IEEE International Ultrasonics Symposium, Rotterdam, The Netherlands, 18–21 September 2005; pp. 419–422.
15. Martinez-Martinez, J.; Benavente, D.; Gomez-Heras, M.; Marco-Castano, L.; Garcia-del-Cura, M.A. Non-linear decay of building stones during freeze-thaw weathering processes. *Constr. Build. Mater.* **2013**, *38*, 443–454. [CrossRef]
16. Kodama, J.; Goto, T.; Fujii, Y.; Hagan, P. The effects of water content, temperature and loading rate on strength and failure process of frozen rocks. *Int. J. Rock Mech. Min. Sci.* **2013**, *62*, 1–13. [CrossRef]
17. Li, X.; Lu, Y.; Weng, Y. Research on Damage Model of Single Jointed Rock Masses under Coupling Action of Freeze-Thaw and Loading. *Chin. J. Rock Mech. Eng.* **2013**, *32*, 2307–2315.
18. Nicholson, D.T.; Nicholson, F.H. Physical deterioration of sedimentary rocks subjected to experimental freeze-thaw weathering. *Earth Surf. Process. Landf.* **2000**, *25*, 1295–1307. [CrossRef]
19. Yamabe, T.; Neaupane, K.M. Determination of some thermo-mechanical properties of Sirahama sandstone under subzero temperature condition. *Int. J. Rock Mech. Min. Sci.* **2001**, *38*, 1029–1034. [CrossRef]
20. Jamshidi, A.; Nikudel, M.R.; Khamehchiyan, M.; Zalooli, A. Statistical Models for Predicting the Mechanical Properties of Travertine Building Stones After Freeze-Thaw Cycles. In Proceedings of the 12th International IAEG Congress, Torino, Italy, 15–19 September 2014; pp. 477–481.
21. Wang, Y.; Song, Z.Y.; Mao, T.Q.; Zhu, C. Macro-Meso Fracture and Instability Behaviors of Hollow-Cylinder Granite Containing Fissures Subjected to Freeze-Thaw-Fatigue Loads. *Rock Mech. Rock Eng.* **2022**, *55*, 4051–4071. [CrossRef]
22. Chen, T.C.; Yeung, M.R.; Mori, N. Effect of water saturation on deterioration of welded tuff due to freeze-thaw action. *Cold Reg. Sci. Technol.* **2004**, *38*, 127–136. [CrossRef]
23. Bayram, F. Predicting mechanical strength loss of natural stones after freeze-thaw in cold regions. *Cold Reg. Sci. Technol.* **2012**, *83–84*, 98–102. [CrossRef]
24. Jamshidi, A. Predicting the Strength of Granitic Stones after Freeze-Thaw Cycles: Considering the Petrographic Characteristics and a New Approach Using Petro-Mechanical Parameter. *Rock Mech. Rock Eng.* **2021**, *54*, 2829–2841. [CrossRef]
25. Ren, J.X.; Yun, M.C.; Cao, X.T.; Zhang, K.; Liang, Y.; Chen, X. Study on the mechanical properties of saturated red sandstone under freeze-thaw conditions. *Environ. Earth Sci.* **2022**, *81*, 376. [CrossRef]
26. Xia, K.; Yao, W. Dynamic rock tests using split Hopkinson (Kolsky) bar system—A review. *J. Rock Mech. Geotech. Eng.* **2015**, *7*, 27–59. [CrossRef]
27. Zhang, Q.B.; Zhao, J. Determination of mechanical properties and full-field strain measurements of rock material under dynamic loads. *Int. J. Rock Mech. Min. Sci.* **2013**, *60*, 423–439. [CrossRef]
28. Dai, F.; Xia, K.W. Loading Rate Dependence of Tensile Strength Anisotropy of Barre Granite. *Pure Appl. Geophys.* **2010**, *167*, 1419–1432. [CrossRef]
29. Ma, Q.Y.; Ma, D.D.; Yao, Z.M. Influence of freeze-thaw cycles on dynamic compressive strength and energy distribution of soft rock specimen. *Cold Reg. Sci. Technol.* **2018**, *153*, 10–17. [CrossRef]
30. Wang, P.; Xu, J.Y.; Liu, S.; Liu, S.H.; Wang, H.Y. A prediction model for the dynamic mechanical degradation of sedimentary rock after a long-term freeze-thaw weathering: Considering the strain-rate effect. *Cold Reg. Sci. Technol.* **2016**, *131*, 16–23. [CrossRef]
31. Weng, L.; Wu, Z.J.; Liu, Q.S. Dynamic Mechanical Properties of Dry and Water-Saturated Siltstones Under Sub-Zero Temperatures. *Rock Mech. Rock Eng.* **2020**, *53*, 4381–4401. [CrossRef]
32. Zhou, K.P.; Li, B.; Li, J.L.; Deng, H.W.; Bin, F. Microscopic damage and dynamic mechanical properties of rock under freeze-thaw environment. *Trans. Nonferrous Met. Soc. China* **2015**, *25*, 1254–1261. [CrossRef]

33. Xie, H.; Ju, Y.; Li, L. Criteria for Strength and Structural Failure of Rocks Based on Energy Dissipation and Energy Release Principles. *Chin. J. Rock Mech. Eng.* **2005**, *24*, 3003–3010.
34. Khazaei, C.; Hazzard, J.; Chalaturnyk, R. Damage quantification of intact rocks using acoustic emission energies recorded during uniaxial compression test and discrete element modeling. *Comput. Geotech.* **2015**, *67*, 94–102. [CrossRef]
35. Wang, P.; Xu, J.Y.; Fang, X.Y.; Wang, P.X. Energy dissipation and damage evolution analyses for the dynamic compression failure process of red-sandstone after freeze-thaw cycles. *Eng. Geol.* **2017**, *221*, 104–113. [CrossRef]
36. Feng, Q.; Jin, J.C.; Zhang, S.; Liu, W.W.; Yang, X.X.; Li, W.T. Study on a Damage Model and Uniaxial Compression Simulation Method of Frozen-Thawed Rock. *Rock Mech. Rock Eng.* **2022**, *55*, 187–211. [CrossRef]
37. Lundberg, B. Split Hopkinson Bar Study of Energy-Absorption in Dynamic Rock Fragmentation. *Int. J. Rock Mech. Min. Sci.* **1976**, *13*, 187–197. [CrossRef]
38. Li, X.B.; Lok, T.S.; Zhao, J. Dynamic characteristics of granite subjected to intermediate loading rate. *Rock Mech. Rock Eng.* **2005**, *38*, 21–39. [CrossRef]
39. Deng, H.W.; Yu, S.T.; Deng, J.R.; Ke, B.; Bin, F. Experimental Investigation on Energy Mechanism of Freezing-Thawing Treated Sandstone under Uniaxial Static Compression. *Ksce J. Civ. Eng.* **2019**, *23*, 2074–2082. [CrossRef]
40. Gong, F.Q.; Si, X.F.; Li, X.B.; Wang, S.Y. Dynamic triaxial compression tests on sandstone at high strain rates and low confining pressures with split Hopkinson pressure bar. *Int. J. Rock Mech. Min. Sci.* **2019**, *113*, 211–219. [CrossRef]
41. Wu, B.; Xia, K. An experimental study of dynamic tensile failure of rocks subjected to hydrostatic confinement. In Proceedings of the 2nd International Conference on Rock Dynamics and Applications (RocDyn), Suzhou, China, 18–19 May 2016; pp. 197–202.
42. Xu, Y.; Yao, W.; Xia, K.W.; Ghaffari, H.O. Experimental Study of the Dynamic Shear Response of Rocks Using a Modified Punch Shear Method. *Rock Mech. Rock Eng.* **2019**, *52*, 2523–2534. [CrossRef]
43. Zhou, Y.X.; Xia, K.; Li, X.B.; Li, H.B.; Ma, G.W.; Zhao, J.; Zhou, Z.L.; Dai, F. Suggested methods for determining the dynamic strength parameters and mode-I fracture toughness of rock materials. *Int. J. Rock Mech. Min. Sci.* **2012**, *49*, 105–112. [CrossRef]
44. Xu, Y.; Chen, B.Q.; Wu, B.B.; Chen, Z.; Yang, L.; Li, P.F. Influence of freeze-thaw cycling on the dynamic compressive failure of rocks subjected to hydrostatic pressure. *Bull. Eng. Geol. Environ.* **2022**, *81*, 276. [CrossRef]
45. Song, B.; Chen, W. Loading and unloading split Hopkinson pressure bar pulse-shaping techniques for dynamic hysteretic loops. *Exp. Mech.* **2004**, *44*, 622–627. [CrossRef]
46. Rong, C.; FangYun, L.U.; YuLiang, L.I.N.; JinGui, Q.I.N. Investigation of the Pre-fixed Gap in Single-Pulse Loading Technique of Hopkinson Pressure Bars. *Chin. J. High Press. Phys.* **2008**, *22*, 187–191.
47. Ping, Q.; Luo, X.; Ma, Q.Y.; Yuan, P. Broken energy dissipation characteristics of sandstone specimens under impact loads. *Chin. J. Rock Mech. Eng.* **2015**, *34*, 4197–4203. (In Chinese) [CrossRef]

minerals

MDPI

Article

Fractal Dimension Analysis of Pores in Coal Reservoir and Their Impact on Petrophysical Properties: A Case Study in the Province of Guizhou, SW China

Wei Fu [1], Wangshui Hu [1,*], Tongsheng Yi [2], Oumar Ibrahima Kane [1], Manting Zhang [1] and Xin Huang [1]

[1] School of Geosciences, Yangtze University, Wuhan 430100, China
[2] Guizhou Engineering Technology Research Center for Coalbed Methane and Shale Gas, Guiyang 550009, China
* Correspondence: hws@yangtzeu.edu.cn

Abstract: Coal is a complex, porous medium with pore structures of various sizes. Therefore, it is difficult to accurately describe the characteristics of pore structure by using the traditional geometry method. The results from the present investigation suggest that the porous media system of the coal reservoir has obvious fractal characteristics at different scales. To study the complexity of the pores in the coal reservoir, 27 coal samples from Guizhou, SW China were studied. The fractal dimensions of coal pores were calculated, and the fractal dimension of a pore in a coal reservoir can be classified into two types: percolation and diffusion. The comprehensive fractal dimension can be obtained using the weighted summation method and the pore volume fraction of different fractal segments as the weight. The percolation fractal dimensions (D_p) of coal samples are between 2.88 and 3.12, the diffusion fractal dimensions (D_d) are between 3.57 and 3.84, and the comprehensive fractal dimensions (D_t) are between 3.05 and 3.63. The D_d values of all coal samples are all larger than the D_p values, which indicates that the random distribution and complexity of diffusion pores in coal are stronger than those of the percolation pores. The percolation fractal dimension decreases as the maturity degree increases, whereas the diffusion and comprehensive fractal dimensions increase. The diffusion pore volume fraction and total pore volume are all highly correlated with the comprehensive and diffusion fractal dimensions, respectively. The correlation between the comprehensive fractal dimension, diffusion pore volume fraction, and coal reservoir porosity is negative exponential, whereas the correlation between the total pore volume and coal reservoir porosity is positive linear. In comparison with the percolation and diffusion fractal dimensions, the comprehensive fractal dimension is better suited for characterizing the permeability of coal reservoirs. The fractal analysis of this paper is beneficial for understanding the relationship between the fractal characteristics of coal pores and properties.

Keywords: Southwestern China; coal pores; mercury injection experiment; fractal dimension; petrophysical properties

Citation: Fu, W.; Hu, W.; Yi, T.; Kane, O.I.; Zhang, M.; Huang, X. Fractal Dimension Analysis of Pores in Coal Reservoir and Their Impact on Petrophysical Properties: A Case Study in the Province of Guizhou, SW China. *Minerals* **2022**, *12*, 1425. https://doi.org/10.3390/min12111425

Academic Editor: Samintha Perera

Received: 16 September 2022
Accepted: 2 November 2022
Published: 10 November 2022

1. Introduction

Coal is a disordered heterogeneous pore medium; the pore structure is directly related to methane adsorption and fluidity [1,2]. The existing research shows that the composition of coal is critical to determine the characteristics of coal reservoirs [3,4]. Southwestern China owns one of the largest coal-producing fields in the country [5,6]. The Longtan Formation of Guizhou province in this region contains abundant coal resources and extremely rich coalbed methane (CBM) resources [7,8]. However, the complex geological conditions in this area seriously restrict the development of coalbed methane production [9,10], with one of the most significant factors being the structure of the pores [11,12]. The pores and fissures of coal reservoirs are important spaces for coalbed methane adsorption, migration,

and accumulation [13,14]. It plays an important role in controlling the adsorption capacity and the permeability of coal [15].

The pore characteristics of coal are highly heterogeneous [16,17]. It is difficult to accurately describe and analyze the complexity of the pore structure by using traditional geometry methods [18,19]. Most scholars tend to use fractal mathematical methods to describe its characteristics [20,21]. Different scholars have carried out significant research on the selection of fractal scale, calculation of fractal dimension, and physical property characterization of fractal geometric parameters of coal reservoir pores, but the conclusions are quite different [22,23]. This is because the pore systems of coal reservoirs in different areas are very different due to differences in metamorphic degree and reservoir physical properties [24,25]. It further indicates the complexity and particularity of the pore system in coal reservoirs [26,27].

In recent years, some scholars have studied pore characteristics, geological controlling factors of coal reservoirs [1,28], and the fractal characteristics of pores on a certain scale in some mining areas of the southwestern part of China [29–31]. However, comprehensive fractal research of the reservoir is seldom involved. This paper uses typical bituminous coal samples and typical anthracite samples in Southwestern China as the research object. In this study, 27 coal samples were collected, and vitrinite reflectance (R_o) measurement, coal proximate analysis, and mercury injection tests were performed on all samples. Based on the results of the equipment, the fractal dimension was calculated. Afterward, the relationship between different parameters was studied. The primary goals of this study were: (1) quantitatively study the fractal characteristics of different pore sizes in coal reservoirs by using fractal theory, and calculate their fractal dimension in sections; (2) discuss the internal relationship of fractal geometric parameters, such as piecewise fractal dimension and comprehensive fractal dimension, and their effects on porosity and permeability. The study is beneficial for better understanding the relationship between fractal characteristics of coal pores and petrophysical properties.

2. Experiment and Methodology

2.1. Equipment and Materials

In order to study the fractal characteristics of coal pores and their impact on the petrophysical properties, a total of 27 coal samples were collected from Longtan Formation in Guizhou province, SW China. A total of 23 of the samples are bituminous coal and 4 are anthracite. The collection process of the samples followed the Standard Method of China (GB/T 19222-2003). All coal samples were obtained from CBM parameter wells or working surfaces of underground mines from 8 coalfields. Some permeability values of coal beds were acquired by an injection test. The statistical results show that the permeability of the coal beds in this region ranges between 0.027×10^{-15} m^2 and 0.573×10^{-15} m^2; this suggests that the permeability of coal beds in this region is relatively low [32–34].

Vitrinite reflectance (R_o) measurement, coal proximate analysis, and mercury injection tests were performed on all samples. The measurement of R_o followed the National Standards of China (GB/T 8899-2013), and the proximate analysis followed the test method ISO 17246-2010. We used the mercury injection method to test the porosity of the samples. The instrument was Autopore IV 9500, which was produced by Micromeritics. To avoid the negative effects of mineral impurity and other factors in the sample preparation process, the samples were crushed into particles with a dimension of 2–4 mm. The mercury injection test shows the distribution of pores diameter in soil based on the fact that non-infiltrative liquid will not flow into solid pores without pressure. The pressure required for liquid injection into the pores is based on the Washburn equation [35,36]. The relationship between the pressure and the volume of mercury was obtained from the mercury injection test. The Washburn equation was used to find the corresponding equivalent diameter through pressure, from which the pore distribution of coal was obtained [37,38]. Additionally, we excluded the results with apertures larger than 100,000 nm as these obviously oversized

pores are not the pore system of the sample itself. The porosity measurements followed the Petroleum Industry Standard of China SY/T 6385-1999.

2.2. Methodology

Fractal dimension is an important parameter to describe fractal characteristics, which describes the complexity or roughness of the coal surface. Many scholars have deduced the fractal dimension of the pore volume of coal reservoirs according to the basic theory of fractals [34,39,40]. Based on the mercury injection data, Equation (1) was utilized to determine the relationship between mercury injection volume and mercury injection pressure in coal samples [11,41,42] and is as follows:

$$\lg\left(\frac{dV_p}{dP}\right) \propto (D-4)\lg P \quad (1)$$

where P is the external pressure in the process of mercury injection, in MPa; dVp is the pore volume increment of the corresponding pressure increment dP, in cm^3/g; and D is the fractal dimension of coal pore volume, and is dimensionless.

Through the scatter plot of $\lg\left(\frac{dV_p}{dP}\right)$ and lg P, one or more linear relationship segments will appear in the graph using the mercury injection test data of coal samples. As long as there is a linear relationship between $\lg\left(\frac{dV_p}{dP}\right)$ and lg P, the pore distribution of the corresponding segment conforms to the fractal characteristics, and the fractal dimension of the segment $D = 4 + K$ can be obtained, where K is the slope of the line [14,18,43].

Existing research on the fractal dimension of coal pores has shown that different pore sizes in coal have fractal dimension characteristics in the subsection [3,23,44]. In the study of mercury injection fractals of coal samples in the Bide–Santang Basin in Southwestern China, it has been found that there are three kinds of pore systems in coal samples, namely large pores between particles of coal matrix >6000 nm, transition pores in coal matrix particles between 6000 and 73 nm, and small/micropores in coal matrix particles <73 nm [45]. They correspond to three different pore fractions, which were called the fractal dimension of large pores (D_1), the fractal dimension of transitional pores (D_2), and the fractal dimension of small/micropores (D_3) [45–47]. The mercury injection fractal of a large number of coal samples collected from different regions of China in different geological times and coal grades were studied based on the diffusion and percolation characteristics of coal bed methane [1,48,49]. The aforementioned study divided the coal pores into percolation pores (>65 nm) and diffusion pores (<65 nm). Jiang et al. studied the samples from West China and found that there are two different fractal characteristics of coal reservoir pores, namely the fractal dimension of the larger pore stage as the percolation fractal dimension and the fractal dimension of the smaller pore stage as the diffusion fractal dimension [50]. The fractal dimension of different pore sizes is related to the physical properties of the reservoir, but it is difficult to represent the complexity of coal [51,52]. Ren et al. proposed a method to study the fractal dimension of coal samples. In this method, the ratio of different pore sizes was taken as the weight coefficient, and the comprehensive fractal dimension was obtained by the weighted summation of the fractal dimension of different pore size segments. The calculation equation is as follows:

$$D_t = \sum D_i T_i \quad (2)$$

where D_t is the comprehensive fractal dimension of coal, D_i is the fractal dimension corresponding to the pore size segment, T_i is the pore size ratio corresponding to the pore size segment, and i is the pore size segment code.

3. Results

3.1. Results of Conventional Coal Test

Table 1 shows the results of the experiments. The R_o of bituminous coals ranges from 1.05% to 1.95%, with an average value of 1.60%. The results of the proximate analysis show that the moisture content (ad), ash yield (Ad), and volatile matter (daf) range from 0.69 to 1.680% (average 0.94%), 12.53 to 28.62% (average 22.18%), and 15.94 to 33.48% (average 22.90%).

Table 1. Test results of coal samples.

Sample Nos.	Coal Rank	R_o/%	Proximate and Ultimate (wt%)			Porosity (%)	Total Pore Volume (mL/g)
			M_{ad}	A_d	V_{daf}		
M_1	Bituminous coal	1.05	0.75	16.28	33.48	4.01	0.0284
M_2		1.07	1.68	25.17	31.80	3.85	0.0236
M_3		1.13	1.52	24.83	30.85	5.68	0.0412
M_4		1.09	0.95	20.48	32.97	3.27	0.0244
M_5		1.10	0.88	19.34	29.18	4.22	0.0251
S_1		1.45	1.06	21.87	23.98	3.74	0.0261
S_2		1.22	1.14	19.27	24.59	4.02	0.0248
S_3		1.43	1.28	21.39	22.48	3.84	0.0257
S_4		1.24	1.85	20.14	24.26	3.97	0..0277
F_1		1.67	0.95	18.67	21.87	3.78	0.0191
F_2		1.54	1.25	19.38	20.89	3.68	0.0210
F_3		1.68	0.85	21.08	18.92	3.47	0.0199
F_4		1.42	1.33	19.38	19.46	4.89	0.0223
W_1		1.81	0.71	26.15	19.38	9.25	0.0538
W_2		1.85	0.98	28.62	18.89	9.14	0.0634
W_3		1.93	0.69	12.53	15.94	5.37	0.0402
W_4		1.84	0.78	18.78	17.96	6.47	0.0581
W_5		1.79	0.93	20.96	20.39	7.25	0.0495
W_6		1.92	0.73	14.69	18.88	5.77	0.0513
G_1		1.95	0.71	28.14	17.82	3.64	0.0227
G_2		1.85	0.92	18.98	20.68	5.87	0.0428
G_3		1.92	1.24	24.39	17.11	4.26	0.0328
G_4		1.89	0.85	20.64	20.47	5.84	0.0215
H_1	Anthracite	3.31	2.30	21.38	7.88	3.41	0.0213
H_2		3.24	2..14	20.86	10.96	4.95	0.0328
H_3		3.14	2.02	17.39	12.85	3.99	0.0259
H_4		3.30	1.95	20.18	8.89	4.12	0.027

R_o: vitrinite reflectance, %; M_{ad}: moisture (air-dried basis); A_d (dry basis); V_{daf} (dry ash-free basis).

3.2. Results of Mercury Intrusion

The mercury intrusion curves for different samples are shown in Figure 1. All curves show the hysteresis loop phenomenon. The mercury injection curves in the study area were divided into two types. Type I (Figure 1a) shows that when low pressure is increased the mercury intake increases significantly with the increase in mercury injection pressure, and shows obvious hysteresis, indicating that there are more open pores and good connectivity.

Type II (Figure 1b) shows that with the continuous increase in mercury intrusion pressure, the mercury intrusion curve continues to rise, indicating that the pores of all scales are distributed and open.

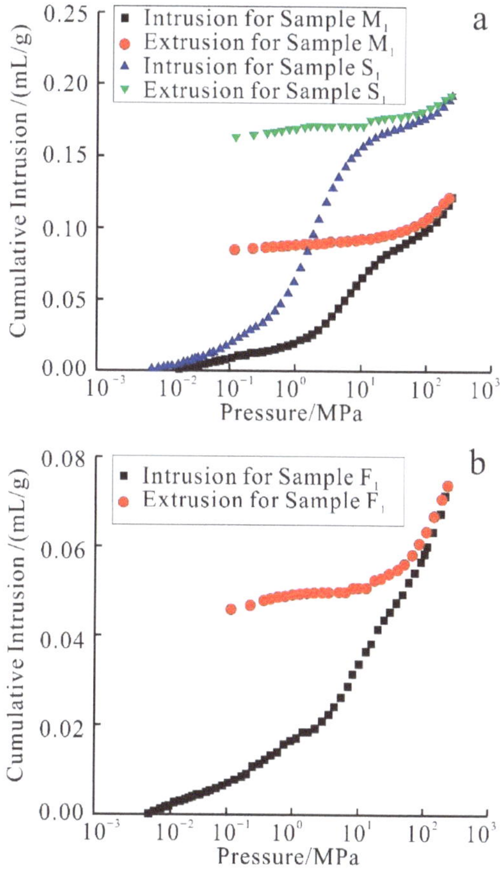

Figure 1. Mercury intrusion curves for different samples. (**a**) Sample M_1 & Sample S_1; (**b**) Sample F_1.

3.3. Results of Fractal Dimension Calculation

According to Equation (1), the relation diagram of $\lg(dV_P/dP)$ and $\lg P$ was drawn from mercury injection data of coal samples (Figure 2). The fractal curve can be divided into two different linear sections: a low-pressure section and a high-pressure section. The linear regression correlation coefficients of each section are all above 0.90. This indicates that there are two different pore systems in the coal samples. The low-pressure section corresponds to a large pore diameter section, and the high-pressure section corresponds to a small pore diameter section. In this paper, the fractal dimension of the larger pore section is called the percolation fractal dimension (D_p), and the fractal dimension of the smaller pore section is called the diffusion fractal dimension (D_d).

It can be observed in Figure 2 that the graph from $\lg P$ and $\lg(dV_P/dP)$ shows two distinct lines and $\lg P$ of the yielding points is between 1.1 and 1.4. The corresponding coal pore radius is between 54 and 85 nm, with an average of 65 nm. This phenomenon is consistent with the study of Fu et al. (2005) which has been shown in the methodology. In this study, we followed the classification scheme of Fu et al. (2005) and set 65 nm as the boundary of percolation pores and diffusion pores.

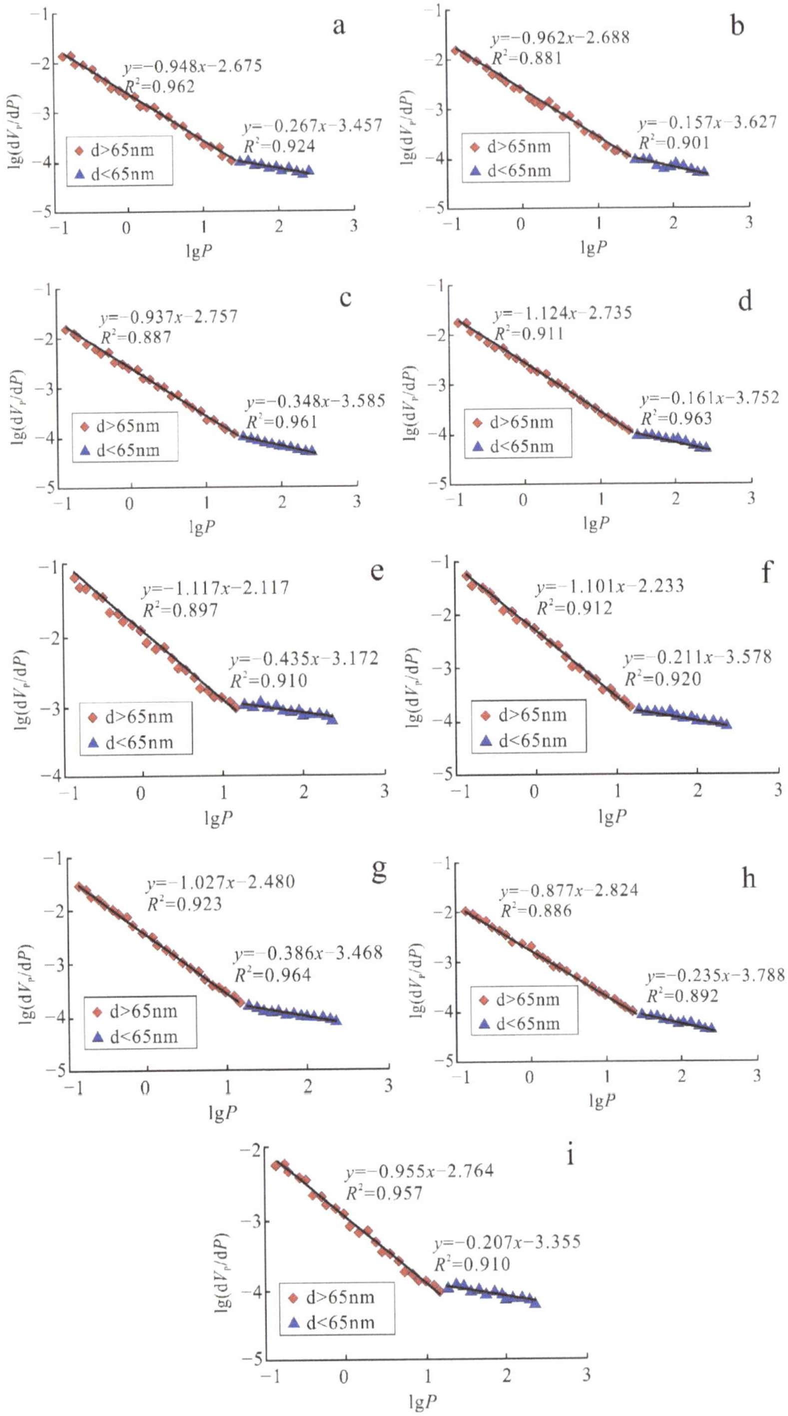

Figure 2. Plots of $\lg(dV_P/dP)$ as a function of $\lg P$ for selected samples (**a**): Sample M_1; (**b**): Sample M_2; (**c**): Sample S_1; (**d**): Sample F_1; (**e**): Sample W_1; (**f**): Sample W_2; (**g**): Sample W_3; (**h**): Sample G_1; (**i**): Sample H_1.

Table 2 lists the fractal dimension, correlation coefficient, and pore size ratio of some coal samples with different pore sizes. According to Equation (2), the comprehensive fractal dimension (D_t) of each coal sample is obtained as a weighted average.

Table 2. Calculation results of fractal dimension of some coal samples.

Sample Nos.	Percolation Pore Segment				Diffusion Pore Segment				D_t
	Linear Fitting Equation	R^2	D_p	Pore Size Ratio/%	Linear Fitting Equation	R^2	D_d	Pore Size Ratio/%	
M_1	$Y = -0.948x - 2.675$	0.96	3.05	44	$Y = -0.267x - 3.457$	0.92	3.73	56	3.43
M_2	$Y = -0.962x - 2.688$	0.88	3.04	35	$Y = -0.157x - 3.627$	0.90	3.84	65	3.56
M_3	$Y = -0.958x - 2.635$	0.81	3.04	41	$Y = -0.257x - 3.659$	0.87	3.74	59	3.46
M_4	$Y = -0.922x - 2.578$	0.84	3.08	39	$Y = -0.195x - 3.561$	0.82	3.81	61	3.52
M_5	$Y = -0.968x - 2.458$	0.80	3.03	42	$Y = -0.281x - 3.486$	0.84	3.72	58	3.43
S_1	$Y = -0.937x - 2.757$	0.89	3.06	37	$Y = -0.348x - 3.585$	0.96	3.65	63	3.43
S_2	$Y = -0.952x - 2.582$	0.81	3.05	38	$Y = -0.385x - 3.518$	0.91	3.62	62	3.40
S_3	$Y = -0.901x - 2.747$	0.85	3.10	35	$Y = -0.315x - 3.685$	0.82	3.69	65	3.48
S_4	$Y = -0.978x - 2.638$	0.88	3.02	41	$Y = -0.341x - 3.725$	0.84	3.66	59	3.40
F_1	$Y = -1.124x - 2.735$	0.91	2.88	32	$Y = -0.161x - 3.752$	0.96	3.84	68	3.53
F_2	$Y = -1.114x - 2.582$	0.88	2.89	39	$Y = -0.158x - 3.857$	0.84	3.84	61	3.47
F_3	$Y = -1.125x - 2.758$	0.81	2.88	35	$Y = -0.181x - 3.581$	0.86	3.82	65	3.49
F_4	$Y = -1.117x - 2.856$	0.84	2.88	37	$Y = -0.178x - 3.893$	0.90	3.82	63	3.47
W_1	$Y = -1.117x - 2.117$	0.90	2.88	76	$Y = -0.435x - 3.172$	0.99	3.57	24	3.05
W_2	$Y = -1.101x - 2.333$	0.91	2.90	69	$Y = -0.211x - 3.578$	0.92	3.79	31	3.17
W_3	$Y = -1.027x - 2.480$	0.92	2.97	48	$Y = -0.386x - 3.468$	0.96	3.61	52	3.31
W_4	$Y = -1.146x - 2.500$	0.90	2.85	65	$Y = -0.582x - 3.573$	0.87	3.42	35	3.05
W_5	$Y = -1.085x - 2.581$	0.88	2.92	58	$Y = -0.379x - 3.458$	0.80	3.62	42	3.21
W_6	$Y = -1.055x - 2.494$	0.82	2.95	49	$Y = -0.375x - 3.483$	0.96	3.63	51	3.29
G_1	$Y = -0.877x - 2.824$	0.89	3.12	34	$Y = -0.235x - 3.788$	0.89	3.77	66	3.55
G_2	$Y = -0.925x - 2.458$	0.87	3.08	38	$Y = -0.285x - 3.581$	0.89	3.72	62	3.47
G_3	$Y = -0.895x - 2.594$	0.81	3.11	42	$Y = -0.246x - 3.673$	0.85	3.75	58	3.48
G_4	$Y = -0.880x - 2.765$	0.79	3.12	36	$Y = -0.249x - 3.657$	0.90	3.75	64	3.52
H_1	$Y = -0.955x - 2.764$	0.96	3.05	22	$Y = -0.207x - 3.355$	0.91	3.79	78	3.63
H_2	$Y = -0.924x - 2.724$	0.90	3.08	38	$Y = -0.182x - 3.278$	0.81	3.82	62	3.54
H_3	$Y = -0.975x - 2.681$	0.85	3.03	32	$Y = -0.214x - 3.281$	0.86	3.79	68	3.54
H_4	$Y = -0.968x - 2.588$	0.80	3.03	29	$Y = -0.193x - 3.589$	0.90	3.81	71	3.58

The results in Table 2 show that the percolation fractal dimensions (D_p) of coal samples are between 2.88 and 3.12, the diffusion fractal dimensions (D_d) are between 3.57 and 3.84, and the comprehensive fractal dimensions (D_t) are between 3.05 and 3.63. The D_d of all coal samples are all larger than D_p, which indicates that the random distribution and complexity of diffusion pores in coal are stronger than that of the percolation pores.

4. Discussion

4.1. The Validity of the Calculation Results

According to the theory of fractal geometry, the fractal dimension of a porous solid should be between 2.0 and 3.0 in general; however, some fractal dimensions in Table 2 exceed 3. Some scholars have found that if the compression deformation of a porous solid occurs under high pressure, the results of the fractal dimension may be larger than 3.0 [51,53].

Yang et al. (2021) and Cai et al. (2018) studied the compressibility of the coal matrix and its impact on the results of MIP. They provided a semi-empirical Tait equation to correct

the results of MIP [54,55]. However, the coal rank and proximate of the samples tested in this work are multiple, which indicated that the compressibility of different samples varies greatly. Therefore, it is difficult to correct the fractal dimension by a unified method. Many scholars have repeatedly obtained the result that the fractal dimension is larger than 3. Therefore, although the calculation results of partial fractal dimension are greater than 3, under the same test conditions, a fractal dimension greater than 3.0 is an effective index to characterize the physical properties of the reservoir.

4.2. Relationship between Fractal Dimensions and Vitrinite Reflectance

By comparing the relationship between the fractal dimension of samples and the degree of maturity, it can be found that the D_p decreases gradually with the increase in the maturity degree (Figure 3), while the D_d and D_t increase with the increase in the maturity degree (Figures 4 and 5). This phenomenon happens because, with the increase in the metamorphic degree, the percolation pores with a larger pore size in the coal are gradually compacted and homogenized, and many irregular diffusive pores with a small pore size are formed during the process of structural compression and compaction of percolation pores. The comprehensive fractal dimension (D_t) increases with the metamorphic degree, which has the same trend as the diffusion fractal dimension (D_d), indicating that the complexity of the coal pores mainly depends on the complexity of diffusion pores (D_d).

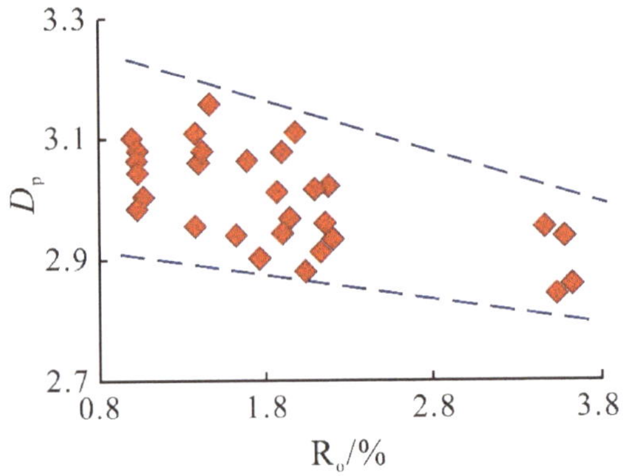

Figure 3. Relationship between percolation fractal dimension (D_p) and vitrinite reflectance (R_o).

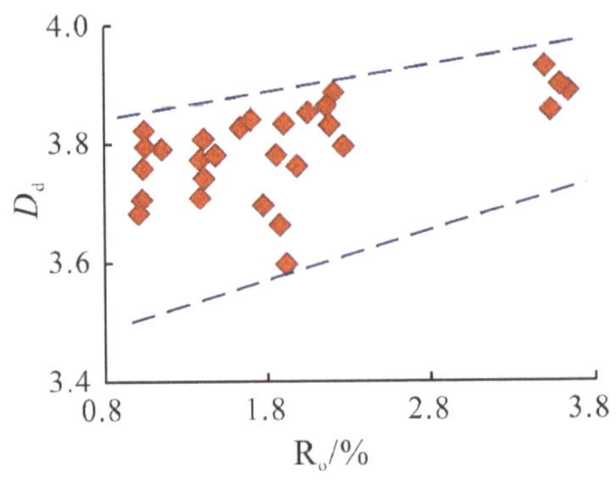

Figure 4. Relationship between diffusion fractal dimension (D_d) and vitrinite reflectance (R_o).

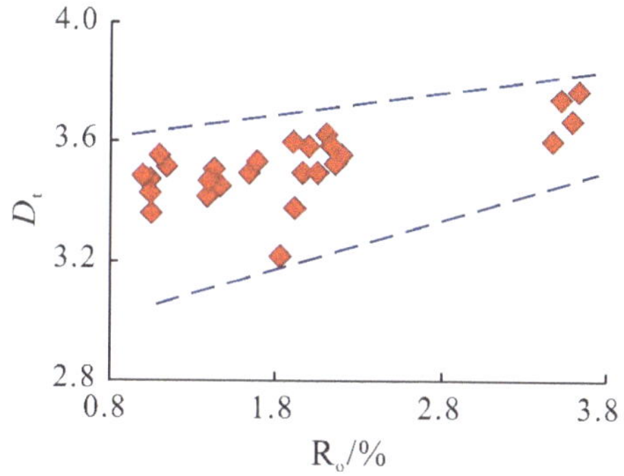

Figure 5. Relationship between comprehensive fractal dimension (D_t) and vitrinite reflectance (R_o).

4.3. Relationship between Volume Fraction of Diffusion Pores and Fractal Dimensions

The investigation of each fractal parameter shows that the diffusion pore volume fractions are poorly correlated with percolation fractal dimension (Figure 6), while it has a good positive linear correlation with diffusion fractal dimension and comprehensive fractal dimension, especially when the linear regression number of diffusion pore volume fraction and comprehensive fractal dimension reaches 0.951 (Figures 7 and 8). It shows that the coal with a smaller volume fraction of diffusion pores has less complexity in the pores. This phenomenon preliminarily shows that the diffusion pore volume fractions are important geometric parameters that indirectly characterize the complexity of coal.

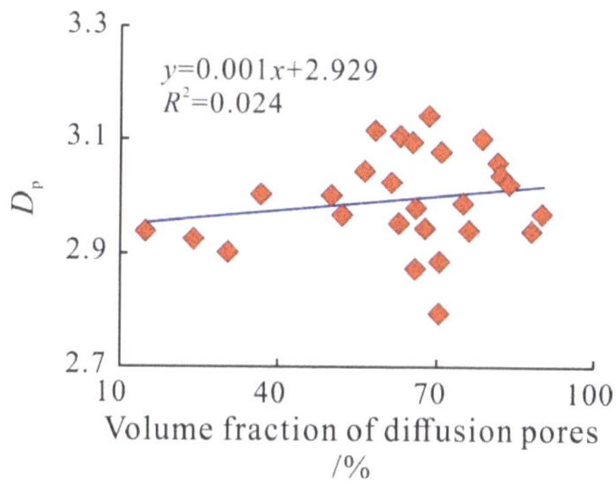

Figure 6. Relationship between volume fraction of diffusion pores and percolation fractal dimension (D_p).

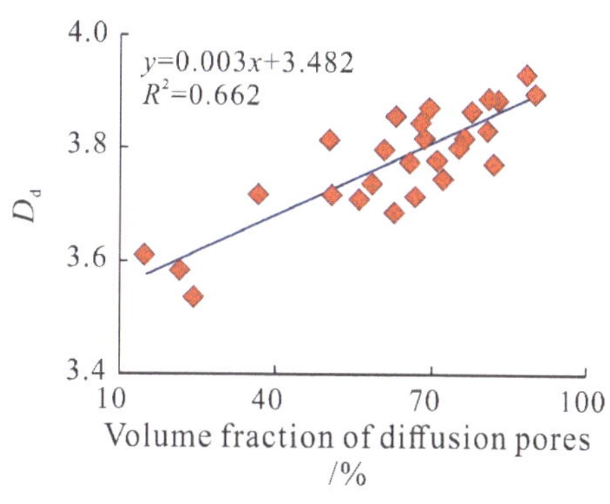

Figure 7. Relationship between volume fraction of diffusion pores and diffusion fractal dimension (D_d).

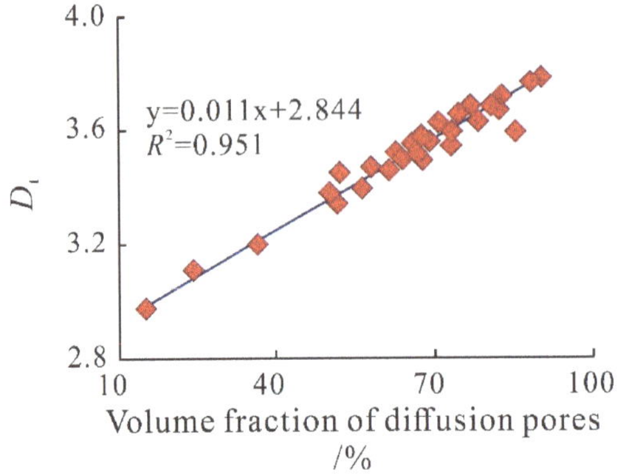

Figure 8. Relationship between volume fraction of diffusion pores and comprehensive fractal dimension (D_t).

4.4. Relationship between Fractal Dimensions and Total Pore Volume

The following figures illustrate that the total pore volume is poorly correlated with the percolation fractal dimension (D_p) (Figure 9), and inversely proportional to the diffusion fractal dimension (D_d) (Figure 10) and comprehensive fractal dimension (D_t) (Figure 11), indicating that increasing total pore volume is beneficial in reducing the complexity of coal pores and making the pore surface of coal, especially the diffusion pore surface, smoother. By comparing the correlation coefficients of the fractal parameters, it is found that the comprehensive fractal dimension has the highest correlation coefficients with each geometrical parameter, the correlation coefficient between diffusion fractal dimension and geometrical parameters is smaller, and the correlation coefficients between percolation fractal dimension and geometrical parameters were very low (less than 0.1) (Figure 9). The correlation between the percolation fractal dimension and the single geometric parameter is not significant. This indicates that the factors affecting the surface roughness of the percolation pore are very complex, and any single geometric parameter is difficult to effectively characterize the complexity of the percolation pore.

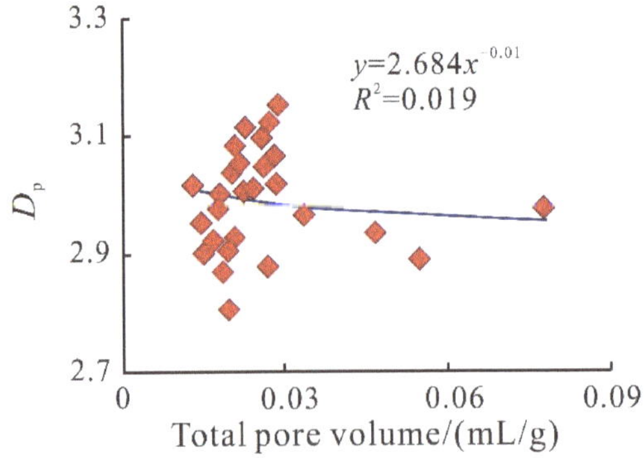

Figure 9. Relationship between total pore volume and percolation fractal dimension (D_p).

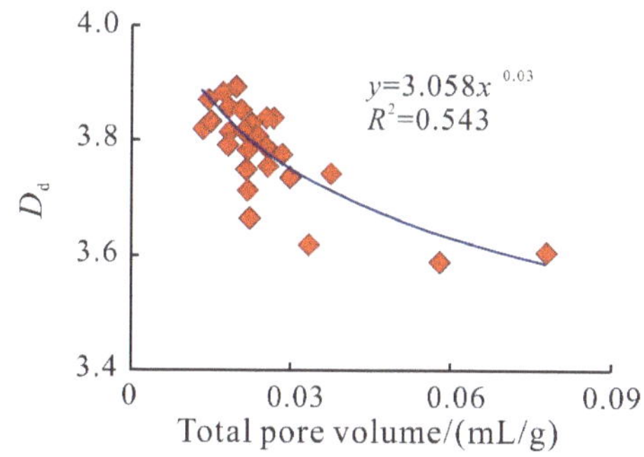

Figure 10. Relationship between total pore volume and diffusion fractal dimension (D_d).

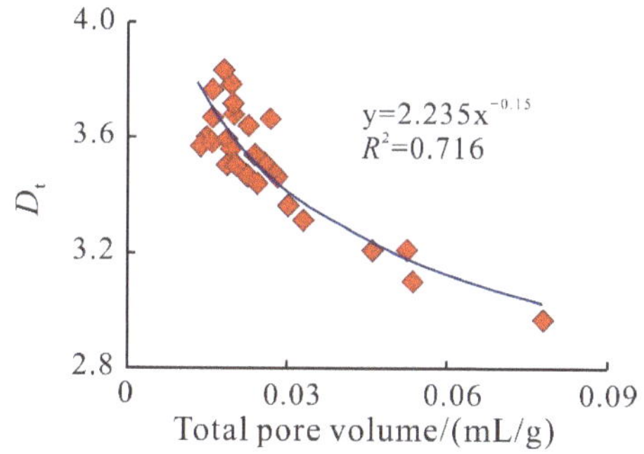

Figure 11. Relationship between total pore volume and comprehensive fractal dimension (D_t).

4.5. Relationship between Fractal Parameters and Porosity

Porosity is an important parameter to describe the petrophysical properties of coal reservoirs. In this study, almost all of the samples were below 5%. The comprehensive fractal dimension of coal samples has a negative power exponential correlation with porosity (Figure 12). This phenomenon can be the result of an increased pore network complexity that leads to poor pore connectivity, which results in lower porosity values. According to the previous analysis, the volume fraction of diffusion pores and total pore volume correlate with the comprehensive fractal dimension. Therefore, they are important fractal geometric parameters in order to characterize the complexity of coal reservoirs. By analyzing the correlation between these three parameters and porosity, the pore volume of fraction diffusion decreases with the increase in porosity (Figure 13). This shows that the comprehensive fractal dimension and the volume fraction of diffusion pores can indirectly characterize coal porosity.

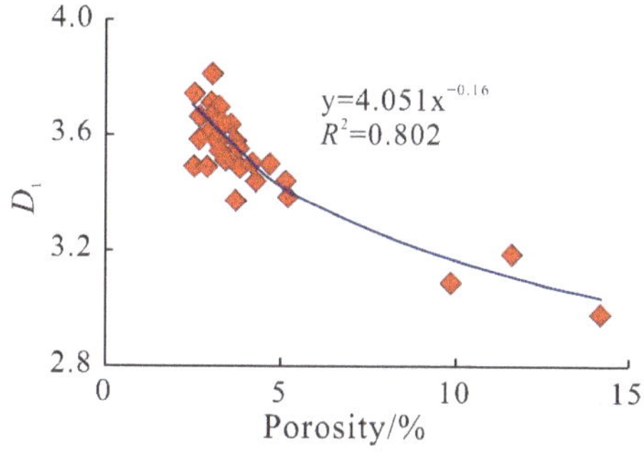

Figure 12. Relationship between porosity and comprehensive fractal dimension (D_t).

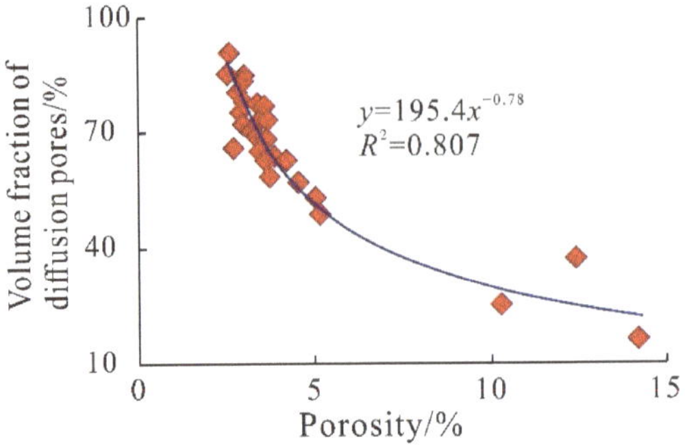

Figure 13. Relationship between porosity and volume fraction of diffusion pores.

4.6. Brief Summary

According to the research above, this paper studied the advantages of the comprehensive fractal dimension of coal pores in detail. The fractal characteristics of coal pores were discussed. Through the establishment of the comprehensive fractal dimension, the petrophysical properties of coal were analyzed.

The result of the research shows that the comprehensive fractal dimension can better reflect the complexity of coal pores. This conclusion is consistent with the conclusions of other scholars [18,20]. Additionally, the result also shows that the diffusion fractal dimension has a greater reflection on the pore complexity of coal pores than the percolation fractal dimension. This phenomenon can also be supported by other scholars [2,8,56].

The shortage of this paper is the lack of study on the permeability, as the complexity of coal pores has an important effect on the permeability of coal reservoirs [27,57,58].

5. Conclusions

Based on the fractal geometry and mercury injection test data, the complexity of pores in collected samples from the Longtan Formation in Guizhou province, SW China, was studied. The pore fractal dimension of the coal reservoir is divided into the percolation fractal dimension and the diffusion fractal dimension. In addition, the comprehensive fractal dimension of the coal sample can be obtained by the weighted summative method.

The percolation fractal dimension decreases as the metamorphic degree increases, whereas the diffusion fractal dimension and comprehensive dimension both increase, and the complexity of diffusion pores has a greater effect on the complexity of pores in coal.

Indirectly, the complexity of coal pores can be determined by the fractal boundary, volume fraction of diffusion pores, and total pore volume. Correlation coefficients between the parameters and the comprehensive fractal dimension are greater than those between the parameter and the diffusion fractal dimension, and there is no obvious correlation between the parameters and the percolation fractal dimension.

The comprehensive fractal dimension, fractal boundary, diffusion pore volume fraction, and total pore volume can all quantitatively reflect the porosity of the coal reservoir. The comprehensive fractal dimension, fractal boundary, and diffusion pore volume fraction all decrease with the increasing porosity of coal. The total pore volume increases linearly in proportion to the coal reservoir porosity.

Author Contributions: Conceptualization, W.H.; methodology, W.F.; validation, W.F., M.Z. and X.H.; formal analysis, W.F.; investigation, W.F.; resources, W.H.; data curation, M.Z.; writing—original draft preparation, W.F.; writing—review and editing, O.I.K.; visualization, M.Z.; supervision, W.H.; project administration, T.Y.; funding acquisition, W.H. All authors have read and agreed to the published version of the manuscript.

Funding: This research received no external funding or This research was funded by National Major Science and Technology Projects of China grant number 2016ZX05034-004 and The APC was funded by Yangtze University.

Data Availability Statement: The data that support the findings of this study are available from the corresponding author upon reasonable request.

Conflicts of Interest: The authors declare no conflict of interest.

References

1. Zhao, S.; Chen, X.; Li, X.; Qi, L.; Zhang, G. Experimental Analysis of the Effect of Temperature on Coal Pore Structure Transformation. *Fuel* **2021**, *305*, 121613. [CrossRef]
2. Nie, A.; Mei, S.; Guan, D.; Wu, P.; Zhang, Z. A Study on the Genetic Relations between Permian Longtan Formation Coal Series Strata and Carlin-Type Gold Deposits, Southwestern Guizhou Province, China. *Chin. J. Geochem.* **2008**, *27*, 291–298. [CrossRef]
3. Zhao, Y.; Sun, Y.; Liu, S.; Wang, K.; Jiang, Y. Pore Structure Characterization of Coal by NMR Cryoporometry. *Fuel* **2017**, *190*, 359–369. [CrossRef]
4. Li, Y.; Yang, J.; Pan, Z.; Tong, W. Nanoscale Pore Structure and Mechanical Property Analysis of Coal: An Insight Combining AFM and SEM Images. *Fuel* **2020**, *260*, 116352. [CrossRef]
5. Li, B.; Zhuang, X.; Li, J.; Querol, X.; Font, O.; Moreno, N. Enrichment and Distribution of Elements in the Late Permian Coals from the Zhina Coalfield, Guizhou Province, Southwest China. *Int. J. Coal Geol.* **2017**, *171*, 111–129. [CrossRef]
6. Li, Z.; Gao, W.; Li, X. Correlations of Permeability and Geological Characteristics Based on Mercury Intrusion Data and Hierarchical Statistical Models: A Case Study. *J. Energy Resour. Technol.* **2020**, *142*, 123004. [CrossRef]
7. Bi, C.; Zhang, J.; Shan, Y.; Hu, Z.; Wang, F.; Chi, H.; Tang, Y.; Yuan, Y.; Liu, Y.; Oil and gas survey, China Geological Survey, Ministry of Natural Resources; et al. Geological Characteristics and Co-Exploration and Co-Production Methods of Upper Permian Longtan Coal Measure Gas in Yangmeishu Syncline, Western Guizhou Province, China. *China Geol.* **2020**, *3*, 38–51. [CrossRef]
8. Li, Z.; Ripepi, N.; Chen, C. Using Pressure Pulse Decay Experiments and a Novel Multi-Physics Shale Transport Model to Study the Role of Klinkenberg Effect and Effective Stress on the Apparent Permeability of Shales. *J. Pet. Sci. Eng.* **2020**, *189*, 107010. [CrossRef]
9. Fu, H.; Tang, D.; Xu, H.; Xu, T.; Chen, B.; Hu, P.; Yin, Z.; Wu, P.; He, G. Geological Characteristics and CBM Exploration Potential Evaluation: A Case Study in the Middle of the Southern Junggar Basin, NW China. *J. Nat. Gas Sci. Eng.* **2016**, *30*, 557–570. [CrossRef]
10. Zhang, S.; Tang, S.; Zhang, J.; Pan, Z. Pore Structure Characteristics of China Sapropelic Coal and Their Development Influence Factors. *J. Nat. Gas Sci. Eng.* **2018**, *53*, 370–384. [CrossRef]
11. Fu, H.; Tang, D.; Xu, T.; Xu, H.; Tao, S.; Li, S.; Yin, Z.; Chen, B.; Zhang, C.; Wang, L. Characteristics of Pore Structure and Fractal Dimension of Low-Rank Coal: A Case Study of Lower Jurassic Xishanyao Coal in the Southern Junggar Basin, NW China. *Fuel* **2017**, *193*, 254–264. [CrossRef]
12. Liu, Z.; Yang, H.; Wang, W.; Cheng, W.; Xin, L. Experimental Study on the Pore Structure Fractals and Seepage Characteristics of a Coal Sample around a Borehole in Coal Seam Water Infusion. *Transp. Porous Med.* **2018**, *125*, 289–309. [CrossRef]
13. Li, Z.; Zhao, Q.; Teng, Y.; Fan, M.; Ripepi, N.; Yin, X.; Chen, C. Experimental Investigation of Non-Monotonic Fracture Conductivity Evolution in Energy Georeservoirs. *J. Pet. Sci. Eng.* **2022**, *211*, 110103. [CrossRef]
14. Zhang, L.; Kou, Z.; Wang, H.; Zhao, Y.; Dejam, M.; Guo, J.; Du, J. Performance Analysis for a Model of a Multi-Wing Hydraulically Fractured Vertical Well in a Coalbed Methane Gas Reservoir. *J. Pet. Sci. Eng.* **2018**, *166*, 104–120. [CrossRef]
15. Luo, Q.; Xiao, Z.; Dong, C.; Ye, X.; Li, H.; Zhang, Y.; Ma, Y.; Ma, L.; Xu, Y. The Geochemical Characteristics and Gas Potential of the Longtan Formation in the Eastern Sichuan Basin, China. *J. Pet. Sci. Eng.* **2019**, *179*, 1102–1113. [CrossRef]
16. Wu, Z.; He, S.; Han, Y.; Zhai, G.; He, X.; Zhou, Z. Effect of Organic Matter Type and Maturity on Organic Matter Pore Formation of Transitional Facies Shales: A Case Study on Upper Permian Longtan and Dalong Shales in Middle Yangtze Region, China. *J. Earth Sci.* **2020**, *31*, 368–384. [CrossRef]
17. He, Q.; Dong, T.; He, S.; Zhai, G. Methane Adsorption Capacity of Marine-Continental Transitional Facies Shales: The Case Study of the Upper Permian Longtan Formation, Northern Guizhou Province, Southwest China. *J. Pet. Sci. Eng.* **2019**, *183*, 106406. [CrossRef]
18. Zhao, L.; Guanhua, N.; Lulu, S.; Qian, S.; Shang, L.; Kai, D.; Jingna, X.; Gang, W. Effect of Ionic Liquid Treatment on Pore Structure and Fractal Characteristics of Low Rank Coal. *Fuel* **2020**, *262*, 116513. [CrossRef]
19. Tsakiroglou, C.D.; Payatakes, A.C. Characterization of the Pore Structure of Reservoir Rocks with the Aid of Serial Sectioning Analysis, Mercury Porosimetry and Network Simulation. *Adv. Water Resour.* **2000**, *23*, 773–789. [CrossRef]
20. Zhang, J.; Li, X.; Wei, Q.; Sun, K.; Zhang, G.; Wang, F. Characterization of Full-Sized Pore Structure and Fractal Characteristics of Marine–Continental Transitional Longtan Formation Shale of Sichuan Basin, South China. *Energy Fuels* **2017**, *31*, 10490–10504. [CrossRef]
21. Wang, H.; Kou, Z.; Guo, J.; Chen, Z. A Semi-Analytical Model for the Transient Pressure Behaviors of a Multiple Fractured Well in a Coal Seam Gas Reservoir. *J. Pet. Sci. Eng.* **2021**, *198*, 108159. [CrossRef]

22. Wood, D.A. Deriving Coal Fractal Dimensions from Low-Pressure Nitrogen Adsorption Isotherms Applying an Integrated Method. *Appl. Geochem.* **2021**, *131*, 105042. [CrossRef]
23. Liu, S.; Tang, S.; Yin, S. Coalbed Methane Recovery from Multilateral Horizontal Wells in Southern Qinshui Basin. *Adv. Geo-Energy Res.* **2018**, *2*, 34–42. [CrossRef]
24. Mahamud, M.M.; Novo, M.F. The Use of Fractal Analysis in the Textural Characterization of Coals. *Fuel* **2008**, *87*, 222–231. [CrossRef]
25. Cai, Y.; Liu, D.; Yao, Y.; Li, J.; Liu, J. Fractal Characteristics of Coal Pores Based on Classic Geometry and Thermodynamics Models. *Acta Geol. Sin.-Engl. Ed.* **2011**, *85*, 1150–1162. [CrossRef]
26. Andhumoudine, A.B.; Nie, X.; Zhou, Q.; Yu, J.; Kane, O.I.; Jin, L.; Djaroun, R.R. Investigation of Coal Elastic Properties Based on Digital Core Technology and Finite Element Method. *Adv. Geo-Energy Res.* **2021**, *5*, 53–63. [CrossRef]
27. Zhang, Q.; Grohmann, S.; Xu, X.; Littke, R. Depositional Environment and Thermal Maturity of the Coal-Bearing Longtan Shale in Southwest Guizhou, China: Implications for Shale Gas Resource Potential. *Int. J. Coal Geol.* **2020**, *231*, 103607. [CrossRef]
28. Kou, Z.; Wang, T.; Chen, Z.; Jiang, J. A Fast and Reliable Methodology to Evaluate Maximum CO_2 Storage Capacity of Depleted Coal Seams: A Case Study. *Energy* **2021**, *231*, 120992. [CrossRef]
29. Pan, J.; Zhang, Z.; Li, M.; Wu, Y.; Wang, K. Characteristics of Multi-Scale Pore Structure of Coal and Its Influence on Permeability. *Nat. Gas Ind. B* **2019**, *6*, 357–365. [CrossRef]
30. Yang, C.; Xiong, Y.; Zhang, J.; Liu, Y.; Chen, C. Comprehensive Understanding of OM-Hosted Pores in Transitional Shale: A Case Study of Permian Longtan Shale in South China Based on Organic Petrographic Analysis, Gas Adsorption, and X-Ray Diffraction Measurements. *Energy Fuels* **2019**, *33*, 8055–8064. [CrossRef]
31. Yang, X.; Ren, T.; He, X. Experimental Study of Coal Burst Risk Prediction Using Fractal Dimension Analysis of AE Spatial Distribution. *J. Appl. Geophys.* **2020**, *177*, 104025. [CrossRef]
32. Cheng, J.; Zhou, J.; Li, Y.; Liu, J.; Cen, K. Effects of Pore Fractal Structures of Ultrafine Coal Water Slurries on Rheological Behaviors and Combustion Dynamics. *Fuel* **2008**, *87*, 2620–2627. [CrossRef]
33. Zhou, Q.; Qin, B.; Wang, J.; Wang, H.; Wang, F. Experimental Investigation on the Changes of the Wettability and Surface Characteristics of Coal Dust with Different Fractal Dimensions. *Colloids Surf. A Physicochem. Eng. Asp.* **2018**, *551*, 148–157. [CrossRef]
34. Wang, Z.; Cheng, Y.; Qi, Y.; Wang, R.; Wang, L.; Jiang, J. Experimental Study of Pore Structure and Fractal Characteristics of Pulverized Intact Coal and Tectonic Coal by Low Temperature Nitrogen Adsorption. *Powder Technol.* **2019**, *350*, 15–25. [CrossRef]
35. Faulon, J.-L.; Mathews, J.P.; Carlson, G.A.; Hatcher, P.G. Correlation between Microporosity and Fractal Dimension of Bituminous Coal Based on Computer-Generated Models. *Energy Fuels* **1994**, *8*, 408–414. [CrossRef]
36. Mondal, D.; Roy, P.N.S.; Behera, P.K. Use of Correlation Fractal Dimension Signatures for Understanding the Overlying Strata Dynamics in Longwall Coal Mines. *Int. J. Rock Mech. Min. Sci.* **2017**, *91*, 210–221. [CrossRef]
37. Reich, M.H.; Snook, I.K.; Wagenfeld, H.K. A Fractal Interpretation of the Effect of Drying on the Pore Structure of Victorian Brown Coal. *Fuel* **1992**, *71*, 669–672. [CrossRef]
38. Mendhe, V.A.; Bannerjee, M.; Varma, A.K.; Kamble, A.D.; Mishra, S.; Singh, B.D. Fractal and Pore Dispositions of Coal Seams with Significance to Coalbed Methane Plays of East Bokaro, Jharkhand, India. *J. Nat. Gas Sci. Eng.* **2017**, *38*, 412–433. [CrossRef]
39. Cai, Y.; Liu, D.; Pan, Z.; Che, Y.; Liu, Z. Investigating the Effects of Seepage-Pores and Fractures on Coal Permeability by Fractal Analysis. *Transp. Porous Med.* **2016**, *111*, 479–497. [CrossRef]
40. Yin, T.; Liu, D.; Cai, Y.; Zhou, Y.; Yao, Y. Size Distribution and Fractal Characteristics of Coal Pores through Nuclear Magnetic Resonance Cryoporometry. *Energy Fuels* **2017**, *31*, 7746–7757. [CrossRef]
41. Huang, F.; Cao, Z.; Guo, J.; Jiang, S.-H.; Li, S.; Guo, Z. Comparisons of Heuristic, General Statistical and Machine Learning Models for Landslide Susceptibility Prediction and Mapping. *CATENA* **2020**, *191*, 104580. [CrossRef]
42. Huang, F.; Chen, J.; Liu, W.; Huang, J.; Hong, H.; Chen, W. Regional Rainfall-Induced Landslide Hazard Warning Based on Landslide Susceptibility Mapping and a Critical Rainfall Threshold. *Geomorphology* **2022**, *408*, 108236. [CrossRef]
43. Li, H.; Shi, S.; Lu, J.; Ye, Q.; Lu, Y.; Zhu, X. Pore Structure and Multifractal Analysis of Coal Subjected to Microwave Heating. *Powder Technol.* **2019**, *346*, 97–108. [CrossRef]
44. Chang, Z.; Du, Z.; Zhang, F.; Huang, F.; Chen, J.; Li, W.; Guo, Z. Landslide Susceptibility Prediction Based on Remote Sensing Images and GIS: Comparisons of Supervised and Unsupervised Machine Learning Models. *Remote Sens.* **2020**, *12*, 502. [CrossRef]
45. Li, Z.; Teng, Y.; Fan, M.; Ripepi, N.; Chen, C. A Novel Multiphysics Multiscale Multiporosity Shale Gas Transport Model for Geomechanics/Flow Coupling in Steady and Transient States. *SPE J.* **2022**, *27*, 452–464. [CrossRef]
46. Deng, E.; Zhang, Q.; Jin, Z.; Zhu, R.; Yan, Z.; Jiang, B.; Littke, R. Non-Overmature Equivalents Confirmed a High Initial Hydrocarbon Generation Potential of the Permian Longtan Shale in Southern China. *Int. J. Coal Geol.* **2022**, *259*, 104043. [CrossRef]
47. Liu, X.; Nie, B. Fractal Characteristics of Coal Samples Utilizing Image Analysis and Gas Adsorption. *Fuel* **2016**, *182*, 314–322. [CrossRef]
48. Fu, X.; Qin, Y.; Zhang, W.; Wei, C.; Zhou, R. Fractal Classification and Natural Classification of Coal Pore Structure Based on Migration of Coal Bed Methane. *Chin. Sci. Bull.* **2005**, *50*, 66–71. [CrossRef]
49. Huang, F.; Cao, Z.; Jiang, S.-H.; Zhou, C.; Huang, J.; Guo, Z. Landslide Susceptibility Prediction Based on a Semi-Supervised Multiple-Layer Perceptron Model. *Landslides* **2020**, *17*, 2919–2930. [CrossRef]

50. Jiang, S.-H.; Huang, J.; Huang, F.; Yang, J.; Yao, C.; Zhou, C.-B. Modelling of Spatial Variability of Soil Undrained Shear Strength by Conditional Random Fields for Slope Reliability Analysis. *Appl. Math. Model.* **2018**, *63*, 374–389. [CrossRef]
51. Huang, F.; Zhang, J.; Zhou, C.; Wang, Y.; Huang, J.; Zhu, L. A Deep Learning Algorithm Using a Fully Connected Sparse Autoencoder Neural Network for Landslide Susceptibility Prediction. *Landslides* **2020**, *17*, 217–229. [CrossRef]
52. Chang, Z.; Catani, F.; Huang, F.; Liu, G.; Meena, S.R.; Huang, J.; Zhou, C. Landslide Susceptibility Prediction Using Slope Unit-Based Machine Learning Models Considering the Heterogeneity of Conditioning Factors. *J. Rock Mech. Geotech. Eng.* **2022**, S1674775522001615. [CrossRef]
53. Ju, Y.; Luxbacher, K.; Li, X.; Wang, G.; Yan, Z.; Wei, M.; Yu, L. Micro-Structural Evolution and Their Effects on Physical Properties in Different Types of Tectonically Deformed Coals. *Int. J. Coal Sci. Technol.* **2014**, *1*, 364–375. [CrossRef]
54. Cai, Y.; Li, Q.; Liu, D.; Zhou, Y.; Lv, D. Insights into Matrix Compressibility of Coals by Mercury Intrusion Porosimetry and N2 Adsorption. *Int. J. Coal Geol.* **2018**, *200*, 199–212. [CrossRef]
55. Yang, Q.; Xue, J.; Li, W.; Du, X.; Ma, Q.; Zhan, K.; Chen, Z. Comprehensive Evaluation and Interpretation of Mercury Intrusion Porosimetry Data of Coals Based on Fractal Theory, Tait Equation and Matrix Compressibility. *Fuel* **2021**, *298*, 120823. [CrossRef]
56. Yao, Y.; Liu, D.; Tang, D.; Tang, S.; Huang, W.; Liu, Z.; Che, Y. Fractal Characterization of Seepage-Pores of Coals from China: An Investigation on Permeability of Coals. *Comput. Geosci.* **2009**, *35*, 1159–1166. [CrossRef]
57. Qin, L.; Zhai, C.; Liu, S.; Xu, J.; Wu, S.; Dong, R. Fractal Dimensions of Low Rank Coal Subjected to Liquid Nitrogen Freeze-Thaw Based on Nuclear Magnetic Resonance Applied for Coalbed Methane Recovery. *Powder Technol.* **2018**, *325*, 11–20. [CrossRef]
58. Liu, L.; Cui, Z.; Wang, J.; Xia, Z.; Duan, L.; Yang, Y.; Li, M.; Li, T. Pore Size Distribution Characteristics of High Rank Coal with Various Grain Sizes. *ACS Omega* **2020**, *5*, 19785–19795. [CrossRef]

minerals

MDPI

Article

A New Shear Constitutive Model Characterized by the Pre-Peak Nonlinear Stage

Shijie Xie [1,2], Hang Lin [2], Zhenyu Han [1,3,*], Hongyu Duan [4], Yifan Chen [2,*] and Diyuan Li [2]

1 School of Civil Engineering, Southeast University, Nanjing 210096, China
2 School of Resources and Safety Engineering, Central South University, Changsha 410083, China
3 Department of Civil Engineering, Monash University, Melbourne, VIC 3800, Australia
4 Discipline of Civil, Surveying & Environmental Engineering, School of Engineering, University of Newcastle, Callaghan, NSW 2308, Australia
* Correspondence: zhenyu_han@seu.edu.cn (Z.H.); 185506005@csu.edu.cn (Y.C.)

Abstract: The pre-peak shear stress-displacement curve is an important part of the study of the shear mechanical behavior of rock joints. Underpinned by the Haldane distribution, a new semi-analytical model for the pre-peak shear deformation of rock joints was established in this paper, the validity of which was verified by laboratory and in situ experimental data. Other existing models were employed to make comparisons. The comparison results show that the model has superior adaptability and is more suitable for convex-type shear constitutive curves than existing models. Besides, only one parameter was introduced to the model, which is more convenient for application. All of these imply that the proposed model is an effective tool to evaluate the pre-peak shear constitutive curves of different rock joints. The research results can provide a reference for further understanding of the shear fracture characteristics of rock materials.

Keywords: fracture characteristics; rock joint; pre-peak shear constitutive curve; shear deformation evolution

Citation: Xie, S.; Lin, H.; Han, Z.; Duan, H.; Chen, Y.; Li, D. A New Shear Constitutive Model Characterized by the Pre-Peak Nonlinear Stage. *Minerals* **2022**, *12*, 1429. https://doi.org/10.3390/min12111429

Academic Editor: Gianvito Scaringi

Received: 10 October 2022
Accepted: 3 November 2022
Published: 10 November 2022

1. Introduction

The intact rock masses in nature are cut by rock joints into rock blocks, forming jointed rock masses as shown in Figure 1 [1,2]. The failure of rock joints dominates more than that of rock blocks during the instability of jointed rock masses [3,4]. Most geotechnical engineering projects are usually constructed in jointed rock masses [5], and the shear mechanical properties of rock joints are one of the main factors controlling the mechanical stability of underground excavations [6]. For example, it has been reported that the collapse of the São Paulo metro station in Brazil may be related to an over-simplified geomechanical model used in the engineering design process [7]. Therefore, understanding the shear deformation behavior of rock joints is very important for the safety assessment of geoengineering structures (such as rock slopes, tunnels, dam foundations, chambers, and waste repositories) [8,9].

Rock materials are a complex mineral aggregate formed under physical and chemical actions after a long geological process [10–12], which makes the link between stress and strain (or displacement) of rock materials has been one of the most pressing issues over the past decades [13]. Extensive theoretical analysis, experimental tests, and numerical simulation studies have been implemented at the laboratory scale on the shear deformation behavior of rock joints, and fruitful results have been obtained [14–20]. It mainly includes the research on shear strength [21,22], shear constitutive models [23,24], surface failure characteristics [25–28], as well as the influencing laws of rock type [29], boundary conditions [30] and loading mode [31], and other factors. Jing [32] and Muralha et al. [33] provide an admirable summary of the suggested method for laboratory shear tests of rock joints and the importance of rock joints in understanding rock mechanics for a variety of

engineering applications. The shear stress/shear displacement curve is one of the most important considerations to evaluate the shear mechanical properties of rock joints [34]. Nonetheless, it is still difficult to establish a constitutive model that can fully reflect the shear stress/shear displacement curve behavior of joint materials, especially considering the complex nonlinear mechanical properties of joint materials [35,36]. Among the aforementioned pioneering studies, it is fully recognized that the shear stress/shear displacement curve of the rock sample subjected to shear tests can be divided by peak points into two critical stages, the pre-peak stage and post-peak stage. Because the shear test requires specific test conditions, the test results themselves have a certain discreteness [37]. In practice, it is difficult to select the most suitable test results to evaluate the shear constitutive relationship [38], and the shear stress/shear displacement curve is essential to describe the shear fracture characteristics of rock joints [39]. According to the classical failure curve of rock joints [40,41], except for peak stress, the stress thresholds of crack compaction, crack initiation and rock damage are all located in the pre-peak stage, which is closely associated with crack development. However, there are rare reports on the pre-peak stress and deformation characteristics [42]. Therefore, it is still necessary to study the theoretical modeling of pre-peak shear stress/shear displacement curves of rock joints.

Figure 1. Example of jointed rock mass(modified after [1]).

Modeling studies to investigate the pre-peak shear stress- displacement curve of rock joints began in the late 1960s and early 1970s. Goodman [43] and Saeb et al. [44] regarded the pre-peak shear stress/shear displacement curve as quasi-linear and used linear functions to characterize the pre-peak shear stress- shear displacement curve of rock joints. Obviously, this oversimplification cannot describe the nonlinear deformation characteristics in the pre-peak stage. Based on a series of direct shear tests, Kulhawy [45] first proposed a hyperbolic function to describe the nonlinear pre-peak shear stress- displacement relationship. Subsequently, referring to the work of Kulhawy [45], Bandis et al. [46] proposed a hyperbolic model to fit the pre-peak shear stress-shear displacement curve and verified the applicability of this model by comparing it with some existing test results. Under the framework of plasticity theory, Desai and Fishman [47] established a pre-peak constitutive model to simulate joints under monotonic loading, unloading and reverse loading. Nassir et al. [48] further generalized the Bandis model [46] by introducing the dilation initiation coefficient to characterize the fraction of peak shear displacement at which dilation initiates, and then calibrated the model based on experimental results. Inspired by the idea of normalization, Ban et al. [49] made the shear stiffness dimensionless and established a hyperbolic shear constitutive considering shear stiffness softening to describe the pre-peak shear stress/shear displacement curve. Kou et al. [50] also studied the pre-peak shear behavior of rock joints contain the triangular-shaped primary and subordinated asperities through pre-peak cyclic shear test. Despite those models inspiring the research on the

characterization of pre-peak shear constitutive curves, they only accord well with curves with specific morphology (i.e., the concave-type curve in Figure 2). Shen et al. [51] pointed out that the pre-peak shear stress curves of many joints presents a convex-type shear constitutive curve (i.e., convex-type curve). However, due to the limitation of hyperbolic function itself, the hyperbolic models mentioned above cannot represent this convex-type curve. In fact, there is currently no model that can simultaneously represent the three types of pre-peak shear curves shown in Figure 2. Therefore, it is necessary to carry out in-depth research.

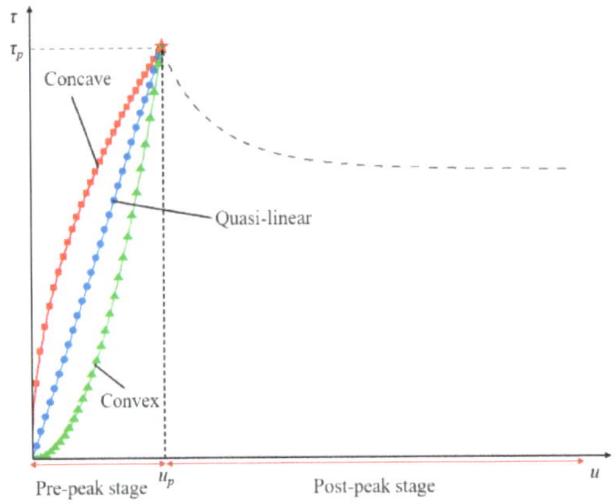

Figure 2. Examples of the pre-peak shear stress-shear displacement curve (modified after [51]).

To remedy this limitation, the semi-analytical model based on the Haldane distribution and exponential correction coefficient was built in this paper, which can represent the shear deformation of three kinds of pre-peak curves. Then the shear test data of different types of rock joints were selected for validity verification. Comprehensive comparisons between the proposed model and other existing models were launched to illustrate the superiority of this model.

2. Model Development

The statistical damage-based approach is widely used to deal with the constitutive response to study the deformation process and failure mechanism of rock materials [52,53]. According to the statistical damage theory, it is assumed that the damage variable obeys a certain distribution function, such as Weibull distribution [54], improved Harris distribution [55], normal (Gaussian) distribution [56], Haldane distribution [57], etc. A comprehensive review of these distribution functions and corresponding constitutive models is provided by Lin et al. [58]. These distribution functions are widely used in many fields, including geophysics, biology, population growth, and economics [59].

The Weibull distribution function is the most commonly used distribution function to numerically model the rock damage, the expression of which $F(x)$ is shown in Equation (1) [60,61].

$$F(x) = 1 - e^{-\left(\frac{x}{x_0}\right)^m} \quad (1)$$

where x is a variable parameter, e is Euler's number, x_0 and m are Weibull distribution parameters without clear physical meanings and mathematical definitions [62].

Although previous studies have proved that the results obtained by the Weibull distribution are statistically acceptable, only focusing on some statistical indicators such as coefficient of determination (R^2) and root mean square error ($RMSE$) (see Equation (A1) in Appendix A) may mask the shortcomings of this distribution, and blind use of this method may cause large errors [63]. Beyond that, it has also been documented [64–67] that there

is not enough evidence showing that the Weibull distribution always takes precedence over other distributions and whether the Weibull distribution is the most appropriate statistical distribution function for quasi-brittle materials (such as rock, ceramics, and concrete) is questionable. As Weibull emphasized in his pioneering paper [54], the Weibull distribution is an empirical distribution function, similar to other distribution functions. The establishment of a constitutive model using the Weibull distribution is also a complicated process in which three unknown variables need to be solved according to at least three sets of data [57]. It can be found that the Weibull function for possible prediction of the stress–strain (σ-ε) relationship may have three unknown constants (λ, m and ε_0):

$$\sigma = \lambda\left[1 - \frac{1}{e^{\left(\frac{\varepsilon}{\varepsilon_0}\right)^m}}\right] \tag{2}$$

It is best to use distribution functions with fewer parameters in modeling. Referring to Palchik [68], the Haldane distribution was used to establish the constitutive model of rock joints. As shown in Equation (3), the Haldane distribution contains only two known parameters (2 and 1/2):

$$F(x) = \frac{1}{2}\left(1 - e^{-2x}\right) \tag{3}$$

It can be deduced that $F(x) = 0$ when $x = 0$, which is consistent with the fact that the shear stress/shear displacement curve initiates from the origin. Additionally, the graph of Equation (3) is in the first quadrant since the shear displacement is nonnegative, which makes the physical significance of Equation (3) more explicit. Therefore, Equation (3) can be applied to establish a joint shear constitutive curve, namely,

$$\tau = \frac{\alpha}{2}\left(1 - e^{-2u}\right) \tag{4}$$

Since the pre-peak curve must pass the peak point (u_p,τ_p), substitute u_p and τ_p into Equation (4) to obtain the expression of α:

$$\alpha = \frac{2\tau_p}{1 - e^{-2u_p}} \tag{5}$$

Substitute Equation (5) into Equation (4):

$$\tau = \tau_p\left(\frac{1 - e^{-2u}}{1 - e^{-2u_p}}\right) \tag{6}$$

To obtain a model with wider applicability, the correction coefficient $\left(\frac{1-e^{-2u}}{1-e^{-2u_p}}\right)^{\gamma-1}$ is adapted to construct an improved model suitable for describing different pre-peak curves:

$$\tau = \tau_p\left(\frac{1 - e^{-2u}}{1 - e^{-2u_p}}\right)\left(\frac{1 - e^{-2u}}{1 - e^{-2u_p}}\right)^{\gamma-1} = \tau_p\left(\frac{1 - e^{-2u}}{1 - e^{-2u_p}}\right)^{\gamma} \tag{7}$$

Equation (7) is the semi-analytical model based on the Haldane distribution function, which is used to describe the pre-peak stage of the shear stress/shear displacement curve. The model parameter γ can be determined using the statistical method of successive approximations. Whether Equation (7) can predict the pre-peak shear constitutive relationship requires a further comparison between the observed and predicted shear stress/shear displacement curves, which is discussed in Section 3.

3. Model Validation

In the first series of verification tests, the shear test results of artificial joints conducted by Bao et al. [69] were adopted. As shown in Figure 3a, the target joints are taken from the face of the Guanshan tunnel in Gansu Province, China, and the lithology of the rock joints

is diorite. The three-dimensional point cloud of target joints was obtained by scanning the collected samples with a handheld 3D laser scanner, and the resin mold was made. Then, three groups of artificial joint samples with different strengths were made according to the mixing ratio shown in Table 1 (numbered as group a, group b and group c). During the shear test, the normal stresses were set as 0.2 MPa, 0.5 MPa, and 1 MPa, respectively. Jointed samples are numbered X-Y-Z, where X represents group number (a, b, c), Y represents the normal stress, and Z represents the shear direction (0°, 90°, 180°, and 270°). The anisotropic shear test results of each joint sample under different normal stresses are shown in Figure 3b.

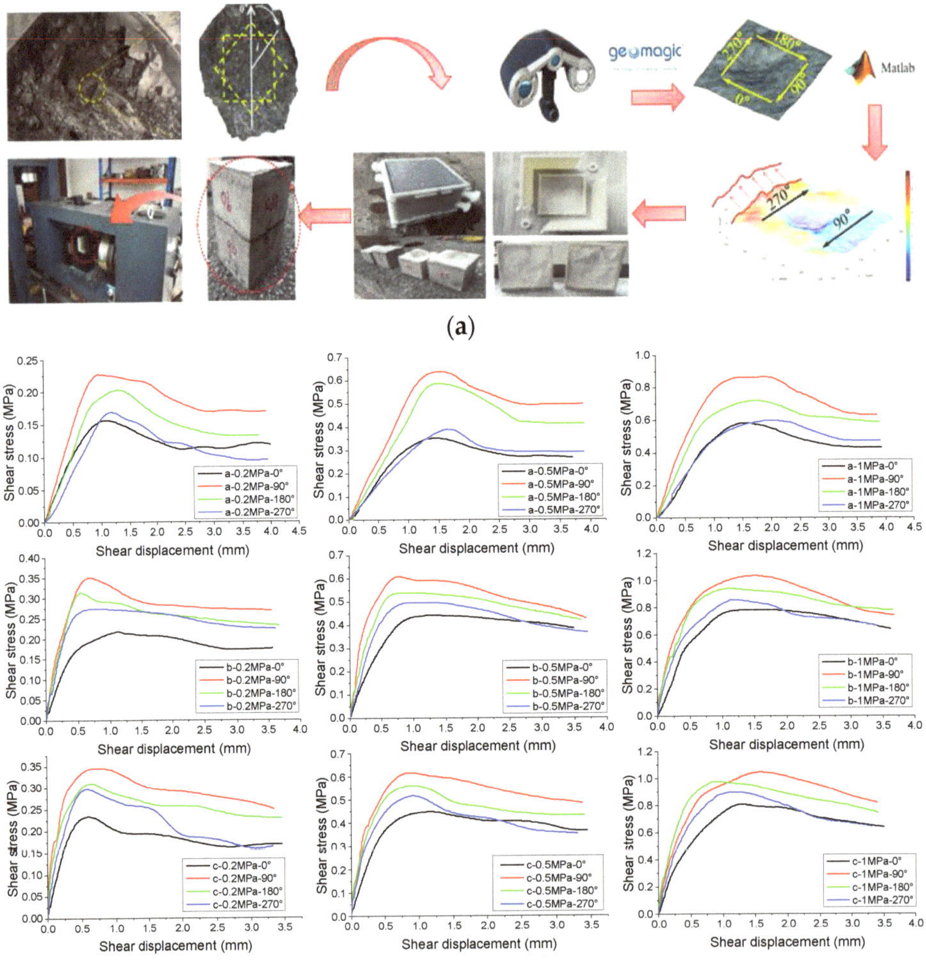

Figure 3. Preparation process and shear test results of joint samples, (**a**) Sample-making process, (**b**) Shear stress-shear displacement curves [69].

Table 1. Ingredient proportions of artificial joints samples [69].

Group	Water	Cement	Sand	Silicon	Water Reducer
a	1	2	2	0	0
b	1	2	3	0.1	0.1
c	1	2	3	0.2	0.2

To facilitate a comparison of model results, the proposed model was compared with Ban et al. [49] model, Bandis et al. [46] model, and Nassir et al. [48] model. The expressions of each model are as follows:

Ban et al. [49] model:

$$\tau = k_i\left[\left(1-\frac{a}{b}\right)u+\frac{au_p}{b^2}\ln\left(\frac{b^2u+bu_p}{bu_p}\right)\right] \tag{8}$$

where parameters a and b are model parameters, and k_i is the initial shear stiffness. All three parameters can be obtained by fitting test data.

Bandis et al. [46] model:

$$\tau = \frac{u}{m+nu} \tag{9}$$

where m represents the reciprocal of the initial shear stiffness and n is the reciprocal of the horizontal asymptote to the hyperbolic curve. Both m and n are positive.

Nassir et al. [48] model:

$$\tau = \frac{k_i u}{1+\delta u} \tag{10}$$

where parameters δ can be obtained by Equation (11):

$$\delta = \frac{1-\frac{\sigma_n\tan(\varphi_b)}{\tau_p\eta}}{u_p\left(\frac{\sigma_n\tan(\varphi_b)}{\tau_p}-1\right)} \tag{11}$$

where η and φ_b are dilation initiation coefficient (ranging from 0 to 1) and basic friction angle, respectively.

According to Refs [70,71], to calibrate the above four models, the expressions of each constitutive model were implemented in Origin software so that we may utilize the Levenberg–Marquardt (L-M) algorithm. The error of each model between the measured stress and the theoretical value is measured by R^2. As shown in Figure 4, the above four models are successively substituted into the experimental results in Figure 3a to solve. Owing to the limited paper length, only the comparison results of experiment b-0.5MPa-0° are given. The R^2 of each theoretical curve fitted by the proposed model and other models was listed in Table 2, which indicates that the theoretical curves agree well with the experimental results. However, it should be noted that the proposed model only contains one unknown parameter γ, while other models contain multiple unknown parameters. Fewer parameters are the advantage of the proposed model.

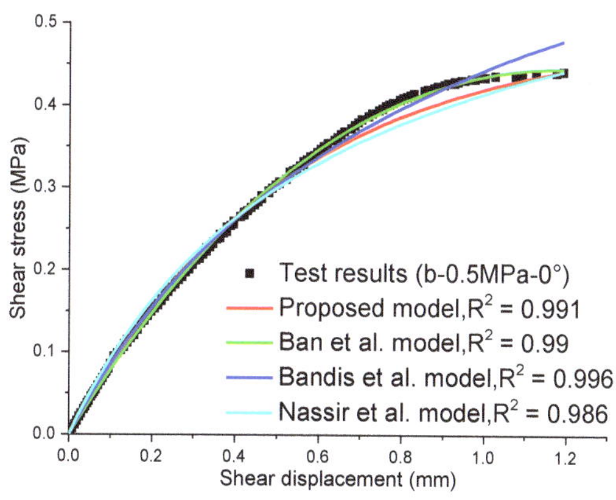

Figure 4. Comparison between model and experimental results of b-0.5MPa-0° [46,48,49].

Table 2. Fitting effect of model and experimental results.

Joint Samples	Model Expression of the Proposed Model	R^2 Proposed Model	Ban et al. [49] Model	Bandis et al. [46] Model	Nassir et al. [48] Model
a-0.2MPa-0°	$\tau = 0.1564\left(\frac{1-e^{-2u}}{1-e^{-2\times 1.03361}}\right)^{1.28}$	0.999	0.991	0.997	0.994
a-0.2MPa-90°	$\tau = 0.2271\left(\frac{1-e^{-2u}}{1-e^{-2\times 0.9496}}\right)^{1.425}$	0.998	0.99	0.98	0.98
a-0.2MPa-180°	$\tau = 0.2036\left(\frac{1-e^{-2u}}{1-e^{-2\times 1.2914}}\right)^{1.559}$	0.983	0.982	0.98	0.98
a-0.2MPa-270°	$\tau = 0.1693\left(\frac{1-e^{-2u}}{1-e^{-2\times 1.174}}\right)^{2.184}$	0.996	/	0.97	0.97
a-0.5MPa-0°	$\tau = 0.352\left(\frac{1-e^{-2u}}{1-e^{-2\times 1.388}}\right)^{1.792}$	0.999	0.994	0.987	0.98
a-0.5MPa-90°	$\tau = 0.6369\left(\frac{1-e^{-2u}}{1-e^{-2\times 1.462}}\right)^{2.208}$	0.982	0.98	0.981	0.972
a-0.5MPa-180°	$\tau = 0.584\left(\frac{1-e^{-2u}}{1-e^{-2\times 1.4486}}\right)^{2.809}$	0.977	/	0.96	0.97
a-0.5MPa-270°	$\tau = 0.3887\left(\frac{1-e^{-2u}}{1-e^{-2\times 1.6379}}\right)^{2.355}$	0.983	0.991	0.983	0.983
a-1.0MPa-0°	$\tau = 0.5807\left(\frac{1-e^{-2u}}{1-e^{-2\times 1.462}}\right)^{2.0407}$	0.999	0.99	0.983	0.974
a-1.0MPa-90°	$\tau = 0.8603\left(\frac{1-e^{-2u}}{1-e^{-2\times 1.421}}\right)^{1.3905}$	0.998	0.992	0.98	0.984
a-1.0MPa-180°	$\tau = 0.7179\left(\frac{1-e^{-2u}}{1-e^{-2\times 1.6919}}\right)^{1.427}$	0.995	0.993	0.985	0.993
a-1.0MPa-270°	$\tau = 0.5947\left(\frac{1-e^{-2u}}{1-e^{-2\times 1.8727}}\right)^{1.846}$	0.996	0.995	0.986	0.99
b-0.2MPa-0°	$\tau = 0.2146\left(\frac{1-e^{-2u}}{1-e^{-2\times 1.0227}}\right)^{0.808}$	0.985	0.99	0.981	0.983
b-0.2MPa-90°	$\tau = 0.35\left(\frac{1-e^{-2u}}{1-e^{-2\times 0.668}}\right)^{0.82}$	0.994	0.989	0.984	0.985
b-0.2MPa-180°	$\tau = 0.3142\left(\frac{1-e^{-2u}}{1-e^{-2\times 0.5537}}\right)^{0.921}$	0.998	0.998	0.988	0.98
b-0.2MPa-270°	$\tau = 0.2731\left(\frac{1-e^{-2u}}{1-e^{-2\times 0.6777}}\right)^{0.7503}$	0.945	0.993	0.975	0.973
b-0.5MPa-0°	$\tau = 0.4419\left(\frac{1-e^{-2u}}{1-e^{-2\times 1.1901}}\right)^{1.0361}$	0.991	0.99	0.996	0.986
b-0.5MPa-90°	$\tau = 0.6062\left(\frac{1-e^{-2u}}{1-e^{-2\times 0.7438}}\right)^{0.72}$	0.979	0.99	0.989	0.988
b-0.5MPa-180°	$\tau = 0.5371\left(\frac{1-e^{-2u}}{1-e^{-2\times 0.8265}}\right)^{0.80}$	0.983	0.979	0.994	0.984
b-0.5MPa-270°	$\tau = 0.4952\left(\frac{1-e^{-2u}}{1-e^{-2\times 0.8926}}\right)^{0.906}$	0.987	0.99	0.991	0.984
b-1.0MPa-0°	$\tau = 0.782\left(\frac{1-e^{-2u}}{1-e^{-2\times 1.2619}}\right)^{1.118}$	0.992	0.99	0.992	0.99
b-1.0MPa-90°	$\tau = 1.0315\left(\frac{1-e^{-2u}}{1-e^{-2\times 1.4763}}\right)^{0.711}$	0.996	0.99	0.99	0.987
b-1.0MPa-180°	$\tau = 0.9371\left(\frac{1-e^{-2u}}{1-e^{-2\times 1.0909}}\right)^{0.764}$	0.976	0.992	0.981	0.979
b-1.0MPa-270°	$\tau = 0.8562\left(\frac{1-e^{-2u}}{1-e^{-2\times 1.1405}}\right)^{0.862}$	0.997	0.99	0.991	0.99
c-0.2MPa-0°	$\tau = 0.2341\left(\frac{1-e^{-2u}}{1-e^{-2\times 0.5905}}\right)^{0.759}$	0.976	0.998	0.976	0.998
c-0.2MPa-90°	$\tau = 0.344\left(\frac{1-e^{-2u}}{1-e^{-2\times 0.6338}}\right)^{0.467}$	0.935	0.96	0.951	0.94
c-0.2MPa-180°	$\tau = 0.3086\left(\frac{1-e^{-2u}}{1-e^{-2\times 0.5918}}\right)^{0.523}$	0.98	0.996	0.97	0.96
c-0.2MPa-270°	$\tau = 0.2981\left(\frac{1-e^{-2u}}{1-e^{-2\times 0.5917}}\right)^{0.683}$	0.985	0.995	0.98	0.982
c-0.5MPa-0°	$\tau = 0.4476\left(\frac{1-e^{-2u}}{1-e^{-2\times 1.128}}\right)^{0.803}$	0.975	0.99	0.971	0.973

Table 2. *Cont.*

Joint Samples	Model Expression of the Proposed Model	R^2 Proposed Model	Ban et al. [49] Model	Bandis et al. [46] Model	Nassir et al. [48] Model
c-0.5MPa-90°	$\tau = 0.6132\left(\frac{1-e^{-2u}}{1-e^{-2\times0.8099}}\right)^{0.6673}$	0.988	0.99	0.964	0.983
c-0.5MPa-180°	$\tau = 0.558\left(\frac{1-e^{-2u}}{1-e^{-2\times0.8316}}\right)^{0.622}$	0.983	0.985	0.979	0.976
c-0.5MPa-270°	$\tau = 0.517\left(\frac{1-e^{-2u}}{1-e^{-2\times0.911}}\right)^{0.661}$	0.99	0.99	0.981	0.98
c-1.0MPa-0°	$\tau = 0.806\left(\frac{1-e^{-2u}}{1-e^{-2\times1.239}}\right)^{1.099}$	0.995	0.99	0.982	0.981
c-1.0MPa-90°	$\tau = 1.0449\left(\frac{1-e^{-2u}}{1-e^{-2\times1.5556}}\right)^{0.828}$	0.997	0.99	0.977	0.983
c-1.0MPa-180°	$\tau = 0.9714\left(\frac{1-e^{-2u}}{1-e^{-2\times0.8282}}\right)^{0.7175}$	0.984	0.996	0.973	0.981
c-1.0MPa-270°	$\tau = 0.8942\left(\frac{1-e^{-2u}}{1-e^{-2\times1.0947}}\right)^{0.8825}$	0.998	0.99	0.981	0.983

Interestingly, in the process of solving, it was found that Ban et al. [49] model, Bandis et al. [46] model, and Nassir et al. [48] model performed poorly for some test results at the initial loading stage, i.e., the prediction results were significantly higher than the test results. On the contrary, the proposed model performed well in the entire loading stage. Taking the test results of a-0.5MPa-0° and a-1MPa-0° as an example, the application of the above four models is presented in Figure 5. It can be found that the Ban et al. [49] model, Bandis et al. [46] model, and Nassir et al. [48] model cannot describe the nonlinear deformation characteristics at the initial loading stage. To facilitate understanding, the shear stress/shear displacement curves of the initial loading stage are excerpted and magnified (i.e., the enlarged view in Figure 5). The R^2 of the above four models in the enlarged view is shown in Figure 6. It can be seen that the fitting effect of the proposed model is still good ($R^2 > 0.9$), while the fitting effect of the other three models is relatively poor (0.539–0.834). In addition, the proposed model guarantees that the model curve passes the peak point, whereas the Ban et al. [49] model and Bandis et al. [46] model do not. All these demonstrate the superiority of the proposed model.

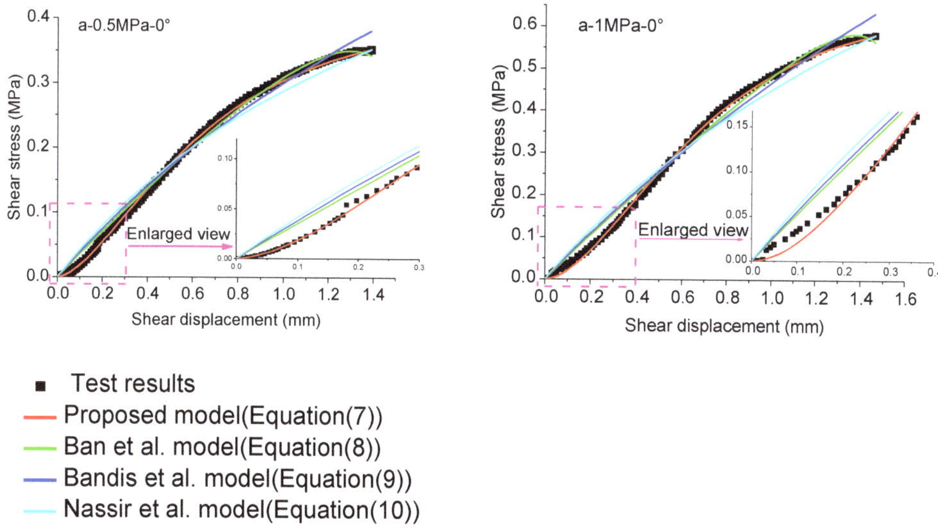

Figure 5. The comparison between each model and experimental results [46,48,49].

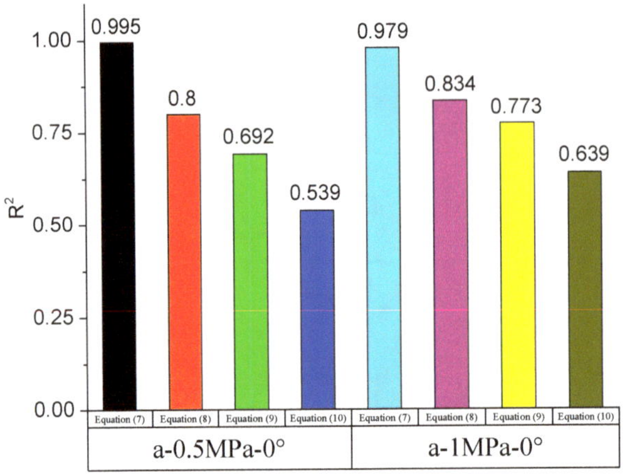

Figure 6. R^2 of each model.

4. Discussion

The shape of the pre-peak shear curve of rock joints depends on roughness, mechanical properties, geological environment, and other perturbation factors. Subjected to certain environmental factors, such as freeze–thaw and immersion, the pre-peak constitutive curve of rock joints is reflected in a convex-type curve. During the initial loading stage, the hangingwall and footwall of rock joints are compressed, and the internal pores and cracks are compacted. This results in a downward concave shear stress/shear displacement curve, and the tangent slope of the curve increases as the shear displacement is applied [72]. There would be a significant error induced by the use of Equations (8)–(10) when describing the pre-peak mechanical behavior of such joints.

The typical convex-type test results in Refs. [73,74] are selected for verification, as shown in Figure 7a,b, which can still be well characterized by the proposed model. However, applying hyperbolic models such as the Ban et al. [49] model, Bandis et al. [46] model, and Nassir et al. [48] model to this validation will result in the parameter fitting being negative (contradictory with each model parameter being greater than 0) or not converging (Figure 8). This is due to the functional nature of the hyperbolic model itself. Under limited conditions (such as model parameters greater than 0), they cannot effectively describe the convex-type curve.

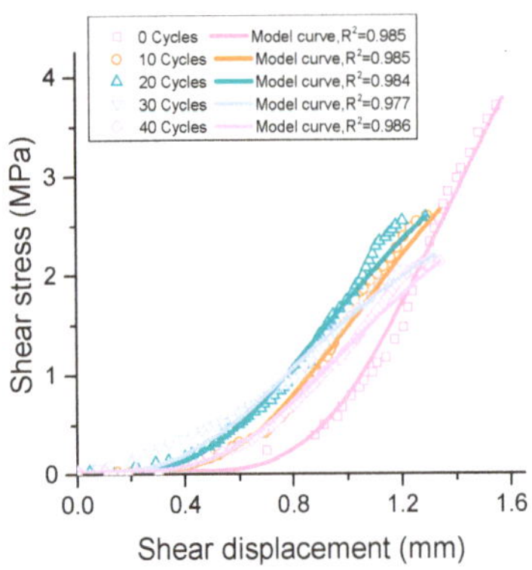

(a) **test results**

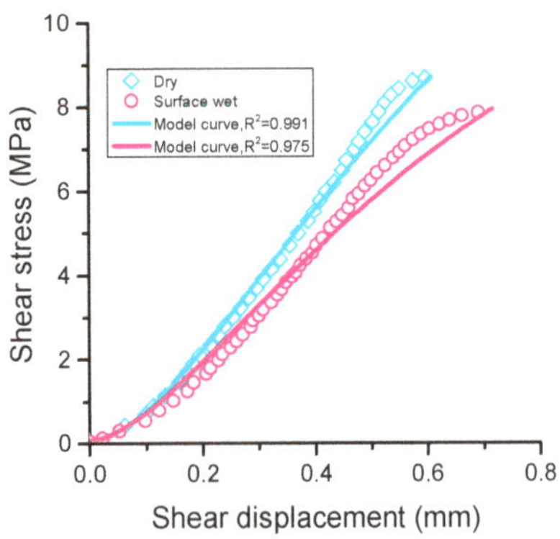

(b) **test results**

Figure 7. Comparison of test results [73,74] with model curves.

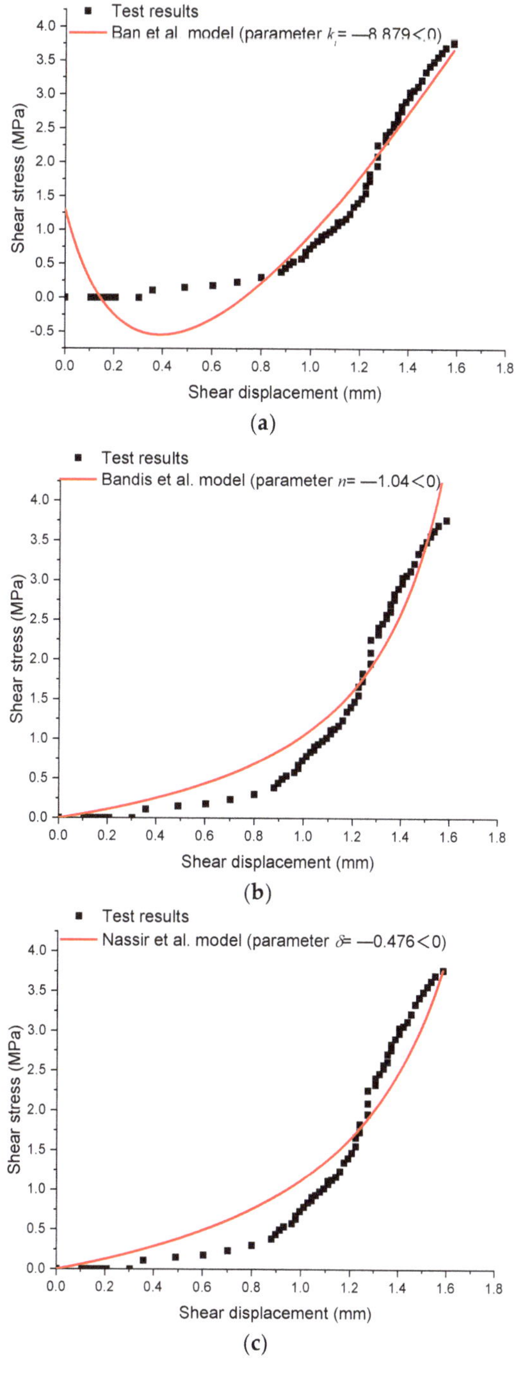

Figure 8. Comparison of test results (0 cycle) [73] with existing models, (**a**) Ban et al. [49] model, (**b**) Bandis et al. [46] model, (**c**) Nassir et al. [48] model.

The experiment is one of the most fundamental and effective research methods. In Section 3, this paper only focuses on the performance of the proposed model in describing

the shear deformation of rock joints in laboratory experiments. Considering that there are many limitations of laboratory experiments, such as difficult field sampling, disturbance to the original state of samples, and defects in laboratory test instruments, the results may not be consistent with reality [75]. Therefore, the in situ shear test, which can obtain the mechanical parameters consistent with engineering practice, was employed.

Taking the test results [76,77] under in-situ conditions as an example, the performance of the model in reproducing in situ rock joint behavior is studied. Figure 9 shows the comparison of shear test data with limestone joints collected from the Lanjiberna limestone mine (located in Odisha, India) with model curves. Figure 10 shows the comparison of shear test results and model curves for sandstone joint samples collected from the upper reservoir of the Azad pump storage power plant project in western Iran. The corresponding fitting curves from the proposed model are well matched with the experimental data in Figures 9 and 10. Obviously, from Figures 9 and 10 (all R^2 values are greater than 0.94), it is not difficult to find that the proposed model can also represent the pre-peak shear deformation behavior of in-situ rock joints to a certain extent. Of course, it is still a complex task to estimate the shear behavior of rock joints under in situ conditions due to various influencing factors presented in the field. We will study the performance of this model in reproducing in situ rock joint behavior in geotechnical engineering features such as rock slopes and tunnels in the future.

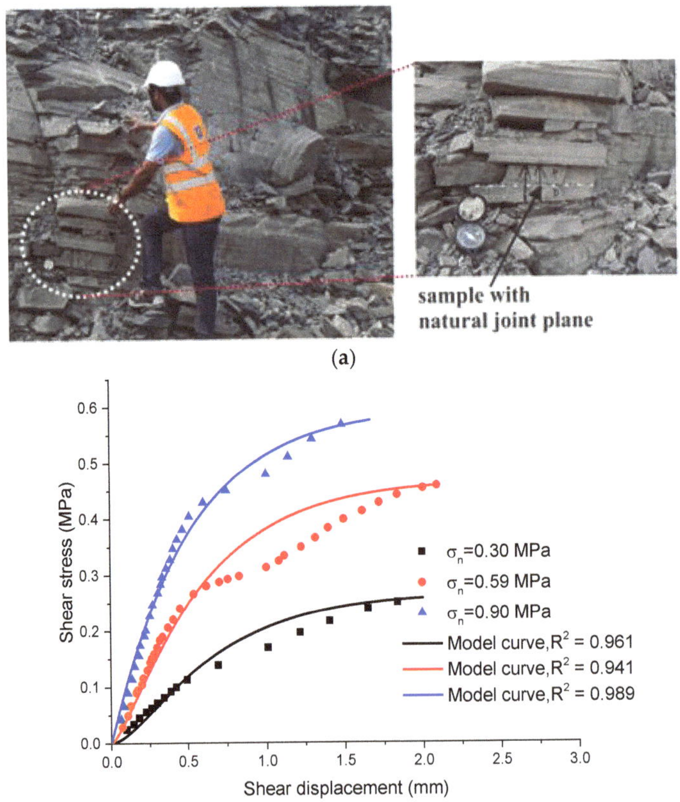

Figure 9. Comparison of in situ test results [76] with model curves, (**a**) sample collection, (**b**) shear stress-shear displacement curve.

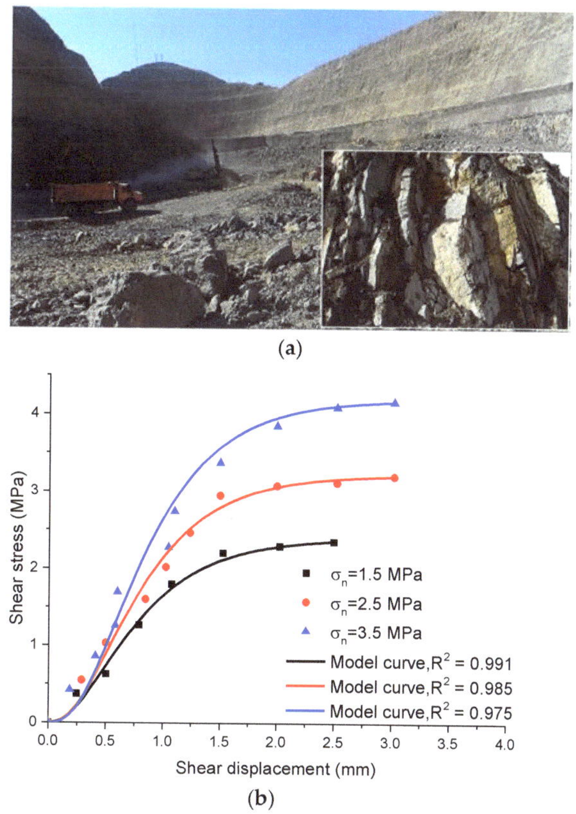

Figure 10. Comparison of in situ test results [77] with model curves, (**a**) sample collection, (**b**) shear stress-shear displacement curve.

5. Conclusions

Based on Haldane distribution, a new semi-analytical model for the pre-peak shear deformation of rock joints is established. The effectiveness of the model was verified by laboratory and in situ experimental data. To better demonstrate the advantages of this model, it is compared with existing models. The results show that the results of the proposed model are similar to those of the existing models for quasilinear-type and concave-type curves. For convex-type curves, the effect of the proposed model is much better than that of the existing models. In addition, compared with existing models that have at least two or more parameters to be fitted, the model in this paper contains only one parameter, which is convenient for application.

Author Contributions: Conceptualization, S.X. and Y.C.; data curation, S.X.; funding acquisition, H.L. and Z.H.; investigation, H.L. and Z.H.; methodology, S.X. and H.D.; writing—original draft, S.X. and Y.C.; writing—review and editing, Z.H. and D.L. All authors have read and agreed to the published version of the manuscript.

Funding: This paper received financial funding from the National Natural Science Foundation of China (No. 52104111 and 42277175) and Postgraduate Research and Practice Innovation Program of Jiangsu Province (No. KYCX21_0119).

Data Availability Statement: Data available on request due to privacy restrictions.

Conflicts of Interest: The authors declare no conflict of interest.

Appendix A

The coefficient of determination (R^2) and root mean square error ($RMSE$) can be used to assess the matching effect of the proposed model against the experimental data, which can be calculated by Equation (A1).

$$\begin{cases} R^2 = 1 - \dfrac{N-1}{N-2} \times \dfrac{\sum_{i=1}^{N}\left(y^i{}_{test} - y^i{}_{cal}\right)^2}{\sum_{i=1}^{N}\left(y^i{}_{test} - y_{ave}\right)^2} \\ RMSE = \sqrt{\dfrac{\sum\limits_{i=1}^{N}\left(y^i{}_{test} - y^i{}_{cal}\right)^2}{N}} \end{cases} \quad \text{(A1)}$$

where N is the number of measured points; y_{test} and y_{cal} are, respectively, the measured stress and the theoretical stress; and y_{ave} is the average of the y_{test}.

References

1. Schreyer, H.; Sulsky, D. Constitutive and numerical framework for modeling joints and faults in rock. *Int. J. Numer. Anal. Met.* **2016**, *40*, 1253–1283. [CrossRef]
2. Luo, X.; Cao, P.; Liu, T.; Zhao, Q.; Meng, G.; Fan, Z.; Xie, W. Mechanical behaviour of anchored rock containing weak interlayer under uniaxial compression: Laboratory test and coupled dem-fem simulation. *Minerals* **2022**, *12*, 492. [CrossRef]
3. Yong, R.; Qin, J.-B.; Huang, M.; Du, S.-G.; Liu, J.; Hu, G.-J. An innovative sampling method for determining the scale effect of rock joints. *Rock Mech. Rock Eng.* **2019**, *52*, 935–946. [CrossRef]
4. Hu, G.J.; Ma, G.; Liu, J.; Qi, K. Size effect of the number of parallel joints on uniaxial compressive strength and characteristic strength. *Minerals* **2022**, *12*, 62. [CrossRef]
5. Lin, Q.; Cao, P.; Meng, J.; Cao, R.; Zhao, Z. Strength and failure characteristics of jointed rock mass with double circular holes under uniaxial compression: Insights from discrete element method modelling. *Theor. Appl. Fract. Mec.* **2020**, *109*, 102692. [CrossRef]
6. Fan, X.; Yu, H.; Deng, Z.; He, Z.; Zhao, Y. Cracking and deformation of cuboidal sandstone with a single nonpenetrating flaw under uniaxial compression. *Theor. Appl. Fract. Mec.* **2022**, *119*, 103284. [CrossRef]
7. Saadat, M.; Taheri, A. A cohesive discrete element based approach to characterizing the shear behavior of cohesive soil and clay-infilled rock joints. *Comput. Geotech.* **2019**, *114*, 103109. [CrossRef]
8. Xie, S.; Han, Z.; Hu, H.; Lin, H. Application of a novel constitutive model to evaluate the shear deformation of discontinuity. *Eng. Geol.* **2022**, *304*, 106693. [CrossRef]
9. Cai, W.; Zhu, H.; Liang, W. Three-dimensional stress rotation and control mechanism of deep tunneling incorporating generalized Zhang–Zhu strength-based forward analysis. *Eng Geol.* **2022**, *308*, 106806. [CrossRef]
10. Yang, Z.; Wu, Y.; Zhou, Y.; Tang, H.; Fu, S. Assessment of machine learning models for the prediction of rate-dependent compressive strength of rocks. *Minerals* **2022**, *12*, 731. [CrossRef]
11. Petrounias, P.; Giannakopoulou, P.P.; Rogkala, A.; Lampropoulou, P.; Koutsopoulou, E.; Papoulis, D.; Tsikouras, B.; Hatzipanagiotou, K. The impact of secondary phyllosilicate minerals on the engineering properties of various igneous aggregates from greece. *Minerals* **2018**, *8*, 329. [CrossRef]
12. Gronen, L.H.; Sindern, S.; Katzmarzyk, J.L.; Bormann, U.; Hellmann, A.; Wotruba, H.; Meyer, F.M. Mineralogical and chemical characterization of zr-ree-nb ores from khalzan buregtei (mongolia)approaches to more efficient extraction of rare metals from alkaline granitoids. *Minerals* **2019**, *9*, 217. [CrossRef]
13. Wang, J.A.; Park, H.D. Fluid permeability of sedimentary rocks in a complete stress-strain process. *Eng. Geol.* **2002**, *63*, 291–300. [CrossRef]
14. Xie, S.; Lin, H.; Cheng, C.; Chen, Y.; Wang, Y.; Zhao, Y.; Yong, W. Shear strength model of joints based on Gaussian smoothing method and macro-micro roughness. *Comput. Geotech.* **2022**, *143*, 104605. [CrossRef]
15. Patel, S.; Martin, C.D. Effect of stress path on the failure envelope of intact crystalline rock at low confining stress. *Minerals* **2020**, *10*, 1119. [CrossRef]
16. Xu, X.Q.; Wu, W.L.; Xu, W.B. Sulfate-dependent shear behavior of cementing fiber-reinforced tailings and rock. *Minerals* **2020**, *10*, 1032. [CrossRef]
17. Han, Z.; Li, D.; Li, X. Dynamic mechanical properties and wave propagation of composite rock-mortar specimens based on SHPB tests. *Int. J. Min. Sci. Technol.* **2022**, *32*, 793–806. [CrossRef]
18. Wu, Y.; Zhou, Y. Hybrid machine learning model and Shapley additive explanations for compressive strength of sustainable concrete. *Constr. Build. Mater.* **2022**, *330*, 127298. [CrossRef]
19. Han, Z.; Li, D.; Li, X. Effects of axial pre-force and loading rate on Mode I fracture behavior of granite. *Int. J. Rock Mech. Min Sci.* **2022**, *157*, 105172. [CrossRef]

20. Pardoen, B.; Seyedi, D.M.; Collin, F. Shear banding modelling in cross-anisotropic rocks. *Int. J. Solids Struct.* **2015**, *72*, 63–87. [CrossRef]
21. Bahaaddini, M.; Hagan, P.C.; Mitra, R.; Hebblewhite, B.K. Scale effect on the shear behaviour of rock joints based on a numerical study. *Eng. Geol.* **2014**, *181*, 212–223. [CrossRef]
22. Asadollahi, P.; Tonon, F. Constitutive model for rock fractures: Revisiting Barton's empirical model. *Eng. Geol.* **2010**, *113*, 11–32. [CrossRef]
23. Mehrishal, S.; Sharifzadeh, M.; Song, J.J.; Shahriar, K. An investigation of the possibility of estimating the residual shear strength of rough joints using planar small joint in limestone rocks. *J. Struct. Geol.* **2021**, *142*, 104220. [CrossRef]
24. Wang, X.Q.; Kang, H.P.; Gao, F.Q. Numerical investigation on the shear behavior of jointed coal mass. *Comput. Geotech.* **2019**, *106*, 274–285. [CrossRef]
25. Karami, A.; Stead, D. Asperity degradation and damage in the direct shear test: A hybrid fem/dem approach. *Rock Mech. Rock Eng.* **2008**, *41*, 229–266. [CrossRef]
26. Asadi, M.S.; Rasouli, V.; Barla, G. A laboratory shear cell used for simulation of shear strength and asperity degradation of rough rock fractures. *Rock Mech. Rock Eng.* **2013**, *46*, 683–699. [CrossRef]
27. Bahaaddini, M.; Hagan, P.C.; Mitra, R.; Khosravi, M.H. Experimental and numerical study of asperity degradation in the direct shear test. *Eng. Geol.* **2016**, *204*, 41–52. [CrossRef]
28. Cai, W.; Zhu, H.; Liang, W. Three-dimensional tunnel face extrusion and reinforcement effects of underground excavations in deep rock masses. *Int. J. Rock Mech. Min. Sci.* **2022**, *150*, 104999. [CrossRef]
29. Meng, F.; Zhou, H.; Li, S.; Zhang, C.; Wang, Z.; Kong, L.; Zhang, L. Shear behaviour and acoustic emission characteristics of different joints under various stress levels. *Rock Mech. Rock Eng.* **2016**, *49*, 4919–4928. [CrossRef]
30. Cui, G.; Zhang, C.; Han, H.; Zeng, Z.; Zhou, H.; Jingjing, L.U. Experiment study on shear behavior of artificial joint under CNL and CNS boundary conditions. *Chin. J. Rock Mech. Eng.* **2019**, *38*, 3384–3392.
31. Fan, W.; Cao, P.; Long, L. Degradation of joint surface morphology, shear behavior and closure characteristics during cyclic loading. *J. Cent South Univ.* **2018**, *25*, 653–661. [CrossRef]
32. Jing, L. A review of techniques, advances and outstanding issues in numerical modelling for rock mechanics and rock engineering. *Int. J. Rock Mech. Min. Sci.* **2003**, *40*, 283–353. [CrossRef]
33. Muralha, J.; Grasselli, G.; Tatone, B.; Blumel, M.; Chryssanthakis, P.; Jiang, Y.J. ISRM Suggested Method for Laboratory Determination of the Shear Strength of Rock Joints: Revised Version. *Rock Mech. Rock Eng.* **2014**, *47*, 291–302. [CrossRef]
34. Xie, S.J.; Lin, H.; Chen, Y.F.; Yong, R.; Xiong, W.; Du, S.G. A damage constitutive model for shear behavior of joints based on determination of the yield point. *Int. J. Rock Mech. Min. Sci.* **2020**, *128*, 104269. [CrossRef]
35. Vakili, A. An improved unified constitutive model for rock material and guidelines for its application in numerical modelling. *Comput. Geotech.* **2016**, *80*, 261–282. [CrossRef]
36. Cai, W.; Zhu, H.; Liang, W.; Wang, X.; Su, C.; Wei, X. A post-peak dilatancy model for soft rock and its application in deep tunnel excavation. *J. Rock Mech. Geotech. Eng.* **2022**. [CrossRef]
37. Cai, W.; Zhu, H.; Liang, W.; Vu, B.; Su, C.; Zhang, K.; Wang, X. Three-dimensional forward analysis and real-time design of deep tunneling based on digital in-situ testing. *Int. J. Mech. Sci.* **2022**, *226*, 107385. [CrossRef]
38. Singh, H.K.; Basu, A. Evaluation of existing criteria in estimating shear strength of natural rock discontinuities. *Eng. Geol.* **2018**, *232*, 171–181. [CrossRef]
39. Develi, K. Computation of direction dependent joint surface parameters through the algorithm of triangular prism surface area method: A theoretical and experimental study. *Int. J. Solids Struct.* **2020**, *202*, 895–911. [CrossRef]
40. Xie, S.J.; Lin, H.; Wang, Y.X.; Chen, Y.F.; Xiong, W.; Zhao, Y.L.; Du, S. A statistical damage constitutive model considering whole joint shear deformation. *Int. J. Damage Mech.* **2020**, *29*, 988–1008. [CrossRef]
41. Xie, S.J.; Han, Z.Y.; Chen, Y.F.; Wang, Y.X.; Zhao, Y.L.; Lin, H. Constitutive modeling of rock materials considering the void compaction characteristics. *Arch. Civ. Mech. Eng.* **2022**, *22*, 60. [CrossRef]
42. Lin, Q.B.; Cao, P.; Wen, G.P.; Meng, J.J.; Cao, R.H.; Zhao, Z.Y. Crack coalescence in rock-like specimens with two dissimilar layers and pre-existing double parallel joints under uniaxial compression. *Int. J. Rock Mech. Min. Sci.* **2021**, *139*, 104621. [CrossRef]
43. Goodman, R.E. *Methods of Geological Engineering in Discontinuous Rocks*; West Publishing, Co.: St. Paul, MN, USA, 1976.
44. Saeb, S.; Amadei, B. Modelling rock joints under shear and normal loading. *Int. J. Rock Mech. Min. Sci.* **1992**, *29*, 267–278. [CrossRef]
45. Kulhawy, F.H. Stress deformation properties of rock and rock discontinuities. *Eng. Geol.* **1975**, *9*, 327–350. [CrossRef]
46. Bandis, S.C.; Lumsden, A.C.; Barton, N.R. Fundamentals of rock joint deformation. *Int. J. Rock Mech. Min. Sci.* **1983**, *20*, 249–268. [CrossRef]
47. Desai, C.S.; Fishman, K.L. Plasticity-based constitutive model with associated testing for joints. *Int. J. Rock Mech. Min. Sci.* **1991**, *28*, 15–26. [CrossRef]
48. Nassir, M.; Settari, A.; Wan, R. Joint stiffness and deformation behaviour of discontinuous rock. *J. Can. Petrol. Technol.* **2010**, *49*, 78–86. [CrossRef]
49. Ban, L.; Qi, C.; Shan, R.; Tao, Z.; Xia, C.; Jiang, K. Pre-peak shear constitutive model considering the softening shear stiffness and its influencing factors. *J. China Coal. Soc.* **2018**, *43*, 2765–2772. (In Chinese)

50. Kou, M.M.; Liu, X.R.; Tang, S.D.; Wang, Y.T. Experimental study of the prepeak cyclic shear mechanical behaviors of artificial rock joints with multiscale asperities. *Soil Dyn. Earthq. Eng.* **2019**, *120*, 58–74. [CrossRef]
51. Shen, H.; Liu, Y.; Liu, B.; Li, H.; Wu, D.; Peng, B. Nonlinear theoretical model for describing shear mechanical behaviors of rock joints. *Chin. J. Rock Mech. Eng.* **2021**, *40*, 2421–2433.
52. Zhao, H.; Zhou, S.; Zhao, M.H.; Shi, C.J. Statistical micromechanics-based modeling for low-porosity rocks under conventional triaxial compression. *Int. J. Geomech.* **2018**, *18*, 04018019. [CrossRef]
53. Li, X.; Cao, W.-G.; Su, Y.-H. A statistical damage constitutive model for softening behavior of rocks. *Eng. Geol.* **2012**, *143*, 1–17. [CrossRef]
54. Weibull, W. A statistical distribution function of wide applicability. *J. Appl. Mech.* **1951**, *18*, 293–297. [CrossRef]
55. Xie, S.J.; Lin, H.; Wang, Y.X.; Cao, R.H.; Yong, R.; Du, S.G.; Li, J. Nonlinear shear constitutive model for peak shear-type joints based on improved Harris damage function. *Arch. Civ. Mech. Eng.* **2020**, *20*, 1–14. [CrossRef]
56. Cao, W.G.; Li, X.; Zhao, H. Damage constitutive model for strain-softening rock based on normal distribution and its parameter determination. *J. Cent. South Univ.* **2007**, *14*, 719–724. [CrossRef]
57. Palchik, V. Stress-strain model for carbonate rocks based on Haldane's distribution function. *Rock Mech. Rock Eng.* **2006**, *39*, 215–232. [CrossRef]
58. Lin, H.; Feng, J.J.; Cao, R.H.; Xie, S.J. Comparative analysis of rock damage models based on different distribution functions. *Geotech. Geol. Eng.* **2021**, *40*, 301–310. [CrossRef]
59. Sun, Z.; Niu, D.; Wang, X.; Zhang, L.; Luo, D. Bond behavior of coral aggregate concrete and corroded Cr alloy steel bar. *J. Build. Eng.* **2022**, *61*, 105294. [CrossRef]
60. Wang, Z.-L.; Li, Y.-C.; Wang, J.G. A damage-softening statistical constitutive model considering rock residual strength. *Comput. Geosci.* **2007**, *33*, 1–9. [CrossRef]
61. Yang, B.C.; Xue, L.; Wang, M.M. Evolution of the shape parameter in the Weibull distribution for brittle rocks under uniaxial compression. *Arab. J. Geosci.* **2018**, *11*, 1–7. [CrossRef]
62. Shen, P.; Tang, H.; Wang, D.; Ning, Y.; Zhang, Y.; Su, X. A statistical damage constitutive model based on unified strength theory for embankment rocks. *Mar. Georesour. Geotec.* **2020**, *38*, 818–829. [CrossRef]
63. Xie, S.J.; Lin, H.; Chen, Y.F.; Wang, Y.X. A new nonlinear empirical strength criterion for rocks under conventional triaxial compression. *J. Cent. South Univ.* **2021**, *28*, 1448–1458. [CrossRef]
64. Deng, J.A.; Gu, D.S. On a statistical damage constitutive model for rock materials. *Comput. Geosci.* **2011**, *37*, 122–128. [CrossRef]
65. Lu, C.S.; Danzer, R.; Fischer, F.D. Fracture statistics of brittle materials: Weibull or normal distribution. *Phys. Rev. E* **2002**, *65*, 067102. [CrossRef] [PubMed]
66. Basu, B.; Tiwari, D.; Kundu, D.; Prasad, R. Is Weibull distribution the most appropriate statistical strength distribution for brittle materials? *Ceram. Int.* **2009**, *35*, 237–246. [CrossRef]
67. Zhou, X.P.; Yang, H.Q. Micromechanical modeling of dynamic compressive responses of mesoscopic heterogenous brittle rock. *Theor. Appl. Fract. Mec.* **2007**, *48*, 1–20. [CrossRef]
68. Palchik, V. Simple stress–strain model of very strong limestones and dolomites for engineering practice. *Geomech. Geophys. Geo-Energy Geo-Resour.* **2019**, *5*, 345–356. [CrossRef]
69. Bao, H.; Xu, X.; Lan, H.; Yan, C.; Xu, J. Stiffness model of rock joint by considering anisotropic morphology. *J. Traffic Transp. Eng.* **2021**, *11*, 1–18. (In Chinese)
70. Wang, S.; Chester, S.A. Experimental characterization and continuum modeling of inelasticity in filled rubber-like materials. *Int. J. Solids Struct.* **2018**, *136*, 125–136. [CrossRef]
71. Wang, S.; Chester, S.A. Modeling thermal recovery of the Mullins effect. *Mech. Mater.* **2018**, *126*, 88–98. [CrossRef]
72. Wang, W.; Yao, Q.; Tang, C.; Li, X.; Chong, Z.; Xu, Q. Experimental study on the shear characteristics and weakening mechanism of water-bearing rock joints. *Bull. Eng. Geol. Environ.* **2021**, *80*, 7653–7668. [CrossRef]
73. Lei, D.X.; Lin, H.; Chen, Y.F.; Cao, R.H.; Wen, Z.J. Effect of cyclic freezing-thawing on the shear mechanical characteristics of nonpersistent joints. *Adv. Mater. Sci. Eng.* **2019**, *2019*, 9867681. [CrossRef]
74. Li, B.; Ye, X.; Dou, Z.; Zhao, Z.; Li, Y.; Yang, Q. Shear strength of rock fractures under dry, surface wet and saturated conditions. *Rock Mech. Rock Eng.* **2020**, *53*, 2605–2622. [CrossRef]
75. Xie, S.; Han, Z.; Lin, H. A quantitative model considering crack closure effect of rock materials. *Int. J. Solids Struct.* **2022**, *251*, 111758. [CrossRef]
76. Basu, A.; Ram, B.K.; Nanda, N.K.; Nayak, S.S. Deterioration of shear strength parameters of limestone joints under simulated acid rain condition. *Int. J. Rock Mech. Min. Sci.* **2020**, *135*, 104508. [CrossRef]
77. Behnia, M.; Nateghpour, B.; Tavakoli, J.; Broujerdi, M.S. Comparison of experimental and empirical methods for estimating the shear strength of rock joints based on the statistical approach. *Environ. Earth Sci.* **2020**, *79*, 361. [CrossRef]

minerals

MDPI

Article

Application of a Hybrid Machine Learning Model for the Prediction of Compressive Strength and Elastic Modulus of Rocks

Xiaoliang Jin [1], Rui Zhao [2,*] and Yulin Ma [1]

[1] School of Architecture and Civil Engineering, Zhengzhou University of Industrial Technology, Zhengzhou 451100, China
[2] School of Architecture and Civil Engineering, Zhengzhou University of Aeronautics, Zhengzhou 451100, China
* Correspondence: civilzz@163.com

Abstract: This paper presents a machine learning-based approach to estimating the compressive strength and elastic modulus of rocks. A hybrid model, GWO-ELM, was built based on an extreme learning machine network optimized by the grey wolf algorithm. The proposed model was carried out on 101 experimental datasets, and four commonly used models were used as benchmarks to evaluate the accuracy of the proposed model. The results showed that the proposed hybrid model can accurately achieve the prediction of elastic modulus and compressive strength with high correlation coefficients and small prediction errors. The prediction performance of the hybrid model is significantly better than the other four original models, and it is an alternative model for predicting the compressive strength and elastic modulus of rocks, which is recommended as an auxiliary tool for real-time prediction of rock mechanical properties.

Keywords: compressive strength; elastic modulus; machine learning; grey wolf algorithm; rock

Citation: Jin, X.; Zhao, R.; Ma, Y. Application of a Hybrid Machine Learning Model for the Prediction of Compressive Strength and Elastic Modulus of Rocks. *Minerals* **2022**, *12*, 1506. https://doi.org/10.3390/min12121506

Academic Editor: Yosoon Choi

Received: 28 October 2022
Accepted: 24 November 2022
Published: 25 November 2022

1. Introduction

When designing and building geotechnical engineering structures, the mechanical properties of rocks are crucial [1–4]. Uniaxial compression strength (UCS), a crucial rock mechanical property metric, determines the material's mechanical strength, which is crucial for assessing the adaptability and resilience of rocks to weathering. The modulus of elasticity (E) is essential to the analysis of deformation and can be used to estimate the bearing capacity of rocks. The slope of the stress–strain curves of unconfined compression tests can be used to assess their uniaxial compressive strength and modulus of elasticity of rocks. The steps and tests for determining these parameters have been standardized by the International Society for Rock Mechanics, American Society for Testing and Materials (ASTM), and the International Society for Rock Mechanics (ISRM) [5]. However, laboratory tests are destructive and are time-consuming and uneconomical to implement for a lot of experimental samples. In addition, the results of the tests are influenced by the quality of the rock samples. Therefore, to some extent, there are some limitations in obtaining these two parameters directly from laboratory experiments. For this reason, it is of positive scientific and engineering interest to develop a nondestructive, cost-effective, fast, efficient, and reliable method for estimating mechanical properties [6–8].

According to the findings of relevant studies, rock qualities and their indices, such as porosity (n), point loading index ($Is_{(50)}$), p-wave velocity (Vp), Schmidt hammer bounce number (Rn), water absorption, mineral composition, etc., have a significant effect on both UCS and E [9–11]. These two mechanical parameters can be determined in a nondirect way from the point of view of the influencing parameters. In this regard, simple models based on statistical regression seem to provide new methods for the prediction of UCS and E. For

the prediction of UCS and E, a number of single regression formulations have been created employing variables including the Schmidt hammer rebound number, p-wave velocity (V_p), and point load strength (PLS) [6,12–15]. In rock engineering, however, the empirical link between a single parameter and UCS or E is not sufficient. Multiple regression analysis was devised to predict UCS, taking into account several criteria related to rock qualities, to overcome this issue [16–19]. It was discovered that the prediction accuracy of the majority of multiple regression equations was not adequate and the predictions given by these models were still subject to large errors. This is mainly attributed to the fact that the regression equations, while being easy to use, also limit the relationship between the input parameters and the output results. In fact, this nonlinear correlation is complex and the established explicit formulas do not adequately reflect this intrinsic relationship. Moreover, statistical prediction methods are deficient for newly available data, i.e., the form of the obtained equation needs to be refitted and updated if the new data differ from the original data [20].

The advent of various soft technologies, such as intelligent models and algorithms [21–24], along with the advancement of computer technology has greatly contributed to new prospects for the regression problem. Predictive models based on intelligent algorithms such as the artificial neural network (ANN) [16,25,26], support vector regression (SVR) [27–32], gene programming (GEP) [6,33–35], random forest (RF) [36,37], long short-term memory (LSTM) [38], and adaptive neuro-fuzzy inference system (ANFIS) [39–41] have received considerable attention and application in the regression analysis of rock properties. For example, Danial et al. [42] employed an adaptive neuro-fuzzy inference system on 45 rock samples to predict compressive strength and elastic modulus, and the model reached better prediction results than multiple regression analysis and an ANN. The R^2 for the prediction of UCS and E were 0.985 and 0.990, respectively. Naseer et al. [43] employed five parameters as input feature vectors to predict the compressive strength and elastic modulus of rocks under different thermal conditions, and the prediction results using multiple linear regression (MLR), ANN, RF, and k-nearest neighbor (KNN) models were compared, where RF achieved the best prediction with R^2 of 0.97 and MAPE of 0.25% for both UCS and E. Yang et al. [44] developed a particle swarm optimization-based SVR model to predict the rate-dependent compressive strength of rocks, and the results showed that the proposed hybrid model predicted better results than the elm and RF models. Matin et al. [36] developed a nonparametric prediction model for UCS and E using the RF model with variable rock properties including porosity (n), point load index ($Is_{(50)}$), p-wave velocity (V_p), and Schmidt hammer rebound number (R_n) The resulting correlation coefficients were 0.93 and 0.91. respectively, which demonstrates that the RF model is a trustworthy and precise method for assessing and estimating complex interactions in rock mechanics. To compare the prediction results of different models for UCS, Arsalan et al. [31] conducted a comparative analysis of model prediction effectiveness on 170 datasets using several models. The results showed that Gaussian process regression (GPR) obtained the best performance with the highest correlation coefficient of 0.9955 and the smallest RMSE of 0.52169. These studies have shown that the predictions using soft computing methods are more accurate than those of traditional statistical models. This is because these models can capture the nonlinear correlation between the input and output in an implicit form instead of an explicit equation.

Although there have been many studies using different intelligent models to predict UCS and elastic modulus, little research has been reported on the use of extreme learning machines (ELMs) in predicting the mechanical property prediction of rock samples [45,46]. In contrast, ELM modes have been widely used in regression analysis such as solving concrete strength prediction [47,48]. Therefore, this paper aims to investigate the predictability of the ELM model for UCS and E. For the ELM network structure, the combined selection of weights and bias has a great effect on the predictive performance of the model. Considering the optimal performance of the GWO algorithm in the selection of hyperparameters of C and γ for the kernel-based ELM model with Gauss kernel function [46], which is an improved algorithm of the ELM model, the GWO algorithm was also employed to optimize

and select the best combination of weights and bias to achieve the accurate prediction of UCS and E in this paper.

2. Methodology

2.1. Extreme Learning Machine Network

The extreme learning machine (ELM) is applicable to the output prediction problem of complex nonlinear systems and has the advantages of fast learning speed, good generalization performance, good scalability, and low computational complexity. The extreme learning machine adopts a three-layer structure, which is shown in Figure 1, with N arbitrary samples, where $x_i \in R^m$ and $y_i \in R^n$. Then, the extreme learning machine model with L hidden layer nodes can be represented as

$$\sum_{j=1}^{L} \beta_j g(\omega_j x_i + b_j) = y_i \tag{1}$$

where β_j is the output weight, $g(x)$ is the activation function, w_j is the input weight vector, b_j is the bias, and y_i is the network output of the sample.

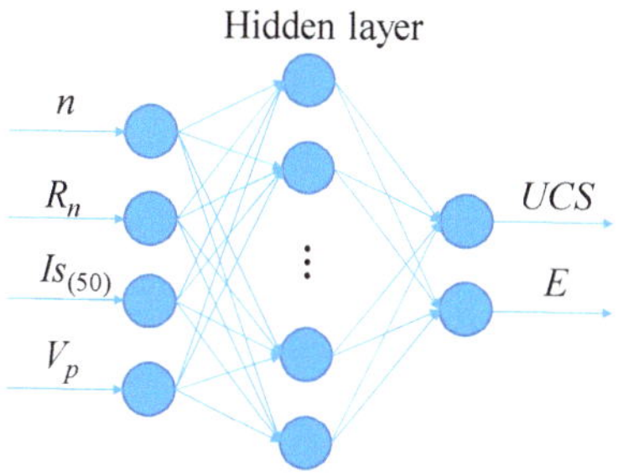

Figure 1. Extreme learning machine network structure.

If the single hidden layer model is able to approximate N samples with zero error, i.e., it satisfies $\sum_{j=1}^{N} \|y_i - t_i\|^2 = 0$, then there exist parameters w_j, b_j, and β_j that satisfy the following conditions.

$$\sum_{j=1}^{L} \beta_j g(\omega_j x_i + b_j) = t_i \tag{2}$$

Transformed into matrix form, the equation can be derived as follows.

$$H\beta = T$$

$$H = \begin{bmatrix} h(x_1) \\ \vdots \\ h(x_N) \end{bmatrix} = \begin{bmatrix} g(\omega_1 x_1 + b_1) & \cdots & g(\omega_L x_1 + b_L) \\ \vdots & & \vdots \\ g(\omega_1 x_N + b_1) & \cdots & g(\omega_L x_N + b_L) \end{bmatrix} \tag{3}$$

Then, the output weight matrix β can be expressed as:

$$\beta = H^{\dagger} T \tag{4}$$

$H^{\dagger}$ is the Moore–Penrose generalized inverse matrix of H.

2.2. Grey Wolf Algorithm Optimizer

The social structure and hunting habits of grey wolf populations served as the inspiration for the development of the wolf optimization algorithm, a new pack intelligence optimization method [49,50]. To find the best solution to the issue at hand, the algorithm mimics the behavior of grey wolves in nature with regard to task division, hunting, and pouncing on prey [51,52]. The wolf hunting process is described as follows.

(1) Social hierarchical stratification. Grey wolves are good at running, like to live in groups, and follow a strict social hierarchy among wolves. As shown in Figure 2, wolves can be divided into 4 grades according to the hierarchy, from high to low, α, β, δ, and ω. Calculate the physical fitness of each grey wolf individual, and mark the 3 grey wolves with the best physical fitness as α, β, and δ in turn, and the remaining grey wolves as ω. The chase is initiated by α, β, and δ, ω tracks and rounds up prey, and the position of the prey corresponds to the global optimal solution of the optimization problem [53].

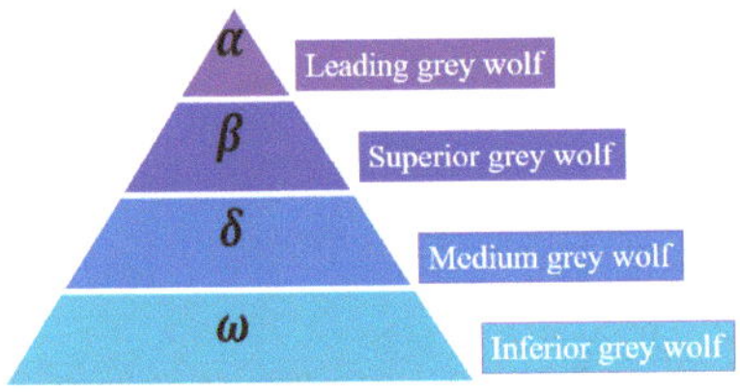

Figure 2. The social hierarchy of the grey wolves.

(2) Surrounding prey. When hunting, grey wolves will first evaluate the distance between themselves and their prey before approaching slowly and then circling.

$$D = |C \cdot X_P(t) - X(t)| \tag{5}$$

$$C = 2r_1 \tag{6}$$

where t denotes the number of current iterations; $X_P(t)$ denotes the position vector of the prey; $X(t)$ denotes the position vector of the grey wolf; C is the coefficient vector; r_1 is a random vector taking values between [0,1]; D denotes the distance between the wolf and the prey. Subsequently, the wolves have to update their positions according to the prey orientation.

$$X(t+1) = X_P(t) - A \cdot D \tag{7}$$

$$A = 2a \cdot r_2 - a \tag{8}$$

where A is a vector of coefficients; r_2 is a random vector taking values between [0,1].

(3) Hunting. When the grey wolf determines the location of the prey, the alpha wolf will lead the beta wolf and the δ wolf in pursuit. α, β, and δ are the most capable among the wolves and are closest to their prey. Use the position relationship of the three to judge the position of the prey, guide the wolf to attack the position of the prey, and finally capture the prey after multiple positioning and movement. The process of position updating is as follows.

$$\begin{cases} D_\alpha = |C_1 \cdot X_\alpha(t) - X(t)| \\ D_\beta = |C_2 \cdot X_\beta(t) - X(t)| \\ D_\delta = |C_3 \cdot X_\delta(t) - X(t)| \\ X_1 = X_\alpha - A_1 \cdot D_\alpha \\ X_2 = X_\beta - A_2 \cdot D_\beta \\ X_3 = X_\delta - A_3 \cdot D_\delta \\ X(t+1) = \frac{X_1+X_2+X_3}{3} \end{cases} \tag{9}$$

where $X(t+1)$ is the updated individual position of the grey wolf. The ω wolves are guided by the 3 head wolves towards the prey position.

To improve the model performance, the grey wolf algorithm is introduced in this paper to optimize the selection of weights w and bias b. The model framework established is presented in Figure 3.

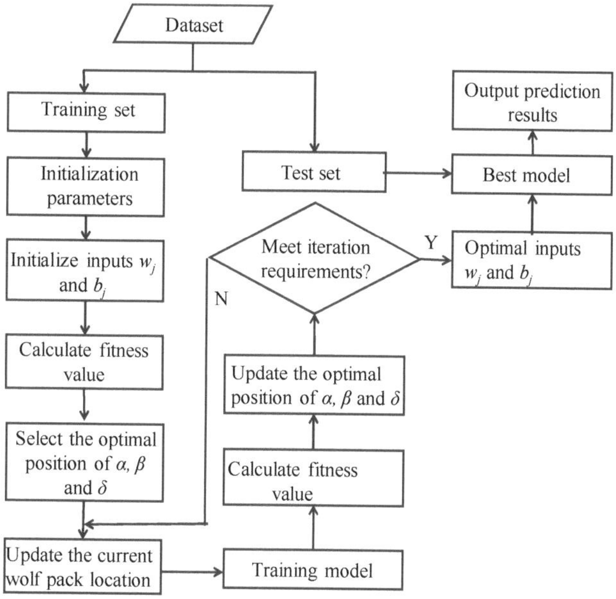

Figure 3. The schematic diagram of the GWO-ELM model structure.

3. Database

Referring to the model input features studied above, in this paper, four effective rock features, namely porosity (n), p-wave velocity (V_p), Schmidt hammer rebound number (R_n), and point loading index ($Is_{(50)}$), are used as input variables of the model to predict UCS and E. For intelligent models, a rich and representative database is necessary. For this purpose, 101 datasets with four input variables and two output variables each were gathered from the literature [16,54], in which 71 granite block samples were collected from the face of the tunnel and 30 Travertine samples were extracted from the mine. More detailed information about the rock samples can be found in References [16,54]. The distribution of these variables and their correlation coefficients with each other are shown in Figures 4 and 5, respectively. The statistical characteristics of these variables are summarized in Table 1. The training process used 10-fold cross-validation [55,56], and the dataset was randomly divided into a training set (80 samples) and a test set (21 samples). For comparison purposes, the dataset was fixed after the first random selection. The training set and the test set are thus the same for different models.

Table 1. Mathematical statistical characteristics of the database.

Variable	n	R_n	V_P	$Is_{(50)}$	UCS	E
Unit	%	-	m/s	MPa	MPa	GPa
Number	101	101	101	101	101	101
Max	10.27	61	7943	7.1	211.9	183.3
Min	0.1	25.63	2823	0.89	22.7	3.05
Mean	2.21	43.13	5517.80	3.28	93.03	63.45
Median	0.46	46	5450	3.16	95.6	66.7
SD	3.24	11.13	930.94	1.28	50.15	47.80
Type	Input	Input	Input	Input	Output	Output

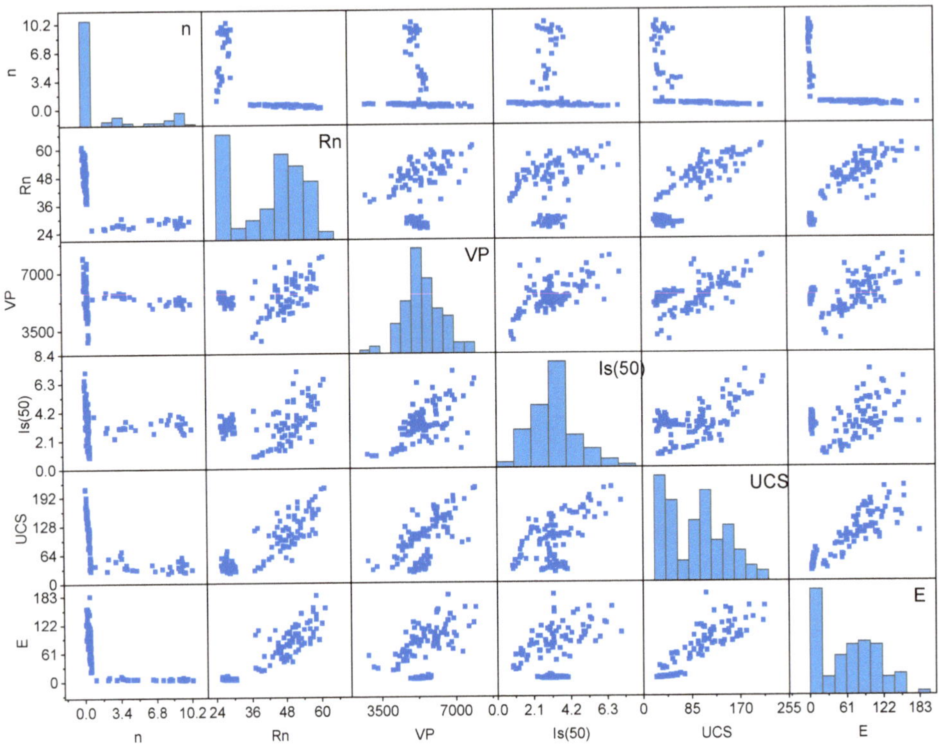

Figure 4. Distribution of the database.

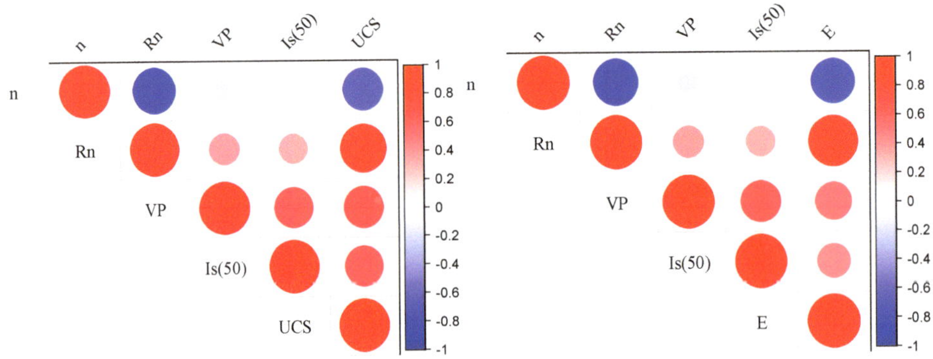

Figure 5. Correlation coefficients between variables.

4. Analysis of Results

Correlation coefficients (R^2), root mean square error (RMSE), and mean absolute percentage error (MAPE) were adopted to measure the prediction outputs of each model in order to quantitatively compare the predictive capabilities of the models [57,58]. Since the results of UCS and E are independent, the two output targets are discussed separately.

4.1. Analysis of Compressive Strength Prediction Results

Figure 6 displays the predictions provided by individual models. Further, the training and testing of the models are presented separately as shown in Figure 7. It can be seen that the suggested GWO-ELM model achieves the maximum value both for the training and

testing phases, with the correlation coefficient R^2 of the prediction outcomes of the machine learning models being greater than those of the MLR model. Additionally, the correlation coefficient shows that the optimized GWO-ELM model performs more predictably than the original ELM model.

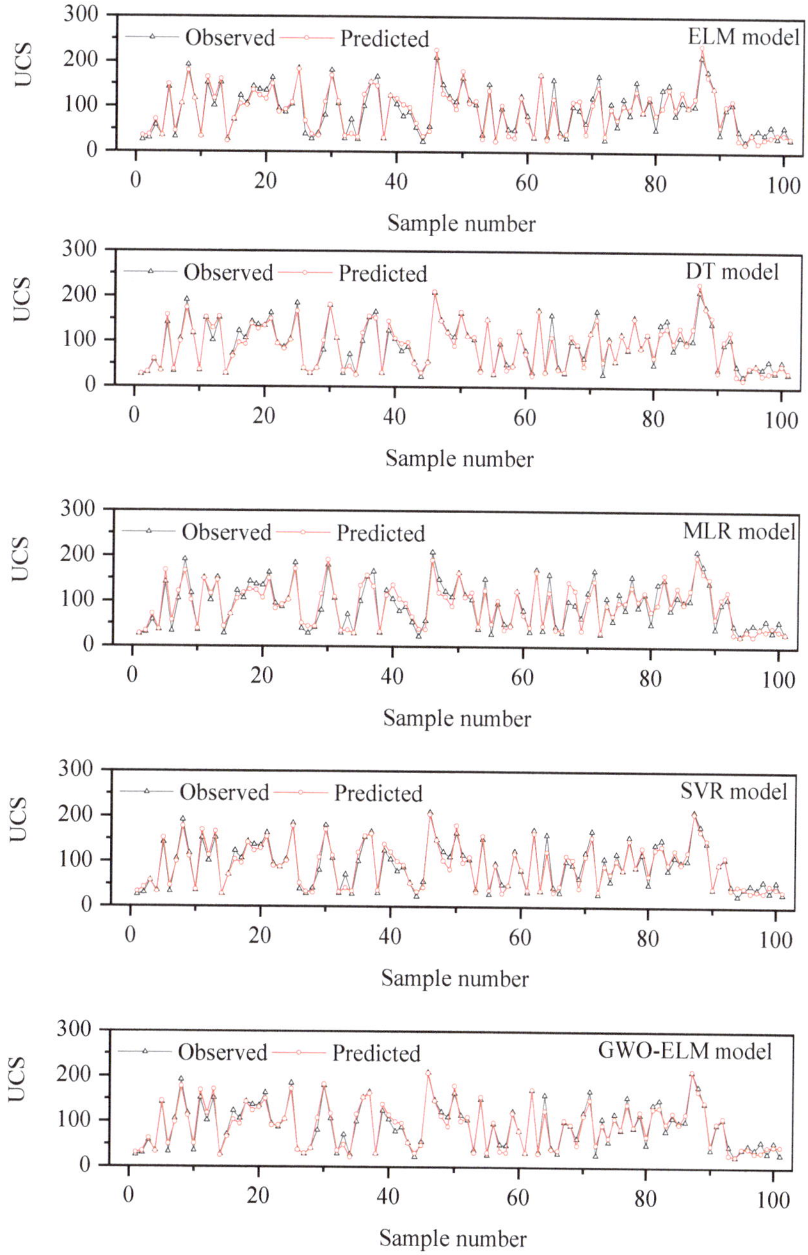

Figure 6. Comparison between prediction results and actual values of different models for UCS.

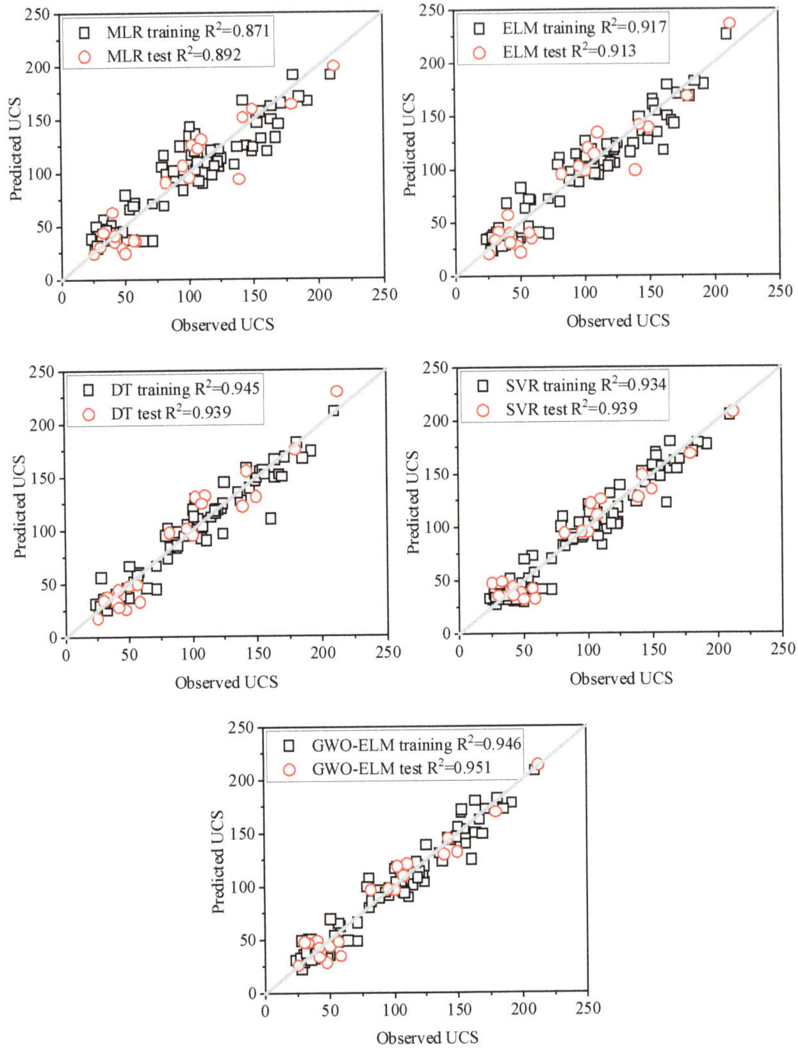

Figure 7. Prediction results of training and test stages for different models for UCS.

Further, the distribution of prediction errors for each model is calculated as shown in Figure 8. Then, the relative error for the test set is calculated and plotted in the box plot shown in Figure 9. In addition, the evaluation metrics are listed in Table 2, where for the test set, the hybrid GWO-ELM model achieves an RMSE of 11.446 and a MAPE of 16.131%. This further demonstrates the excellent prediction accuracy of the optimization model suggested in this paper and its ability to capture the nonlinear relationship between the UCS and the four input variables.

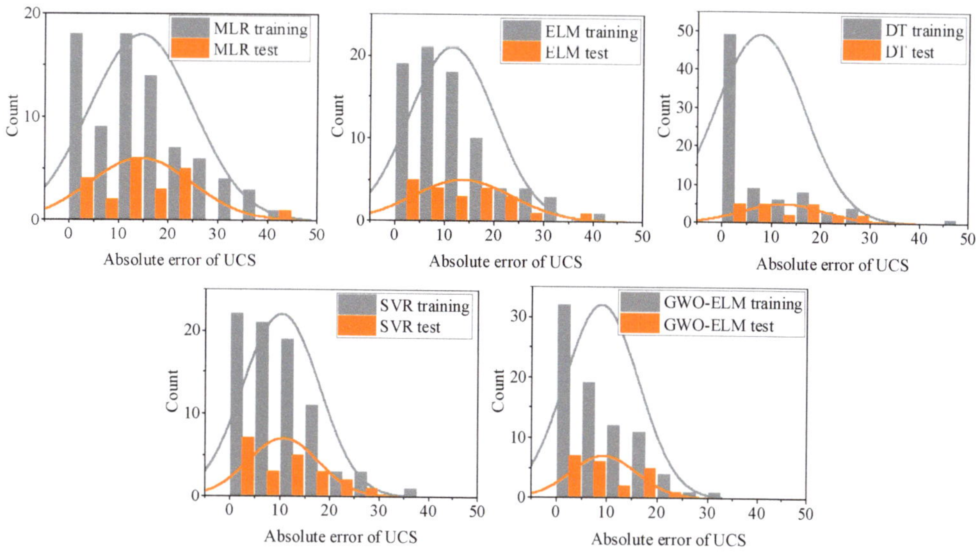

Figure 8. Prediction error distribution of different models for UCS.

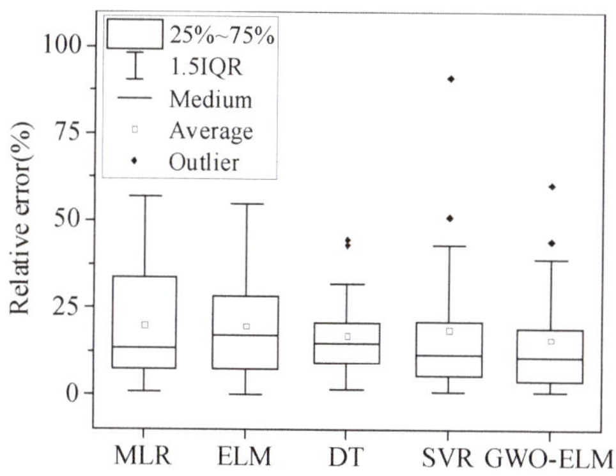

Figure 9. Box plot of relative error for UCS.

Table 2. Evaluation index of different models for UCS.

Model	Training			Test		
	R^2	RMSE	MAPE	R^2	RMSE	MAPE
MLR	0.871	17.914	19.444%	0.891	17.711	19.960%
ELM	0.917	14.334	15.890%	0.913	16.914	19.855%
DT	0.945	11.673	10.039%	0.939	14.817	17.191%
SVR	0.934	12.750	14.379%	0.939	12.754	18.844%
GWO-ELM	0.946	11.519	12.424%	0.951	11.446	16.131%

4.2. Analysis of Elastic Modulus Results

Similarly, the comparison between the predicted results of the different models for the elastic modulus and the actual values is shown in Figure 10. Based on the test set's prediction findings, which are shown in Figure 11, the correlation coefficient of the GWO-ELM model achieved 0.87, which is still a small value compared to the prediction of compressive strength, although it is the maximum among the five models. From the error analysis in Figures 12 and 13, it is clear that the optimized model improved the prediction

accuracy compared to that before optimization. Among the evaluation indexes of the model prediction results listed in Table 3, the optimized model achieved the smallest RMSE of 17.571 and MAPE of 23.649%, which were better than the prediction results of the other four models.

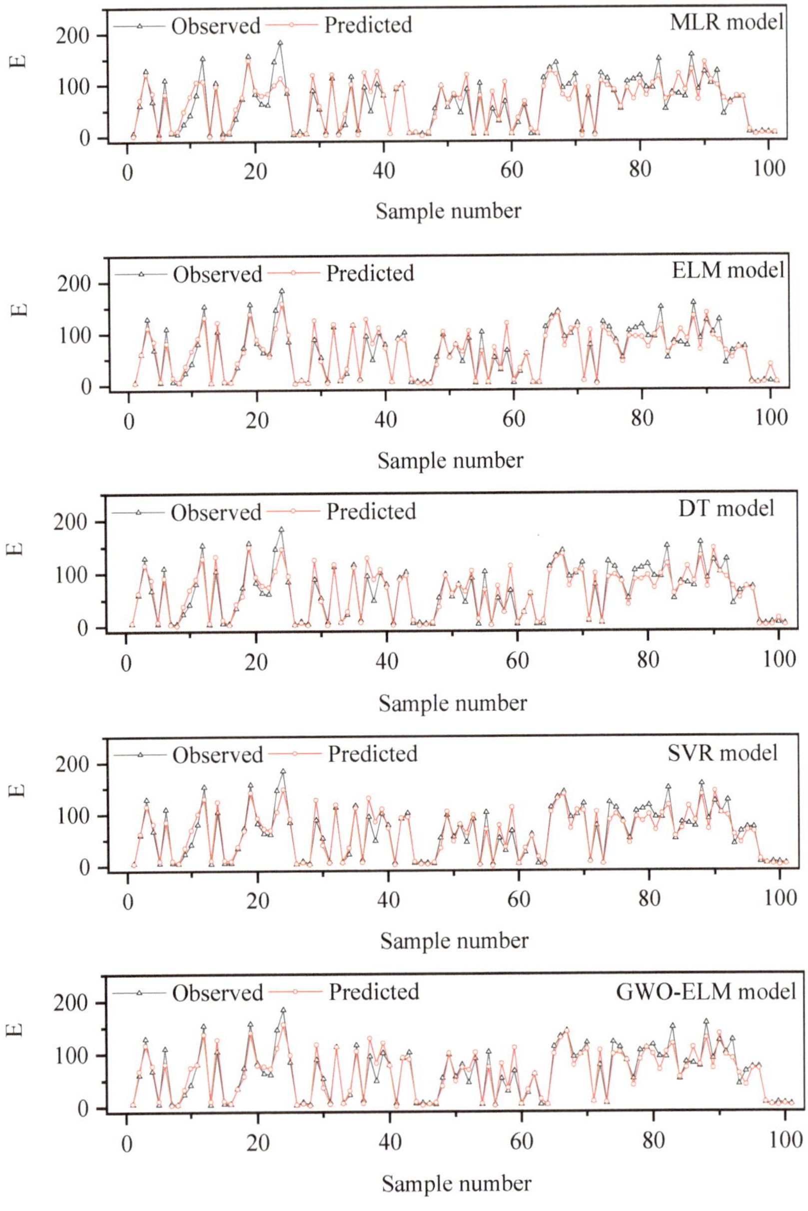

Figure 10. Comparison between prediction results and actual values of different models for E.

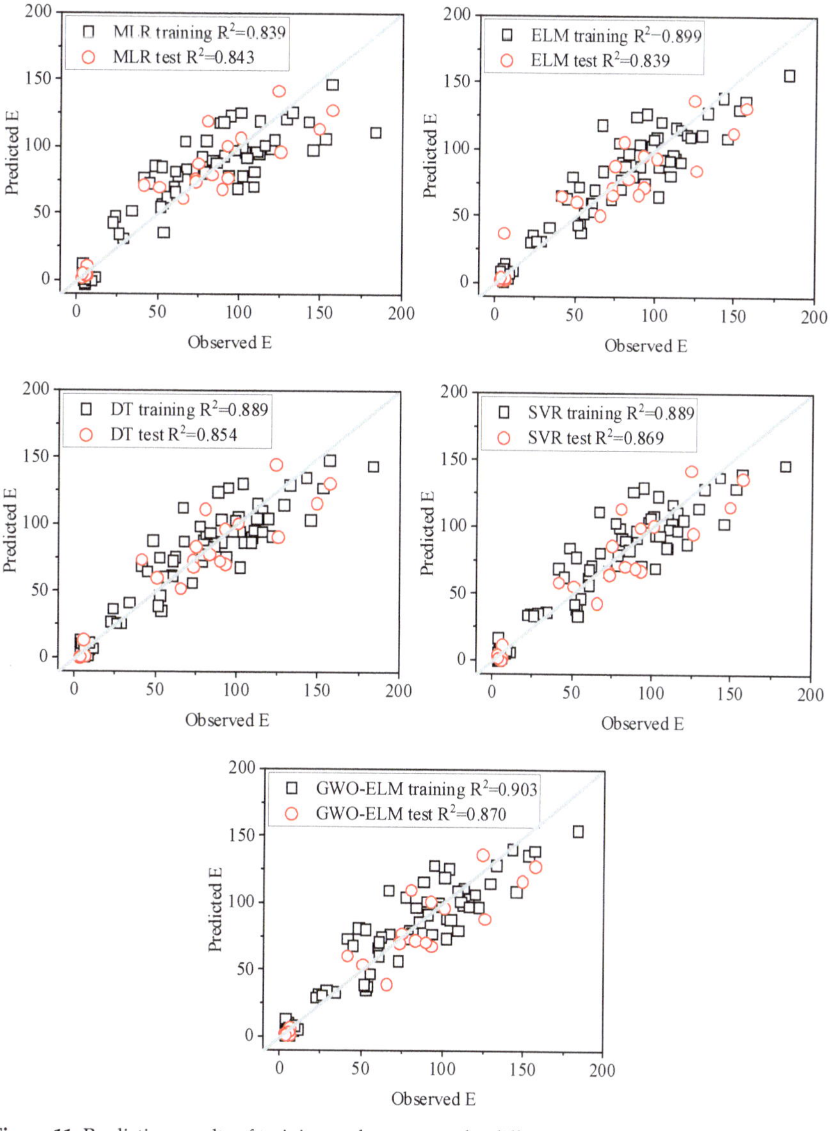

Figure 11. Prediction results of training and test stages for different models for E.

Table 3. Evaluation index of different models for E.

Model	Training			Test		
	R^2	RMSE	MAPE	R^2	RMSE	MAPE
MLR	0.839	19.269	42.293%	0.843	18.275	26.298%
ELM	0.899	15.204	31.114%	0.839	18.927	48.527%
DT	0.889	16.014	36.778%	0.854	17.801	37.133%
SVR	0.889	16.005	35.212%	0.869	17.203	30.282%
GWO-ELM	0.903	14.949	26.781%	0.870	17.571	23.649%

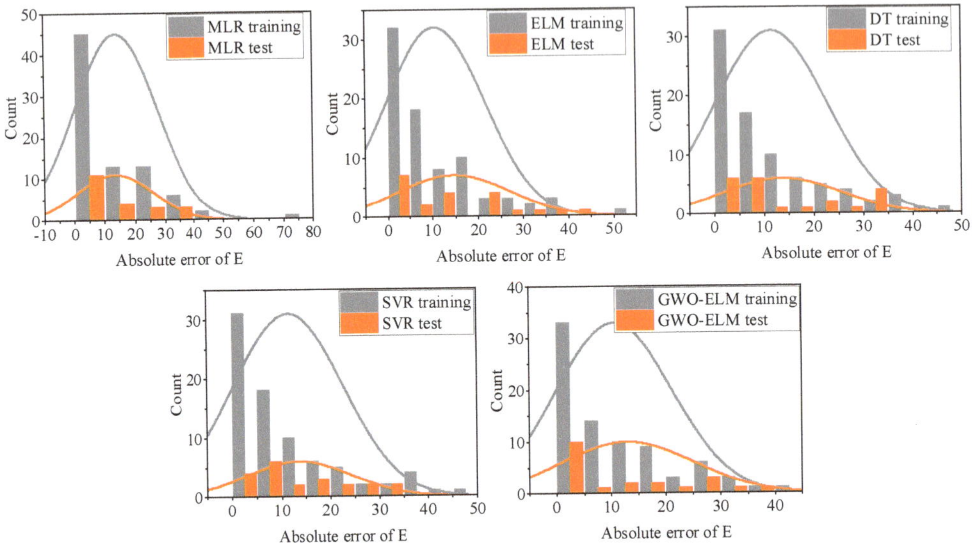

Figure 12. Prediction error distribution of different models for E.

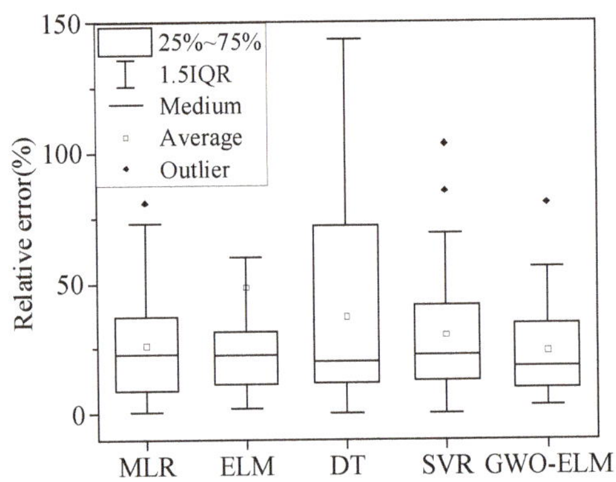

Figure 13. Box plot of relative error for E.

5. Discussion

Comparing the prediction results of different models, it can be found that the other four models all have better prediction results than the MLR model, which is mainly due to the fact that the MLR model restricts the mapping relationship between multiple input parameters and outputs, and this inherent mapping relationship is often nonlinear. Compared with the other three original machine learning models, the hybrid model proposed in this paper improves the predictive performance of the model due to the optimal combination of hyperparameters (weights and bias) obtained by using the GWO algorithm.

As far as the prediction results of compressive strength and elastic modulus are compared, the intrinsic correlation between the four input features used in this paper and compressive strength is stronger, while that with elastic modulus is weaker. Therefore, future research can consider input feature diversity and introduce more input parameters, which have a stronger intrinsic correlation with compressive strength and elastic modulus, to build the model. Moreover, the number of data samples in this study is limited and increasing the number and diversity of samples as much as possible can help to improve the prediction accuracy and generalization ability of the hybrid model.

6. Conclusions

In this paper, a hybrid GWO-ELM model was developed to predict the UCS and E of rocks. The main findings are as follows.

(1) In terms of evaluation metrics of prediction results, the proposed hybrid model reached the R^2 of 0.951, RMSE of 11.446, MAPE of 16.131% for UCS, and R^2 of 0.870, RMSE of 17.571, MAPE of 23.649% for E. The model achieved a better predictive performance than the other four models with the highest correlation coefficient R^2 and the lowest error indicator.

(2) Compared with the prediction results of the original ELM model, the GWO algorithm was validated as an efficient optimization algorithm that can improve the prediction performance of the ELM model by selecting the best combination of weights and bias.

(3) The proposed GWO-ELM model could be recommended as a prediction auxiliary tool for mechanical properties to provide support and assistance for real-time analysis of rock mechanical properties.

(4) The input feature diversity and the number of dataset samples considered in this paper limit the prediction performance of the model to some extent and increasing the effective input parameters and the number of dataset samples as much as possible can help improve the prediction performance of the hybrid model.

Author Contributions: X.J.: Writing—original draft, Methodology. R.Z.: Software, Resources. Y.M.: Data curation, Validation. All authors have read and agreed to the published version of the manuscript.

Funding: We gratefully acknowledge the financial supports provided by the Key Science and Technology Program of Henan Province, China (No. 222102320201).

Data Availability Statement: Not applicable.

Conflicts of Interest: The authors declare no conflict of interest.

References

1. Ersoy, H.; Kanik, D. Multicriteria decision-making analysis based methodology for predicting carbonate rocks' uniaxial compressive strength. *Earth Sci. Res. J.* **2012**, *16*, 65–74.
2. Gao, H.; Wang, Q.; Jiang, B.; Zhang, P.; Jiang, Z.; Wang, Y. Relationship between rock uniaxial compressive strength and digital core drilling parameters and its forecast method. *Int. J. Coal Sci. Technol.* **2021**, *8*, 605–613. [CrossRef]
3. Xie, S.; Han, Z.; Hu, H.; Lin, H. Application of a novel constitutive model to evaluate the shear deformation of discontinuity. *Eng. Geol.* **2022**, *304*, 106693. [CrossRef]
4. Xie, S.; Lin, H.; Chen, Y. New constitutive model based on disturbed state concept for shear deformation of rock joints. *Arch. Civ. Mech. Eng.* **2022**, *23*, 26. [CrossRef]
5. Ying, J.; Han, Z.; Shen, L.; Li, W. Influence of Parent Concrete Properties on Compressive Strength and Chloride Diffusion Coefficient of Concrete with Strengthened Recycled Aggregates. *Materials* **2020**, *13*, 4631. [CrossRef]
6. Beiki, M.; Majdi, A.; Givshad, A.D. Application of genetic programming to predict the uniaxial compressive strength and elastic modulus of carbonate rocks. *Int. J. Rock Mech. Min. Sci.* **2013**, *63*, 159–169. [CrossRef]
7. Xie, S.-j.; Lin, H.; Chen, Y.-f.; Wang, Y.-X. A new nonlinear empirical strength criterion for rocks under conventional triaxial compression. *J. Cent. South Univ.* **2021**, *28*, 1448–1458. [CrossRef]
8. Xie, S.; Lin, H.; Chen, Y.; Yong, R.; Xiong, W.; Du, S. A damage constitutive model for shear behavior of joints based on determination of the yield point. *Int. J. Rock Mech. Min. Sci.* **2020**, *128*, 104269. [CrossRef]
9. Jahed Armaghani, D.; Tonnizam Mohamad, E.; Hajihassani, M.; Yagiz, S.; Motaghedi, H. Application of several non-linear prediction tools for estimating uniaxial compressive strength of granitic rocks and comparison of their performances. *Eng. Comput.* **2016**, *32*, 189–206. [CrossRef]
10. Jamshidi, A.; Nikudel, M.R.; Khamehchiyan, M.; Sahamieh, R.Z. The effect of specimen diameter size on uniaxial compressive strength, P-wave velocity and the correlation between them. *Geomech. Geoengin.* **2016**, *11*, 13–19. [CrossRef]
11. Sousa, L.M.O. Petrophysical properties and durability of granites employed as building stone: A comprehensive evaluation. *Bull. Eng. Geol. Environ.* **2014**, *73*, 569–588. [CrossRef]
12. Kahraman, S. Evaluation of simple methods for assessing the uniaxial compressive strength of rock. *Int. J. Rock Mech. Min. Sci.* **2001**, *38*, 981–994. [CrossRef]
13. Sharma, P.K.; Singh, T.N. A correlation between P-wave velocity, impact strength index, slake durability index and uniaxial compressive strength. *Bull. Eng. Geol. Environ.* **2008**, *67*, 17–22. [CrossRef]

14. Kahraman, S. The determination of uniaxial compressive strength from point load strength for pyroclastic rocks. *Eng. Geol.* **2014**, *170*, 33–42. [CrossRef]
15. Jahed Armaghani, D.; Hajihassani, M.; Yazdani Bejarbaneh, B.; Marto, A.; Tonnizam Mohamad, E. Indirect measure of shale shear strength parameters by means of rock index tests through an optimized artificial neural network. *Measurement* **2014**, *55*, 487–498. [CrossRef]
16. Nasiri, H.; Homafar, A.; Chelgani, S.C. Prediction of uniaxial compressive strength and modulus of elasticity for Travertine samples using an explainable artificial intelligence. *Results Geophys. Sci.* **2021**, *8*, 100034. [CrossRef]
17. Madhubabu, N.; Singh, P.K.; Kainthola, A.; Mahanta, B.; Tripathy, A.; Singh, T.N. Prediction of compressive strength and elastic modulus of carbonate rocks. *Measurement* **2016**, *88*, 202–213. [CrossRef]
18. Mishra, D.A.; Basu, A. Estimation of uniaxial compressive strength of rock materials by index tests using regression analysis and fuzzy inference system. *Eng. Geol.* **2013**, *160*, 54–68. [CrossRef]
19. Tiryaki, B. Predicting intact rock strength for mechanical excavation using multivariate statistics, artificial neural networks, and regression trees. *Eng. Geol.* **2008**, *99*, 51–60. [CrossRef]
20. Rezaei, M.; Majdi, A.; Monjezi, M. An intelligent approach to predict unconfined compressive strength of rock surrounding access tunnels in longwall coal mining. *Neural Comput. Appl.* **2014**, *24*, 233–241. [CrossRef]
21. Zhu, H.; Wu, X.; Luo, Y.; Jia, Y.; Wang, C.; Fang, Z.; Zhuang, X.; Zhou, S. Prediction of Early Compressive Strength of Ultrahigh-Performance Concrete Using Machine Learning Methods. *Int. J. Comput. Methods* **2022**, 2141023. [CrossRef]
22. Zhuang, X.Y.; Zhou, S. The Prediction of Self-Healing Capacity of Bacteria-Based Concrete Using Machine Learning Approaches. *CMC Comput. Mater. Contin.* **2019**, *59*, 57–77. [CrossRef]
23. Wu, Y.; Li, S. Damage degree evaluation of masonry using optimized SVM-based acoustic emission monitoring and rate process theory. *Measurement* **2022**, *190*, 110729. [CrossRef]
24. Wu, Y.; Zhou, Y. Splitting tensile strength prediction of sustainable high-performance concrete using machine learning techniques. *Environ. Sci. Pollut. Res.* **2022**, *29*, 89198–89209. [CrossRef]
25. Mohamad, E.T.; Jahed Armaghani, D.; Momeni, E.; Alavi Nezhad Khalil Abad, S.V. Prediction of the unconfined compressive strength of soft rocks: A PSO-based ANN approach. *Bull. Eng. Geol. Environ.* **2015**, *74*, 745–757. [CrossRef]
26. Yılmaz, I.; Yuksek, A.G. An Example of Artificial Neural Network (ANN) Application for Indirect Estimation of Rock Parameters. *Rock Mech. Rock Eng.* **2008**, *41*, 781–795. [CrossRef]
27. Ren, Q.; Wang, G.; Li, M.; Han, S. Prediction of Rock Compressive Strength Using Machine Learning Algorithms Based on Spectrum Analysis of Geological Hammer. *Geotech. Geol. Eng.* **2019**, *37*, 475–489. [CrossRef]
28. Ceryan, N. Application of support vector machines and relevance vector machines in predicting uniaxial compressive strength of volcanic rocks. *J. Afr. Earth Sci.* **2014**, *100*, 634–644. [CrossRef]
29. Barzegar, R.; Sattarpour, M.; Nikudel, M.R.; Moghaddam, A.A. Comparative evaluation of artificial intelligence models for prediction of uniaxial compressive strength of travertine rocks, Case study: Azarshahr area, NW Iran. *Model. Earth Syst. Environ.* **2016**, *2*, 76. [CrossRef]
30. Çelik, S.B. Prediction of uniaxial compressive strength of carbonate rocks from nondestructive tests using multivariate regression and LS-SVM methods. *Arab. J. Geosci.* **2019**, *12*, 193. [CrossRef]
31. Mahmoodzadeh, A.; Mohammadi, M.; Hashim Ibrahim, H.; Nariman Abdulhamid, S.; Ghafoor Salim, S.; Farid Hama Ali, H.; Kamal Majeed, M. Artificial intelligence forecasting models of uniaxial compressive strength. *Transp. Geotech.* **2021**, *27*, 100499. [CrossRef]
32. Hanzelik, P.P.; Gergely, S.; Gáspár, C.; Győry, L. Machine learning methods to predict solubilities of rock samples. *J. Chemom.* **2020**, *34*, e3198. [CrossRef]
33. Baykasoğlu, A.; Güllü, H.; Çanakçı, H.; Özbakır, L. Prediction of compressive and tensile strength of limestone via genetic programming. *Expert Syst. Appl.* **2008**, *35*, 111–123. [CrossRef]
34. Çanakcı, H.; Baykasoğlu, A.; Güllü, H. Prediction of compressive and tensile strength of Gaziantep basalts via neural networks and gene expression programming. *Neural Comput. Appl.* **2008**, *18*, 1031. [CrossRef]
35. Ozbek, A.; Unsal, M.; Dikec, A. Estimating uniaxial compressive strength of rocks using genetic expression programming. *J. Rock Mech. Geotech. Eng.* **2013**, *5*, 325–329. [CrossRef]
36. Matin, S.S.; Farahzadi, L.; Makaremi, S.; Chelgani, S.C.; Sattari, G. Variable selection and prediction of uniaxial compressive strength and modulus of elasticity by random forest. *Appl. Soft Comput.* **2018**, *70*, 980–987. [CrossRef]
37. Barzegar, R.; Sattarpour, M.; Deo, R.; Fijani, E.; Adamowski, J. An ensemble tree-based machine learning model for predicting the uniaxial compressive strength of travertine rocks. *Neural Comput. Appl.* **2020**, *32*, 9065–9080. [CrossRef]
38. Mahmoodzadeh, A.; Mohammadi, M.; Ghafoor Salim, S.; Farid Hama Ali, H.; Hashim Ibrahim, H.; Nariman Abdulhamid, S.; Nejati, H.R.; Rashidi, S. Machine Learning Techniques to Predict Rock Strength Parameters. *Rock Mech. Rock Eng.* **2022**, *55*, 1721–1741. [CrossRef]
39. Yilmaz, I.; Yuksek, G. Prediction of the strength and elasticity modulus of gypsum using multiple regression, ANN, and ANFIS models. *Int. J. Rock Mech. Min. Sci.* **2009**, *46*, 803–810. [CrossRef]
40. Yesiloglu-Gultekin, N.; Sezer, E.A.; Gokceoglu, C.; Bayhan, H. An application of adaptive neuro fuzzy inference system for estimating the uniaxial compressive strength of certain granitic rocks from their mineral contents. *Expert Syst. Appl.* **2013**, *40*, 921–928. [CrossRef]

41. Chentout, M.; Alloul, B.; Rezouk, A.; Belhai, D. Experimental study to evaluate the effect of travertine structure on the physical and mechanical properties of the material. *Arab. J. Geosci.* **2015**, *8*, 8975–8985. [CrossRef]
42. Jahed Armaghani, D.; Tonnizam Mohamad, E.; Momeni, E.; Narayanasamy, M.S.; Mohd Amin, M.F. An adaptive neuro-fuzzy inference system for predicting unconfined compressive strength and Young's modulus: A study on Main Range granite. *Bull. Eng. Geol. Environ.* **2015**, *74*, 1301–1319. [CrossRef]
43. Khan, N.M.; Cao, K.; Yuan, Q.; Bin Mohd Hashim, M.H.; Rehman, H.; Hussain, S.; Emad, M.Z.; Ullah, B.; Shah, K.S.; Khan, S. Application of Machine Learning and Multivariate Statistics to Predict Uniaxial Compressive Strength and Static Young's Modulus Using Physical Properties under Different Thermal Conditions. *Sustainability* **2022**, *14*, 9901. [CrossRef]
44. Yang, Z.; Wu, Y.; Zhou, Y.; Tang, H.; Fu, S. Assessment of Machine Learning Models for the Prediction of Rate-Dependent Compressive Strength of Rocks. *Minerals* **2022**, *12*, 731. [CrossRef]
45. Liu, Z.; Shao, J.; Xu, W.; Wu, Q. Indirect estimation of unconfined compressive strength of carbonate rocks using extreme learning machine. *Acta Geotech.* **2015**, *10*, 651–663. [CrossRef]
46. Li, C.; Zhou, J.; Dias, D.; Gui, Y. A Kernel Extreme Learning Machine-Grey Wolf Optimizer (KELM-GWO) Model to Predict Uniaxial Compressive Strength of Rock. *Appl. Sci.* **2022**, *12*, 8468. [CrossRef]
47. Al-Shamiri, A.K.; Kim, J.H.; Yuan, T.-F.; Yoon, Y.S. Modeling the compressive strength of high-strength concrete: An extreme learning approach. *Constr. Build. Mater.* **2019**, *208*, 204–219. [CrossRef]
48. Han, B.; Wu, Y.; Liu, L. Prediction and uncertainty quantification of compressive strength of high-strength concrete using optimized machine learning algorithms. *Struct. Concr.* **2022**, 1–14. [CrossRef]
49. Mirjalili, S.; Mirjalili, S.M.; Lewis, A. Grey Wolf Optimizer. *Adv. Eng. Softw.* **2014**, *69*, 46–61. [CrossRef]
50. Subudhi, U.; Dash, S. Detection and classification of power quality disturbances using GWO ELM. *J. Ind. Inf. Integr.* **2021**, *22*, 100204. [CrossRef]
51. Shariati, M.; Mafipour, M.S.; Ghahremani, B.; Azarhomayun, F.; Ahmadi, M.; Trung, N.T.; Shariati, A. A novel hybrid extreme learning machine–grey wolf optimizer (ELM-GWO) model to predict compressive strength of concrete with partial replacements for cement. *Eng. Comput.* **2022**, *38*, 757–779. [CrossRef]
52. Muro, C.; Escobedo, R.; Spector, L.; Coppinger, R.P. Wolf-pack (*Canis lupus*) hunting strategies emerge from simple rules in computational simulations. *Behav. Process.* **2011**, *88*, 192–197. [CrossRef] [PubMed]
53. Roushangar, K.; Shahnazi, S.; Azamathulla, H.M. Partitioning strategy for investigating the prediction capability of bed load transport under varied hydraulic conditions: Application of robust GWO-kernel-based ELM approach. *Flow Meas. Instrum.* **2022**, *84*, 102136. [CrossRef]
54. Armaghani, D.J.; Tonnizam Mohamad, E.; Momeni, E.; Monjezi, M.; Sundaram Narayanasamy, M. Prediction of the strength and elasticity modulus of granite through an expert artificial neural network. *Arab. J. Geosci.* **2015**, *9*, 48. [CrossRef]
55. Wu, Y.; Zhou, Y. Hybrid machine learning model and Shapley additive explanations for compressive strength of sustainable concrete. *Constr. Build. Mater.* **2022**, *330*, 127298. [CrossRef]
56. Wu, Y.; Zhou, Y. Prediction and feature analysis of punching shear strength of two-way reinforced concrete slabs using optimized machine learning algorithm and Shapley additive explanations. *Mech. Adv. Mater. Struct.* **2022**, 1–11. [CrossRef]
57. Chen, H.; Li, X.; Wu, Y.; Zuo, L.; Lu, M.; Zhou, Y. Compressive Strength Prediction of High-Strength Concrete Using Long Short-Term Memory and Machine Learning Algorithms. *Buildings* **2022**, *12*, 302. [CrossRef]
58. Tang, F.; Wu, Y.; Zhou, Y. Hybridizing Grid Search and Support Vector Regression to Predict the Compressive Strength of Fly Ash Concrete. *Adv. Civ. Eng.* **2022**, *2022*, 3601914. [CrossRef]

minerals

MDPI

Article

A Multi-Scale Study on the Property Degradation of High-Temperature Treated Beishan Granite

Xiang Zhang, Manke Wei *, Zhen Lei and Ying Chen

School of Architecture and Planning, Yunnan University, Kunming 650091, China
* Correspondence: weimanke@mail.ynu.edu.cn

Abstract: Granite is the main host rock for the underground storage of nuclear waste in Beishan, China. Heat is continuously generated during the long-term disposal of nuclear waste; therefore, it is important to investigate the influence of high temperature on the physical and mechanical properties and the constitutive relation of Beishan granite. In this study, laboratory experiments on heat-treated (25 °C, 200 °C, 400 °C, 600 °C and 800 °C) Beishan granite samples were performed in combination with nuclear magnetic resonance (NMR) analysis and regular physical-mechanical tests. The results show that the elastic modulus tends to decline faster at the temperature ranges of 25–200 °C and 600–800 °C by approximately 26.767% and 66.996%, respectively. Compared with the results at 25 °C, the peak stress decreases by 72.664% at 800 °C. The peak strain increases gradually from 25 °C to 600 °C and abruptly from 600 °C to 800 °C. The peak strain at 800 °C is 2.303× greater than that at 25 °C. Based on the damage theory, the Weibull distribution, the rock damage threshold point, and the residual strength, this study corrected the Drucker–Prager (D–P) criterion to consider the damage stress and then to establish the constitutive model of thermally damaged Beishan granite. The parameters required for the model are conventional mechanical parameters that can be calculated from the uniaxial test results, thus making the model convenient to apply. Meanwhile, the mechanical behavior of thermally damaged Beishan granite under uniaxial compression was simulated using the Particle Flow Code (PFC) to explore the development of cracks from the microscopic scale. The research results can provide theoretical support for the calculation and numerical simulation related to the mechanics of high-temperature treated rocks.

Keywords: Beishan granite; constitutive model; high-temperature; thermal damage; PFC

Citation: Zhang, X.; Wei, M.;Lei, Z.; Chen, Y. A Multi-Scale Study on the Property Degradation of High-Temperature Treated Beishan Granite. *Minerals* **2023**, *13*, 27. https://doi.org/10.3390/min13010027

Academic Editor: Mamadou Fall

Received: 16 November 2022
Revised: 10 December 2022
Accepted: 21 December 2022
Published: 23 December 2022

1. Introduction

The long-term safe disposal of high-level radioactive waste is a major issue for the sustainable development of nuclear energy and environmental protection [1]. Currently, deep geological disposal is considered to be one of the best options for the storage of high-level radioactive waste [2]. The preferential host rocks include salt beds [3,4], shale [5,6] and granite [7,8]. Because of its good integrity, homogeneity, and low porosity and permeability [9–11], the host rock formation of granite has been used in many countries [12], such as the crystalline Archean granite–gneisses at Yeniseisky in Russia [13], the granite formation at the Grimsel test site in Switzerland [14], and the Lac du Bonnet granite formation in southeastern Manitoba in Canada [15]. In recent years, the Chinese nuclear power industry has developed rapidly. According to the data of the World Nuclear Association, as of July 2022, China has 20 nuclear reactors under construction, with the largest net capacity in the world [16]. A deep granite formation located in the Beishan region of Gansu Province in China is currently considered the preferred site for the geological storage of high-level radioactive waste [17]. Nuclear waste will continue to release heat during disposal, and the duration of the nuclear heat pulse is 10^3 to 10^4 years [18]. Due to the complexity of the mineral composition of granite and the different coefficients of thermal expansion, the mineral expansion that occurs at a high temperature can exert significant thermal stresses

between minerals, resulting in the formation of microcracks and fractures in rock [19,20]. This effect will significantly influence the mechanical properties of rock, including its elastic modulus, peak stress, and peak strain [21], which will in turn affect the long-term safety and stability of the repository. Therefore, thermal damage to granite is a key scientific research issue in the construction and development of nuclear waste underground repositories.

To date, researchers have investigated the physical and mechanical properties of heat-treated granite using a variety of experimental and numerical methods. Zhao et al. [22] studied the thermal conductivity of heat-damaged Beishan granite under uniaxial compression with the transient plane source method using a servo-controlled compression machine and an acoustic emission (AE) monitoring system. Their results show that with increasing temperature, the cumulative AE hits of the specimens at the crack damage stress and the peak stress increase nonlinearly, the initial and peak lateral thermal conductivities of the specimens decrease, and their differences increase. In addition, the high-temperature treated specimens experienced greater shrinkage deformation to reach the peak thermal conductivity. Chen et al. [23] used high-temperature uniaxial compression and a three-point bending method to determine the macroscopic mechanical properties of Fujian granite and performed scanning electron microscopy (SEM) and optical microscopy to study the microstructure of the damaged samples. The results show that the uniaxial compressive strength, Young's modulus, and fracture toughness of granite decrease with increasing temperature. In addition, the connection between minerals becomes weaker, and the main failure mode is intergranular cracking when the temperature is below 800 °C and transgranular cracking when the temperature is close to 1000 °C. Gautam et al. [24] experimentally investigated the effect of temperature on the physical and mechanical properties of Jalore granite in India using SEM. The results show that thermogenic cracks have a long densification phase and exhibit a rapid increase in crack density due to the complexity of their thermal expansion rate, with the expansion of quartz as the most significant effect. The compressive and tensile strengths increase gradually for temperatures below 300 °C and decrease sharply above 300 °C. In the range of 300–600 °C, the stress–strain curves exhibit plastic behavior, reflecting the brittle-tough transition. Tomas et al. [25] performed physical and mechanical tests on granite specimens from northern Portugal and compared the results with a large number of published case studies, showing that the variations in porosity, unit weight, ultrasonic wave velocity, elastic modulus, and uniaxial compressive stress are mainly related to the induced thermal stress from the anisotropic thermal expansion of mineral particles. Wu et al. [26] used uniaxial compression tests, ultrasonic velocity analysis, AE monitoring, and SEM to study the effect of thermal damage at different temperatures on the physical and mechanical properties of Beishan granite, and the results show that the peak strength, P-wave velocity, and mass loss rate decreased sharply with increasing temperature. Cracks increased and gradually formed fracture zones that eventually lead to the destruction of the sample.

The constitutive model is mainly used to study the relationship between stress and strain. Based on the Weibull distribution and Lemaitre strain equivalence principle, Xu et al. [27] used the Drucker–Prager (D–P) criterion and analyzed thermal damage at high temperature and mechanical damage at high pressure, proposed nonlinear coupled total damage parameters, and then established a thermomechanical coupled damage model of granite. The theoretical results obtained using the model are in good agreement with the experimental values. Liu et al. [28] used the D–P criterion and the Weibull distribution to develop a statistical constitutive model of high-temperature granite that considered the damage threshold, the residual stress, and the thermal damage. This model enables the evaluation of the rock toughness and brittleness based on the damage evolution and analyzes the effect of the Weibull parameter and temperature on the damage evolution. Wang et al. [29] derived the temperature dependence based on the variations in the parameters such as the tensile strength, Young's modulus, Poisson's ratio, and coefficient of thermal expansion of granite at 1000 °C and then combined it with the constitutive model based on the classical Mohr–Coulomb model with strain softening and tension cutoff. Then,

a new constitutive law was verified by performing a uniaxial compression test. Wang et al. [30] combined the 3D D–P criterion, the compaction coefficient, the damage variable correction factor, and the thermal damage variable to establish a new whole-process thermomechanical constitutive model considering the rock strain softening caused by the compaction stage and thermal shock in the uniaxial compression test. The model can be used to estimate the overall evolution of rock damage after high-temperature treatment. Xie et al. [31] determined the yield point based on the stress difference method, modified the Weibull distribution, and proposed a piecewise constitutive model bounded by the yield point, which could reflect the complete stress-strain curves of the rock.

For the microstructure and microcracks of thermally damaged granite, Zhao et al. [32] used the mercury intrusion technique to study the pore characteristics of granite under different thermal cycling conditions. The results show that the number of thermal cycles has little effect on the pore structure, while the temperature has a greater effect, and decreases the complexity of the pore structure for pores with sizes from 0.01 to 10 μm and increases the complexity of pores with sizes from 10 to 200 μm. Miao et al. [33] combined AE, digital image correlation, and optical microscopic observations to reveal the key mechanism of the mechanical and fracture behavior of heat-treated Beishan granite. Li et al. [34] investigated the effect of cyclic heating and cooling on the pore structure of granite by performing nuclear magnetic resonance (NMR) and magnetic resonance imaging (MRI). The results show that the porosity increases significantly and the ultrasonic velocity decreases sharply after cyclic heating and cooling. It was confirmed by MRI observations that the expansion and deepening of microcracks in granite during thermal cycling were due to the combined effects of the inhomogeneous expansion and contraction of particles, the transformation of mineral crystal structure, the fatigue damage, and the thermal shock caused by cooling. Wang F et al. [35] investigated the effect of low and high heating rates on the thermal cracking of granite using a continuous medium-based numerical method. The results show that at a low heating rate, the thermal stress variation is mainly controlled by the coefficient of nonhomogeneous thermal expansion of the mineral grains. At a high heating rate, larger cracks and newly induced microcracks form due to a wider range of concentrated tensile stresses. Different heating rates induce different cracking behaviors, affecting the strength and final damage form of the granite. Liu et al. [36] performed discrete element modeling of mineral structures to study the stress–strain, AE, and crystal-scale crack extension patterns of inhomogeneous rocks. Based on the effective medium theory, Xie et al. [37] proposed an exponential model to describe the influence of the crack closure effect on the constitutive relation, which was verified by the uniaxial compression test. The results show that the microcrack volume ratio is positively correlated with the crack density parameters, and the material damage caused by microcrack can be identified and quantified by it.

This work is a multi-scale study that combines the theoretical study, the physical-mechanical tests, and the numerical simulation to investigate the degradation effect of high temperature on granite properties. Heat treatment, NMR, physical, and mechanical experiments were carried out. Based on the Lemaitre strain equivalence principle and the D–P criterion, the Weibull distribution was introduced to describe the microelement strength distribution of rock and to build the constitutive and high-temperature thermal damage constitutive laws of Beishan granite. The parameters required for the damage constitutive model are conventional mechanical parameters that can be calculated from the uniaxial test results, thus making the model convenient to apply. The model can reflect the stress-strain relationship of high-temperature treated granite. The simulation of the uniaxial compression mechanical behavior of heat-treated granite was completed by using PFC, and the fracture mechanism was revealed from the perspective of micro-crack development. The research results can provide theoretical support for the calculation and numerical simulation related to the mechanics of high-temperature treated rocks.

2. Experiments

This section is about the high-temperature treatment test, and the physical and mechanical properties tests on Beishan granite.

2.1. Beishan Granite Sample

The samples were obtained from the granite stratum in the preselection area of Beishan, Gansu. The cores were drilled horizontally from the wall of the tunnel after excavation at the pit exploration facility. The core is granitic in structure and is light gray with black spots, and the main minerals include plagioclase, alkali feldspar, quartz, and black mica. The mineral composition was obtained by X-ray diffraction analysis (SmartLab, Rigaku, Japan) as 67.90% feldspar, 27.40% quartz, and 4.70% mica. The preparation method and processing accuracy met the relevant standards. The specimens of φ 50 mm $\times$ 25 mm and φ 50 mm $\times$ 100 mm were made and treated at different temperature levels for the uniaxial compression test and Brazilian splitting test, respectively, as shown Figure 1.

(a) (b) (c)

Figure 1. Beishan granite specimens. (**a**) Beishan granite specimens; (**b**) φ 50 mm $\times$ 25 mm; (**c**) φ 50 mm $\times$ 100 mm.

2.2. Experimental Studies

2.2.1. High-Temperature Treatment Test

A KSL-1200X muffle furnace (maximum temperature of 1200 °C, heating rate $\leq$ 10 °C/min, temperature control accuracy $\pm$1 °C, HF-Kejing, Hefei, China) was used for the high-temperature treatment at 25 °C, 200 °C, 400 °C, 600 °C and 800 °C for a total of five test groups. The experimental work of Chen [38] has indicated that 5 °C/min is usually the threshold of the heating rate; above this threshold the damage mechanism of granite changes. To avoid thermal shock to the specimens, this study fixed a heating rate of 4 °C/min. The preset temperature of the muffle furnace was maintained for 1h to allow the rock specimen to be fully heated. When the heating process was completed, the specimen naturally cooled to room temperature before being removed. As shown in Figure 2, with increasing temperature, the specimens gradually became more reddish brown and was more susceptible to damage with smaller forces, and the specimens of the 800 °C test group made a sharp sound after a slight collision.

Figure 2. Heat-treated granite specimens.

2.2.2. Physical Property Tests

Electronic scales, Vernier calipers, and wave velocity testers (RSM-SY5(T), SINOROCK, Wuhan, China) were used to measure the variations in mass, volume, density, and wave velocity of the heat-treated specimens.

Figure 3 shows the relationship of some physical properties with increasing temperature. The mass, density, and wave velocity of the specimens gradually decrease with increasing temperature, while the volume shows the opposite trend, and all four parameters subsequently vary. In particular, the mass of the specimen decreased by 0.116% and 0.118% at the temperature intervals of 25 °C to 200 °C and 400 °C to 600 °C, respectively, showing faster change compared with the result obtained at 25 °C. The former decrease is due to the evaporation of free and bound water inside the specimens during the heating process. In the latter temperature interval, particularly between 300 °C and 500 °C, the contents of quartz, feldspar, hornblende, and mica fluctuate with increasing temperature, indicating that heating promotes physical and chemical reactions in granite components [39]. Meanwhile, the volume, density, and wave velocity change significantly after 400 °C, and a change-rate inflection point is present between 400 and 600 °C. It indicates that with increasing temperature, particularly above 400 °C, the internal physicochemical reaction of granite is more intense, and the degree of thermal damage is more important.

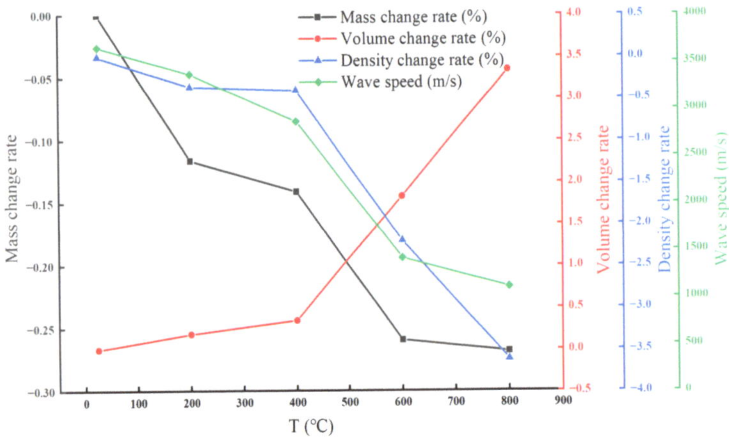

Figure 3. Effect of heat treatment on the physical properties of Beishan granite.

A MesoMR23-060H-I NMR analyzer (magnetic field strength of 0.52 ± 0.05 T, magnetic field stability of ≤300 Hz/h, maximum sampling width/kHz of 2000 kHz, Niumai, Suzhou, China) was used to perform NMR experiments on heat-treated specimens, and the pore size distribution characteristics were analyzed by T_2 spectroscopy (MesoMR23-060H-I, Niumai, Suzhou, China) to reveal the development of microcracks.

NMR was used to detect the relaxation response of ^1H protons in rock samples by performing Carr–Purcell–Meiboom–Gil pulse sequence tests on fully saturated rock specimens to obtain the transverse relaxation time T_2 [40]. T_2 is mainly influenced by the ratio of the surface area to the volume of the pores, which is proportional to the pore radius. Therefore, the pore size distribution of rock can be characterized by the NMR signal intensity of the T_2 curve (amplitude, peak area, and continuity) [41]. Figure 4 shows the T_2 profiles of the specimens obtained after different temperature treatments. The distribution characteristics of the microcrack of the specimens can be analyzed according to the T_2 spectrum amplitude. The T_2 amplitudes show a double-peak feature. The left peak corresponds to the fine pores, and the right peak corresponds to the microcracks. The left peak is lower than the right peak.

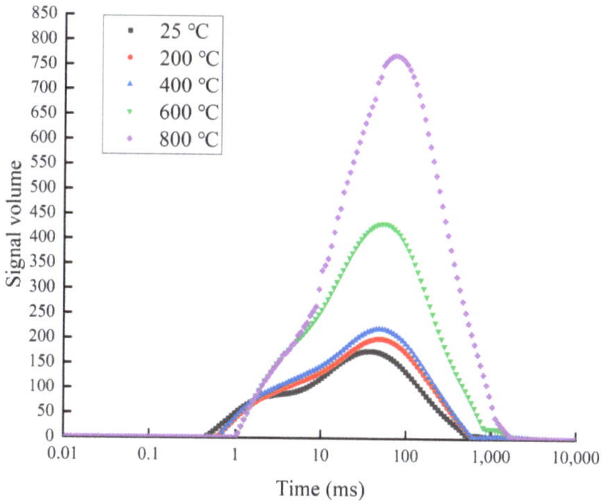

Figure 4. T_2 spectra from NMR on heat-treated Beishan granite.

Overall, the high temperature has a significant effect on the T_2 values of the Beishan granite. In the range of 25–400 °C, there is no significant difference in the left peak T_2 amplitude with increasing temperature, and the right peak T_2 amplitude increases slightly, indicating that there is no significant expansion of granite pores in this temperature range. In the range of 400–800 °C, the left peak T_2 amplitude increases with increasing temperature, and the right peak and the corresponding area under the curve increase significantly. Based on the observation of the microstructure, Sun Z et al. [42] found that boundary cracks were generated between feldspar crystals and between feldspar grains and quartz grains at 500 °C. At 600 °C, the quartz α/β phase transition point was exceeded, and large transparent transgranular cracks were generated in quartz particles, accompanied by an obvious transgranular network. These results show that a high temperature leads to the phase transformation of quartz, the considerable development of internal cracks, and increases in the size and volume of the pores. A large number of microcracks develop as new pores are formed.

Figure 5 shows the MRI results from NMR on the specimens of 25 °C, 200 °C and 600 °C. Figure 5d shows the relative intensity range of ^1H in the internal pores, and the red color shows the stronger ^1H signal that indicates greater fluid intrusion into the specimen and greater crack development. In the image, the black part is the imaging background, and the middle-colored part is the proton density-weighted image.

It can be deduced from the NMR images that the proton density distribution is uniform at 25 °C, and there is no obvious concentration of proton density, indicating that the specimen is mainly dominated by fine pores at this temperature. With increasing temperature, the rock specimen undergoes more thermal damage, the proton density exhibits a more clustered distribution, the microcracks gradually initiate, develop and expand, and the volume gradually increases. There is a distribution of high-proton-density (red) regions at 600 °C. Meanwhile, green pixel clusters are commonly produced by the aggregation of microcrack interactions, indicating that the crack development of Beishan granite is particularly pronounced at this temperature. The T_2 pattern and imaging analysis show that with increasing temperature, the moisture inside the specimen evaporates through the microcracks, and the resulting pressure leads to a slight expansion of the cracks. When the temperature exceeds the phase change point of quartz, the temperature gives rise to cracks through the crystal as well as microcrack expansion between the grains and more mineral physicochemical reactions. The fine cracks inside the specimen gradually develop to form a larger connected area and agglomerate into a crack network, leading to the further deterioration of the physical properties of the rock.

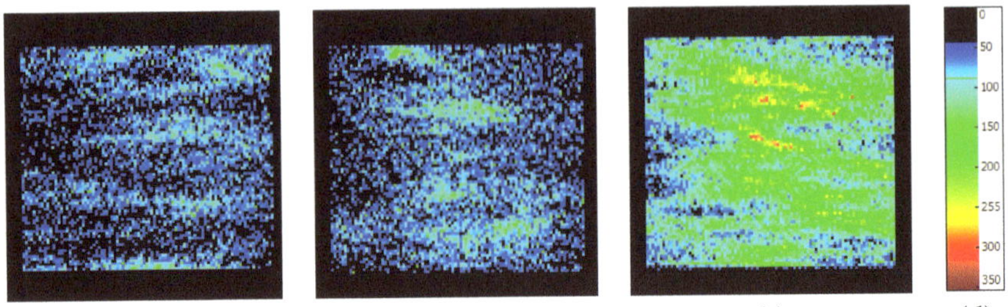

(a) (b) (c) (d)

Figure 5. MRI of heat-treated Beishan granite. (**a**) 25 °C (**b**) 200 °C (**c**) 600 °C (**d**) legend.

2.2.3. Mechanical Property Tests

The uniaxial compression test (φ 50 mm × 25 mm) and Brazilian splitting test (φ 50 mm × 100 mm) were conducted on the heat-treated rock samples using a HUT305A type 300 kN microcomputer-controlled electrohydraulic servo universal testing machine (as shown in Figure 6, WANCE, Shenzhen, China) to study the effect of high temperature on the mechanical properties of rock.

Figure 6. HUT305A type 300 kN microcomputer-controlled electrohydraulic servo universal testing machine.

Figure 7 shows the damage modes of rock after different temperature treatments. With increasing temperature, the damage at the peak strength increased, and the specimens were gradually powdered. In the Brazilian splitting test, when the stress reached the peak strength, a primary fracture expanded along the loading direction, and with increasing temperature, more secondary cracks were produced around the primary fracture, with a denser distribution.

Figure 8 shows the results of the mechanical tests. In the uniaxial compression test, with the increase in temperature, the peak of the stress–strain curve decreases significantly and gradually levels off. The modulus of elasticity and the peak stress gradually decreases, and the peak strain shows an overall gradual increasing trend. In particular, the elastic modulus tends to decline faster at the temperature ranges of 25–200 °C and 600–800 °C by approximately 26.767% and 66.996%, respectively. The former is due to the evaporation of water vapor that leads to a slight expansion of the microcracks. The latter is due to the thermal stress generated by the thermal expansion between the grains, leading to the rapid expansion of the crack and the rapid deterioration of mechanical properties. Comparing with the results at 25 °C, the peak stress decreases by 72.664% at 800 °C. The peak strain increases gradually from 25 °C to 600 °C and abruptly from 600 °C to 800 °C. The peak strain at 800 °C is 2.303× greater than that at 25 °C. This indicates that the granite gradually changes from brittle to ductile after reaching a certain temperature. It has been shown that the temperature range from 400 °C to 600 °C can be considered the critical temperature range for the brittle-to-ductile transition [43].

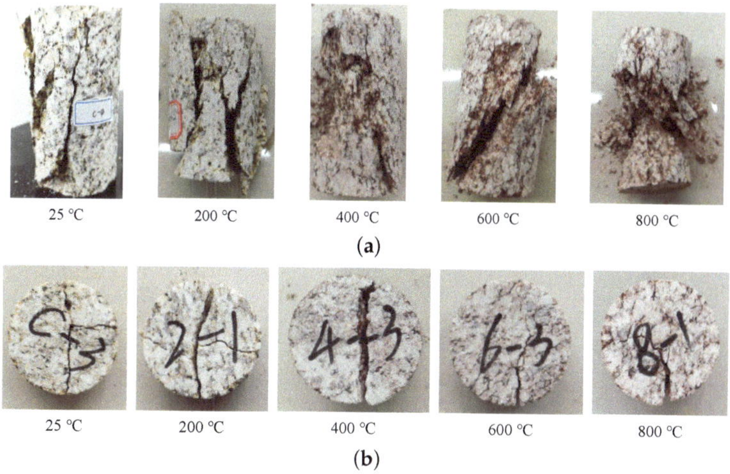

Figure 7. Specimens after damage by uniaxial compression and Brazilian splitting tests. (**a**) Uniaxial compression test; (**b**) Brazilian splitting test.

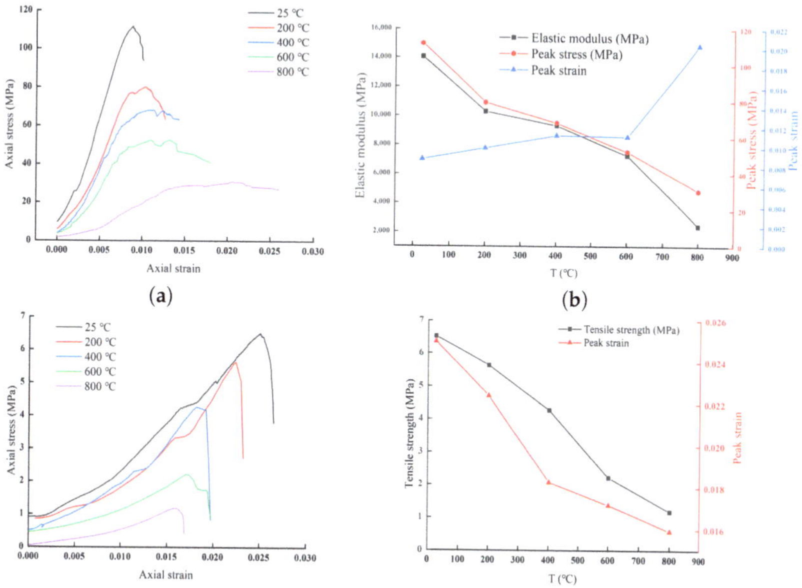

Figure 8. Effect of temperature on the mechanical properties of Beishan granite. (**a**) Uniaxial compression test: stress–strain curve; (**b**) Uniaxial compression test: mechanical properties; (**c**) Brazilian splitting test: stress–strain curve; (**d**) Brazilian splitting test: mechanical properties.

3. Constitutive Model

Based on the results of uniaxial compression tests, this section introduces the establishment of the damage constitutive model under normal temperature loading and the deduction of the high-temperature damage constitutive model. The proposed mechanical model consists of a Kelvin body and a damaged body in parallel, and the D–P criterion is modified.

3.1. Damage Constitutive Model under Normal Temperature Loading

Many natural pores and tectonic fractures are distributed inside rock; they develop under loading and increasing temperature and eventually cause damage and failure. Among various mechanical components, the Calvin body can be used to characterize the viscoelastic properties of materials [44], and can be used to establish constitutive models for the mechanical response of materials such as rocks. However, since such models do not include damage properties to consider the influence of thermal damage of granite, this paper assumes that the mechanical model of granite consists of a Calvin body and a damaged body in parallel, as shown in Figure 9.

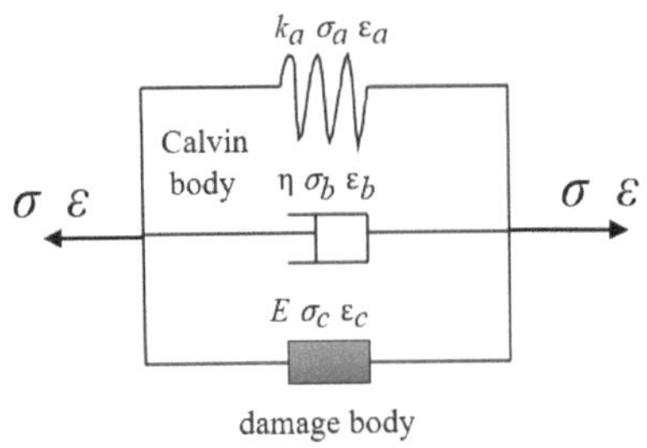

Figure 9. Damage intrinsic mechanics model.

According to the damage theory proposed by Liu et al. [45], assuming that the number of microelements contained inside the granite is N, the number of microelements completely destroyed inside the rock due to external loading is n, and the damage variable is denoted by D, Then, the damage variable D_σ caused by the loading is given by:

$$D_\sigma = \frac{n}{N}. \tag{1}$$

Considering that the rock microelement strength obeys the Weibull distribution [46], the microelement strength probability density function $f(x)$ of granite can be computed as:

$$f(x) = \begin{cases} 0, x \geqslant 0 \\ \frac{m}{K}\left(\frac{x}{K}\right)^{m-1} \exp\left[-\left(\frac{x}{K}\right)^m\right], x < 0. \end{cases} \tag{2}$$

The distribution function $F(x)$ is obtained as:

$$F(x) = 1 - \exp\left[-\left(\frac{x}{K}\right)^m\right], \tag{3}$$

where x is a Weibull distributed random variable and m and K are parameters in the Weibull distribution that represent the shape parameter and the size parameter, respectively.

A threshold exists in the process of deformation and damage. When the stress is lower than the threshold, no damage occurs [47]. When the external load exceeds the threshold and reaches a certain value, the number of microelements that undergo damage inside the rock can be expressed as:

$$n = N\int_0^x f(\sigma)d\sigma. \tag{4}$$

Combining Equation (3) and substituting Equation (4) into Equation (1), the following equation is obtained:

$$D_\sigma = \frac{n}{N} = \frac{N\int_0^x f(\sigma)d\sigma}{N} = \int_0^x f(\sigma)d\sigma = 1 - \exp\left[-\left(\frac{f(\sigma)}{K}\right)^m\right]. \tag{5}$$

Using the D–P criterion to measure the strength of the rock microelement, we obtain:

$$f(\sigma) = \alpha I_1 + \sqrt{J_2}, \tag{6}$$

where $\alpha = \frac{\sin\varphi}{\sqrt{9+3(\sin\varphi)^2}}$, φ is the angle of internal friction, and I_1 is the first invariant of the stress tensor and J_2 is the second invariant of the deviatoric stress tensor that are expressed as:

$$\begin{cases} I_1 = \sigma_1^* + \sigma_2^* + \sigma_3^*, \\ J_2 = \frac{1}{6}\left[(\sigma_1^* - \sigma_2^*)^2 + (\sigma_2^* - \sigma_3^*)^2 + (\sigma_3^* - \sigma_1^*)^2\right]. \end{cases} \tag{7}$$

Although the D–P criterion is widely used, it is used as an ideal elastic–plastic criterion that assumes that damage occurs when the rock is subjected to a certain load. Since there exists a threshold for rock damage, its influence needs to be considered. Therefore, the microelement intensity can be corrected as follows:

$$f(\sigma)' = f(\sigma) - C, \tag{8}$$

where C is the strength at the threshold (threshold strength), and its corresponding strain is the threshold strain ε_c. The threshold stress $C = 40\%\ \sigma_D$ (σ_D is the peak strength) was set according to the experimental study of Martin et al. [48] on Lac du Bonnet granite.

When the load is less than C, no damage or very little damage occurs, and its damage variable can be considered as 0 [49]. Equation (8) can then be substituted into Equation (6) to obtain the expression for the load-related damage variable.

$$D_\sigma = \begin{cases} 0, \ \varepsilon < \varepsilon_c, \\ 1 - exp\left[-\left(\frac{\alpha I_1 + \sqrt{J_2} - C}{K}\right)^m\right], \varepsilon \geqslant \varepsilon_c. \end{cases} \tag{9}$$

Based on the relationship between the nominal stress σ_i and the effective stress σ_i^* in the strain equivalence assumption proposed by Lemaitre et al. [50], the following equation is obtained:

$$\sigma_i = \sigma_i^*(1 - D), i = 1, 2, 3 \tag{10}$$

In addition, the rock will possess a certain residual strength after damage. Considering this, the damage variable correction factor β [51] in Equation (10) is introduced as:

$$\sigma_i = \sigma_i^*(1 - \beta D), i = 1, 2, 3, \tag{11}$$

where $\beta = \sqrt{\frac{\sigma_r}{\sigma_D}}$, and σ_r is the residual strength.

When the rock is in the elastic phase, it obeys Hooke's law:

$$\varepsilon_1 = \frac{1}{E}[\sigma_1^* - \nu(\sigma_2^* + \sigma_3^*)], \tag{12}$$

where E is the modulus of elasticity and ν is the Poisson ratio.

Substitution of Equation (11) into Equation (12) yields:

$$\varepsilon_1 = \frac{1}{E(1 - \beta D)}[\sigma_1 - \nu(\sigma_2 + \sigma_3)]. \tag{13}$$

According to Equation (13), Equation (11) can be expressed as:

$$\begin{cases} \sigma_1^* = \dfrac{\sigma_1}{(1-\beta D)} = \dfrac{\sigma_1 E\varepsilon_1}{\sigma_1 - \nu(\sigma_2+\sigma_3)}, \\ \sigma_2^* = \dfrac{\sigma_2}{(1-\beta D)} = \dfrac{\sigma_2 E\varepsilon_1}{\sigma_1 - \nu(\sigma_2+\sigma_3)}, \\ \sigma_3^* = \dfrac{\sigma_3}{(1-\beta D)} = \dfrac{\sigma_3 E\varepsilon_1}{\sigma_1 - \nu(\sigma_2+\sigma_3)}. \end{cases} \tag{14}$$

Substitution of Equation (14) into Equation (7) yields the expressions for I_1 and J_2 as:

$$\begin{cases} I_1 = \sigma_1^* + \sigma_2^* + \sigma_3^* = \dfrac{E\varepsilon_1(\sigma_1+\sigma_2+\sigma_3)}{\sigma_1 - \nu(\sigma_2+\sigma_3)}, \\ J_2 = \dfrac{1}{6}\left\{ \dfrac{E^2{\varepsilon_1}^2\left[(\sigma_1-\sigma_2)^2 + (\sigma_2-\sigma_3)^2 + (\sigma_3-\sigma_1)^2\right]}{[\sigma_1 - \nu(\sigma_2+\sigma_3)]^2} \right\}. \end{cases} \tag{15}$$

In the uniaxial compression test, $\sigma_2 = \sigma_3 = 0$, then $I_1 = E\varepsilon$, $J_2 = \frac{1}{3}E^2\varepsilon^2$, so that the damage body of the load damage variable D_σ is obtained as follows:

$$D_\sigma = \begin{cases} 0, \varepsilon < \varepsilon_c, \\ 1 - \exp\left[-\left(\dfrac{\alpha E\varepsilon + \frac{1}{\sqrt{3}}E\varepsilon - C}{K} \right)^m \right], \varepsilon \geqslant \varepsilon_c. \end{cases} \tag{16}$$

Therefore, when the axial stress exceeds C, the damage constitutive model of the damaged body is obtained as:

$$\sigma_c = E\varepsilon\left\{ 1 - \beta + \beta\exp\left[-\left(\dfrac{\alpha E\varepsilon + \frac{1}{\sqrt{3}}E\varepsilon - C}{K} \right)^m \right] \right\}. \tag{17}$$

In this study, the Kelvin body is parallel with the damaged body, and for the Kelvin body we obtain:

$$\begin{cases} \sigma_k = \sigma_a + \sigma_b, \\ \sigma_a = k_a\varepsilon_a, \\ \sigma_b = \eta\,\dot{\varepsilon_b}\,. \end{cases} \tag{18}$$

In the combinatorial model, the parallel relationship leads to:

$$\begin{cases} \sigma - \sigma_k + \sigma_c, \\ \varepsilon = \varepsilon_k = \varepsilon_a = \varepsilon_b. \end{cases} \tag{19}$$

Since the uniaxial compressive stress–strain curve crosses the origin of the coordinate system, when $\varepsilon < \varepsilon_c$, the stress–strain function is assumed to be $\sigma = A\varepsilon^2 + B\varepsilon$ Meanwhile, by substituting Equations (17) and (18) into Equation (19), the damage constitutive model under the normal temperature loading of Beishan granite can be expressed as follows:

$$\sigma = \begin{cases} A\varepsilon^2 + B\varepsilon, \varepsilon < \varepsilon_c, \\ k_a\varepsilon + \eta\dot{\varepsilon} + E\varepsilon\left\{ 1 - \beta + \beta\exp\left[-\left(\dfrac{\alpha E\varepsilon + \frac{1}{\sqrt{3}}E\varepsilon - C}{K} \right)^m \right] \right\}, \varepsilon \geqslant \varepsilon_c. \end{cases} \tag{20}$$

3.2. High-Temperature Damage Constitutive Model

It is observed from Figure 8b, that the uniaxial compressive strength and the modulus of elasticity gradually decrease with increasing temperature, and the thermal damage variable D_T is defined as:

$$D_T = 1 - \frac{E_T}{E_0}, \tag{21}$$

where E_T and E_0 are the modulus of elasticity after heat treatment and at 25 °C, respectively.

Temperature affects the parameters K and m of the Weibull distribution [52], assuming that the parameters are:

$$\begin{cases} K(T) = K_0(1 - D_T), \\ m(T) = m_0(1 - D_T). \end{cases} \tag{22}$$

Meanwhile, the viscosity η and the angle of internal friction φ are bound to vary with temperature [53–55]. Substituting Equations (21) and (22) into Substituting Equation (20), the constitutive equation of the thermally damaged Beishan granite is obtained as:

$$\sigma_T = \begin{cases} A\varepsilon^2 + B\varepsilon, \varepsilon < \varepsilon_c, \\ k_a\varepsilon(1 - D_T) + \eta_T\dot{\varepsilon} + E_T\varepsilon\left\{1 - \beta + \beta\exp\left[-\left(\dfrac{\alpha_T E_T\varepsilon + \frac{1}{\sqrt{3}}E_T\varepsilon - C}{K(T)}\right)^{m(T)}\right]\right\}, \varepsilon \geqslant \varepsilon_c. \end{cases} \tag{23}$$

where $\alpha_T = \frac{\sin\varphi_T}{\sqrt{9+3(\sin\varphi_T)^2}}$. And φ_T, E_T, and η_T are the internal friction angle, the elastic modulus, and the viscosity coefficient, respectively, at different temperatures.

4. Solution of Constitutive Equation

The proposed constitutive equation of thermally damaged Beishan granite is relatively complex and involves several parameters. Based on the threshold point and peak point of the uniaxial compressive stress–strain curve, some of the parameters can be determined. The viscosity coefficient η_T and internal friction angle φ_T can be determined by referring to the relevant experimental results.

4.1. K and m

According to the constitutive equation, when $\varepsilon \geq \varepsilon_c$, the following equation can be obtained at the peak point $(\varepsilon_D, \sigma_D)$.

$$\sigma_D = k_a\varepsilon_D(1 - D_T) + \eta_T\dot{\varepsilon}_D + E_T\varepsilon_D\left\{1 - \beta + \beta\exp\left[-\left(\frac{\alpha_T E_T\varepsilon_D + \frac{1}{\sqrt{3}}E_T\varepsilon_D - C}{K(T)}\right)^{m(T)}\right]\right\} \tag{24}$$

Additionally, the peak point satisfies $\left.\frac{\partial\sigma}{\partial\varepsilon}\right|_{\substack{\sigma=\sigma_D \\ \varepsilon=\varepsilon_D}} = 0$, so that:

$$\begin{aligned} \left.\frac{\partial\sigma}{\partial\varepsilon}\right|_{\substack{\sigma=\sigma_D \\ \varepsilon=\varepsilon_D}} &= k_a(1 - D_T) + E_T\left\{1 - \beta + \beta\exp\left[-\left(\frac{\alpha_T E_T\varepsilon_D + \frac{1}{\sqrt{3}}E_T\varepsilon_D - C}{K(T)}\right)^{m(T)}\right]\right\} - \\ & E_T\varepsilon_D\beta\exp\left[-\left(\frac{\alpha_T E_T\varepsilon_D + \frac{1}{\sqrt{3}}E_T\varepsilon_D - C}{K(T)}\right)^{m(T)}\right] \cdot m(T)\left(\frac{\alpha_T E_T\varepsilon_D + \frac{1}{\sqrt{3}}E_T\varepsilon_D - C}{K(T)}\right)^{m(T)-1} \\ & \cdot \frac{\alpha_T E_T + \frac{1}{\sqrt{3}}E_T}{K(T)} = 0 \end{aligned} \tag{25}$$

The values of K and m at different temperatures can be obtained using Equations (24) and (25) at different temperatures.

4.2. A and B

A and B are the fitted coefficients of the quadratic function, and since the function is continuous at the segment point, the threshold point (ε_C, C) satisfies:

$$A\varepsilon_C{}^2 + B\varepsilon_C = k_a\varepsilon_C(1 - D_T) + \eta_T\dot{\varepsilon_C} + E_T\varepsilon_C\left\{1 - \beta + \beta\exp\left[-\left(\frac{\alpha_T E_T\varepsilon_C + \frac{1}{\sqrt{3}}E_T\varepsilon_C - C}{K(T)}\right)^{m(T)}\right]\right\} \tag{26}$$

$$\left.\frac{\partial\sigma}{\partial\varepsilon}\right|_{\substack{\sigma = C\\ \varepsilon = \varepsilon_C}} = k_a(1 - D_T) + E_T\left\{1 - \beta + \beta\exp\left[-\left(\frac{\alpha_T E_T\varepsilon_C + \frac{1}{\sqrt{3}}E_T\varepsilon_C - C}{K(T)}\right)^{m(T)}\right]\right\} - E_T\varepsilon_C\beta\exp\left[-\left(\frac{\alpha_T E_T\varepsilon_C + \frac{1}{\sqrt{3}}E_T\varepsilon_C - C}{K(T)}\right)^{m(T)}\right]\cdot m(T)\left(\frac{\alpha_T E_T\varepsilon_C + \frac{1}{\sqrt{3}}E_T\varepsilon_C - C}{K(T)}\right)^{m(T)-1}\cdot \frac{\alpha_T E_T + \frac{1}{\sqrt{3}}E_T}{K(T)} = 2A\varepsilon_C + B \tag{27}$$

The values of A and B at different temperatures can be obtained using Equations (26) and (27).

5. The Evolution of Thermal Damage in Beishan Granites

To verify the applicability of the model of the thermally damaged Beishan granite, uniaxial compression tests were conducted on Beishan granite following treatment at different temperatures (25 °C, 200 °C, 400 °C, 600 °C and 800 °C). The results are shown in Table 1.

Table 1. Parameters of the uniaxial compression experiment on Beishan granite with different temperature treatments.

$T/°C$	E/GPa	σ_D/MPa	ε_D	C/MPa	ε_C	η	φ
25	14,051.58	111.8710	0.00882	44.7484	0.00376	2.8669	47.36
200	10,290.00	79.4000	0.00995	31.7600	0.00417	1.5620	52.86
400	9322.24	68.1520	0.01122	27.2608	0.00572	1.1271	52.20
600	7280.79	52.3744	0.01109	20.9498	0.00651	0.8726	53.29
800	2402.92	30.5807	0.02031	12.2323	0.01252	0.6921	50.87

The model was validated by comparison to the above-mentioned solution and by uniaxial compression experimental results. Through multiple fitting tests, the damage correction coefficient β was determined to be 0.9, and the values of the relevant parameters are presented in Table 2. Figure 10 shows the comparison between the theoretical and the experimental values.

Table 2. Parameters of the constitutive model of thermally damage granite.

$T/°C$	D_T	K	m	A	B
25	0.00000	38.1773	2.709271	1,490,249.16	6318.17
200	0.26770	75.7117	2.328035	1,045,909.61	5939.64
400	0.33657	94.5893	1.671817	337,823.69	6882.43
600	0.48185	47.8525	1.380419	140,391.48	5815.79
800	0.82899	37.5851	1.484007	17,505.03	1881.04

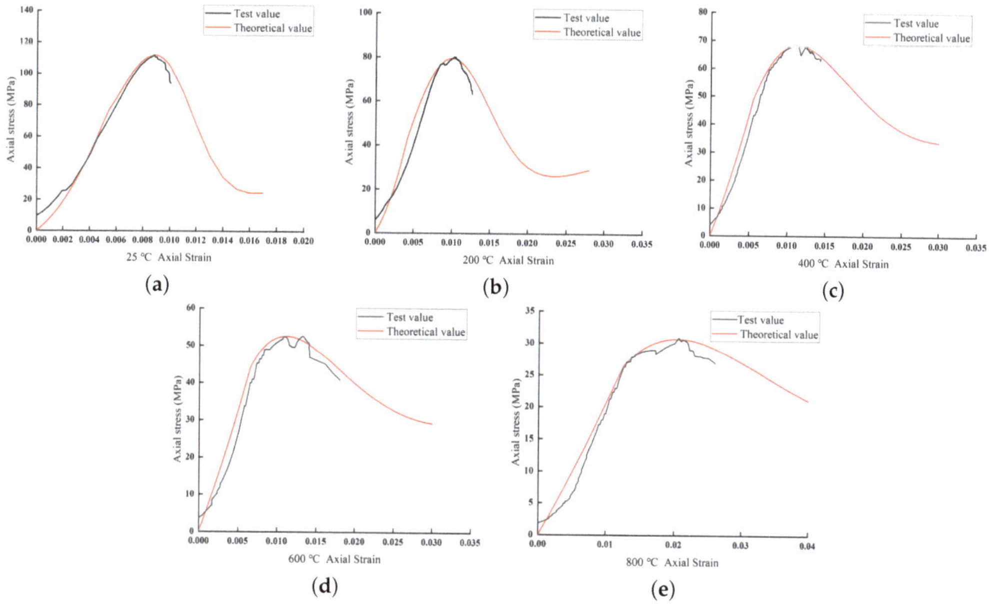

Figure 10. Comparison between the theoretical value and the experimental value. (**a**) T = 25 °C; (**b**) T = 200 °C; (**c**) T = 400 °C; (**d**) T = 600 °C; (**e**) T = 800 °C.

It is observed from Figure 10 that as the temperature rises, the theoretical values are greater than the experimental values except for the peak point and its nearby points. The main reasons for this are as follows: first, the theoretical derivation assumes the ideal state. Second, the microelement strength criterion of rock is the D–P criterion, and the calculated damage strength is usually high [56]. However, the correction in this study is desirable and should be further adjusted based on experimental results to make it closer to the actual situation. Third, in the thermal damage analysis, the temperature only causes damage to accommodate several rock microelements, but in fact, the strength of other microelements will also be influenced and then affect the strength and stability of the rock. Overall, this study optimizes the mechanical model and criterion and has certain applicability and research significance.

6. PFC-Based High-Temperature Damage Simulation

Particle flow code (PFC) is one of the most important tools for investigations of the mechanical properties of rocks [57]. PFC uses particles as the basic unit to study the complex deformation properties of materials. It treats the material as a collection of tiny spherical or disc particles, which are connected to each other by "adhesions". The mechanical properties are determined by both the physical and mechanical properties of the particles and the contact. The advantage of PFC is that it can simulate the material transport and stress transfer inside the specimen, as well as the crack opening, extension, and elongation processes. It can monitor the location and number of microcracks generated during the damage of the specimen, the stress-strain state, the morphology of the crack when it penetrates, and the strength of the specimen. Therefore, it can simulate the mechanical properties of materials at the microscopic scale. In this study, PFC^{2D} was applied to simulate the high-temperature damage of Beishan granite. The simulation process was divided into four steps: sample formation, heat treatment, prepressure, and loading. The flow chart of the modeling is shown in Figure 11.

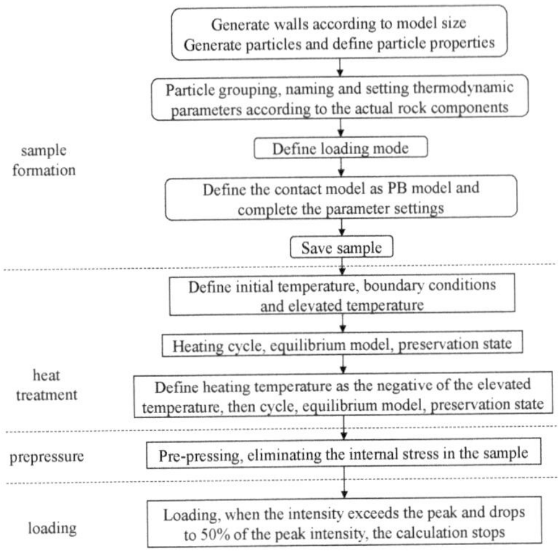

Figure 11. Flow chart of PFC simulation.

First, the model was generated using the FISH function of PFC^{2D}. An outer wall with a height of 100 mm and width of 50 mm was built and filled with particles. The minimum particle size was 0.6 mm, and the ratio of the maximum particle size to the minimum particle size was 1.66. A total of 9017 particles were filled, and the initial establishment of the model was completed after equilibrating the system. Considering that the components of Beishan granite are 67.90% feldspar, 27.40% quartz, and 4.70% mica, 6123 particles were randomly selected and designated as the feldspar group. Similarly, 2471 particles were in the quartz group, and 423 particles were in the mica group. Figure 12 describes the generated particle flow model.

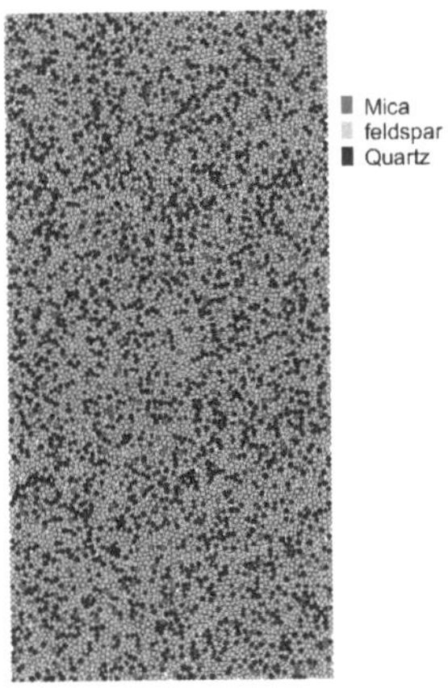

Figure 12. Particle flow model of Beishan granite.

Since the parallel bonding model has successfully reproduced the mechanical behavior of rock [58], it was chosen as the interparticle cementation model in this study. When calibrating the related parameters, the PFC^{2D} manual recommends using the trial-and-error method to obtain simulation results that are consistent with the experimental results [59]. After trial calculations and simulation tests, the parameters of the particle and the contact were determined and are shown in Table 3. Following the parameter calibration, the thermal microparameters of the granite minerals were defined and are shown in Table 4.

Table 3. Parameters of the particle and the contact.

Category	Parameter	Unit	Value
Particle	Minimum particle size	m	6×10^{-4}
	Max/min particle size ratio	-	1.66
	Density	kg/cm^3	2.564
	Elastic modulus	Pa	1.4×10^{10}
	Friction coefficient	-	0.5
	Stiffness ratio	-	1
Contact	Bonding tensile strength	Pa	1×10^8
	Gluing adhesion	Pa	1×10^8
	Gluing friction angle	°	47.36
	Friction coefficient	-	0.5
	Radius multiplier	-	1
	Normal to tangential stiffness ratio	-	1
	Stiffness ratio	-	1

Table 4. Thermal microparameters of granite minerals.

Mineral	Linear Coefficient of Thermal Expansion /10^{-6}·C^{-1}	Thermal Conductivity /W·Mk^{-1}	Specific Heat Capacity /J·kg °C
Feldspar	15	2×10^{-5}	800
Quartz	10	2×10^{-5}	800
Mica	3.5	2×10^{-5}	800

Following the definition of the parameters, the model was heat-treated using the thermal module. It was set as the first module, and the force module was set as the second module because they cannot be calculated simultaneously in PFC^{2D}. The calculation modes can be modified at any time during a PFC analysis by issuing the set thermal and set mechanical commands with the on/off keywords. To perform the thermal calculation, the velocity and displacement of both the ball and the wall are specified to be zero, the initial and boundary conditions are specified, the temperature is applied to the particle with the given thermodynamic parameters, and the thermal step is executed by means of a loop command. When the particle reaches the defined temperature (200 °C, 400 °C, 600 °C and 800 °C) and reaches a steady state, the computational state is saved. Further, the specimen is cooled down to room temperature by the same operation. The distribution of thermal cracks is shown in Figure 13.

Figure 13 shows that 10 tensile cracks are produced at 200 °C. With increasing temperature, the number of cracks gradually increases. Shear cracks appear at 400 °C, and tensile cracks account for 72.93% of the total cracks. With a further increase in the temperature, the crack initiation, expansion and development accelerate, producing 7967 tensile cracks and 7704 shear cracks at 600 °C. The numbers of the two types of cracks are approximately equal. When the temperature reaches 800 °C, the thermal cracks increase rapidly and spread throughout the model, 6281 tensile cracks and 8046 shear cracks are formed, and the number of tensile cracks accounts for 42.08% of the total cracks.

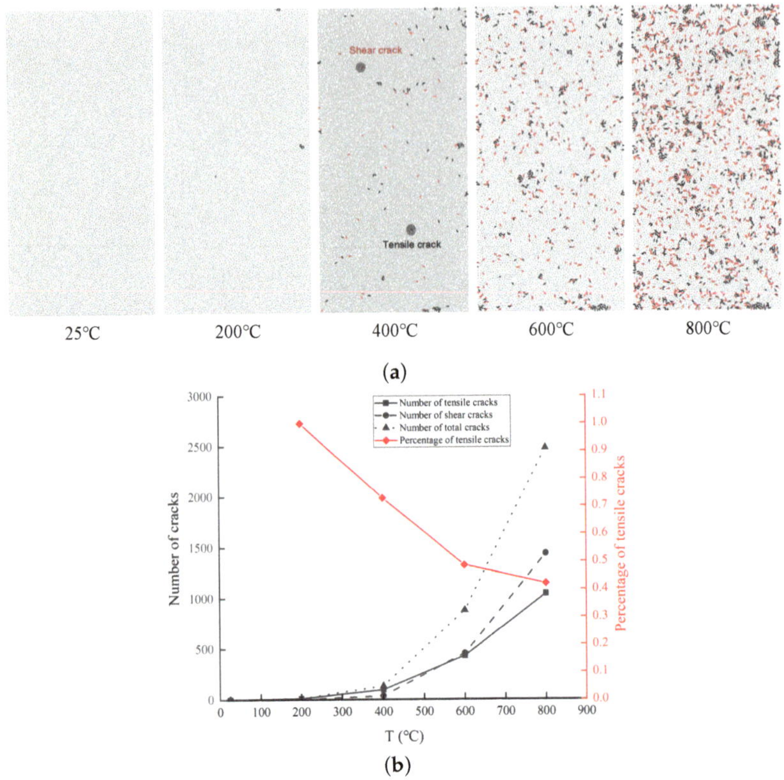

Figure 13. Thermal crack development. (**a**) Thermal crack distribution; (**b**) The number of cracks.

After the cooling step, the thermal module was closed, and the compressive stage was initiated. Considering that the prestress was generated by the particle contact in the model generation step, the model was precompressed at an axial strain rate of 0.2 s^{-1}. The uniaxial compression test was performed at the same rate after precompression. When the load reached the peak strength and then decreased 50%, the calculation stopped. Figure 14a shows the distribution of cracks after model failure.

Figure 14b shows that 829 tensile cracks and 126 shear cracks form after reaching the damage strength at 25 °C, the tensile cracks account for 86.81% of the total number of cracks, and the damage is mainly due to tensile damage. From the appearance of the sample, the distribution of cracks is concentrated, and is mainly located in the upper part of the model, and the cracks develop from the middle of the upper surface to the middle of the left and right sides. The shear zone gradually develops into a rupture surface. With increasing temperature, the number of tensile cracks gradually increases, while the number of shear cracks increases faster, and the latter exceeds the former at 800 °C. The trend of the total number of cracks is similar to that of shear cracks, indicating that the specimen is significantly affected by shear damage. A total of 1489 tensile cracks and 1890 shear cracks form after reaching the damage strength at 800 °C. The shear cracks account for 55.93% of the total number of cracks, which is 15× that at 25 °C, indicating that the damage is changed from mainly tensile cracks to mainly shear cracks with increasing temperature. From the images it is observed that the cracks are more scattered at 800 °C than at 25 °C; they develop from the top of the upper surface to the middle of the right side while also expanding from the middle of the left side to the lower surface, forming two rupture surfaces. A higher temperature corresponds to a higher degree of damage.

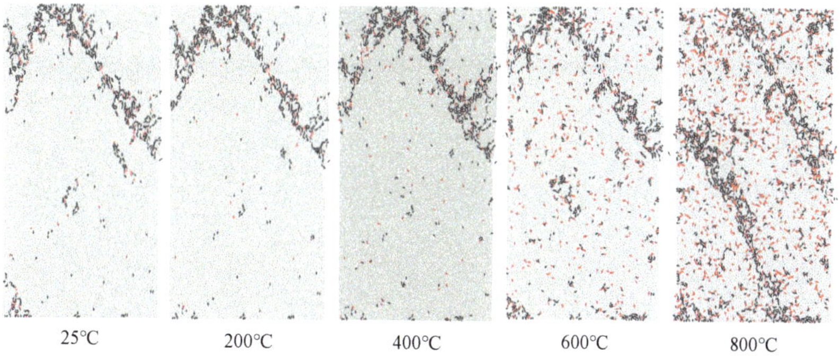

(**a**)

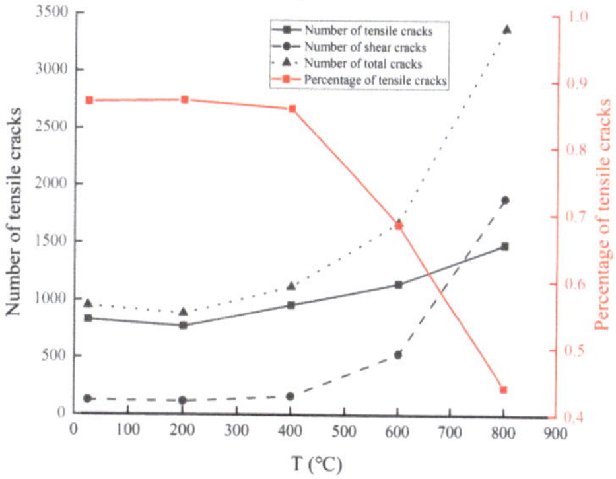

(**b**)

Figure 14. Crack development following uniaxial compression. (**a**) Distribution of crack of model failure; (**b**) The number of cracks.

7. Conclusions

In this study, heat treatment, NMR, uniaxial compression, and Brazilian splitting tests were carried out to construct a constitutive model of thermally damaged Beishan granite. Moreover, PFC^{2D} was used to simulate the development of fractures in heat-treated rock models under uniaxial compression. The following conclusions are obtained:

1. The mass, volume, density, wave velocity, elastic modulus, peak stress, and other experimental results of Beishan granite were obtained. They indicate that temperature has a significant effect on physical and mechanical properties. The high temperature not only gives rise to microcracks between grains but also penetration cracks when the phase change point of quartz is reached. Among these, the microcracks initiate and expand rapidly. High temperature reduces the strength and brittleness of rock.
2. The mechanical model is assumed to consist of a Kelvin body and a damaged body in parallel. Based on the damage theory, the Weibull distribution, the Lemaitre strain equivalence assumption, the damage threshold, and the residual strength, this study innovatively modified the D–P criterion and damage stress. The constitutive model of thermally damaged Beishan granite. The solution of the parameters in the constitutive model was proposed. It was found that the Weibull distribution parameters K and m have significant influences on the curve trend. The theoretical study has produced results essentially in agreement with the experimental ones. The parameters required for the model are conventional mechanical parameters that can be calculated from the uniaxial test results, thus making the model convenient to apply.
3. In fact, the variation of stress-strain curves of rocks under different treatment temperatures is complex. Since rock is non-homogeneous material, there will be certain

differences in the properties of specimens used for different temperature levels, which may cause dispersion of the experimental results. Parameters such as K, m, A, and B in the proposed constitute model highly depend on the experimental results. Therefore, it is difficult to define temperature-dependent expressions for them. A constitutive equation that can accurately describe the stress-strain relationship of rock with temperature as the unique variable will be an important research direction. The following work will be conducted from the aspects of specimen preparation, selection, test conditions, test methods, and equation derivation.

4. According to the simulation of PFC^{2D}, there are two main types of rock cracks: shear cracks and tensile cracks. The cracks change from mainly tensile cracks to mainly shear cracks with increasing temperature. A higher temperature implies a more significant effect of shear cracking and a higher degree of damage.

Author Contributions: Conceptualization, X.Z. and M.W.; methodology, X.Z. and M.W.; software, X.Z., M.W. and Y.C.; validation, X.Z., M.W. and Y.C.; formal analysis, X.Z., M.W. and Z.L.; investigation, X.Z., M.W. and Y.C.; resources, X.Z., M.W. and Z.L.; data curation, X.Z., M.W., Z.L. and Y.C.; writing—original draft, X.Z. and M.W.; writing—review and editing, X.Z., M.W. and Z.L.; visualization, X.Z. and M.W.; supervision, X.Z., M.W., Z.L. and Y.C.; project administration, X.Z. and M.W.; funding acquisition, X.Z. and M.W. All authors have read and agreed to the published version of the manuscript.

Funding: This research was funded by the National Natural Science Foundation of China [No. 52268038], The Applied Basic Research Project of Yunnan Province [No. 202201AT070159], The 1st Professional Degree Postgraduate Practice Innovation Project of Yunnan University [No. 2021Y061].

Institutional Review Board Statement: Not applicable.

Informed Consent Statement: Not applicable.

Data Availability Statement: The data that support the findings of this study are available from the corresponding author upon reasonable request.

Acknowledgments: The authors would like to thank all of the cited authors and the anonymous reviewers in this article for their helpful suggestions.

Conflicts of Interest: The authors declare no conflict of interest.

References

1. Wang, J. Progress of Geological Disposal of High–level Radioactive Waste in China in the 21th Century. *At. Energy Sci. Technol.* **2019** , *53*, 2072–2082.
2. Wang, R.; Zhuo, Z.; Zhou, H.W.; Liu, J.F. A fractal derivative constitutive model for three stages in granite creep. *Results Phys.* **2017**, *7*, 2632–2638. [CrossRef]
3. Rechard, R.P. Historical background on performance assessment for the Waste Isolation Pilot Plant. *Reliab. Eng. Syst. Saf.* **2000**, *69*, 5–16. [CrossRef]
4. Goldstein, J. How to build a better sepulcher: Lessons from New Mexico's Waste Isolation Pilot Plant. *Bull. At. Sci.* **2011**, *67*, 77–88. [CrossRef]
5. Sasaki, T.; Rutqvist, J. Effects of time-dependent deformation of shale on the integrity of a geological nuclear waste repository. *Int. J. Rock Mech. Min. Sci.* **2022**, *158*, 105206. [CrossRef]
6. Chang, K.W.; Nole, M.; Stein, E.R. Reduced-order modeling of near-field THMC coupled processes for nuclear waste repositories in shale. *Comput. Geotech.* **2021**, *138*, 104326. [CrossRef]
7. Li, L.; Shu, X.; Cheng, Y.; Wen, M.; Zhou, H.; Lu, X.; Lu, Y.; Chen, J.; Dong, F. High immobilizing capacity of natural granite as glass-ceramic matrix to simulated trivalent actinide waste. *Radiat. Phys. Chem.* **2022**, *195*, 110067. [CrossRef]
8. Gautam, P.K.; Verma, A.K.; Singh, T.N.; Hu, W.; Singh, K.H. Experimental investigations on the thermal properties of Jalore granitic rocks for nuclear waste repository. *Thermochim. Acta* **2019**, *681*, 178381. [CrossRef]
9. Hudson, J.A.; Bäckström, A.; Rutqvist, J.; Jing, L.; Backers, T.; Chijimatsu, M.; Christiansson, R.; Feng, X.T.; Kobayashi, A.; Koyama, T.; et al. Characterising and modelling the excavation damaged zone in crystalline rock in the context of radioactive waste disposal. *Environ. Geol.* **2009**, *57*, 1275–1297. [CrossRef]
10. Tsang, C.F.; Bernier, F.; Davies, C. Geohydromechanical processes in the Excavation Damaged Zone in crystalline rock, rock salt, and indurated and plastic clays—In the context of radioactive waste disposal. *Int. J. Rock Mech. Min. Sci.* **2005**, *42*, 109–125. [CrossRef]

11. Rutqvist, J.; Börgesson, L.; Chijimatsu, M.; Hernelind, J.; Jing, L.; Kobayashi, A.; Nguyen, S. Modeling of damage, permeability changes and pressure responses during excavation of the TSX tunnel in granitic rock at URL, Canada. *Environ. Geol.* **2009**, *57*, 1263–1274. [CrossRef]
12. Cai, Y.; Liu, Y. Geometry Parameter Analysis on the Fracture Nterconnectivity of Rock Mass Surrounding Nuclear Waste Repositories. *J. Geol. Hazards Environ. Preserv.* **2001**, *33*, 39–47.
13. Laverov, N.P.; Yudintsev, S.V.; Kochkin, B.T.; Malkovsky, V.I. The Russian Strategy of using Crystalline Rock as a Repository for Nuclear Waste. *Elements* **2016**, *12*, 253–256. [CrossRef]
14. Hadermann, J.; Heer, W. The Grimsel (Switzerland) migration experiment: Integrating field experiments, laboratory investigations and modelling. *J. Contam. Hydrol.* **1996**, *21*, 87–100. [CrossRef]
15. Fairhurst, C. Nuclear waste disposal and rock mechanics: Contributions of the Underground Research Laboratory (URL), Pinawa, Manitoba, Canada. *Int. J. Rock Mech. Min. Sci.* **2004**, *41*, 1221–1227. [CrossRef]
16. WNA. World Nuclear Performance Report 2022. Available online: https://www.world-nuclear.org/world-nuclear-performance-report.aspx (accessed on 1 November 2022).
17. Chen, L.; Wang, C.P.; Liu, J.F.; Li, Y.; Liu, J.; Wang, J. Effects of temperature and stress on the time-dependent behavior of Beishan granite. *Int. J. Rock Mech. Min. Sci.* **2017**, *93*, 316–323. [CrossRef]
18. Heierli, J. A comparative study of engineering options for the disposal of high-level radioactive waste with regard to thermal effects and chemical degradation. *J. Nucl. Sci. Technol.* **2016**, *53*, 1276–1295. [CrossRef]
19. Takarli, M.; Prince, W.; Siddique, R. Damage in granite under heating/cooling cycles and water freeze–thaw condition. *Int. J. Rock Mech. Min. Sci.* **2008**, *45*, 1164–1175. [CrossRef]
20. Qin, Y.; Tian, H.; Xu, N.X.; Chen, Y. Physical and Mechanical Properties of Granite After High-Temperature Treatment. *Rock Mech. Rock Eng.* **2020**, *53*, 305–322. [CrossRef]
21. Zhang, W.; Sun, Q.; Hao, S.; Geng, J.; Lv, C. Experimental study on the variation of physical and mechanical properties of rock after high temperature treatment. *Appl. Therm. Eng.* **2016**, *98*, 1297–1304. [CrossRef]
22. Zhao, X.G.; Xu, H.R.; Zhao, Z.; Guo, Z.; Cai, M.; Wang, J. Thermal conductivity of thermally damaged Beishan granite under uniaxial compression. *Int. J. Rock Mech. Min. Sci.* **2019**, *115*, 121–136. [CrossRef]
23. Chen, Y.L.; Wang, S.R.; Ni, J.; Azzam, R.; Fernández-steeger, T.M. An experimental study of the mechanical properties of granite after high temperature exposure based on mineral characteristics. *Eng. Geol.* **2017**, *220*, 234–242. [CrossRef]
24. Gautam, P.K.; Verma, A.K.; Jha, M.K.; Sharma, P.; Singh, T.N. Effect of high temperature on physical and mechanical properties of Jalore granite. *J. Appl. Geophys.* **2018**, *159*, 460–474. [CrossRef]
25. Tomás, R.; Cano, M.; Pulgarín, L.F.; Brotóns, V.; Benavente, D.; Miranda, T.; Vasconcelos, G. Thermal effect of high temperatures on the physical and mechanical properties of a granite used in UNESCO World Heritage sites in north Portugal. *J. Build. Eng.* **2021**, *43*, 102823. [CrossRef]
26. Wu, Y.; Li, X.Z.; Huang, Z.; Xue, S. Effect of temperature on physical, mechanical and acoustic emission properties of Beishan granite, Gansu Province, China. *Nat. Hazards* **2021**, *107*, 1577–1592. [CrossRef]
27. Xu, X.L.; Karakus, M. A coupled thermo-mechanical damage model for granite. *Int. J. Rock Mech. Min. Sci.* **2018**, *103*, 195–204. [CrossRef]
28. Liu, W.; Dan, Z.; Jia, Y.; Zhu, X. On the Statistical Damage Constitutive Model and Damage Evolution of Hard Rock at High-Temperature. *Geotech. Geol. Eng.* **2020**, *38*, 4307–4318. [CrossRef]
29. Wang, F.; Konietzky, H. Thermo-Mechanical Properties of Granite at Elevated Temperatures and Numerical Simulation of Thermal Cracking. *Rock Mech. Rock Eng.* **2019**, *52*, 3737–3755. [CrossRef]
30. Wang, S.; Liao, H.; Chen, Y.; Fernández-Steeger, T.M.; Du, X.; Xiong, M.; Liao, S. Damage Evolution Constitutive Behavior of Rock in Thermo-Mechanical Coupling Processes. *Materials* **2021**, *14*, 7840. [CrossRef]
31. Xie, S.; Han, Z.; Chen, Y.; Wang, Y.; Zhao, Y.; Lin, H. Constitutive modeling of rock materials considering the void compaction characteristics. *Arch. Civ. Mech. Eng.* **2022**, *22*, 60. [CrossRef]
32. Zhao, F.; Sun, Q.; Zhang, W. Fractal analysis of pore structure of granite after variable thermal cycles. *Environ. Earth Sci.* **2019**, *78*, 677. [CrossRef]
33. Miao, S.; Pan, P.Z.; Zhao, X.; Shao, C.; Yu, P. Experimental Study on Damage and Fracture Characteristics of Beishan Granite Subjected to High-temperature Treatment with DIC and AE Techniques. *Rock Mech. Rock Eng.* **2021**, *54*, 721–743. [CrossRef]
34. Li, Q.; Li, X.; Yin, T. Factors affecting pore structure of granite under cyclic heating and cooling: A nuclear magnetic resonance investigation. *Geothermics* **2021**, *96*, 102198. [CrossRef]
35. Wang, F.; Konietzky, H. Thermal damage evolution of granite under slow and high-speed heating conditions. *Comput. Geotech.* **2020**, *123*, 103590. [CrossRef]
36. Liu, L.; Li, H.; Li, X.; Zhang, G.; Wu, R. Research on mechanical properties of heterogeneous rocks using grain-based model under uniaxial compression. *Chin. J. Geotech. Eng.* **2020**, *42*, 542–550.
37. Xie, S.; Han, Z.; Lin, H. A quantitative model considering crack closure effect of rock materials. *Int. J. Solids Struct.* **2022**, *251*, 111758. [CrossRef]
38. Chen, S.; Yang, C.; Wang, G. Evolution of thermal damage and permeability of Beishan granite. *Appl. Therm. Eng.* **2017**, *110*, 1533–1542. [CrossRef]

39. Sun, Q.; Hu, J. Effects of heating on some physical properties of granite, Shandong, China. *J. Appl. Geophys.* **2021**, *193*, 104410. [CrossRef]
40. Wu, Y.; Peng, K.; Zou, Q.; Long, K.; Wang, Y. Tensile Properties and Damage Evolution Laws of Granite After High- and Low-Temperature Cycles. *Nat. Resour. Res.* **2022**, *31*, 1289–1306. [CrossRef]
41. Fang, W.; Jiang, H.; Li, J.; Li, W.; Li, J.; Zhao, L.; Feng, X. A new experimental methodology to investigate formation damage in clay-bearing reservoirs. *J. Pet. Sci. Eng.* **2016**, *143*, 226–234. [CrossRef]
42. Sun, Z.; Jiang, D.; Xie, K.; Wang, K.; Li, L.; Jiang, X. Thermal damage study of Beishan granite based on low field magnetic resonance. *J. China Coal Soc.* **2020**, *45*, 1081–1088. [CrossRef]
43. Wu, Y.; Xi, B.; Wang, L.; Niu, X.; Wang, S.; Zhao, Y. Experimental study on physico-mechanical properties of granite after high temperature. *J. Cent. South Univ. Sci. Technol.* **2020**, *51*, 193–203.
44. Zhao, X.; Meng, W.; Zhou, L. Indentation rolling resistance based on a three-parameter Kelvin solid model. *Adv. Mech. Eng.* **2019**, *11*, 1687814019839517. [CrossRef]
45. Liu, Q.; Xu, X. Damage Analysis of Brittle Rock at High Temperature. *Chin. J. Rock Mech. Eng.* **2000**, 408–411.
46. Wang, F.; Konietzky, H.; Herbst, M. Influence of heterogeneity on thermo-mechanical behaviour of rocks. *Comput. Geotech.* **2019**, *116*, 103184. [CrossRef]
47. Jiang, H.; Jiang, A.; Yang, X. Statistical damage constitutive model of high temperature rock based on Weibull istribution and its verification. *Rock Soil Mech.* **2021**, *42*, 1894–1902. [CrossRef]
48. Martin, C.D.; Chandler, N.A. The progressive fracture of Lac du Bonnet granite. *Int. J. Rock Mech. Min. Sci. Geomech. Abstr.* **1994**, *31*, 643–659. [CrossRef]
49. Cao, W.; Zhao, H.; Zhang, Y.; Zhang, L. Strain softening and hardening damage constitutive model for rock considering effect of volume change and its parameters determination method. *Rock Soil Mech.* **2011**, *32*, 647–654.
50. Lemaitre, J.; Sermage, J.; Desmorat, R. A two scale damage concept applied to fatigue. *Int. J. Fract.* **1999**, *97*, 67. [CrossRef]
51. Cao, R.; He, S.; Wei, J.; Wang, F. Study of modified statistical damage softening constitutive model for rock considering residual strength. *Rock Soil Mech.* **2013**, *34*, 1652–1660+1667. [CrossRef]
52. Li, T.; Gao, M.; Chen, G.; Ma, C.; Xu, Z.; Yin, H.; Chen, C.; Men, L. A thermal-mechanical-damage constitutive model for hard brittle rocks and its preliminary application. *Chin. J. Geotech. Eng.* **2017**, *39*, 1477–1484.
53. Wang, C.; Chen, L.; Liang, J.; Liu, J.; Liu, Y.; Liu, J.; Wang, J.; Zhou, H. Creep constitutive model for full creep process of granite considering thermal effect. *Rock Soil Mech.* **2014**, *35*, 2493–2500+2506. [CrossRef]
54. Zhu, Z.; Jiang, G.; Tian, H.; Wu, W.; Liang, R.; Dou, B. Study on statistical thermal damage constitutive model of rock based on normal distribution. *J. Cent. South Univ. Sci. Technol.* **2019**, *50*, 1411–1418.
55. Xi, B.; Zhao, Y.; Wan, Z.; Zhao, J.; Wang, Y. Study of Constitutive Equation of Granite Rheological Model with Thermo-Mechanical Coupling Effects. *Chin. J. Rock Mech. Eng.* **2009**, *28*, 956–967.
56. Xu, W.; Zhao, G.; Meng, X.; Cheng, X. Modified D–P criterion based on the double-fold reduction method of the true triaxial single-side unloading strength. *Chin. J. Rock Mech. Eng.* **2018**, *37*, 1813–1822. [CrossRef]
57. Chen, P.; Kong, Y.; Yu, H. Research on the Calibration Method of Microparameters of a Uniaxial Compression PFC2D Model for Rock. *Chin. J. Undergr. Space Eng.* **2018**, *14*, 1240–1249.
58. Huang, Y.H.; Yang, S.Q.; Ranjith, P.G.; Zhao, J. Strength failure behavior and crack evolution mechanism of granite containing pre-existing non-coplanar holes: Experimental study and particle flow modeling. *Comput. Geotech.* **2017**, *88*, 182–198. [CrossRef]
59. Liu, X.; Li, D.; Wang, Z.; Zhang, L. The effect of dry-wet cycles with acidic wetting fluid on strength deterioration of shaly sandstone. *Chin. J. Rock Mech. Eng.* **2016**, *35*, 1543–1554. [CrossRef]

minerals

MDPI

Article

Comparative Analysis of Tunnel Plastic Zone Calculation and Engineering Measurement Based on Various Methods

Zhihong Dong [1,2], Qingwen Yan [3], Qifeng Guo [1,2,*], Xun Xi [1,2], Xianquan Lei [4] and Wenhui Tan [1,2,*]

1 School of Civil and Resource Engineering, University of Science and Technology Beijing, Beijing 100083, China
2 Key Laboratory of High-Efficient Mine and Safety of Metal Mines, University of Science and Technology Beijing, Beijing 100083, China
3 Huize Mining Company, Yunnan Chihong Zinc and Germanium Co., Ltd., Qujing 654200, China
4 CINF Engineering Co., Ltd., Changsha 410019, China
* Correspondence: guoqifeng@ustb.edu.cn (Q.G.); wenhui.t@163.com (W.T.)

Abstract: The characteristics of plastic zone are a critical basis for the control and stability analysis of the surrounding rock of roadways. This paper aims to investigate the rationality and applicability of the numerical methods for the plastic zone analysis of deep jointed rock roadways. Based on the detailed investigation and experiments, The plastic zone distribution of roadway surrounding rock under different GSI values and different buried depths was analyzed by analytical methods, parameter reduction, and equivalent rock mass technology, and then the acoustic wave measurement method was used to carry out the field measurement and was compared with the simulation results. The results show that when GSI is large, the difference between the results is not obvious. When GSI is small, the results obtained by the parameter reduction method and the analytical method show a more drastic increase and the discreteness increases. The results obtained by the equivalent rock mass technique are generally close to the measured values, and the growth rate is more uniform. According to the convenience of the calculation parameters and the accuracy of the calculation, the suitable calculation methods for different working conditions were suggested.

Keywords: plastic zone; parameters reduction method; synthetic rock mass technique; loose circle; surrounding rock grades

Citation: Dong, Z.; Yan, Q.; Guo, Q.; Xi, X.; Lei, X.; Tan, W. Comparative Analysis of Tunnel Plastic Zone Calculation and Engineering Measurement Based on Various Methods. *Minerals* **2023**, *13*, 141. https://doi.org/10.3390/min13020141

Academic Editor: Mamadou Fall

Received: 8 November 2022
Revised: 4 January 2023
Accepted: 6 January 2023
Published: 18 January 2023

1. Introduction

The volume and distribution of surrounding rock plastic zone are important parameters for roadway stability analysis, support and reinforcement design, monitoring, and early warning of surrounding rock safety control technology. In engineering, the physical detection method is often used to conduct field measurement of surrounding loose rock zone. Zhang et al. [1] used rock mass ultrasonic detection and a digital panoramic drilling camera to obtain the range of surrounding rock loosening zone in deep roadway. Xu Kun et al. [2] used the single-hole acoustic test method to test the depth of the fracture zone of the surrounding rock. Jing Hongwen et al. [3] took the circularity of cracks measured from borehole images as the criterion for judging surrounding rock loosening zone. Guo Liang et al. [4] measured the range of loose circle by GPR. However, the operation of field measurement of surrounding rock plastic zone is difficult and is easily limited by field environment and construction conditions, and the cost is high, so it is impossible to give prediction guidance before roadway excavation.

Analytical methods based on rock elastic-plastic theory and damage theory can well explain the formation and evolution mechanism of surrounding rock [5–9]. However, when the rock section is complex, it is difficult to obtain an accurate analytical solution. At the same time, there are structural planes of different sizes in rock mass, so it is difficult for the theoretical analysis method to consider the real weak plane of rock mass structure. At the same time, there are structural planes of different sizes in rock mass, so it is difficult

for the theoretical analysis method to consider the real weak plane of rock mass structure. The numerical simulation method can efficiently solve the plastic zone distribution of surrounding rock, solve the complex section and engineering geology problems, and has been more widely used. Xiao Ming et al. [10] determined the range of surrounding rock loosening zone under excavation blasting through numerical calculation. Zhou Zhihua [11] used ANSYS software to establish a model to simulate the loose zone of surrounding rock of soft rock roadway and identified the distribution of plastic zone of roadway through the stress distribution characteristics of roadway. Xiong Liangxiao et al. [12] used FLAC3D to analyze the stress path and the change process of safety factor caused by deep tunnel excavation and also made certain research achievements in dynamic damage and environmental impact [13,14]. However, most studies regard surrounding rock as homogenate, but from the actual situation, roadway surrounding rock is usually anisotropic and discontinuous, composed of discontinuous surface and structure. The existence of discontinuities makes the rock mass appear as a collection of broken blocks, which reduces the stability of the rock mass [15–17]. Compared with the intact rock, various parameters of the jointed rock mass can be obtained through reduction or equivalent calculation and then simulated and analyzed. Hoek-Brown [18] criterion is widely used as a commonly used method for rock mass parameter reduction. Massimiliano Fraldi et al. [19–21] compared the differences between different calculation methods based on limit analysis approaches when the plastic zone is calculated and proposed a straight characterization, which can help to evaluate the risk of a tunnel collapse with respect to the depth of excavation. Wu Shunchuan et al. [22] proposed equivalent proximity and compared and analyzed the rationality of various reduction paths. Hu Mingsheng et al. [23] comprehensively considered the characteristics of the structural plane and the number of joint volumes and used the Hoek-Brown criterion to reduce the rock mass parameters after classifying the rock mass based on the GSI system. Su Yonghua et al. [24] tracked the progress of the methods adopted by Hoek-Brown criterion to estimate the mechanical parameters of rock mass by using the mechanical parameters of rock blocks and various improvement measures.

Equivalent rock mass technology is a reliable mean of numerical calculation in terms of large-scale rock mass stability analysis in recent years. The method by adding discrete fracture network (DFN) numerical model to build realistic rock joints and fissures distribution model, and the calculation results compared with the calculation model and homogenization of continuous medium joint calculation model is more reliable. Wu Shunchuan et al. [25–27] constructed an equivalent rock mass model that could fully reflect the distribution characteristics of engineering rock mass joints and verified the suitability and reliability of equivalent rock mass technology in the study of mechanical properties of discontinuous jointed rock masses. Zhu Wancheng et al. [28–30] conducted numerical simulation tests on equivalent rock mass to determine the rock mass characterization element and then studied the mechanical properties of engineering scale rock mass.

In engineering, when different rock mass parameters of different quality levels are converted by different methods, the difficulty of obtaining various basic properties and necessary parameters is different. The applicability, difference, and accuracy of parameter reduction method and equivalent rock mass technique in plastic zone analysis for different grades of surrounding rock and in situ stress conditions are still unclear. Relying on a mine in Yunnan province, this paper is based on the detailed engineering geological investigation, rock mechanics test, in situ stress test, equivalent rock mass parameters, which were used respectively to fold the subtraction and technical analysis of the plastic zone of surrounding rock, programming to extract the plastic zone size calculation, using the method of acoustic measurement of roadway surrounding rock loose circle of field measurement analysis, comparing simulation and the measured results, comparing the difference between the two numerical methods, and the applicability of the two methods to calculate the plastic zone under different surrounding rock grades and depths is also analyzed.

2. Analysis of Plastic Zone of Roadway

2.1. Analytical Solution Analysis of Plastic Zone in Roadway Surrounding Rock

According to the existing research results [31,32], based on the plane strain mechanical model of circular holes shown in Figure 1, the implicit equation of plastic zone boundary of non-isobaric circular roadway surrounding rock with respect to r, θ is obtained.

$$\begin{aligned} f(r,\theta) = &\left\{(P+\lambda P)(R_0^2/r^2) - \cos 2\theta(\lambda P - P)[1+3(R_0^2/r^2)^2 - 2(R_0^2/r^2)]\right\}^2 + \\ &\left\{\sin 2\theta(\lambda P - P)[1-3(R_0^2/r^2)^2 + 2(R_0^2/r^2)]\right\}^2 - \\ &\left\{[(P+\lambda P) - 2\cos 2\theta(\lambda P - P)(R_0^2/r^2)]^2 - 4c^2\right\}\frac{(1-\cos 2\varphi)}{2} - \\ &4c^2 - 2c\sin 2\varphi[(P+\lambda P) - 2\cos 2\theta(\lambda P - P)(R_0^2/r^2)] \end{aligned} \tag{1}$$

where r and θ are the polar coordinates of any point on the boundary of plastic zone of roadway surrounding rock; P is vertical principal stress; R_0 is the roadway radius; λ is the lateral pressure coefficient; c is the cohesion of surrounding rock; and φ is the friction angle in the surrounding rock. This method is abbreviated as AN in the following.

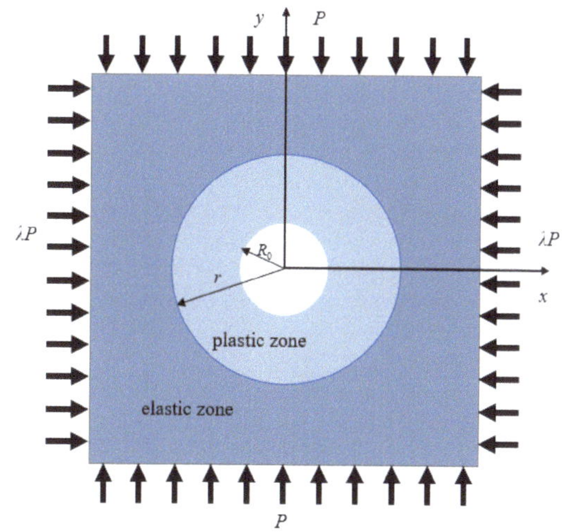

Figure 1. Circular roadway surrounding rock force model in the inhomogeneous stress field.

2.2. Analysis of Plastic Zone of Roadway Based on Parameter Reduction Method

Hoek-Brown criterion is widely used in slope engineering, tunnel chamber, and hydraulic engineering to calculate the rock mass parameters. The GSI value of roadway surrounding rock is obtained through joint and fracture investigation, combined with complete rock physical and mechanical parameters, and according to the latest Hoek-Brown criterion [18], the strength parameters of rock mass under different quality levels can be calculated, as shown in the Figure 2. And then the reduced relevant parameters are assigned to the numerical model for calculation, and the settlement result of roadway plastic zone can be obtained. This method is abbreviated as RE in the following.

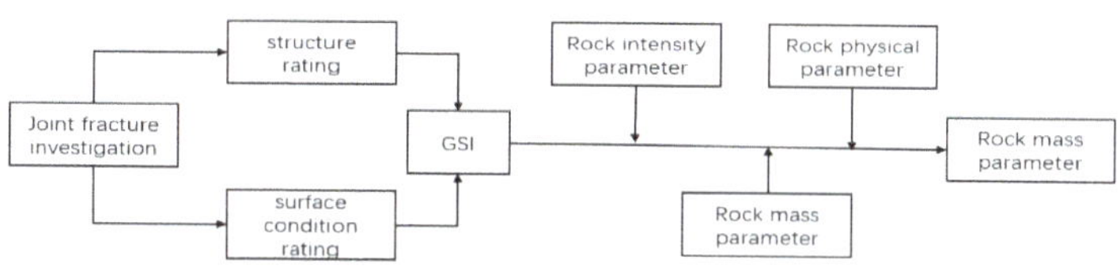

Figure 2. Process of parameter reduction method to obtain rock mass parameters.

2.3. Analysis of Plastic Zone of Roadway Based on Equivalent Rock Mass Technology

By combing discrete fracture network (DFN) with discrete element numerical software (3DEC, PFC, etc.), the equivalent rock mass model is formed by block model, particle model, and smooth joint model, as shown in the Figure 3. In this paper, a calculation model close to the real rock mass is constructed by combining 3DEC and DFN models, which can be obtained based on the joint structure information of roadway surrounding rock. Through the smooth joint model, the parameters, such as joint density, trace length, and occurrence information, can be more truly reflected in the calculation model. Combined with different boundary conditions and mechanical properties of intact rock, the plastic zone of the roadway of jointed rock mass under different working conditions can be calculated and analyzed. This method is abbreviated as EQ in the following.

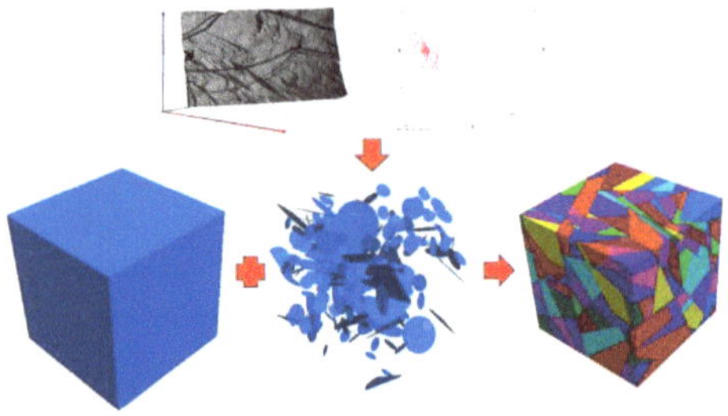

Figure 3. Main process of equivalent rock mass technology.

3. Calculation Parameter Acquisition of Plastic Zone of Jointed Rock Mass Roadway

3.1. Determination of Mechanical Properties of Surrounding Rock Foundation in Deep Roadway

The rock mechanics experiments of uniaxial compression deformation, Brazilian splitting, and variable angle shear were carried out after the rock samples were processed. The basic parameters obtained from the rock mechanics experiments are as shown in the Table 1.

Table 1. Mechanical properties of rocks at different depths.

Depth/m	Uniaxial Compressive Strength/MPa	Uniaxial Tensile Strength/MPa	Cohesion/MPa	Internal Friction Angle/°	Modulus of Elasticity/GPa	Poisson Ratio
1100	69	8.2	19.6	46	86.3	0.25
1300	65	6.8	17.5	48	90.2	0.26
1500	58	4.3	17.2	45	98.7	0.25

3.2. Structural Plane Survey and GSI Grading

The joint survey of typical engineering surrounding rocks at different depths was carried out by using manual measurement and SIROVISION three-dimensional rock mass measurement technology. There were two measurement points for each depth and a total of six measurement points. The distribution law of joints and fractures was counted at each measurement point, and the occurrence patterns of dominant joints at each measurement point were projected by the stereographic projection, as shown in Figure 4. In the figure, the arc is the projection of the dominant joint group on the equatorial plane, the square is the average vector of the dominant joint group, and the triangle is the vector of a single joint within the dominant joint group.

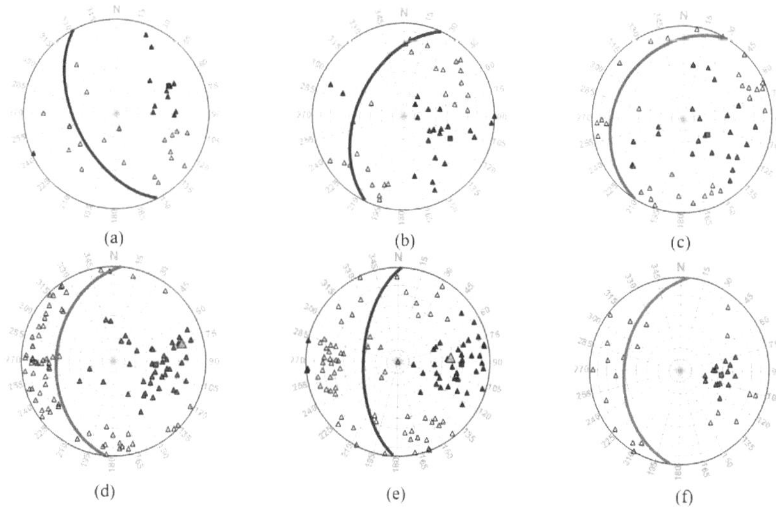

Figure 4. Stereographic projection of the occurrence of dominant joint groups at each measurement point. (a) point 1 (b) point 2 (c) point 3 (d) point 4 (e) point 5 (f) point 6.

The rock mass quality of the scanned site is graded according to the Geological Strength Index (GSI) classification method, and the results are shown in Table 2.

Table 2. Test information and joint occurrence statistics of each measurement point.

Measuring Point Number	Measuring Point Depth/m	GSI	Dominant Dip of Joint/°	Dominant Dip Angle of Joint/°	Joint Trace Length/m	Average Joint Spacing/m
1	1100	78	62.1	62.6	2.3–5.7	2.21
2	1100	52	119.6	59.2	1.3–3.4	1.38
3	1300	72	126.9	28.6	2.0–3.9	2.39
4	1300	41	92.4	49.4	0.5–2.2	0.82
5	1500	59	93.2	69.6	0.8–4.8	1.89
6	1500	44	91.9	39.9	0.9–3.3	0.94

3.3. Distribution of Ground Stress in Deep Roadway

In order to simulate the actual stress state of deep jointed rock mass under high stress, static boundary conditions should be applied in the simulation process to make the simulation as close to the actual working conditions as possible. According to the stress relief method of casing hole, the stress of the original rock at different depths of the mining area is measured, and the deep crustal stress value of the mining area is obtained, as shown in the Figure 5, Table 3.

Table 3. In situ stress distribution.

Depth/m	Maximum Principal Stress			Minimum Principal Stress			Intermediate Principal Stress		
	Value/MPa	Direction/°	Dip Angle/°	Value/MPa	Direction/°	Dip Angle/°	Value/MPa	Direction/°	Dip Angle/°
1000	25.74	330.27	29.50	16.36	76.72	26.59	22.83	203.03	48.27
1100	45.95	4.14	8.88	15.30	95.6	9.47	19.11	231.72	76.96
1300	38.05	33	−10	26.25	72	76	33.24	124	82

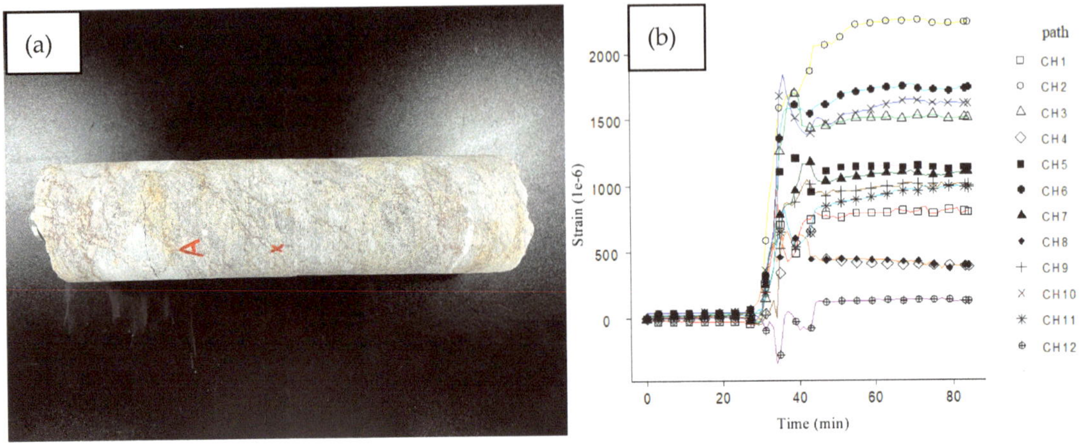

Figure 5. (**a**) Core; (**b**) Stress relief curve.

The regression analysis of the variation of ground stress with buried depth under different depths can be obtained:

$$\sigma_{h\max} = 0.06H + 3.11 \tag{2}$$

$$\sigma_{h\min} = 0.036H - 21.57 \tag{3}$$

$$\sigma_Z = 0.04H - 20.1 \tag{4}$$

where $\sigma_{h\max}$, $\sigma_{h\min}$, and σ_Z are, respectively, the maximum, minimum principal stress, and vertical stress, and the unit is MPa. *H* represents buried depth.

4. Calculation of Plastic Zone of Roadway Surrounding Rock

4.1. Calculation Results of Parameter Reduction Method

Based on the rule of parameter reduction, the mechanical parameters of rock mass with different buried depths and different GSI values were obtained. Simulation results are used in order to improve the coverage of integrity, and the field investigation in the process of different depth of surrounding rock of roadway GSI values are not covered all in interval, so on the basis of GSI scale, added interpolation GSI values 90, 80, 60, and 40 are complete, with 20 representatives constituting surrounding rock, thus representing a more complete, more broken surrounding rock.. The rock mass parameters after reduction are shown in Table 4.

The numerical model is shown in Figure 6, the size of which is 100 m × 50 m × 100 m, and the bottom boundary of the model is fixed, and the normal displacement is fixed for all four boundaries in the horizontal direction. In order to improve the calculation accuracy and efficiency as much as possible, the grid is partitioned. The closer the grids were to the surrounding roadway, the denser the grid was, and the size of grids closed to the boundary was increased appropriately. The displacement boundary conditions defined by the normal displacements of the front, and back and left surfaces were applied to the model, and the stress boundary conditions were applied according to the actual measured in situ stress data.

According to the parameters obtained by parameter reduction in Table 4, the model is assigned to simulate excavation calculation. The distribution and accumulation results of plastic zone are shown in Figure 7 and Table 5.

It can be seen from the calculation results that, with the decrease of GSI value of rock mass, the plastic zone of roadway surrounding rock gradually increases, and with the increase in roadway buried depth, the plastic zone volume also increases.

Table 4. Strength parameters of rock mass under different buried depths and different surrounding rock grades.

Depth	GSI	Rock Uniaxial Compressive Strength/MPa	Rock Uniaxial Tensile Strength/MPa	Cohesion/MPa	Internal Friction Angle/°	Modulus of Elasticity /GPa	Poisson Ratio
1100	90	32.21	2.08	5.17	42.47	36.08	0.25
	78	16.92	0.66	2.91	40.18	26.22	0.25
	52	5.82	0.05	1.20	37.97	6.47	0.26
	40	3.60	0.01	0.81	32.18	3.44	0.28
	20	1.34	0.002	0.34	24.27	2.01	0.32
1300	90	30.71	1.86	4.97	42.55	37.71	0.25
	72	11.47	0.27	2.13	41.68	19.31	0.26
	60	7.68	0.10	1.54	38.16	10.98	0.28
	41	3.48	0.015	0.82	33.16	3.60	0.32
	20	1.30	0.002	0.32	26.28	2.10	0.35
1500	90	28.29	1.46	5.26	44.80	41.27	0.25
	80	10.84	0.21	2.51	43.66	21.13	0.25
	59	7.29	0.084	1.84	40.22	12.02	0.27
	44	3.32	0.012	0.97	38.67	3.94	0.30
	20	1.25	0.001	0.41	34.33	2.30	0.33

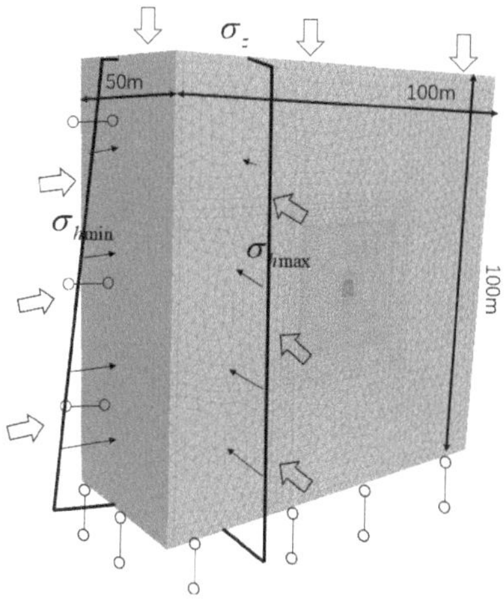

Figure 6. Numerical model.

Table 5. Plastic zone volume of roadway under different rock mass grades.

GSI	Plastic Zone Volume/m^3		
	1100	1300 m	1500 m
90	491.6	551.5	552.2
80	616.9	698.8	734.3
60	905.3	1031.1	1175.1
40	1279.1	1425.9	1673.2
20	1430.4	1545.3	1814.7

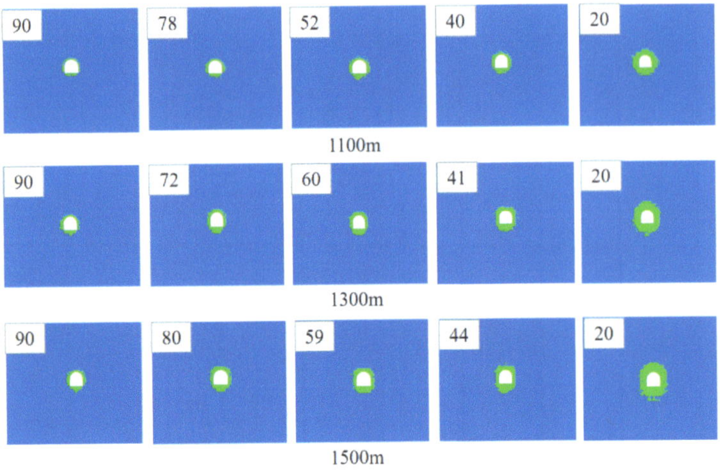

Figure 7. Plastic zone distribution of surrounding rock.

4.2. Equivalent Rock Mass Technical Simulation Results

In the EQ, the size and boundary conditions of the numerical model are the same as those in the parameter reduction method, and the physical and mechanical parameters of the model are consistent with the intact rock to avoid secondary reduction. The fracture stochastic network is constructed, which is close to the actual working condition, as shown in the Figure 8. The joint formation can be controlled by the key words "dipdirlimit" and "diplimit", and the joint length is controlled by the command "size". The surrounding rocks with different GSI values are mainly reflected by the joint spacing, that is, the joint density is converted and controlled in the fish statement. For the surrounding rocks with higher GSI values without measured data, the field measured values are fitted to obtain the joint spacing under different GSI values:

$$d = 0.042 \times \text{GSI} - 1.03 \tag{5}$$

where d is the joint spacing.

Figure 8. Numerical model of equivalent rock mass with different GSI values.

It can be seen from the calculation results, as shown in the Figure 9, similar to the results obtained by the RE, with the decrease in the GSI value of rock mass, the plastic zone of roadway surrounding rock gradually increases, but the overall increase is relatively gentle, and the calculation results are smaller than that of the parameter reduction method when the quality of rock mass is poor.

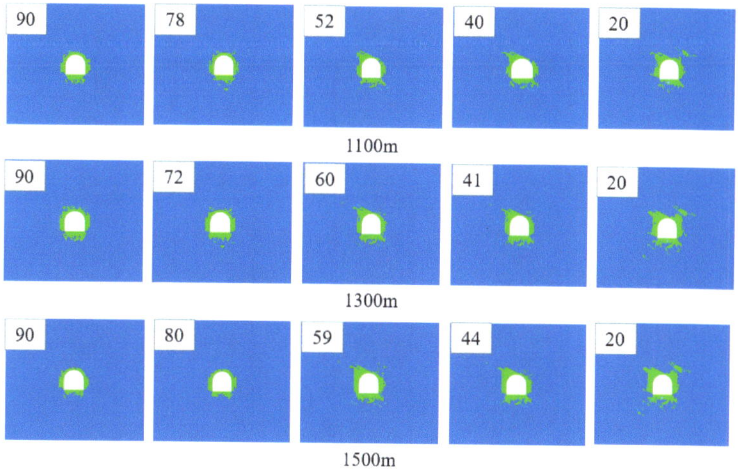

Figure 9. Plastic zone distribution of surrounding rock.

4.3. Analytical Calculation Results

Substitute the parameters in Table 4 into Equation (1). P is the size of the vertical stress on the upper boundary of the model. λ is the ratio of half of the sum of the horizontal principal stresses to the vertical stress, which can be calculated according to the content in Section 3.3, and R_0 is 2.5m. MATLAB software was used to calculate and draw the contour map of plastic zone boundary of Equation (1), as shown in the Figure 10, and then the plastic zone area enclosed by the contour line was integrated and multiplied by the axial length of the roadway model to obtain the volume of plastic zone, as shown in the Table 6.

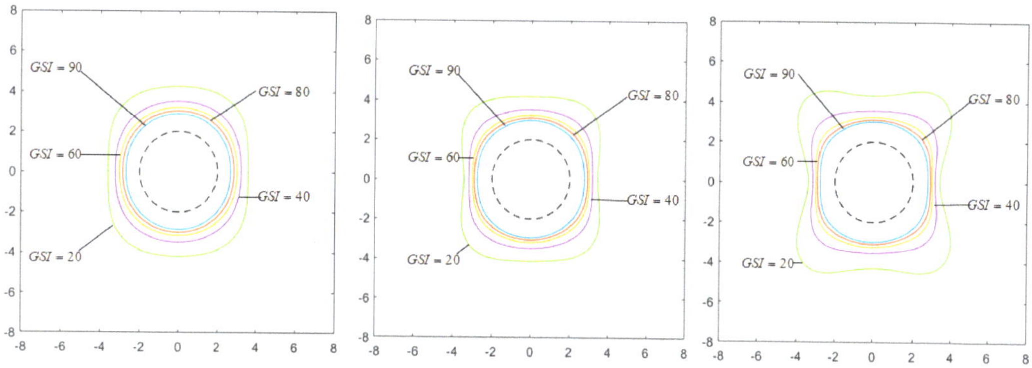

Figure 10. Plastic zone distribution of surrounding rock.

Table 6. Plastic zone volume of roadway under different rock mass grades.

GSI	Plastic Zone Volume/m^3		
	1100	1300 m	1500 m
90	479.9	549.9	605.5
80	603.8	667.6	721.1
60	743.7	781.4	854.9
40	991.7	1042.2	1182.4
20	1662.1	1680.8	2135.1

5. The Site Observation of Plastic Zone in Roadway

The acoustic wave measurement method was used to measure the loosening circle at the survey site of the structural plane. The measuring device was the RSM-SY5 acoustic wave detector produced by Wuhan Zhongke Intellectual Innovation (Wuhan, China), and it is the matching one, with a double receiver probe.

The steps of field measurement are as follows. First, 7655 drilling was used to drill holes, and a measuring borehole was set at 1.7 m from the side wall of the roadway to the depth of 3 m, with a diameter of 60 mm and a depth of 3 m, and a certain downward angle was maintained to facilitate water injection, and the borehole was cleaned to remove gravel and mud. Then, the probe of the ultrasonic instrument was placed at the bottom of the hole, water was injected into the hole as a coupling agent, and the probe was pulled outward each time when it was full of water and measured every 20 cm until it reached the hole, as shown in the Figure 11.

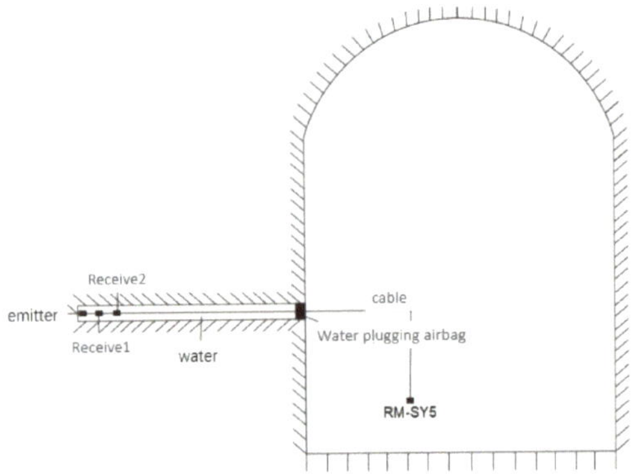

Figure 11. Schematic diagram of the field measurement in the plastic zone of the roadway.

The wave velocity values of rock mass at different depths from the surface of surrounding rock were measured by acoustic detection method, and the size range of surrounding rock loose zone was deduced according to the variation rule of the obtained wave velocity. The acoustic detection results are shown in Figure 12.

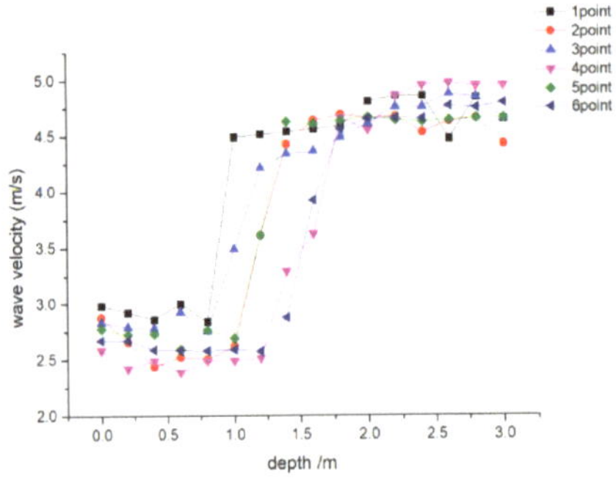

Figure 12. Acoustic wave distribution of each measuring point.

The points with obvious changes in acoustic wave velocity were identified as the boundary of the loose circle, and the areas with small wave velocity were identified as the loose circle. Then, the distance between the loose circle of measuring points 1 to 6, and the side wall of the roadway was about 1 m, 1.5 m, 1.1 m, 1.8 m, 1.5 m, and 1.9 m, respectively.

Combined with the radius of the roadway, the radius of the loose ring is converted into the radius of the plastic zone, according to the formula proposed in the literature [2].

$$R = \left(\frac{1}{1+3\alpha}\right)^{\frac{1-3\alpha}{6\alpha}} R_P \tag{6}$$

$$\alpha = \frac{\sin\varphi}{\sqrt{3(3+\sin^2\varphi)}} \tag{7}$$

where φ is the internal friction angle and R_P is the radius of the tunnel plastic zone. After conversion, the radius of the plastic zone of the roadway of the three surrounding rock grades is 3.09 m, 3.49 m, 3.09 m, 3.68 m, 3.79 m, and 3.78 m, respectively. Based on the calculation method of the cylinder volume and combined with the tunnel axial length used in the numerical model, the measured values of the plastic zone of surrounding rock in the three surrounding rock grades are 697.1, 1027.1, 697.1, 1119.5, 1302.3, and 1294.8 m^3.

6. Volume Contrast Analysis of Plastic Zone

The plastic zone volumes obtained by different methods are compared. In the Figure 13, ME represents measured method.

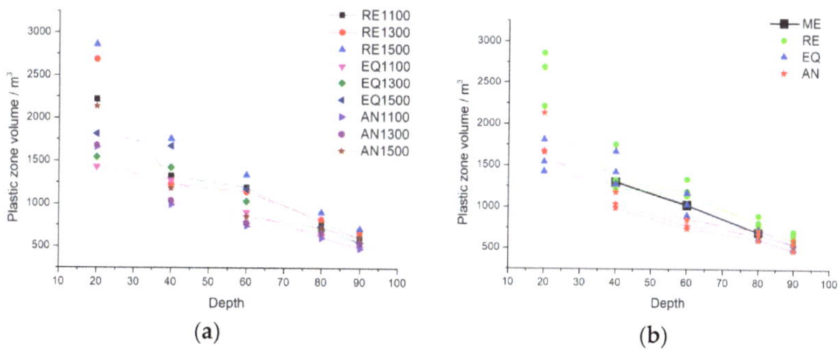

Figure 13. (**a**) Comparison between results of two methods and measured values; (**b**) Comparison between measurement and calculation.

When the GSI value of rock mass is greater than 60, the difference between the results obtained by the three calculation methods is small. When the GSI value is less than 60, the calculation results begin to show a big difference. With the decrease of GSI value, the quality of rock mass deteriorates and the degree of fragmentation increases. The change rate of the result obtained by parameter reduction method becomes larger and larger. When the GSI value is less than 40, the trend of surge occurs, and the data is generally larger than the measured value. The variation trend of the calculated results obtained by equivalent rock mass technique is the most gentle, and the error between the calculated results and the measured values is small. The results obtained by analytical method are generally smaller than the measured values.

The ratio of the difference between the calculated value and the measured value is defined as the error rate.

$$v_i = \left|\frac{V_c - V_r}{V_r}\right| \tag{8}$$

where V_c is the calculated value, V_r is the measured value, and v_i is the error rate.

The parameters required for calculation by different methods and the error rates are sorted out in the following Figure 14.

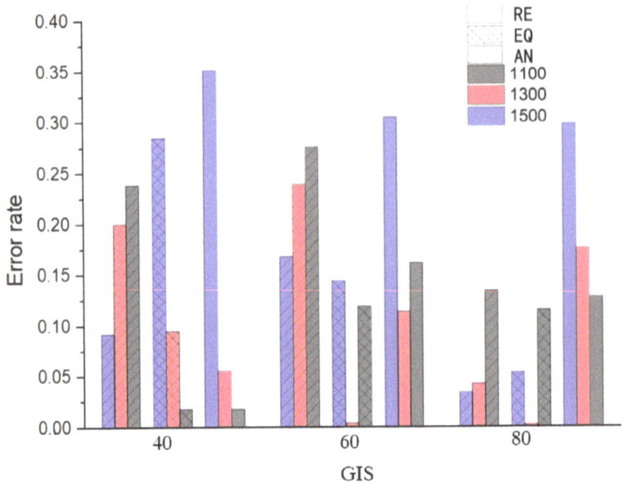

Figure 14. Error rates distribution of plastic zone under different methods, different GSI values, and different depths.

As can be seen from Figure 14, the error of RE method is larger among the three methods. The smaller the GSI value is and the deeper the depth is, the larger the error is and the more uneven the error distribution between different depths is. The error of EQ method is the smallest among the three methods. With the decrease of GSI value, the position error of deeper buried depth is gradually larger than that of shallower buried depth. The error of AN method is more uniform in depth, and the deeper the depth, the smaller the error.

The calculation parameters required by different methods are summarized in Table 7.

Table 7. Calculation parameters to be obtained by various methods.

Methods	GSI	Elasticity Modulus	Uniaxial Compressive Strength	Uniaxial Tensile Strength	Density	Internal Friction Angle	Cohesion	Side Pressure Coefficient	Frequency of Joints	Joint Trace Length	The Joint Stiffness
RE	√	√	√	√	√	√	√	√			
EQ	√	√	√	√	√	√	√	√	√	√	√
AN	√					√	√	√			

It can be seen from Table 7 that the equivalent EQ method requires the most parameters, which not only requires the physical and mechanical parameters of rocks, but also needs to pay attention to the mechanical properties of joints. The RE method requires complete rock parameters for reduction calculation, and the selection of some empirical coefficients in the reduction process is subjective. The analytical method requires the least parameters, and only some parameters of two-dimensional calculation are concerned.

7. Conclusions

(1) There is a high correlation between the plastic zone volume calculated by the three methods and the measured value, and they all increase with the decrease of GSI value, and the smaller the GSI value, the more significant the increase. When the GSI value is greater than 40, the error between the results of the three calculation methods and the measured values is small. When the GSI value is less than 40, the results obtained by the parameter reduction method and the analytical method show a more drastic increase, and the discreteness increases. The results obtained by the equivalent rock mass technique are generally close to the measured values, and the growth rate is more uniform.

(2) There is a certain error between the results obtained by the three calculation method and the measured values. In general, the error of equivalent rock mass technology i the smallest. With the decrease of GSI value, its error rate at different depths graduall increases. The error of the parameter reduction method is the largest, and it show great discreteness in different depths. The error of analytical method increases mor evenly under different GSI values, and there is little difference between differen depths.
(3) Considering the convenience of obtaining calculation parameters and the accuracy o calculation, analytical method can be used to calculate when the GSI is high, the deptl is shallow, and the approximate plastic zone range needs to be determined quickly When the GSI is high and the depth is shallow, the parameter reduction method car be used for calculation and analysis. On the premise that GSI is small, the depth is deep, and the joint mechanical properties can be obtained, and the equivalent rock mass technology can be used to calculate the plastic zone of roadway more accurately so as to guide the support and stability analysis of deep roadway.

In general, there are some errors in the three methods compared with the measured results, but the measured values are not completely accurate, so different methods can be selected to calculate the plastic zone of the roadway according to the GSI value of the surrounding rock, the distribution interval of the buried depth, and the convenience of parameter acquisition.

Author Contributions: Conceptualization, Z.D.; methodology, Z.D.; software, Z.D.; validation, X.X. X.L.; formal analysis, Z.D.; investigation, Z.D.; resources, Q.Y.; data curation, Z.D.; writing—original draft preparation, Z.D.; writing—review and editing, Q.G. and W.T.; visualization, Z.D.; supervision Z.D.; project administration, Z.D.; funding acquisition, Z.D. All authors have read and agreed to the published version of the manuscript.

Funding: This research received no external funding.

Data Availability Statement: Data available in a publicly accessible repository.

Conflicts of Interest: The authors declare no conflict of interest.

References

1. Zhang, Z.H.; An, L.; Wu, D.W.; Wu, C. Field detection of roadway surrounding rock loose zone and support technology of broken rock mass in jiaojia gold mine. *Met. Mine* **2019**, *12*, 56–61. [CrossRef]
2. Xu, K.; Wang, Z.J.; Meng, X.L. Research on detection technology for deep tunnel surrounding rock loose circle and numerical simulation analysis. *Rock Soil Mech.* **2013**, *34*, 464–470. [CrossRef]
3. Jing, H.W.; Li, Y.H.; Liang, J.Q.; Yu, D.C. Borehole camera technology for measuring the relaxation zone of surrounding rock: Mechanism and application. *J. China Univ. Min. Technol.* **2009**, *38*, 645–649, 669.
4. Guo, L.; Li, J.C.; Zhang, Z.C.; Chen, S.J. Research on surrounding rock loose zone of tunnel under unsymmrtrical loading with ground penetrating radar ang its application. *Chin. J. Rock Mech. Eng.* **2011**, *30*, 3009–3015.
5. Yang, K.; Wu, S.K.; Wu, Q.L.; Gao, Y.T.; Wang, Y.Q. Support optimization design of a deep-buried tunnel in consideration of the broken rock zone support. *Chin. J. Eng.* **2015**, *37*, 839–843. [CrossRef]
6. Wang, L.; Gao, Q. Elastoplastic model of damaged rock based on the strength theory. *J. Univ. Sci. Technol. Beijing* **2008**, *30*, 461–467.
7. Cai, W.; Zhu, H.; Liang, W.; Wang, X.; Su, C.; Wei, X. A post-peak dilatancy model for soft rock and its application in deep tunnel excavation. *J. Rock Mech. Geotech. Eng.* **2022**, *in press*. [CrossRef]
8. Cai, W.Q.; Zhu, H.H.; Liang, W.H. Three-dimensional stress rotation and control mechanism of deep tunneling incorporating generalized Zhang–Zhu strength-based forward analysis. *Eng. Geol.* **2022**, *308*, 106806. [CrossRef]
9. Cai, W.; Zhu, H.; Liang, W.; Vu, B.; Su, C.; Zhang, K.; Wang, X. Three-dimensional forward analysis and real-time design of deep tunneling based on digital in-situ testing. *Int. J. Mech. Sci.* **2022**, *226*, 107385. [CrossRef]
10. Xiao, M.; Zhang, Y.T.; Chen, J.T.; Tian, H. Numerical analysis of excavation damaged zone of underground caverns include by excavation blasting. *Rock Soil Mech.* **2010**, *31*, 2613–2618. [CrossRef]
11. Zhou, Z.H.; Yang, L.D.; Duan, Y. Numerical simulation of the broken zone in soft rock roads. *China Min. Mag.* **2009**, *18*, 97–99.
12. Xiong, L.X.; Yang, D.L. Stress evolution of deep cavern included by excavating. *J. Cent. South Univ. (Sci. Technol.)* **2009**, *40*, 236–242. [CrossRef]
13. Zuo, S.Y.; Xiao, M.; Xu, J.K. Numerical simulation of dynamic damage effect of surrounding rock for tunnels by blasting excavation. *Rock Soil Mech.* **2011**, *32*, 3171–3176, 3184. [CrossRef]

14. Sun, L.H.; Ji, H.G.; Zeng, P.; Jiang, H. Variability characteristics simulation of roadway plastic zone in different side pressure coefficients. *Saf. Coal Mines* **2015**, *46*, 227–230. [CrossRef]
15. Xu, W.Y.; Zhang, G.K. Study on orthotropic equivalent strength parameters of jointed rock mass. *Chin. J. Geotech. Eng.* **2007**, *29*, 806–810.
16. Fan, W.C.; Cao, P. Degradation of joint surface morphology, shear behavior and closure characteristics during cyclic loading. *J. Cent. South Univ.* **2018**, *25*, 653–661. [CrossRef]
17. Liu, C.; Bai, S.W. Fuzzy mathematics method for analysis of the weathering degree of rocks. *Chin. J. Rock Mech. Eng.* **2005**, *26*, 252–256.
18. Hoek, E.; Brown, E.T. The Hoek–Brown failure criterion and GSI—2018 edition. *J. Rock Mech. Geotech. Eng.* **2019**, *11*, 445–463. [CrossRef]
19. Fraldi, M.; Cavuoto, R.; Cutolo, A.; Guarracino, F. Stability of tunnels according to depth and variability of rock mass parameters. *Int. J. Rock Mech. Min. Sci.* **2019**, *119*, 222–229. [CrossRef]
20. Fraldi, M.; Guarracino, F. Limit analysis of progressive tunnel failure of tunnels in Hoek-Brown rock masses. *Int. J. Rock Mech. Min. Sci.* **2012**, *50*, 170–173. [CrossRef]
21. Fraldi, M.; Guarracino, F. Evaluation of impending collapse in circular tunnels by analytical and numerical approaches. *Tunn. Undergr. Space Technol.* **2011**, *26*, 507–516. [CrossRef]
22. Gao, Y.T.; Fan, G.L.; Wu, S.C.; Han, H.L.; Gao, L.L.; Zhou, Y. Stability analysis of tunnel surrounding rock based on different strength reduction paths. *J. Univ. Sci. Technol. Beijing* **2013**, *35*, 393–401.
23. Hu, S.M.; Hu, X.W. Estimation of rock mass parameters based on quantitative GSI system and Hoke-Brown criterion. *Rock Soil Mech.* **2011**, *32*, 861–866. [CrossRef]
24. Su, Y.H.; Feng, L.Z.; Li, Z.Y.; Zhao, M. Quantification of elements for geological strength index in Hoke-Brown criterion. *Chin. J. Rock Mech. Eng.* **2009**, *28*, 679–686.
25. Wu, S.C.; Zhou, Y.; Gao, Y.T.; Misra, A. Research on construction method of stochastic joints 3d-network model of equivalent rock mass. *Chin. J. Rock Mech. Eng.* **2012**, *31*, 3082–3090.
26. Zhou, Y.; Wu, S.C.; Wang, L.; Yan, Q.; Zhao, W.; Zhang, X.P. Application of equivalent rock mass technique to mesoscopic analysis of fracture mechanism of rock specimen containing two intermittent joint. *Rock Soil Mech.* **2013**, *34*, 2801–2809. [CrossRef]
27. Yang, Z.M.; Gao, Y.T.; Wu, S.C.; Chengzi, B.; Jin, A. Study of influence of joint parameters on rock mass strength based on equivalent rock mass technology. *J. China Univ. Min. Technol.* **2018**, *47*, 979–986. [CrossRef]
28. Zhu, W.C.; Zhang, M.S.; Zhang, H.X.; Guo, X.Q.; Guan, K. Numerical simulation for determining the size of representative element volume (REV) of jointed rock mass. *Chin. J. Geotech. Eng.* **2013**, *35*, 1121–1127.
29. Wu, S.C.; Zhou, Y.; Gao, L.L.; Zhang, X.P. Application of equivalent rock mass technique to rock mass engineering. *Chin. J. Rock Mech. Eng.* **2010**, *29*, 1435–1441.
30. Wang, X.M.; Zheng, Y.H. Review of advances in investigation of representative elementary volume and scale effect of fractured rock masses. *Rock Soil Mech.* **2015**, *36*, 3456–3464.
31. Zhao, Z.Q. Study on Deformation Failure Mechanism and Control Method of Surrounding Rock in Large Deformation Mining Roadway. Ph.D. Thesis, China University of Mining & Technology, Beijing, China, 2014.
32. Yuan, C. Study on Deformation Failure Mechanism and Stability Control Principle of Deep Roadway Surrounding Rock. Ph.D. Thesis, China University of Mining & Technology, Beijing, China, 2014.

MDPI AG
Grosspeteranlage 5
4052 Basel
Switzerland
Tel.: +41 61 683 77 34

Minerals Editorial Office
E-mail: minerals@mdpi.com
www.mdpi.com/journal/minerals

www.ingramcontent.com/pod-product-compliance
Lightning Source LLC
Chambersburg PA
CBHW041430210926
51850CB00013B/3396
* 9 7 8 3 7 2 5 8 3 2 4 1 5 *